Lecture Notes in Artificial Intelligence 8725

Subseries of Lecture Notes in Computer Science

LNAI Series Editors

Randy Goebel
University of Alberta, Edmonton, Canada
Yuzuru Tanaka
Hokkaido University, Sapporo, Japan
Wolfgang Wahlster
DFKI and Saarland University, Saarbrücken, Germany

LNAI Founding Series Editor

Joerg Siekmann
DFKI and Saarland University, Saarbrücken, Germany

Toon Calders   Floriana Esposito
Eyke Hüllermeier   Rosa Meo (Eds.)

Machine Learning and Knowledge Discovery in Databases

European Conference, ECML PKDD 2014
Nancy, France, September 15-19, 2014
Proceedings, Part II

 Springer

Volume Editors

Toon Calders
Université Libre de Bruxelles, Faculty of Applied Sciences
Department of Computer and Decision Engineering
Av. F. Roosevelt, CP 165/15, 1050 Brussels, Belgium
E-mail: toon.calders@ulb.ac.be

Floriana Esposito
Università degli Studi "Aldo Moro", Dipartimento di Informatica
via Orabona 4, 70125 Bari, Italy
E-mail: floriana.esposito@uniba.it

Eyke Hüllermeier
Universität Paderborn, Department of Computer Science
Warburger Str. 100, 33098 Paderborn, Germany
E-mail: eyke@upb.de

Rosa Meo
Università degli Studi di Torino, Dipartimento di Informatica
Corso Svizzera 185, 10149 Torino, Italy
E-mail: meo@di.unito.it

ISSN 0302-9743   e-ISSN 1611-3349
ISBN 978-3-662-44850-2   e-ISBN 978-3-662-44851-9
DOI 10.1007/978-3-662-44851-9
Springer Heidelberg New York Dordrecht London

Library of Congress Control Number: 2014948041

LNCS Sublibrary: SL 7 – Artificial Intelligence

© Springer-Verlag Berlin Heidelberg 2014

This work is subject to copyright. All rights are reserved by the Publisher, whether the whole or part of the material is concerned, specifically the rights of translation, reprinting, reuse of illustrations, recitation, broadcasting, reproduction on microfilms or in any other physical way, and transmission or information storage and retrieval, electronic adaptation, computer software, or by similar or dissimilar methodology now known or hereafter developed. Exempted from this legal reservation are brief excerpts in connection with reviews or scholarly analysis or material supplied specifically for the purpose of being entered and executed on a computer system, for exclusive use by the purchaser of the work. Duplication of this publication or parts thereof is permitted only under the provisions of the Copyright Law of the Publisher's location, in ist current version, and permission for use must always be obtained from Springer. Permissions for use may be obtained through RightsLink at the Copyright Clearance Center. Violations are liable to prosecution under the respective Copyright Law.
The use of general descriptive names, registered names, trademarks, service marks, etc. in this publication does not imply, even in the absence of a specific statement, that such names are exempt from the relevant protective laws and regulations and therefore free for general use.
While the advice and information in this book are believed to be true and accurate at the date of publication, neither the authors nor the editors nor the publisher can accept any legal responsibility for any errors or omissions that may be made. The publisher makes no warranty, express or implied, with respect to the material contained herein.

Typesetting: Camera-ready by author, data conversion by Scientific Publishing Services, Chennai, India

Printed on acid-free paper

Springer is part of Springer Science+Business Media (www.springer.com)

Preface

The European Conferences on Machine Learning (ECML) and on Principles and Practice of Knowledge Discovery in Data Bases (PKDD) have been organized jointly since 2001, after some years of mutual independence. Going one step further, the two conferences were merged into a single one in 2008, and these are the proceedings of the 2014 edition of ECML/PKDD. Today, this conference is a world-wide leading scientific event. It aims at further exploiting the synergies between the two scientific fields, focusing on the development and employment of methods and tools capable of solving real-life problems.

ECML PKDD 2014 was held in Nancy, France, during September 15–19, co-located with ILP 2014, the premier international forum on logic-based and relational learning. The two conferences were organized by Inria Nancy Grand Est with support from LORIA, a joint research unit of CNRS, Inria, and Université de Lorraine.

Continuing the tradition, ECML/PKDD 2014 combined an extensive technical program with a demo track and an industrial track. Recently, the so-called Nectar track was added, focusing on the latest high-quality interdisciplinary research results in all areas related to machine learning and knowledge discovery in databases. Moreover, the conference program included a discovery challenge, a variety of workshops, and many tutorials.

The main technical program included five plenary talks by invited speakers, namely, Charu Aggarwal, Francis Bach, Lise Getoor, Tie-Yan Liu, and Raymond Ng, while four invited speakers contributed to the industrial track: George Hébrail (EDF Lab), Alexandre Cotarmanac'h (Twenga), Arthur Von Eschen (Activision Publishing Inc.) and Mike Bodkin (Evotec Ltd.).

The discovery challenge focused on "Neural Connectomics and on Predictive Web Analytics" this year. Fifteen workshops were held, providing an opportunity to discuss current topics in a small and interactive atmosphere: Dynamic Networks and Knowledge Discovery, Interactions Between Data Mining and Natural Language Processing, Mining Ubiquitous and Social Environments, Statistically Sound Data Mining, Machine Learning for Urban Sensor Data, Multi-Target Prediction, Representation Learning, Neural Connectomics: From Imaging to Connectivity, Data Analytics for Renewable Energy Integration, Linked Data for Knowledge Discovery, New Frontiers in Mining Complex Patterns, Experimental Economics and Machine Learning, Learning with Multiple Views: Applications to Computer Vision and Multimedia Mining, Generalization and Reuse of Machine Learning Models over Multiple Contexts, and Predictive Web Analytics.

Nine tutorials were included in the conference program, providing a comprehensive introduction to core techniques and areas of interest for the scientific community: Medical Mining for Clinical Knowledge Discovery, Patterns in Noisy and Multidimensional Relations and Graphs, The Pervasiveness of

Machine Learning in Omics Science, Conformal Predictions for Reliable Machine Learning, The Lunch Is Never Free: How Information Theory, MDL, and Statistics are Connected, Information Theoretic Methods in Data Mining, Machine Learning with Analogical Proportions, Preference Learning Problems, and Deep Learning.

The main track received 481 paper submissions, of which 115 were accepted. Such a high volume of scientific work required a tremendous effort by the area chairs, Program Committee members, and many additional reviewers. We managed to collect three highly qualified independent reviews per paper and one additional overall input from one of the area chairs. Papers were evaluated on the basis of their relevance to the conference, their scientific contribution, rigor and correctness, the quality of presentation and reproducibility of experiments. As a separate organization, the demo track received 24 and the Nectar track 23 paper submissions.

For the second time, the conference used a double submission model: next to the regular conference track, papers submitted to the Springer journals *Machine Learning* (MACH) and *Data Mining and Knowledge Discovery* (DAMI) were considered for presentation in the conference. These papers were submitted to the ECML/PKDD 2014 special issue of the respective journals, and underwent the normal editorial process of these journals. Those papers accepted for the of these journals were assigned a presentation slot at the ECML/PKDD 2014 conference. A total of 107 original manuscripts were submitted to the journal track, 15 were accepted in DAMI or MACH and were scheduled for presentation at the conference. Overall, this resulted in a number of 588 submissions, of which 130 were selected for presentation at the conference, making an overall acceptance rate of about 22%.

These proceedings of the ECML/PKDD 2014 conference contain the full papers of the contributions presented in the main technical track, abstracts of the invited talks and short papers describing the demonstrations, and the Nectar papers. First of all, we would like to express our gratitude to the general chairs of the conference, Amedeo Napoli and Chedy Raïssi, as well as to all members of the Organizing Committee, for managing this event in a very competent and professional way. In particular, we thank the demo, workshop, industrial, and Nectar track chairs. Special thanks go to the proceedings chairs, Élisa Fromont, Stefano Ferilli and Pascal Poncelet, for the hard work of putting these proceedings together. We thank the tutorial chairs, the Discovery Challenge organizers and all the people involved in the conference, who worked hard for its success. Last but not least, we would like to sincerely thank the authors for submitting their work to the conference and the reviewers and area chairs for their tremendous effort in guaranteeing the quality of the reviewing process, thereby improving the quality of these proceedings.

July 2014

Toon Calders
Floriana Esposito
Eyke Hüllermeier
Rosa Meo

Organization

ECML/PKDD 2014 Organization

Conference Co-chairs

Amedeo Napoli   Inria Nancy Grand Est/LORIA, France
Chedy Raïssi   Inria Nancy Grand Est/LORIA, France

Program Co-chairs

Toon Calders   Université Libre de Bruxelles, Belgium
Floriana Esposito   University of Bari, Italy
Eyke Hüllermeier   University of Paderborn, Germany
Rosa Meo   University of Turin, Italy

Local Organization Co-chairs

Anne-Lise Charbonnier   Inria Nancy Grand Est, France
Louisa Touioùi   Inria Nancy Grand Est, France

Awards Committee Chairs

Johannes Fürnkranz   Technical University of Darmstadt, Germany
Katharina Morik   University of Dortmund, Germany

Workshop Chairs

Bettina Berendt   KU Leuven, Belgium
Patrick Gallinari   LIP6 Paris, France

Tutorial Chairs

Céline Rouveirol   University of Paris-Nord, France
Céline Robardet   University of Lyon, France

Demonstration Chairs

Ricard Gavaldà   UPC Barcelona, Spain
Myra Spiliopoulou   University of Magdeburg, Germany

Publicity Chairs

Stefano Ferilli   University of Bari, Italy
Pauli Miettinen   Max-Planck-Institut, Germany

Panel Chairs

Jose Balcazar   UPC Barcelona, Spain
Sergei O. Kuznetsov   HSE Moscow, Russia

Industrial Chairs

Michael Berthold   University of Konstanz, Germany
Marc Boullé   Orange Labs, France

PhD Chairs

Bruno Crémilleux   University of Caen, France
Radim Belohlavek   University of Olomouc, Czech Republic

Nectar Track Chairs

Evimaria Terzi   Boston University, USA
Pierre Geurts   University of Liège, Belgium

Sponsorship Chairs

Francesco Bonchi   Yahoo ! Research Barcelona, Spain
Jilles Vreeken   Saarland University/Max-Planck-Institut,
                                   Germany

Proceedings Chairs

Pascal Poncelet   University of Montpellier, France
Élisa Fromont   University of Saint Etienne, France
Stefano Ferilli   University of Bari, Italy

EMCL PKDD Steering Committee

Fosca Giannotti   University of Pisa, Italy
Michèle Sebag   Université Paris Sud, France
Francesco Bonchi   Yahoo! Research Barcelona, Spain
Hendrik Blockeel   KU Leuven, Belgium and Leiden University,
                                   The Netherlands

Katharina Morik   University of Dortmund, Germany
Tobias Scheffer   University of Potsdam, Germany
Arno Siebes   Utrecht University, The Netherlands
Dimitrios Gunopulos   University of Athens, Greece
Michalis Vazirgiannis   École Polytechnique, France
Donato Malerba   University of Bari, Italy
Peter Flach   University of Bristol, UK
Tijl De Bie   University of Bristol, UK
Nello Cristianini   University of Bristol, UK
Filip Železný   Czech Technical University in Prague,
    Czech Republic
Siegfried Nijssen   LIACS, Leiden University, The Netherlands
Kristian Kersting   Technical University of Dortmund, Germany

Area Chairs

Hendrik Blockeel   KU Leuven, Belgium
Henrik Boström   Stockholm University, Sweden
Ian Davidson   University of California, Davis, USA
Luc De Raedt   KU Leuven, Belgium
Janez Demšar   University of Ljubljana, Slovenia
Alan Fern   Oregon State University, USA
Peter Flach   University of Bristol, UK
Johannes Fürnkranz   TU Darmstadt, Germany
Thomas Gärtner   University of Bonn and Fraunhofer IAIS,
    Germany
João Gama   University of Porto, Portugal
Aristides Gionis   Aalto University, Finland
Bart Goethals   University of Antwerp, Belgium
Andreas Hotho   University of Würzburg, Germany
Manfred Jaeger   Aalborg University, Denmark
Thorsten Joachims   Cornell University, USA
Kristian Kersting   Technical University of Dortmund, Germany
Stefan Kramer   University of Mainz, Germany
Donato Malerba   University of Bari, Italy
Stan Matwin   Dalhousie University, Canada
Pauli Miettinen   Max-Planck-Institut, Germany
Dunja Mladenić   Jozef Stefan Institute, Slovenia
Marie-Francine Moens   KU Leuven, Belgium
Bernhard Pfahringer   University of Waikato, New Zealand
Thomas Seidl   RWTH Aachen University, Germany
Arno Siebes   Utrecht University, The Netherlands
Myra Spiliopoulou   Magdeburg University, Germany
Jean-Philippe Vert   Mines ParisTech, France
Jilles Vreeken   Max-Planck-Institut and Saarland University,
    Germany

Marco Wiering   University of Groningen, The Netherlands
Stefan Wrobel   University of Bonn & Fraunhofer IAIS,
                               Germany

Program Committee

Foto Afrati	Wray Buntine	Wei Ding
Leman Akoglu	Robert Busa-Fekete	Ying Ding
Mehmet Sabih Aksoy	Toon Calders	Stephan Doerfel
Mohammad Al Hasan	Rui Camacho	Janardhan Rao Doppa
Omar Alonso	Longbing Cao	Chris Drummond
Aijun An	Andre Carvalho	Devdatt Dubhashi
Aris Anagnostopoulos	Francisco Casacuberta	Ines Dutra
Annalisa Appice	Michelangelo Ceci	Sašo Džeroski
Marta Arias	Loic Cerf	Tapio Elomaa
Hiroki Arimura	Tania Cerquitelli	Roberto Esposito
Ira Assent	Sharma Chakravarthy	Ines Faerber
Martin Atzmüller	Keith Chan	Hadi Fanaee-Tork
Chloe-Agathe Azencott	Duen Horng Chau	Nicola Fanizzi
Antonio Bahamonde	Sanjay Chawla	Elaine Faria
James Bailey	Keke Chen	Fabio Fassetti
Elena Baralis	Ling Chen	Hakan
Daniel Barbara'	Weiwei Cheng	Ferhatosmanoglou
Christian Bauckhage	Silvia Chiusano	Stefano Ferilli
Roberto Bayardo	Vassilis Christophides	Carlos Ferreira
Aurelien Bellet	Frans Coenen	Cèsar Ferri
Radim Belohlavek	Fabrizio Costa	Jose Fonollosa
Andras Benczur	Bruno Cremilleux	Eibe Frank
Klaus Berberich	Tom Croonenborghs	Antonino Freno
Bettina Berendt	Boris Cule	Élisa Fromont
Michele Berlingerio	Tomaz Curk	Fabio Fumarola
Indrajit Bhattacharya	James Cussens	Patrick Gallinari
Marenglen Biba	Maria Damiani	Jing Gao
Albert Bifet	Jesse Davis	Byron Gao
Enrico Blanzieri	Martine De Cock	Roman Garnett
Konstantinos Blekas	Jeroen De Knijf	Paolo Garza
Francesco Bonchi	Colin de la Higuera	Eric Gaussier
Gianluca Bontempi	Gerard de Melo	Floris Geerts
Christian Borgelt	Juan del Coz	Pierre Geurts
Marco Botta	Krzysztof Dembczyński	Rayid Ghani
Jean-François Boulicaut	François Denis	Fosca Giannotti
Marc Boullé	Anne Denton	Aris Gkoulalas-Divanis
Kendrick Boyd	Mohamed Dermouche	Vibhav Gogate
Pavel Brazdil	Christian Desrosiers	Marco Gori
Ulf Brefeld	Luigi Di Caro	Michael Granitzer
Björn Bringmann	Jana Diesner	Oded Green

Tias Guns
Maria Halkidi
Jiawei Han
Daniel Hernandez
  Lobato
José Hernández-Orallo
Frank Hoeppner
Jaakko Hollmén
Geoff Holmes
Arjen Hommersom
Vasant Honavar
Xiaohua Hu
Minlie Huang
Eyke Hüllermeier
Dino Ienco
Robert Jäschke
Frederik Janssen
Nathalie Japkowicz
Szymon Jaroszewicz
Ulf Johansson
Alipio Jorge
Kshitij Judah
Tobias Jung
Hachem Kadri
Theodore Kalamboukis
Alexandros Kalousis
Pallika Kanani
U Kang
Panagiotis Karras
Andreas Karwath
Hisashi Kashima
Ioannis Katakis
John Keane
Latifur Khan
Levente Kocsis
Yun Sing Koh
Alek Kolcz
Igor Kononenko
Irena Koprinska
Nitish Korula
Petr Kosina
Walter Kosters
Georg Krempl
Konstantin Kutzkov
Sergei Kuznetsov

Nicolas Lachiche
Pedro Larranaga
Silvio Lattanzi
Niklas Lavesson
Nada Lavrač
Gregor Leban
Sangkyun Lee
Wang Lee
Carson Leung
Jiuyong Li
Lei Li
Tao Li
Rui Li
Ping Li
Juanzi Li
Lei Li
Edo Liberty
Jefrey Lijffijt
shou-de Lin
Jessica Lin
Hsuan-Tien Lin
Francesca Lisi
Yan Liu
Huan Liu
Corrado Loglisci
Eneldo Loza Mencia
Chang-Tien Lu
Panagis Magdalinos
Giuseppe Manco
Yannis Manolopoulos
Enrique Martinez
Dimitrios Mavroeidis
Mike Mayo
Wannes Meert
Gabor Melli
Ernestina Menasalvas
Roser Morante
João Moreira
Emmanuel Müller
Mohamed Nadif
Mirco Nanni
Alex Nanopoulos
Balakrishnan
  Narayanaswamy
Sriraam Natarajan

Benjamin Nguyen
Thomas Niebler
Thomas Nielsen
Siegfried Nijssen
Xia Ning
Richard Nock
Niklas Noren
Kjetil Nørvåg
Eirini Ntoutsi
Andreas Nürnberger
Salvatore Orlando
Gerhard Paass
George Paliouras
Spiros Papadimitriou
Apostolos Papadopoulos
Panagiotis Papapetrou
Stelios Paparizos
Ioannis Partalas
Andrea Passerini
Vladimir Pavlovic
Mykola Pechenizkiy
Dino Pedreschi
Nikos Pelekis
Jing Peng
Ruggero Pensa
Fabio Pinelli
Marc Plantevit
Pascal Poncelet
George Potamias
Aditya Prakash
Doina Precup
Kai Puolamaki
Buyue Qian
Chedy Raïssi
Liva Ralaivola
Karthik Raman
Jan Ramon
Huzefa Rangwala
Zbigniew Raś
Chotirat
  Ratanamahatana
Jan Rauch
Soumya Ray
Steffen Rendle
Achim Rettinger

Fabrizio Riguzzi
Céline Robardet
Marko Robnik Sikonja
Pedro Rodrigues
Juan Rodriguez
Irene Rodriguez-Lujan
Fabrice Rossi
Juho Rousu
Céline Rouveirol
Stefan Rüping
Salvatore Ruggieri
Yvan Saeys
Alan Said
Lorenza Saitta
Ansaf Salleb-Aouissi
Scott Sanner
Vítor Santos Costa
Raul Santos-Rodriguez
Sam Sarjant
Claudio Sartori
Taisuke Sato
Lars Schmidt-Thieme
Christoph Schommer
Matthias Schubert
Giovanni Semeraro
Junming Shao
Junming Shao
Pannaga Shivaswamy
Andrzej Skowron
Kevin Small
Padhraic Smyth
Carlos Soares
Yangqiu Song
Mauro Sozio
Alessandro Sperduti
Eirini Spyropoulou
Jerzy Stefanowski
Jean Steyaert
Daniela Stojanova
Markus Strohmaier
Mahito Sugiyama

Johan Suykens
Einoshin Suzuki
Panagiotis Symeonidis
Sandor Szedmak
Andrea Tagarelli
Domenico Talia
Pang Tan
Letizia Tanca
Dacheng Tao
Nikolaj Tatti
Maguelonne Teisseire
Evimaria Terzi
Martin Theobald
Jilei Tian
Ljupco Todorovski
Luis Torgo
Vicenç Torra
Ivor Tsang
Panagiotis Tsaparas
Vincent Tseng
Grigorios Tsoumakas
Theodoros Tzouramanis
Antti Ukkonen
Takeaki Uno
Athina Vakali
Giorgio Valentini
Guy Van den Broeck
Peter van der Putten
Matthijs van Leeuwen
Maarten van Someren
Joaquin Vanschoren
Iraklis Varlamis
Michalis Vazirgiannis
Julien Velcin
Shankar Vembu
Sicco Verwer
Vassilios Verykios
Herna Viktor
Christel Vrain
Willem Waegeman
Byron Wallace

Fei Wang
Jianyong Wang
Xiang Wang
Yang Wang
Takashi Washio
Geoff Webb
Jörg Wicker
Hui Xiong
Jieping Ye
Jeffrey Yu
Philip Yu
Chun-Nam Yu
Jure Zabkar
Bianca Zadrozny
Gerson Zaverucha
Demetris Zeinalipour
Filip Železný
Bernard Zenko
Min-Ling Zhang
Nan Zhang
Zhongfei Zhang
Junping Zhang
Lei Zhang
Changshui Zhang
Kai Zhang
Kun Zhang
Shichao Zhang
Ying Zhao
Elena Zheleva
Zhi-Hua Zhou
Bin Zhou
Xingquan Zhu
Xiaofeng Zhu
Kenny Zhu
Djamel Zighed
Arthur Zimek
Albrecht Zimmermann
Indre Zliobaite
Blaz Zupan

Demo Track Program Committee

Martin Atzmueller	Jaakko Hollmén	Mykola Pechenizkiy
Bettina Berendt	Andreas Hotho	Bernhard Pfahringer
Albert Bifet	Mark Last	Pedro Rodrigues
Antoine Bordes	Vincent Lemaire	Jerzy Stefanowski
Christian Borgelt	Ernestina Menasalvas	Grigorios Tsoumakas
Ulf Brefeld	Kjetil Nørvåg	Alice Zheng
Blaz Fortuna	Themis Palpanas	

Nectar Track Program Committee

Donato Malerba	George Karypis	Rosa Meo
Dora Erdos	Louis Wehenkel	Myra Spiliopoulou
Yiannis Koutis	Leman Akoglu	Toon Calders

Additional Reviewers

Argimiro Arratia	Elad Liebman	Kiumars Soltani
Rossella Cancelliere	Babak Loni	Ricardo Sousa
Antonio Corral	Emmanouil Magkos	Eleftherios
Joana Côrte-Real	Adolfo Martínez-Usó	Spyromitros-Xioufis
Giso Dal	Dimitrios Mavroeidis	Jiliang Tang
Giacomo Domeniconi	Steffen Michels	Eleftherios Tiakas
Roberto Esposito	Pasquale Minervini	Andrei Tolstikov email
Pedro Ferreira	Fatemeh Mirrashed	Tiago Vinhoza
Asmelash Teka Hadgu	Fred Morstatter	Xing Wang
Isaac Jones	Tsuyoshi Murata	Lorenz Weizsäcker
Dimitris Kalles	Jinseok Nam	Sean Wilner
Yoshitaka Kameya	Rasaq Otunba	Christian Wirth
Eamonn Keogh	Roberto Pasolini	Lin Wu
Kristian Kersting	Tommaso Pirini	Jinfeng Yi
Rohan Khade	Maria-Jose	Cangzhou Yuan
Shamanth Kumar	Ramirez-Quintana	Jing Zhang
Hongfei Li	Irma Ravkic	

Sponsors

Gold Sponsor
Winton   http://www.wintoncapital.com

Silver Sponsors

Deloitte	http://www.deloitte.com
Xerox Research Centre Europe	http://www.xrce.xerox.com

Bronze Sponsors

EDF	http://www.edf.com
Orange	http://www.orange.com
Technicolor	http://www.technicolor.com
Yahoo! Labs	http://labs.yahoo.com

Additional Supporters

Harmonic Pharma	http://www.harmonicpharma.com
Deloitte	http://www.deloitte.com

Lanyard

Knime	http://www.knime.org

Prize

Deloitte	http://www.deloitte.com
Data Mining and Knowledge Discovery	http://link.springer.com/journal/10618
Machine Learning	http://link.springer.com/journal/10994

Organizing Institutions

Inria	http://www.inria.fr
CNRS	http://www.cnrs.fr
LORIA	http://www.loria.fr

Invited Talks Abstracts

Scalable Collective Reasoning Using Probabilistic Soft Logic

Lise Getoor

University of California, Santa Cruz
Santa Cruz, CA, USA
getoor@cs.umd.edu

Abstract. One of the challenges in big data analytics is to efficiently learn and reason collectively about extremely large, heterogeneous, incomplete, noisy interlinked data. Collective reasoning requires the ability to exploit both the logical and relational structure in the data and the probabilistic dependencies. In this talk I will overview our recent work on probabilistic soft logic (PSL), a framework for collective, probabilistic reasoning in relational domains. PSL is able to reason holistically about both entity attributes and relationships among the entities. The underlying mathematical framework, which we refer to as a hinge-loss Markov random field, supports extremely efficient, exact inference. This family of graphical models captures logic-like dependencies with convex hinge-loss potentials. I will survey applications of PSL to diverse problems ranging from information extraction to computational social science. Our recent results show that by building on state-of-the-art optimization methods in a distributed implementation, we can solve large-scale problems with millions of random variables orders of magnitude faster than existing approaches.

Bio. In 1995, Lise Getoor decided to return to school to get her PhD in Computer Science at Stanford University. She received a National Physical Sciences Consortium fellowship, which in addition to supporting her for six years, supported a summer internship at Xerox PARC, where she worked with Markus Fromherz and his group. Daphne Koller was her PhD advisor; in addition, she worked closely with Nir Friedman, and many other members of the DAGS group, including Avi Pfeffer, Mehran Sahami, Ben Taskar, Carlos Guestrin, Uri Lerner, Ron Parr, Eran Segal, Simon Tong.

In 2001, Lise Getoor joined the Computer Science Department at the University of Maryland, College Park.

Network Analysis in the Big Data Age: Mining Graph and Social Streams

Charu Aggarwal

IBM T.J. Watson Research Center, New York
Yorktown, NY, USA
charu@us.ibm.com

Abstract. The advent of large interaction-based communication and social networks has led to challenging streaming scenarios in graph and social stream analysis. The graphs that result from such interactions are large, transient, and very often cannot even be stored on disk. In such cases, even simple frequency-based aggregation operations become challenging, whereas traditional mining operations are far more complex. When the graph cannot be explicitly stored on disk, mining algorithms must work with a limited knowledge of the network structure. Social streams add yet another layer of complexity, wherein the streaming content associated with the nodes and edges needs to be incorporated into the mining process. A significant gap exists between the problems that need to be solved, and the techniques that are available for streaming graph analysis. In spite of these challenges, recent years have seen some advances in which carefully chosen synopses of the graph and social streams are leveraged for approximate analysis. This talk will focus on several recent advances in this direction.

Bio. Charu Aggarwal is a Research Scientist at the IBM T. J. Watson Research Center in Yorktown Heights, New York. He completed his B.S. from IIT Kanpur in 1993 and his Ph.D. from Massachusetts Institute of Technology in 1996. His research interest during his Ph.D. years was in combinatorial optimization (network flow algorithms), and his thesis advisor was Professor James B. Orlin. He has since worked in the field of data mining, with particular interests in data streams, privacy, uncertain data and social network analysis. He has published over 200 papers in refereed venues, and has applied for or been granted over 80 patents. Because of the commercial value of the above-mentioned patents, he has received several invention achievement awards and has thrice been designated a Master Inventor at IBM. He is a recipient of an IBM Corporate Award (2003) for his work on bio-terrorist threat detection in data streams, a recipient of the IBM Outstanding Innovation Award (2008) for his scientific contributions to privacy technology, and a recipient of an IBM Research Division Award (2008) for his scientific contributions to data stream research. He has served on the program committees of most major database/data mining conferences, and served as program vice-chairs of the SIAM Conference on Data Mining, 2007, the IEEE ICDM Conference, 2007, the WWW Conference 2009, and the IEEE ICDM Conference, 2009. He served as an associate editor of the IEEE Transactions on Knowledge and Data Engineering Journal from 2004 to 2008. He is an associate editor of the ACM TKDD Journal, an action editor of the Data Mining and Knowledge Discovery Journal, an associate editor of the ACM SIGKDD Explorations, and an associate editor of the Knowledge and Information Systems Journal. He is a fellow of the ACM (2013) and the IEEE (2010) for contributions to knowledge discovery and data mining techniques.

Big Data for Personalized Medicine: A Case Study of Biomarker Discovery

Raymond Ng

University of British Columbia
Vancouver, B.C., Canada
mg@cs.ubc.ca

Abstract. Personalized medicine has been hailed as one of the main frontiers for medical research in this century. In the first half of the talk, we will give an overview on our projects that use gene expression, proteomics, DNA and clinical features for biomarker discovery. In the second half of the talk, we will describe some of the challenges involved in biomarker discovery. One of the challenges is the lack of quality assessment tools for data generated by ever-evolving genomics platforms. We will conclude the talk by giving an overview of some of the techniques we have developed on data cleansing and pre-processing.

Bio. Dr. Raymond Ng is a professor in Computer Science at the University of British Columbia. His main research area for the past two decades is on data mining, with a specific focus on health informatics and text mining. He has published over 180 peer-reviewed publications on data clustering, outlier detection, OLAP processing, health informatics and text mining. He is the recipient of two best paper awards from 2001 ACM SIGKDD conference, which is the premier data mining conference worldwide, and the 2005 ACM SIGMOD conference, which is one of the top database conferences worldwide. He was one of the program co-chairs of the 2009 International conference on Data Engineering, and one of the program co-chairs of the 2002 ACM SIGKDD conference. He was also one of the general co-chairs of the 2008 ACM SIGMOD conference. For the past decade, Dr. Ng has co-led several large scale genomic projects, funded by Genome Canada, Genome BC and industrial collaborators. The total amount of funding of those projects well exceeded $40 million Canadian dollars. He now holds the Chief Informatics Officer position of the PROOF Centre of Excellence, which focuses on biomarker development for end-stage organ failures.

Machine Learning for Search Ranking and Ad Auction

Tie-Yan Liu

Microsoft Research Asia
Beijing, P.R. China
tyliu@microsoft.com

Abstract. In the era of information explosion, search has become an important tool for people to retrieve useful information. Every day, billions of search queries are submitted to commercial search engines. In response to a query, search engines return a list of relevant documents according to a ranking model. In addition, they also return some ads to users, and extract revenue by running an auction among advertisers if users click on these ads. This "search + ads" paradigm has become a key business model in today's Internet industry, and has incubated a few hundred-billion-dollar companies. Recently, machine learning has been widely adopted in search and advertising, mainly due to the availability of huge amount of interaction data between users, advertisers, and search engines. In this talk, we discuss how to use machine learning to build effective ranking models (which we call learning to rank) and to optimize auction mechanisms. (i) The difficulty of learning to rank lies in the interdependency between documents in the ranked list. To tackle it, we propose the so-called listwise ranking algorithms, whose loss functions are defined on the permutations of documents, instead of individual documents or document pairs. We prove the effectiveness of these algorithms by analyzing their generalization ability and statistical consistency, based on the assumption of a two-layer probabilistic sampling procedure for queries and documents, and the characterization of the relationship between their loss functions and the evaluation measures used by search engines (e.g., NDCG and MAP). (ii) The difficulty of learning the optimal auction mechanism lies in that advertisers' behavior data are strategically generated in response to the auction mechanism, but not randomly sampled in an i.i.d. manner. To tackle this challenge, we propose a game-theoretic learning method, which first models the strategic behaviors of advertisers, and then optimizes the auction mechanism by assuming the advertisers to respond to new auction mechanisms according to the learned behavior model. We prove the effectiveness of the proposed method by analyzing the generalization bounds for both behavior learning and auction mechanism learning based on a novel Markov framework.

Bio. Tie-Yan Liu is a senior researcher and research manager at Microsoft Research. His research interests include machine learning (learning to rank, online learning, statistical learning theory, and deep learning), algorithmic game theory, and computational economics. He is well known for his work on learning to rank for information retrieval. He has authored the first book in this area, and published tens of highly-cited papers on both algorithms and theorems of learning to rank. He has also published extensively on other related topics. In particular, his paper won the best student paper award of SIGIR (2008), and the most cited paper award of the Journal of Visual Communication and Image Representation (2004-2006); his group won the research break-through award of Microsoft Research Asia (2012). Tie-Yan is very active in serving the research community. He is a program committee co-chair of ACML (2015), WINE (2014), AIRS (2013), and RIAO (2010), a local co-chair of ICML 2014, a tutorial co-chair of WWW 2014, a demo/exhibit co-chair of KDD (2012), and an area/track chair of many conferences including ACML (2014), SIGIR (2008-2011), AIRS (2009-2011), and WWW (2011). He is an associate editor of ACM Transactions on Information System (TOIS), an editorial board member of Information Retrieval Journal and Foundations and Trends in Information Retrieval. He has given keynote speeches at CCML (2013), CCIR (2011), and PCM (2010), and tutorials at SIGIR (2008, 2010, 2012), WWW (2008, 2009, 2011), and KDD (2012). He is a senior member of the IEEE and the ACM.

Beyond Stochastic Gradient Descent for Large-Scale Machine Learning

Francis Bach

INRIA, Paris
Laboratoire d'Informatique de l'Ecole Normale Superieure
Paris, France
francis.bach@inria.fr

Abstract. Many machine learning and signal processing problems are traditionally cast as convex optimization problems. A common difficulty in solving these problems is the size of the data, where there are many observations ("large n") and each of these is large ("large p"). In this setting, online algorithms such as stochastic gradient descent which pass over the data only once, are usually preferred over batch algorithms, which require multiple passes over the data. In this talk, I will show how the smoothness of loss functions may be used to design novel algorithms with improved behavior, both in theory and practice: in the ideal infinite-data setting, an efficient novel Newton-based stochastic approximation algorithm leads to a convergence rate of $O(1/n)$ without strong convexity assumptions, while in the practical finite-data setting, an appropriate combination of batch and online algorithms leads to unexpected behaviors, such as a linear convergence rate for strongly convex problems, with an iteration cost similar to stochastic gradient descent.
(joint work with Nicolas Le Roux, Eric Moulines and Mark Schmidt)

Bio. Francis Bach is a researcher at INRIA, leading since 2011 the SIERRA project-team, which is part of the Computer Science Laboratory at Ecole Normale Superieure. He completed his Ph.D. in Computer Science at U.C. Berkeley, working with Professor Michael Jordan, and spent two years in the Mathematical Morphology group at Ecole des Mines de Paris, then he joined the WILLOW project-team at INRIA/Ecole Normale Superieure from 2007 to 2010. Francis Bach is interested in statistical machine learning, and especially in graphical models, sparse methods, kernel-based learning, convex optimization vision and signal processing.

Industrial Invited Talks Abstracts

Making Smart Metering Smarter
by Applying Data Analytics

Georges Hébrail

EDF Lab
CLAMART, France
georges.hebrail@edf.fr

Abstract. New data is being collected from electric smart meters which are deployed in many countries. Electric power meters measure and transmit to a central information system electric power consumption from every individual household or enterprise. The sampling rate may vary from 10 minutes to 24 hours and the latency to reach the central information system may vary from a few minutes to 24h. This generates a large amount of - possibly streaming - data if we consider customers from an entire country (ex. 35 millions in France). This data is collected firstly for billing purposes but can be processed with data analytics tools with several other goals. The first part of the talk will recall the structure of electric power smart metering data and review the different applications which are considered today for applying data analytics to such data. In a second part of the talk, we will focus on a specific problem: spatio-temporal estimation of aggregated electric power consumption from incomplete metering data.

Bio. Georges Hébrail is a senior researcher at EDF Lab, the research centre of Electricité de France, one of the world's leading electric utility. His background is in Business Intelligence covering many aspects from data storage and querying to data analytics. From 2002 to 2010, he was a professor of computer science at Telecom ParisTech, teaching and doing research in the field of information systems and business intelligence, with a focus on time series management, stream processing and mining. His current research interest is on distributed and privacy-preserving data mining on electric power related data.

Ads That Matter

Alexandre Cotarmanac'h

VP Platform & Distribution
Twenga
alexandre.cotarmanach@twenga.com

Abstract. The advent of realtime bidding and online ad-exchanges has created a new and fast-growing competitive marketplace. In this new setting, media-buyers can make fine-grained decisions for each of the impressions being auctioned taking into account information from the context, the user and his/her past behavior. This new landscape is particularly interesting for online e-commerce players where user actions can also be measured online and thus allow for a complete measure of return on ad-spend.

Despite those benefits, new challenges need to be addressed such as:

- the design of a real-time bidding architecture handling high volumes of queries at low latencies,
- the exploration of a sparse and volatile high-dimensional space,
- as well as several statistical modeling problems (e.g. pricing, offer and creative selection).

In this talk, I will present an approach to realtime media buying for online e-commerce from our experience working in the field. I will review the aforementioned challenges and discuss open problems for serving ads that matter.

Bio. Alexandre Cotarmanac'h is Vice-President Distribution & Platform for Twenga.

Twenga is a services and solutions provider generating high value-added leads to online merchants that was founded in 2006.

Originally hired to help launch Twenga's second generation search engine and to manage the optimization of revenue, he launched in 2011 the affinitAD line of business and Twenga's publisher network. Thanks to the advanced contextual analysis which allows for targeting the right audience according to their desire to buy e-commerce goods whilst keeping in line with the content offered, affinitAD brings Twenga's e-commerce expertise to web publishers. Alexandre also oversees Twenga's merchant programme and strives to offer Twenga's merchants new services and solutions to improve their acquisition of customers.

With over 14 years of experience, Alexandre has held a succession of increasingly responsible positions focusing on advertising and web development. Prior to joining Twenga, he was responsible for the development of Search and Advertising at Orange. Alexandre graduated from Ecole polytechnique.

Machine Learning and Data Mining in Call of Duty

Arthur Von Eschen

Activision Publishing Inc.
Santa Monica, CA, USA
Arthur.VonEschen@activision.com

Abstract. Data science is relatively new to the video game industry, but it has quickly emerged as one of the main resources for ensuring game quality. At Activision, we leverage data science to analyze the behavior of our games and our players to improve in-game algorithms and the player experience. We use machine learning and data mining techniques to influence creative decisions and help inform the game design process. We also build analytic services that support the game in real-time; one example is a cheating detection system which is very similar to fraud detection systems used for credit cards and insurance. This talk will focus on our data science work for Call of Duty, one of the bestselling video games in the world.

Bio. Arthur Von Eschen is Senior Director of Game Analytics at Activision. He and his team are responsible for analytics work that supports video game design on franchises such as Call of Duty and Skylanders. In addition to holding a PhD in Operations Research, Arthur has over 15 years of experience in analytics consulting and R&D with the U.S. Fortune 500. His work has spanned across industries such as banking, financial services, insurance, retail, CPG and now interactive entertainment (video games). Prior to Activision he worked at Fair Isaac Corporation (FICO). Before FICO he ran his own analytics consulting firm for six years.

Algorithms, Evolution and Network-Based Approaches in Molecular Discovery

Mike Bodkin

Evotec Ltd.
Oxfordshire, UK
Mike.Bodkin@evotec.com

Abstract. Drug research generates huge quantities of data around targets, compounds and their effects. Network modelling can be used to describe such relationships with the aim to couple our understanding of disease networks with the changes in small molecule properties. This talk will build off of the data that is routinely captured in drug discovery and describe the methods and tools that we have developed for compound design using predictive modelling, evolutionary algorithms and network-based mining.

Bio. Mike did his PhD in protein de-novo design for Nobel laureate sir James Black before taking up a fellowship in computational drug design at Cambridge University. He moved to AstraZeneca as a computational chemist before joining Eli Lilly in 2000. As head of the computational drug discovery group at Lilly since 2003 he recently jumped ship to Evotec to work as the VP for computational chemistry and cheminformatics. His research aims are to continue to develop new algorithms and software in the fields of drug discovery and systems informatics and to deliver and apply current and novel methods as tools for use in drug research.

Table of Contents – Part II

Main Track Contributions

Robust Distributed Training of Linear Classifiers Based on Divergence Minimization Principle

Junpei Komiyama, Hidekazu Oiwa, and Hiroshi Nakagawa

The University of Tokyo,
7-3-1, Hongo, Bunkyo-ku, Tokyo, 113-0033, Japan
junpei@komiyama.info
oiwa@r.dl.itc.u-tokyo.ac.jp
nakagawa@dl.itc.u-tokyo.ac.jp

Abstract. We study a distributed training of a linear classifier in which the data is separated into many shards and each worker only has access to its own shard. The goal of this distributed training is to utilize the data of all shards to obtain a well-performing linear classifier. The iterative parameter mixture (IPM) framework (Mann et al., 2009) is a state-of-the-art distributed learning framework that has a strong theoretical guarantee when the data is clean. However, contamination on shards, which sometimes arises in real world environments, largely deteriorates the performances of the distributed training. To remedy the negative effect of the contamination, we propose a divergence minimization principle for the weight determination in IPM. From this principle, we can naturally derive the Beta-IPM scheme, which leverages the power of robust estimation based on the beta divergence. A mistake/loss bound analysis indicates the advantage of our Beta-IPM in contaminated environments. Experiments with various datasets revealed that, even when 80% of the shards are contaminated, Beta-IPM can suppress the influence of the contamination.

1 Introduction

A linear classifier is one of the most fundamental concepts in the field of machine learning. Online learning algorithms [20,5,6] are able to train linear classifiers effectively. An online algorithm sequentially processes data points, and thus, it requires all data to be accessible from a single machine. While the training on a single machine is of its own importance, training in distributed environments has attracted increasing interest [1,10,16]. In such environments, data is divided up into disjoint sets of shards and each worker has access to only one shard.

Iterative Parameter Mixture (IPM) [16,17] is a state-of-the-art distributed training framework, which involves a master node and worker nodes. Advantages of IPM lies in its communication efficiency and simplicity: in each epoch, each worker trains a model in parallel on his own shard, and the master mixes the training results (Fig. 1).

IPM implicitly assumes that each shard is noiseless. However, it is not always the case: there can be some adversarially or randomly labelled data in some distributed learning scenarios. For example, web mail systems possibly involve some users who

T. Calders et al. (Eds.): ECML PKDD 2014, Part II, LNCS 8725, pp. 1–17, 2014.
© Springer-Verlag Berlin Heidelberg 2014

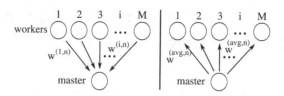

Fig. 1. Illustration of the communication in IPM

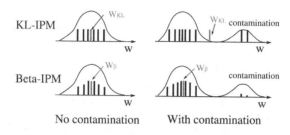

No contamination   With contamination

Fig. 2. Illustrative example of KL-IPM and Beta-IPM. Each horizontal line represents a parameter space, and each vertical line represents a parameter returned by a worker, the height of which is proportional to the mixing weight. While KL-IPM equally weights all the parameter vectors, Beta-IPM adaptively weights each parameter as described later.

adversarially labels spams and non-spams, or incorrect data formats [9] lead to corrupted classification result. Let us call such flawed data contamination. We verified that, the performance of the trained classifier is deteriorated by contamination.

Meanwhile, IPM has freedom in how to weight the individual workers' results. If some shards are known to be contaminated, we can avoid the effect of these shards by setting their weights to zero. However, it is unlikely that there will be prior knowledge about which shards are contaminated, and thus, the strategy we should take is to weight seemingly contaminated results less on the basis of their statistical anomalousness.

With this in mind, we propose a weight determination principle based on a divergence minimization criterion. This criterion reinterprets the most straightforward choice, which is to weight each worker equally, as the minimization of the Kullback-Leibler divergence (KL-IPM). On the other hand, the beta divergence, which is the extension of the KL divergence, provides robust inference against contamination. We propose the weight determination formula by minimizing the beta divergence (Beta-IPM). Moreover, We prove a mistake/loss bound of IPM. This theoretical result shows that, by weighting less heavily to contaminated shards with Beta-IPM we can suppress the upper bound of losses over training. The difference between KL-IPM and Beta-IPM is illustrated in Fig. 2. Finally, an empirical evaluation on various datasets confirms that Beta-IPM remedies the effect of contamination.

2 Related Work

2.1 Distributed Training of Linear Models

Distributed training frameworks for linear models have been studied in the literature. Asynchronous updates are sets of models in which all workers simultaneously operate on shared parameters. There is a long line of work related to asynchronous updates from the 1980s [22] onwards [24,12]. A problem of the shared memory resource it that, it does not scale with many shards because the communication cost is proportional to the number of updates.

The distributed gradient method [4] is a distributed extension of the gradient descent method, which optimizes some smooth function by taking steps proportional to the negative gradient. In the distributed gradient method, individual workers compute partial gradients based on their shards, which are then summed to make an update.

In the IPM method, each worker operates independently and shares parameters after each worker finishes an iteration. A master mixes the parameters of the workers with weights whose sum is normalized. IPM is used in many linear and regression models, such as logistic regression [16], structured perceptron [17], etc. McDonald et al. [17] proved that IPM with perceptron has mistake bound in a linearly separable case (i.e., the case in which every data point is correctly classified by some classifier). Moreover, Hall et al. [14] empirically compared the asynchronous updates, the distributed gradient method and IPM using large-scale click-through data. The results show that IPM combined with the perceptron [20] performed the best.

2.2 Robust Training Against Flawed Data

Detection of spam and malicious activities is an important problem in the highly distributed web industry [18,7]. In addition, poorly formatted data [19,9] causes a considerable problem in distributed training. Despite significant efforts made at removing such flawed data, there still is a need for robust models. Robust models are used in many fields, including multi-task learning [13,23] and sensor networks [3].

The problem of contamination in distributed data can be broken down into two cases: the first case is when the contamination is scattered across every shard, and the second case is when some shards are clean while others are contaminated. Studies on the robustness of online learning algorithms [5,6,15] have mainly dealt with the first case, which considers a single data repository affected by noise. In this case, the clean data are hard to distinguish from the noise. Instead, we consider the second case and show that a significant improvement is possible by putting less importance on statistically extraordinary shards which are likely to be wrongly labeled or corrupted.

We note that, Daumé III et al. [8] proposed a distributed learning algorithm with adversarially distributed data. Their definition of adversarially distributed data is different from our adversarial noise: while they considered separable data with an adversary who can generate an arbitrary imbalance among shards, we consider an adversarial attacker that can harm the separability assumption by maliciously labelling.

Algorithm 1. Iterative Parameter Mixture (IPM)

1: Shards: $\mathcal{T}_1, ..., \mathcal{T}_M$, $\mathbf{w}^{(\mathrm{avg},0)} = \mathbf{0}$
2: **for** $n = 1, ..., N$ **do**
3:    $\mathbf{w}^{(i,n)} = \mathrm{SingleIterationTrain}(\mathcal{T}_i, \mathbf{w}^{(\mathrm{avg},n-1)})$
4:    $\mathbf{w}^{(\mathrm{avg},n)} = \sum_i \alpha_{i,n} \mathbf{w}^{(i,n)}$
5: **end for**

---

3 Problem Setup

We consider a binary classification. Let $\mathcal{X} \in \mathbb{R}^d$ be the input space and $\mathcal{Y} = \{-1, 1\}$ be the output space. A data point is defined as an input-output tuple $(\mathbf{x}, y) \in \mathcal{X} \times \mathcal{Y}$. A linear classifier with parameter vector $\mathbf{w}$ predicts an output as $\hat{y} = \mathrm{sign}(\mathbf{w} \cdot \mathbf{x})$. The goal of our distributed training framework is to find the parameter vector $\mathbf{w}$ which explains the whole data most.

In distributed training, the training data is divided into M non-overlapping subsets (shards), and a shard is assigned to each worker. There also is a master who integrates the results of workers. Training based on IPM (Algorithm 1) goes as follows. In each epoch $n = 1, 2, ..., N$, each worker i independently does a single iteration of training its own parameters $\mathbf{w}^{(i,n)}$, which are then sent to the master. The master waits until all workers finish their training before it computes a mixed parameter $\mathbf{w}^{(\mathrm{avg},n)}$, which is a weighted sum of the trainers' parameters. The weight $\alpha_{i,n}$ of each worker i in each epoch n can be chosen arbitrarily as long as it is normalized (i.e. $\sum_{i=1}^{M} \alpha_{i,n} = 1$). Later in Section 4 we propose weight determination formulas that we call KL-IPM and Beta-IPM. At the end of the epoch, the mixed parameters $\mathbf{w}^{(\mathrm{avg},n)}$ are sent back to the workers, who in the next epoch start the new single iteration training based on the received mixed parameter. After N epochs have been completed, the master outputs the final parameter vector (a linear classifier).

3.1 IPM Combined with Online Algorithms

We use online learning algorithms in single iteration training ("SingleIterationTrain" in Algorithm 1). We specifically deal with the perceptron [20] and the Passive Aggressive (PA) method [5]. Section 5 describes that, IPM combined with perceptron and PA is able to extend the theoretical guarantee of these single-machine online algorithms.

4 Divergence Minimization Principle

In this section, we describe our main proposal, which is how to determine the weights based on the divergence minimization principle. Section 4.1 describes our statistical assumptions and the divergence minimization principle. Section 4.2 describes the KL and beta divergences, and Section 4.3 shows KL-IPM and Beta-IPM formulas. Section 4.4 demonstrates the behavior of KL-IPM and Beta-IPM with a simple example.

4.1   Statistical Assumption And Divergence Minimization Principle

The statistical assumption is as follows: the parameters returned by the workers in each epoch n should be drawn from a Gaussian distribution Q_n. Our proposal is that, in each epoch n the mixed parameter vector $\mathbf{w}^{(\text{avg},n)}$ is determined to be the mean of the Gaussian Q_n. However, the parameters are actually drawn from P_n, which possibly contains contamination. Namely, $\mathbf{w}^{(i,n)} \sim P_n$. The mean $\boldsymbol{\mu}$ and covariance $\boldsymbol{\Sigma}$ of Q_n are determined in such a way as to minimize the divergence:

$$\arg\min_{\boldsymbol{\mu},\boldsymbol{\Sigma}} D(P_n \| Q_n(\boldsymbol{\mu}, \boldsymbol{\Sigma})), \tag{1}$$

where D is the divergence between P_n and Q_n. If we use a robust divergence that suppresses the influence of contamination, we are able to estimate the true $\boldsymbol{\mu}$ and $\boldsymbol{\Sigma}$.

4.2   KL and Beta Divergences

The KL divergence is the most basic measure that indicates the deviation of a distribution from another distribution. The KL divergence between two probability distributions P and Q on $\mathbb{R}^d$ is defined as

$$D_{KL}(P\|Q) = \int P(\mathbf{w}) \log \frac{P(\mathbf{w})}{Q(\mathbf{w})} d\mathbf{w}, \tag{2}$$

which is non-negative and equal to zero if and only if $P = Q$ almost everywhere. While the KL divergence is of fundamental importance in information theory, it is not robust to the contamination of outliers.

The beta divergence, which was introduced by [2] and [11], is parameterized by a real parameter $\beta > 0$. The beta divergence between P and Q is defined as

$$D_\beta(P\|Q) = \int \left\{ P(\mathbf{w}) \frac{P^\beta(\mathbf{w}) - Q^\beta(\mathbf{w})}{\beta} \right\} - \frac{P^{\beta+1}(\mathbf{w}) - Q^{\beta+1}(\mathbf{w})}{\beta+1} d\mathbf{w}. \tag{3}$$

When $\beta \to 0$, the beta divergence is consistently defined as $\lim_{\beta \to 0} D_\beta(P\|Q) = D_{KL}(P\|Q)$. Therefore, the beta divergence can be considered as an extension of the KL divergence. One of the main motivations of investigating the beta divergence is to devise a robust inference against contamination. That is, the beta divergence between two distributions P and Q remains undisturbed by some fraction of the contamination in P. β is a trade-off parameter. The bigger β is, the more robust and less computationally effective the divergence becomes.

4.3   KL-IPM and Beta-IPM

KL-IPM: KL-IPM is a weight determination formula in IPM that equally weights each worker. Namely,

$$\alpha_{i,n} = \frac{1}{M}. \tag{4}$$

However, if flawed data contaminate some of the shards, the performance of KL-IPM deteriorates. To remedy this problem, we derive Beta-IPM that minimizes the beta divergence.

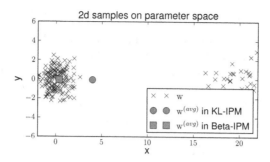

Fig. 3. Illustration of a two-dimensional example. Each of the 100 crosses represents the parameters $\mathbf{w} \sim P$. The 80 crosses are from the true distribution (Gaussian with $\boldsymbol{\mu} = (0,0)^\top$ and $\boldsymbol{\Sigma} = \mathrm{diag}(1,1)$). The other 20 crosses are contamination and generated from a false distribution (Gaussian with $\boldsymbol{\mu} = (20,0)^\top$ and $\boldsymbol{\Sigma} = \mathrm{diag}(2,2)$). The large red circle is the simple mean of all parameters, determined by KL-IPM (Equation (4)). The large blue square is the mixed parameter determined by Beta-IPM with $\beta = 0.1$ (Equation (6)).

Beta-IPM: Let $\boldsymbol{\mu}_c$ and $\boldsymbol{\Sigma}_c$ are respectively the empirical mean and covariance of the parameter vectors $\{\mathbf{w}^{(i,n)}\}$ defined as

$$\boldsymbol{\mu}_c = \frac{1}{M}\sum_{i=1}^{M}\mathbf{w}^{(i,n)}, \text{ and } \boldsymbol{\Sigma}_c = \frac{1}{M}\sum_{i=1}^{M}(\mathbf{w}^{(i,n)} - \boldsymbol{\mu}_c)(\mathbf{w}^{(i,n)} - \boldsymbol{\mu}_c)^\top. \quad (5)$$

Beta-IPM is defined as a weight determination formula in IPM that in each epoch n chooses weight $\alpha_{i,n}$ as follows:

$$\alpha_{i,n} = \frac{\exp S(\mathbf{w}^{(i,n)}|\boldsymbol{\mu}_c, \frac{1}{\beta}\boldsymbol{\Sigma}_c)}{\sum_{j=1}^{M}\exp S(\mathbf{w}^{(j,n)}|\boldsymbol{\mu}_c, \frac{1}{\beta}\boldsymbol{\Sigma}_c)}, \quad (6)$$

where $S(\mathbf{w}^{(i,n)}|\boldsymbol{\mu}, \boldsymbol{\Sigma}) = -(1/2)(\mathbf{w}^{(i,n)} - \boldsymbol{\mu})^\top \Sigma^{-1}(\mathbf{w}^{(i,n)} - \boldsymbol{\mu})$ is the exponent part of the Gaussian. Namely, each weight of a shard is determined by the distance of the parameter vector from the mean. Beta-IPM is parameterized by $\beta \geq 0$ and is equivalent to KL-IPM when $\beta \to 0$ because the covariance $(1/\beta)\boldsymbol{\Sigma}_c$ in (6) becomes infinitely large. The KL-IPM and Beta-IPM formulas above are derived in Appendix A.1. Note that the problem of minimizing the beta divergence is non-convex, so we have made some approximations in order to derive (6).

4.4   Example of KL-IPM and Beta-IPM

Fig. 3 is a two-dimensional example that displays the behaviors of KL-IPM and Beta-IPM. While KL-IPM equally weights each parameter vector, Beta-IPM weights the vector farther from the mean less, and in this way it suppresses the influence of contamination. As a result, the mixed parameter vector chosen by Beta-IPM is closer to the true center ($=(0,0)^\top$) than that by KL-IPM.

5 Mistake / Loss Bound in IPM

This section provides a theoretical viewpoint for the weight determination by Beta-IPM. We first discuss the separable mistake bound of IPM with a single iteration perceptron (IPM-perceptron) in Section 5.1, then goes to the corresponding loss bound of IPM with a single iteration PA (IPM-PA) in Section 5.2. With these bounds, we discuss about Beta-IPM as a suppressor of weights in contaminated shards in Section 5.3.

5.1 Mistake bound of IPM-perceptron

The following theorem, which is proven by McDonald et al. [17], is an extension of the well-known mistake bound of the single machine perceptron to IPM,

Theorem 1. (Mistake bound of IPM-perceptron in the separable case) [Theorem 3 in [17]] *Assume all the training data is separable by a margin γ. Suppose that $\|\mathbf{x}\| \leq R$ holds for any training input $\mathbf{x}$, and let $k_{i,n}$ be the number of mistakes in shard i during the nth epoch of training. For any number of epochs N, the number of mistakes during the training in the IPM-perceptron is bounded as*

$$\sum_{n=1}^{N}\sum_{i=1}^{M}\alpha_{i,n}k_{i,n} \leq \frac{R^2}{\gamma^2}. \tag{7}$$

Theorem 1 states that the IPM-perceptron with separable data has a finite number of misses, which guarantees it converges to parameters that correctly classify the entire data.

In contrast, when some fraction of the dataset is non-separable, there are no parameters that perfectly classify all the data. Yet even in this case, we can bound the mistake in terms of the loss of the possible best classifier (parameter vector) $\mathbf{u}$.

Theorem 2. (Mistake bound of IPM-perceptron in the non-separable case) *Let $k_{i,n}$ be the number of mistakes in shard i during the nth epoch of training. Furthermore, let $\mathbf{u}$ be an arbitrary normalized parameter vector $\mathbf{u} \in \mathbb{R}^n(\|\mathbf{u}\| = 1)$ Let $\xi = \max\{0, \gamma - y(\mathbf{u} \cdot \mathbf{x})\}$ and $\Xi_i = \sum_{t'} \xi$, where the index t' runs over all data points in shard i. For any number of epochs N and any $\gamma \geq 0$, the following inequality holds:*

$$\sum_{n=1}^{N}\sum_{i=1}^{M}\alpha_{i,n}k_{i,n} \leq \frac{R^2}{\gamma^2} + \frac{2}{\gamma}\sum_{n=1}^{N}\sum_{i=1}^{M}\alpha_{i,n}\Xi_i. \tag{8}$$

Theorem 2 is proven by the combination of the technique for the IPM loss bound [17] and an ordinary technique for the non-separable mistake bound of perceptron. The proof of Theorem 2 is in a full version of this paper. Notice that, ξ is the distance from the margin with a data point $(\mathbf{x}, y)$, which indicates how the classification with a classifier $\mathbf{u}$ fails for this data point. Therefore, Ξ_i, the sum of ξ over shard i, can be considered as a cumulative loss if $\mathbf{u}$ is run on shard i. From inequality (8), the number of mistakes of IPM-perceptron is bounded in terms of the cumulative loss of an arbitrary vector $\mathbf{u}$.

5.2 Loss Bound of IPM-PA

Here, we describe the loss bound of IPM with the Passive Aggressive algorithm (IPM-PA). As in the case of IPM-perceptron, we can obtain separable and non-separable loss bounds. The proofs of the bounds are in Appendix A.2.

Theorem 3. (Loss bound of IPM-PA in the separable case)
 Let there be a parameter vector $\mathbf{u}$ *that suffers no loss for any data point* $(\mathbf{x}, y)$ *in the training data set. Suppose that* $||\mathbf{x}|| \leq R$ *holds for any input* $\mathbf{x}$. *Then,*

$$\sum_{n=1}^{N} \sum_{i=1}^{M} \alpha_{i,n} L_{i,n} \leq ||\mathbf{u}||^2 R^2, \tag{9}$$

where $L_{i,n}$ *is the cumulative squared loss which worker* i *suffers in epoch* n.

Theorem 4. (Loss bound of IPM-PA in the non-separable case)
 Assume $||\mathbf{x}|| = 1$ *holds for any data point. Then, for any parameter vector* $\mathbf{u}$,

$$\sum_{n=1}^{N} \sum_{i=1}^{M} \alpha_{i,n} L_{i,n} \leq \left(||\mathbf{u}|| + 2 \sqrt{\sum_{n=1}^{N} \sum_{i=1}^{M} \alpha_{i,n} L_i^*} \right)^2, \tag{10}$$

where L_i^* *is the cumulative squared loss of parameter vector* $\mathbf{u}$ *with data on shard* i.

5.3 Superiority of Beta-IPM from a Theoretical Perspective

The cumulative loss in (8) is weighted by $\alpha_{i,n}$. Suppose the shards are divided into two categories: separable shards $i = 1, ..., m$ which can be classified by $\mathbf{u}$ and non-separable shards $i = m+1, ..., M$ with no vector to correctly classify them. The smaller the weights of the non-separable shards $\alpha_{m+1}, ..., \alpha_M$ are, the smaller the weighted cumulative loss $\sum_{i=1}^{M} \alpha_{i,n} \Xi_i$ we can obtain, and this means that it is very important to reduce the weights corresponding to contaminated shards. The same argument goes with PA. In general, Beta-IPM suppresses the weights of non-separable shards as described in Section 4, and thus Beta-IPM is expected to have a smaller mistake count than KL-IPM.

6 Empirical Evaluation

We conducted an evaluation with various datasets. The overall goal of these experiments was to study how KL-IPM and Beta-IPM behave in contaminated environments.

6.1 Setup

Our experiments involved 16 datasets (Table 1). Zeta and ocr datasets are from the Pascal large-scale learning challenge[1], and the imdb and citeseer datasets are from Paul Komarek's webpage[2]. The other datasets are from the LIBSVM dataset repository[3].

---

[1] http://largescale.ml.tu-berlin.de/
[2] http://komarix.org/ac/ds/
[3] http://www.csie.ntu.edu.tw/~cjlin/libsvmtools/datasets/binary.html

Table 1. List of the binary classification datasets evaluated. The tasks of the datasets are CI (census income prediction), DC (document categorization), HA (human answer prediction), IP (involvement prediction of person to some contents), IR (image recognition), MD (malware / suspicious contents detection), S (synthetically created problem), TC (toxicity categorization), or TD (text decoding).

	# of features	# of data points	task		# of features	# of data points	task
ijcnn1.tr	22	49,990	TD	rcv1	47,236	20,242	DC
mushrooms	112	8,124	TC	citeseer	105,354	181,395	IP
a8a	123	22,696	CI	imdb	685,569	167,773	IP
ocr	1,156	3,500,000	IR	news20	1,355,191	19,996	DC
epsilon	2,000	400,000	S	url	3,231,961	2,396,130	MD
zeta	2,000	500,000	S	webspam	16,609,143	350,000	MD
gisette	5,000	6,000	IR	kdda	20,216,830	8,407,752	HA
real-sim	20,958	72,309	DC	kddb	29,890,095	19,264,097	HA

Data shards: For each dataset, we used 80% of the data for training and 20% for testing. Then, the training dataset is divided into 100 shards associated with workers. Algorithms are trained with the training data and evaluated in terms of the classification accuracy of the test data.

To study the proposed algorithms' robustness against contamination, we studied the clean setting (i.e. no contamination) and two following contamination settings. Note that the contaminations are only on the training shards (the test data is always clean).

Setting 1 - adversarial labels: In this setting, 30 out of 100 shards are adversarial data: the labels of the data are reversed. This setting models situations where the data in some shards are maliciously labeled.

Setting 2 - random labels: In this setting, 80 out of 100 shards are assigned random labels. Each data point in these randomly labeled shards is labeled $y_t \sim \text{Bernoulli}(p)$, regardless of its true label. The ratio of positive labels, p, varies from 0.1 to 0.9 among workers. This setting models situations where data in most shards are corrupted.

Algorithms: We compared four algorithms: KL-IPM with a single-iteration perceptron or Passive Aggressive (KL-IPM-perceptron and KL-IPM-PA, respectively) and Beta-IPM with a single-iteration perceptron or Passive Aggressive (Beta-IPM-perceptron and Beta-IPM-PA, respectively). All values of β were the best among $\{10^{-1}, 10^{-2}, ..., 10^{-8}\}$. Since our research includes high-dimensional datasets, we assumed that the Gaussian in Beta-IPM was diagonal. The features with zero-variances were ignored in the weight calculation. We normalized the parameter vector of each worker by using the $l2$-norm when calculating the weights in Beta-IPM.

6.2   Results

The results for all datasets are shown in Table 2. The results of KL-IPM in the clean setting can be considered to be the possible best performance of linear classifiers in our distributed setting. As aforementioned, the performance of IPM is degraded by contamination. Note that our main interest in these experiments is the extent to which Beta-IPM can remedy the effects of contamination. First, let us compare the results of KL-IPM in

Table 2. Accuracy of the algorithms in the clean/adversarial/random settings at the 50th epoch. Boldface entries in the contamination settings are the best among the individual datasets.

	Clean Setting							
	ijcnn1.tr	mushrooms	a8a	ocr	epsilon	zeta	gisette	real-sim
KL-IPM-perceptron	0.913	0.999	0.845	0.763	0.899	0.628	0.947	0.968
KL-IPM-PA	0.912	0.999	0.845	0.762	0.899	0.694	0.958	0.975
	rcv1	citeseer	imdb	news20	url	webspam	kdda	kddb
KL-IPM-perceptron	0.960	0.976	0.981	0.953	0.986	0.990	0.881	0.886
KL-IPM-PA	0.966	0.977	0.985	0.958	0.986	0.990	0.882	0.887

	Contamination Setting 1: Adversarial Labels							
	ijcnn1.tr	mushrooms	a8a	ocr	epsilon	zeta	gisette	real-sim
KL-IPM-perceptron	**0.908**	0.937	0.838	0.760	0.881	0.582	0.854	0.824
KL-IPM-PA	**0.908**	0.983	0.837	0.760	0.886	0.651	0.912	0.904
Beta-IPM-perceptron	**0.908**	**0.998**	**0.846**	**0.763**	**0.898**	**0.665**	0.935	0.961
Beta-IPM-PA	**0.908**	0.989	**0.846**	0.762	**0.898**	0.663	**0.957**	**0.972**
	rcv1	citeseer	imdb	news20	url	webspam	kdda	kddb
KL-IPM-perceptron	0.762	0.976	0.980	0.703	0.983	0.987	0.743	0.759
KL-IPM-PA	0.871	**0.977**	**0.984**	0.844	0.983	0.987	0.689	0.712
Beta-IPM-perceptron	0.955	0.976	0.981	0.945	**0.986**	**0.991**	**0.876**	**0.882**
Beta-IPM-PA	**0.962**	**0.977**	**0.984**	**0.950**	**0.986**	**0.991**	0.676	0.693

	Contamination Setting 2: Random Labels							
	ijcnn1.tr	mushrooms	a8a	ocr	epsilon	zeta	gisette	real-sim
KL-IPM-perceptron	0.886	0.858	0.820	0.750	0.737	0.516	0.698	0.680
KL-IPM-PA	0.855	0.942	0.817	0.674	0.758	0.545	0.827	0.741
Beta-IPM-perceptron	0.911	0.980	0.825	**0.755**	0.886	**0.642**	0.888	0.948
Beta-IPM-PA	**0.913**	**0.999**	**0.830**	0.723	**0.890**	0.624	**0.942**	**0.958**
	rcv1	citeseer	imdb	news20	url	webspam	kdda	kddb
KL-IPM-perceptron	0.600	0.657	0.611	0.644	0.971	0.951	0.739	0.734
KL-IPM-PA	0.701	0.685	0.684	**0.730**	0.971	0.960	0.761	0.757
Beta-IPM-perceptron	**0.919**	0.836	0.826	0.717	0.981	**0.986**	0.853	0.833
Beta-IPM-PA	0.910	**0.916**	**0.943**	**0.730**	**0.985**	0.985	**0.868**	**0.868**

the adversarial/random settings with those in the clean setting. The contamination negatively affected the results on almost all datasets. On the random setting, where 80% of the shards are contaminated, the damage to the results tended to be more severe than that on the adversarial setting. Second, let us compare the performances of Beta-IPM and KL-IPM in the adversarial/random settings. One can see that Beta-IPM outperformed KL-IPM on almost all datasets. Indeed, one many datasets Beta-IPM performed almost as well as KL-IPM under the clean setting; this confirms that Beta-IPM can remove the influence of contamination.

Fig. 4 shows the classification results of Beta-IPM with a single iteration perceptron for several values of β. The optimal value with this dataset was $\beta = 10^{-5}$, and the accuracy with this β value showed a steady rise in epochs. The accuracy after epoch 50 was nearly 95%. With a β value smaller than the optimal one ($\beta = 10^{-6}$) and

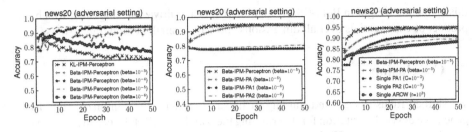

Fig. 4. Classification accuracy of Beta-IPM with a single iteration perceptron for several values of beta. The algorithms were run with news20 in the adversarial setting.

Fig. 5. Classification accuracies of Beta-IPM with various algorithms. The algorithms were run with news20 in the adversarial setting.

Fig. 6. Classification accuracy of Beta-IPM and single machine PA-I/PA-II/AROW. The algorithms were run with news20 in the adversarial setting.

with no beta (KL-IPM-perceptron), the algorithm failed to suppress the influence of the adversarial workers. Conversely, with β values bigger than optimal ($\beta = 10^{-3}$), the regularization was so strong that even the influence of some of the correct workers was suppressed. As a result, the learning rate with this beta value was very slow.

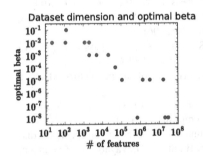

Fig. 7. Number of features and optimal value of beta in the adversarial setting. Each point corresponds to a dataset.

Fig. 5 compares several different algorithms with the best beta values. Given a proper value of β, Beta-IPM with a perceptron or PA successfully learned the parameter vectors. However, PA-I and PA-II[4], the noise-tolerant version of PA, did not perform well. These results indicate that robustness in a distributed environment is essentially different from that of single machine online learning: while we assume some fraction of the data is clean and the rest is contaminated, robust learning in a single machine aims to learn in environments where the clean and contaminated data are mixed. A possible hypothesis is that, the regularization of the learning rate in PA-I and PA-II obscured the difference between clean and contaminated shards, which made the accuracies of IPM with PA-I and PA-II poor.

Fig. 6 compares Beta-IPM-perceptron with a single machine PA-I or PA-II and AROW [6]. The hyper-parameter C in PA-I and PA-II and r in AROW were optimized in $\{10^{-4}, 10^{-3}, ..., 10^4\}$. The data of all 100 shards were put into a single shard in the single machine experiments. The two Beta-IPM algorithms performed better than the single machine algorithms. These results are empirical evidence that Beta-IPM can reduce the weights of adversarial shards.

Fig. 7 shows the optimal value of beta as a function of the number of features. Overall, in high-dimensional datasets, the value of beta tends to be small. The reason for

---

[4] The parameter C in PA-I and PA-II was set to be 0.001.

this is that the weight in Beta-IPM (Equation (6)) is a multivariate Gaussian, which is a product of exponentials over all dimensions and thus is small at high dimensions.

7 Conclusion

We studied robust distributed training of linear classifiers. By minimizing the divergence, we devised a criterion for determining the weights in IPM. Experiments revealed that the performance of IPM is significantly recovered on many contaminated datasets by determining the weights based on the beta divergence. An interesting direction of future work is to remove the statistiscal assumption of Gaussian distribution, by allowing more wider class of distributions, or non-parametric models.

References

1. Aberdeen, D., Pacovsky, O., Slater, A.: The learning behind gmail priority inbox. In: LCCC: NIPS 2010 Workshop on Learning on Cores, Clusters and Clouds (2010)
2. Basu, A., Harris, I.R., Hjort, N.L., Jones, M.C.: Robust and efficient estimation by minimising a density power divergence. Biometrika 85(3), 549–559 (1998)
3. Chouvardas, S., Slavakis, K., Theodoridis, S.: Adaptive robust distributed learning in diffusion sensor networks. IEEE Transactions on Signal Processing 59(10), 4692–4707 (2011)
4. Chu, C.T., Kim, S.K., Lin, Y.A., Yu, Y., Bradski, G.R., Ng, A.Y., Olukotun, K.: Map-reduce for machine learning on multicore. In: NIPS, pp. 281–288 (2006)
5. Crammer, K., Dekel, O., Keshet, J., Shalev-Shwartz, S., Singer, Y.: Online passive-aggressive algorithms. Journal of Machine Learning Research 7, 551–585 (2006)
6. Crammer, K., Kulesza, A., Dredze, M.: Adaptive regularization of weight vectors. Machine Learning 91(2), 155–187 (2013)
7. Curtsinger, C., Livshits, B., Zorn, B.G., Seifert, C.: Zozzle: Fast and precise in-browser javascript malware detection. In: USENIX Security Symposium (2011)
8. Daumé III, H., Phillips, J.M., Saha, A., Venkatasubramanian, S.: Efficient protocols for distributed classification and optimization. In: Bshouty, N.H., Stoltz, G., Vayatis, N., Zeugmann, T. (eds.) ALT 2012. LNCS, vol. 7568, pp. 154–168. Springer, Heidelberg (2012)
9. Dekel, O., Shamir, O., Xiao, L.: Learning to classify with missing and corrupted features. Mach. Learn. 81(2), 149–178 (2010)
10. Djuric, N., Grbovic, M., Vucetic, S.: Distributed confidence-weighted classification on mapreduce. In: IEEE Bigdata (2013)
11. Eguchi, S., Kano, Y.: Robustifying maximum likelihood estimation. Technical report, Institute of Statistical Mathematics (June 2001)
12. Gimpel, K., Das, D., Smith, N.A.: Distributed asynchronous online learning for natural language processing. In: CoNLL 2010, pp. 213–222 (2010)
13. Gong, P., Ye, J., Zhang, C.: Robust multi-task feature learning. In: KDD, pp. 895–903 (2012)
14. Hall, K.B., Inc, G., Gilpin, S., Mann, G.: Mapreduce/bigtable for distributed optimization. In: LCCC: NIPS 2010 Workshop on Learning on Cores, Clusters and Clouds (2010)
15. Hoi, S.C.H., Wang, J., Zhao, P.: Exact soft confidence-weighted learning. In: ICML (2012)
16. Mann, G., McDonald, R.T., Mohri, M., Silberman, N., Walker, D.: Efficient large-scale distributed training of conditional maximum entropy models. In: NIPS, pp. 1231–1239 (2009)
17. McDonald, R., Hall, K., Mann, G.: Distributed training strategies for the structured perceptron. In: NAACL, HLT 2010, pp. 456–464 (2010)

18. Meyer, T.A., Whateley, B.: Spambayes: Effective open-source, bayesian based, email classi-fication system. In: CEAS (2004)
19. Rahm, E., Do, H.H.: Data cleaning: Problems and current approaches. IEEE Data Engineer-ing Bulletin 23, 2000 (2000)
20. Rosenblatt, F.: The perceptron: A probabilistic model for information storage and organiza-tion in the brain. Psychological Review 65(6), 386–408 (1958)
21. Runnalls, A.R.: A kullback-leibler approach to gaussian mixture reduction. IEEE Trans. Aerosp. Electron. Syst., 989–999 (2007)
22. Tsitsiklis, J., Bertsekas, D., Athans, M.: Distributed asynchronous deterministic and stochas-tic gradient optimization algorithms. IEEE Transactions on Automatic Control 31(9), 803–812 (1986)
23. Xu, H., Leng, C.: Robust multi-task regression with grossly corrupted observations. Journal of Machine Learning Research - Proceedings Track 22, 1341–1349 (2012)
24. Zinkevich, M., Smola, A.J., Langford, J.: Slow learners are fast. In: NIPS, pp. 2331–2339 (2009)

A Appendix

A.1 Derivation of KL-IPM and Beta-IPM

Derivation of KL-IPM. We want to show that a mixed weight based on KL-IPM mini-mizes the KL-divergence between P and Q based on the parameter vectors $\{\mathbf{w}^{(1,n)}, ..., \mathbf{w}^{(M,n)}\}$. The following lemma states that KL divergence minimization based on Gaus-sian distributions preserves the mean and covariance.

Lemma 5. [Theorem 3.2 in [21]] *Let P be an arbitrary probability distribution on $\mathbb{R}^d$ with a well-defined mean $\boldsymbol{\mu}^*$ and covariance matrix $\boldsymbol{\Sigma}^*$, where $\boldsymbol{\Sigma}^*$ is strictly positive-definite. Let Q be a Gaussian distribution with mean $\boldsymbol{\mu}$ and covariance matrix $\boldsymbol{\Sigma}$. The unique minimum value of $D_{KL}(P||Q)$ is achieved when $\boldsymbol{\mu} = \boldsymbol{\mu}^*$ and $\boldsymbol{\Sigma} = \boldsymbol{\Sigma}^*$.*

The inequality (4) follows by using Lemma 5 and the fact that the empirical mean of P on the parameter vectors is $(1/M) \sum_i \mathbf{w}^{(i,n)}$.

Derivation of Beta-IPM. Let the parameters of the workers be $\{\mathbf{w}^{(1,n)}, ..., \mathbf{w}^{(M,n)}\}$, which is generated from a distribution P, and $Q(\boldsymbol{\mu}, \boldsymbol{\Sigma})$ be a Gaussian distribution. We would like to minimize the beta divergence, namely, $\mathbf{w}^{(\mathrm{avg},n)} = \arg\min_{\boldsymbol{\mu}} D_\beta(P||Q(\boldsymbol{\mu}, \boldsymbol{\Sigma}))$. Then,

$$D_\beta(P||Q(\boldsymbol{\mu}, \boldsymbol{\Sigma})) \tag{11}$$

$$= -\frac{1}{\beta} \int P(\mathbf{w}) Q^\beta(\mathbf{w}|\boldsymbol{\mu}, \boldsymbol{\Sigma}) d\mathbf{w} + \frac{1}{\beta+1} \int Q^{\beta+1}(\mathbf{w}|\boldsymbol{\mu}, \boldsymbol{\Sigma}) d\mathbf{w} + \text{Const.} \tag{12}$$

$$= -\frac{1}{\beta} E_{P(\mathbf{w})}[Q^\beta(\mathbf{w}|\boldsymbol{\mu}, \boldsymbol{\Sigma})] + \frac{1}{\beta+1} \int Q^{\beta+1}(\mathbf{w}|\boldsymbol{\mu}, \boldsymbol{\Sigma}) d\mathbf{w} + \text{Const.}, \tag{13}$$

where Const. is a term independent of $\boldsymbol{\mu}$ and $\boldsymbol{\Sigma}$, and $E_{P(\mathbf{w})}$ is the expectation un-der the assumption that $\mathbf{w}$ follows the probability distribution $P(\mathbf{w})$. Replacing the expectation of the first term with an empirical expectation over the parameter vectors $\{\mathbf{w}^{(1,n)}, ..., \mathbf{w}^{(M,n)}\}$ yields

$$(13) = -\frac{1}{\beta}\sum_{i=1}^{M}\frac{1}{M}\left[Q^{\beta}(\mathbf{w}^{(i,n)}|\boldsymbol{\mu}, \boldsymbol{\Sigma})\right] + \frac{1}{\beta+1}\int Q^{\beta+1}(\mathbf{w}|\boldsymbol{\mu}, \boldsymbol{\Sigma})d\mathbf{w}. \quad (14)$$

The multivariate Gaussian Q is explicitly written as $Q(\mathbf{w}|\boldsymbol{\mu}, \boldsymbol{\Sigma}) = Z(\boldsymbol{\Sigma})$ $\exp S(\mathbf{w}|\boldsymbol{\mu}, \boldsymbol{\Sigma})$, where $Z(\boldsymbol{\Sigma}) = 1/\sqrt{(2\pi)^d|\boldsymbol{\Sigma}|}$ and $S(\mathbf{w}|\boldsymbol{\mu}, \boldsymbol{\Sigma}) = -\frac{1}{2}(\mathbf{w} - \boldsymbol{\mu})^{\top}\boldsymbol{\Sigma}^{-1}$ $(\mathbf{w} - \boldsymbol{\mu})$. The second term in (14) is, from the property of multivariate Gaussian distribution,

$$\frac{1}{\beta+1}\int Q^{\beta+1}(\mathbf{w}|\boldsymbol{\mu}, \boldsymbol{\Sigma})d\mathbf{w} = Z(\boldsymbol{\Sigma})^{\beta}(\beta+1)^{-1-d/2}, \quad (15)$$

which is independent on $\boldsymbol{\mu}$. By using these facts, the first derivative of $D_{\beta}(P\|Q(\boldsymbol{\mu}, \boldsymbol{\Sigma}))$ is equivalent to the one of the first term in the RHS of (14), which is transformed as,

$$\frac{d}{d\boldsymbol{\mu}}\left\{-\frac{1}{\beta}\sum_{i=1}^{M}\frac{1}{M}\left[Q^{\beta}(\mathbf{w}^{(i,n)}|\boldsymbol{\mu}, \boldsymbol{\Sigma})\right]\right\} = \frac{Z(\boldsymbol{\Sigma})^{\beta}}{\beta M}\left\{-\frac{d}{d\boldsymbol{\mu}}\sum_{i=1}^{M}\exp S(\mathbf{w}^{(i,n)}, \boldsymbol{\mu}, \frac{1}{\beta}\boldsymbol{\Sigma})\right\}.$$

$$= \frac{Z(\boldsymbol{\Sigma})^{\beta}}{\beta M}\left\{\sum_{i=1}^{M}\exp S(\mathbf{w}^{(i,n)}, \boldsymbol{\mu}, \frac{1}{\beta}\boldsymbol{\Sigma})(\beta\boldsymbol{\Sigma}^{-1})(\mathbf{w}^{(i,n)} - \boldsymbol{\mu})\right\}$$

$$= \frac{Z(\boldsymbol{\Sigma})^{\beta}}{\beta M}\left\{\beta\boldsymbol{\Sigma}^{-1}\sum_{i=1}^{M}\exp S(\mathbf{w}^{(i,n)}, \boldsymbol{\mu}, \frac{1}{\beta}\boldsymbol{\Sigma})(\mathbf{w}^{(i,n)} - \boldsymbol{\mu})\right\}. \quad (16)$$

Therefore, we obtain

$$\frac{d}{d\boldsymbol{\mu}}\left\{-\frac{1}{\beta}\sum_{i=1}^{M}\frac{1}{M}\left[Q^{\beta}(\mathbf{w}^{(i,n)}|\boldsymbol{\mu}, \boldsymbol{\Sigma})\right]\right\} = 0 \Leftrightarrow \sum_{i=1}^{M}\exp S(\mathbf{w}^{(i,n)}, \boldsymbol{\mu}, \frac{1}{\beta}\boldsymbol{\Sigma})(\mathbf{w}^{(i,n)} - \boldsymbol{\mu}) = 0$$

$$\Leftrightarrow \boldsymbol{\mu} = \frac{\sum_{i=1}^{M}\exp S(\mathbf{w}^{(i,n)}, \boldsymbol{\mu}, \frac{1}{\beta}\boldsymbol{\Sigma})\mathbf{w}^{(i,n)}}{\sum_{j=1}^{M}\exp S(\mathbf{w}^{(j,n)}, \boldsymbol{\mu}, \frac{1}{\beta}\boldsymbol{\Sigma})}. \quad (17)$$

RHS of (17) states that $\boldsymbol{\mu}$ that minimizes $D_{\beta}(P\|Q)$ is a weighted mean of each $\mathbf{w}^{(i,n)}$ with weight $\exp S(\mathbf{w}^{(i,n)}, \boldsymbol{\mu}, \frac{1}{\beta}\boldsymbol{\Sigma})$. Unfortunately, the weight $\exp S(\mathbf{w}^{(i,n)}, \boldsymbol{\mu}, \frac{1}{\beta}\boldsymbol{\Sigma})$ on the RHS includes $\boldsymbol{\mu}$, and thus, an exact solution is unattainable. To get a reasonable solution, we can approximate $\boldsymbol{\mu}$ and $\boldsymbol{\Sigma}$ on the RHS of (17) by the mean and covariance of the samples $\{\mathbf{w}^{(1,n)}, ..., \mathbf{w}^{(M,n)}\}$, which finally yields (6).

A.2   Proof of Theorem 3 and 4

The crux in the mistake/loss bound proofs in online classifiers is to find some value that can be bounded from both the upper and lower side: in the case of PA we bound the value $\Delta_n = \|\mathbf{w}^{(\mathrm{avg},n-1)} - \mathbf{u}\| - \|\mathbf{w}^{(\mathrm{avg},n)} - \mathbf{u}\|$. By using these lower and upper bounds we obtain Lemma 6, which leads to the proofs of Theorem 3 and 4.

Algorithm 2. Single iteration Passive Aggressive

1: $T = \{(\mathbf{x}_t, y_t)\}$, $\mathbf{w}$
2: **for** $t = 1, ..., |T|$ **do**
3:     $\hat{y}_t \leftarrow \text{sign}(\mathbf{w} \cdot \mathbf{x}_t)$
4:     $l_t \leftarrow \max(0, 1 - y_t(\mathbf{w} \cdot \mathbf{x}_t))$
5:     $\tau_t \leftarrow l_t / ||\mathbf{x}_t||^2$
6:     $\mathbf{w} \leftarrow \mathbf{w} + \tau_t y_t \mathbf{x}_t$
7: **end for**

Lemma 6. *Let the index $t = 1, ..., k_{i,n}$ denotes the data points on shard i that the worker suffered non-zero losses, and $(\mathbf{x}_{i,t}, y_{i,t})$ be the data point at that round. Moreover, let $l_{i,t}$ be the corresponding loss of the worker, and $\tau_{i,t} = l_{i,t} / ||\mathbf{x}_{i,t}||^2$, and $l_{i,t}^*$ be the loss of any constant classifier $\mathbf{u}$ with the data point. Then,*

$$\sum_{n=1}^{N} \left\{ \sum_{i=1}^{M} \alpha_{i,n} \sum_{t=1}^{k_{i,n}} \left\{ \tau_{i,t}(2l_{i,t} - \tau_{i,t}||\mathbf{x}_{i,t}||^2 - 2l_{i,t}^*) \right\} \right\} \leq ||\mathbf{u}||. \tag{18}$$

Proof (Lemma 6). Consider shard i in epoch n. Let $\Delta_{i,n} = ||\mathbf{w}^{(\text{avg},n-1)} - \mathbf{u}||^2 - ||\mathbf{w}^{(i,n)} - \mathbf{u}||^2$. Notice that the parameter vector is updated only when the loss is non-zero, and For $1 \leq t \leq k_{i,n}$, let $\mathbf{w}^{([i,n]+t)}$ be the parameter vector on shard i in epoch n in the round just before the t-th loss occurred. Also, for $t = k_{i,n} + 1$, let $\mathbf{w}^{([i,n]+t)} = \mathbf{w}^{(i,n)}$. Notice that $\mathbf{w}^{([i,n]+1)} = \mathbf{w}^{(\text{avg},n-1)}$. The update per single loss is,

$$||\mathbf{w}^{([i,n]+t)} - \mathbf{u}||^2 - ||\mathbf{w}^{([i,n]+(t+1))} - \mathbf{u}||^2$$
$$= ||\mathbf{w}^{([i,n]+t)} - \mathbf{u}||^2 - ||\mathbf{w}^{([i,n]+t)} - \mathbf{u} + y_{i,t}\tau_{i,t}\mathbf{x}_{i,t}||^2$$
$$= ||\mathbf{w}^{([i,n]+t))} - \mathbf{u}||^2 - \left\{ ||\mathbf{w}^{([i,n]+t)} - \mathbf{u}||^2 + 2y_{i,t}\tau_{i,t}(\mathbf{w}^{([i,n]+t)} - \mathbf{u}) \cdot \mathbf{x}_{i,t} + \tau_{i,t}^2 ||\mathbf{x}_{i,t}||^2 \right\}$$
$$= -2y_{i,t}\tau_{i,t}(\mathbf{w}^{([i,n]+t)} - \mathbf{u}) \cdot \mathbf{x}_{i,t} - \tau_{i,t}^2 ||\mathbf{x}_{i,t}||^2. \tag{19}$$

Since we assumed $l_{i,t} > 0$ with this data point, $y_{i,t}(\mathbf{w}^{([i,n]+t)} \cdot \mathbf{x}_{i,t}) = 1 - l_{i,t}$ and $l_{i,t}^* \geq 1 - y_{i,t}(\mathbf{u} \cdot \mathbf{x}_{i,t})$ always holds. Thus, (19) can be bounded as,

$$(19) \geq 2\tau_{i,t}((1 - l_{i,t}^*) - (1 - l_{i,t})) - \tau_{i,t}^2 ||\mathbf{x}_{i,t}||^2 = \tau_{i,t}(2l_{i,t} - \tau_{i,t}||\mathbf{x}_{i,t}||^2 - 2l_{i,t}^*). \tag{20}$$

By using (20), $\Delta_{i,n}$ is bounded as,

$$\Delta_{i,n} = ||\mathbf{w}^{(\text{avg},n-1)} - \mathbf{u}||^2 - ||\mathbf{w}^{(i,n)} - \mathbf{u}||^2$$
$$= \sum_{t=1}^{k_{i,n}} \left(||\mathbf{w}^{([i,n]+t)} - \mathbf{u}||^2 - ||\mathbf{w}^{([i,n]+(t+1))} - \mathbf{u}||^2 \right)$$
$$\geq \sum_{t=1}^{k_{i,n}} \left\{ \tau_{i,t}(2l_{i,t} - \tau_{i,t}||\mathbf{x}_{i,t}||^2 - 2l_{i,t}^*) \right\}. \tag{21}$$

We now lower-bound Δ_n as follows:

$$\Delta_n = ||\mathbf{w}^{(\mathrm{avg},n-1)}- \mathbf{u}||^2 - ||\mathbf{w}^{(\mathrm{avg},n)}- \mathbf{u}||^2 = ||\mathbf{w}^{(\mathrm{avg},n-1)}- \mathbf{u}||^2 - ||\sum_i \alpha_{i,n}(\mathbf{w}^{(i,n)}- \mathbf{u})||^2$$

$$\geq \sum_{i=1}^{M} \alpha_{i,n}\left(||\mathbf{w}^{(\mathrm{avg},n-1)} - \mathbf{u}||^2 - ||\mathbf{w}^{(i,n)} - \mathbf{u}||^2\right) = \sum_{i=1}^{M} \alpha_{i,n}\Delta_{i,n}$$

$$\geq \sum_{i=1}^{M} \alpha_{i,n} \sum_{t=1}^{k_{i,n}} \left\{\tau_{i,t}(2l_{i,t} - \tau_{i,t}||\mathbf{x}_{i,t}||^2 - 2l_{i,t}^*)\right\}, \tag{22}$$

where we have used $\sum_i \alpha_{i,n} = 1$ in going between the first and second line, and used (21) at the last transformation.

On the other hand, the sum of Δ_n is upper-bounded as follows:

$$\sum_{n=1}^{N} \Delta_n = \sum_{n=1}^{N} \left(||\mathbf{w}^{(\mathrm{avg},n-1)} - \mathbf{u}||^2 - ||\mathbf{w}^{(\mathrm{avg},n)} - \mathbf{u}||^2\right)$$

$$= ||\mathbf{w}^{(\mathrm{avg},0)} - \mathbf{u}||^2 - ||\mathbf{w}^{(\mathrm{avg},N)} - \mathbf{u}||^2 \leq ||\mathbf{u}||, \tag{23}$$

where the last inequality follows from the fact that the initial parameter vector is the zero vector and $||\mathbf{w}^{(\mathrm{avg},N)} - \mathbf{u}||^2 \geq 0$. Using (22) and (23) yields (18).

$\square$

Proof (Theorem 3). By using the fact that $l_{i,t}^* = 0$, $l_{i,t} = \tau_{i,t}||\mathbf{x}_{i,t}||^2$, Lemma 6 is transformed as follows:

$$\sum_{n=1}^{N} \left\{\sum_{i=1}^{M} \alpha_{i,n} \sum_{t=1}^{k_{i,n}} \frac{(l_{i,t})^2}{||\mathbf{x}_{i,t}||^2}\right\} \leq ||\mathbf{u}||^2. \tag{24}$$

With the fact that $||\mathbf{x}_{i,t}|| < R$, we finally obtain

$$\sum_{n=1}^{N} \left\{\sum_{i=1}^{M} \alpha_{i,n} \sum_{t=1}^{k_{i,n}} (l_{i,t})^2\right\} \leq ||\mathbf{u}||^2 R^2. \tag{25}$$

$L_{i,n}$, the cumulative squared loss the worker i suffers during epoch n, corresponds to $\sum_{t=1}^{k_{i,n}} (l_{i,t})^2$. Therefore, the inequality (25) is equivalent to (9).

$\square$

Proof (Theorem 4). Next, we consider the case where $l_{i,t}^*$ is not necessarily zero. Let us assume $||\mathbf{x}_{i,t}|| = 1$. Notice that $\tau_{i,t} = l_{i,t}/||\mathbf{x}_{i,t}||^2 = l_{i,t}$. By these facts and Lemma 6,

$$\sum_{n=1}^{N} \left\{\sum_{i=1}^{M} \alpha_{i,n} \sum_{t=1}^{k_{i,n}} \left\{(l_{i,t})^2 - 2l_{i,t}l_{i,t}^*\right\}\right\} \leq ||\mathbf{u}||^2. \tag{26}$$

Let

$$X_N = \sqrt{\sum_{n=1}^{N} \sum_{i=1}^{M} \alpha_{i,n} \sum_{t=1}^{k_{i,n}} (l_{i,t})^2}, \text{ and } Y_N = \sqrt{\sum_{n=1}^{N} \sum_{i=1}^{M} \alpha_{i,n} \sum_{t=1}^{k_{i,n}} (l_{i,t}^*)^2}. \tag{27}$$

Using the Cauchy-Schwarz inequality on the LHS of (26), we obtain $X_N^2 - 2X_N Y_N \leq ||\mathbf{u}||^2$, which is a quadratic inequality of X_N, and thus $X_N \leq Y_N + \sqrt{Y_N^2 + ||\mathbf{u}||^2} \leq ||\mathbf{u}|| + 2Y_N$, where we used the fact that $\sqrt{a+b} \leq \sqrt{a} + \sqrt{b}$ for $a, b \geq 0$. By explicitly writing X_N and Y_N we obtain

$$\sqrt{\sum_{n=1}^{N} \sum_{i=1}^{M} \alpha_{i,n} \sum_{t=1}^{k_{i,n}} (l_{i,t})^2} \leq ||\mathbf{u}|| + 2\sqrt{\sum_{n=1}^{N} \sum_{i=1}^{M} \alpha_{i,n} \sum_{t=1}^{k_{i,n}} (l_{i,t}^*)^2}. \qquad (28)$$

The cumulative squared loss the worker i suffers in epoch n is $L_{i,n} = \sum_{t=1}^{k_{i,n}} (l_{i,t})^2$. Moreover, $L_i^* \geq \sum_{t=1}^{k_{i,n}} (l_{i,t}^*)^2$ holds because the index t runs the subset of data on shard i. Taking these into consideration, we finally obtain (10).

$\square$

Reliability Maps: A Tool to Enhance Probability Estimates and Improve Classification Accuracy

Meelis Kull and Peter Flach

Intelligent Systems Laboratory, University of Bristol, United Kingdom
{Meelis.Kull,Peter.Flach}@bristol.ac.uk

Abstract. We propose a general method to assess the reliability of two-class probabilities in an instance-wise manner. This is relevant, for instance, for obtaining calibrated multi-class probabilities from two-class probability scores. The LS-ECOC method approaches this by performing least-squares fitting over a suitable error-correcting output code matrix, where the optimisation resolves potential conflicts in the input probabilities. While this gives all input probabilities equal weight, we would like to spend less effort fitting unreliable probability estimates. We introduce the concept of a reliability map to accompany the more conventional notion of calibration map; and LS-ECOC-R which modifies LS-ECOC to take reliability into account. We demonstrate on synthetic data that this gets us closer to the Bayes-optimal classifier, even if the base classifiers are linear and hence have high bias. Results on UCI data sets demonstrate that multi-class accuracy also improves.

1 Introduction

Classification problems can be approached using a range of machine learning models. Some of these models, including decision trees, naive Bayes and nearest neighbour, deal naturally with more than two classes. Others – most notably linear models and their kernelised variants – are essentially two-class or binary. In order to solve a multi-class problem with binary models we need to decompose the multi-class problem into a set of binary subproblems, train a classifier on each subproblem and aggregate the predicted classes or scores obtained on each subproblem into an overall multi-class prediction or score vector. In the most common scenarios these subproblems are either pairwise (one class against another class) or one-vs-rest (one class against all other classes), which in matrix form could be described as follows:

$$\mathbf{M} = \begin{pmatrix} +1 & +1 & 0 \\ -1 & 0 & +1 \\ 0 & -1 & -1 \end{pmatrix} \quad \mathbf{N} = \begin{pmatrix} +1 & -1 & -1 \\ -1 & +1 & -1 \\ -1 & -1 & +1 \end{pmatrix}$$

These are known as code matrices, with binary subproblems in columns and classes in rows. $\mathbf{M}$ encodes pairwise subproblems and $\mathbf{N}$ encodes one-vs-rest. A vector of outputs from the binary classifiers can often be traced back to one of the classes: e.g., if we receive $(+1, +1, -1)$ in the pairwise case we can construe this as two votes for the first class and one vote for the third class.

T. Calders et al. (Eds.): ECML PKDD 2014, Part II, LNCS 8725, pp. 18–33, 2014.
© Springer-Verlag Berlin Heidelberg 2014

The general approach of error-correcting output codes was pioneered by [4] and later refined to take classifier scores into account [8,1]. Error-correcting capability is achieved by building redundancy into the code matrix: for instance, we can use two different binary classifiers for each subproblem, leading to two copies of each column in the code matrix (in fact, ensemble methods can be represented by a code matrix with repeated columns). More generally, ECOC can be described as an approach to combining the opinions of many different experts. Each expert has its own group of positive and negative classes (not necessarily covering all classes) and is trained to decide whether an unlabelled example falls in the positive or negative group. Expert opinions may disagree, in which case we need to figure out how to tweak the opinions to agree, and possibly which experts to trust more than others. Therefore, it is important to know how reliable or confident each expert is. For example, a properly Bayesian expert would output a posterior probability distribution over all possible opinions, from which we can infer a confidence level (e.g., expressed as a variance). However, while most machine learning models can be made to output a (more or less calibrated) probability score, they rarely give information about their confidence, and so a model-independent method needs to learn the reliability of these scores. Note that a calibrated probability score quantifies the expert's uncertainty in the class value, but here we are after the uncertainty in that probability estimate. That is, a weather forecaster can be very certain that the chance of rain is 50%; or her best estimate at 20% might be very uncertain due to lack of data.

This paper proposes a practical method to learn the reliability of probability scores output by experts in the above scenario, in an instance-wise manner. Being able to assess the reliability of a probability score for each instance is much more powerful than assigning an aggregate reliability score to each expert, independent of the instance to be classified. For example, we show later that an ECOC-based method that takes instance-wise reliability into account allows us to learn non-linear decision boundaries even when employing linear base models. As such the method can be seen as reducing the bias of the base classifier. But the basic method has applicability beyond ECOC. For example, in comparison with another bias-reducing technique, boosting [12], which uses a single confidence factor per base classifier, our method offers the possibility to generalise this to instance-wise confidence which should result in a better model. The advantage of having calibrated probability estimates in a cost-based scenario is that we can better minimise expected overall cost by predicting the class that minimises the cost averaged over all possible true classes. Taking reliability of the probability scores into account gives us the choice of choosing a non-minimising class if it has the benefit of less uncertainty. This would be useful in the presence of hard constraints of the form 'the probability that the cost exceeds budget B must be less than 5%' which may be true for a class even if it does not minimise expected cost.

The outline of the paper is as follows. Section 2 introduces reliability maps and their relation to squared bias of probability estimates from the respective true posterior probabilities. Section 3 develops an algorithm to learn reliability maps from class-labelled data, without access to true posterior probabilities. Section 4 introduces LS-ECOC-R, a reliability-weighted version of the LS-ECOC method to obtain multi-class probability scores. In Section 5 we present two kinds of experiments: we investigate how far our estimates are from the truth on synthetic data, and we investigate the effect of using reliabilities on the quality of multi-class predictions and probability scores. Section 6 discusses related work, and Section 7 concludes.

2   Calibration and Reliability

Let X, Y be the random variables representing the unknown true model of our binary classification task. That is, X is a random variable over the instance space $\mathscr{X}$ and Y is a binary random variable with 1 and 0 standing for positive and negative classes, respectively. Ideally, we would like to know the true positive class posterior $q(x)$ for each possible $x \in \mathscr{X}$:

$$q(x) = P(Y=1|X=x). \tag{1}$$

In reality, we use training data to learn a model $g : \mathscr{X} \to [0,1]$ such that $g(x)$ is approximating $q(x)$. If the output of the model is $g(x) = s$, what can we say about the true value $q(x)$? Let $\mu_g(s)$ be the expected proportion of positives among all instances x with the same value $g(x) = s$:

$$\mu_g(s) = \mathbb{E}[q(X)|g(X)=s] = P(Y=1|g(X)=s). \tag{2}$$

The function μ_g is known as the (true) *calibration map* of the probability estimator g. If the estimator g is perfectly calibrated, then μ_g is the identity function. If not, then there are many methods of calibration which can be applied to learn an estimated calibration map $\hat{\mu}_g$ such that $\hat{\mu}_g(g(x))$ is approximately equal to $\mu_g(g(x))$. However, for individual instances x the expected proportion of positives $q(x)$ can deviate from the mean proportion $\mu_g(g(x))$, i.e. the following variance is non-zero:

$$\sigma_g^2(s) = \mathrm{var}[q(X)|g(X)=s] = \mathbb{E}[(q(X) - \mu_g(s))^2|g(X)=s]. \tag{3}$$

The magnitude of $\sigma_g^2(s)$ across the estimates s from the model g actually determines how useful g is for estimating q. For instance, a constant probability estimator $g(x) = P(Y=1)$ is perfectly calibrated, but has high $\sigma_g^2(s)$ for its constant estimate $s = P(Y=1)$. The perfect estimator $g(x) = q(x)$ has $\sigma_g^2(s) = 0$ for all s. The variance σ_g^2 is bounded from above by $\sigma_g^2(s) \le \mu_g(s) \cdot (1 - \mu_g(s))$, where the equality holds when $q(x)$ is either 0 or 1 for each x with $g(x) = s$.[1] This leads to our following definition of the *reliability map* r_g of the probability estimator g.

Definition 1. *Let $g : \mathscr{X} \to [0,1]$ be an estimator of probability $q(x) = P(Y=1|X=x)$. Then the* reliability map r_g *of the probability estimator g is defined as follows:*

$$r_g(s) = 1 - \frac{\sigma_g^2(s)}{\mu_g(s) \cdot (1 - \mu_g(s))}, \tag{4}$$

where the calibration map $\mu_g(s)$ and variance $\sigma_g^2(s)$ are defined by (2) and (3). For an estimate s from the model g, we refer to the value $r_g(s)$ as local reliability *of g at s.*

---

[1] This can be seen as follows: $\sigma_g^2(s) = \mathbb{E}[(q(X))^2|g(X) = s] - (\mu_g(s))^2 \le \mathbb{E}[q(X)|g(X) = s] - (\mu_g(s))^2 = \mu_g(s) - (\mu_g(s))^2 = \mu_g(s) \cdot (1 - \mu_g(s))$. The equality holds when $P[(q(X))^2 = q(X)|g(X) = s] = P[q(X) \in \{0,1\}|g(X) = s] = 1$.

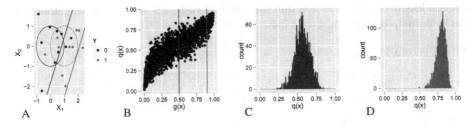

Fig. 1. Synthetic 2-class data of Gaussians centered at (0,0) and (1,0). (A) Standard deviation circles of the two Gaussians, 20 training instances and lines $g(x) = 0.5$ (blue) and $g(x) = 0.9$ (red) where g is the learned logistic regression model; (B) scatterplot of $q(x)$ and $g(x)$ for 2000 test points; (C) histogram of $q(x)$ for the 1530 test points out of 200000 which have $g(x)$ in range $[0.495, 0.505]$; (D) histogram of $q(x)$ for the 1727 test points out of 200000 with $g(x)$ in $[0.895, 0.905]$.

Minimum and maximum values 0 and 1 for the local reliability mean respectively that $q(x) \in \{0, 1\}$ and $q(x) = g(x)$ for all x with $g(x) = s$. We call it local reliability so that we can still talk about the (global) reliability of the probability estimator as a whole. Figure 3 presents the calibration and reliability maps for a synthetic dataset described in Section 5.

Example 1. To illustrate the above notions consider a synthetic two-class generative model with uniform class distribution $P(Y = 1) = P(Y = 0) = 1/2$ and X distributed as a standard 2-dimensional normal distribution centered at $(1,0)$ for class $Y = 1$ and at $(0,0)$ for class $Y = 0$. We generated 20 training instances from this generative model and learned a logistic regression model $g : \mathbb{R}^2 \to [0, 1]$ to estimate posterior class probabilities $q(x) = P(Y = 1 | X = x)$, see Fig. 1A. In this experiment our logistic regression learner resulted in the model $g(x) = 1/(1 + \exp(1.37 - 1.68x_1 + 0.76x_2))$ whereas the true model is $q(x) = P(Y = 1 | X = x) = 1/(1 + \exp(0.5 - x_1))$. This implies that for any instance x the learned estimate $g(x)$ can deviate slightly from the true value $q(x)$, see Fig. 1B with 2000 test points drawn randomly from the same generative model with two Gaussians.

Consider now the group of all instances with $g(x) = 0.5$, located on the blue line in Fig. 1AB. The histogram of $q(x)$ for a sample of these points is given in Fig. 1C with mean $\mu_g(0.5) = 0.5675$ and estimated variance $\sigma_g^2(0.5) = 0.0101$, leading to a reliability value of $r_g(s) = 0.9589$. What this demonstrates is that at predicted score 0.5 there is little variation in the true probabilities, even though the estimator is not perfectly calibrated at that score. For the group $g(x) = 0.9$ shown in red in Fig. 1AB and with a histogram of $q(x)$ in Fig. 1D the mean and variance are $\mu_g(x) = 0.7979$ and $\sigma_g^2(x) = 0.0042$, resulting again in a high reliability of $r_g(s) = 0.9740$.

The estimated $g(x) = s$ can differ from the true $q(x)$ for one or both of the following two reasons. First, if $\mu_g(s) \neq s$ then there is a bias in $g(x)$ from the average $q(x)$ of the group of instances with the same estimate s. Second, if $\sigma_g^2(s) > 0$ then there is variance in $q(x)$ for the group of instances with the same estimate s. In fact, the instance-wise squared loss between g and q within the group of instances with the same estimate s can be decomposed into these two losses:

$$\mathbb{E}[(g(X)-q(X))^2|g(X)=s] =$$
$$= \mathbb{E}[(s-q(X))^2|g(X)=s] = \mathbb{E}[(s-\mu_g(s)+\mu_g(s)-q(X))^2|g(X)=s] =$$
$$= (s-\mu_g(s))^2 + 2(s-\mu_g(s))\mathbb{E}[\mu_g(s)-q(X)|g(X)=s] + \mathbb{E}[(\mu_g(s)-q(X))^2|g(X)=s] =$$
$$= (s-\mu_g(s))^2 + \sigma_g^2(s),$$

where the last equality holds because $\mathbb{E}[\mu_g(s)-q(X)|g(X)=s] = 0$. This decomposition can be averaged over the whole instance space, resulting in the following decomposition:

$$\mathbb{E}[(g(X)-q(X))^2] = \mathbb{E}[(g(X)-\mu_g(g(X)))^2] + \mathbb{E}[(\mu_g(g(X))-q(X))^2].$$

We will refer to these three quantities as *instance-wise calibration loss*, *group-wise calibration loss*, and *grouping loss*.[2] Any calibration procedure which transforms values of $g(x)$ with a calibration map can decrease the group-wise calibration loss but not the grouping loss, which is inherent to the model. Grouping loss arises from the model's decision to group certain instances together with the same probability estimate whereas the true probabilities are different. The quantity $\sigma_g^2(s)$ can be interpreted as the local grouping loss for one group of instances with the same estimate $g(x) = s$ and the total grouping loss is the average $\sigma_g^2(s)$ across all groups s:

$$\mathbb{E}[(\mu_g(g(X))-q(X))^2] = \mathbb{E}[\sigma_g^2(g(X))].$$

Example 1 (continued). In the example of Fig. 1, the group $g(x) = 0.5$ suffers instance-wise calibration loss equal to 0.0147 decomposing into group-wise calibration loss of $(0.5675-0.5)^2 = 0.0046$ and grouping loss of $\sigma_g^2(s) = 0.0101$. Calibration of g can decrease the group-wise calibration loss, but the grouping loss remains irreducible, unless a new model is trained instead of g.

3   Learning Calibration and Reliability Maps

Learning calibration maps is a task that has been solved earlier with various methods. One simple approach that we revisit below is to view the binary label Y as a dependent variable and the probability estimate $S = g(X)$ as the independent variable and apply any standard regression learning algorithm. The training data for such approach is a list of pairs (S_i, Y_i). Although each individual instance S_i is far from the true calibration map, the expected value of Y given a fixed estimate S lies at the calibration map (see (6) below). In other words, Y is an unbiased estimator of the calibrated probability. Assuming that the true calibration map is continuous, this allows to estimate it with regression.

---

[2] The instance-wise calibration loss bears similarity to the calibration loss which is obtained by decomposing the Brier score [10], but the difference is that there the comparison is made with the empirical probability rather than the true probability.

For learning reliability maps it is also possible to use regression, but the challenge is to come up with a suitable unbiased estimator. Since the Bernoulli distribution of the binary label Y given an estimate S has only one parameter determining the calibrated probability, it does not contain information about the reliability map. For each instance X we need more information about the true probability $q(X)$ than just a single binary label Y. Our solution is to gather a small group of similar instances $X_1, \ldots, X_m$ with approximately the same estimate $g(X_i) \approx S$ and approximately the same posterior $q(X_i) \approx Q$. We obtain such groups of instances by splitting the training instances into clusters of equal size m according to some distance measure in the instance space. The clustering method that we have used in the experiments to obtain clusters of size $m = 10$ is described in Section 3.1 below. The reason for building clusters is that the variance in the number of positives $\sum_{i=1}^{m} Y_i$ in a cluster contains information about the variance in the posterior, $\sigma_g^2(S)$. As an estimator of local reliability of g at S we use $R^{(m)}$, defined as follows:

$$R^{(m)} = 1 + \frac{1}{m-1} - \frac{(\sum_{i=1}^{m} Y_i - m\mu_g(S))^2}{m(m-1)\mu_g(S)(1-\mu_g(S))}. \tag{5}$$

Theorem 1 proves by equality (7) that this estimator is unbiased if the instances within the cluster have equal g and equal q.

Theorem 1. *Let* $g : \mathcal{X} \to [0,1]$ *be a fixed probability estimator and let* (X_i, Y_i) *for* $i = 1, \ldots, m$ *with* $m \geq 2$ *be an i.i.d. random sample distributed identically to* (X,Y) *where* X *is a random variable over* $\mathcal{X}$ *and* Y *is a binary random variable. Additionally, let* $\mathcal{C}$ *stand for the condition where* $g(X_i) = g(X)$ *and* $q(X_i) = q(X)$ *for* $i = 1, \ldots, m$, *where* q *is defined as in (1). Then the following two equalities hold:*

$$\mu_g(s) = \mathbb{E}[Y | S{=}s] \tag{6}$$
$$r_g(s) = \mathbb{E}[R^{(m)} | S{=}s, \mathcal{C}] \tag{7}$$

where $S = g(X)$ *and* $\mu_g, r_g, R^{(m)}$ *are defined above respectively in (2), (4) and (5).*

Proof. Equation (6) can easily be proved by denoting $Q = q(X)$ and applying the law of total expectation:

$$\mathbb{E}[Y | S{=}s] = \mathbb{E}[\mathbb{E}[Y | Q, S{=}s] | S{=}s] = \mathbb{E}[Q | S{=}s] = \mu_g(s).$$

Let us denote $Z = \sum_{i=1}^{m} Y_i$. As Y_i are independent given $\mathcal{C}$ then $\mathbb{E}[Z | S{=}s, \mathcal{C}] = m\mu_g(s)$. Therefore,

$$\mathbb{E}[R^{(m)} | S{=}s, \mathcal{C}] = 1 + \frac{1}{m-1} - \frac{\text{var}[Z | S{=}s, \mathcal{C}]}{m(m-1)\mu_g(s)(1-\mu_g(s))}.$$

Due to (4) it now remains to prove that

$$\frac{\text{var}[Z | S{=}s, \mathcal{C}]}{m(m-1)\mu_g(s)(1-\mu_g(s))} = \frac{\sigma_g^2(s)}{\mu_g(s)(1-\mu_g(s))} + \frac{1}{m-1},$$

or equivalently, that

$$\mathrm{var}[Z|S{=}s,\mathscr{C}] = m(m-1)\sigma_g^2(s) + m\mu_g(s)(1-\mu_g(s)).$$

Let us denote $Q = q(X_i)$. As Z is binomially distributed given Q, $S{=}s$ and $\mathscr{C}$, we have $\mathbb{E}[Z|Q,S{=}s,\mathscr{C}] = mQ$ and $\mathrm{var}[Z|Q,S{=}s,\mathscr{C}] = mQ(1-Q)$. Also, $\mathbb{E}[Q|S{=}s,\mathscr{C}] = \mu_g(s)$ and $\mathrm{var}[Q|S{=}s,\mathscr{C}] = \sigma_g^2(s)$ and $\mathbb{E}[Q^2|S{=}s,\mathscr{C}] = \mathrm{var}[Q|S{=}s,\mathscr{C}] + (\mathbb{E}[Q|S{=}s,\mathscr{C}])^2 = \sigma_g^2(s) + \mu_g^2(s)$. Using this and the law of total variance (and algebraic manipulations) we obtain the following:

$$\begin{aligned}
\mathrm{var}[Z|S{=}s,\mathscr{C}] &= \mathbb{E}[\mathrm{var}[Z|Q,S{=}s,\mathscr{C}]] + \mathrm{var}[\mathbb{E}[Z|Q,S{=}s,\mathscr{C}]] = \\
&= \mathbb{E}[mQ(1-Q)|S{=}s,\mathscr{C}] + \mathrm{var}[mQ|S{=}s,\mathscr{C}] = \\
&= m\mathbb{E}[Q|S{=}s,\mathscr{C}] - m\mathbb{E}[Q^2|S{=}s,\mathscr{C}] + m^2\mathrm{var}[Q|S{=}s,\mathscr{C}] = \\
&= m\mu_g(s) - m\sigma_g^2(s) - m\mu_g^2(s) + m^2\sigma_g^2(s) = \\
&= m\mu_g(s)(1-\mu_g(s)) + m(m-1)\sigma_g^2(s),
\end{aligned}$$

which completes the proof.   □

In practice, the estimates g and true probabilities q are equal for a cluster only approximately, so the equality (7) also holds only approximately. Due to clustering the number of training instances for regression is m times smaller than for the original problem, so learning the reliability map is harder than learning the calibration map. However, the experiments show that with a training set of 2000 instances the learned reliability map can be already accurate enough to improve multi-class probability estimation and classification. Next we describe what regression and clustering methods we are using to achieve this.

3.1   Regression and Clustering Methods for Learning the Maps

First let us stress that there is a wide variety of regression and clustering methods and many could be used for learning calibration and reliability maps. The choice has certainly implications on the performance of multi-class probability estimation and classification, but the comparison of different methods remains as future work. Here we describe the methods we have chosen.

For regression we use local linear regression with the Epanechnikov kernel and fixed bandwidth. For learning the calibration map we have the training pairs (S_i, Y_i) for $i = 1, \ldots n$. The regression estimate $\hat{\mu}_g(s)$ for a target point s is calculated as follows:

$$\hat{\mu}_g(s) = \alpha(s) + \beta(s) \cdot s,$$

$$\alpha(s), \beta(s) = \underset{\alpha,\beta \in \mathbb{R}}{\mathrm{argmin}} \sum_{i=1}^{n} K_\lambda(s, S_i) \cdot (Y_i - \alpha - \beta \cdot S_i)^2$$

where $\lambda > 0$ is the fixed bandwidth of the Epanechnikov kernel K_λ defined as follows:

$$K_\lambda(s, S_i) = \begin{cases} \frac{3}{4}(1 - (s - S_i)^2) & \text{if } |s - S_i| \le 1; \\ 0 & \text{otherwise.} \end{cases}$$

For learning the reliability map we use the same method, that is:

$$\hat{r}_g(s) = \alpha'(s) + \beta'(s) \cdot s,$$

$$\alpha'(s), \beta'(s) = \underset{\alpha', \beta' \in \mathbb{R}}{\operatorname{argmin}} \sum_{i=1}^{n} K_{\lambda'}(s, S_i) \cdot (R_i^{(m)} - \alpha' - \beta' \cdot S_i)^2.$$

Local linear regression can produce estimates outside of our range $[0,1]$ and this problem needs to be addressed. Actually, the extreme values 0 and 1 are also undesired, because they present an over-confident statement about the probabilities. In the experiments we used 0.001 and 0.999 as the lower and upper bound for regression and all estimates outside of this range were changed to these values.

For the clustering method we set the following requirements:

(a) the resulting clusters must all be of fixed size m;
(b) within each cluster the estimate g should be approximately equal;
(c) within each cluster the true posterior q should be approximately equal.

Our first step is to order the instances by g and to cut the ordered list into *super-clusters* of size $k \cdot m$, in the experiments we used $k = 20$ and $m = 10$. With $n \gg k \cdot m$ every super-cluster satisfies requirement (b) to some extent. We then cluster each super-cluster into k clusters of size m according to some distance measure between the instances, in the experiments we used the Euclidean distance. Depending on how smooth q is and how tightly together the instances are, the resulting clusters can satisfy the requirement (c) to some extent. A few instances can remain unclustered to satisfy the requirement (a).

To cluster $k \cdot m$ instances of a super-cluster into k clusters of size m according to some distance measure we modify the DIANA clustering algorithm for this purpose [7]. DIANA is a divisive algorithm which splits at each step one of the existing clusters into two. The splitting is initialised by creating an empty new cluster besides the existing one. Then the algorithm iterates and in each iteration reassigns one instance from the old cluster to the new one. For reassignment it chooses the instance with the largest value for the sum of distances to the instances of the old cluster minus to the new cluster. The original version of the algorithm stops reassignments when the respective value becomes negative, we stop when the size of the new cluster is divisible by m and differs from the size of the old cluster by at most m. The original DIANA has to decide which cluster to split next, for us the order does not matter because of the required fixed size m. Our algorithm ends when all clusters are of size m, except one can be smaller. The smaller cluster is discarded from learning the reliability map.

4   LS-ECOC-R: Multi-class Probability Estimation with Reliabilities

Next we show that the learned calibration and reliability maps $\hat{\mu}_g$ and $\hat{r}_g$ can be used for multi-class probability estimation with ECOC. The ECOC decomposition of a K-class task into L binary tasks is represented as a code matrix $M \in \{-1, 0, +1\}^{K \times L}$. The binary task represented by column l aims at discriminating between the positive group of classes $\mathscr{C}_l^+ = \{k | M_{k,l} = +1\}$ and the negative group of classes $\mathscr{C}_l^- = \{k | M_{k,l} = -1\}$.

The neutral group of classes $\mathcal{C}_l^0 = \{k | M_{k,l} = 0\}$ is excluded from the training set for the l-th binary model. Suppose that for a given coding matrix M we have trained L binary probability estimators $g_l : \mathcal{X} \to [0,1]$ for tasks $l = 1, \ldots, L$, and learned their calibration maps $\hat{\mu}_{g_l}$ and reliability maps $\hat{r}_{g_l}$. LS-ECOC [8] estimates multi-class posterior probabilities by combining the calibrated probability estimates $\hat{\mu}_{g_l}(g_l(x))$ only.

First denote by $q_l(x)$ the true posterior of the positive group given that instance x is not neutral:

$$q_l(x) = P(Y \in \mathcal{C}_l^+ | Y \in \mathcal{C}_l^{\pm}, X = x) = \frac{\sum_{k \in \mathcal{C}_l^+} p_k}{\sum_{k \in \mathcal{C}_l^{\pm}} p_k}$$

where $\mathcal{C}_l^{\pm} = \mathcal{C}_l^+ \cup \mathcal{C}_l^-$ and $p_k = P(Y = k | X = x)$. Let ε_l be the error of $\hat{\mu}_{g_l}(g_l(x))$ in estimating the true $q_l(x)$:

$$\varepsilon_l = q_l(x) - \hat{\mu}_{g_l}(g_l(x))$$

The idea of LS-ECOC is to estimate the posterior probabilities p_k such that the total squared error $\sum_{l=1}^L \varepsilon_l^2$ is minimised:

$$\hat{p} = \underset{\substack{p_k \geq 0 \\ \sum p_k = 1}}{\operatorname{argmin}} \sum_{l=1}^L \varepsilon_l^2 = \underset{\substack{p_k \geq 0 \\ \sum p_k = 1}}{\operatorname{argmin}} \sum_{l=1}^L \left(\frac{\sum_{k \in \mathcal{C}_l^+} p_k}{\sum_{k \in \mathcal{C}_l^{\pm}} p_k} - \hat{\mu}_{g_l}(g_l(x)) \right)^2$$

If $\sum_{k \in \mathcal{C}_l^{\pm}} p_k = 1$ for each l, that is if the coding matrix is actually binary, then this is a straightforward least-squares optimisation with linear constraints which is convex and can easily be solved to estimate $\hat{p}$. The optimisation for ternary coding matrices can in general be non-convex.

Effectively, LS-ECOC assumes that ε_l is normally distributed around 0 with the same variance for all l. Therefore, LS-ECOC is equally confident in each value of $\hat{\mu}_{g_l}(g_l(x))$ regardless of which binary model it is resulting from and what the value of the estimate is. For example, the calibrated probability estimates 0.01 from model g_1 and 0.5 from model g_2 are equally likely to be off by 0.1 according to LS-ECOC.

This is where we can benefit from the learned reliability map $\hat{r}_{g_l}$. We propose a variant of LS-ECOC which we denote LS-ECOC-R (for LS-ECOC with reliability estimates). LS-ECOC-R assumes that ε_l is normally distributed around 0 with variance $\hat{\sigma}_g^2(g_l(x))$ where $\hat{\sigma}_g^2$ is calculated due to (4) as follows:

$$\hat{\sigma}_g^2(s) = (1 - \hat{r}_g(s)) \cdot \hat{\mu}_g(s) \cdot (1 - \hat{\mu}_g(s)). \tag{8}$$

So there is potentially a different level of confidence in each probability estimate for each instance. The multi-class probability estimates with LS-ECOC-R are obtained as follows:

$$\hat{p} = \underset{\substack{p_k \geq 0 \\ \sum p_k = 1}}{\operatorname{argmin}} \sum_{l=1}^L \frac{\varepsilon_l^2}{\hat{\sigma}_g^2(g_l(x))} = \underset{\substack{p_k \geq 0 \\ \sum p_k = 1}}{\operatorname{argmin}} \sum_{l=1}^L \left(\frac{\sum_{k \in \mathcal{C}_l^+} p_k}{\hat{\sigma}_{g_l}(g_l(x)) \sum_{k \in \mathcal{C}_l^{\pm}} p_k} - \frac{\hat{\mu}_{g_l}(g_l(x))}{\hat{\sigma}_{g_l}(g_l(x))} \right)^2 .$$

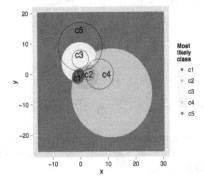

Fig. 2. Generative model for the synthetic dataset with 5 Gaussians with black circles centered at the means μ_c and radius equal to respective σ_c. Bayes-optimal decision regions are coloured by classes.

5   Experimental Evaluation

In the experiments we have three objectives. First, we demonstrate that the proposed learning methods can indeed provide good estimates of the calibration and reliability maps. For this we use a synthetic dataset where we know exactly the true calibration and reliability maps and can therefore compare our estimates to ground truth. Second, we show on the same dataset that LS-ECOC-R using the estimated reliability map outperforms LS-ECOC on probability estimation and classification. Finally, we show that LS-ECOC-R outperforms LS-ECOC also on 6 real datasets.

5.1   Experiments on Synthetic Data

As we need to know the true posterior distribution for an in-depth evaluation, we generate synthetic data with a probabilistic generative model. We use the same model that has earlier been used in several papers relating to multi-class probability estimation [16,15,18]. This generative model has 5 equiprobable classes and the instances of each class are distributed as a 2-dimensional normal distribution with the following parameters:

	class 1	class 2	class 3	class 4	class 5
μ_c	(0,0)	(3,0)	(0,5)	(7,0)	(0,9)
σ_c^2	1	4	9	25	64

where μ_c is the mean and the covariance matrix is the unit matrix multiplied by σ_c^2. Figure 2 shows in colours the Bayes-optimal decision regions of this probabilistic model and the black circles are centered at the means of the Gaussians and have radius equal to the respective σ_c.

In order to evaluate our calibration and reliability learning algorithms we consider binary base estimators for which we can calculate the true calibration and reliability

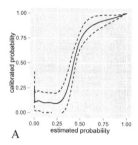
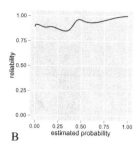

Fig. 3. Solid lines are the true (A) calibration map μ_{g_l} and (B) reliability map r_{g_l} for the logistic regression model g_l comparing classes 3 and 5 against others for the synthetic dataset. The dashed lines in (A) mark $\mu_{g_l} - \sigma_{g_l}$ and $\mu_{g_l} + \sigma_{g_l}$.

maps. This is possible whenever the contours of equal probability estimate are straight lines. Along any straight line the instances of each class have univariate Gaussian distribution, and we can analytically derive the class proportions and the parameters of the respective Gaussians. The true calibration maps can be determined using the class proportions and the reliability maps by numerical integration of the product of a density ratio and mixture density across the line. We choose logistic regression as our base model for the synthetic task, as it is used often for probability estimation and has linear contours.

For a 5-class problem there are 15 different ways to split the classes into two groups, which correspond to 5 one-vs-four (one-vs-rest) and 10 two-vs-three (pair-vs-rest) tasks. We first generate $n = 400$ (and repeat the same with 2000 and 10000) training instances and train a logistic regression model g_l for each of these tasks, $l = 1, 2, \ldots, 15$. Now we can calculate the true calibration maps μ_{g_l} and reliability maps r_{g_l}. For the model comparing classes 3 and 5 against others these are plotted in Fig. 3. We next learn the calibration maps $\hat{\mu}_{g_l}$ and reliability maps $\hat{r}_{g_l}$ using our method described in previous sections. For the clustering method we use super-cluster size 200 and cluster size 10. For the regression task in learning calibration maps we test bandwidth values 0.005, 0.01, 0.02 and 0.05 and find 0.01 as the best performer. For reliability we use then bandwidth 0.1 as there are $m = 10$ times less instances for training the regression model.

Ultimately we are going to use the estimated calibration and reliability maps for multi-class probability estimation. Therefore, we need the estimated and true distribution of $q(X)$ given $g(X)$ to be maximally alike. As LS-ECOC and LS-ECOC-R both assume Gaussian distribution, we assess how close to each other are $\mathcal{N}(\hat{\mu}_{g_l}, \hat{\sigma}_{g_l}^2)$ and $\mathcal{N}(\mu_{g_l}, \sigma_{g_l}^2)$, averaged over all instances, where $\sigma_{g_l}^2$ and $\hat{\sigma}_{g_l}^2$ can be calculated as in (8) from the true and estimated calibration and reliability maps, respectively. As a distance measure we use Cramér distance, which is half of the energy distance [14]. Intuitively, it measures how much work has to be done carrying pixels from one density plot to another. The advantage over Kullback-Leibler divergence is that for equal variance it measures the distance of means. Therefore its value is easier to interpret and perhaps ultimately more relevant for multi-class probability estimation and classification.

With few training instances the regression learner for the reliability map can have a large variance. Therefore, we consider in addition to our local linear regression method an averaging regression method which produces constant reliability across all values of s. The method calculates the average of $R^{(m)}$ over all training instances. Table 1 shows the results of the comparison between the estimated and true calibration and probability maps for various training set sizes and the two regression methods. Cramér distance is averaged over results on 10 independently generated training and test sets, the test set size is 10000. For bandwidths we use 0.01 for calibration and 0.1 for reliability, as these are the best according to the results shown later. The results indicate that we are able to learn reliability maps that are better than the average line for already a training set of size 400.

Next we proceed to evaluation of multi-class probabilities that can be obtained using LS-ECOC-R from the learned calibration and reliability maps. For evaluating the probability estimates we use the following measures: root mean square error (RMSE), mean absolute error (MAE), Pearson correlation (Pearson) and Brier score (Brier). For evaluating classification performance we use Error rate (Error) and Error compared to the Bayes-optimal class (ErrVsOpt). Zhou et al. have recently compared 9 ECOC decoding algorithms with regards to classification performance and LS-ECOC out-performed other methods for 7 out of 8 datasets [18]. We use the same datasets (leaving out the two smallest), therefore we compare LS-ECOC-R only against LS-ECOC.

On the synthetic dataset we consider three coding matrices — 'one-vs-rest' (5 columns), 'pair-vs-rest' (10 columns) and 'all' (15 columns). First we study which bandwidth is best for calibration, in order to ensure that we use LS-ECOC at its best. The RMSE between the true and LS-ECOC estimates of multi-class probabilities is presented in Table 2 for three matrices and training set sizes 400, 2000, 10000. The results are averaged over 10 runs over independently generated datasets. Relying on these results we have decided to use bandwidth 0.01 for calibration throughout the paper. As reliability maps are learned on 10 times fewer values because of the clustering, we have chosen 0.1 as the bandwidth for learning reliability.

Table 1. Comparison of the standard (REL) and averaging (R-AVE) regression method for learning reliabilities, assessed by Cramér distance and averaged over 10 runs. Bandwidths for calibration and regression are 0.01 and 0.1, respectively.

n	method	1vsR	2vsR	3vsR	4vsR	5vsR	12vsR	13vsR	14vsR	15vsR	23vsR	24vsR	25vsR	34vsR	35vsR	45vsR
400	R-AVE	.0208	.0253	.0271	.0589	.0567	.0321	.0336	.0577	.0483	.0386	.0678	.0453	.0430	.0684	.0358
400	REL	.0175	.0233	.0258	.0588	.0564	.0303	.0327	.0564	.0481	.0382	.0672	.0451	.0431	.0683	.0354
2000	R-AVE	.0081	.0080	.0093	.0209	.0162	.0101	.0102	.0165	.0137	.0143	.0242	.0127	.0096	.0228	.0096
2000	REL	.0046	.0063	.0079	.0203	.0160	.0087	.0097	.0159	.0136	.0139	.0235	.0128	.0095	.0227	.0095
10000	R-AVE	.0049	.0038	.0046	.0076	.0043	.0036	.0029	.0054	.0034	.0108	.0072	.0039	.0031	.0073	.0025
10000	REL	.0018	.0023	.0035	.0071	.0041	.0025	.0025	.0048	.0032	.0103	.0067	.0037	.0027	.0072	.0024

Table 2. RMSE of multi-class probability estimates obtained with LS-ECOC on three different ECOC matrices and training set sizes 400, 2000, 10000. Results are averaged over 10 runs.

matrix	1vsR			2vsR			all		
λ_{cal} n	400	2000	10000	400	2000	10000	400	2000	10000
.005	.1933	.1621	.1563	.1643	.1478	.1449	.1567	.1421	.1392
.01	.1729	.1594	.1562	.1563	.1480	.1464	.1494	.1425	.1406
.02	.1656	.1597	.1578	.1566	.1515	.1507	.1503	.1461	.1450
.05	.1708	.1673	.1667	.1664	.1635	.1632	.1616	.1588	.1584

Table 3 compares the performance of LS-ECOC, LS-ECOC-R-ave and LS-ECOC-R across three matrices and training set sizes 400, 2000, 10000. Both RMSE and error rate are averaged over 10 runs. We also provide the standard deviation estimates for these values, calculated as the sample standard deviation over square root of 10 (the number of runs). The results indicate that both LS-ECOC-R methods outperform LS-ECOC on all cases with $n \geq 2000$ and in some cases with $n = 400$. The full version of LS-ECOC-R performs better than the averaged, with some exceptions for $n = 400$.

It is in principle possible to improve LS-ECOC by improving the calibration map learning method. The following results show that LS-ECOC-R remains superior, with sufficient data given. For this we apply LS-ECOC on the true calibration map (not the estimated one) and consider this as the performance-bound for LS-ECOC. Table 4 compares the results of LS-ECOC, LS-ECOC-R-ave and LS-ECOC-R with the results obtained on the true calibration and reliability maps, averaged over 10 runs. The results indicate that each following method is better than the upper bound for the previous method. Therefore, on this dataset LS-ECOC-R remains superior to LS-ECOC even if calibration is perfect.

5.2 Experiments on Real Data

Finally we show that LS-ECOC-R outperforms LS-ECOC on some real datasets. For this purpose we use 6 UCI datasets shown in Table 5. As the binary base model we use logistic regression and support vector machines with polynomial kernel. Zhou et al. used also the polynomial kernel and published the best degree for it according to cross-validation results [18]. We use the same degree, for easier comparison with their work.

Table 3. Comparison of LS-ECOC, LS-ECOC-R-ave and LS-ECOC-R across three matrices and three training set sizes. RMSE and error rate are both averaged over 10 runs and standard deviation is calculated as the sample standard deviation over square root of 10.

n	method	RMSE 1vsR	RMSE 2vsR	RMSE all	Error 1vsR	Error 2vsR	Error all
400	LS-ECOC	.1729 ± .0015	.1563 ± .0012	.1494 ± .0009	.3860 ± .0065	.3450 ± .0049	.3265 ± .0044
400	LS-ECOC-R-ave	.1774 ± .0023	.1526 ± .0020	.1456 ± .0025	.3872 ± .0056	.3572 ± .0051	.3366 ± .0049
400	LS-ECOC-R	.1788 ± .0025	.1527 ± .0021	.1475 ± .0025	.3852 ± .0058	.3464 ± .0055	.3325 ± .0059
2000	LS-ECOC	.1594 ± .0006	.1480 ± .0004	.1425 ± .0004	.3588 ± .0032	.3392 ± .0027	.3184 ± .0031
2000	LS-ECOC-R-ave	.1487 ± .0006	.1314 ± .0005	.1135 ± .0011	.3539 ± .0029	.3367 ± .0025	.3034 ± .0030
2000	LS-ECOC-R	.1468 ± .0006	.1243 ± .0007	.1102 ± .0011	.3493 ± .0026	.3224 ± .0030	.3004 ± .0030
10000	LS-ECOC	.1562 ± .0004	.1464 ± .0003	.1406 ± .0003	.3509 ± .0020	.3350 ± .0017	.3143 ± .0018
10000	LS-ECOC-R-ave	.1448 ± .0005	.1290 ± .0003	.1111 ± .0004	.3490 ± .0018	.3344 ± .0016	.2996 ± .0019
10000	LS-ECOC-R	.1429 ± .0005	.1227 ± .0002	.1079 ± .0003	.3449 ± .0018	.3214 ± .0016	.2987 ± .0017

Table 4. Comparison of LS-ECOC, LS-ECOC-R-ave and LS-ECOC-R with the respective methods which use true calibration and reliability maps (bound). The results were obtained on the full ECOC matrix 'all', on 10000 training instances, and are averaged over 10 runs.

method	RMSE	MAE	Pearson	Brier	Error	ErrVsOpt
LS-ECOC	.1406	.1046	.8801	.2499	.3143	.1327
LS-ECOC bound	.1356	.0976	.8843	.2463	.3108	.1239
LS-ECOC-R-ave	.1111	.0745	.9253	.2314	.2996	.0943
LS-ECOC-R-ave bound	.1093	.0727	.9269	.2303	.2988	.0929
LS-ECOC-R	.1079	.0712	.9316	.2297	.2987	.0914
LS-ECOC-R bound	.1018	.0666	.9411	.2264	.2928	.0810

There are many possible choices for the ECOC matrix, some even domain-specific, in the sense that each column is chosen based on the performance of the models for the previous columns. In the LS-ECOC experiments of Zhou et al. the equidistant code matrices performed best among the domain-independent matrices for 5 of our 6 datasets [18]. We therefore use equidistant code matrices, which we create using BCH codes. We have built a 15x15 binary matrix which was obtained by creating a BCH code with code-length 15 and aligning all 15 code-words with exactly seven 1's as columns of the matrix. The Hamming distance between each pair of columns and each pair of rows is exactly 8, making it equidistant. Another nice property is that any top k rows with $k \geq 5$ have all columns splitting the classes differently into two groups. For a k-class problem we use this matrix for our experiments if $k \geq 5$ and for $k = 4$ we use the matrix with all 7 different splits of classes into two.

We perform 10-fold cross-validation and use error-rate as the evaluation measure. The sample mean and variance of the training set are used to normalize each input feature individually. Before calibration, the scores of SVM are transformed using the standard logistic map. The calibration and reliability maps are learned with bandwidths 0.01 and 0.1, respectively — the same which performed best on the synthetic data.

Table 5 lists the error rate for methods LS-ECOC, LS-ECOC-R-ave, LS-ECOC-R on the 6 datasets with two different base learners. We first note that LS-ECOC results in our experiments are superior to the LS-ECOC results by Zhou et al. [18], probably due to differences in normalization and calibration. To assess the differences between LS-ECOC and the two variants of LS-ECOC-R we have performed significance tests with t-test on confidence level 95%. The stars in the table indicate which errors of LS-ECOC-R are significantly lower than the respective error for LS-ECOC. To conclude, LS-ECOC-R outperforms LS-ECOC on four larger datasets ($n \geq 2000$), with significance in 3 out of the 4.

6   Related Work

The reliability of model predictions has been studied before but mostly in the context of regression, where it is known as conditional variance estimation [5]. Conformal

Table 5. The comparison of 10-fold cross-validated error rate of classification for LS-ECOC, LS-ECOC-R-ave and LS-ECOC-R on 6 real datasets with n instances, a attributes and k classes. The stars in the table indicate which errors of LS-ECOC-R are significantly lower than respective error for LS-ECOC according to t-test at confidence level 95%.

dataset	n	a	k	model	LS-ECOC	LS-ECOC-R-ave	LS-ECOC-R
shuttle	14500	9	7	LR	.0383± .0016	.0259± .0014 *	.0323± .0017 *
				SVM	.0914± .0015	.0888± .0017	.0859± .0018 *
sat	6435	36	6	LR	.1713± .0021	.1514± .0028 *	.1489± .0027 *
				SVM	.1737± .0026	.1554± .0021 *	.1610± .0026 *
page-blocks	5473	10	5	LR	.0453± .0025	.0420± .0034	.0426± .0035
				SVM	.0426± .0021	.0429± .0026	.0411± .0028
segment	2310	19	7	LR	.0887± .0040	.0775± .0036	.0753± .0046 *
				SVM	.0987± .0041	.0788± .0036 *	.0771± .0031 *
yeast	1481	8	10	LR	.4327± .0134	.4279± .0124	.4314± .0165
				SVM	.4084± .0147	.4246± .0145	.4198± .0132
vehicle	846	18	4	LR	.2174± .0088	.2258± .0085	.2081± .0136
				SVM	.2316± .0125	.2375± .0113	.2553± .0136

prediction is a general approach that can be applied to both regression and classification in an on-line setting. It outputs a so-called region prediction which might be a confidence interval (in regression) or a set of possible values (in classification) that contains the true value with a certain level of confidence [13]. This is different from the approach in this paper where we try to assess the uncertainty associated with a point estimate.

In the area of multi-class classification and probability scores, the original error-correcting output codes method is due to [4]. The least-squares method for obtaining multi-class posterior probabilities was developed not much later by [8], but does not appear to be widely known. Better-known is the loss-based decoding method by [1], which takes classifier margins rather than probabilities as input and outputs classes rather than posterior probabilities. A review on combinations of binary classifiers in a multi-class setting is given by [9]. [16] study coding and decoding strategies in ECOC, and also originated the synthetic 5-class data set we used in this paper. The same dataset was used by [15] and [18] to study the behaviour of LS-ECOC.

Calibration of multi-class posterior probabilities is often studied in a cost-sensitive setting [17,11]. The effect of calibration in classifier combination is studied by [2]. Perhaps closest in spirit to our work in this paper is the work by [6] who propose methods to identify and remove unreliable classifiers in a one-vs-one setting. Also related is the work on neighborhood-based local sensitivity by [3].

7   Concluding Remarks

Assessing the reliability of probability scores in classification is clearly an important task if we want to combine scores from different classifiers. If we want to combine two scores of 0.5 and 0.3, say, it makes a difference if one of them is deemed much more reliable than the other. Yet the problem of estimating this reliability in an instance-wise manner appears not to have been widely studied in the machine learning literature. In this paper we present a theoretically well-founded and practically feasible approach to the problem. We demonstrate the quality of the reliability estimates both in comparison with the true values on synthetic data, and in obtaining well-calibrated multi-class probability scores through the improved LS-ECOC-R method.

The paper opens many avenues for further work. We are particularly interested in developing cost models in cost-sensitive classification that can take these reliability estimates into account. Incorporating instance-wise confidence ratings into boosting also appears a fruitful research direction.

Acknowledgments We would like to thank Tijl De Bie and Thomas Gärtner for helpful discussions. This work is supported by the REFRAME project granted by the European Coordinated Research on Long-term Challenges in Information and Communication Sciences & Technologies ERA-Net (CHIST-ERA), and funded by the Engineering and Physical Sciences Research Council in the UK under grant EP/K018728/1.

References

1. Allwein, E., Schapire, R., Singer, Y.: Reducing multiclass to binary: A unifying approach for margin classifiers. Journal of Machine Learning Research 1, 113–141 (2001)
2. Bella, A., Ferri, C., Hernández-Orallo, J., Ramírez-Quintana, M.J.: On the effect of calibration in classifier combination. Applied Intelligence 38(4), 566–585 (2012)
3. Bennett, P.N.: Neighborhood-based local sensitivity. In: Kok, J.N., Koronacki, J., Lopez de Mantaras, R., Matwin, S., Mladenič, D., Skowron, A. (eds.) ECML 2007. LNCS (LNAI), vol. 4701, pp. 30–41. Springer, Heidelberg (2007)
4. Dietterich, T., Bakiri, G.: Solving Multiclass Learning Problems via Error-Correcting Output Codes. Journal of Artificial Intelligence Research 2, 263–286 (1995)
5. Fan, J., Yao, Q.: Efficient estimation of conditional variance functions in stochastic regression. Biometrika 85(3), 645–660 (1998)
6. Galar, M., Fernández, A., Barrenechea, E., Bustince, H., Herrera, F.: Dynamic classifier selection for one-vs-one strategy. Pattern Recognition 46, 3412–3424 (2013)
7. Kaufman, L., Rousseeuw, P.J.: Finding groups in data: an introduction to cluster analysis, vol. 344. John Wiley & Sons (2009)
8. Kong, E.B., Dietterich, T.: Probability estimation via error-correcting output coding. In: International Conference of Artificial Intelligence and Soft Computing, Banff, Canada (1997)
9. Lorena, A.C., Carvalho, A., Gama, J.: A review on the combination of binary classifiers in multiclass problems. Artificial Intelligence Review 30, 19–37 (2009)
10. Murphy, A.H.: A new vector partition of the probability score. Journal of Applied Meteorology 12(4), 595–600 (1973)
11. O'Brien, D., Gupta, M., Gray, R.: Cost-sensitive multi-class classification from probability estimates. In: Proceedings of the 25th International Conference on Machine Learning, Helsinki, Finland, pp. 712–719 (2008)
12. Schapire, R.E., Singer, Y.: Improved boosting algorithms using confidence-rated predictions. Machine Learning 37(3), 297–336 (1999)
13. Shafer, G., Vovk, V.: A tutorial on conformal prediction. Journal of Machine Learning Research 9, 371–421 (2008)
14. Székely, G.J., Rizzo, M.L.: A new test for multivariate normality. Journal of Multivariate Analysis 93(1), 58–80 (2005)
15. Wang, X., Zhou, J.: Research on the characteristic of the probabilistic outputs via LS-ECOC. In: Eighth International Conference on Fuzzy Systems and Knowledge Discovery, vol. 2, pp. 1330–1334. IEEE (2011)
16. Windeatt, T., Ghaderi, R.: Coding and decoding strategies for multi-class learning problems. Information Fusion 4(1), 11–21 (2003)
17. Zadrozny, B., Elkan, C.: Transforming classifier scores into accurate multiclass probability estimates. In: Eighth ACM SIGKDD International Conference, pp. 694–699. ACM Press, New York (2002)
18. Zhou, J.D., Wang, X.D., Song, H.: Research on the unbiased probability estimation of error-correcting output coding. Pattern Recognition 44(7), 1552–1565 (2011)

Causal Clustering for 2-Factor Measurement Models

Erich Kummerfeld, Joe Ramsey, Renjie Yang, Peter Spirtes,
and Richard Scheines*

Department of Philosophy, Carnegie Mellon University

Abstract. Many social scientists are interested in inferring causal relations between "latent" variables that they cannot directly measure. One strategy commonly used to make such inferences is to use the values of variables that can be measured directly that are thought to be "indicators" of the latent variables of interest, together with a hypothesized causal graph relating the latent variables to their indicators. To use the data on the indicators to draw inferences about the causal relations between the latent variables (known as the *structural model*), it is necessary to hypothesize causal relations between the indicators and the latents that they are intended to indirectly measure, (known as the *measurement model*). The problem addressed in this paper is how to reliably infer the measurement model given measurements of the indicators, without knowing anything about the structural model, which is ultimately the question of interest. In this paper, we develop the *Find-TwoFactorClusters* (*FTFC*) algorithm, a search algorithm that, when compared to existing algorithms based on vanishing tetrad constraints, also works for a more complex class of measurement models, and does not assume that the model describing the causal relations between the latent variables is linear or acyclic.

1   Introduction

Social scientists are interested in inferring causal relations between "latent" variables that they cannot directly measure. For example, Bongjae Lee conducted a study in which the question of interest was the causal relationships between *Stress*, *Depression*, and (religious) *Coping*. One strategy commonly used to make such inferences is to use the values of variables that can be measured directly (e.g. answers to questions in surveys) that are thought to be "indicators" of the latent variables of interest, together with a hypothesized causal graph relating the latent variables to their indicators. A model in which each latent variable of interest is measured by multiple indicators (which may also be caused by other latents of interest as well as by an error variable) is called a *multiple indicator model* [1]. Lee administered a questionnaire to 127 students containing questions

* We thank NSF for providing funding through grant 1317428 for this work. We thank anonymous referees for helpful suggestions.

T. Calders et al. (Eds.): ECML PKDD 2014, Part II, LNCS 8725, pp. 34–49, 2014.
© Springer-Verlag Berlin Heidelberg 2014

whose answers were intended to be indicators of *Stress*, *Depression*, and *Coping*. There were 21 questions relating to *Stress* (such as meeting with faculty, etc.) which students were asked to rate on a seven point scale, and similar questions for the other latents [2].

To use the data on the indicators to draw inferences about the causal relations between the latents (known as the *structural model*), it is necessary to hypothesize causal relations between the indicators and the latents that they are intended to indirectly measure (i.e. the subgraph containing all of the vertices, and all of the edges except for the edges between the latent variables, known as the *measurement model*). Given the measurement model, there are well known algorithms for making inferences about the structural model [2]. The problem addressed in this paper is how to reliably infer the measurement model given sample values of the indicators, without knowing anything about the structural model. In [2], Silva et al. developed an algorithm that reliably finds certain kinds of measurement models without knowing anything about the structural model other than its linearity and acyclicity. Their method first employs a clustering method to identify "pure" measurement sub-models (discussed below). (Note that in this context, *variables* rather than *individuals* are being clustered.) In this paper, we develop the *FindTwoFactorClusters* (*FTFC*) algorithm, an algorithm for reliably generating pure measurement submodels on a much wider class of measurement models, and does not assume that the model describing the causal relations between the latent variables is linear or acyclic.

1.1   Structural Equation Models (SEMs)

We represent causal structures as structural equation models (SEMs). In what follows, random variables are in italics, and sets of random variables are in boldface. Linear structural equation models are described in detail in [3]. In a structural equation model (SEM) the random variables are divided into two disjoint sets, the *substantive variables* (typically the variables of interest) and the *error variables* (summarizing all other variables that have a causal influence on the substantive variables) [3]. Corresponding to each substantive random variable V is a unique error term ϵ_V. A *fixed parameter SEM S* has two parts $\langle \phi, \theta \rangle$, where ϕ is a set of equations in which each substantive random variable V is written as a function of other substantive random variables and a unique error variable, together with θ, the joint distributions over the error variables. Together ϕ and θ determine a joint distribution over the substantive variables in S, which will be referred to as the distribution entailed by S. A *free parameter linear SEM model* replaces some of the real numbers in the equations in ϕ with real-valued variables and a set of possible values for those variables, e.g. $X = a_{X,L}L + \epsilon_X$, where $a_{X,L}$ can take on any real value. In addition, a free parameter SEM can replace the particular distribution over ϵ_X and ϵ_L with a parametric family of distributions, e.g. the bi-variate Gaussian distributions with zero covariance. The free parameter SEM also has two parts $\langle \Phi, \Theta \rangle$, where Φ contains the set of equations with free parameters and the set of values the free parameters are allowed to take, and Θ is a family of distributions over the error variables. In general, we will assume that there is a finite set of free parameters, and all allowed values of the free parameters lead to fixed parameter SEMs that have a reduced form (i.e. each substantive variable X can be expressed as a function of the error variables of X and the error variables of its ancestors), all variances and partial variances among the substantive variables are finite and positive, and there are no deterministic relations among the measured variables.

The *path diagram* (or *causal graph*) of a SEM with is a directed graph, written with the conventions that it contains an edge $B \to A$ if and only if B is a non–trivial argument of the equation for A. The error variables are not included in the path diagram unless they are correlated, in which case they are included and a double-headed arrow is placed between them. A fixed-parameter acyclic structural equation model (without double-headed arrows) is an instance of a Bayesian Network $\langle G, P(V) \rangle$, where the path diagram is G, and $P(V)$ is the joint distribution over the variables in G entailed by the set of equations and the joint distribution over the error variables, which in this case is just the product of the marginal distribution over the error variables [4]. A polynomial equation Q where the variables represent covariances is *entailed* by a free parameter SEM when all values of the free parameters entail covariance matrices that are solutions to Q. For example, a vanishing tetrad difference holds among $\{X, W\}$ and $\{Y, Z\}$, iff $cov(X, Y)cov(Z, W) - cov(X, Z)cov(Y, W) = 0$, and is entailed by a free parameter linear SEM S in which X, Y, Z, and W are all children of just one latent variable L.

1.2   Pure 2-Factor Measurement Models

In *1–factor measurement models* and *2–factor measurement models* each indicator has the specified number of latent parents in addition to its "error" variable. There is often no guarantee, however, that the indicators do not have unwanted additional latent common causes, or that none of the indicators are causally influenced by any other indicators. However, pure measurement models (defined below) have properties described below that make them easy to find, regardless of the structural models, and for that reason the strategy we will adopt in this paper is to search for a subset of variables that form a pure measurement model. In what follows, we will assume that no measured variable (indicator) causes a latent variable.

A set of variables $\mathbf{V}$ is *minimally causally sufficient* when every cause of any two variables in $\mathbf{V}$ is also in $\mathbf{V}$, and no proper subset of $\mathbf{V}$ is causally sufficient. If $\mathbf{O}$ is a set of indicators, and $\mathbf{V}$ is a minimally causally sufficient set of variables containing $\mathbf{O}$, then an *n-factor model* for $\mathbf{V}$ is a model in which there is a partition P of the indicators, and where each element of the partition is a set of indicators, all of which have exactly n latent parents, and that share the same n latent parents; if in addition there are no other edges (either directed, or bidirected representing correlated errors) into or out of any of the indicators the measurement model is said to be *pure*. We will refer to any n-factor model whose measurement model is pure as a *pure n-factor model*. Figure 1 is not a pure 2-factor measurement model. There are three reasons for this: X_1 causes

X_9, X_{15} has three latent direct causes, L_2, L_3, and L_4, and there is a latent cause L_5 of X_8 and L_1. However, note that the sub-model that does not contain the vertices X_1, X_8, X_9 and X_{15} is a 2-pure measurement model, because when those variables are not included, there are no edges out of any indicator, and the only edges into each indicator are from their two latent parents.

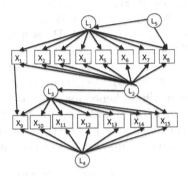

Fig. 1. Impure 2-factor model

Given a measurement model, any subset $\mathbf{S}$ of $\mathbf{O}$ for which every member of $\mathbf{S}$ is a child of the same n latent parents (and has no other parents), is adjacent to no other member of $\mathbf{O}$, and has a correlated error with no other member of $\mathbf{V}$, is an n-*pure* subset. In Figure 1, $\{X_2, X_3, X_4, X_5, X_6, X_7\}$ is a 2-pure sextet, but $\{X_{10}, X_{11}, X_{12}, X_{13}, X_{14}, X_{15}\}$ and $\{X_2, X_3, X_4, X_{10}, X_{11}, X_{12}\}$ are not.

2   Trek Separation

This section describes the terminology used in this paper. A *simple trek* in directed graph G from i to j is an ordered pair of directed paths $(P_1; P_2)$ where P_1 has sink i, P_2 has sink j, and both P_1 and P_2 have the same source k, and the only common vertex among P_1 and P_2 is the common source k. One or both of P_1 and P_2 may consist of a single vertex, i.e., a path with no edges. There is a trek between a set of variables $\mathbf{V_1}$ and a set of variables $\mathbf{V_2}$ iff there is a trek between any member of $\mathbf{V_1}$ and any member of $\mathbf{V_2}$. Let $\mathbf{A}$, $\mathbf{B}$, be two disjoint subsets of vertices $\mathbf{V}$ in G, each with two vertices as members. Let $\mathbf{S(A, B)}$ denote the sets of all simple treks from a member of $\mathbf{A}$ to a member of $\mathbf{B}$.

Let $\mathbf{A}$, $\mathbf{B}$, $\mathbf{C_A}$, and $\mathbf{C_B}$ be four (not necessarily disjoint) subsets of the set $\mathbf{V}$ of vertices in G. The pair $(\mathbf{C_A}; \mathbf{C_B})$ *t-separates* $\mathbf{A}$ from $\mathbf{B}$ if for every trek $(P_1; P_2)$ from a vertex in $\mathbf{A}$ to a vertex in $\mathbf{B}$, either P_1 contains a vertex in $\mathbf{C_A}$ or P_2 contains a vertex in $\mathbf{C_B}$; $\mathbf{C_A}$ and $\mathbf{C_B}$ are *choke sets* for $\mathbf{A}$ and $\mathbf{B}$ [6]. Let $\#\mathbf{C}$ be the number of vertices in $\mathbf{C}$. For a choke set $(\mathbf{C_A}; \mathbf{C_B})$, $\#\mathbf{C_A} + \#\mathbf{C_B}$ is the *size of the choke set*. We will say that a vertex X is in a choke set $(\mathbf{C_A}; \mathbf{C_B})$ if $X \in \mathbf{C_A} \cup \mathbf{C_B}$.

The exact definition of linear acyclicity (or LA for short) below a choke set is somewhat complex and is described in detail in [6]; for the purposes of this paper it suffices to note that roughly speaking a directed graphical model is *LA below sets* (C_A; C_B) for A and B respectively, if there are no directed cycles between C_A and A or C_B and B, and A is a linear function with additive noise of C_A, and similarly for B and C_B.

For two sets of variables A and B, and a covariance matrix over a set of variables V containing A and B, let $cov(A, B)$ be the sub-matrix of the covariance matrix that contains the rows in A and columns in B. In the case where A and B both have 3 members, if the rank of the $cov(A, B)$ is less than or equal to 2, the determinant of $cov(A, B) = 0$. In that case the matrix is said to satisfy a *vanishing sextad constraint* since there are six members of $A \cup B$ if A and B are disjoint. For any given set of six variables, there are 10 different ways of partitioning them into two sets of three; hence for a given sextet of variables there are 10 distinct possible vanishing sextad constraints. The following two theorems [6] (extensions of the theorems in [5]) relate the structure of the causal graph to the rank of the determinant of sub-matrices of the covariance matrix.

Theorem 1. *(Extended Trek Separation Theorem): Suppose G is a directed graph containing C_A, A, C_B, and B, and (C_A;C_B) t-separates A and B in G. Then for all covariance matrices entailed by a fixed parameter structural equation model S with path diagram G that is LA below the sets C_A and C_B for A and B, $rank(cov(A, B)) \leq \#C_A + \#C_B$.*

Theorem 2. *For all directed graphs G, if there does not exist a pair of sets C'_A, C'_B, such that (C'_A; C'_B) t-separates A and B and $\#C'_A + \#C'_B \leq r$, then for any C_A, C_B there is a fixed parameter structural equation model S with path diagram G that is LA below the choke sets (C_A; C_B) for A and B that entails $rank(cov(A, B)) > r$.*

Theorem 1 guarantees that trek separation entails the corresponding vanishing sextad for all values of the free parameters, and Theorem 2 guarantees that if the trek separation does not hold, it is not the case that the corresponding vanishing sextad will hold for all values of the free parameters. It is still possible that if the vanishing sextad does not hold for all values of the free parameters, it will hold for some values of the free parameters. See [6].

3   Algorithm

Before stating the *sample* version of the algorithm (described below), we will motivate the intuitions behind it by an example (Figure 1). Let a *vanishing sextet* be a set of 6 indicators in which all ten sextads among the six variables are entailed to vanish by the Extended Trek Separation Theorem. In general, 2-pure sets of 5 variables (henceforth referred to as *pure pentads*) can be distinguished from non-2-pure sets of 5 variables (henceforth referred to as *mixed pentads*) by the following property: A pentad is 2-pure only if adding each of the other variable in O to the pentad creates a vanishing sextet. For example, in Figure 1, $S_1 = \{X_3, X_4, X_5, X_6, X_7\}$ is a 2-pure pentad. Adding any other variable to

$\mathbf{S_1}$ creates a sextet of variables which, no matter how they are partitioned, will have one side t-separated from the other side by a choke set ($\{L_1, L_2\} : \emptyset$). In contrast, $\mathbf{S_2} = \{X_1, X_4, X_5, X_6, X_7\}$ is not pure, and when X_9 is added to $\mathbf{S_2}$, the resulting sextet is not a vanishing sextet, since when X_1 and X_9 are on different sides of a partition, at least 3 variables (including L_1, L_2, and X_1 or X_9) are needed to t-separate the treks between the variables in the two sides of the partition.

The first stage of the algorithm calls *FindPureClusters*, which tests each pentad to see if it has the property that adding any other member of **O** creates a vanishing sextet; if it does have the property it is added to the list *PureList* of 2-pure pentads. *FindPureClusters* tests whether a given sextet of variables is a vanishing sextet by calling *PassesTest*, which takes as input a sextet of variables, a sample covariance matrix, and the search parameter alpha that the user inputs to *FTFC*. *PassesTest* is implemented with an asymptotically distribution-free statistical test of sets of vanishing sextad constraints that is a modification of a test devised by Bollen and Ting [7]. The list of 2-pure pentads at this point of the algorithm is $\{X_{10}, X_{11}, X_{12}, X_{13}, X_{14}\}$ and every subset of X_2 through X_7 of size 5. X_1, X_9, and X_{15} do not appear in any 2-pure pentad. X_8 is also not in any pure sub-cluster, but *FTFC* is unable to detect that it is impure. This is the only kind of impurity *FTFC* cannot detect. See the explanation in Section 4 for why this is the case, and why this kind of mistake is not important. *GrowClusters* then initializes the *ClusterList* to *PureList*.

If any of the 2-pure sets of variables overlap, their union is also 2-pure. So *FTFC* calls *GrowClusters* to see if any of the 2-pure sextets in *PureClusters* can be combined into a larger 2-pure set. Theoretically, *GrowClusters* could simply check whether any two subsets on *PureClusters* overlap, in which case they could be combined into a larger 2-pure set. In practice, however, in order to determine whether a given variable o can be added to a cluster **C** in *ClusterList*, *GrowClusters* checks whether a given fraction (determined by the parameter *GrowParameter*) of the sub-clusters of size 5 containing 4 members of **C** and o are on *PureList*. If they are not, then *GrowClusters* tries another possible expansion of clusters on *ClusterList*; if they are, then *GrowClusters* adds o to **C** in *ClusterList*, and deletes all subsets of the expanded cluster of size 5 from *PureList*. *GrowClusters* continues until it runs out of possible expansions to examine.

Finally, when *GrowClusters* is done, *SelectClusters* goes through *ClusterList*, outputting the largest cluster **C** still on *ClusterList*, and deleting any other clusters on *ClusterList* that intersect **C** (including **C** itself).

Algorithm 1: FindTwoFactor Clusters (FTFC)

Data: *Data, V, GrowParameter, α*
Result: *SelectedClusters*
$\langle Purelist, \mathbf{V} \rangle = FindPureClusters(Data, \mathbf{V}, \alpha)$
$Clusterlist = GrowClusters(Purelist, \mathbf{V})$
$SelectedClusters = SelectClusters(Clusterlist)$

Algorithm 2: FindPureClusters

Data: $\mathbf{V}, \mathbf{Data}, \alpha$
Result: $PureList$
$PureList = \varnothing$
for $\mathbf{S} \subseteq \mathbf{V}, |\mathbf{S}| = 5$ **do**
    $Impure = FALSE$
    for $v \in \mathbf{V} \setminus \mathbf{S}$ **do**
        if $PassesTest(\mathbf{S} \cup \{v\}, Data, \alpha) = FALSE$ **then**
            $Impure = TRUE$
            break

    if $Impure = FALSE$ **then**
        $PureList = c(\mathbf{S}, PureList)$

$\mathbf{V} = \bigcup_{i \in PureList} i$
return($\langle PureList, \mathbf{V} \rangle$)

Algorithm 3: GrowClusters

Data: $PureList, \mathbf{V}$
Result: $Clusterlist$
$Clusterlist = PureList$
for $cluster \in Clusterlist$ **do**
    for $sub \subset \mathbf{cluster}, |\mathbf{sub}| = 4$ **do**
        for $o \in \mathbf{V} \setminus \mathbf{cluster}$ **do**
            $testcluster = \mathbf{sub} \cup \{o\}$
            if $testcluster \in PureList$ **then**
                $accepted + +$
            else
                $rejected + +$

        if $accepted \div (rejected + accepted) \geq GrowParameter$ **then**
            $Clusterlist = c(Clusterlist, \mathbf{cluster} \cup \{o\})$
            for $s \subset \mathbf{cluster} \cup \{o\}, \mathbf{s} \in Clusterlist$ **do**
                $Purelist = Purelist \setminus \{\mathbf{s}\}$

The complexity of the algorithm is dominated by *FindPureClusters*, which in the worst case requires testing n choose 6 sets of variables, and for each sextet requires testing five of the ten possible vanishing sextad constraints in order to determine if they all vanish. In practice, we have found that it can be easily applied to about 30 measured variables at a time, but not 60 measured variables. http://www.phil.cmu.edu/projects/tetrad_download/launchers/ contains an implementation available by downloading tetrad-5.0.0-15-experimental.jnlp, creating a "Search" box, selecting "BPC" from the list of searches, and then setting "Test" to "TETRAD-DELTA", and "Algorithm" to "FIND_TWO_FACTORS_CLUSTER".

Algorithm 4: SelectClusters

Data: *Clusterlist*
Result: *Selectedlist*
Selectedlist = ∅
while *Clusterlist* ≠ ∅ **do**
  Choose a largest C
  Selectedlist = *Selectedlist* ∪ {C}
  for $s ∈$ *Clusterlist*, $s ∩ C ≠ ∅$ **do**
    $\lfloor$ *Clusterlist* = *Clusterlist* \ {s}

---

4   Correctness of Algorithm

In what follows, we will assume that if there is not a trek between some pair of indicators, or if there are entailed vanishing partial correlations among the observed indicators, or if there are rank constraints of size 1 on the relevant sub-matrices that the relevant variables are removed in a pre-processing phase. We will make the assumption that sextad constraints vanish only when they are entailed to vanish for all values of the free parameters (i.e vanishing sextad constraints that hold in the population are entailed to hold by the structure of the graph, not the particular values of the free parameters). In the linear case and other natural cases, the set of parameters that violates this assumption is Lebesgue measure 0 [6]. This still leaves the question of whether there are common "almost" violations of rank faithfulness that could only be discovered with enormous sample sizes (i.e. the relevant determinants are very close to zero), which we will address through simulation studies.

There is also a *population* version of the *FTFC* algorithm that differs from the sample algorithm described above in two respects. First, in *PassesTest* it takes as input a sextet of variables and a population covariance matrix, and tests whether all ten possible vanishing sextad constraints among a sextet of variables hold exactly. Second, in *GrowClusters* it sets *GrowParameter* to 1 (whereas in the simulation tests *GrowParameter* was set to 0.5.)

In a 2-factor model, two variables *belong to the same cluster* if they share the same two latent parents. A $5×1$ sextad contains a sextet of variables, 5 of which belong to one cluster, and 1 of which belongs to a different cluster. For a given variable X, $L_1(X)$ is one of the two latent parents of X, and $L_2(X)$ is a second latent parent of X not equal to $L_1(X)$. An indicator X is *impure* if there is an edge into or out of X other than $L_1(X)$ or $L_2(X)$. Define **L** as the set of latent variables L such that $L = L_1(X)$ or $L_2(X)$ for some indicator X. (Latent variables not in **L** might be included in the graph if there are more than two common causes of a pair of indicators, or common causes of an indicator or a member of **L**, e.g. L_5 in Figure 1.

Theorem 3 states that given a measurement model that has a large enough pure sub-model, the output of the *FTFC* algorithm is correct in the sense that the variables in the same output cluster share the same pair of latent parents, and that the only impure indicators X in the output are impure because there is a latent variable not in $\mathbf{L}$ that is a parent of X and $L_1(X)$ or $L_2(X)$ (e.g. L_5 in Figure 1 is a parent of L_1 and X_8). This kind of impurity is not detectible by the algorithm, but is also not important, because it does not affect the estimate of the value of the latent parent from the indicators. In addition, in the output, no single latent parent in $\mathbf{L}$ can be on two treks between latent variables not in $\mathbf{L}$ and an impure indicator; e.g. there cannot be two latent common causes of L_1 and two distinct indicators.

Theorem 3 assumes that the relationships between the indicators and their latent parents is linear. This assumption does not in general entail the model is LA below the choke sets for any arbitrary sextad, since in some cases the latent variables that are in a choke set are not the parents of the indicators in the sextad, in which case it is possible that non-linear relationships between the latent variables will lead to a non-linear relationship between the indicators and the latent variables in the choke set. However, for the particular kind of sextads that the *FTFC* algorithm relies on (i.e. 5×1 sextads) all of the choke sets contain parents of the indicators in the sextad. Hence, linear relationships between the indicators and their latent parents do entail the structure is LA below the choke sets for the sextads that the *FTFC* algorithm relies on for determining the structure of the output clustering,

Theorem 3. *If a SEM S is a 2-factor model that has a 2-pure measurement sub-model in which each indicator X is a linear function of $L_1(X)$ and $L_2(X)$, S has at least six indicators, and at least 5 indicators in each cluster, then the population FTFC algorithm outputs a clustering in which any two variables in the same output cluster have the same pair of latent parents. In addition, each output cluster contains no more than two impure indicators X_1 and X_2, one of which is on a trek whose source is a common cause of $L_1(X_1)$ and X_1, and the other of which is on a trek whose source is a common cause of $L_2(X_1)$ and X_2.*

Proof. First we will show that pure clusters of variables in the true causal graph appear clustered together in the output. Suppose $\mathbf{C} = \{X_1, X_2, X_3, X_4, X_5\}$ belong to a single pure cluster with latent variables L_a and L_b. For any sixth variable Y, and any partition of $\{X_1, X_2, X_3, X_4, X_5, Y\}$ into two sets of size 3, $\{X_a, X_b, X_c\}$ and $\{X_d, X_e, Y\}$, $\{X_a, X_b, X_c\}$ is trek-separated from $\{X_d, X_e, Y\}$ by a choke set containing just $\{L_a$ and $L_b\}$ since there are no other edges into or out of $\{X_a, X_b, X_c\}$ except for those from L_a and L_b. Hence $\mathbf{C}$ is correctly added to *PureList*.

Next we show that variables from different pure clusters in the true causal graph are not clustered together in the output. Suppose that two of the variables in $\mathbf{C}$ belong to different clusters. There are two cases. Either every member of $\mathbf{C}$ belongs to a different cluster or some pair of variables in $\mathbf{C}$ belong to the same cluster. Suppose first two members of $\mathbf{C}$, say X_1 and X_2, belong to a single cluster with latents L_a and L_b, and X_3 belongs to a different cluster with latent L_c. In that case, for any sixth variable Y from the same cluster as X_3, the partitions $\{X_1, X_3, X_4\}$ and $\{X_2, X_5, Y\}$ are not trek-separated by any choke set $\mathbf{S}$ of size 2, since L_a, and L_b would both have to be in $\mathbf{S}$ in order for $\mathbf{S}$ to trek-separate X_1 and $\{X_2, X_5, Y\}$, and L_c would have to be in $\mathbf{S}$ in order for $\mathbf{S}$ to trek-separate X_3 and $\{X_2, X_5, Y\}$. Hence $\mathbf{C}$ will correctly not be added to *PureList*. If, on the other hand, every member of $\mathbf{C}$ belongs to a different cluster, then choose a trek T between two variables in $\mathbf{C}$ such that there is no shorter trek between any two members of $\mathbf{C}$ than T. Suppose without loss of generality that these two variables in $\mathbf{C}$ are X_1 and X_2. Because the clusters are pure by assumption, every trek between X_1 and X_2 contains some pair of latent parents L_1 (a parent of X_1) and L_2 (a parent of X_2). The subtrek of T between L_1 and L_2 does not contain any latent parent of any member of $\mathbf{C} \setminus \{X_1, X_2\}$ since otherwise there would be a trek between two members of $\mathbf{C}$ shorter than T. By the assumption that the model has a pure 2-factor model measurement model, there is a third variable X_3 in $\mathbf{C}$ that is not equal to X_1 or X_2, and some other variable Y that belongs to the same cluster as X_3. X_3 and Y have two latent parents, L_{3a} and L_{3b} that do not lie on T. Consider the sextad $\mathrm{cov}(\{X_1, X_3, X_4\}, \{X_2, X_5, Y\})$. Then in order to trek-separate X_1 from the 3 variables in the side of the partition containing X_2, some latent not equal to L_{3a} or L_{3b} is required to be in the choke set. In order to trek-separate X_3 from the side of the partition containing Y, both L_{3a} and L_{3b} are required to be in the choke set. It follows that no choke set of size 2 trek-separates $\{X_1, X_3, X_4\}$ and $\{X_2, X_5, Y\}$, and $\mathbf{C}$ will not be added to *PureList*. Similarly, if two variables are from different impure clusters, they will not both be added to $\mathbf{C}$, since impurities imply the existence of even more treks, and hence choke sets that are at least as large as in the pure case.

Now we will show that only one kind of impure vertex can occur in an output cluster. Suppose that X is in cluster $\mathbf{C}$, but impure. By definition, there is either an edge E into or out of X that is not from $L_1(X)$ or $L_2(X)$. If E is out of X, then by the assumption that none of the measured indicators cause any of the latent variables in G, E is into some indicator Y. If $(\mathbf{S_1} : \mathbf{S_2})$ t-separates X from Y, and $\mathbf{S} = \mathbf{S_1} \cup \mathbf{S_2}$, then $\mathbf{S}$ contains either X or Y. Consider the sextad $cov(\{X, X_a, X_b\}, \{X_c, X_d, Y\})$, where X_a, X_b, X_c, X_d all belong to $\mathbf{C}$. In order to trek-separate X from X_c, $L_1(X)$ and $L_2(X)$ must be in choke set $\mathbf{S}$. Hence in order to separate both sets in the partition from each other, $\mathbf{S}$ must contain at least 3 elements ($L_1(X)$, $L_2(X)$, and X or Y), and there is a 5×1 sextad that is not entailed to vanish, so X is not clustered with the other variables by *FTFC*.

Suppose E is into X. If the tail of E is a measured indicator Y, then by the same argument as above, there is a 5×1 sextad that is not entailed to vanish, so X is not clustered with the other variables by *FTFC*. If the tail of E is $L_1(Y)$ or $L_2(Y)$ for some Y that is a measured indicator but not in $\mathbf{C}$, consider the sextad $cov(\{X, X_a, X_b\}, \{X_c, X_d, Y\})$, where X_a, X_b, X_c, X_d all belong to $\mathbf{C}$. In order to trek-separate X from X_c, $L_1(X)$ and $L_2(X)$ must be in choke set $\mathbf{S}$. Hence in order to separate both sets in the partition from each other, $\mathbf{S}$ must contain at least 3 elements ($L_1(X)$, $L_2(X)$, and $L_1(Y)$ or $L_2(Y)$). So there is a 5×1 sextad that is not entailed to vanish, and X is not clustered with the other variables by *FTFC*. If the tail of E is a latent variable L that is not equal to $L_1(Y)$ or $L_2(Y)$ for any Y that is a measured indicator but not in $\mathbf{C}$, then there is a shortest trek T between L and some latent parent $L_1(Y)$ of a measured indicator Y. If T contains a measured indicator, then this reduces to one of the previous cases. If Y is not in $\mathbf{C}$ then any trek-separating set of $\mathbf{S}$ must contain at least 3 elements ($L_1(X)$, $L_2(X)$, and some vertex along T that is not equal to $L_1(X)$ or $L_2(X)$). Hence there is a 5×1 sextad that is not entailed to vanish, and X is not clustered with the other variables by *FTFC*.

Finally, consider the case where there are two indicators X_1 and X_2 in $\mathbf{C}$ such that there is a latent common cause M_1 of X_1 and $L_1(X_1)$ and a latent common cause M_2 of X_2 and $L_1(X_1)$, or there is a latent common causes M_1 of X_1 and $L_2(X_1)$ and a latent common cause M_2 of X_2 and $L_2(X_1)$. Suppose without loss of generality that it is the former. If $M_1 = M_2$, then this reduces to one of the previous cases. Otherwise, there are treks T_1 and T_2 between X_1 and X_2 whose sources are M_1 and M_2 respectively. Because X_1 and X_2 are in the same cluster $\mathbf{C}$, in order to trek-separate X_1 and X_2 with a choke set ($\mathbf{S_1}$:$\mathbf{S_2}$), $\mathbf{S_1}$ or $\mathbf{S_2}$ must contain $L_2(X_1)$. In order to separate T_1 and T_2, $L_1(X_1)$ must be in both $\mathbf{S_1}$ and $\mathbf{S_2}$ since $L_1(X_1)$ occurs on the X_2 side of T_1 and the X_1 side of T_2. It follows that $\mathbf{S_1} \cup \mathbf{S_2}$ contains at least 3 elements. Hence there is a 5×1 sextad that is not entailed to vanish, and X_1 and X_2 are not both clustered with the other variables by *FTFC*.

So after the first stage of the algorithm, *PureList* is correct, and hence *ClusterList* is correct (up to the kinds of impurities just described) before it is subsequently modifed.

Now we will show that each stage of modifying *ClusterList* and *PureList* is correct. For a given cluster $\mathbf{C}$, if a variable o belongs to the same cluster, then for every subset of $\mathbf{C} \cup \{o\}$ of size 5, a choke set that contains $L_a(\mathbf{C})$ and $L_b(\mathbf{C})$ t-separates any two members of $\mathbf{C} \cup \{o\}$. Hence $\mathbf{C} \cup \{o\}$ will have passed the purity test, and be found on *PureList*; hence *GrowClusters* will correctly add o to $\mathbf{C}$, and subsets of $\mathbf{C} \cup \{o\}$ will be correctly deleted from *PureList*. If on the other hand o does not belong to the same cluster as $\mathbf{C}$, then some subsets of $\mathbf{C} \cup \{o\}$ of size 5 are not pure, and will not appear in *PureList*. Hence $\mathbf{C} \cup \{o\}$ will not be added to *ClusterList*. Finally, the same argument showing the kinds of impurities that could occur on *PureList* can be applied to *ClusterList*. □

This theorem entails that if there is a 2-factor model with a 2-pure measurement model with sufficiently many variables and a large enough sample size, then *FTFC* will detect it and output the correct clustering. Unfortunately the converse is not true — there are models that do not contain 2-pure measurement sub-models that entail exactly the same set vanishing sextad differences over the measured variables (i.e. are *sextad-equivalent*)[5]; for those alternative models, *FTFC* will output clusters anyway. However, for linear models, it is possible to perform a chi-squared test of whether the measurement model is 2-pure, using structural equation modeling programs such as *EQS*, or *sem* in *R*, or the tests in *TETRAD IV*. In practice, a pure 2-factor model will be rejected by a chi-squared test given data generated by all of the known models sextad-equivalent to a 2-factor model (because of differences between the models in inequality constraints). For this reason, in ideal circumstances, the *FTFC* algorithm would be one part of a larger generate (with $FTFC$) and test (with structural equation modeling estimating and testing) algorithm. See [6] for details.

5 Tests

We tested the $FTFC$ algorithm on simulated and real data sets. We did not directly compare it to other algorithms for the non-linear cases, since to our knowledge there are no other algorithms that can handle non-linearities and/or cyclic relations among the latent variables, impurities in the measurement model, and multiple factors for each cluster. Factor analysis has been used to cluster variables, but has not proved successful even in cases where each cluster has a single latent common cause but impurities [2]. The BuildPureClusters Algorithm uses vanishing tetrad constraints, instead of vanishing sextad constraints to cluster variables, but assumes that each cluster has at most one latent common cause [2]. In the linear, acyclic case, we did compare $FTFC$ to a semi-automated search for a special case of two-factor linear acyclic models, as described in the section on Linear Acyclic models.

5.1 Simulations

The first directed graph we used to generate data has 3 clusters of 10 measured variables each, with each cluster having two latent variables as causes of each measured variable in the cluster, and one of each pair of latent variables for the second cluster causing one of each pair of latent variables in the first cluster, and one of each pair of latent variables in the third cluster. The second directed acyclic graph we used to generate data in addition contained 7 impurities: X_1 is a parent of X_2 and X_3, X_2 is a parent of X_3, L_1 is a parent of X_{11} and X_{21}, X_{20} is a parent of X_{21}, and L_4 is a parent of X_{30}.

For each graph, we generated data at three different sample sizes, $n = 100$, 500, and 1000. The $FTFC$ algorithm was run with significance level (for the vanishing sextad tests) of 0.1 for sample sizes 500 and 1000, and 0.4 for sample size 100. Theoretically, non-linearity among the latent-latent connections should not negatively affect the performance of the algorithm, as long as the sample size is large enough that the asymptotic normality assumed by the sextad test that we employed is a good approximation. Theoretically, non-linearity among the latent-observed connections should negatively affect the performance of the algorithm, since if there are non-linear latent-observed connections, the Extended Trek Separation Theorem generally does not apply. For each graph and each sample size we generated four kinds of models, with each possible combination of linear or non-linear latent-latent connections and linear or non-linear latent-observed connections. In all cases, the non-linearities replace linear relationships with a convex combination of linear and cubic relationships. For example, in the pure model with non-linear latent-latent connections + non-linear latent-measured connections, each variable X was set to the sum over the parents of $0.5 * c_1 * P + 0.5 * d_1 * (.5 * P)^3$ plus an error, where P is one of the parents of X,

c_1 and d_1 were chosen randomly from a Uniform(.35,1.35) distribution, and each error variable was a Gaussian with mean zero, and a variance chosen randomly from a Uniform(2,3) distribution. We tested a few of the simulated data sets with a White test in R for non-linearity, and they rejected the null hypothesis of linearity quite strongly.

In many applications of multiple indicator models, the indicators are deliberately chosen so that the correlations are fairly large (greater than 0.1 in most cases), and all positive; in addition, there are relatively few correlations greater than 0.9. In order to produce correlation matrices with these properties, we had to adjust some of the parameters of the various models we considered according to the type of model (i.e. whether the latent-latent connections were linear or not, whether the latent-measured connections were linear or not, and whether the model was pure of not). We did not however, adjust the model parameters according to the results of the algorithm.

We calculated the precision for each cluster output, and the sensitivity for each cluster output. We then evaluated the output of the algorithm by the number of clusters found, and for each run, the average of the sensitivities and the average of the precisions over the clusters.

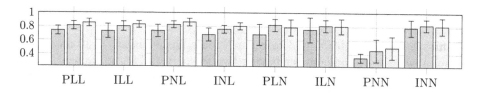

Fig. 2. Average Precision of The Output

The correct number of clusters in each case is 3, and the average number of clusters output ranged between 2.7 and 3.1 for each kind of model and sample size, except for PNN. As expected, non-linearities for the latent-observed connections degraded the performance, and the number of clusters for PNN at sample sizes 100, 500, and 1000 were1.05, 1.38, and 1.54 respectively.

Figure 2 shows the mean (over 50 runs) of the average precision of the clustering output for each simulation case. The error bars shows the standard deviation of the average precision. Figure 3 shows the mean (over 50 runs) average sensitivity of the clustering output for each simulation case. The error bars shows the standard deviation of the average sensitivity. The blue, red and green bars represent cases with 100, 500, and 1000 sample size respectively. In the three-letter lable for every group of three bars, the first letter refers to the purity of the generative model, with "P" being "Pure" and "I" being "Impure". The second letter refers to the linearity of the latent-latent connection, with "L" representing linear connections and "N" representing non-linear connections. The third letter refers to the linearity of the latent-measured connection, the letter "L" and "N" have the same meaning as the case of the second letter. For example, "PNL" represent the case in which the generative model is pure, with non-linear latent-latent connections, and linear latent-observed connections. We generated 50 models of each kind, except that due to time limitations, the sample size 100 PNN case has 40 runs, and the sample size 500 PLN case has 10 runs. The run times varied between 44 and 1328 seconds.

In general, as expected, the result is better as the sample size increases, and is worse when there are impurities adding to the graphical model. The non-linear latent-latent connections does not have an obvious effect upon the clustering output. However, as expected, when the non-linear latent-observed connections are added to the generative model, the mean value of the purity is lower than the corresponding linear cases, and the standard deviation of the two measures starts to increase. Most notably, in the case of "PNN", the interaction of the two kinds of non-linearities renders most of the clustering result being very large clusters (as indicated by the small number of clusters output). That is why the average precision becomes very small while the average sensitivity is relatively large.

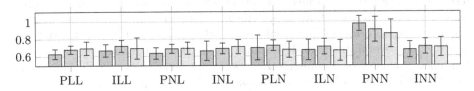

Fig. 3. Average Sensitivity of The Output

5.2 Real Data

We applied *FTFC* to six data sets in R for which there are published bifactor models (see the "Bechthold" help page in R). We ran *FTFC* at 5 significance levels 0.05, 0.1, 0.2, 0.3, and 0.4 and chose the best model. In some cases where there were multiple clusters which together did not pass a chi-squared test, we chose the best individual cluster. In Table 1, p is the number of variables, n is the sample size, *indicators* is the number of indicators in the output, *clusters* is the number of clusters in the output, and $p - value$ is the p-value of the best model. Because we did not have access to the original raw data (just the correlation matrices), we could not divide the data into a training set and a test set, leading to somewhat higher p-values than we would expect if we calculated the p-value on a separate test set.

Table 1. Results of Application of *FTFC* to *R* data sets

Data Set	p	n	*indicators*	*clusters*	$p-value$
Thurstone	9	213	6	1	0.96
Thurstone.33	9	417	5	1	0.52
Holzinger	14	355	7	1	0.23
Holzinger.9	9	145	6	1	0.82
Bechtholdt.1	17	212	8	1	0.59
Reise	16	1000	13	2	0.32

We also applied *FTFC* to the depression data. Lee's model fails a chi-square test: $p = 0$. Although the depression data set contained too many variables to test how well *FTFC* performed overall, we did use it to test whether it could remove impure variables from given clusters (formed from background knowledge), leading to a model that would pass a chi-squared test. Using the output of *FTFC* at several different significance levels, the best model that we found contained a cluster of 9 coping variables, 8 stress variables, and 8 depression variables (all latent variables directly connected) with a p-value of 0.28.

5.3   The Linear Acyclic Case

A bifactor model is a model in which there is a single general factor that is a cause of all of the indicators, and a set of "specific" factors that are causes of some of the indicators. It is a special case of a two-factor model. The *schmid* function in *R* takes as input a correlation matrix and (at least 3 specific) factors, and outputs a bifactor model; it first performs an ordinary factor analysis and then transforms the output into a bifactor model (which is a proper supermodel of one-factor models). We compare *FTFC* algorithm to a *FTFC-schmid* algorithm hybrid on real and simulated data.

We turned the two-factor model described in the previous set of simulations into a linear bifactor model by collapsing three of the latent variables from different clusters into a single variable. We did not find any functions for reliably automatically estimating the number of factors in a bifactor model, so we compared the *FTFC* algorithm to a *FTFC-schmid* hybrid, in which *FTFC* provided the number of factors input to *schmid*. The hybrid *FTFC-schmid* algorithm removed 1.6% of the intra-cluster impurities (e.g. X_1, X_2, X_3), and 48% of the inter-cluster impurities (e.g. X_{11}, X_{20}, X_{21}, X_{30}) while removing 8% of the pure variables . In contrast, *FTFC* removed 61% of the intra-cluster impurities, and 58% of inter-cluster impurities, while also removing 30% of the pure variables. While *FTFC* incorrectly removed many more pure variables than the hybrid *FTFC-schmid*, for the purposes of finding submodels that pass chi-squared tests, this is far less important than its superiority in removing far more of the impure variables

We then compared *schmid* to *FTFC* on the Reise data. The published bi-factor model [8], the output of the *schmid* function in *R* with 5 specific factors (as in the published model), and versions of both of these models that removed the same variables that *FTFC* algorithm did, all failed chi-squared tests and had p-values of 0. The output of *FTFC* removed 3 of the 16 variables, and combined the five specific factors into two specific factors (with the exception of one variable.) We turned the resulting two-factor graph into a bifactor graph, and It passed a chi-squared test with a p-value of 0.32.

6   Future Research

Further research into making the output of *FTFC* more reliable and more stable is needed. It would also be useful to automate the use of chi-squared tests of the output models and to combine the strengths of the *schmid* and *FTFC* algorithms. The ultimate goal of the clustering is to find causal relations among the latent variables; when clusters have multiple latent common causes, some edges become unidentifiable (i.e. the parameters associated with the edge are not a function of the covariance matrix among the measured variables.) Computationally feasible necessary and sufficient conditions for identifiability of linear models are not known, and the possibility that the relations among the latents are non-linear complicates these issues further.

References

[1] Bartholomew, D.J., Steele, F., Moustaki, I., Galbraith, J.I.: The Analysis and Interpretation of Multivariate Data for Social Scientists. Texts in Statistical Science Series. Chapman & Hall/CRC (2002)

[2] Silva, R., Glymour, C., Scheines, R., Spirtes, P.: Learning the structure of latent linear structure models. Journal of Machine Learning Research 7, 191–246 (2006)

[3] Bollen, K.A.: Structural Equations with Latent Variables. Wiley-Interscience (1989)

[4] Spirtes, P., Glymour, C., Scheines, R.: Causation, Prediction, and Search, 2nd edn. Adaptive Computation and Machine Learning. The MIT Press (2001)

[5] Sullivant, S., Talaska, K., Draisma, J.: Trek Separation for Gaussian Graphical Models. Ann. Stat. 38(3), 1665–1685 (2010)

[6] Spirtes, P.: Calculation of Entailed Rank Constraints in Partially Non-Linear and Cyclic Models. In: Proceedings of the Twenty-Ninth Conference Annual Conference on Uncertainty in Artificial Intelligence (UAI 2013), pp. 606–615 (2013)

[7] Bollen, K., Ting, K.: Confirmatory tetrad analysis. Sociological Methodology 23, 147–175 (1993)

[8] Reise, S., Morizot, J., Hays, R.: The role of the bifactor model in resolving dimensionality issues in health outcomes measures. Quality of Life Research 16, 19–31 (2007)

Support Vector Machines for Differential Prediction

Finn Kuusisto[1], Vitor Santos Costa[2], Houssam Nassif[3], Elizabeth Burnside[1], David Page[1], and Jude Shavlik[1]

[1] University of Wisconsin - Madison, Madison, WI, USA
[2] University of Porto, Porto, Portugal
[3] Amazon, Seattle, WA, USA

Abstract. Machine learning is continually being applied to a growing set of fields, including the social sciences, business, and medicine. Some fields present problems that are not easily addressed using standard machine learning approaches and, in particular, there is growing interest in *differential prediction*. In this type of task we are interested in producing a classifier that specifically characterizes a subgroup of interest by maximizing the difference in predictive performance for some outcome between subgroups in a population. We discuss adapting maximum margin classifiers for differential prediction. We first introduce multiple approaches that do not affect the key properties of maximum margin classifiers, but which also do not directly attempt to optimize a standard measure of differential prediction. We next propose a model that directly optimizes a standard measure in this field, the *uplift* measure. We evaluate our models on real data from two medical applications and show excellent results.

Keywords: support vector machine, uplift modeling.

1 Introduction

Recent years have seen increased interest in machine learning, with novel applications in a growing set of fields, such as social sciences, business, and medicine. Often, these applications reduce to familiar tasks, such as classification or regression. However, there are important problems that challenge the state-of-the-art.

One such task, *differential prediction*, is motivated by studies where one submits two different subgroups from some population to stimuli. The goal is then to gain insight on the different reactions by producing, or simply identifying, a classifier that demonstrates significantly better predictive performance on one subgroup (often called the *target* subgroup) over another (the *control* subgroup). Examples include:

- Seminal work in sociology and psychology used regression to study the factors accounting for differences in the academic performance of students from different backgrounds [5,15,26].

T. Calders et al. (Eds.): ECML PKDD 2014, Part II, LNCS 8725, pp. 50–65, 2014.
© Springer-Verlag Berlin Heidelberg 2014

- Uplift modeling is a popular technique in marketing studies. It measures the impact of a campaign by comparing the purchases made by a subgroup that was targeted by some marketing activity versus a control subgroup [16,9,20].
- Medical studies often evaluate the effect of a drug by comparing patients who have taken the drug against patients who have not [7,4].
- Also within the medical domain, breast cancer is a major disease that often develops slower in older patients. Insight on the differences between older and younger patients is thus crucial in determining whether treatment is immediately necessary [19,18].

Differential prediction has broad and important applications across a range of domains and, as specific motivating applications, we will consider two medical tasks. One is a task in which we want to specifically identify older patients with breast cancer who are good candidates for "watchful waiting" as opposed to treatment. The other is a task in which we want to specifically identify patients who are most susceptible to adverse effects of COX-2 inhibitors, and thus not prescribe such drugs for these patients.

The adverse drug event task alone is of major worldwide significance, and the significance of the breast cancer task cannot be overstated. Finding a model that is predictive of an adverse event for people on a drug versus not could help in isolating the key causal relationship of the drug to the event, and using machine learning to uncover causal relationships from observational data is a big topic in current research. Similarly, finding a model that can identify patients with breast cancer that may not be threatening enough in their lifetime to require treatment could greatly reduce overtreatment and costs in healthcare as a whole.

Progress in differential prediction requires the ability to measure differences in classifier performance between two subgroups. The standard measure of differential prediction is the *uplift* curve [23,22], which is defined as the difference between the *lift* curves for the two subgroups. Several classification and regression algorithms have been proposed and evaluated according to this measure [22,23,19,10]. These models were designed to improve uplift, but do not directly optimize it. We show that indeed it is possible to directly optimize uplift and we propose and implement the SVM^{upl} model, which does so. This model is constructed by showing that optimizing uplift can be reduced to optimizing a linear combination of a weighted combination of features, thus allowing us to apply Joachims' work on the optimization of multivariate measures [13]. We evaluate all models on our motivating applications and SVM^{upl} shows the best performance in differential prediction in most cases.

The paper is organized as follows. Section 2 presents our motivating applications in greater detail. In Section 3 we introduce uplift modeling and the uplift measure that we will use to evaluate our models. We also present results on a synthetic dataset in this section to give further insight in the task. We discuss multiple possible approaches to differential prediction that do not directly optimize uplift in Section 4. Section 5 discusses previous work on SVMs that optimize for multi-variate measures, and Section 6 presents how to extend this work to optimize uplift directly. We discuss methodology in Section 7 and evaluate all of the proposed models on our motivating applications in Section 8. Finally, Section 9 presents conclusions and future work.

2   Medical Applications

To illustrate the value of differential prediction in our motivating applications we first discuss both in further detail.

Breast cancer is the most common cancer among women [2] and has two basic stages: an earlier *in situ* stage where cancer cells are still localized, and a subsequent *invasive* stage where cancer cells infiltrate surrounding tissue. Nearly all in situ cases can be cured [1], thus current practice is to treat in situ occurrences in order to avoid progression into invasive tumors [2]. Treatment, however, is costly and may produce undesirable side-effects. Moreover, an in situ tumor may never progress to invasive stage in the patient's lifetime, increasing the possibility that treatment may not have been necessary. In fact, younger women tend to have more aggressive cancers that rapidly proliferate, whereas older women tend to have more indolent cancers [8,11]. Because of this, younger women with in situ cancer should be treated due to a greater potential time-span for progression. Likewise, it makes sense to treat older women who have in situ cancer that is similar in characteristics to in situ cancer in younger women since the more aggressive nature of cancer in younger patients may be related to those features. However, older women with in situ cancer that is significantly different from that of younger women may be less likely to experience rapid proliferation, making them good candidates for "watchful waiting" instead of treatment. For this particular problem, predicting in situ cancer that is specific to older patients is the appropriate task.

COX-2 inhibitors are a family of non-steroidal anti-inflammatory drugs (NSAIDs) used to treat inflammation and pain by directly targeting the COX-2 enzyme. This is a desirable property as it significantly reduces the occurrence of various adverse gastrointestinal effects common to other NSAIDs. As such, some early COX-2 inhibitors enjoyed rapid and widespread acceptance in the medical community. Unfortunately, clinical trial data later showed that the use of COX-2 inhibitors also came with a significant increase in the rate of myocardial infarction (MI), or "heart attack" [14]. As a result, physicians must be much more careful when prescribing these drugs. In particular, physicians want to avoid prescribing COX-2 inhibitors to patients who may be more susceptible to the adverse effects that they entail. For this problem, predicting MI that is specific to patients who have taken COX-2 inhibitors, versus those who did not, is the appropriate task to identify the at-risk patients.

3   Uplift Modeling

The fundamental property of differential prediction is the ability to quantify the difference between the classification of subgroups in a population, and much of the reference work in this area originates from the marketing domain. Therefore, we first give a brief overview of differential prediction as it relates to marketing.

In marketing, customers can be broken into four categories [21]:

Persuadables Customers who respond positively (e.g. buy a product) when targeted by marketing activity.

Sure Things Customers who respond positively regardless of being targeted.

Lost Causes Customers who do not respond (e.g. not buy a product) regardless of being targeted or not.

Sleeping Dogs Customers who do not respond as a result of being targeted.

Thus, targeting *Persuadables* increases the value produced by the marketing activity, targeting *Sleeping Dogs* decreases it, and targeting customers in either of the other groups has no effect, but is a waste of money. Ideally then, a marketing team would only target the *Persuadables* and avoid targeting *Sleeping Dogs* whenever possible. Unfortunately, the group to which a particular individual belongs is unknown and is not readily observable. An individual cannot be both targeted and not targeted to determine their response to marketing activity directly. Only the customer response and whether they were in the target or control group can be observed experimentally (see Table 1).

Table 1. Customer groups and their expected responses based on targeting. Only the shaded region can be observed experimentally.

Target		Control	
Response	No Response	Response	No Response
Persuadables, Sure Things	Sleeping Dogs, Lost Causes	Sleeping Dogs, Sure Things	Persuadables, Lost Causes

In this scenario, since we cannot observe customer groups beforehand, standard classifiers appear less than ideal. For example, training a standard classifier to predict response, ignoring that the target and control subgroups exist, is likely to result in a classifier that identifies *Persuadables*, *Sure Things*, and *Sleeping Dogs* as they represent the responders when the target and control subgroups are combined. Recall, however, that targeting *Sure Things* is a waste of money, and targeting *Sleeping Dogs* is harmful. Even training on just the target subgroup is likely to produce a classifier that identifies both *Persuadables* and *Sure Things*. The point of differential prediction in this domain is then to quantify the difference between the target and control subgroups. While it may be simple and intuitive to simply learn two separate models and subtract the output of the control model from the target model, recent work suggests that this is less effective than modeling the difference directly [22]. Thus, the goal is to produce a single classifier that maximizes predictive performance on the target subgroup over the control subgroup. The idea is that such a classifier characterizes properties that are specific to the target subgroup, thereby making it effective at identifying *Persuadables*. That is, such a classifier will produce a larger output for customers who are more likely to respond positively as a direct result of targeting, and a smaller output for those who are unaffected or are more likely to respond negatively. The classifier could then be used in subsequent marketing campaigns to select who should be targeted and who should not.

There are many possible measures that could be used to quantify the difference in predictive performance between the target and control subgroups. In marketing, the uplift measure is often used to quantify this difference as well as to evaluate the performance of classifiers designed for differential prediction. Thus, this task is often referred to as *uplift modeling*.

3.1   Uplift

In this work, we will consider two subgroups, which we will refer to as A and B, representing *target* and *control* subgroups respectively, and where subgroup A is the subgroup of most interest.

The *lift* curve [24] reports the total percentage examples that a classifier must label as positive (x-axis) in order to obtain a certain recall (y-axis), expressed as a count of true positives instead of a rate. As usual, we can compute the corresponding area under the lift curve (AUL). Note that the definition of the lift curve is very similar to that of an ROC curve.

Uplift is the difference in lift produced by a classifier between subgroups A and B, at a particular threshold percentage of all examples. We can compute the area under the uplift curve (AUU) by subtracting their respective AULs:

$$AUU = AUL_A - AUL_B \tag{1}$$

Notice that, because uplift is simply a difference in lift at a particular threshold, uplift curves always start at zero and end at the difference in the total number of positive examples between subgroups. Higher AUU indicates an overall stronger differentiation of subgroup A from B, and an uplift curve that is skewed more to the left suggests a more pronounced ranking of positives from subgroup A ahead of those from subgroup B.

3.2   Simulated Customer Experiments

To demonstrate that uplift modeling does help to produce classifiers that can specifically identify *Persuadables*, we generated a synthetic population of customers and simulated marketing activity to produce a dataset for which we knew the ground truth customer groups. We present results on this synthetic dataset, but save algorithmic details for later sections.

To generate a customer population, we first generated a random Bayesian network with 20 nodes and 30 edges. We then randomly selected one node with four possible values to be the customer group feature. Next, we drew 10,000 samples from this network. This left us with a population of customers for which one feature defined the group they belonged to and the rest represented observable features.

We then subjected this population to a simulated marketing activity. We randomly selected roughly 50% of the entire population to be part of the target subgroup. Next, we produced a response for each customer based on their customer group and whether or not they were chosen to be targeted. For this demonstration, we determined each response based on the strongest stereotypical interpretation of each customer group. That is, *Persuadables* always responded when targeted and never responded when not. *Sleeping Dogs* never responded when targeted and always responded when not. *Sure Things* and *Lost Causes* always and never responded respectively.

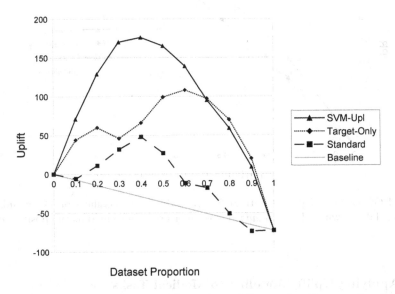

Fig. 1. Uplift curves (higher is better) for three different classifiers on the simulated customer dataset

We removed the customer group feature from the training set and trained three different classifiers to demonstrate performance. First, we trained a standard SVM classifier on the entire dataset with a positive response as the positive class. Next, we trained a standard SVM on just the target subgroup. Finally, we trained an SVM designed to maximize uplift, about which we will go into greater detail later.

We evaluated the results using 10-fold cross-validation and used internal cross-validation to select parameters in the same way that we will show later on our medical datasets.

Figure 1 shows the uplift curves on the synthetic customer dataset. As expected, the SVM designed to maximize uplift produces the highest uplift curve, while the standard SVM trained on the entire dataset produces the lowest. Figure 2 shows ROC curves on the synthetic customer dataset when the *Persuadable* customers are considered to be the positive class. Recall that this feature was unobserved at training time, but identifying *Persuadables* is the real goal in the marketing domain. As hoped, the SVM that maximizes uplift has the highest ROC curve whereas the standard SVM trained on the entire dataset hovers around the diagonal.

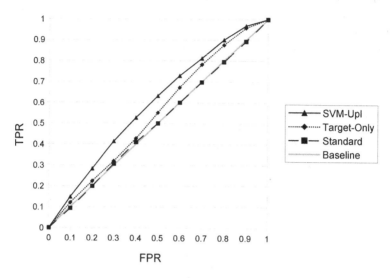

Fig. 2. ROC curves (higher is better) for three different classifiers on the simulated customer dataset when the *Persuadable* customer group is treated as the positive class

3.3   Applying Uplift Modeling to Medical Tasks

In this work, we propose that the task addressed in the marketing domain can be mapped to our motivating medical tasks, suggesting that the uplift measure is a reasonable measure for evaluation of our models.

In the COX-2 inhibitor task, variability in response to the drug suggests that there will be some people at increased risk of MI as a result of taking the drug, some who are at increased risk of MI regardless of treatment, some who are at decreased risk regardless, and perhaps even some who are at decreased risk as a result of taking the drug. Just like in the marketing task, which group an individual belongs to cannot be directly observed. An individual cannot both take the drug and not take the drug to determine its effect. Only the MI outcome and whether or not the individual took the drug can be observed experimentally. We propose that training a classifier to identify individuals for which taking a COX-2 inhibitor increases their risk of MI is analogous to identifying *Persuadables*.

In the breast cancer task, the analogy is not as obvious, but we know that younger patients often have aggressive cancers while older patients have both aggressive and indolent cancers. Again, which type of cancer a patient has is not directly observable and it is unreasonable to not treat patients in an attempt to determine which have less aggressive varieties. We propose that training a classifier to identify less aggressive varieties of cancer (seen in older patients) is also analogous to identifying *Persuadables*.

4   Uplift-Agnostic Models

There are many different possible approaches to learning a classifier that is differentially predictive and we have reviewed how this task is approached and evaluated in the marketing domain. We first introduce a number of possibilities in this section that do not directly optimize the uplift measure at training time.

4.1   Standard SVM

To better understand the problem, we start from the standard maximum margin classifier [25]. This classifier minimizes:

$$\frac{1}{2}||\mathbf{w}||^2 + C \sum_{i=1}^{N} \xi_i \tag{2}$$

subject to $\xi_i \geq 1 - y_i \langle \mathbf{x_i}, \mathbf{w} \rangle, \xi_i \geq 0$, and where is $(\mathbf{x}, y)$ feature vector and label pair notation representing examples. The formulation tries to minimize the norm of the weight vector, $\mathbf{w}$, and hence maximize the margin, while softly allowing a number of errors ξ_i whose cost depends on the parameter C.

For the sake of comparison, we evaluate the ability of a standard linear SVM model to produce uplift in our applications of interest. In this case we simply ignore the fact that the examples fall into two subgroups.

4.2   Subgroup-Only SVM

Another intuitive possible approach to achieving differential prediction, without modifying the original optimization, is to only train on the subgroup of most interest. In this way, the classifier should perform well on the subgroup used to train it, whereas it should not perform as well on the other subgroup. In our applications, that would mean only training on the data for the older subgroup of breast cancer patients, or the subgroup of MI patients who have been prescribed COX-2 inhibitors.

4.3   Flipped-Label SVM

Jaśkowski and Jaroszewicz [10] propose a general method for adapting standard models to be differentially predictive. This is accomplished by flipping the classification labels in the secondary subgroup during training. In this way, the classifier is trained to correctly predict the positive class on the subgroup of interest, subgroup A, whereas it is trained to predict the negative class in the secondary subgroup, subgroup B. The resulting classifier should then, ideally, perform much better on subgroup A than subgroup B.

4.4   Two-Cost SVM

Another possibility is to simply treat the errors on the different subgroups differently. In the case of the SVM optimization, we would clearly like the cost to be different for the two subgroups. Specifically, we would like to maximize the cost difference between the two, but that problem is ill-defined, suggesting the following adaptation of the standard minimization problem:

$$\frac{1}{2}||\mathbf{w}||^2 + C_A \sum_{i=1}^{|A|} \xi_i + C_B \sum_{j=1}^{|B|} \xi_j \tag{3}$$

subject to $\xi_i \geq 1 - y_i\langle \mathbf{x_i}, \mathbf{w}\rangle, \xi_j \geq 1 - y_j\langle \mathbf{x_j}, \mathbf{w}\rangle, \xi_i \geq 0, \xi_j \geq 0$. As a first step, we assume $C_A \geq 0$ and $C_B \geq 0$, so we continue penalizing errors on subgroup B. We call this method the two-cost model, and although this problem is similar to addressing class weight, there is an important difference. When addressing class skew, the ratio between C^+ and C^- can be estimated from the class skew in the data. On the other hand, a natural ratio between C_A and C_B may not be known beforehand: if $C_A \approx C_B$, there will be little differential classification, but if $C_A \gg C_B$ the errors may be captured by set B only, leading to over-fitting.

5   Multivariate Performance Measures

Our goal is to find the parameters $\mathbf{w}$ that are optimal for a specific measure of uplift performance, such as AUU. Similar to AUC [13,28,17], AUL depends on the rankings between pairs of examples. We next, we focus on the SVM^{perf} approach [13]. This approach hypothesizes that we want to find the h that minimizes the area of a generic loss function Δ over an unseen set of examples S':

$$R^\Delta(h) = \int \Delta((h(\mathbf{x_1'}), \dots, h(\mathbf{x_{n'}'})), (y_1', \dots, y_{n'}'))dPr(S') \tag{4}$$

Note that we use a $(\mathbf{x}, y)$ feature vector and label pair notation to represent examples throughout. Also, in practice we cannot use equation (4), we can only use the training data:

$$\hat{R}^\Delta(h) = \Delta((h(\mathbf{x_1}), \dots, h(\mathbf{x_n})), (y_1, \dots, y_n)) \tag{5}$$

Let tuples $\bar{y} = (y_1, \dots, y_n)$ and $\bar{y}'$ be assignments over the n examples, $\bar{\mathcal{Y}}$ is the set of all possible assignments. $\Psi(\mathbf{x}, y)$ is a measure-specific combination of features of inputs and outputs in our problem, such that one wants to maximize $\mathbf{w}^T\Psi$:

$$\underset{\bar{y}' \in \bar{\mathcal{Y}}}{\operatorname{argmax}} \left\{\mathbf{w}^T \Psi(\bar{\mathbf{x}}, \bar{y}')\right\} \tag{6}$$

Then the problem reduces to:

$$\min_{\mathbf{w}, \xi \geq 0} \frac{1}{2}||\mathbf{w}||^2 + C\xi \tag{7}$$

given the constraints:

$$\forall \bar{y}' \in \bar{\mathcal{Y}} \setminus \bar{y} : \mathbf{w}^T [\Psi(\bar{\mathbf{x}}, \bar{y}) - \Psi(\bar{\mathbf{x}}, \bar{y}')] \geq \Delta(\bar{y}', \bar{y}) - \xi \qquad (8)$$

which is a quadratic optimization problem and can be solved by a cutting plane solver [12], even if it involves many constraints (one per element in $\bar{\mathcal{Y}} \setminus \bar{y}$).

The formulation applies to the AUC by defining it as $1 - \frac{BadPairs}{N \times P}$, where N is the number of negative examples, P is the number of positive examples, and BadPairs is the number of pairs (i, j) such that $y_i = 1, y_j = -1$, and $y_i' < y_j'$. Joachims thus addresses the optimization problem in terms of pairs y_{ij}', where y_{ij}' is 1 if $y_i' > y_j'$, and -1 otherwise. The loss is the number of swapped pairs:

$$\Delta_{AUC}(\bar{y}', \bar{y}) = \sum_{i=1}^{P} \sum_{j=1}^{N} \frac{1}{2}(1 - y_{ij}') \qquad (9)$$

The combination of features Ψ should be symmetric to the loss, giving:

$$\mathbf{w}^T \Psi(\bar{\mathbf{x}}, \bar{y}') = \frac{1}{2} \sum_{i=1}^{P} \sum_{j=1}^{N} y_{ij}'(\mathbf{w}^T \mathbf{x}_i - \mathbf{w}^T \mathbf{x}_j) = \mathbf{w}^T \frac{1}{2} \sum_{i=1}^{P} \sum_{j=1}^{N} y_{ij}'(\mathbf{x}_i - \mathbf{x}_j) \qquad (10)$$

The optimization algorithm [12] finds the most violated constraint in equation (8). This corresponds to finding the y_{ij}^* that minimize $\mathbf{w}^T[\Psi(\bar{\mathbf{x}}, \bar{y}) - \Psi(\bar{\mathbf{x}}, \bar{y}')] - \Delta_{AUC}(\bar{y}', \bar{y})$, or, given that $\Psi(\bar{\mathbf{x}}, \bar{y})$ is fixed, that maximize:

$$\mathbf{w}^T \Psi(\bar{\mathbf{x}}, \bar{y}') + \Delta_{AUC}(\bar{y}', \bar{y})$$

Expanding this sum resumes into independently finding the y_{ij}^* such that:

$$y_{ij}^* = \operatorname*{argmax}_{y_{ij}' \in \{1, -1\}} \ y_{ij}'((\mathbf{w}^T \mathbf{x}_i - \frac{1}{2}) - (\mathbf{w}^T \mathbf{x}_j + \frac{1}{2})) \qquad (11)$$

Joachims' algorithm then sorts the $\mathbf{w}^T \mathbf{x}_i - \frac{1}{2}$ and $\mathbf{w}^T \mathbf{x}_j + \frac{1}{2}$, and generates labels from this total order.

6   Maximizing Uplift

Recall from Section 3.1 the similarity between lift and ROC. The two are actually closely related. As shown in Tufféry [24], and assuming that we are given the skew $\pi = \frac{P}{P+N}$, the AUL is related to the AUC by:

$$AUL = P\left(\frac{\pi}{2} + (1 - \pi)AUC\right) \qquad (12)$$

Expanding equation (1) with equation (12):

$$AUU = P_A\left(\frac{\pi_A}{2} + (1 - \pi_A)AUC_A\right) - P_B\left(\frac{\pi_B}{2} + (1 - \pi_B)AUC_B\right) \qquad (13)$$

P_A, P_B, π_A, and π_B are properties of the two subgroups, and thus independent of the classifier. Removing constant terms we see that maximizing uplift is equivalent to:

$$max(AUU) \equiv max(P_A(1 - \pi_A)AUC_A - P_B(1 - \pi_B)AUC_B)$$
$$\propto max\left(AUC_A - \frac{P_B(1 - \pi_B)}{P_A(1 - \pi_A)}AUC_B\right) \quad (14)$$

Defining $\lambda = \frac{P_B(1-\pi_B)}{P_A(1-\pi_A)}$ we have:

$$max(AUU) \equiv max(AUC_A - \lambda AUC_B) \quad (15)$$

Therefore, maximizing AUU is equivalent to maximizing a weighted difference between two AUCs.

Equation (15) suggests that we can use the AUC formulation to optimize AUU. First, we make it a double maximization problem by switching labels in subgroup B:

$$max(AUU) \equiv max(AUC_A - \lambda(1 - AUC_B^-))$$
$$\equiv max(AUC_A + \lambda AUC_B^-) \quad (16)$$

The new formulation reverses positives with negatives making it a sum of separate sets.

At this point, we can encode our problem using Joachims' formulation of the AUC. In this case, we have two AUCs. One, as before, is obtained from the y_{ij} where the i, j pairs range over A. The second corresponds to pairs y_{kl} where the k, l pairs range over B. On switching the labels, we must consider y_{lk} where k ranges over the positives in B, and l over the negatives in B.

After switching labels, we can expand equation (9) to obtain our new loss Δ_{AUU} as the weighted sum of two losses:

$$\Delta_{AUU}(\bar{y}', \bar{y}) = \sum_{i=1}^{P_A} \sum_{j=1}^{N_A} \frac{1}{2}(1 - y'_{ij}) + \lambda \sum_{k=1}^{P_B} \sum_{j=1}^{N_B} \frac{1}{2}(1 - y'_{lk}) \quad (17)$$

From equation (10) we construct a corresponding weighted sum as the new Ψ:

$$\Psi(\bar{x}, \bar{y}') = \frac{1}{2} \sum_{i=1}^{P_A} \sum_{j=1}^{N_A} y'_{ij}(x_i - x_j) + \lambda \frac{1}{2} \sum_{k=1}^{P_B} \sum_{l=1}^{N_B} y'_{lk}(x_l - x_k) \quad (18)$$

The two sets are separate, so optimizing the y_{ij} does not change from equation (11), as their maximization does not depend on the y_{lk}. Optimizing the y_{lk} follows similar reasoning to the y_{ij} and gives:

$$y^*_{lk} = \operatorname*{argmax}_{y'_{lk} \in \{1, -1\}} y'_{lk}((w^T x_l - \frac{1}{2}) - (w^T x_k + \frac{1}{2})) \quad (19)$$

Thus, we now have two independent rankings: one between the labels for the examples in A, and the other between the labels for the examples in B. We can sort them together or separately, but we simply have to label the sets independently to obtain the $\bar{y}^*$ of the most violated constraint. Note that the computation of the $\bar{y}^*$ in this setting is independent of λ, but λ still affects the solutions found by the cutting-plane solver through Δ and Ψ.

7   Experiments

We implemented our SVM^{Upl} method using the SVM^{perf} codebase, version 3.00[1]. We implemented the two-cost model using the LIBSVM codebase [3], version 3.17[2]. All other uplift-agnostic approaches were run using LIBSVM, but required no changes to the code.

Recall that our motivating applications are to produce a differential older-specific classifier for in situ breast cancer, and produce a differential COX-2 specific classifier for myocardial infarction (MI). We apply all of the proposed approaches to the breast cancer data used in Nassif et al. [19] and the MI data used in Davis et al. [6]. Their composition is shown in Table 2.

The breast cancer data consists of two cohorts: patients younger than 50 years old form the *younger* cohort, while patients aged 65 and above form the *older* cohort. The older cohort has 132 in situ and 401 invasive cases, while the younger one has 110 in situ and 264 invasive.

Table 2. Composition of the breast cancer and MI datasets for our motivating applications. In the breast cancer dataset the older subgroup is the target subgroup, and in situ breast cancer is the positive class. In the MI dataset the COX-2 inhibitor subgroup is the target subgroup, and MI is the positive class.

Older		Younger		COX-2 Inhibitors		No COX-2 Inhibitors	
In Situ	Invasive	In Situ	Invasive	MI	No MI	MI	No MI
132	401	110	264	184	1,776	184	1,776

The MI data consists of patients separated into two equally-sized subgroups: patients who have been prescribed COX-2 inhibitors and those who have not. The group prescribed COX-2 inhibitors has 184 patients who had MI, and 1776 who did not. The subgroup not prescribed COX-2 inhibitors has the same number of patients for each outcome.

We use 10-fold cross-validation for evaluation. Final results were produced by concatenating the output test results for each fold. Cost parameters were selected for each fold using 9-fold internal cross-validation. For all approaches, the cost parameter was selected from $\{10.0, 1.0, 0.1, 0.01, 0.001, 0.0001, 0.00001\}$. For the two-cost model, C_A and C_B were selected from all combinations of values from

---

[1] http://www.cs.cornell.edu/people/tj/svm_light/svm_perf.html
[2] http://www.csie.ntu.edu.tw/~cjlin/libsvm

Table 3. 10-fold cross-validated performance for all proposed approaches on the breast cancer dataset (* indicates significance)

Model	Older AUL	Younger AUL	AUU	Per-fold AUU μ	Per-fold AUU σ	SVM^{Upl} p-value
SVM^{Upl}	64.26	45.05	19.21	1.93	0.78	-
Two-Cost	74.30	60.76	13.54	1.45	1.18	0.432
Older-Only	67.70	61.85	5.85	1.03	1.15	0.037 *
Standard	75.35	64.34	11.01	1.26	0.38	0.049 *
Flipped	53.90	49.08	4.82	0.77	0.58	0.020 *
Baseline	66.00	55.00	11.00	1.10	0.21	0.004 *

Table 4. 10-fold cross-validated performance for all proposed approaches on the MI dataset (* indicates significance)

Model	COX-2 AUL	No COX-2 AUL	AUU	Per-fold AUU μ	Per-fold AUU σ	SVM^{Upl} p-value
SVM^{Upl}	123.38	72.70	50.68	5.07	2.04	-
Two-Cost	126.23	106.25	19.99	2.43	1.54	0.004 *
COX-2-Only	151.50	137.70	13.80	1.18	1.52	0.002 *
Standard	147.69	146.49	1.20	-0.16	1.25	0.002 *
Flipped	102.15	73.63	28.52	2.97	1.35	0.037 *
Baseline	0.00	0.00	0.00	0.00	0.00	0.002 *

the set such that $C_A > C_B$. We plot the final uplift curves for each approach along with the uplift for a baseline random classifier in Figures 3 and 4.

Tables 3 and 4 compare SVM^{Upl} with every other approach proposed as well as a fixed baseline random classifier. We use the Mann-Whitney test at the 95% confidence level to compare approaches based on per-fold AUU. We show the per-fold mean, standard deviation, and p-value of the 10-fold AUU paired Mann-Whitney of each method as compared to SVM^{Upl} (* indicates significance).

8   Evaluation

The results on the breast cancer dataset in Table 3 show that SVM^{Upl} produces significantly greater uplift than all proposed approaches, except for the two-cost model. This exception may be a result of the higher variance of the model on this particular dataset. The results on the MI dataset in Table 4 show that SVM^{Upl} produces the greatest uplift in all cases.

Figure 3 shows SVM^{Upl} with an uplift curve that dominates the rest of the approaches until around the 0.7 threshold on the breast cancer dataset. Most other approaches produce curves that sit around or below the baseline.

Figure 4 tells a similar story, with SVM^{Upl} dominating all other methods across the entire space on the MI dataset. In this dataset, however, only the standard SVM approach consistently performs below the baseline, whereas all other methods appear to produce at least modest uplift.

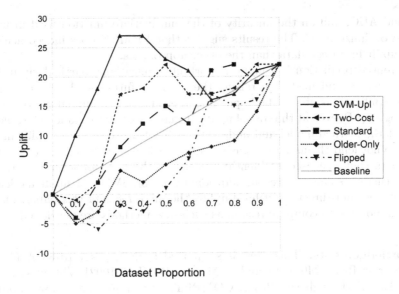

Fig. 3. Uplift curves (higher is better) for all approaches on the breast cancer dataset

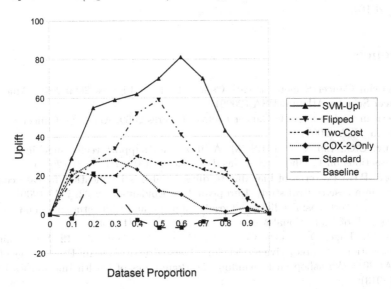

Fig. 4. Uplift curves (higher is better) for all approaches on the MI dataset. Note that the baseline uplift lies on the x-axis due to the equal number of patients with MI in each subgroup.

9   Conclusions and Future Work

We introduced a support vector model directed toward differential prediction. The SVM^{Upl} approach optimizes uplift by relying on the relationship between

AUL and AUC, and on the linearity of the multivariate function used in prior work to optimize AUC. The results suggest that SVM^{Upl} does indeed achieve better uplift in unseen data than the other approaches.

Differential prediction has many important applications, particularly in the human sciences and medicine, raising the need for future work. For example, in some applications, it may be important to ensure some minimal performance over subgroup B, even at the cost of uplift. It may also be important to be able to interpret the learned model and understand what features improve uplift most. SVMs do not lend themselves as easily to this task as some models, but feature coefficients could be used to identify which are the most or least important. Finally, there is some very recent additional work on SVMs for uplift modeling [27] that does not directly optimize uplift, the main focus of this paper, but it will be important to compare results as such new methods are developed.

Acknowledgements. This work is supported by NIH grant R01-CA165229, NIGMS grant R01-GM097618, and NLM grant R01-LM010921. VSC was funded by the ERDF through the Progr. COMPETE, the Portuguese Gov. through FCT, proj. ABLe ref. PTDC/EEI-SII/2094/2012, ADE (PTDC/ EIA-EIA/ 121686/2010).

References

1. American Cancer Society: Breast Cancer Facts & Figures 2009-2010. American Cancer Society, Atlanta, USA (2009)
2. American Cancer Society: Cancer Facts & Figures 2009. American Cancer Society, Atlanta, USA (2009)
3. Chang, C.C., Lin, C.J.: LIBSVM: A library for support vector machines. ACM Transactions on Intelligent Systems Technology 2(3), 27:1–27:27 (2011)
4. Chyou, P.H.: Patterns of bias due to differential misclassification by case control status in a case control study. European J. of Epidemiology 22, 7–17 (2007)
5. Cleary, T.: Test bias: Prediction of grades of negro and white students in integrated colleges. J. of Educational Measurement 5(2), 115–124 (1968)
6. Davis, J., Page, D., Santos Costa, V., Peissig, P., Caldwell, M.: A preliminary investigation into predictive models for adverse drug events. In: Proceedings of the AAAI 2013 Workshop on Expanding the Boundaries of Health Informatics Using AI (2013)
7. Flegal, K., Keyl, P., Nieto, F.: Differential misclassification arising from nondifferential errors in exposure measurement. Am. J. of Epidemiology 134(10), 1233–1244 (1991)
8. Fowble, B., Schultz, D., Overmoyer, B., Solin, L., Fox, K., Jardines, L., Orel, S., Glick, J.: The influence of young age on outcome in early stage breast cancer. Intl. J. of Radiation Oncology Biology Physics 30(1), 23–33 (1994)
9. Hansotia, B., Rukstales, B.: Incremental value modeling. J. of Interactive Marketing 16(3), 35–46 (2002)
10. Jaśkowski, M., Jaroszewicz, S.: Uplift modeling for clinical trial data. In: ICML 2012 Workshop on Clinical Data Analysis (2012)

11. Jayasinghe, U., Taylor, R., Boyages, J.: Is age at diagnosis an independent prognostic factor for survival following breast cancer? ANZ J. of Surgery 75(9), 762–767 (2005)
12. Joachims, T., Finley, T., Yu, C.N.: Cutting-plane training of structural SVMs. Machine Learning 77(1), 27–59 (2009)
13. Joachims, T.: A support vector method for multivariate performance measures. In: Proceedings of the 22nd International Conference on Machine Learning, pp. 377–384 (2005)
14. Kearney, P., Baigent, C., Godwin, J., Halls, H., Emberson, J., Patrono, C.: Do selective cyclo-oxygenase-2 inhibitors and traditional non-steroidal anti-inflammatory drugs increase the risk of atherothrombosis? meta-analysis of randomised trials. BMJ 332(7553), 1302–1308 (2006)
15. Linn, R.: Single-group validity, differential validity, and differential prediction. J. of Applied Psychology 63, 507–512 (1978)
16. Lo, V.: The true lift model - a novel data mining approach to response modeling in database marketing. SIGKDD Explorations 4(2), 78–86 (2002)
17. Narasimhan, H., Agarwal, S.: A structural SVM based approach for optimizing partial AUC. In: Dasgupta, S., Mcallester, D. (eds.) Proceedings of the 30th International Conference on Machine Learning, pp. 516–524 (2013)
18. Nassif, H., Kuusisto, F., Burnside, E.S., Page, D., Shavlik, J., Santos Costa, V.: Score as you lift (SAYL): A statistical relational learning approach to uplift modeling. In: Blockeel, H., Kersting, K., Nijssen, S., Železný, F. (eds.) ECML PKDD 2013, Part III. LNCS, vol. 8190, pp. 595–611. Springer, Heidelberg (2013)
19. Nassif, H., Santos Costa, V., Burnside, E.S., Page, D.: Relational differential prediction. In: Flach, P.A., De Bie, T., Cristianini, N. (eds.) ECML PKDD 2012, Part I. LNCS, vol. 7523, pp. 617–632. Springer, Heidelberg (2012)
20. Radcliffe, N.: Using control groups to target on predicted lift: Building and assessing uplift models. Direct Marketing J. 1, 14–21 (2007)
21. Radcliffe, N., Simpson, R.: Identifying who can be saved and who will be driven away by retention activity. J. of Telecommunications Management 1(2), 168–176 (2008)
22. Radcliffe, N., Surry, P.: Real-world uplift modelling with significance-based uplift trees. White Paper TR-2011-1, Stochastic Solutions (2011)
23. Rzepakowski, P., Jaroszewicz, S.: Decision trees for uplift modeling with single and multiple treatments. Knowledge and Information Systems 32, 303–327 (2012)
24. Tufféry, S.: Data Mining and Statistics for Decision Making, 2nd edn. John Wiley & Sons (2011)
25. Vapnik, V.: Statistical Learning Theory. John Wiley & Sons (1998)
26. Young, J.: Differential validity, differential prediction, and college admissions testing: A comprehensive review and analysis. Research Report 2001-6, The College Board (2001)
27. Zaniewicz, L., Jaroszewicz, S.: Support vector machines for uplift modeling. In: IEEE ICDM Workshop on Causal Discovery, CD 2013 (2013)
28. Zhang, S., Hossain, M., Hassan, M., Bailey, J., Ramamohanarao, K.: Feature weighted SVMs using receiver operating characteristics. In: Proceedings of the SIAM International Conference on Data Mining, pp. 497–508 (2009)

Fast LSTD Using Stochastic Approximation: Finite Time Analysis and Application to Traffic Control

L.A. Prashanth[1], Nathaniel Korda[2], and Rémi Munos[1]

[1] INRIA Lille - Nord Europe, Team SequeL, France
{prashanth.la, remi.munos}@inria.fr
[2] Oxford University, United Kingdom
nathaniel.korda@eng.ox.ac.uk

Abstract. We propose a stochastic approximation based method with randomisation of samples for policy evaluation using the least squares temporal difference (LSTD) algorithm. Our method results in an $O(d)$ improvement in complexity in comparison to regular LSTD, where d is the dimension of the data. We provide convergence rate results for our proposed method, both in high probability and in expectation. Moreover, we also establish that using our scheme in place of LSTD does not impact the rate of convergence of the approximate value function to the true value function. This result coupled with the low complexity of our method makes it attractive for implementation in *big data* settings, where d is large. Further, we also analyse a similar low-complexity alternative for least squares regression and provide finite-time bounds there. We demonstrate the practicality of our method for LSTD empirically by combining it with the LSPI algorithm in a traffic signal control application.

1 Introduction

Several machine learning problems involve solving a linear system of equations from a given set of training data. In this paper we consider the problem of policy evaluation in reinforcement learning (RL) using the method of temporal differences (TD). Given a fixed training data set, one popular temporal difference algorithm for policy evaluation is LSTD [4]. However, LSTD is computationally expensive as it requires $O(d^2)$ computations. We propose a stochastic approximation (SA) based algorithm that draws data samples from a uniform distribution on the training set. From the finite time analyses that we provide, we observe our algorithm converges at the optimal rate, in high probability as well as in expectation. Moreover, using our scheme in place of LSTD does not impact the rate of convergence of the approximate value function to the true value function. This finding coupled with the significant decrease in the computational cost of our algorithm, makes it appealing in the canonical *big data* settings.

The problem considered here is to estimate the value function V^π of a given policy π. Temporal difference (TD) methods are well-known in this context, and they are known to converge to the fixed point $V^\pi = \mathcal{T}^\pi(V^\pi)$, where $\mathcal{T}^\pi$ is the Bellman operator (see Section 3.1 for a precise definition). A popular approach to overcome the curse of dimensionality associated with large state spaces is to parameterize the value function using a linear function approximation architecture. For every s in the state space $\mathcal{S}$,

T. Calders et al. (Eds.): ECML PKDD 2014, Part II, LNCS 8725, pp. 66–81, 2014.
© Springer-Verlag Berlin Heidelberg 2014

we approximate $V^\pi(s) \approx \theta^\intercal \phi(s)$, where $\phi(\cdot)$ is a d-dimensional feature vector with $d << |\mathcal{S}|$, and θ is a tunable parameter. The function approximation variant of TD [23] is known to converge to the fixed point of $\Phi\theta = \Pi\mathcal{T}^\pi(\Phi\theta)$, where Π is the orthogonal projection onto the space within which we approximate the value function, and Φ is the feature matrix that characterises this space.

LSTD estimates the fixed point of $\Pi\mathcal{T}^\pi$ using empirical data $\mathcal{D} := \{(s_i, r_i, s'_i), i = 1, \dots, T)\}$ obtained by simulating the Markov decision process (MDP) with the underlying policy π. For every $i = 1, \dots, T$, the 3-tuple (s_i, r_i, s'_i) corresponds to a transition from state s_i to s'_i under action $\pi(s_i)$ and the resulting reward is denoted by r_i. The LSTD estimate is given as the solution to $\hat{\theta}_T = \bar{A}_T^{-1}\bar{b}_T$, where $\bar{A}_T = \frac{1}{T}\sum_{i=1}^T \phi(s_i)(\phi(s_i) - \beta\phi(s'_i))^\intercal$, and $\bar{b}_T = \frac{1}{T}\sum_{i=1}^T r_i\phi(s_i)$.

Computing the inverse of the matrix $\bar{A}_T$ is computationally expensive, especially when d is large. Indeed, assuming that the features $\phi(s_i)$ evolve in a compact subset of $\mathbb{R}^d$, the complexity of the above approach is $O(d^2 T)$, where $\bar{A}_T^{-1}$ is computed iteratively using the Sherman-Morrison lemma. On the other hand, if we employ the Strassen algorithm or the Coppersmith-Winograd algorithm for computing $\bar{A}_T^{-1}$, the complexity is of the order $O(d^{2.807})$ and $O(d^{2.375})$, respectively, in addition to $O(d^2 T)$ complexity for computing $\bar{A}_T$.

A common trick, in practice, to alleviate this problem in high dimensions, is to replace the inversion of the $\bar{A}_T$ matrix by an iterative procedure that performs a fixed point iteration. From a theoretical standpoint, this comes under the purview of stochastic approximation (SA), and one requires that the samples be chosen randomly to ensure convergence. In this paper, we analyse such an SA based scheme and show that it converges to the LSTD solution. The advantage is that the SA based scheme incurs lower computational cost in comparison to the approaches mentioned above. We also analyse a similar low-complexity alternative for the classic least squares parameter estimation problem.

We provide convergence rate results for our proposed method, both in high probability and in expectation. In particular, we show that, with probability $1 - \delta$, the SA based scheme constructs an ϵ-approximation of the corresponding LSTD solution with $O(d\ln(1/\delta)/\epsilon^2)$ complexity, irrespective of the number of samples T. Moreover, we also establish that using the SA based scheme in place of LSTD does not impact the rate of convergence of the approximate value function to the true value function (see Theorem 2).

The rate results coupled with the low complexity of our scheme make it more amenable to practical implementation in the canonical *big data* settings, where both d and T are large. Further, we provide explicit constants in the high probability bounds and we believe this opens several avenues for the use of SA based low complexity alternatives in higher level decision making procedures, for instance, least squares policy iteration (LSPI) [11] and linear bandit [5] algorithms. We demonstrate the practicality of our solution scheme for LSTD empirically by using it as a subroutine in the LSPI algorithm for adaptive traffic signal control[1]. In particular, for the experiments we

---

[1] See [16] for another set of experiments that combines the SA based low-complexity variant for least squares regression with the LinUCB algorithm for contextual bandits, using the large scale news recommendation dataset from Yahoo [24].

employ step-sizes that were used to derive the finite-time bounds (see Corollary 1). We demonstrate that this choice results in rapid convergence of our SA based scheme in the experiments and also that the performance of the SA variant of LSPI is comparable to that of LSPI.

The rest of the paper is organized as follows: In Section 2, we review relevant previous works and relevant literature. In Section 3 we present the fast LSTD algorithm based on stochastic approximation and in Section 4 we provide the non-asymptotic bounds for this algorithm. In Section 5, we outline the variants of our algorithm to incorporate regularization and iterate averaging, while in Section 7, we provide extensions to solve the problem of least squares regression. Next, in Section 6, we provide outlines for the proof and derivation of rates. In Section 8, we provide experiments on a traffic signal control application. Finally, in Section 9 we provide the concluding remarks.

2   Literature Review

Our algorithms are based on the well-known stochastic approximation technique, originally proposed for finding zeroes of a nonlinear function in [17]. The reader is referred to [10] for a textbook introduction to SA. Iterate averaging is a standard approach to accelerate the convergence of SA schemes and was proposed independently in [18] and [13]. Non asymptotic bounds for Robbins Monro schemes have been provided in [7] and extended to incorporate iterate averaging in [6].

In the context of the problem of prediction in RL, temporal difference (TD) learning is a well-known algorithm. See [3,20] for a textbook introduction and [23] for an asymptotic analysis. LSTD [4] is a popular batch algorithm that converges asymptotically to the TD solution. Finite time analysis of LSTD is provided in [12] and we extend it to the case when LSTD solution is replaced by a SA iterate.

A popular line of research in RL is on improving the complexity of TD-like algorithms (cf. GTD [21], GTD2 [22], iLSTD [8] and the references therein). The popular Computer Go with dimension $d = 10^6$ [19] and several practical applications (e.g. transportation, networks) involve high-feature dimensions. Moreover, considering that linear function approximation is effective with a large number of features, our $O(d)$ improvement in complexity of LSTD by employing SA is meaningful.

In comparison to previous work, we would like to point out that there is no finite time analysis of GTD-type algorithms. While iLSTD is an efficient approximation to LSTD, analysis in [8] requires that the feature matrix be sparse. In contrast, we provide finite-time bounds and do not make any sparsity assumption. To the best of our knowledge, efficient SA algorithms that approximate LSTD without impacting its rate of convergence to true value function, have not been proposed before in the literature. The high probability bounds that we derive for the SA based scheme do not directly follow from earlier work on LSTD algorithms. Further, unlike [7], we provide explicit constants in the bounds that we derive (see Corollary 1) and we employ these in our experiments as well.

Stochastic gradient descent (SGD) is a well-known method for optimising a function given only noisy observations. In the context of machine learning, finite time analysis of such methods have been provided in [1]. While the bounds in [1] are given in expectation, many machine learning applications require high probability bounds, which we provide for our case. Regret bounds for online SGD techniques have been given iin [25,9]: the gradient descent algorithm in [25] is in the setting of optimising the average of convex loss functions whose gradients are available, while that in [9] is for strongly convex loss functions.

In comparison to previous work w.r.t. least squares regression, we highlight the following differences: (i) Earlier works on least squares regression (cf. [9]) require the knowledge of the strong convexity constant in deciding the step-size, while we average the iterates to get rid of this dependency. (ii) Our analysis is much simpler (since we work directly with least squares problems) and we make all the constants explicit for the problems considered.

3   Fast LSTD Using Stochastic Approximation (fLSTD-SA)

We propose here a stochastic approximation variant of the least squares temporal difference (LSTD) algorithm, whose iterates converge to the same fixed point as the regular LSTD algorithm, while incurring much smaller overall computational cost.

The algorithm, which we call fast LSTD through Stochastic Approximation (fLSTD-SA), is a simple stochastic approximation scheme with randomised samples. The results that we present establish that fLSTD-SA computes an ϵ-approximation to the LSTD solution $\hat{\theta}_T$ with probability $1 - \delta$, while incurring a complexity of the order $O(d \ln(1/\delta)/\epsilon^2)$, irrespective of the number of samples T. In turn, this enables us to give a performance bound for the approximate value function computed by fLSTD-SA. A schema of fLSTD-SA is given in Figure 1.

Although our analysis for fLSTD-SA depends on a strong convexity assumption that may not hold in all situations, we present also a variant of fLSTD-SA employing iterate averaging for which error bounds can be given without resorting to a strong convexity assumption.

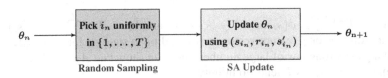

Fig. 1. Overall flow of the fLSTD-SA algorithm

3.1   Background for LSTD

Consider an MDP with (finite) state space $\mathcal{S}$, (finite) action space $\mathcal{A}$ and transition probabilities $p(s, a, s')$, $s, s' \in \mathcal{S}$, $a \in \mathcal{A}$. For a given stationary policy $\pi : \mathcal{S} \to \mathcal{A}$, the value function V^π is defined by

$$V^\pi(s) := E\left[\sum_{t=0}^{\infty} \gamma^t r(s_t, \pi(s_t)) \mid s_0 = s\right], \tag{1}$$

where s_t denotes the state of the MDP at time t, $\beta \in (0,1)$ is the discount factor, and $r(s, a)$ denotes the instantaneous rewards obtained in state s with action a. The value function V^π can be expressed as the fixed point of the Bellman operator $\mathcal{T}^\pi$ defined by

$$\mathcal{T}^\pi(V)(s) := r(s, \pi(s)) + \beta \sum_{s'} p(s, \pi(s), s') V(s'), \tag{2}$$

When the cardinality of $\mathcal{S}$ is huge and in the absence of knowledge of the transition dynamics, a popular approach is to parameterize the value function using a linear function approximation architecture, i.e., for every $s \in \mathcal{S}$, we approximate $V^\pi(s) \approx \phi(s)^\mathsf{T}\theta$, where $\phi(s)$ is a d-dimensional feature vector with $d << |\mathcal{S}|$, and θ is a tunable parameter. The well-known TD learning algorithm [3] attempts to find the fixed point of the operator $\Pi\mathcal{T}^\pi$ given by

$$\Phi\theta = \Pi\mathcal{T}^\pi(\Phi\theta), \tag{3}$$

where $\mathcal{B} = \{\Phi\theta \mid \theta \in \mathbb{R}^d\}$ is the space within which we want to approximate the value function V^π, Π is the orthogonal projection onto $\mathcal{B}$, and Φ is the feature matrix with rows $\phi(s)^\mathsf{T}, \forall s \in \mathcal{S}$ denoting the features corresponding to state $s \in \mathcal{S}$. Let θ^* denote the solution to (3), P the transition probability matrix with components $p(s, \pi(s), s')$ and Ψ the stationary distribution (assuming it exists) of the Markov chain for the underlying policy π. Then, θ^* can be written as the solution to the following system of equations (cf. [2, Section 6.3])

$$A\theta^* = b, \text{ where } A = \Phi^\mathsf{T}\Psi(I - \beta P)\Phi \text{ and } b = \Phi^\mathsf{T}\Psi r. \tag{4}$$

The LSTD approach is to approximate A and b using T samples $\{(s_i, r_i, s_i'), i = 1, \dots, T)\}$ obtained by simulating the MDP with the underlying policy π.

An approximate solution to (4) is constructed as follows:

$$\hat{\theta}_T = \bar{A}_T^{-1}\bar{b}_T \tag{5}$$

where $\bar{A}_T = T^{-1}\sum_{i=1}^{T} \phi(s_i)(\phi(s_i) - \beta\phi(s_i'))^\mathsf{T}$, and $\bar{b}_T = T^{-1}\sum_{i=1}^{T} r_i\phi(s_i)$. Here $\phi(s_i)$ is a d-dimensional feature vector corresponding to state s_i, for all $i = 1, \dots, T$. By invoking the strong law of large numbers, one can show that $\bar{A}_T \to A$ and $\bar{b}_T \to b$ as the number of samples T tends to infinity.

3.2   Update Rule for Flstd-SA

Starting with an arbitrary θ_0, we update the parameter θ_n as follows:

$$\theta_n = \theta_{n-1} + \gamma_n \left(r_{i_n} + \beta\theta_{n-1}^\mathsf{T}\phi(s_{i_n}') - \theta_{n-1}^\mathsf{T}\phi(s_{i_n})\right)\phi(s_{i_n}), \tag{6}$$

where each i_n is chosen uniformly randomly from the set $\{1, \dots, T\}$. In other words, we pick a sample with uniform probability $1/T$ from the set $\mathcal{D} := \{(s_i, r_i, s_i'), i = 1, \dots, T)\}$ and use it to perform a fixed point iteration in (6). The quantities γ_n above are *step sizes* that are chosen in advance and satisfy standard stochastic approximation conditions (see (A1) below). Notice that the above update is the usual TD update, except that the samples are drawn uniformly randomly from the sample set $\mathcal{D}$.

4   Main Results

4.1   Error Bounds

We make the following assumptions for the analysis fLSTD-SA:

(A1) The step sizes γ_n satisfy $\sum_n \gamma_n = \infty$, and $\sum_n \gamma_n^2 < \infty$.

(A2) Bounded features: $\|\phi(s_i)\|_2 \leq 1$, for $i = 1, \dots, T$.

(A3) Bounded rewards: $|r_i| \leq R_{\max} < \infty$ for $i = 1, \dots, T$ and bounded linear space, i.e., $-V_{\max} \leq \Phi\theta \leq V_{\max} < \infty$.

(A4) Strong Convexity: Writing $\Phi_T \triangleq (\phi(s_1)^\mathsf{T}; \dots; \phi(s_T)^\mathsf{T})$, the covariance matrix $\frac{1}{T}\Phi_T^\mathsf{T}\Phi_T$ is positive definite and its smallest (positive) eigenvalue is at least μ.

By working in a bounded linear space along with bounded rewards and features, along with step sizes that satisfy standard stochastic approximation conditions, we ensure that the parameter θ remains stable, and hence that (6) converges.

To obtain high probability bounds on the error we consider separately the deviation of z_n from its mean (see (7) in Theorem 1), and the size of its mean itself (see (8) in Theorem 1). In this way the first quantity can be directly decomposed as a sum of martingale differences, and then a standard martingale concentration argument applied, while the second quantity can be analyzed by directly unrolling iteration (6) (a proof outline is provided in Section 6, while the detailed proofs are available in [16]).

Theorem 1. *Under (A1)-(A4), we have* $\forall \epsilon > 0$,

$$P(\|\theta_n - \hat{\theta}_T\|_2 - \mathbb{E}\|\theta_n - \hat{\theta}_T\|_2 \geq \epsilon) \leq \exp\left(-\epsilon^2 / (2\sum_{i=1}^{n} L_i^2)\right), \tag{7}$$

$$\mathbb{E}\|\theta_n - \hat{\theta}_T\|_2 \leq \underbrace{\exp(-(1-\beta)\mu\Gamma_n)\|\theta_0 - \hat{\theta}_T\|_2}_{\textit{initial error}}$$

$$+ \underbrace{\left(\sum_{k=1}^{n-1} H_\beta^2 \gamma_{k+1}^2 \exp(-2(1-\beta)\mu(\Gamma_n - \Gamma_{k+1}))\right)^{\frac{1}{2}}}_{\textit{sampling error}}, \tag{8}$$

where $L_i := \gamma_i \prod_{j=i}^{n-1}(1 - 2\gamma_{j+1}\mu((1-\beta) - \beta(2-\beta)\gamma_{j+1}))^{1/2}$, $\Gamma_n := \sum_{i=1}^{n}\gamma_i$ *and* $H_\beta^2 := R_{\max}(R_{\max} + 2) + (1+\beta)^2 V_{\max}^2$.

The initial error depends on the initial point θ_0 of the algorithm. The sampling error arises out of a martingale difference sequence that depends on the random deviation of the stochastic update from the standard fixed point iteration, and is the dominant term in (8). Under a suitable choice of step-sizes (see Corollary 1), it can be shown that the initial error is forgotten faster than the sampling error.

The above theorem assumes no specific form for the step-sizes γ_n. Specifying the step-size sequence, we can merge the two claims above to deduce the following bounds on the approximation error z_n with explicit constants:

Corollary 1 (Error Bound for iterates of fLSTD-SA). *Under (A2)-(A4), choosing* $\gamma_n = \frac{(1-\beta)c}{2(c+n)}$ *and c such that* $(1-\beta)^2 \mu c \in (1.33, 2)$, *we have, for any* $\delta > 0$,

$$\mathbb{E}\|\theta_n - \hat{\theta}_T\|_2 \leq \frac{K_1(n)}{\sqrt{n+c}} \text{ and } P\left(\|\theta_n - \hat{\theta}_T\|_2 \leq \frac{K_2(n)}{\sqrt{n+c}}\right) \geq 1 - \delta, \qquad (9)$$

where $K_1(n)$ *and* $K_2(n)$ *are functions of order* $O(1)$, *defined by:*

$$K_1(n) = \frac{\sqrt{c}\|\theta_0 - \hat{\theta}_T\|_2}{n^{((1-\beta)^2\mu c - 1)/2}} + \frac{(1-\beta)cH_\beta}{2}, \; K_2(n) = \frac{(1-\beta)c\sqrt{\log \delta^{-1}}}{2\sqrt{\left(\frac{4}{3}(1-\beta)^2\mu c - 1\right)}} + K_1(n).$$

Remark 1. We note that setting c such that $(1-\beta)^2\mu c = \eta \in (1.33, 2)$ we can rewrite the constants in Corollary 1 as:

$$K_1(n) = \frac{\|\theta_0 - \hat{\theta}_T\|_2}{(1-\beta)\sqrt{\mu n^{(\eta-1)}}} + \frac{H_\beta}{2(1-\beta)\mu}, \; K_2(n) = \frac{\sqrt{\log \delta^{-1}}}{2(1-\beta)\mu\sqrt{\left(\frac{4}{3}\eta - 1\right)}} + K_1(n).$$

So both the bounds in expectation and high probability have a linear dependence on the inverse of $(1-\beta)\mu$.

4.2 Performance Bound

Let $\tilde{v}_T := \Phi\theta_T$ denote the approximate value function obtained from T steps of fLSTD-SA, and let v denote the true value function, evaluated at the states $s_1, \ldots, s_T$. Then the following lower bound on the performance of $\tilde{v}_T$ can be deduced from Corollary 1 in conjunction with Theorem 1 of [12]:

Theorem 2. *Under conditions of Corollary 1, for any* $\delta > 0$, *with probability* $1 - \delta$,

$$\|v - \tilde{v}_T\|_T \leq \underbrace{\frac{\|v - \Pi v\|_T}{\sqrt{1-\beta^2}}}_{\text{residual error}} + \underbrace{O\left(\sqrt{\frac{d}{(1-\beta)^2\mu T}}\right)}_{\text{estimation error}} + \underbrace{O\left(\sqrt{\frac{1}{(1-\beta)\mu T}} \ln\frac{1}{\delta}\right)}_{\text{approximation error}},$$

where $\|f\|_T^2 := T^{-1}\sum_{i=1}^{T} f(s_i)^2$, *for any function f.*

The residual and estimation errors (first and second terms in the RHS above) are artifacts of function approximation and least squares methods, respectively. The third term, of order $O(1/\sqrt{T})$, is a consequence of using fLSTD-SA in place of the LSTD. From the above theorem, we observe that using our scheme in place of LSTD does not impact the rate of convergence of the approximate value function $\tilde{v}_T$ to the true value function v. This finding coupled with the fact that our scheme is of low complexity makes it attractive for implementation in *big data* settings, where the feature dimension d is large.

5   Variants

To obtain the best performance from fLSTD-SA we need to know the value of μ. However with minor adjustments to the analysis we can provide two variants of fLSTD-SA for which it is not necessary to know the value of μ to obtain the (optimal) approximation error of order $O(n^{-1/2})$ and explicit constants.

5.1   Regularization

A popular approach is to search not for the LSTD solution, but instead for a regularized LSTD solution defined as follows:

$$\hat{\theta}_T^{reg} = (\bar{A}_T + \mu I)^{-1} \bar{b}_T \tag{10}$$

where μ is now a constant set in advance. The update rule for this variant is

$$\theta_n^{reg} = (1 - \gamma_n \mu)\theta_{n-1} + \gamma_n \left(r_{i_n} + \beta \theta_{n-1}^{\mathsf{T}} \phi(s'_{i_n}) - \theta_{n-1}^{\mathsf{T}} \phi(s_{i_n}) \right) \phi(s_{i_n}). \tag{11}$$

This algorithm retains all the properties of the non-regularized fLSTD-SA algorithm, except that it converges to the solution of (10) rather than to that of (5). In particular the conclusions of Theorem 1, and of Corollary 1 hold without requiring assumption (A4), but measuring $\theta_n - \hat{\theta}_T^{reg}$, the error to the regularized fixed point $\hat{\theta}_T^{reg}$.

5.2   Iterate Averaging

Another well-known approach is to employ the Polyak-Ruppert scheme of averaging the iterates, together with choosing larger step-sizes. In particular, we fix the step-size $\gamma_n := \frac{(1-\beta)}{2} \left(\frac{c}{c+n} \right)^\alpha$, and then use the averaged iterate $\bar{\theta}_{n+1} := (\theta_1 + \ldots + \theta_n)/n$ to approximate the LSTD solution. Here the quantities θ_n are just the iterates of the fLSTD-SA presented earlier. An analogue of Corollary 1 for iterate averaging is as follows (see [16] for a detailed proof):

Corollary 2. *Under (A2)-(A3), choosing* $\gamma_n = \frac{(1-\beta)}{2} \left(\frac{c}{c+n} \right)^\alpha$, *with* $\alpha \in (1/2, 1)$ *and* $c \in (1.33, 2)$, *we have, for any* $\delta > 0$,

$$\mathbb{E}\|\bar{\theta}_n - \hat{\theta}_T\|_2 \leq \frac{K_1^{IA}(n)}{(n+c)^{\alpha/2}} \text{ and } P\left(\|\bar{\theta}_n - \hat{\theta}_T\|_2 \leq \frac{K_2^{IA}(n)}{(n+c)^{\alpha/2}} \right) \geq 1 - \delta, \tag{12}$$

where, writing $C = \sum_{n=1}^{\infty} \exp(-\mu c n^{1-\alpha})(< \infty)$,

$$K_1^{IA}(n) := \frac{C\|\theta_0 - \hat{\theta}_T\|_2}{(n+c)^{(1-\alpha)/2}} + \frac{H_\beta c^\alpha (1-\beta)}{(\mu c^\alpha (1-\beta)^2)^\alpha \frac{1+2\alpha}{2(1-\alpha)}}, \text{ and}$$

$$K_2^{IA}(n) := \frac{\sqrt{\log \delta^{-1}}}{\mu(1-\beta)} \left[3^\alpha + \left[\frac{2\alpha}{\mu c^\alpha (1-\beta)^2} + \frac{2^\alpha}{\alpha} \right]^2 \right] \frac{1}{(n+c)^{(1-\alpha)/2}} + K_1^{IA}(n).$$

Thus, it is possible to remove the dependency on the knowledge of μ for the choice of c through averaging of the iterates, at the cost of $(1 - \alpha)/2$ in the rate. However, choosing α close to 1 causes a sampling error blowup. As suggested by earlier works on stochastic approximation, it is preferred to average after a few iterations since the initial error is not forgotten exponentially faster than the sampling error with averaging.

6 Outline of Analysis

In this section we give outline proofs of the main results concerning the fLSTD-SA algorithm. We split these into two sections: first, we sketch the martingale analysis that leads to the proof of Theorem 1 and which forms the template for the proof for extension to least squares regression (see Appendix C in [16]) and the regularized and iterate averaged variants of fLSTD-SA (see Corollary 2); second, we give the derivation of the rates when the step sizes a chosen in specific forms.

6.1 Outline of Theorem 1 Proof

Denote the approximation error by $z_n := \theta_n - \hat{\theta}_T$. Recall that Theorem 1 decomposes the problem of bounding z_n into bounding the deviation from its mean in high probability and then the mean of z_n itself. In the following, we first provide a sketch of the proof of high probability bound and later outline the proof for the bound in expectation. For the former, we employ a proof technique similar to that used in [7]. However, our analysis is much simpler and we make all the constants explicit for the problem at hand. Moreover, in order to eliminate a possible exponential dependence of the constants in the resulting bound on the inverse of $(1 - \beta)\mu$, we depart from the argument in [7].

Proof (**High probability bound.**). *(Sketch)* Recall that $z_n := \theta_n - \hat{\theta}_T$. We rewrite $\|z_n\|_2^2 - \mathbb{E}\|z_n\|_2^2$ as a telescoping sum of martingale differences:

$$\|z_n\|_2 - \mathbb{E}\|z_n\|_2 = \sum_{i=1}^{n} g_i - \mathbb{E}[g_i \,|\, \mathcal{F}_{i-1}] = \sum_{i=1}^{n} D_i,$$

where $D_i := g_i - \mathbb{E}[g_i \,|\, \mathcal{F}_{i-1}]$, $g_i := \mathbb{E}[\|z_n\|_2 \,|\, \theta_i]$, and $\mathcal{F}_i$ denotes the sigma algebra generated by the random variables $\{i_1, \ldots, i_n\}$.

The next step is to show that the functions g_i are Lipschitz continuous in the rewards, with Lipschitz constants L_i. In order to obtain constants with no exponential dependence on the inverse of $(1 - \beta)\mu$ we depart from the general scheme of [7], and use our knowledge of the form of the update function f_i to eliminate the noise due to the rewards between time $i + 1$ and time n. Specifically, letting $\Theta_j^i(\theta)$ denote the mapping that returns the value of the iterate θ_j at instant j, given that $\theta_i = \theta$, we show that

$$\mathbb{E}\left[\|\Theta_n^i(\theta) - \Theta_n^i(\theta')\|_2^2\right] = \mathbb{E}\left[\mathbb{E}\left([I - \gamma_n[\phi(s_{i_n})\phi(s_{i_n})^\mathsf{T} - \beta\phi(s_{i_n})\phi(s_{i_n}')^\mathsf{T}]]\right)\right.$$
$$\left. \cdot(\Theta_{n-1}^i(\theta) - \Theta_{n-1}^i(\theta')) \,|\, \Theta_{n-1}^i(\theta), \Theta_{n-1}^i(\theta'))\right]$$
$$\leq (1 - \gamma_n\mu(1 - \beta - \gamma_n\beta(2 - \beta)))\mathbb{E}\left[\|\Theta_{n-1}^i(\theta) - \Theta_{n-1}^i(\theta')\|_2^2\right],$$

where we used the specific form of f_i in obtaining the equality, and have applied assumption (A4) to obtain the inequality. Unrolling this iteration then yields the new Lipschitz constants.

Now we can invoke a standard martingale concentration bound: Using the L_i-Lipschitz property of the g_i functions and the assumption (A3) we find that

$$P(\|z_n\|_2 - \mathbb{E}\|z_n\|_2 \geq \epsilon) = P\left(\sum_{i=1}^{n} D_i \geq \epsilon\right) \leq \exp(-\lambda\epsilon)\exp\left(\frac{\alpha\lambda^2}{2}\sum_{i=1}^{n} L_i^2\right).$$

The claim follows by optimizing the above over λ. The full proof is available in [16].

Proof (**Bound in expectation.**). (*Sketch*) First we extract a martingale difference from the update rule (6): Recall that $z_n := \theta_n - \hat{\theta}_T$. Let $f_n(\theta) := (\theta^\mathsf{T} x_{i_n} - (r_{i_n} + \beta \theta^\mathsf{T} x'_{i_n})) x_{i_n}$ and let $F(\theta) := \mathbb{E}_{i_n}(f_n(\theta))$. Then, we have

$$z_n = \theta_n - \hat{\theta}_T = \theta_{n-1} - \hat{\theta}_T - \gamma_n \left(F(\theta_{n-1}) - \Delta M_n \right),$$

where $\Delta M_{n+1}(\theta) = F_n(\theta) - f_n(\theta)$ is a martingale difference. Now since $\hat{\theta}_T$ is the LSTD solution, $F(\hat{\theta}_T)) = 0$. Moreover, $F(\cdot)$ is linear, and so we obtain

$$z_n = z_{n-1} - \gamma_n \left(z_{n-1} \bar{A}_n - \Delta M_n \right) = \Pi_n z_0 - \sum_{k=1}^{n} \gamma_k \Pi_n \Pi_k^{-1} \Delta M_k,$$

where $\bar{A}_n = \dfrac{1}{n} \sum_{i=1}^{n} x_i (x_i - \beta x'_i)^\mathsf{T}$ and $\Pi_n := \prod_{k=1}^{n} \left(I - \gamma_k \bar{A}_k \right)$.
By Jensen's inequality, we obtain

$$\mathbb{E}(\|z_n\|_2) \leq (\mathbb{E}(\langle z_n, z_n \rangle))^{\frac{1}{2}} = \left(\mathbb{E}\|\Pi_n z_0\|_2^2 + \sum_{k=1}^{n} \gamma_k^2 \mathbb{E}\|\Pi_n \Pi_k^{-1} \Delta M_k\|_2^2 \right)^{\frac{1}{2}} \quad (13)$$

The rest of the proof amounts to bounding the martingale difference ΔM_n as follows:

$$\mathbb{E}[\|\Delta M_n\|_2^2] \leq \mathbb{E}_{i_t} \langle f_{i_t}(\theta_{t-1}), f_{i_t}(\theta_{t-1}) \rangle \leq R_{\max}(R_{\max} + 2) + (1 + \beta)^2 \|\theta_{t-1}\|_2^2 \leq H_\beta^2.$$

6.2 Derivation of Rates

Now we give the proof of Corollary 1, which gives explicitly the rate of convergence of the approximation error in high probability for the specific choice of step sizes:

Proof (**Proof of Corollary 1:**). Note that when $\gamma_n = \dfrac{(1-\beta)c}{2(c+n)}$,

$$\sum_{i=1}^{n} L_i^2 = \sum_{i=1}^{n} \frac{(1-\beta)^2 c^2}{4(c+i)^2} \prod_{j=i}^{n} \left(1 - 2\mu \frac{(1-\beta)c}{2(c+n)} ((1-\beta) - \beta(2-\beta) \frac{(1-\beta)c}{2(c+n)}) \right)$$

$$\leq \sum_{i=1}^{n} \frac{(1-\beta)^2 c^2}{4(c+i)^2} \exp \left(-\frac{3}{4}(1-\beta)^2 \mu c \sum_{j=i}^{n} \frac{1}{(c+n)} \right)$$

$$\leq \frac{(1-\beta)^2 c^2}{4(n+c)^{\frac{3}{4}(1-\beta)^2 \mu c}} \sum_{i=1}^{n} (i+c)^{-(2-\frac{3}{4}(1-\beta)^2 \mu c)}.$$

We now find three regimes for the rate of convergence, based on the choice of c:

(i) $\sum_{i=1}^{n} L_i^2 = O\left((n+c)^{\frac{3}{4}(1-\beta)^2 \mu c} \right)$ when $\frac{3}{4}(1-\beta)^2 \mu c \in (0,1)$,

(ii) $\sum_{i=1}^{n} L_i^2 = O\left(n^{-1} \ln n \right)$ when $\frac{3}{4}(1-\beta)^2 \mu c = 1$, and

(iii) $\sum_{i=1}^{n} L_i^2 = \dfrac{(1-\beta)^2 c^2}{4(\frac{3}{4}(1-\beta)^2 \mu c - 1)} (n+c)^{-1}$ when $\frac{3}{4}(1-\beta)^2 \mu c \in (1,2)$.

(We have used comparisons with integrals to bound the summations.) Thus, setting $2/((1-\beta)^2\mu) > c > 1/((1-\beta)^2\mu)$, the high probability bound from Theorem 1 gives

$$P(\|\theta_n - \hat{\theta}_T\|_2 - \mathbb{E}\|\theta_n - \hat{\theta}_T\|_2 \geq \epsilon) \leq \exp\left(-\frac{\epsilon^2(n+c)}{2K_{\mu,c,\beta}}\right) \qquad (14)$$

where $K_{\mu,c,\beta} := \frac{(1-\beta)^2c^2}{4((1-\beta)^2\mu c - 1)}$.

Under the same choice of step-size, the bound in expectation in Theorem 1 we have:

$$\sum_{k=1}^{n-1} H_\beta^2 \gamma_{k+1}^2 \exp(-2(1-\beta)\mu(\Gamma_n - \Gamma_{k+1}))$$

$$\leq \frac{(1-\beta)^2c^2H_\beta^2}{4(n+c)^{(1-\beta)^2\mu c}} \sum_{k=1}^{n} (c+k)^{-(2-(1-\beta)^2\mu c)} \leq \frac{(1-\beta)^2c^2H_\beta^2}{4(n+c)}$$

we in the last inequality we have again compared the sum with an integral. Similarly

$$\exp(-(1-\beta)\mu\Gamma_n) \leq \left(\frac{c}{n+c}\right)^{\frac{(1-\beta)^2\mu c}{2}} \leq \left(\frac{c}{n+c}\right)^{\frac{1}{2}}.$$

So we have

$$\mathbb{E}\|\theta_n - \hat{\theta}_T\|_2 \leq \left(\sqrt{c}\|\theta_0 - \theta^*\|_2 + \frac{(1-\beta)cH_\beta}{2}\right)(c+n)^{-\frac{1}{2}}, \qquad (15)$$

and the result now follows.

7 Extension to Least Squares Regression

In this section, we describe the classic parameter estimation problem using the method of least squares, the standard approach to solve this problem and a low-complexity alternative using stochastic approximation.

In this setting, we are given a set of samples $\mathcal{D} := \{(x_i, y_i), i = 1, \ldots, T\}$ with the underlying observation model $y_i = x_i^\top\theta^* + \xi_i$ (ξ_i is zero mean and variance bounded by $\sigma < \infty$, and θ^* is an unknown parameter). The least squares estimate $\hat{\theta}_T$ minimizes $\sum_{i=1}^{T}(y_i - \theta^\top x_i)^2$. It can be shown that $\hat{\theta}_T = \bar{A}_T^{-1}\bar{b}_T$, where $\bar{A}_T = T^{-1}\sum_{i=1}^{T} x_i x_i^\top$ and $\bar{b}_T = T^{-1}\sum_{i=1}^{T} x_i y_i$.

Notice that, unlike the RL setting, $\hat{\theta}_T$ here is the minimizer of an empirical loss function. However, as in the case of LSTD, the computational cost for a Sherman-Morrison lemma based approach for solving the above would be of the order $O(d^2T)$. Similarly to the case of the fLSTD-SA algorithm, we update the iterate θ_n using a SA scheme as follows (starting with an arbitrary θ_0),

$$\theta_n = \theta_{n-1} + \gamma_n(y_{i_n} - \theta_{n-1}^\top x_{i_n})x_{i_n}, \qquad (16)$$

where, as before, each i_n is chosen uniformly randomly from the sample set $\mathcal{D}$ and γ_n are step-sizes.

Unlike fLSTD-SA which is a fixed point iteration, the above is a stochastic gradient descent procedure. Nevertheless, using the same proof template as for fLSTD-SA earlier, we can derive bounds on the approximation error, i.e., the distance between θ_n and least squares solution $\hat{\theta}_T$, both in high probability as well as expectation.

Results. As in the case of fLSTD-SA, we assume that the features are bounded, the noise is i.i.d, zero-mean and bounded and the matrix $\bar{A}_T$ is positive definite, with smallest eigenvalue at least $\mu > 0$. An analogue of Corollary 1 for this setting is as follows (See Appendix C in [16] for a detailed proof.):

Corollary 3. *Choosing* $\gamma_n = \frac{c}{2(c+n)}$ *and c such that* $\mu c \in (1.33, 2)$, *for any* $\delta > 0$,

$$\mathbb{E}\|\theta_n - \hat{\theta}_T\|_2 \leq \frac{K_1^{LS}}{\sqrt{n+c}} \text{ and } P\left(\|\theta_n - \hat{\theta}_T\|_2 \leq \frac{K_2^{LS}}{\sqrt{n+c}}\right) \geq 1 - \delta,$$

where, defining $h(n) := c\left[\left(\sigma + 2\|\theta_0 - \hat{\theta}_T\|_2^2\right) + 4\|\theta_0 - \hat{\theta}_T\|_2 \ln n + 2\ln^2 n\right]$,

$$K_1^{LS}(n) := \frac{\sqrt{c}\|\theta_0 - \hat{\theta}_T\|_2}{(n+c)^{(\mu c - 1)/2}} + \frac{h(n)}{2}, \quad K_2^{LS}(n) := \frac{\sqrt{c}}{\sqrt{((\mu c)/2 - 1)}}\sqrt{\log\frac{1}{\delta}} + K_1(n).$$

8   Traffic Control Application

LSPI [11] is a well-known algorithm for control based on the policy iteration procedure for MDPs. It performs policy evaluation and policy improvement in tandem. For the purpose of policy evaluation, LSPI uses a LSTD-like algorithm called LSTDQ, which learns the state-action value function. In contrast, LSTD learns the state value function.

We now briefly describe LSTDQ and its fast SA variant fLSTDQ-SA: We are given a set of samples $\mathcal{D} := \{(s_i, a_i, r_i, s_i'), i = 1, \ldots, T)\}$, where each sample i denotes a one-step transition of the MDP from state s_i to s_i' under action a_i, while resulting in a reward r_i. LSTDQ attempts to approximate the Q-value function for any policy π by solving the linear system $\hat{\theta}_T = \bar{A}_T^{-1}\bar{b}_T$, where $\bar{A}_T = T^{-1}\sum_{i=1}^{T}\phi(s_i, a_i)(\phi(s_i, a_i) - \beta\phi(s_i', \pi(s_i')))^\mathsf{T}$, and $\bar{b}_T = T^{-1}\sum_{i=1}^{T} r_i\phi(s_i, a_i)$. fLSTDQ-SA approximates LSTDQ by an iterative update scheme as follows (starting with an arbitrary θ_0):

$$\theta_k = \theta_{k-1} + \gamma_k\left(r_{i_k} + \beta\theta_{k-1}^\mathsf{T}\phi(s_{i_k}', \pi_n(s_{i_k}')) - \theta_{k-1}^\mathsf{T}\phi(s_{i_k}, a_{i_k})\right)\phi(s_{i_k}, a_{i_k}) \quad (17)$$

From Section 3, it is evident that the claims in Theorem 1 and Corollary 1 hold for the above scheme as well.

The idea behind the experimental setup is to study both LSPI and a variant of LSPI, referred to as fLSPI-SA, where we use fLSTDQ-SA as a subroutine to approximate the LSTDQ solution. Algorithm 1 provides the pseudo-code for the latter algorithm.

We consider a traffic signal control application for conducting the experiments. The problem here is to adaptively choose the sign configurations for the signalized intersections in the road network considered, in order to maximize the traffic flow in the long run. Let L be the total number of lanes in the road network considered. Further, let $q_i(t), i = 1, \ldots, L$ denote the queue lengths and $t_i(t), i = 1, \ldots, L$ the elapsed time (since signal turned to red) on the individual lanes of the road network. Following [14], the traffic signal control MDP is formulated as follows:

Algorithm 1. fLSPI-SA

Input: Sample set $D := \{s_i, a_i, r_i, s_i'\}_{i=1}^{T}$, obtained from an initial (arbitrary) policy
Initialisation: ϵ, τ, step-sizes $\{\gamma_k\}_{k=1}^{T}$, initial policy π_0 (given as θ_0)
$\pi \leftarrow \pi_0, \theta \leftarrow \theta_0$
repeat
    Policy Evaluation
        Approximate LSTDQ(D, π) using fLSTDQ-SA(D, π) as follows:
        for $k = 1 \ldots \tau$ **do**
            Get random sample index: $i_k \sim U(\{1, \ldots, T\})$
            Update fLSTD-SA iterate θ_k using (17)
        end for
    $\theta' \leftarrow \theta_\tau, \Delta = \|\theta - \theta'\|_2$
    Policy Improvement
        Obtain a greedy policy π' as follows: $\pi'(s) = \arg\max_{a \in \mathcal{A}} \theta'^{\mathsf{T}} \phi(s, a)$
    $\theta \leftarrow \theta', \pi \leftarrow \pi'$
until $\Delta < \epsilon$

State $s_t = \big(q_1(t), \ldots, q_L(t), t_1(t), \ldots, t_L(t)\big)$,
Action a_t belongs to the set of feasible sign configurations,
Single-stage cost $h(s_t) = u_1 \left[\sum_{i \in I_p} u_2 \cdot q_i(t) + \sum_{i \notin I_p} w_2 \cdot q_i(t) \right] + w_1 \left[\sum_{i \in I_p} u_2 \cdot t_i(t) + \sum_{i \notin I_p} w_2 \cdot t_i(t) \right]$, where $u_i, w_i \geq 0$ such that $u_i + w_i = 1$ for $i = 1, 2$ and $u_2 > w_2$. Here, the set I_p is the set of prioritized lanes.

Table 1. Feature selection

State	Action	Feature $\phi_i(s, a)$
$q_i < \mathcal{L}_1$ and $t_i < \mathcal{T}_1$	RED	0.01
	GREEN	0.06
$q_i < \mathcal{L}_1$ and $t_i \geq \mathcal{T}_1$	RED	0.02
	GREEN	0.05
$\mathcal{L}_1 \leq q_i < \mathcal{L}_2$ and $t_i < \mathcal{T}_1$	RED	0.03
	GREEN	0.04
$\mathcal{L}_1 \leq q_i < \mathcal{L}_2$ and $t_i \geq \mathcal{T}_1$	RED	0.04
	GREEN	0.03
$q_i \geq \mathcal{L}_2$ and $t_i < \mathcal{T}_1$	RED	0.05
	GREEN	0.02
$q_i \geq \mathcal{L}_2$ and $t_i \geq \mathcal{T}_1$	RED	0.06
	GREEN	0.01

Function approximation is a standard technique employed to handle high-dimensional state spaces (as is the case with the traffic signal control MDP on large road networks). We employ the feature selection scheme from [15], which is briefly described in the

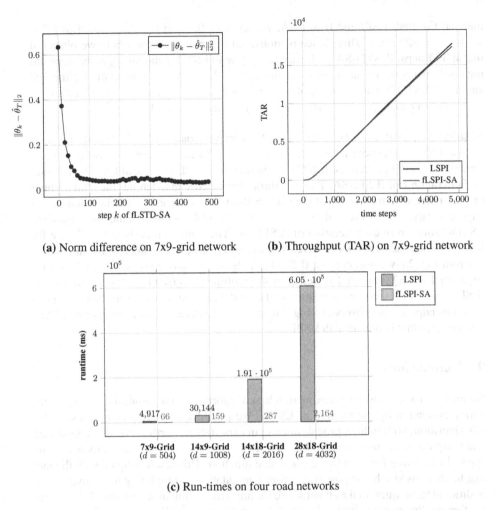

(a) Norm difference on 7x9-grid network   **(b)** Throughput (TAR) on 7x9-grid network

(c) Run-times on four road networks

Fig. 2. Norm difference, throughput and runtime performance of LSPI and fLSPI-SA

following: The features $\phi(s, a)$ corresponding to any state-action tuple (s, a) is a L-dimensional vector, with one bit for each line in the road network. The feature value $\phi_i(s, a), i = 1, \ldots, L$ corresponding to lane i is chosen as described in Table. 1, with q_i and t_i denoting the queue length and elapsed times for lane i. Thus, as the size of the network increases, the feature dimension scales in a linear fashion.

Note that the above feature selection scheme depends on certain thresholds $\mathcal{L}_1$ and $\mathcal{L}_2$ on the queue length and $\mathcal{T}_1$ on the elapsed times. The motivation for using such graded thresholds is owing to the fact that queue lengths are difficult to measure precisely in practice. We set $(\mathcal{L}_1, \mathcal{L}_2, \mathcal{T}_1) = (6, 14, 130)$ in all our experiments and this choice has been used, for instance, in [15].

We implement both LSPI as well as fLSPI-SA for the above problem. We collect $T = 10000$ samples from a exploratory policy that picks the actions in a uniformly random manner. For both LSPI and fLSPI-SA, we set $\beta = 0.9$ and $\epsilon = 0.1$. For fLSPI-SA, we set $\tau = 500$ steps. This choice is motivated by an experiment where we observed that at 500 steps, fLSTD-SA is already very close to LSTDQ and taking more steps did not result in any significant improvements for fLSPI-SA. We implement the regularized variant of LSTDQ, with regularization constant μ set to 1. Motivated by Corollary 1, we set the step-size $\gamma_k = (1 - \beta)c/(2(c + k))$, with $c = 1.33(1 - \beta)^{-2}$.

Results We report the norm differences, total arrived road users (TAR) and run-times obtained from our experimental runs in Figs. 2a–2c. Norm difference measures the distance in ℓ^2 norm between the fLSTD-SA iterate θ_k, $k = 1, \ldots, \tau$ and LSTDQ solution $\hat{\theta}_T$ in iteration 1 of fLSPI-SA. TAR is a throughput metric that denotes the total number of road users who have reached their destination. The choice 1 of the iteration in Fig 2a is arbitrary, as we observed that fLSTD-SA iterate θ_τ is close to the corresponding LSTDQ solution in each iteration of fLSPI-SA. The runtime reports in Fig. 2c are for four different road networks of increasing size and hence, increasing feature dimension.

From Fig. 2a, we observe that fLSTD-SA algorithm converges rapidly to the corresponding LSTDQ solution. Further, from the runtime plots (see Fig. 2c), we notice that fLSPI-SA is several orders of magnitude faster than regular LSPI. From a traffic application standpoint, we observe in Fig. 2b that fLSPI-SA results in a throughput (TAR) performance that is on par with LSPI.

9   Conclusions

We analysed a stochastic approximation based algorithm with randomised samples for policy evaluation by the method of LSTD. We provided convergence rate results for this algorithm, both in high probability and in expectation. Further, we also established that using this scheme in place of LSTD does not impact the rate of convergence of the approximate value function to the true value function. This result coupled with the fact that the SA based scheme possesses lower computational complexity in comparison to traditional techniques, makes it attractive for implementation in *big data* settings, where the feature dimension is large. On a traffic signal control application, we demonstrated the practicality of a low-complexity alternative to LSPI that uses our SA based scheme in place of LSTDQ for policy evaluation.

Acknowledgments. The first and third authors would like to thank the European Community's Seventh Framework Programme (FP7/2007 − 2013) under grant agreement n° 270327 for funding the research leading to these results. The second author was gratefully supported by the EPSRC Autonomous Intelligent Systems project EP/I011587.

References

1. Bach, F., Moulines, E.: Non-asymptotic analysis of stochastic approximation algorithms for machine learning. In: NIPS (2011)
2. Bertsekas, D.P.: Dynamic Programming and Optimal Control, 4th edn. Approximate Dynamic Programming, vol. II (2012)

3. Bertsekas, D.P., Tsitsiklis, J.N.: Neuro-Dynamic Programming. Optimization and Neural Computation Series 3, vol. 7. Athena Scientific (1996)
4. Bradtke, S., Barto, A.: Linear least-squares algorithms for temporal difference learning. Machine Learning 22, 33–57 (1996)
5. Dani, V., Hayes, T.P., Kakade, S.M.: Stochastic linear optimization under bandit feedback. In: COLT, pp. 355–366 (2008)
6. Fathi, M., Frikha, N.: Transport-entropy inequalities and deviation estimates for stochastic approximation schemes. arXiv preprint arXiv:1301.7740 (2013)
7. Frikha, N., Menozzi, S.: Concentration Bounds for Stochastic Approximations. Electron. Commun. Probab. 17(47), 1–15 (2012)
8. Geramifard, A., Bowling, M., Zinkevich, M., Sutton, R.S.: iLSTD: Eligibility traces and convergence analysis. In: NIPS, vol. 19, p. 441 (2007)
9. Hazan, E., Kale, S.: Beyond the regret minimization barrier: an optimal algorithm for stochastic strongly-convex optimization, pp. 421–436 (2011)
10. Kushner, H.J., Yin, G.: Stochastic approximation and recursive algorithms and applications, vol. 35. Springer (2003)
11. Lagoudakis, M.G., Parr, R.: Least-squares policy iteration. The Journal of Machine Learning Research 4, 1107–1149 (2003)
12. Lazaric, A., Ghavamzadeh, M., Munos, R.: Finite-sample analysis of least-squares policy iteration. Journal of Machine Learning Research 13, 3041–3074 (2012)
13. Polyak, B.T., Juditsky, A.B.: Acceleration of stochastic approximation by averaging. SIAM Journal on Control and Optimization 30(4), 838–855 (1992)
14. Prashanth, L., Bhatnagar, S.: Reinforcement Learning with Function Approximation for Traffic Signal Control. IEEE Transactions on Intelligent Transportation Systems 12(2), 412–421 (2011)
15. Prashanth, L., Bhatnagar, S.: Threshold Tuning using Stochastic Optimization for Graded Signal Control. IEEE Transactions on Vehicular Technology 61(9), 3865–3880 (2012)
16. Prashanth, L., Korda, N., Munos, R.: Fast LSTD using stochastic approximation: Finite time analysis and application to traffic control. arXiv preprint arXiv:1306.2557v4 (2014)
17. Robbins, H., Monro, S.: A stochastic approximation method. In: The Annals of Mathematical Statistics, pp. 400–407 (1951)
18. Ruppert, D.: Stochastic approximation. In: Handbook of Sequential Analysis, pp. 503–529 (1991)
19. Silver, D., Sutton, R.S., Müller, M.: Reinforcement Learning of Local Shape in the Game of Go. In: IJCAI, vol. 7, pp. 1053–1058 (2007)
20. Sutton, R.S., Barto, A.G.: Reinforcement learning: An introduction, vol. 1. Cambridge Univ. Press (1998)
21. Sutton, R.S., Szepesvári, C., Maei, H.R.: A convergent O(n) algorithm for off-policy temporal-difference learning with linear function approximation, pp. 1609–1616 (2009)
22. Sutton, R.S., et al.: Fast gradient-descent methods for temporal-difference learning with linear function approximation. In: ICML, pp. 993–1000. ACM (2009)
23. Tsitsiklis, J.N., Van Roy, B.: An analysis of temporal-difference learning with function approximation. IEEE Transactions on Automatic Control 42(5), 674–690 (1997)
24. Webscope, Y.: Yahoo! Webscope dataset ydata-frontpage-todaymodule-clicks-v2_0 (2011), "http://research.yahoo.com/Academic_Relations"
25. Zinkevich, M.: Online convex programming and generalized infinitesimal gradient ascent. In: ICML, pp. 928–925 (2003)

Mining Top-K Largest Tiles in a Data Stream

Hoang Thanh Lam[1,3], Wenjie Pei[1], Adriana Prado[2], Baptiste Jeudy[2],
and Élisa Fromont[2]

[1] Technische Universiteit Eindhoven, 5612 AZ, Eindhoven, The Netherlands
[2] H. Curien Lab, UMR CNRS 5516, Université de St-Etienne, Université de Lyon,
France
[3] IBM Research Lab, Damastown, Dublin, Ireland

Abstract. Large tiles in a database are itemsets with the largest *area* which is defined as the itemset frequency in the database multiplied by its size. Mining these large tiles is an important pattern mining problem since tiles with a large area describe a large part of the database. In this paper, we introduce the problem of mining top-k largest tiles in a data stream under the sliding window model. We propose a candidate-based approach which summarizes the data stream and produces the top-k largest tiles efficiently for moderate window size. We also propose an approximation algorithm with theoretical bounds on the error rate to cope with large size windows. In the experiments with two real-life datasets, the approximation algorithm is up to hundred times faster than the candidate-based solution and the baseline algorithms based on the state-of-the-art solutions. We also investigate an application of large tile mining in computer vision and in emerging search topics monitoring.

1 Introduction

Mining frequent patterns is an important research topic in data mining. However, instead of focusing on exhaustive search to find all possible frequent patterns, many works are now focusing on designing methods that are not only efficient in the context of very big data but also limit, e.g. with constraints or with new interestingness measures, the number of patterns output by these algorithms.

Area is a measure of pattern interestingness defined as a pattern's frequency in the database multiplied by its size. It has been shown that in some applications such as in role mining [10,16] where the idea is, given a set of users and a set of permissions, to find a minimum set of roles such that all users will be assigned a role for which some permissions will be granted, mining roles is equivalent to a variant of mining itemsets with large area in a database.

Recent applications produce a large variety of transactional data streams, such as text stream from *twitter*[1] or video stream in which video frames can be converted into transactions [6]. In the context of data streams, data usually arrive continuously with high speed, hence requiring efficient mining techniques for summarizing the data stream and keeping track of important patterns.

---

[1] www.twitter.com

T. Calders et al. (Eds.): ECML PKDD 2014, Part II, LNCS 8725, pp. 82–97, 2014.
© Springer-Verlag Berlin Heidelberg 2014

In this paper we tackle the problem of mining the top-k largest tiles, i.e. the k closed itemsets with the largest area in a stream of itemsets. The problem of mining the largest tile in a database is well-known to be NP-hard and in-approximable [7]. Therefore, the straightforward approach that recalculates the set of top-k largest tiles from scratch every time a sliding window is updated is not efficient. To deal with this situation, we first introduce in Section 4 a candidate-based approach which incrementally summarizes the stream by keeping the itemsets that can be the top-k largest tiles in some future windows. For each candidate, an *upper-bound* and a *lower-bound* of the area in any future window are kept. These bounds are used to prune the candidates when they cannot be the top-k largest tiles in any future window. In doing so, the candidate-based algorithm is more efficient than the straightforward approach because updating is cheaper than recalculating the top-k tiles from scratch.

However, when the widow size is large, the candidate-based algorithm is inefficient because the summary grows very quickly. Therefore, we introduce an approximation algorithm with theoretical bounds on the error rate. In the experiments with two real-life datasets presented in Section 6, the approximation algorithm is two to three orders of magnitude faster and more accurate than the candidate-based solution and the baseline algorithms created based on the state-of-the-art solutions for the top-k largest tiles mining. We also discuss potential applications of mining large tiles for object tracking in video and emerging search topics monitoring problems.

2   Related Works

Recent works [12,9,17] propose approaches to solve the redundancy and trivial patterns issues in frequent pattern mining. For instance, the first two works focus on finding a relevant or concise representation of sets of frequent patterns. Meanwhile, the last work solves the aforementioned issues by proposing a MDL-based approach that finds patterns compressing the database well. In all these cases as well as in ours, the purpose is to limit the output of the algorithms to a set of useful patterns.

The problem of mining large tiles has already been tackled in [10,16,13]. The authors of the two first papers show that the problem of finding roles is equivalent to variants of the tiling database problem. The last paper shows how tiles can be used to understand complex proteins by identifying their subunits. The authors of [4] also showed that the dense rectangles in a binary matrix seem to correspond to interesting concepts. Therefore, mining large tiles may help to identify those interesting concepts from the databases. In [15] the authors investigate how to output the set of tiles in a tree representation which is easily interpretable by the users. Tiles are also used to identify and characterize anomalies in a database in [14].

In many applications, data arrives in a streaming fashion with high speed [1]. Besides the popularity of data streams in many applications, the temporal aspect of the patterns such as the evolution of patterns overtime [1] provides

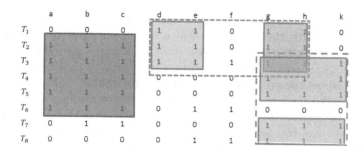

Fig. 1. An example of large tiles in a window with $w = 8$ transactions

useful insights about the data for the users. Despite the importance of data stream paradigm, there is no work yet addressing the problem of mining large tiles in a data stream. The algorithms introduced in this paper are, to the best of our knowledge, the first to solve the problem of mining the top-k largest tiles in a stream of itemsets under the sliding windows model.

3   Problem Definition

Let Σ be an alphabet of items, a transaction T is an itemset $T \subseteq \Sigma$. Let $S = T_1 T_2 \cdots T_n$ be a stream of transactions where each transaction T_t is associated with a timestamp t indicating the order of transaction generation.

In this work we consider the sliding window model in which a window of the w most recent transactions is monitored. A sliding window of size w at time point t denoted by W_t is a sequence of w consecutive transactions in the data stream, i.e. $W_t = T_{t-w+1} \cdots T_{t-1}T_t$.

For any given itemset $I \subseteq \Sigma$ the frequency of I in a sliding window W_t, denoted by $f_t(I)$, is defined as the number of transactions in W_t that are the supersets of I. Let $|I|$ be the cardinality of the set I, the area of I in the window W_t denoted by $A_t(I)$ is defined as $A_t(I) = f_t(I) * |I|$.

An itemset I is closed in a window W_t if there is no superset of I with the same frequency as I in the window W_t. Such itemsets are usually called large tiles in the literature [7]. Itemsets that are not closed correspond to sub-tiles of large tiles and are thus not interesting for our problem. From now on we use the term tiles to refer to closed itemsets.

Example 1 (Large tiles). Figure 1 shows a window with 8 transactions represented as rows of a binary matrix. In each row, an element is equal to 1 if the corresponding item belongs to the transaction. In this figure, three large tiles abc, $degh$ and ghk with respective area 15, 12 and 15 are highlighted. Large tiles are the maximal sub-matrices containing only 1.

The problem of stream tiling can be formulated as :

Definition 1 (Stream Tiling). *Given a data stream of itemset transactions S, a parameter k and a window size w, the stream tiling problem consists in computing the k tiles with the largest area in every window of size w.*

4   Algorithms

It was proven that the problem of mining the largest tile in a database is NP-Complete and is even inapproximable [7]. Therefore, recalculation of the top-k largest tiles from scratch every time the window is updated is very time-demanding. In this section we discuss efficient solutions for the given problem under the streaming context.

4.1   Candidate-Based Algorithm

We first discuss an exact algorithm named $cTile$ which maintains a summary of the sliding window containing candidate itemsets that potentially can become top-k largest tiles in any future window. This summary is designed such that it can avoid expensive recalculation of the set of largest tiles from scratch when the window is sliding.

The general idea of the algorithm is as follows: at every transaction T_i we keep a candidate list C_i of all closed subsets of the given transaction which can become a top-k tile in a future sliding window. In order to identify these closed itemsets, we keep a *lower-bound* and an *upper-bound* on the area of these itemsets in any future sliding window which contains T_i. These bounds will be used to infer which itemsets can be top-k tiles and which sets for sure cannot be top-k tiles in any future window (thus can be removed from the summary).

$$W_w$$
$$T_1 \, T_2 \, \cdots \cdots \cdots T_i \cdots \cdots \cdots T_{w-1} \, T_w \qquad i-1 \text{ transactions}$$
$$W_{w+i-1}$$
$$J^- = \min A_{w+i-1}(J) \qquad J^+ = \max A_{w+i-1}(J)$$

Given a window $W_w = T_1 \cdots T_{w-1}T_w$, for every closed itemset J in the candidate list $C_i (1 \le i \le w)$, the *lower-bound* and the *upper-bound* of the area of this itemset in any future sliding window containing T_i are denoted J^- and J^+. The lower-bound J^- at time point w is calculated as the area of J in the transactions $T_i, T_{i+1}, \cdots, T_w$ and the *upper-bound* J^+ is calculated as

$$J^+ = J^- + (i-1) \times |J| \tag{1}$$

Proposition 1. J^- and J^+ are the correct lower-bound and upper-bound on the area of J in any future window W_t of size w, $w \le t$, containing the transaction T_i.

Example 2 (Bounds). Figure 2 (upper-part) shows an example of a sliding window with size $w = 5$. Each candidate J in a candidate list C_i is associated with two numbers corresponding to the lower-bound and the upper-bound on the area of the candidate. For instance, for the candidate $J = ac$ associated with the

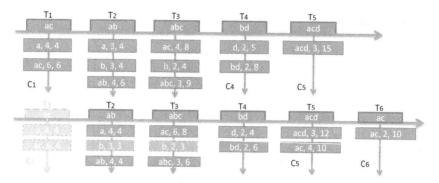

Fig. 2. An example of the summary maintained by the *cTile* algorithm for sliding windows with size $w = 5$. The summary contains candidate itemsets which can become a top-k tile in the future. Every candidate is associated with a *lower-bound* and an *upper-bound* on the area which are used to prune the candidate list.

candidate list C_3 the lower-bound is $J^- = 4$ because the area of ac in the set of currently observed transactions $T_3, \cdots, T_5$ is $2 * 2 = 4$. For any future window in which the transaction T_3 is not expired, the area of ac is at least as large as 4. Meanwhile $J^+ = 8$ because the area of ac in any future window that contains T_3 is at most $4 + 2 * 2 = 8$. The area is equal to J^+ only when the new transactions T_6 and T_7 both contain ac.

Algorithm 1 incrementally maintains the candidate lists C_i in the summary. When a new transaction is appended to the end of the window, the bounds J^+ and J^- are updated accordingly (line 10-13). Then the new transaction is intersected with every element in the summary and the intersections are added to the corresponding candidate list (line 8-9). If there exists another candidate $B \in C_j$ kept at a younger transaction T_j ($j > i$), such that B^- is ranked k^{th} in all transactions younger than T_i (including itself) and meanwhile $B^- > J^+$, then J will never be among top-k tiles in any future sliding window, hence J is removed from the summary. When a transaction is expired, it is removed from the window along with the candidate list stored at the given transaction.

Example 3 (cTile). Figure 2 (bottom part) shows one update step of Algorithm 1 when a new transaction $T_6 = \{a, c\}$ is appended to the window and T_1 together with C_1 are removed from the summary. First, T_6 is intersected with all the existing candidates in $C_2, C_3, \cdots, C_5$ to add new closed sets to the candidate lists. After that all the lower-bounds and the upper-bounds are updated accordingly.

Assume that we want to get top-3 largest tiles in the sliding window, i.e. $k = 3$. Since $(b, 3, 3)$ in C_2 has the *upper-bound* equal to 3. It is removed from the summary because the itemset ranked at the third position in the candidates lists of younger transactions (including T_2) are $(a, 4, 4)$, $(ac, 4, 10)$, $(ab, 4, 4)$ whose *lower-bound* is 4. The same pruning operation can be applied for $(b, 2, 3)$ in C_3.

Theorem 1. *Given k, w and a stream of transactions, using the summary in Algorithm 1 we can answer the top-k largest tiles exactly.*

Algorithm 1. cTile(S_t)

1: **Input**: A stream of transactions $T_1, T_2, \cdots, T_\infty$, a sliding window size w and a parameter k
2: **Output**: Summary C for calculating top-k largest tiles
3: $C \longleftarrow \{C_1, C_2, \cdots, C_w\}$ //candidate lists
4: **for** $t = w \to \infty$ **do**
5:    $C \leftarrow C \setminus \{C_{t-w+1}\}$
6:    $C_{t+1} \leftarrow \{T_{t+1}\}$
7:    $C \leftarrow C \cup \{C_{t+1}\}$
8:    **for** $i = t \to t - w + 2$ **do**
9:       $C_i \longleftarrow C_i \bigcup \text{intersect}(C_i, T_{t+1})$
10:      **for** $J \in C_i$ **do**
11:         $J^- = |\{i \leq l \leq t | J \subseteq T_l\}| * |J|$
12:         $J^+ = J^- + (w - t + i - 1)|J|$
13:      **end for**
14:      Pruning(C_i)
15:   **end for**
16: **end for**
17: Return C

Proof. Assume that there exists a top-k largest tile I in a window W_t not recognized as a top-k largest tile (false negative). The first transaction of W_t containing I is denoted by T_i. False negative can only occur if at time t, $I \notin C_i$. Therefore either I is directly pruned from C_i by pruning criteria or indirectly pruned because its closed supersets are pruned.

Since the bounds are exact the former case cannot happen. The latter case happens when I is not a closed itemset at the moment its closed supersets are pruned. In such case, the upper-bound of I is always less than the upper-bound of its closed supersets which were used to prune the supersets. Therefore, I is not a top-k tile in W_t. Both cases lead to contradiction.

4.2   Approximation Algorithm

The size of the summary maintained by *cTile* grows quickly when the window size increases making it inefficient for monitoring large windows. The main reason is that each time a new transaction arrives, it has to be intersected with a large amount of candidates, which is a time-consuming operation. Therefore, in this section we discuss an approximation algorithm named *aTile* that approximates the set of largest tiles efficiently.

The main process of the approximation algorithm is almost the same as Algorithm 1. The only difference of these two algorithms lies in the method of candidate pruning. Instead of using an *upper-bound* on the area, *aTile* tries to approximate the future area of a tile. This estimate is used to prune the candidates in the same way as *cTile* does.

Let us denote the probability of observing an itemset J as a subset of a transaction in the data stream by μ_J. Therefore, the expectation of the area of the itemset J in a window of size w is $\mu_J.|J|.w$.

Given a window $W_w = T_1 \cdots T_{w-1} T_w$, assume that J is kept in the candidate list C_i of transaction T_i. The *lower-bound* J^- is calculated as in the $cTile$ algorithm. Instead of using an upper bound J^+ on the area of J, we compute an *estimate* of the area of J denoted by J^*.

Since μ_J is unknown, we cannot directly define J^* as $\mu_J.|J|.w$. The probability μ_J is thus estimated in the sub-window $T_i \cdots T_w$. However, if this sub-window is too small, the estimation of μ_J is not accurate. Therefore, we introduced a threshold $L \leq w$: if the sub-window is smaller than L, we do not use this estimation and fall back on using the upper bound J^+ for J^*:

$$J^* = \begin{cases} J^- + (i-1) \times |J| & \text{if } w - i + 1 \leq L \ (\text{i.e., } J^* = J^+) \\ \frac{J^-}{(w-i+1)} \times w & \text{if } w - i + 1 > L \end{cases} \tag{2}$$

All the steps of the of the $aTile$ algorithm are very similar to the $cTile$ algorithm except that it uses J^* instead of J^+ for candidate pruning. In the experiment section we empirically show that the $aTile$ algorithm is more efficient than the $cTile$ algorithm because its candidate set is more concise. An important property of the $aTile$ algorithm is that the error rate on the accuracy of the result is bounded if we assume that the transactions are i.i.d. in a window of size $2w$.

5   Theoretical Analysis

In this section, we show theoretical bounds on the probability of errors induced by the $aTile$ algorithm. We mainly show that the error rate is extremely low when L is large enough. We consider two types of error event: *False negative (FN)*: in a window W_{t^*} for some t^*, there is a true top-k largest tile that is not present in the results returned by the $aTile$ algorithm and *False positive (FP)*: in a window W_{t^*} for some t^*, there is a non top-k largest tile that is present in the result returned by the $aTile$ algorithm.

5.1   False Negative Bound

Let us assume that I is a true top-k largest tile in the window W_{t^*} and $I_k^{t^*}$ is ranked at position k in the list of a true top-k largest tile of the window W_{t^*}. Let $\triangle \geq 0$ be the difference between the average area of I and $I_k^{t^*}$ in the window W_{t^*}, i.e. $\triangle = \frac{A_{t^*}(I) - A_{t^*}(I_k^{t^*})}{w}$. The following lemma show the relationship between the probability of a false negative, L and $\triangle$:

Lemma 1 *If the transactions are independent and identically distributed (i.i.d) in a window of size $2w$, the probability that a random top-*k* tile is not reported, is bounded as follows:*

$$Pr(FN) < 4w * e^{-\frac{L\triangle^2}{2|I|^2}}$$

Proof. A false negative happens in the window W_{t^*} with respect to an itemset I if its area is underestimated in that window. This only happens if there exists at least one moment $t < t^*$ such that I is pruned from the candidate list C_i $(t - w < i \leq t)$.

Given a time t, let W_0 be a window containing transactions $T_i, T_{i+1}, \cdots, T_t$ where $|W_0| = w_o > L$ and let $f_0(I)$ be the frequency of I in the window W_0. When t is given, the event "I is removed from C_i" (denoted by R_t) only happens when the estimate of the *upper-bound* on the area of I is less than the area of $I_k^{t^*}$, i.e. $f_{W_0}(I)|I|\frac{w}{w_0} < A_{t^*}(I_k^{t^*})$. Therefore, we have:

$$Pr(R_t) < Pr\left(f_0(I)|I|\frac{w}{w_0} \leq A_{t^*}(I_k^{t^*})\right) < Pr\left(f_0(I)|I|\frac{w}{w_0} < A_{t^*}(I) - w\triangle\right)$$

$$< Pr\left(\frac{f_0(I)}{w_0} < \frac{f_{t^*}(I)}{w} - \frac{\triangle}{|I|}\right) < Pr\left(\left|\frac{f_0(I)}{w_0} - \frac{f_{t^*}(I)}{w}\right| > \frac{\triangle}{|I|}\right)$$

$$< Pr\left(\left|\frac{f_0(I)}{w_0} - \mu_I\right| + \left|\frac{f_{t^*}(I)}{w} - \mu_I\right| > \frac{\triangle}{|I|}\right)$$

$$< Pr\left(\left|\frac{f_0(I)}{w_0} - \mu_I\right| > \frac{\triangle_0}{w_0}\right) + Pr\left(\left|\frac{f_{t^*}(I)}{w} - \mu_I\right| > \frac{\triangle_1}{w}\right)$$

$$< Pr\left(|f_0(I) - \mu_I w_0| > \triangle_0\right) + Pr\left(|f_{t^*}(I) - \mu_I w| > \triangle_1\right)$$

Where $\triangle_0 = \frac{\triangle w_0}{2|I|}$ and $\triangle_1 = \frac{w\triangle}{2|I|}$. It is important to notice that $w_0\mu_I$ and $w\mu_I$ are the expectation of the frequency of I in the window W_0 and the window W_{t^*} respectively. Since the transactions are independent to each other, according to the Hoeffding inequality [11] we have:

$$Pr\left(|f_0(I) - w_0\mu_I| > \triangle_0\right) < 2e^{-\frac{2\triangle_0^2}{w_0}} < 2e^{-w_0\frac{\triangle^2}{2|I|^2}} < 2e^{-L\frac{\triangle^2}{2|I|^2}} \qquad (3)$$

A similar inequality can be obtained for bounding the second term as follows:

$$Pr\left(|f_{t^*}(I) - w\mu_I| > \triangle_1\right) < 2e^{-\frac{2\triangle_1^2}{w}} < 2e^{-w\frac{\triangle^2}{2|I|^2}} < 2e^{-L\frac{\triangle^2}{2|I|^2}} \qquad (4)$$

Moreover, since FN happens only when there at least one moment t such that R_t happens, therefore:

$$Pr(FN) < Pr(\cup_{t^*-w<t\leq t^*} R_t) < \sum_{t^*-w<t\leq t^*} Pr(R_t) \qquad (5)$$

Inequalities 3, 4 and 5 prove the lemma.

A direct corollary of Lemma 1 is shown in the following theorem:

Theorem 2. *If I is strictly more important than $I_k^{t^*}$, i.e. the expectation of the area of I in a transaction is strictly greater than the expectation of the area of $I_k^{t^*}$ in a transaction and $L = O(w)$ then:* $\lim_{L\longrightarrow\infty} Pr(FN) = 0$

Proof. Let $E(A(I))$ be the expectation of the area of I in a transaction and $E(A(I_k^{t^*}))$ be the expectation of the area of $I_k^{t^*}$ in a transaction. Since $\triangle = \frac{A_{t^*}(I) - A_{t^*}(I_k^{t^*})}{w}$ we can imply that:

$$\frac{\triangle}{|I|} = \frac{A_{t^*}(I) - A_{t^*}(I_k^{t^*})}{w|I|} \simeq \frac{E(A(I)) - E(A(I_k^{t^*}))}{|I|} \tag{6}$$

The last equation is the result of the law of large number when L goes to ∞. From the last equation we can imply that $\lim\limits_{L \to \infty} 4we^{-L\frac{\triangle^2}{2|I|^2}} = 0$ from which the theorem is proved.

5.2   False Positive Bound

In this subsection we prove bound for false positive error which can be obtained in a similar way as the bound of false negative.

Let J be an itemset that is not a true top-k largest tile in the window W_{t^*} but returned by the $aTile$ algorithm as a false positive tile. An itemset ranked at position k in the list of a true top-k largest tile of the window W_{t^*} is denoted by $I_k^{t^*}$. Let $\triangle \geq 0$ be the difference between the area of J and $I_k^{t^*}$ in the window W_{t^*}, i.e. $\triangle = \frac{A_{t^*}(I_k^{t^*}) - A_{t^*}(J)}{w}$.

The following lemma show the relationship between the probability of a false positive and L, $\triangle$ (proof is similar to Lemma 1):

Lemma 2 *If the transactions are i.i.d. in a window of size $2w$ the probability of false positive, i.e. the event that a random non top-k tile is reported, is bounded as follows:*

$$Pr(FP) < 4w * e^{-\frac{L\triangle^2}{2|I|^2}}$$

A corollary of Lemma 2 is shown as follows (proof is similar to Theorem 2):

Theorem 3. *If tile size is bounded and if J is strictly less important than $I_k^{t^*}$, i.e. the expectation of the area of J is strictly less than the expectation of the area of $I_k^{t^*}$ in a transaction and $L = O(w)$ then:* $\lim\limits_{L \to \infty} Pr(FP) = 0$

Theorem 3 and Theorem 2 show an interesting result that the probability of false positive and false negative decrease exponentially with $L = O(w)$. It is important to notice that the condition $L = O(w)$ can be replaced by a weaker assumption $L = k \log w$ for some constant values k. If k is large enough the bound is also closed to zero. Although the bounds are not tight, in experiments, we empirically show that false negative rate and false positive rate are negligible even when L is set to a small value.

5.3   Long Lasting Tiles

The two previous Theorems give probabilistic results. We can also show a deterministic one: if an itemset stays in the top-k largest tiles for more than w consecutive windows and its area is at least twice the area of the k-th tile, then the $aTile$ algorithm finds it. An important point of this theorem is that it does not depends on the value of L. Therefore, even if the probability of false negative or false positive is higher with a small L, the algorithm is still able to mine these tiles that we call the long lasting tiles.

Theorem 4. *Let z be a time and J be an itemset. If the area of J in every windows W_t with $z \leq t < z+w$ is larger than two time the area of the k-th largest tile in this window (i.e., $A_t(J) \geq 2A_t(I_k^t)$), then there is a time $z-w < t^* < z+w$ such that J (or one of its supersets) is in C_{t^*} and is never pruned by aTile.*

Proof. We define hypothesis H as: t^* *does not exist*. We will show that if H is true, it leads to a contradiction and thus the theorem must be true. We denote by $[x, y]$ the subsequence of transactions $T_x \cdots T_y$. The occurrence times of J in the subsequence $[z-w+1, z+w-1]$ are denoted by $z-w < t_1 < \ldots < t_k < z+w$.

If H is true, then for every t_i, itemset J and its supersets must be pruned from C_{t_i} at some time t'_i no later than $t_i + w - 1$, i.e., $t'_i < t_i + w$.

Let S_i be the subsequence $[t_i, t'_i] = T_{t_i} \cdots T_{t'_i}$. If for all $z - w < t_i \leq z$, we have $t'_i \leq z$ then we define $W_0 = W_z$. Otherwise, we take $W_0 = W_{t'_{max}}$ where $t'_{max} = \max\{t'_i \mid t_i \leq z\}$.

We now construct a set S of these subsequences S_i such that every subsequence of S is included in W_0, every occurrence of J is in at least one of these subsequences, and at most two of these subsequences intersect at any given time. We start with $S = \emptyset$ if $W_0 = W_z$ or with $S = \{[t_{max}, t'_{max}]\}$ otherwise. We scan the window W_0 from left to right. If there is an occurrence of J at time t_i not already in a subsequence of S, then we add S_i in S. All the added subsequences are disjoint by construction, only the last added one may intersect with (t_{max}, t'_{max}).

For every $S_i \in S$, let A_i be the area of the k-th largest tile of S_i used to prune J, i.e., $J^* < A_i$. Since by Eq. 2 $J^* \geq wA_{S_i}(J)/(t'_i - t_i + 1)$, we have $wA_{S_i}(J)/(t'_i - t_i + 1) < A_i$. If we define $A_m = \max A_i$, then $A_{S_i}(J) < A_m(t'_i - t_i + 1)/w$ and by summation for all $S_i \in S$: $\sum_i A_{S_i}(J) < A_m \sum_i(t'_i - t_i + 1)/w$. Since all occurrences of J are covered by at least one S_i, $\sum_i A_{S_i}(J) > A_{W_0}(J)$ and since at most two subsequences S_i from S intersect at any given time, $\sum_i(t'_i - t_i + 1) \leq 2w$ and thus $A_{W_0}(J) < 2A_m$. Since A_m is the size of the k-th tile of one of the $S_i \subseteq W_0$, it is less than the size of the k-th tile in the whole window W_0. Finally, $A_{W_0}(J)$ is strictly less than two time the area of the k-th tile in W_0 which is a contradiction.

6   Experiments

In this section, we perform experiments with two real-life data streams to compare the proposed algorithms to the baseline approaches with respect to the efficiency and the accuracy of the results. The two datasets are:

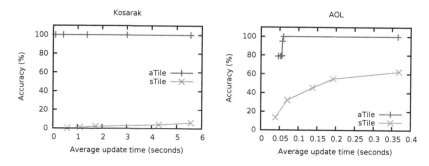

Fig. 3. Accuracy versus average update time in seconds of the *aTile* algorithm and the *sTile* algorithm when L and the number samples increases

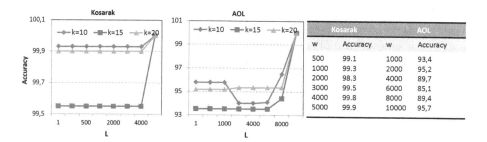

Fig. 4. The average accuracy of the *aTile* algorithm when L is varied. In the table: the average accuracy of the top-*10* largest tiles returned by *aTile* when the window size is varied and $L = 0.1w$

- Kosarak: 1M transactions of click log by the users of a website. Item are a pages in the website. The largest transaction contains 500 items.
- AOL: 100K search queries by the users of the AOL search engine. Each item is a keyword in the search query. Most queries are short and the longest query contains only 26 keywords.

The datasets and the source codes of *aTile* and *cTile* in C++ are available for download[2]. The baseline algorithms for comparison are as follows:

- *Tile*: the original implementation of the tiling algorithm for static database [7]. In order to adopt this algorithm for a data stream, we use *Tile* to recalculate the top-k largest tiles from scratch whenever the window is sliding.
- *sTile*: a sampling based technique proposed in [2]. The *sTile* algorithm samples N itemsets from the window such that each itemset is sampled with probability proportional to the area of the itemset. The top-k largest tiles are extracted from the samples every time the window is sliding.

---

[2] http://www.win.tue.nl/~lamthuy/tile.htm

	sTile		aTile
# samples	Sum of area	L	Sum of area
5000	2652	1000	20384
10000	2753	2000	20384
20000	2822	3000	20384
40000	2918	4000	20384
80000	2992	5000	20388

Fig. 5. The average sum (larger is better) of the area of the top-10 tiles returned by the *aTile* and the *sTile* algorithms

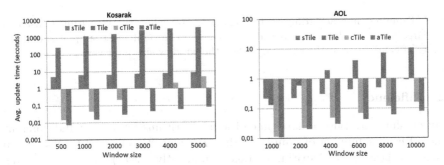

Fig. 6. Average time per update according to the window size

6.1   Accuracy

We performed experiments to demonstrate the effectiveness of the *aTile* algorithm in term of accuracy. In the first experiment the window size is set to $w = 5000$ for the Kosarak dataset and $w = 10000$ for the AOL dataset. Figure 4 shows the average accuracy calculated as the precision of the top-k largest tiles returned by the *aTile* algorithm when L is varied. Even for very small L, the accuracy is very high, e.g. 99% in the Kosarak dataset and 93% in the AOL dataset. The accuracy increases and reaches 100% accuracy when L is increased. In the same figure, we can also see that the results are similar when k is varied.

In order to compare to the baseline algorithm *sTile*, we plot the accuracy (y-axis) versus average update time (x-axis) in Figure 3. We varied both L for the *aTile* algorithm and the number of samples of the *sTile* algorithm. Recall that the *sTile* algorithm depends on the number of samples it collects from the window. When the number of samples increases, so do the accuracy and the update time. This fact is illustrated in Figure 3 in which the *sTile* algorithm is significantly less accurate than the *aTile* algorithm given about the same average update time.

The accuracy of *sTile* may be negatively influenced by the fact that *sTile* has a very low probability to find each top-k tile. When we calculated the average sum of the area of the top-*10* tiles in the Kosarak dataset for $w = 5000$, we observed that it was significantly lower for *sTile* than for the *aTile* algorithm (Figure 5). This confirmed that *sTile* is not able to find the large tiles with a reasonable number of samples.

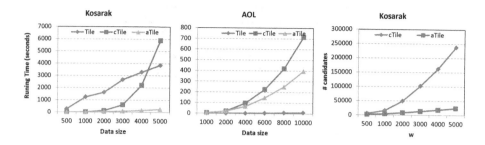

Fig. 7. Running time according to the database size when the algorithm is used to mine large tiles in a database instead of a stream. The right most subplot shows the number of candidates in the summary when the window size increases.

6.2 Efficiency

Figure 6 shows average update time of four algorithms for $k = 10$ and different values of w. The number of samples in the $sTile$ algorithm was set to $N = 80000$ (when N is larger, the algorithm become significantly slower). We set $L = \frac{w}{10}$ (as shown in Sect. 6.1, the accuracy is then always above 85%). In term of update time, $aTile$ is up to an order of magnitude (50x) faster than the $cTile$ algorithm and about two to three orders of magnitude faster than the $sTile$ (100x) and the $Tile$ (1000x) algorithms. The speed-up increases with the window size.

In Figure 7, the last plot shows the number of candidates kept in the summary of the $aTile$ and the $cTile$ algorithm for varying values of w (the result of the AOL dataset is omitted because it is very similar to the results of the Kosarak dataset in this experiment). The $aTile$ algorithm is more memory efficient than the $cTile$ algorithm as the number of candidates it keeps in the summary is much lower.

Finally, in Figure 7 we show the running time when the $aTile$, $cTile$ and $Tile$ algorithms are used to find the top-10 largest tiles in a static corpus with varying size. The purpose of the experiment is to see whether the $aTile$ and the $cTile$ algorithm can find the large tiles in a static corpus more efficiently than the $Tile$ algorithm. For the AOL dataset, when most of transactions are small, the $Tile$ algorithm is the fastest one. However, for the Kosarak dataset, when the average transaction size is larger, $aTile$ outperforms both $Tile$ and $cTile$. Therefore, the $aTile$ algorithm can not only be used for mining large tiles from a data stream but also to mine large tiles from a database efficiently.

6.3 Application to Topics Monitoring in the AOL Query Stream

In order to show a potential application of the work we created a demo video[3] to visualize the top-k largest tiles of the AOL query stream. Each snapshot of the video corresponds to a list of the largest tiles extracted from a sliding window.

---

[3] http://www.youtube.com/watch?v=3UCjs9d91_g

Fig. 8. Top-10 largest tiles in windows (ordered by timestamps) from the AOL query stream. Larger words correspond to tiles with bigger area. The tile "high school" emerges in the second window related to searches for the "high school musical movies".

Figure 8 shows example snapshots taken from the demo video. Three snapshots of the demo (ordered by timestamps) show the evolution of the largest tiles overtime. Each snapshot is visualized by the *wordcloud* tool in *R*. Larger words correspond to tiles with larger area.

For example, in the first snapshot the keyword *"real estate"* is the most important tile. However, it becomes less important in the second and the third snapshot. Meanwhile, new important tiles such as *"high school"* is emerging as search for the "high school musical movies" increases in 2006. With this demo, users can track the dynamic of important search topics online.

6.4   Application to Tracking in Videos

We investigate how mining top-k largest tiles in a data stream can be useful for analysis of videos and, in particular, for tracking. A more detailed description of this application can be found in [3]. We worked on a real video made of 5619 frames. This video is shot from a car while following another car (the main object). The main object is present in almost all the frames of the video. If we are able to represent the video as a data stream and use our tile mining algorithm on it, it should discover the car as a large tile or a set of large tiles.

We used the segmentation algorithm[4] [8] to generate a stream of graphs (one graph per frame). Each graph is a Region Adjacency Graph (RAG) where each node is a region (a set of adjacent pixels) and two nodes are connected if their regions are adjacent. This algorithm performs a temporal segmentation, which means that a given region (and thus, the corresponding node) is present in several successive frames. However, a single region is not enough to track an object in the video: due to change in the object pose or illumination, some regions will split, or merge or disappear and so do the corresponding nodes. We then build a transaction consisting in the set of nodes of each RAG.

Mining tiles on this data stream results in tiles containing regions spread all over the frames. Moreover, the top-k tiles are generally very similar (only differing by one or two regions). Indeed, by removing the information about the

---

[4] http://www.videosegmentation.com

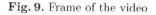

Fig. 9. Frame of the video

Fig. 10. top-30 star-shaped tiles in the segmented frame

adjacency of the regions, it was not possible to find meaningful tiles. The main problem is that the regions composing the car (or more generally, an object to be tracked) must be "close" to each other and this was not taken into account in the original setting. Therefore, we added a spatial constraint to the problem. We chose to mine tiles that are star subgraphs of the RAGs. A star subgraph is a graph with nodes $i_0, i_1, \ldots i_n$ where $i_1, \ldots, i_n$ are nodes adjacent to i_0 in the RAG. This formalization ensures that the items in a tile are adjacent to a "center" item i_0. Moreover, it is easy to integrate this constraint into our algorithms: the intersection of two star graphs with the same center is just the intersection of their set of nodes. The algorithm is otherwise unchanged.

We extracted the top-30 tiles with a window of 300 frames (see an example on Fig 10). Then we checked if the extracted tiles could be useful to track the main object (car). We manually drawn the bounding box containing the car in each fifth frame. Then, we selected, for each frame, the tile with the best precision. We end up with a set of 69 tiles with an overall precision of 0.92 and recall of 0.74 on the whole video (and even a recall of 1 on the 2000 first frames). This means that the tiles can actually be used as a high level feature for tracking. Of course, in a real application, the bounding box is not known. But techniques like those described in [5] can be used to build tracks from the extracted tiles.

7 Conclusions and Future Works

In this paper, we proposed two algorithms for mining the top-k largest tiles from a data stream with a sliding window model. The first candidate-based algorithm *cTile* solves the problem exactly but the update time becomes important when the window size increases. The second one is an approximation algorithm with theoretical bounds on the error rate. Experiments with two real-life datasets show that the approximation algorithm can find large tiles with high accuracy while being an order of magnitude faster than the candidate based algorithm and two to three order of magnitude faster than the other baseline algorithms.

We also show potential applications of large tiles mining in monitoring emerging popular topics in a search engine query log or in a tracking problem. A possible extension for future work is to consider mining different types of tiles with constraints for meaningful applications.

Acknowledgements. This work is supported by the Netherlands Organization for Scientific Research (NWO) through the project *Mining Complex Patterns in Stream* (COMPASS) and the French ANR project SoLStiCe (ANR-13-BS02-0002-01). We would like to thank Dr. Bart Goethals for providing us the implementation of the *Tile* algorithm and Pr. Toon Calders for his precious comments.

References

1. Aggarwal, C.C. (ed.): Data Streams - Models and Algorithms. Advances in Database Systems, vol. 31. Springer (2007)
2. Boley, M., Lucchese, C., Paurat, D., Gärtner, T.: Direct local pattern sampling by efficient two-step random procedures. In: KDD, pp. 582–590 (2011)
3. Calders, T., Fromont, É., Jeudy, B., Lam, H.T.: Analysis of videos using tile mining. In: Real-World Challenges for Data Stream Mining Workshop (2013)
4. Cerf, L., Besson, J., Nguyen, K.N., Boulicaut, J.F.: Closed and noise-tolerant patterns in n-ary relations. Data Min. Knowl. Discov. 26(3), 574–619 (2013)
5. Diot, F., Fromont, E., Jeudy, B., Marilly, E., Martinot, O.: Graph mining for object tracking in videos. In: Flach, P.A., De Bie, T., Cristianini, N. (eds.) ECML PKDD 2012, Part I. LNCS, vol. 7523, pp. 394–409. Springer, Heidelberg (2012)
6. Fernando, B., Fromont, E., Tuytelaars, T.: Effective use of frequent itemset mining for image classification. In: Fitzgibbon, A., Lazebnik, S., Perona, P., Sato, Y., Schmid, C. (eds.) ECCV 2012, Part I. LNCS, vol. 7572, pp. 214–227. Springer, Heidelberg (2012)
7. Geerts, F., Goethals, B., Mielikäinen, T.: Tiling databases. In: Suzuki, E., Arikawa, S. (eds.) DS 2004. LNCS (LNAI), vol. 3245, pp. 278–289. Springer, Heidelberg (2004)
8. Grundmann, M., Kwatra, V., Han, M., Essa, I.: Efficient hierarchical graph-based video segmentation. In: CVPR (2010)
9. van Leeuwen, M., Knobbe, A.J.: Diverse subgroup set discovery. Data Min. Knowl. Discov. 25(2), 208–242 (2012)
10. Lu, H., Vaidya, J., Atluri, V.: Optimal boolean matrix decomposition: Application to role engineering. In: ICDE, pp. 297–306 (2008)
11. Motwani, R., Raghavan, P.: Randomized algorithms. Cambridge University Press, USA (1995)
12. Pasquier, N., Bastide, Y., Taouil, R., Lakhal, L.: Discovering frequent closed itemsets for association rules. In: Beeri, C., Bruneman, P. (eds.) ICDT 1999. LNCS, vol. 1540, pp. 398–416. Springer, Heidelberg (1998)
13. Remmerie, N., Vijlder, T.D., Valkenborg, D., Laukens, K., Smets, K., Vreeken, J., Mertens, I., Carpentier, S.C., Panis, B., Jaeger, G.D., Blust, R., Prinsen, E., Witters, E.: Unraveling tobacco by-2 protein complexes with {BN} page/lcms/ms and clustering methods. Journal of Proteomics 74(8), 1201–1217 (2011)
14. Smets, K., Vreeken, J.: The odd one out: Identifying and characterising anomalies. In: SDM, pp. 804–815 (2011)
15. Tatti, N., Vreeken, J.: Discovering descriptive tile trees by mining optimal geometric subtiles. In: Flach, P.A., De Bie, T., Cristianini, N. (eds.) ECML PKDD 2012, Part I. LNCS, vol. 7523, pp. 9–24. Springer, Heidelberg (2012)
16. Vaidya, J., Atluri, V., Guo, Q.: The role mining problem: A formal perspective. ACM Trans. Inf. Syst. Secur. 13(3) (2010)
17. Vreeken, J., van Leeuwen, M., Siebes, A.: Krimp: mining itemsets that compress. Data Min. Knowl. Discov. 23(1), 169–214 (2011)

Ranked Tiling

Thanh Le Van[1], Matthijs van Leeuwen[1], Siegfried Nijssen[1,2],
Ana Carolina Fierro[3], Kathleen Marchal[3,4,5], and Luc De Raedt[1]

[1] Department of Computer Science, KU Leuven, Belgium
`firstname.lastname@cs.kuleuven.be`
[2] Leiden Institute for Advanced Computer Science, Universiteit Leiden,
The Netherlands
[3] Department of Microbial and Molecular Systems, KU Leuven, Belgium
`carolina.fierro@biw.kuleuven.be`
[4] Department of Plant Biotechnology and Bioinformatics, Ghent University, Belgium
[5] Department of Information Technology, iMinds, Ghent University, Belgium
`kathleen.marchal@ugent.be`

Abstract. Tiling is a well-known pattern mining technique. Traditionally, it discovers large areas of ones in binary databases or matrices,
where an area is defined by a set of rows and a set of columns. In this
paper, we introduce the novel problem of *ranked* tiling, which is concerned with finding interesting areas in ranked data. In this data, each
transaction defines a complete ranking of the columns. Ranked data occurs naturally in applications like sports or other competitions. It is also
a useful abstraction when dealing with numeric data in which the rows
are incomparable.

We introduce a scoring function for ranked tiling, as well as an algorithm using constraint programming and optimization principles. We empirically evaluate the approach on both synthetic and real-life datasets,
and demonstrate the applicability of the framework in several case studies. One case study involves a heterogeneous dataset concerning the discovery of biomarkers for different subtypes of breast cancer patients. An
analysis of the tiles by a domain expert shows that our approach can
lead to the discovery of novel insights.

Keywords: tiling, ranked data, numerical data, pattern mining.

1 Introduction

The problem of tiling was introduced by Geerts et al. [1]. It is a popular pattern
mining technique that searches for a set of tiles (that is, a *tiling*) in a 0/1 matrix.
Such matrices often represent transactional data, where each transaction specifies the presence or absence of a set of items in the transaction. A tile is then
a subset of the rows and columns of the matrix, for which the corresponding
submatrix contains all 1s. Tilings are interesting as they provide groupings of
both the rows and the columns that may give new insights in the data.

In this paper, we extend tiling towards a setting which is not binary. That
is, we introduce the problem of *ranked tiling*, in which each transaction in the

T. Calders et al. (Eds.): ECML PKDD 2014, Part II, LNCS 8725, pp. 98–113, 2014.
© Springer-Verlag Berlin Heidelberg 2014

data is a ranking of all available items. This type of data naturally occurs in many situations of interest. Consider, for instance, cycling competitions where the items could be the cyclists and each transaction would correspond to a race, or consider a business context, where the items could be companies and the transactions specify the rank of their quotation for a particular service. Ranking is also a natural abstraction for purely numeric data, which often arises in practice and may be noisy or imprecise. Numeric data is hard to analyse with many existing pattern mining approaches.

One real-life example that we shall use is concerned with the discovery of biomarkers to group cancer patients into *subtypes*. Finding a set of biomarkers that characterise different cancer subtypes is clearly important. Thanks to advances in genome sequencing and high throughput technologies, a lot of data is becoming available about patients. However, different types of data may be obtained with different technologies, for example, data concerning mRNA, miRNA, copy number variations, or proteins. This means that we are given a number of data matrices, each of which corresponds to one data type or experiment, and each of which is measured on a different scale. Still we would expect to find a tiling in which the same set of patients is shared across the different data matrices. We shall show that by using a ranked version of the data and ranked tiling, this is feasible.

To illustrate the problem of ranked tiling, let us consider the toy example in Figure 1. It depicts a rank matrix containing five rows and ten columns. Assuming no ties, each row contains each of the numbers one to ten exactly once. In this paper, we assume that a desirable high rank is indicated by a high number, i.e., in this case the highest possible rank is ten. Now, we are intuitively interested in rectangular areas in the matrix that have relatively high values, as these correspond to columns and rows which are highly ranked. In this particular example, the maximal ranked tile that we would like to find consists of five rows and three columns, i.e., the area defined by $\{R1, R2, R3, R5\}$ and $\{C1, C2, C3\}$.

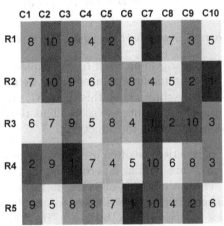

Fig. 1. Example rank matrix, with maximal ranked tile $B = (\{R1, R2, R3, R5\}, \{C1, C2, C3\})$

The key contributions of our paper are 1) the introduction of the problem of ranked tiling, 2) the introduction of an optimisation model for ranked tiling and its implementation in a constraint programming solver, and 3) an empirical evaluation on synthetic and real-life datasets that shows the ability of our ranked tiler to discover interesting tiles, and shows the promise of the approach in practical discovery tasks such as that concerned with breast cancer.

2   Ranked Tiling

In this section, we formally define ranked data and introduce the problems that we consider: maximal ranked tile mining and ranked tiling.

Definition 1 (Rank matrix). *Let $\mathcal{M}$ be a matrix consisting of m rows and n columns. Let $\mathcal{R} = \{1, ..., m\}$, $\mathcal{C} = \{1, ..., n\}$ be index sets for rows and for columns respectively. The matrix $\mathcal{M}$ is a rank matrix iff:*

$$\forall r \in \mathcal{R} : \cup_{c \in \mathcal{C}} \mathcal{M}_{r,c} \subseteq \sigma, \tag{1}$$

where $\sigma = \{1, 2, ..., n\}$.

Note that the values of a row can only be a (strict) subset of σ iff the row contains ties, otherwise the set of values must be exactly equal to σ. Given such a rank matrix, we would like to find a maximal ranked tile, i.e., a tile with relatively highly ranked values. Formally, we have the following problem.

Problem 1 (Maximal ranked tile mining) *Given a rank matrix $\mathcal{M} \in \sigma^{m \times n}$ and a threshold θ, find the ranked tile $B = (R^*, C^*)$, with $R^* \subseteq \mathcal{R}$ and $C^* \subseteq \mathcal{C}$, such that:*

$$B = (R^*, C^*) = \operatorname*{argmax}_{R,C} \sum_{r \in R, c \in C} (\mathcal{M}_{r,c} - \theta). \tag{2}$$

where θ is an absolute-valued threshold.

Example 1. Going back to the example rank matrix in Figure 1 and choosing $\theta = 5$, the maximal ranked tile is defined by $R = \{1, 2, 3, 5\}$ and $C = \{1, 2, 3\}$. The score obtained by this tile is 37, and no more columns or rows can be added without decreasing the score. This result matches the desired outcome that we described in the introduction.

In practice, we often use a relative instead of an absolute threshold. We denote such a threshold as a percentage, i.e., $\theta = a\%$ implies $\theta = a\% \times n$.

The optimisation objective in Equation 2 rewards cells in a tile having values higher than θ, and vice versa for cells having lower values. Since we look for tiles that maximise this score, this threshold θ plays an important role. That is, higher values for θ result in smaller tiles with larger ranks. This implies that the threshold can be used to influence both 1) the size of the mined tiles, and 2) the extent to which the ranks deviate from the mean rank. An alternative interpretation is that the threshold can be used by the analyst to express her prior belief about how high the ranks should be to make a tile interesting.

In practice, it happens quite often that we have numerical data, either discrete or continuous, that we would like to analyse. In gene expression analysis, for example, we are given a matrix with continuous data, whose columns represent patients and rows represent genes. A value in a cell is then the expression level of the gene for a specific patient.

Fortunately, converting a matrix with numeric data to a rank matrix is straightforward. Given any ranking function, i.e., a function that sorts a set of values and assigns ranks based on the resulting order, each row of a numeric matrix can be transformed. In practice, ties may occur: the same value may occur more than once within a single row. Such ties can be broken using either average rank scores or minimum rank scores (in this paper, we use the former unless noted otherwise).

The maximal ranked tiling problem aims to find a single tile, but we are of course interested in finding a set of such tiles. In other words, we would like to discover a ranked tiling.

Problem 2 (Ranked tiling) *Given a rank matrix* $\mathcal{M}$*, a number* k*, a threshold* θ*, and a penalty term* P*, the ranked tiling problem is to find a set of ranked tiles* $B_i = (R_i, C_i)$*,* $i = 1 \ldots k$*, such that they together maximise the following objective function:*

$$\underset{R_i, C_i}{\arg\max} \sum_{r \in \mathcal{R}, c \in \mathcal{C}} \mathbb{1}_{(t_{r,c} \geq 1)}((\mathcal{M}_{r,c} - \theta) - (t_{r,c} - 1)P) \tag{3}$$

where $t_{r,c} = |\{i \in \{1, \ldots, k\} \mid r \in R_i, c \in C_i\}|$ *indicates the number of tiles that cover a cell, and* $\mathbb{1}_\varphi$ *is an indicator function that returns 1 if the test* φ *is true, and 0 otherwise.* P *indicates a penalty that is assigned when tiles overlap.*

For the remainder of this paper we will fix P to $\theta - 1$, i.e., we penalise overlap with the lowest possible rank minus the score θ.

Before we describe our approach to solving the aforementioned problems, we introduce two variations that allow the data analyst to do *query-based* tiling. By allowing the data analyst to provide queries to the system, it becomes possible to explicitly search for interesting patterns in specific areas of the data.

Problem 3 (Maximal query-based ranked tile mining) *Given a query* Q*, a rank matrix* $\mathcal{M} \in \sigma^{m \times n}$*, and a threshold* θ*, find the maximal ranked tile that satisfies the following additional constraint:*

$$\forall q \in Q : q \in C \tag{4}$$

In other words, a query-based tile is a maximal ranked tile whose columns contains those specified in the user-defined query.

Finally, an alternative method can be used to avoid overlap between tiles.

Problem 4 (Diverse ranked tiling) *Given a number* k*, the problem of diverse ranked tiling is to find a set of* k *ranked tiles,* $B_i = (R_i, C_i), i = 1 \ldots k$*, such that the following objective function is optimised:*

$$\underset{R_i, C_i}{\arg\max} \sum_{r \in \mathcal{R}, c \in \mathcal{C}} \mathbb{1}_{(t_{r,c} \geq 1)}(\mathcal{M}_{r,c} - \theta), \tag{5}$$

under the constraint that:

$$R_i \cap R_j = \varnothing, \forall i \neq j, i, j = 1 \ldots k, \tag{6}$$

i.e., no two tiles share the same row.

The diverse ranked tiling and maximal query-based ranked tile mining problems can be combined to find a *diverse query-based ranked tiling*.

3 Ranked Tiling Using Constraint Programming

In this section, we present the techniques that we propose to solve the problems introduced in the previous section. That is, we introduce a constraint-based model equivalent to Equation 2, but add two constraints to make solving more efficient without affecting the results.

The technique we use to solve the optimisation problem is constraint programming (CP). We follow the approach that was originally introduced by De Raedt et al. [2]. The idea is to formalise the problems as *constraint satisfaction problems* and then use existing solvers to find solutions. There are a number of advantages to this solving paradigm. First, it is a declarative approach, meaning that the data analyst can focus on modelling the problem rather than on complex procedural implementations. Second, CP is very flexible. As we will demonstrate, it is easy to implement small variations of a problem by adding or modifying constraints.

3.1 Constraint-Based Model

To speed up the search process, we add two redundant constraints to the optimisation problem of Equation 2. That is, we require that the average values in rows and columns in the selected submatrix $\mathcal{M}_{R,C}$ are higher than the threshold θ. The resulting constraint-based model is as follows:

$$\text{argmax}_{R,C} \sum_{r \in R, c \in C} (\mathcal{M}_{r,c} - \theta) \tag{7}$$

subject to

$$\forall r \in \mathcal{R} : r \in R \leftrightarrow \frac{\sum_{c \in C} \mathcal{M}_{r,c}}{|C|} \geq \theta \tag{8}$$

$$\forall c \in \mathcal{C} : c \in C \leftrightarrow \frac{\sum_{r \in R} \mathcal{M}_{r,c}}{|R|} \geq \theta \tag{9}$$

Theorem 1 (Model equivalence). *The constraint-based optimisation model in Equations 7–9 is equivalent to the optimisation model in Equation 2.*

Proof. The proof is given in the Appendix.

3.2 Problem Formalisation Using CP

The formalisation in Equations 7 − 9 is defined over set variables. Unfortunately, in earlier work it was shown [2] that set variables do not lead to good performance in CP and a reformalization in terms of boolean variables is necessary.

We therefore introduce two Boolean decision vectors: $T = (T_1, T_2, ..., T_m)$, with $T_i \in \{0, 1\}$, for rows and $I = (I_1, I_2, ..., I_n)$, with $I_i \in \{0, 1\}$, for columns. An assignment to the Boolean vectors T and I corresponds to an indication of rows and columns belonging to a tile. Given this, we have the following theorem.

Theorem 2 (Highly ranked rows constraint). *If the following constraint is satisfied:*

$$\forall t \in \mathcal{R} : T_t = 1 \leftrightarrow \sum_{i \in \mathcal{C}} (\mathcal{M}_{t,i} - \theta) * I_i \geq 0 \tag{10}$$

then rows that satisfy the inequality of Equation 8 are identified by:

$$\{r \in \mathcal{R} \mid T_r = 1\} \tag{11}$$

Proof.

$$\forall t \in \mathcal{R} : T_t = 1 \leftrightarrow \sum_{i \in \mathcal{C}} (\mathcal{M}_{t,i} - \theta) * I_i \geq 0 \leftrightarrow \sum_{i \in \mathcal{C}} \mathcal{M}_{t,i} * I_i \geq \theta * \sum_{i \in \mathcal{C}} I_i$$

$$\leftrightarrow \frac{\sum_{i \in \mathcal{C}} \mathcal{M}_{t,i} * I_i}{\sum_{i \in \mathcal{C}} I_i} \geq \theta \quad \leftrightarrow \quad \frac{\sum_{i \in \mathcal{C}} \mathcal{M}_{i,c}}{|\mathcal{C}|} \geq \theta$$

Here we assume that $\sum_{i \in \mathcal{C}} I_i \geq 1$. Overall, $T_t = 1 \leftrightarrow t \in R$, which concludes the proof.

A similar property can be obtained for the column constraint. This leads to a CP model which is equivalent to the constraint-based model in Equations 7 – 9:

$$\underset{T,I}{\text{argmax}} \sum_{t \in \mathcal{R}} T_t * \left(\sum_{i \in \mathcal{C}} (\mathcal{M}_{t,i} - \theta) * I_i \right) \tag{12}$$

subject to

$$\forall t \in \mathcal{R} : T_t = 1 \leftrightarrow \sum_{i \in \mathcal{C}} (\mathcal{M}_{t,i} - \theta) * I_i \geq 0 \tag{13}$$

$$\forall i \in \mathcal{C} : I_i = 1 \leftrightarrow \sum_{t \in \mathcal{R}} (\mathcal{M}_{t,i} - \theta) * T_t \geq 0 \tag{14}$$

The constraints in this formalisation are similar to those used to mine frequent itemsets in [2]. The main difference is that we also have an objective function.

3.3   Mining Maximal Ranked Tiles Using CP

Mining a single ranked tile is equivalent to finding an assignment to vectors T, I such that T and I satisfy constraints 13 – 14 and maximise objective function 12. We solve this constrained optimisation problem using constraint programming.

Solving a problem using CP is done in two phases: 1) modelling, and 2) solving. Equations 12 – 14 can be written down as a model in any CP solver. As we will see, however, the problem of ranked tiling is not easy to solve, and finding exact solutions is difficult. To allow for finding approximate solutions, we choose the OscaR[1] solver, which is an open source CP solver written in Scala. A distinguishing feature is that it provides good support for both exhaustive and heuristic search methods, which we will use for our approach.

---

[1] https://bitbucket.org/oscarlib/oscar/wiki/Home.

When the CP solver is asked to perform exhaustive search, it essentially builds a search tree. The key idea here is that it uses the constraints to remove inconsistent values while searching. This removal of inconsistent values is called propagation and can reduce the search space significantly.

Variable and value ordering heuristic. The order in which variables are considered for branching, as well as the order in which values are assigned to the variables, determine the shape of the search tree and the effectiveness of constraint propagation. We use the following heuristic. We order column variables I_c, $c = 1 \ldots n$, by their total sum scores, $\sum_{r \in \mathcal{R}} (\mathcal{M}_{r,c} - \theta)$, in ascending order. That is, variables that have lower scores will be branched on first, and the value zero will be assigned to a variable before the value one. Using this heuristic, CP has a higher chance to add variables having high scores to the solution when backtracking. Consequently, CP needs less backtracks to find a first valid assignment as variables having higher scores have higher probability of satisfying constraint 14.

Large neighbourhood search. Equation 13 shows that vector T can be completely determined given a complete assignment to I. Hence, the size of the search space is $O(2^n)$, where n is the number of columns. Even taking into account propagation, this search space is in practice often still too large to be traversed completely and hence we use a form of local search to speed up the search.

Large Neighbourhood Search (LNS) is a hybridization of local search and exhaustive search in CP. Local search refers to the idea that one solution can be transformed into another by changing the assignment of a number of variables. While traditional local search methods only change a limited number of variables (e.g., one variable at a time), LNS selects a relatively large subset of the variables in a problem (e.g., a random subset of half of the variables) and performs complete search over these variables while fixing the remaining variables. Two main questions involved with LNS include a) which variables should be selected to search over, and b) how to search on these variables? In our implementation, we use stochastic variable selection and an exhaustive search approach. The stochastic variable selection uniformly selects half of the column variables to search over.

3.4   Ranked Tiling

Ranked tiling was introduced in Problem 2. We propose to approximate the optimal solution by using a greedy approach, as is common for this type of pattern set mining problem. The first tile is found by solving the optimisation problem in Equations 12–14. Next, we remove that tile by setting all cells in the matrix that are covered to the lowest rank (or another value, depending on parameter P). Then, we search in the resulting matrix for the second tile. This process is repeated until the number of desired tiles is found. The sum of the scores of all discovered tiles will correspond to the score of Equation 3 for this solution. However, as the search is greedy, the solution is not necessarily optimal.

4   Experiments on Synthetic Datasets

In this section, we set up a number of experiments on synthetic data sets to evaluate the proposed approach. The results will show: 1) the gain in runtime by adding the redundant constraints; 2) the accuracy of the discovered tiles in two settings: ranked tiling and query-based ranked tiling; 3) the robustness of the algorithm with respect to noise thresholds and variations of the local search; 4) a comparison to constant-row bi-clustering algorithms.

4.1   Data Generation

By generating data in which rows are incomparable due to different scales, our aim is to show that our technique finds the relevant patterns in such data, whereas methods for bi-clustering cannot. Since bi-clustering methods only work on numeric data, we use a simple generative model to generate synthetic, continuous data. This numeric data is then transformed to a rank matrix to apply ranked tiling. For bi-clustering, we choose the constant-row setting, as there are many bi-clustering algorithms specifically designed for this type of pattern.

To generate synthetic datasets, we first generate background data, and then implant a number of constant-row bi-clusters with higher average values.

Background information is generated such that each row has a potentially different scale than the others. Values within each row are sampled from two distributions, with certain probability: one for modelling background noise, the other for interfering with the implanted patterns. First, for each row, we uniformly sample μ_r^1, μ_r^2 from two ranges:

$$\mu_r^1 \sim U(0,3), \forall r \in \mathcal{R} \tag{15}$$

$$\mu_r^2 \sim U(3,5), \forall r \in \mathcal{R} \tag{16}$$

Second, for every cell in a row, we sample a latent binary variable $X_{r,c}$ from a Bernoulli distribution $Bin(p, 1-p)$, given some p. Depending on the value of this latent variable, the background data is sampled either from the low-average or high-average distribution:

$$\mathcal{D}_{r,c} \sim \begin{cases} N(\mu_r^1, 1) & \text{if } X_{r,c} = 1 \\ N(\mu_r^2, 1) & \text{otherwise} \end{cases} \tag{17}$$

To plant a constant-row bi-cluster in a submatrix $\mathcal{D}_{R,C}$, we use the following two equations:

$$\forall r \in R, \quad \mu_r \sim U(3,5) \tag{18}$$

$$\forall r \in R, \quad \mathcal{D}_{r,c} \sim N(\mu_r, 1) \tag{19}$$

Equation 18 is used to sample the mean value for every row in a bi-cluster. The mean value is uniformly sampled from the range $[3 \ldots 5]$, which is higher than the main sampling range for the background ($[0 \ldots 3]$).

4.2   Evaluation

Performance Gain by Adding Constraints. To evaluate how the redundant constraints affect the time needed for search, we compare the runtime of the optimisation model in Equation 2, which does not have any constraints, to the runtime of the constraint-based model shown in Equations 7–9. To make the comparison fair, we perform exhaustive search in both cases.

To this aim, we generate a number of datasets with varying sizes using the procedure described in Section 4.1. All datasets have the same number of columns, i.e. 20, and a varying number of rows: $\{100, 200, 300, 400, 500, 600\}$, with $p = 0.1$. They have the

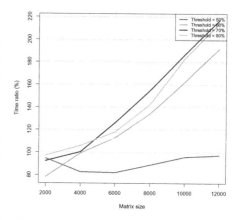

Fig. 2. Speed up achieved by adding the two redundant constraints. The y-axis indicates the ratio between the runtime needed with and without the constraints, and is computed for varying threshold θ values and matrix sizes.

same two non-overlapping tiles: one is 10×5 and the other is 30×10. We execute the mentioned models on these datasets to find the maximal ranked tile. All experiments are executed single threaded on a desktop computer (Intel i7-2600 CPU @ 3.40GHz, 16GB RAM).

Figure 2 shows the ratio between the time needed to solve the problem with and without the extra constraints, for varying θ thresholds. We can see that adding the extra constraints reduces the search time when $\theta > 50\%$. In particular for larger datasets and values of the threshold, the model with the added constraints always outperforms the optimisation-only model. This demonstrates that adding the constraints results in better propagation and hence more efficient search. The absolute time needed to find the optimal solution on these datasets ranges from 1ms to 4h 37m 26s.

Accuracy of the recovered tiles. We now evaluate the ability of the algorithm to recover the implanted ranked tiles. We do this by measuring *recall* and *precision*, using the implanted tiles as ground truth. Overall performance is quantified by the *F1* measure, which is the average of the two scores.

We generate seven 1000 rows × 100 columns datasets for different p, i.e., $p \in \{0.10, 0.15, 0.20, 0.25, 0.30, 0.35, 0.40\}$, using the same procedure as before. In each dataset, we implant five ranked tiles. Figure 3a shows a heatmap of the numerical dataset for $p = 10\%$. Figure 3b shows its corresponding rank matrix, using the same color coding as in Figure 1 (blue = low rank, red = high rank).

We varied the threshold θ; for each value of θ and each dataset, we performed ranked tiling five times, each time mining $k = 5$ tiles. Then, we calculated average precision, recall and F1 score over these five runs. Figure 3c summarises the results obtained with and without using the variable ordering heuristic (see Section 3.3). It can be seen that the heuristic contributes to improved performance.

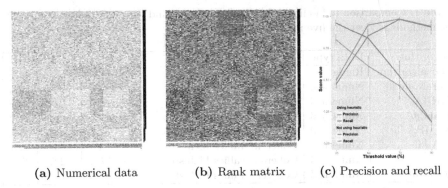

(a) Numerical data   (b) Rank matrix   (c) Precision and recall

Fig. 3. Recovering five implanted ranked tiles from synthetic data

When the threshold θ is around 60%, the algorithm achieves high prediction accuracy (average F1 = 86%). At lower thresholds, it has low precision, while higher thresholds result in lower recall. This completely matches our expectation, since higher thresholds result in smaller tiles with higher values.

Robustness of the local search. The error bars in the precision and recall curves shown in the Figure 3c show the robustness of the large neighbourhood search on the synthetic datasets. The variation is typically low with respect to the number of times the local search is repeated and the different noise levels.

Comparison to bi-clustering. In this experiment, we compare to several bi-clustering algorithms. SAMBA [3] was designed for coherent evolution bi-clusters, in which there is coherence of the signs of values, i.e., up or down. The other methods discover coherent-valued bi-clusters, of which a constant-row bi-cluster is a special case. CC [4], Spectral [5], and Plaid [6] are implemented in the R biclust[2] package. FABIA[3] [7] and SAMBA[4] are downloaded from their website. ISA [8] is from the R isa2 package[5].

Since large noise levels may conversely affect the performance of the algorithms, we use a dataset also used for the previous experiments, with $p = 0.20$ (average noise level). We ran all algorithms on this dataset and took the first five tiles/bi-clusters they produced. For most of the benchmarked algorithms, we used their default values. For CoreNode, we use $msr = 1.0$ and $overlap = 0.5$, as preliminary experiments showed that this combination produced the best result. For ISA, we applied its built-in normalised method before running the algorithm.

The results in Table 1 show that our algorithm achieves much higher precision and recall on this task than the bi-clustering methods. This indicates that when the rows in a numerical matrix are incomparable, converting the data to a ranked matrix and applying ranked tiling is a better solution than applying bi-clustering.

---

[2] http://cran.r-project.org/web/packages/biclust/
[3] http://www.bioinf.jku.at/software/fabia/fabia.html
[4] http://acgt.cs.tau.ac.il/expander/
[5] http://cran.r-project.org/web/packages/isa2/

Table 1. Comparison to bi-clustering. Precision, recall and F1 quantify how accurately the methods recover the five implanted tiles. $k = 5$.

Algorithm	Data type	Pattern	Precision	Recall	F1
Our algorithm	Ranks	Ranked tile	88%	83%	86%
CoreNode [9]	Numerical	Coherent values bicluster	43%	72%	58%
FABIA [7]	Numerical	Coherent values bicluster	40%	24%	32%
Plaid [6]	Numerical	Coherent values bicluster	90%	6%	48%
SAMBA [3]	Numerical	Coherent evolution bicluster	67%	3%	35%
ISA [8]	Numerical	Coherent values bicluster	64%	44%	54%
CC [4]	Numerical	Coherent values bicluster	35%	22%	29%
Spectral [5]	Numerical	Coherent values bicluster	-	-	-

(a) A query of 2 columns.   (b) The two mined tiles.

Fig. 4. Query-based ranked tiling on a synthetic dataset; $\theta = 70\%$

Diverse query-based ranked tiling Finally, we use the same dataset, with $p = 0.2$, to illustrate diverse query-based ranked tiling. Figure 4a shows a query consisting of two columns, and Figure 4b shows the two discovered ranked tiles given that query. The rows and columns of each tile are marked by coloured bars (tile 1 = red, tile 2 = green). The results demonstrate the flexibility of our approach based on constraint programming: adding a few constraints results in ranked tiles for a given query. Both discovered tiles contain the query, but are also clearly present in the generated data. This allows the analyst to focus the search on specific areas of the data and still obtain high-quality results.

5   Real-world Case Studies

In this section, we present two applications of ranked tiling: 1) voting pattern discovery in Eurovision Song Contest data, and 2) biomarker discovery for different subtypes from a heterogeneous genomic dataset.

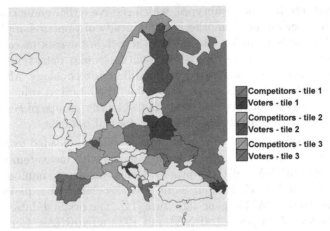

Fig. 5. Voting patterns in the Eurovision Song Contest data

5.1   European Song Contests

The Eurovision Song Contest (ESC) has been held annually since 1956. Each participating country gives voting scores, which are a combination of televoting and jury voting, to competing countries. Scores are in the range of $1 \ldots 8$, 10 and 12. Each country awards 12 points to their most favourite country, 10 points to the second favourite, and $8 \ldots 1$ to the third...tenth favourites respectively. The data can be represented by a matrix, in which rows correspond to voting countries, columns correspond to competing countries, and values are the scores.

We collected voting data for the final rounds of the period 2010 – 2013. We filtered out countries participating in fewer than 3 years. Data is aggregated by calculating average scores that voting countries award to competing countries during the period considered. After the pre-processing step, the data consists of 44 rows and 37 columns. The pre-processed data is then transformed to ranked data, using minimum ranks in case of ties.

We ran the algorithm on the ESC dataset and obtained 10 ranked tiles with the threshold θ set to 80%. The first tile shows that Azerbaijan and Sweden were highly voted for by other 19 countries located in many regions in Europe. Other tiles reveal that there are local voting patterns in the contests: countries tend to distribute high scores to their neighbours. Figure 5 shows three representative tiles: one for eastern countries, one for western and northern countries, and one for (mostly) southern countries. In general, the discovered tiles confirm that countries give high scores to their neighbours, which matches our expectations.

5.2   Biomarker Discovery for Breast Cancer Subtypes

Breast cancer is known to be a heterogeneous disease that can be categorised in clinical and molecular subtypes. Assignment of patients to such subtypes is crucial to give adapted treatments to patients. Currently, breast cancer patients are categorised into 3 clinical subtypes, which receive either endocrine therapy, targeted HER2 therapy, or chemotherapy. The study of tumour samples at multiple molecular levels helps to understand the driving events in cancer. Most studies, however, focus on the analysis of each molecular data type separately [10], since each data type is measured with a different technology. Although raw data between molecular levels are incomparable, their ranked values can easily be compared. The goal of this case study is to identify groups of genes or copy number regions that are highly specific to a subset of patients.

Data pre-processing. A breast cancer dataset was downloaded from the Cancer Genome Atlas[6]. A subset of 94 tumour samples with measurements at four molecular levels (mRNA, miRNA, protein levels, and copy number variation (CNV)) was selected. Each tumour sample is associated to a molecular subtype according to the PAM50 gene signature [11]. The original dataset contains measurements for 17814 genes (mRNA) and 1222 microRNAs. Genes and microRNAs were selected based on their potential subtype-specific activity and their differential expression relative to normal (non-tumour) samples.

To capture subtype-specific expression changes, we evaluated the 5- and 95-percentile of the tumour samples. Genes in which the p-value for these percentiles was below 0.001 and their log-fold change relative to the mean normal expression was at least 2.5 were selected. The final dataset contains 1761 genes and 138 microRNAs. Segmented files with copy number alterations were pre-processed and filtered by germline aberrations as described by [10]. Significant copy number regions were identified with GISTIC2.0 [12]. No selection was done for protein levels and copy number regions. Each data level was converted to rank scores and combined into a single matrix, which consists of 2211 rows and 94 columns.

Ranked tiling analysis. The ten ranked tiles discovered by the algorithm are shown in Figure 6 ($\theta = 65\%$, 20 LNS repeats, in 3h 45m 9s). A first important observation is that each tile contains features from all four different data types. All PAM50 subtypes (LuminalA, LuminalB, Basal and HER2 subtypes) are covered by the tiles. Some discrepancies between the tiles and PAM50 subtypes are expected, since PAM50 is based on a 50-genes classifier derived from expression data only. We find that tile 1 captures the largest group of genes and samples, but due to its large size functional annotation did not lead to interesting insights.

Tiles 2 and 3 match a known basal subtype, with tile 3 containing most basal samples. Among the genes specific for tile 3, we observe MYC at the protein and mRNA levels, which has been previously suggested as a basal characteristic pattern [10]. Both tiles 2 and 3 largely overlap in terms of genes also with tile 5 (LuminalB). The genes in common to these 3 tiles are enriched for cell cycle and cell division (Gene Ontology and Reactome enrichment), consistent with the high proliferation present in LuminalB, Basal and Her2 subtypes.

Tile 6 only contains samples from the known HER2 subtype. The HER2 molecular subtype is known to over-express the ERBB2 gene and to contain copy number alterations for the same gene. Tile 6 captures all the molecular levels related to the gene, as it contains the amplified region of ERBB2, the protein

---

[6] https://tcga-data.nci.nih.gov/tcga

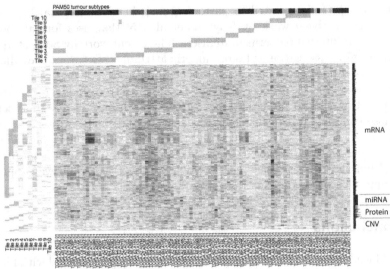

Fig. 6. Ranked tiling on heterogeneous breast cancer data. The rows correspond to mRNA, miRNA, protein and CNV levels, the columns correspond to breast cancer samples. The left and upper gray bars indicate the ten tiles. PAM50 subtypes are indicated at the top (LuminalA=blue, LuminalB=light blue, Basal=red and HER2=pink).

and the mRNA gene. LuminalA and LuminalB PAM50 subtypes correspond to the estrogen positive (ER+) clinical subtype which is the most common one and thus they are present in many tiles. Among them, tile 10 contains the estrogen receptor (ESR1) and other genes related to the pathway, suggesting that the pathway activity might be higher for the patients in tile 10.

Overall, we conclude that ranked tiling successfully identifies known subtypes and includes different data types in tiles, as desired. Such an integrated analysis of heterogeneous data has large potential for this type of application.

6   Related Work

Ranked tiling is related to bi-clustering, but is different because of the type of regularities it aims to find. The literature describes four types of bi-clusters: constant-valued, constant-row, constant-column and coherent [13]. In ranked tiling, the absolute values within the specified areas matter, i.e., the values must be higher than a given threshold. This is clearly different from the objectives of bi-clustering, as also demonstrated by the results presented in Section 4.

Calders et al. [14] use rank measures to find interesting patterns in numeric data. However, they rank values on individual columns and use frequent itemset mining-based techniques to find correlated sets of columns (items). Here, we consider ranks in rows and use optimisation-based techniques to find sets of rows and columns (tiles) in a matrix for which the ranks are relatively high.

Ranked tiling is also related to pattern mining in numeric data. In this direction, there has been work by Kaytoue et al. [15] that uses formal concept analysis to find interval patterns of itemsets. The recent work by Song et al. [16] proposes to mine association rules of numeric attributes. However, they did not consider ranked data and tilings on this type of data.

Kontonasios et al. [17] propose to use a Maximum Entropy model to iteratively mine interesting tiles in numeric data. This approach also aims to find sets of tiles whose content are contrasted to the background model. However, we do not contrast a tile against expected values given some prior beliefs, but consider absolute values relative to a given threshold. Apart from this, they do not provide an algorithm that directly searches for high-scoring tiles, while we propose a mining approach based on constraint programming.

7   Conclusions

We introduced the novel problem of ranked tiling, which is concerned with finding areas in ranked data in which the ranks are relatively high. Ranked data occurs naturally in many applications, but is also a useful abstraction when dealing with numeric data in which the rows are incomparable.

We presented an optimisation-based approach to solving the ranked tiling problem using constraint programming, and demonstrated the flexibility of this approach by extending it to query-based ranked tiling. The experiments on both synthetic and real data show that our approach finds high-quality ranked tiles that can lead to the discovery of novel insights in heterogeneous data.

Acknowledgements. This research was supported by the DBOF 10/044 Project, the Natural and Artifical Genetic Variation in Microbes project, Post-doctoral Fellowships of the Research Foundation Flanders (FWO) for Siegfried Nijssen and Matthijs van Leeuwen, and the EU FET Open project Inductive Constraint Programming.

References

1. Geerts, F., Goethals, B., Mielikäinen, T.: Tiling Databases. In: Suzuki, E., Arikawa, S. (eds.) DS 2004. LNCS (LNAI), vol. 3245, pp. 278–289. Springer, Heidelberg (2004)
2. De Raedt, L., Guns, T., Nijssen, S.: Constraint programming for itemset mining. In: KDD, pp. 204–212 (2008)
3. Tanay, A., Sharan, R., Shamir, R.: Discovering statistically significant biclusters in gene expression data. Bioinformatics 18(suppl. 1), S136–S144 (2002)
4. Cheng, Y., Church, G.M.: Biclustering of expression data. In: The 8th International Conference on Intelligent Systems for Molecular Biology, vol. 8, pp. 93–103 (2000)
5. Kluger, Y., Basri, R., Chang, J.T., Gerstein, M.: Spectral Biclustering of Microarray Data: Coclustering Genes and Conditions. Genome Research 13, 703–716 (2003)

6. Turner, H., Bailey, T., Krzanowski, W.: Improved biclustering of microarray data demonstrated through systematic performance tests. Computational Statistics & Data Analysis 48(2), 235–254 (2005)
7. Hochreiter, S., Bodenhofer, U., Heusel, M., Mayr, A., Mitterecker, A., Kasim, A., Khamiakova, T., Van Sanden, S., Lin, D., Talloen, W., Bijnens, L., Göhlmann, H.W.H., Shkedy, Z., Clevert, D.A.: FABIA: Factor analysis for bicluster acquisition. Bioinformatics 26(12), 1520–1527 (2010)
8. Ihmels, J., Friedlander, G., Bergmann, S., Sarig, O., Ziv, Y., Barkai, N.: Revealing modular organization in the yeast transcriptional network. Nature Genetics 31(4), 370–377 (2002)
9. Truong, D.T., Battiti, R., Brunato, M.: Discovering Non-redundant Overlapping Biclusters on Gene Expression Data. In: ICDM 2013, pp. 747–756. IEEE (2013)
10. The Cancer Genome Atlas Network: Comprehensive molecular portraits of human breast tumours. Nature 490(7418), 61–70 (October 2012)
11. Parker, J.S., Mullins, M., Cheang, M.C.U., Leung, S., Voduc, D., Vickery, T., Davies, S., Fauron, C., He, X., Hu, Z., Quackenbush, J.F., Stijleman, I.J., Palazzo, J., Marron, J.S., Nobel, A.B., Mardis, E., Nielsen, T.O., Ellis, M.J., Perou, C.M., Bernard, P.S.: Supervised risk predictor of breast cancer based on intrinsic subtypes. Journal of Clinical Oncology 27(8), 1160–1167 (2009)
12. Mermel, C.H., Schumacher, S.E., Hill, B., Meyerson, M.L., Beroukhim, R., Getz, G.: GISTIC2.0 facilitates sensitive and confident localization of the targets of focal somatic copy-number alteration in human cancers. Genome Biology 12(4) (2011)
13. Madeira, S.C., Oliveira, A.L.: Biclustering algorithms for biological data analysis: A survey. IEEE/ACM Transactions on Computational Biology and Bioinformatics 1(1), 24–45 (2004)
14. Calders, T., Goethals, B., Jaroszewicz, S.: Mining rank-correlated sets of numerical attributes. In: KDD 2006, pp. 96–105. ACM, New York (2006)
15. Kaytoue, M., Kuznetsov, S.O., Napoli, A.: Revisiting Numerical Pattern Mining with Formal Concept Analysis. In: IJCAI, pp. 1342–1347 (2011)
16. Song, C., Ge, T.: Discovering and managing quantitative association rules. In: CIKM 2013, pp. 2429–2434. ACM, New York (2013)
17. Kontonasios, K.-N., Vreeken, J., De Bie, T.: Maximum entropy models for iteratively identifying subjectively interesting structure in real-valued data. In: Blockeel, H., Kersting, K., Nijssen, S., Železný, F. (eds.) ECML PKDD 2013, Part II. LNCS, vol. 8189, pp. 256–271. Springer, Heidelberg (2013)

Proof Theorem 1. Let us assume that (R, C) is the optimum solution found without additional constraints. Without loss of generality we can assume that (R, C) is maximal in both rows and columns, i.e., there is no row nor column that can be added to obtain the same score or a better score.

Then this optimal solution must also satisfy the constraint:

$$\forall r \in \mathcal{R} : r \in R \leftrightarrow \frac{\sum_{c \in C} \mathcal{M}_{r,c}}{|C|} \geq \theta \leftrightarrow (\sum_{c \in C} \mathcal{M}_{r,c} - \theta) \geq 0. \qquad (20)$$

Indeed, assume that $r \in R$ while (R, C) is optimal, then $(\sum_{c \in C} \mathcal{M}_{r,c} - \theta) \geq 0$ must hold, as otherwise the score $\sum_{c \in C, r \in R}(\mathcal{M}_{r,c} - \theta)$ could be improved by removing r from R while keeping C fixed. Conversely, if $r \notin R$ while (R, C) is optimal, it can only be the case that $(\sum_{c \in C} \mathcal{M}_{r,c} - \theta) < 0$, as otherwise the score of the tile could be improved by adding r to R while keeping C fixed.

Fast Estimation of the Pattern Frequency Spectrum

Matthijs van Leeuwen[1] and Antti Ukkonen[2]

[1] Department of Computer Science, KU Leuven, Belgium
[2] Helsinki Institute for Information Technology HIIT, Aalto University, Finland
matthijs.vanleeuwen@cs.kuleuven.be, antti.ukkonen@aalto.fi

Abstract. Both exact and approximate counting of the number of frequent patterns for a given frequency threshold are hard problems. Still, having even coarse prior estimates of the number of patterns is useful, as these can be used to appropriately set the threshold and avoid waiting endlessly for an unmanageable number of patterns. Moreover, we argue that the number of patterns for different thresholds is an interesting summary statistic of the data: the *pattern frequency spectrum*.

To enable fast estimation of the number of frequent patterns, we adapt the classical algorithm by Knuth for estimating the size of a search tree. Although the method is known to be theoretically suboptimal, we demonstrate that in practice it not only produces very accurate estimates, but is also very efficient. Moreover, we introduce a small variation that can be used to estimate the number of patterns under constraints for which the Apriori property does not hold. The empirical evaluation shows that this approach obtains good estimates for closed itemsets.

Finally, we show how the method, together with isotonic regression, can be used to quickly and accurately estimate the frequency pattern spectrum: the curve that shows the number of patterns for every possible value of the frequency threshold. Comparing such a spectrum to one that was constructed using a random data model immediately reveals whether the dataset contains any structure of interest.

1 Introduction

Pattern mining aims to enable the discovery of patterns from data. As such, it is one of the most-studied problems in exploratory data mining. A *pattern* is a description of some structure that occurs locally in the data. That is, a pattern is an element of a given pattern language $\mathcal{L}$ that describes a subset of a dataset $\mathcal{D}$. The most commonly used formalisation is theory mining, where the goal is to find the theory $Th(\mathcal{L}; \mathcal{D}; q) = \{X \in \mathcal{L} \mid q(X, \mathcal{D}) = \texttt{true}\}$, with q a selection predicate that returns true iff X satisfies the imposed constraints on $\mathcal{D}$.

The best-known instance of pattern mining is frequent itemset mining [1], which discovers sets of items that frequently occur together in transactional data. Given a minimum support threshold σ, the theory to be mined consists of all itemsets that occur at least σ times in the data. That is, $Th(\mathcal{L}; \mathcal{D}; q) = \{X \in \mathcal{L} \mid \text{freq}(X, \mathcal{D}) \geq \sigma\}$, where $\text{freq}(X, \mathcal{D})$ denotes the *frequency* of X in $\mathcal{D}$,

T. Calders et al. (Eds.): ECML PKDD 2014, Part II, LNCS 8725, pp. 114–129, 2014.
© Springer-Verlag Berlin Heidelberg 2014

i.e., the number of transactions in which the pattern occurs. In general, frequent pattern mining techniques have been developed for quite some data types and corresponding pattern types, e.g., for sequences [2], and for graphs [18].

A major problem in frequent pattern mining is that choosing low values for the minimum frequency threshold results in vast amounts of patterns – the infamous *pattern explosion*. One may try to avoid this by choosing the threshold such that the number of patterns is still manageable. Parameter tuning can be a tricky business though, because small changes in σ often have a large impact on the number of patterns. For that reason, it would be beneficial to know how many patterns to expect without having to actually mine them.

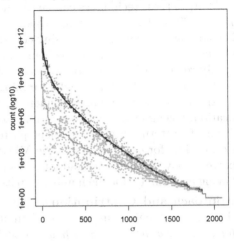

Fig. 1. Exact (black line) and estimated (blue line) numbers of patterns for all possible frequency thresholds in *Mammals*. The green line shows the expected number of patterns in random data having the same column marginals, the grey dots represent individual path sample estimates (see Alg. 1)

Unfortunately, both exact and approximate counting of the number of frequent patterns for a given frequency threshold are hard problems [11,6,19]. Nevertheless, having even coarse prior estimates of the number of patterns is useful, as these can be used to appropriately set the threshold σ and avoid waiting endlessly for an unmanageable number of patterns. In this paper we introduce methods for the fast estimation of the number of frequent patterns in a dataset, both for individual thresholds and the complete frequency spectrum.

Figure 1 illustrates both the pattern explosion and the accuracy of our method. Note that the counts on the y-axis are in logarithmic scale ($\log_{10}$). It took only 11 seconds to accurately estimate the curve for the complete frequency range, whereas the exact curve took over 12 hours to compute and still does not go lower than a frequency threshold of 10. A quick comparison to the expected curve, which is computed under the assumption that all items are independent, shows that the dataset contains quite some structure: there are many more itemsets than can be explained by this simple model.

Using Frequent Patterns for Knowledge Discovery. In practice, frequent pattern mining is seldom used as final step in the KDD process, because interpretation of a large amount of patterns by a domain expert is impracticable. Nevertheless, frequent pattern mining and hence estimating the number of patterns are important problems.

The most important reason is that frequent patterns are often used as input for some other algorithm, as part of the KDD process. Pattern-based approaches to data mining are popular, in particular for exploratory purposes. Using patterns has clear advantages, the most obvious one being that patterns are interpretable representations and can thus provide explanations.

Many of these techniques can be captured under the umbrella term *pattern set mining* [7], a class of solutions proposed to address the pattern explosion. This is commonly achieved by imposing constraints and/or an optimisation criterion on the complete set of patterns being mined. Pattern set mining methods commonly require a large number of frequent patterns as input. Examples include KRIMP [17] and the iterative data mining framework by De Bie [4].

Another context in which frequent patterns are used is pattern-based classification: Cheng et al. [9], for example, construct a classifier from a large set of frequent patterns. Finally, frequent patterns can serve as input for interactive exploration, for example, using the MIME tool [10]. In all these cases, frequent patterns have to be mined and a frequency threshold needs to be chosen. Hence, having an estimate for the number of patterns given a certain threshold is useful.

Approach and Contributions. The first main contribution of this paper is the FASTEST algorithm, for Fast Estimation. It is *a fast and accurate method for estimating the number of frequent patterns for a given dataset and frequency threshold.* For this, we adapt the classical algorithm by Knuth [14] for estimating the size of a search tree. We demonstrate that the method is fast and produces very accurate estimates in practice. In particular, we focus on frequent itemsets and rely on the Apriori property, which states that any subset of a frequent itemset must also be frequent. Our method can be easily applied to other types of data and patterns, as long as the Apriori, or monotonicity, property holds.

We also introduce a small variation of the method that can be used to estimate the number of patterns under constraints for which the Apriori property does not hold. The variation empirically adjusts the total number of estimated frequent patterns for the considered constraints.

The second main contribution is *a method for efficiently estimating the total number of frequent patterns for all possible frequency thresholds*: the pattern frequency spectrum. The algorithm, dubbed SPECTRA, uses isotonic regression [3] to compute a complete spectrum from point estimates obtained with FASTEST. The resulting curves are extremely accurate and provide useful summary statistics of the data. To demonstrate this, *we investigate spectra constructed for random data*, i.e., assuming that all items in a dataset are independent. Comparing an actual to a randomized spectrum immediately reveals whether the dataset contains any structure of interest (see Fig. 1).

The remainder of this paper is organised as follows. First, we discuss related work in Section 2. Next, in Section 3 we describe the classical algorithm by Knuth upon which we base our method. Sections 4 and 5 introduce our techniques for estimating the number of patterns in a dataset for individual thresholds and threshold ranges respectively. We present the empirical evaluation in Section 6, after which we round up with discussion and conclusions in Sections 7 and 8.

2   Related Work

We briefly discuss two categories of related work that are relevant to our work: 1) frequent pattern counting, and 2) estimating the size of a search tree.

Frequent Pattern Counting. Exactly counting the number of frequent patterns is #P-complete [11], as is counting the number of maximal frequent itemsets [19]. For that reason, methods designed specifically for this task usually compute approximate counts. Boley and Grosskreutz [6] devised an approximate counting algorithm for frequent itemsets that is based on MCMC simulation [12]. Although technically solid, the use of MCMC simulation makes the method rather complex. One of the aims of this paper is to develop a much simpler method that is therefore easier to implement and use. We will compare our algorithm to the MCMC-based method in Section 6. Later, Boley et al. [5] proposed a similar method for sampling and counting closed itemsets.

The alternative approach is to perform exact counting by adapting existing pattern mining techniques for this purpose. Highly optimised frequent itemset mining implementations have been available since the FIMI workshops[1] in 2003 and 2004. One such very efficient implementation is AFOPT [15], which we will use as a baseline in our experiments.

Estimating Search Tree Sizes. Knuth's original algorithm [14] was proposed for the problem of studying algorithm performance on a given input by estimating the size of the associated search tree. Purdom suggested a modification that incorporates partial backtrack into the algorithm [16]. This will lead to better estimates when the trees have some long and "thin" branches. Kilby et al. [13] addressed the same problem using a slightly different technique, but focused on binary trees. Chen [8] proposed a generalisation of Knuth's method that is based on the idea of stratified sampling. The algorithm we propose for estimating the number of closed patterns has some similarities with this method.

3   Preliminaries

This section provides a short recap on the tree size estimation algorithm by Knuth [14] that we will use as foundation for our algorithms.

Most combinatorial problems can be framed in terms of finding an assignment of values to a set of variables, so that the assignment satisfies some constraints and optimizes an objective function. This is also true for pattern mining problems, except that one aims to find all assignments that satisfy the constraints.

Backtracking algorithms perform depth-first search over feasible assignments and can be characterised in terms of a *search tree*, where the root contains the "empty" assignment, and nodes correspond to (partial or complete) assignments. The size of the search tree is a practical measure of the hardness of the problem instance, because a simple backtracking algorithm must enumerate all feasible assignments. However, counting the number of nodes is a hard problem, and can

---

[1] http://fimi.ua.ac.be/

usually be solved exactly only by an exhaustive traversal of the search tree. That is, knowing the hardness of a problem instance requires us to solve it!

Knuth proposed an algorithm [14] that computes an *estimate* of the size of the search tree without exhaustive traversal. The intuition of the algorithm is the following: If a search tree is perfectly regular (every internal node has the same outdegree), we can compute its exact size by summing the sizes of every level of the tree. The size of a level is given by the *product of the outdegrees* observed on a path from the root node that ends just above the level.

Example 1. Consider a complete binary tree with h levels, including the root at level 1. The size of every level l, $2 \leq l \leq h$, is $\prod_{i=1}^{l-1} 2 = 2^{l-1}$, and summing these yields $1 + \sum_{l=2}^{h} 2^{l-1} = \sum_{l=0}^{h-1} 2^l = 2^h - 1$, that is, the number of nodes in a complete binary tree.

Since search trees that arise in practice are rarely (if ever) regular, determining the size by only considering a single path from the root to a leaf is not going to produce the correct size. However, we can consider a number of *random paths*, and compute a *path estimate* for each. In detail, let $(v_1, v_2, \ldots, v_h)$ denote a random path from the root v_1 to a leaf v_h in a search tree, and let $d(v)$ denote the outdegree of node v. The associated path estimate is given by the sum

$$1 + \sum_{i=2}^{h} \prod_{l=1}^{i-1} d(v_l), \tag{1}$$

where the product $\prod_{l=1}^{i-1} d(v_l)$ is the estimate associated with the ith level of the tree. Notice that the path estimate is a sum of such levelwise estimates.

To compute a single path estimate, the algorithm starts from the root, selects one child at random at every level until it reaches a leaf, and then applies Eq. 1. The final estimate is defined as *the average of the path estimates* from a number of random paths. Depending on their number, this process only considers a very small part of the search tree, but can in practice obtain an accurate and *unbiased* estimate of the tree size [14].

4   Estimating the Number of Patterns

Let a database $\mathcal{D}$ be a bag of transactions over a set of items $\mathcal{I}$, where a transaction t is a subset of $\mathcal{I}$, i.e., $t \subseteq \mathcal{I}$. Furthermore, a pattern X is an itemset, $X \subseteq \mathcal{I}$, and pattern language $\mathcal{L}$ is the set of all such possible patterns, $\mathcal{L} = 2^{\mathcal{I}}$. An itemset X occurs in a transaction t iff $X \subseteq t$, and its frequency is defined as the number of transactions in $\mathcal{D}$ which it occurs, i.e., $\text{freq}(X, \mathcal{D}) = |\{t \subseteq D \mid X \subseteq t\}|$. A pattern X is said to be frequent iff its frequency exceeds the minimum frequency threshold σ. The frequent itemset mining problem is to find all frequent patterns, i.e., all $X \in \mathcal{L}$ for which $\text{freq}(X, \mathcal{D}) \geq \sigma$. Frequent itemsets can be mined efficiently due to monotonicity of the frequency constraint with respect to set inclusion, which is also known as the Apriori property [1]. This property states that for any frequent itemset X, all itemsets $Y \subseteq X$ must also be frequent.

4.1   Frequent Patterns and the FASTEST Algorithm

Next we describe a modification to Knuth's algorithm [14] and use it to estimate the number of frequent itemsets[2]. The core question we must address is how to turn the task of counting frequent itemsets to that of estimating the size of a tree. By constructing a tree where every node corresponds to a frequent itemset, we can use Knuth's method. The following discussion focuses on frequent itemsets, but the same approach can be applied also to other patterns that are constructed "piece-by-piece" from elements of some language.

For any itemset X, let $X \cup u$, where u is some item not in X, denote an *expansion* of X. Consider a tree T rooted at $\emptyset$, where each node is a frequent itemset. The children of a node X are all of its frequent expansions. More formally,

$$\text{children}(X) = \{X \cup u \mid u \in \{\mathcal{I} \setminus X\} \land \text{freq}(X \cup u) \geq \sigma\}.$$

That is, the root $\emptyset$ has all singleton items having a high enough frequency as children, these have all frequent pairs as children, and so on. The leaves of the tree correspond to *maximal* frequent itemsets, i.e., those that cannot be expanded without violating the frequency constraint. Note that there cannot be any itemset that is frequent but *not* in T, because of the monotonicity of the frequency constraint: if itemset X is not frequent, no $Y \supset X$ can be frequent either.

Now we could use Knuth's algorithm "as is" to estimate the size of T. However, observe that T contains multiple copies of the frequent itemsets. The tree T is in fact an "unfolded" *itemset lattice*. Indeed, every frequent itemset X of size $|X|$ is contained in T exactly $|X|!$ times. This is because every node of T is connected to the root $\emptyset$ by a path where the items in X are added one by one in some particular order, and this happens in as many ways as there are permutations of $|X|$ items.

To obtain a proper estimate of the number of frequent itemsets, we need to correct for this property of T. As noted earlier, Eq. 1 is in fact a sum over all levels of the tree. In the tree T, the level i (with root $\emptyset$ on level 0) contains $i!$ copies of the same itemset. We must thus replace Equation 1 with

$$1 + \sum_{i=1}^{l} \frac{1}{i!} \prod_{j=0}^{i-1} d(v_j), \tag{2}$$

where $\frac{1}{i!}$ corrects the sizes of the levels so that each itemset is counted only once.

Pseudocode of the full FASTEST algorithm is shown in Algorithm 1. In short, this is Knuth's algorithm applied on the frequent pattern lattice combined with the modified path estimate equation. In practice we do not materialise the tree T, but only sample paths through it using the SAMPLEPATH subroutine. On every step it finds the set E of extensions to the current pattern P that are still frequent (lines 2 and 8), and the size of E gives the outdegree of P (line 5).

---

[2] Or any other type of pattern for which the Apriori / monotonicity property w.r.t. pattern inclusion holds.

Algorithm 1. The FASTEST algorithm
___

1: Sample a number of paths using the SAMPLEPATH subroutine.
2: Use Equation 2 to compute the path-specific estimates.
3: **Return** the average of these as the final estimate.

1: SAMPLEPATH:
2: $P \leftarrow \emptyset, i \leftarrow 0$
3: $E \leftarrow \{x \in \mathcal{I} \mid \text{freq}(P \cup x) \geq \sigma\}$
4: **while** $|E| > 0$ **do**
5:     $i \leftarrow i + 1, d_i \leftarrow |E|$
6:     $e \leftarrow$ random element of E
7:     $P \leftarrow P \cup e, E \leftarrow E \setminus e$
8:     $E \leftarrow \{x \in E \mid \text{freq}(P \cup x) \geq \sigma\}$
9: **return** $(d_1, \ldots, d_i)$

The algorithm proceeds until it hits a maximal frequent itemset, after which it returns the sequence of encountered outdegrees.

The main bottleneck when running FASTEST are the support computations in SAMPLEPATH. These can be made very efficient by using a vertical representation of the database where we have a list of transaction identifiers for each item. As SAMPLEPATH proceeds deeper into the tree, we simply maintain a list of transaction identifiers for the current node P, and intersect this with the lists for other items when computing the support for each extension (line 8).

Remark: Algorithms for pattern mining often perform search by considering a depth-first tree of the patterns. Applying Knuth's method directly on this tree is a bad idea, however. This is because the DFS tree is by construction imbalanced, and therefore the random paths have very different lengths. Most paths will underestimate the tree size, while few paths blow up the estimate. This can, and will, to some extent also happen with the tree T we defined, but since an itemset can be reached along several paths, we expect T to be less imbalanced. Also, the DFS tree has a different structure depending on the order in which the items are considered; T is not dependent on any such ordering.

4.2   A Non-monotonic Constraint: Closed Patterns

We conclude the section by discussing a simple approach that extends our estimation algorithm for non-monotonic constraints. The example that we focus on are *closed* patterns, i.e., patterns that are frequent and cannot be extended without decreasing the support. More formally, a pattern X is closed iff $\text{freq}(X) > \text{freq}(X \cup u)$ for every possible u.

As described in [14], Knuth's algorithm can be modified to count only those nodes of the tree that satisfy a given property. The first idea is thus to use this approach for closed patterns, as closedness is simply a property of the pattern associated with a node. However, since closed patterns can be rare, this approach leads to poor estimates. The high variance of individual path estimates implies that a very large number of samples are needed to produce reasonable estimates. Instead, we propose a method somewhat related to [8]. This estimates the fractions of closed patterns on each level of the tree T, and corrects the final estimate with these.

In more detail, when sampling a path through T, we can collect statistics on the number of patterns we observe on every level. We maintain two vectors of counters that have as many elements as there are levels in T. The first, denoted q_p, counts the number of all patterns found. The second, denoted q_c, keeps track of the number of closed patterns we observe. Then, we estimate the fraction of closed patterns on level l by computing $q_c(l)/q_p(l)$. Given the output of Algorithm 1, we can also compute independent estimates for the sizes of every level of T. (Recall that Eq. 2 is just the sum of these.) By multiplying these with $q_c(l)/q_p(l)$, we obtain estimates of the numbers of closed patterns on every level. The path-estimate is simply the sum of these, and the final estimate is again the average over a number of random paths.

5 The Pattern Frequency Spectrum

The pattern frequency spectrum shows the number of frequent patterns in data $\mathcal{D}$ as a function of σ. More formally, we define the spectrum as

$$f(\sigma, \mathcal{D}) = |\{X \in \mathcal{L} \mid \text{freq}(X, \mathcal{D}) \geq \sigma\}|.$$

Below we write $f(\sigma)$ for short, unless $\mathcal{D}$ is not clear from the context.

5.1 The SPECTRA Algorithm

A simple method to estimate $f(\sigma)$ is to run the FASTEST algorithm for a number of fixed values of σ, and construct $f(\sigma)$ by interpolating from these point estimates. The problem with this approach is that determining the values of σ to use is not easy. By using a too coarse grid, we will miss some of the structure present in the frequency spectrum. On the other hand, using many values of σ may be very slow. Instead, we will use an approach where we obtain a number of estimates for *random values* of σ using the algorithm from the previous section, and then fit a nonlinear regression line through these.

In more detail, we propose the following algorithm called SPECTRA:

1. Compute the set of points $S = \{(\sigma_1, g(\sigma_1)), \ldots, (\sigma_N, g(\sigma_N))\}$, where every σ_i is drawn uniformly at random from some predefined interval, and $g(\sigma_i)$ is a single path estimate given by SAMPLEPATH (Alg. 1).
2. We fit a nonlinear, nonincreasing regression line through S and use this as our estimate of the pattern frequency spectrum.

This method has the advantage that we can simultaneously estimate $f(\sigma)$ for all values of σ, rather than interpolate from point estimates at predefined locations.

We know $f(\sigma)$ must be monotonically nonincreasing as σ increases. A suitable method for step 2 is thus *isotonic regression* [3]. The task is to find a strictly nonincreasing function that minimises the squared error to the input points in S. More formally, we find an estimate $\hat{f}$ by solving

$$\min \sum_{(\sigma, g(\sigma)) \in S} \left(\hat{f}(\sigma) - g(\sigma)\right)^2,$$
$$\text{s.t.} \quad \hat{f}(\sigma_i) \geq \hat{f}(\sigma_j) \ \forall \sigma_i \leq \sigma_j.$$

In this paper we use the `isoreg` function of GNU R that represents $\hat{f}$ by a *piecewise constant* function (a step function). However, any other algorithm for finding monotonically non-increasing functions subject to squared error can be applied just as well. We point out that solving the regression problem is in general orders of magnitude faster than obtaining the set S, and is thus not a critical component from a complexity point of view.

5.2   Frequency Curves in Random Data

We now study what frequency spectra look like in data that has no real structure. They can be used as a kind of 'null hypothesis' to compare real curves to: is there any structure in the dataset or not? For this we consider two types of random data: 1) constant background, i.e., each value occurs with fixed probability, 2) variable column marginals, i.e., each value in each column occurs with a given probability.

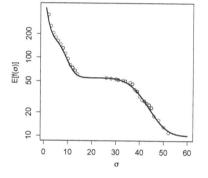

Uniformly Random Data with Constant Background. We derive an analytic expression for the expected frequency spectrum for data that is uniformly random. Turns out that even under this simple model $f(\sigma)$ has non-trivial structure.

Fig. 2. Expected frequency curve in uniformly random data; $n = 1000$, $m = 10$, and $p = 0.2$. The points show observed frequencies in one instance of random data

Let $\mathcal{D}$ denote a random binary dataset, with n rows and m attributes, where every item appears with probability p in every row. The expected value of $f(\sigma, \mathcal{D})$ can be written as

$$\mathbb{E}_{\mathcal{D}}[f(\sigma, \mathcal{D})] = \sum_{l=1}^{m} \binom{m}{l} \sum_{k=\sigma}^{n} \text{Binomial}(k; n, p^l). \tag{3}$$

The above equation follows from taking the expected value of $\sum_{X \subseteq \mathcal{I}} \mathrm{I}\{\text{freq}(X) \geq \sigma\}$, and observing that the probability of the indicator function is given by the tail of a Binomial distribution with parameters n and p^l.

Figure 2 shows an example of the expected frequency spectrum for uniformly random data with parameters $n = 1000$, $m = 100$, and $p = 0.2$. Perhaps somewhat surprisingly, the plot shows a clear "staircase" pattern, despite there not being any structure in the data. The points in Fig. 2 are exact frequencies computed from a single instance of random data that matches the parameters. We can see that the expected curve closely matches the observed points.

Uniformly Random Data with Variable Column Marginals. Next, we consider the case when the data are generated by a model where the items are still all independent, but every item i has its own occurrence probability p_i. In this case it is no longer straightforward to derive a closed form expression for $E[f(\sigma)]$. However, we can in fact use the SPECTRA algorithm to estimate the expected frequency spectrum under this model as well. We point out that in this case, SPECTRA becomes a heuristic that seems to give useful results, but it does not necessarily converge to the correct expected pattern spectrum.

The algorithm is exactly the same as for real data, but we replace the support computation in SAMPLEPATH with *expected supports* defined for itemset X as

$$E_{\mathcal{D}}[\text{freq}(X)] = \sum_{\mathcal{D}} \text{freq}(X \mid \mathcal{D}) \Pr[\mathcal{D}] = n \prod_{i \in X} p_i,$$

where the expectation is taken over all possible datasets. Under the considered model this becomes the expected value of a Binomial distribution. For a given σ, the algorithm again starts from the empty itemset, and proceeds to sample a path by determining whether the expected support of an extension is larger than σ. As described in Section 5.1, we run SAMPLEPATH for a number of randomly selected values of σ, and fit an isotonic regression line through the points obtained.

6   Experiments

In this section we empirically evaluate the FASTEST and SPECTRA algorithms.

Datasets: For the experiments, we selected 14 moderately-sized datasets for which exact frequent itemset counting is still possible for reasonable frequency thresholds. The datasets were taken from the FIMI[3] (*Accidents*, *Kosarak*, and *Pumsbstar*) and LUCS-KDD[4] dataset repositories. Table 2 lists the dataset dimensions: the number of transactions and the total number of items.

Evaluation criteria: Primary evaluation criteria are 1) accuracy of the estimated counts, and 2) runtime. For purposes of presentation, all counts are given in logarithmic scale, i.e., $\log_{10}$. An additional reason is that we are primarily interested in correctly estimating the order of magnitude; in practice the difference between mining and processing 1 million or 1.1 million itemsets is negligible.

Implementation: FASTEST was implemented in C++ and is publicly available[5]. The implementation of the MCMC-based method was kindly provided by the authors of [6]. For exact counting, also used by the MCMC-based method, we use the original AFOPT implementation [15] taken from the FIMI repository. All experiments were executed single-threaded on a regular desktop computer (Intel i5-2500 @ 3.3GHz, 8GB RAM).

---

[3] http://fimi.ua.ac.be/data/
[4] http://cgi.csc.liv.ac.uk/~frans/KDD/Software/
[5] http://patternsthatmatter.org/implementations/

Table 1. Itemset counts for a fixed frequency threshold σ: comparing exact counting, FASTEST (FE), and the MCMC-based method. Estimated counts are given with their 95% confidence intervals. The last column shows the fraction of frequency computations required by FASTEST when compared to exact counting.

Dataset	σ	Count [conf. interval] ($\log_{10}$)			Time (sec)			Frac.
		Exact	FE	MCMC	Exact	FE	MCMC	FE
Adult	1	7.8	7.7 [7.5,7.9]	7.8 [7.8,7.8]	3	18	27	0.0245
Anneal	1	6.6	6.6 [6.4,6.8]	6.8 [6.7,6.8]	1	1	5	0.2291
Hepatitis	1	7.8	7.8 [7.7,7.9]	7.9 [7.8,7.9]	2	1	2	0.0146
Letrecog	1	8.8	8.7 [8.6,9.0]	8.8 [8.8,8.9]	23	12	51	0.0029
Mushroom	1	9.7	9.7 [9.6,9.9]	10.0 [9.9,10.0]	91	8	67	0.0005
Pendigits	1	8.7	8.7 [8.6,8.8]	8.7 [8.6,8.7]	21	8	39	0.0030
Waveform	1	10.1	10.1 [9.9,10.3]	10.2 [10.1,10.3]	331	5	53	0.0002
Accidents	4000	9.5	9.2 [8.5,9.7]	-	496	475	-	0.0006
Chess	300	9.8	9.8 [9.5,10.1]	9.8 [9.7,9.9]	344	7	28	0.0002
Connect	7500	10.6	10.5 [10.2,10.8]	10.7 [10.5,10.7]	753	250	117	0.0000
Ionosphere	10	10.7	10.6 [10.5,10.8]	11.5 [11.4,11.5]	1231	1	146	0.0000
Kosarak	800	7.6	7.5 [6.7,7.9]	-	15	206	-	0.1938
Mammals	10	12.2	12.3 [11.8,12.7]	12.1 [12.0,12.2]	43841	3	45	0.0000
Pumsbstar	7500	10.7	10.7 [10.6,10.8]	8.5 [8.4,8.5]	1042	320	554	0.0000

6.1   Estimating the Number of Frequent Itemsets

We first evaluate how well the FASTEST algorithm performs when estimating the number of frequent itemsets for a fixed threshold σ. We compare our method to both exact counting and the existing MCMC-based approach for approximate counting. Table 1 shows the results obtained on all 14 datasets. The results are split into two groups, according to the used frequency threshold: $\sigma = 1$ was used for the upper seven datasets, higher thresholds were used for the lower seven datasets to make sure that exact counting finished in reasonable time. Note that lower thresholds are no problem for FASTEST, as we will see later.

The estimated counts are evaluated and compared using the following procedure. For FASTEST, we first obtain a large population of 10000 path samples. We then compute the expected estimate and 95% confidence interval for 1000 path samples by subsampling the large pool 100 times. For MCMC we use a similar approach: the procedure is executed 100 times, and expected estimates and confidence intervals are based on taking a single sample from that pool. The presented runtimes for FASTEST and MCMC match this procedure, that is, they are based on computing 1000 and 1 sample(s) respectively.

Looking at the upper half of the table first, we observe that the estimates by our method are spot on in expectation. There is clearly some variance between the estimates when using 1000 samples, but the 95% confidence intervals indicate that they are always in the same order of magnitude. The expected estimates obtained by MCMC are sometimes slightly off, but the variance is smaller. Except for *Adult* and *Anneal*, FASTEST is always the fastest of the three methods.

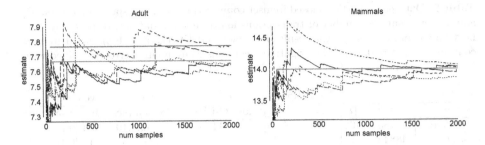

Fig. 3. Estimation stabilisation: the running estimate is updated after each new sample. Five independent runs of 2000 samples each, for *Adult* (left, $\sigma = 1$) and *Mammals* (right, $\sigma = 2$). Also indicated are the expected estimates (green) and the actual counts (red, only for *Adult*).

The results for the larger datasets in the lower half of the table show similar behaviour. The results for MCMC are missing for *Accidents* and *Kosarak* due to the implementation running out of memory. In general, the estimates by both FE and MCMC are accurate, although those for *Ionosphere* and *Pumbstar* by MCMC are clearly off. In terms of runtime, FASTEST is the fastest in five out of seven cases. The gain in runtime is particularly large for *Chess*, *Ionosphere*, and *Mammals*, with attained speed-ups of 25-1230x.

Finally, the rightmost column of Table 1 shows the ratio of the number of the frequency computations required by FE, relative to the number needed by exact counting with Apriori [1]. Computing the frequency of an itemset is the expensive part of the algorithm, and requiring fewer such computations results in lower runtimes. The generally very low ratios confirm that FASTEST needs very few frequency computations to obtain accurate estimates.

An important question we have not addressed so far is how many samples are needed to obtain accurate estimates. In the previous, we have shown that 1000 samples are generally enough, but it could be possible that fewer would be sufficient as well. To investigate this, consider the stabilisation plots in Figure 3. Although some small jumps in the estimates remain, they stabilise rather quickly. Taking into account the scales of the y-axes, we conclude that 1000 samples is more than enough to get at the correct order of magnitude. We observed similar behaviour for other datasets, and Table 1 also confirms this finding.

As a side note: computing the 10000 samples required for the *Mammals* plot, with $\sigma = 2$, took only 33 seconds. This demonstrates that FASTEST can easily deal with lower frequency thresholds. Unfortunately, we were not able to produce the exact count for this threshold; we killed the process after it ran for two days.

Closed Frequent Itemsets. We now present results obtained with our adaptation of FASTEST for estimating the number of closed frequent itemsets, here denoted by C-FE. The results, all for $\sigma = 0.05|\mathcal{D}|$, are shown in the 'Count' and 'Time' columns in Table 2. We do not include confidence intervals for reasons of space, but observed similar intervals as previously. The exact results for *Chess* are missing because the AFOPT miner crashed.

Table 2. Dataset properties, closed itemset count estimation, and spectrum errors. $|\mathcal{D}|$ and $|\mathcal{I}|$ represent the number of transactions and items in a dataset, respectively. The next four columns contain the results for exact and FASTEST (C-FE) closed itemset counting, with $\sigma = 0.05|\mathcal{D}|$. The rightmost columns contain curve errors obtained with SPECTRA and 1000 and 10000 samples (lower error is better).

Dataset			Count ($\log_{10}$)		Time (s)		SPECTRA curve error	
Name	$|\mathcal{D}|$	$|\mathcal{I}|$	Exact	C-FE	Exact	C-FE	1000 samples	10000 samples
Accidents	340183	468	7.81	7.49	27665	1168	0.68 [0.42,1.10]	0.34 [0.24,0.48]
Adult	48842	15	4.41	4.35	0	71	0.23 [0.16,0.35]	0.12 [0.10,0.16]
Anneal	898	71	3.55	3.91	0	2	0.24 [0.15,0.36]	0.14 [0.10,0.18]
Chess	3196	37	-	8.64	-	12	0.57 [0.38,0.81]	0.30 [0.23,0.39]
Connect	67557	43	7.45	10.09	3324	1301	0.54 [0.37,0.79]	0.29 [0.22,0.42]
Hepatitis	155	20	5.18	5.17	2	1	0.28 [0.20,0.40]	0.14 [0.11,0.18]
Ionosphere	351	35	7.45	8.42	40974	4	0.74 [0.50,1.13]	0.40 [0.30,0.53]
Kosarak	990002	41270	1.52	1.52	0	51	0.11 [0.07,0.15]	0.05 [0.04,0.07]
Letrecog	20000	17	4.48	4.54	1	14	0.35 [0.22,0.58]	0.15 [0.11,0.22]
Mammals	2183	121	7.61	7.71	5722	3	0.47 [0.34,0.65]	0.23 [0.18,0.28]
Mushroom	8124	23	4.11	4.53	0	17	0.39 [0.28,0.55]	0.20 [0.15,0.26]
Pendigits	10992	17	3.76	3.74	0	8	0.25 [0.18,0.37]	0.13 [0.11,0.15]
Pumsbstar	49046	2088	6.97	10.63	1609	805	0.44 [0.30,0.64]	0.22 [0.18,0.27]
Waveform	5000	22	5.59	5.53	3	5	0.42 [0.28,0.64]	0.22 [0.16,0.27]

The estimates are pretty accurate for most datasets, but there are some exceptions. Of these, *Connect* is the most obvious: the estimate is three orders of magnitude off. Investigating this in more detail, it turns out that this is due to the extreme ratio between the number of frequent and closed itemsets: with only 1000 samples, the estimated correction coefficient cannot be reliable if it is much lower than 0.1. If we increase the number of samples for *Connect* to 2000, for example, the estimate becomes 9.2 – already one order of magnitude better.

Closed frequent itemset mining is a much harder problem than frequent itemset mining, as is also reflected by the runtimes. The exact miner is faster in eight cases, but those are the relatively easy datasets. The exact counting runtimes explode for the more difficult datasets, whereas C-FE is relatively fast.

6.2   Estimating the Pattern Frequency Spectrum

We now turn our focus to the SPECTRA algorithm. That is, we estimate the number of frequent itemsets for the complete frequency ranges. For each dataset, we obtain a curve using both 1000 and 10000 samples, and compare the resulting curves to the exact curves as far as they are available, i.e., for the frequency range $[\sigma, |\mathcal{D}|]$ (for values of σ mentioned in Table 1).

We initially computed the mean error over the complete curve, but since the spectra are quite accurate these were hardly informative. We therefore turned to another measure, i.e., the curve errors presented in the rightmost columns of Table 2. These numbers are the 95% quantiles of the errors: 95% of the curve has an error that is at most as large as indicated. For all datasets, 1000 samples are

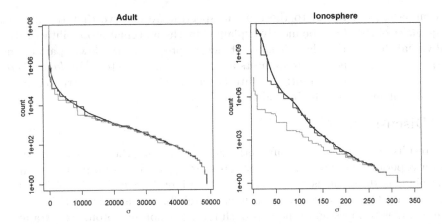

Fig. 4. Frequency curves obtained with SPECTRA for *Adult* and *Ionosphere*: estimated on actual data (blue) and expected in random data (green). For comparison, the exact curve (black) is also drawn.

enough to estimate at least 95% of the *full spectrum* with an error of less than one order of magnitude, and often much less. With 10000 samples, 95% of the curve is often only 0.1 - 0.2 away from the actual counts. Of course, the estimates are inevitably less accurate than when using the same number of samples for a single σ, but nonetheless SPECTRA succeeds in obtaining accurate estimates for the *whole frequency range* with a limited number of samples.

The runtimes of SPECTRA are not listed, but they are even shorter than when estimating the number of itemsets for a single, low threshold. The reason is that samples are drawn for arbitrary values of σ, and runtimes are shorter for higher values. When 1000 samples are used, computing a curve takes only seconds for the smaller datasets, and a few minutes for the larger ones. The longest runtime was measured for *Accidents*: 401 seconds.

Figure 4 shows example frequency spectra obtained for *Adult* and *Ionosphere*. The estimated curves (in blue) clearly match the actual curves (in black) very well, although there is a slight tendency towards underestimation for the lower frequency ranges. It is very easy to read out ballpark pattern counts and use these to tune the frequency threshold. For *Ionosphere*, for example, choosing $\sigma = 80$ would result in a modest 10^6 frequent itemsets, whereas anything lower will quickly get you vast amounts of patterns.

Comparing Against Random Frequency Spectra. The plots in Figure 4 also show expected spectra obtained with the procedure described in Section 5.2. That is, we consider the model where the items are independent but variable column marginals are given. The number of transactions and item probabilities of the random data are equal to those of the real dataset, and high-confidence random frequency spectra are obtained by obtaining 10 samples for each possible value of σ; this procedure is extremely fast and can be done in one or two seconds for any dataset considered in this paper.

Comparing the actual to the random spectra, we observe that the itemset frequencies of *Adult* can be mostly explained by the marginal probabilities of the individual items. Still, there is some structure present in the lower part of the frequency range, where more itemsets are found than expected. For *Ionosphere*, the picture is rather different: this dataset has clearly much more structure than can be explained from the individual item probabilities.

7 Discussion

The results demonstrate that our methods for estimating the number of frequent patterns perform very well. Still, there are many possibilities for future research. One obvious direction is to investigate the adaptation for closed itemsets in more detail. One task would be to make it reliable for any dataset, but more interesting is to investigate the approach for other non-monotonic constraints.

The estimated frequency spectra can be reliably used to choose a minimum support threshold for frequent itemset mining; the order of magnitude is always right. Furthermore, these spectra are potentially useful as 'fingerprints' of the data. For example, we witnessed rather different shapes and curves, and we can imagine that it might be possible to cluster datasets based on their frequency spectra. Also, more advanced models could be used for the generation of expected curves, to see whether the actual curves match those.

We only considered frequency spectra for complete datasets, but one could also consider subsets of the pattern language. For example, by only considering those patterns that are supersets of a given 'query' pattern. This would give query-based frequency spectra, which could inform us about the local structure for a given query. The sampling procedure would remain almost unchanged: given a query, each path is sampled starting from that query instead of the empty set.

Note that the FASTEST algorithm provides estimates for each individual depth in the search tree, i.e., for each itemset size. This implies that our method could be used to estimate, e.g., the number of n-grams in a document dataset.

Finally, our approach can be easily parallelised to make it run efficiently on very large datasets. When sampling a path, we must compute the frequency of a pattern several times. We can easily adapt the algorithm to make it an Apriori-style algorithm, which has the advantage of being database friendly. Then, the database scan can be efficiently implemented on, for example, relational databases or distributed platforms for large-scale data processing.

8 Conclusions

We introduced two methods for approximate counting of the number of frequent patterns in a given dataset. Based on Knuth's classical algorithm for estimating the size of a search tree, the FASTEST algorithm estimates the number of frequent itemsets for a given frequency threshold. The SPECTRA algorithm combines FASTEST with isotonic regression to estimate the complete pattern frequency spectrum for a given dataset.

The experiments showed that both methods are very accurate and efficient, when compared to exact counting and an existing MCMC-based estimator. The adaptation for closed itemsets gives good estimates in most cases. Finally, we showed how pattern frequency spectra provide interesting summary statistics that can be compared to expected curves generated for random data models.

Acknowledgements. Matthijs van Leeuwen is supported by a Post-doctoral Fellowship of the Research Foundation Flanders (FWO).

References

1. Agrawal, R., Srikant, R.: Fast algorithms for mining association rules in large databases. In: Proceedings of VLDB 1994, pp. 487–499 (1994)
2. Agrawal, R., Srikant, R.: Mining sequential patterns. In: Proceedings of ICDE 1995, pp. 3–14 (1995)
3. Barlow, R.E., Brunk, H.D.: The isotonic regression problem and its dual. Journal of the American Statistical Association 67(337), 140–147
4. De Bie, T.: Maximum entropy models and subjective interestingness: An application to tiles in binary databases. Data Min. Knowl. Discov. 23(3), 407–446 (2011)
5. Boley, M., Gärtner, T., Grosskreutz, H.: Formal concept sampling for counting and threshold-free local pattern mining. In: Proc. of SDM 2010, pp. 177–188 (2010)
6. Boley, M., Grosskreutz, H.: A randomized approach for approximating the number of frequent sets. In: Proceedings of ICDM 2008, pp. 43–52 (2008)
7. Bringmann, B., Nijssen, S., Tatti, N., Vreeken, J., Zimmermann, A.: Mining sets of patterns: Next generation pattern mining. In: Tutorial at ICDM 2011 (2011)
8. Chen, P.C.: Heuristic sampling: A method for predicting the performance of tree searching programs. SIAM Journal on Computing 21(2), 295–315 (1992)
9. Cheng, H., Yan, X., Han, J., Hsu, C.-W.: Discriminative frequent pattern analysis for effective classification. In: Proceedings of the ICDE, pp. 716–725 (2007)
10. Goethals, B., Moens, S., Vreeken, J.: MIME: A framework for interactive visual pattern mining. In: Proceedings of KDD 2011, pp. 757–760 (2011)
11. Gunopulos, D., Khardon, R., Mannila, H., Saluja, S., Toivonen, H., Sharm, R.S.: Discovering all most specific sentences. ACM Trans. Database Syst. 28(2), 140–174 (2003)
12. Jerrum, M., Sinclair, A.: The markov chain monte carlo method: An approach to approximate counting and integration. Approximation Algorithms for NP-hard Problems, 482–520 (1996)
13. Kilby, P., Slaney, J.K., Thiébaux, S., Walsh, T.: Estimating search tree size. In: Proceedings of AAAI 2006, pp. 1014–1019 (2006)
14. Knuth, D.E.: Estimating the efficiency of backtrack programs. Mathematics of Computation 29(129), 122–136 (1975)
15. Liu, G., Lu, H., Yu, J.X., Wang, W., Xiao, X.: AFOPT: An efficient implementation of pattern growth approach. In: Proc. of FIMI at ICDM 2003 (2003)
16. Purdom, P.W.: Tree size by partial backtracking. SIAM Journal on Computing 7(4), 481–491 (1978)
17. Vreeken, J., van Leeuwen, M., Siebes, A.: Krimp: mining itemsets that compress. Data Min. Knowl. Discov. 23(1), 169–214 (2011)
18. Yan, X., Han, J.: gSpan: Graph-based substructure pattern mining. In: Proceedings of ICDM 2002, pp. 721–724 (2002)
19. Yang, G.: The complexity of mining maximal frequent itemsets and maximal frequent patterns. In: Proceedings of KDD 2004, pp. 344–353 (2004)

Recurrent Greedy Parsing with Neural Networks

Joël Legrand[1,2] and Ronan Collobert[1]

[1] Idiap Research Institute,
Rue Marconi 19, Martigny, Switzerland
[2] Ecole Polytechnique Fédérale de Lausanne (EPFL)
Lausanne, Switzerland
`joel.legrand@idiap.ch`,
`ronan@collobert.com`

Abstract. In this paper, we propose a bottom-up greedy and purely discriminative syntactic parsing approach that relies only on a few simple features. The core of the architecture is a simple neural network architecture, trained with an objective function similar to that of a Conditional Random Field. This parser leverages continuous word vector representations to model the conditional distributions of context-aware syntactic rules. The learned distribution rules are naturally smoothed, thanks to the continuous nature of the input features and the model. Generalization accuracy compares favorably to existing generative or discriminative (non-reranking) parsers (despite the greedy nature of our approach), while the prediction speed is very fast.

Keywords: Syntactic Parsing, Natural Language Processing, Neural Networks.

1 Introduction

While discriminative methods are at the core of most state-of-the-art approaches in Natural Language Processing (NLP), historically the task of syntactic parsing has been mainly solved with generative approaches. A major contribution in the parsing field is certainly probabilistic context-free grammar (PCFGs)-based parsers [1,2,3]. These types of parsers model the syntactic grammar by computing statistics of simple grammar rules occurring in a training corpus. Inference is then achieved with a simple bottom-up chart parser. These methods face a classical learning dilemma: on one hand PCFG rules have to be refined enough to avoid any ambiguities in the prediction. On the other hand, too much refinement in these rules implies lower occurrences in the training set, and thus a possible generalization issue. PCFGs-based parsers are thus judiciously composing with carefully chosen PCFG rules and clever regularization tricks.

Given the success of discriminative methods for various NLP tasks, similar methods have been attempted for the syntactic parsing task. One of the first successful discriminative parsers [4] was based on MaxEnt classifiers (trained over a large number of different features) and a greedy shift-reduce strategy. However, it did not perform on par with the best generative parsers of the time. Costa *et al.* [5] introduced a left-to-right incremental parser which used a recursive neural network to re-rank possible phrase attachments. They showed that RNN was able to capture enough information to make

T. Calders et al. (Eds.): ECML PKDD 2014, Part II, LNCS 8725, pp. 130–144, 2014.
© Springer-Verlag Berlin Heidelberg 2014

correct parsing decisions. Their system was, however, only tested on a subset of 2000 sentences. One had to wait a few more years before discriminative parsers could match Collins' parser performance. To this extent , Taskar *et al.* [6] proposed an approach which discriminates the entire space of parse trees, with a max margin criterion applied to Context Free Grammars. Other discriminative approaches [7,8] also outperformed standard PCFG-based generative parsers, but only by discriminatively re-ranking the K-best predicted trees coming from a generative parser.

Turian and Melamed [9] later proposed a bottom-up greedy algorithm to construct the parse tree, using a feature boosting approach. The parsing is performed following a left-to-right or a right-to left strategy. The greedy decisions regarding the tree construction are made using decision tree classifiers. However, both of these parsers were limited to sentences of less than 15 words, due to a training time growing exponentially with the size of the input.

McClosky *et al.* [10] successfully leveraged unlabeled data to train a parser using a self-training technique. In this approach, a re-ranker is trained over a generative model. The re-ranker is used to generate "labels" over a large unlabeled corpus. These "labels" are then used to retrain the original generative model. This work is currently considered the state-of-the-art in syntactic parsing.

Most recent discriminative parsers [11,12] rely on Conditional Random Fields (CRFs) with PCFG-like features. Carreras *et al.* [13] used a global-linear model (instead of a CRF) with PCFGs and various new advanced features.

While PCFG-based parsers are widely used, other approaches do exist. In [14], the proposed parser relies on continuous word vector representations, and a discriminative model to predict "levels" of the syntactic tree. Socher *et al.* [15] also relies on continuous word vector representations, which are "compressed" in a pairwise manner to form higher level chunk representations. Their approach is used as a re-ranker of the Stanford Parser [16].

Finally, it is worth noting that generative parsers are still evolving. PCFGs with latent-variables [17] have been used in various ways to improve the performance of classical PCFG as in [18].

In this paper, we propose a greedy and purely discriminative parsing approach. In contrast with most existing methods, it relies on a few simple features. The core of our architecture is a simple neural network which is fed with continuous word vector representations (as in [19,15]). It models the conditional distributions of *context-aware* syntactic rules. The learned distribution rules are naturally smoothed, thanks to the continuous nature of the input features.

Section 2 introduces our algorithm and relates it to PCFG-based parsers. Section 3 describes the classification model at the core of our architecture. Section 4 reports experimental comparisons with existing approaches. We conclude in Section 5.

2 A Greedy Discriminative Parser

2.1 Smoothed Context Rule Learning

PCFG-based parsers rely on the statistical modeling of rules of the form $A \to B, C$, where A, B and C are tree nodes. The context-free grammar is always normalized in the Chomsky Normal Form (CNF) to make the global tree inference practical (with a dynamic programming like CYK or similar). In general a tree node is represented as several features, including for example its own parsing tags and head word (for non-terminal nodes) or word and Part Of Speech (POS) tag (for terminal nodes) [2]. State-of-the-art parsers rely on a judicious blending of carefully chosen features and regularization: adding features in PCFG rules might resolve some ambiguities, but at the cost of sparser occurrences of those rules. In that respect, the learned distributions must be carefully smoothed so that the model can generalize on unseen data. Some parsers also leverage other types of features (such as bigram or trigram dependencies between words [13]) to capture additional regularities in the data.

In contrast, our system models non-CNF rules of the form $A \rightarrow B_1, ..., B_K$. The score of each rule is determined by looking at a large context of tree nodes. More formally, we learn a classifier of the form:

$$f(C_{left}, B_1, ..., B_K, C_{right}) = (s_1, ..., s_{|\mathcal{T}|}) \tag{1}$$

where the B_k are either terminal or non-terminal nodes, K is the size of the right part of the rule, C_{left} and C_{right} are context terminals or non-terminals and s_t is the score for the parsing tag $t \in \mathcal{T}$. Each possible rule $A_i \rightarrow B_1, ..., B_K$ is thus assigned a score s_i by the classifier (with $A_i \in \mathcal{T}$). These scores can be interpreted as probabilities by performing a softmax operation. We used a Multi Layer Perceptron (MLP) as classifier. Formal details will be given in Section 3.2.

The only tree node features considered in our system are parsing tags (or POS tags for terminals), as well as the headword (or words for terminals). We overcome the problem of data sparsity which occurs in most classical parsers by leveraging *continuous vector representations* for all features associated to each tree node. In particular, word (or headword) representations are derived from recent distributed representations computed on large unlabeled corpora (such as [19,20]). Thanks to this approach, our system can naturally generalize a rule like $NP \rightarrow a, clever, guy$ to a possibly unseen rule like $NP \rightarrow a, smart, guy$, as the vector representation of $smart$ and $clever$ are close to each other, given that they are semantically and syntactically related.

Several works leveraging continuous vector representations have been previously proposed for syntactic parsing. [14] introduced a neural network-based approach, iteratively tagging "levels" of the parse tree where the full sentence was seen at each level. A complex pooling approach was introduced to capture long-range dependencies, and performance only matched early lexicalized parsers. [21] introduced a recursive approach, where representations are "compressed" two by two to form higher-level representations. However, the system was limited to bracketing, and did not produce parsing tags. The authors later proposed an improved version in [15], where their approach was used to re-rank the output of the Stanford Parser, approximately reaching state-of-the-art performance. In contrast, our approach does not rely on CNF grammars and does not re-rank an external generative parser.

2.2 Greedy Recurrent Algorithm

Our parser follows a bottom-up iterative approach: the tree is constructed starting from the terminal nodes (sentence words). Assuming that a part of the tree has been already

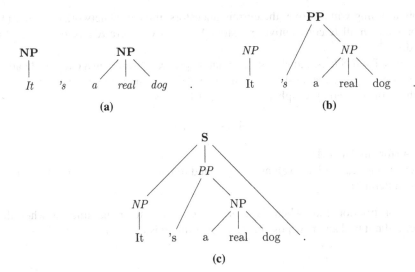

Fig. 1. Illustration of our greedy algorithm: at each iteration (a)→(b)→(c), the classifier sees only the previous tree heads (ancestors), shown here in italics. It predicts new nodes (here in bold). New tree heads become the ancestors at the next iteration. All other previously discovered tree nodes (shown in regular black here) will remain unchanged and ignored in subsequent iterations.

predicted (see Figure 1), the next iteration of our algorithm looks for all possible new tree nodes which combine ancestors (i.e., heads of the trees predicted so far). New nodes are found by maximizing the score of our context-rule classifier (1), constrained in such a way so that two new nodes cannot overlap, thanks to a dynamic programming approach. The system is recurrent, in the sense that new predicted parsing labels are used in the next iteration of our algorithm.

For each iteration, assuming N ancestors

$$X = [X_1, ..., X_N],$$

finding all possible new nodes with K ancestors would require to apply

$$f(C_{left}, B_1, ..., B_K, C_{right})$$

over all possible windows of K ancestors in X. One would also have to vary K from 1 to N, to discover new nodes of all possible sizes. Obviously, this could quickly become time consuming for large sentence sizes. This problem of finding nodes with a various number of ancestors can be viewed as the classical NLP problem of finding "chunks" of various sizes. This problem is typically transformed into a tagging task: finding the chunk with label A in the rule $A \rightarrow X_i, X_{i+1}, ..., X_j$ can simply be viewed as tagging the ancestors with $B\text{-}A, I\text{-}A, ... E\text{-}A$, where we use the standard BIOES label prefixing (Begin, Intermediate, Other, End, Single). See Table 1 for a concrete example. The classifier outputs the "Other" tag, when the considered ancestors do not correspond to any possible rule.

In the end, our approach can be summarized as the following iterative algorithm:

1. Apply a sliding window over the current ancestors: the neural network classifier (1) is applied over all K consecutive ancestors $X_1, ..., X_N$, where K has to be carefully tuned.
2. Aggregate BIOES tags into chunks: a dynamic program (based on a CRF, as detailed in Section 3.3) finds the most likely sequence of BIOES parsing tags. The new nodes are then constructed by simply aggregating BIES tags

$$B\text{-}A, I\text{-}A, \dots E\text{-}A$$

into A (for any label A).
3. Ancestors tagged as O, as well as newly found tree nodes are passed as ancestors to the next iteration.

The tree construction ends when there is only one ancestor remaining, or when the classifier did not find any new possible rule (everything is tagged as O).

Table 1. A simple example of a grammar rule extracted from the sentence "*It's a real dog .*", and its corresponding BIOES grammar. In both cases, we include a left and right context of size 1. The middle column shows the required classifier evaluations. The right column shows the type of scores produced by the classifier.

GRAMMAR	CLASSIFIER EVALUATIONS	SCORES
NP → 'S A REAL DOG .	$f($'S, A, REAL, DOG, .$)$	$s_{NP}, ..., s_{VP}, s_O$
B-NP → 'S A REAL	$f($'S, A, REAL$)$	
I-NP → A REAL DOG	$f($ A, REAL, DOG$)$	$s_{B\text{-}NP}, ..., s_{E\text{-}VP}, s_O$
E-NP → REAL DOG .	$f($ REAL, DOG, .$)$	

3 Architecture

In this section, we formally introduce the classification architecture used to find new tree nodes at each iteration of our greedy recurrent approach. A simple two-layer neural network is at the core of the system. It leverages continuous vector word representations. In this respect, the network is clearly inspired by the work of [22] in the context of language modeling, and later re-introduced in [23] for various NLP tagging tasks.

Given an input sequence of N tree node ancestors $X_1, ..., X_N$ (as defined in Section 2.2), our model outputs a BIOES-prefixed parsing tag for each ancestor X_i, by applying a sliding window approach. These scores are then fed as input to a properly constrained graph on which we apply the Viterbi algorithm to infer the best sequence of parsing tags. The whole architecture (including transition scores in the graph) is trained in an end-to-end manner by maximizing the graph likelihood. The end-to-end neural network training approach was first introduced in [24]. The system can be also viewed as a particular Graph Transformer Network [25], or a particular *non-linear* Conditional Random Field (CRF) for sequences [26]. Each layer of the architecture is presented in detail in the following paragraphs. The objective function will be introduced in Section 3.4.

3.1   Words Embeddings

Our system relies on *raw words*, following the idea of [19]. Each word is mapped into a continuous vector space. For efficiency, words are fed into our architecture as indices taken from a finite dictionary $\mathcal{W}$. Word vector representations, as other network parameters, are trained by back-propagation.

More formally, given a sentence of N words, $w_1, w_2, ..., w_N$, each word $w_n \in \mathcal{W}$ is first embedded in a D-dimensional vector space by applying a lookup-table operation:

$$LT_W(w_n) = W_{w_n} \, ,$$

where the matrix $W \in \mathbb{R}^{D \times |\mathcal{W}|}$ represents the parameters to be trained in this lookup layer. Each column $W_n \in \mathbb{R}^D$ corresponds to the vector embedding of the nth word in our dictionary $\mathcal{W}$.

These types of architectures allow us to take advantage of word vector representations trained on large unlabeled corpora, by simply initializing the word lookup table with a pre-trained representation [19]. In this paper, we chose to use the representations from [27], obtained by a simple PCA on a matrix of word co-occurrences. As shown in [14] for various NLP tasks, we will see that these representations can provide a great boost in parsing performance.

In practice, it is common to give several features (for each tree node) as input to the network. This can be easily done by adding a different lookup table for each discrete feature. The input becomes the concatenation of the outputs of all these lookup-tables:

$$LT_{W_1,...,W_K}(w_n) = (LT_{W_1}(w_n))^T,$$
$$...,$$
$$(LT_{W_{|\mathcal{F}|}}(w_n))^T$$

where $|\mathcal{F}|$ is the number of features. For simplicity, we consider only one lookup-table in the rest of the architecture description.

3.2   Sliding Window BIOES Tagger

The second module of our architecture is a simple neural network which applies a sliding window over the output of the lookup tables, as shown in Figure 2. The nth window is defined as

$$u_n = [LT(X_{n-\frac{K-1}{2}}), ..., LT(X_n), ..., LT(X_{n+\frac{K-1}{2}})] \, ,$$

where K is the size of window. The module outputs a vector of scores $s(u_n) = [s_1, ..., s_{|\mathcal{T}|}]$ (where s_t is the score of the BIOES-prefixed parsing tag $t \in \mathcal{T}$ for the ancestor X_n). The ancestors with indices exceeding the input boundaries ($n - (K-1)/2 < 1$ or $n + (K-1)/2 > N$) are mapped to a special padding vector (which is also learned). As any classical two-layer neural network, our architecture performs several matrix-vector operations on its inputs, interleaved with some non-linear transfer function $h(\cdot)$,

$$s(u_n) = M_2 \, h(M_1 \, u_n) \, ,$$

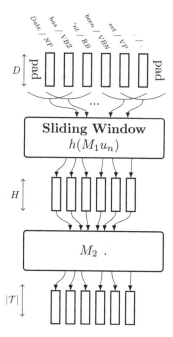

Fig. 2. Sliding window tagger. Given the concatenated output of lookup tables (here the ancestor words/headwords and ancestor tags), the tagger outputs a BIOES-prefixed parsing tag for each ancestor node. The neural network itself is a standard two-layer neural network.

where the matrices $M_1 \in \mathbb{R}^{H \times K|D|}$ and $M_1 \in \mathbb{R}^{|\mathcal{T}| \times H}$ are the trained parameters of the network. The number of hidden units H is a hyper-parameter to be tuned.

As transfer function, we chose in our experiments a (fast) "hard" version of the hyperbolic tangent:

$$
h(x) = \begin{cases} -1 \text{ if } & x < -1 \\ x \text{ if } & -1 <= x <= 1 \\ 1 \text{ if } & x > 1 \end{cases} \tag{2}
$$

3.3   Aggregating BIOES Predictions

The scores obtained from the previous module of our architecture are in BIOES format. The next module in our system aggregates these tags and finds the new tree nodes at each iteration of our greedy recurrent approach. We introduce a graph G of scores as shown in Figure 3: each node of the graph corresponds to a BIOES score produced for each ancestor by the neural network module. This graph is constrained in such a way that only feasible sequences of tags are possible (e.g. B-A tags can only be followed by I-A tags, for any parsing label A). Our graph also includes a duration model: on each edge, we add a transition score $A_{tt'}$ for jumping from tag $t \in \mathcal{T}$ to $t' \in \mathcal{T}$.

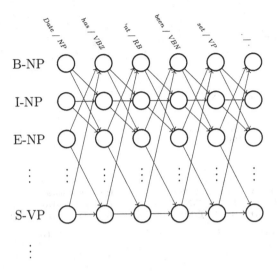

Fig. 3. Constrained graph for tag inference. Only feasible sequences of tags are considered. The nodes of the graph are assigned a score from the tagger shown in Figure 2. Edges of the graph are assigned a transition score which is learned similarly to other parameters in the architecture.

A score for a sequence of tags $[t]_1^N$ in the lattice G is obtained as the sum of scores along $[t]_1^N$ in G:

$$S([t]_1^N, [u]_1^N, \theta) = \sum_{n=1}^{N} \left(A_{t_{n-1}t_n} + s(u_n)_{t_n} \right),$$

where θ represents all the trainable parameters of the complete architecture. The sequence of tags $[t^*]_1^N$ for the input sequence of tree node ancestors $X_1, \ldots, X_N$ is then inferred by finding the path which leads to the maximal score:

$$[t^*]_1^N = \underset{[t]_1^N \in \mathcal{T}^N}{\operatorname{argmax}} \, S([t]_1^N, [u]_1^N, \theta)$$

The Viterbi algorithm is the natural choice for this inference. From this optimal BIOES tag sequence, we extract sub-sequences $B\text{-}A, \ldots, E\text{-}A$ and $S\text{-}A$ as new nodes for the tree. O tags are simply ignored. See Section 2.2 for more details.

3.4   Training Likelihood

Our architecture sees sequences of ancestor tree nodes, and outputs new possible syntactic tree nodes only from this history. Technically speaking, the training set can be prepared by iterating over each tree in the training corpus, removing all possible leaves in an iterative process so that all training rules are uncovered (see Figure 4).

The neural network is trained by maximizing a likelihood over the training data, using stochastic gradient ascent. The score for a path can be interpreted as a conditional

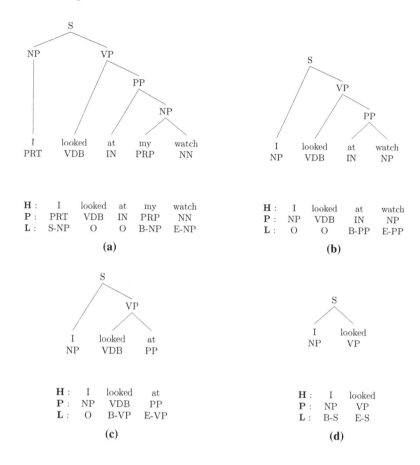

Fig. 4. Iterative procedure (a)→(b)→(c)→(d) to generate the training data, which involves cutting out all tree leaves at each step. The data fed to our network architecture is then easily uncovered (H: ancestor headwords/words, P: ancestor POS/parsing tags, L: parsing labels to be predicted).

probability over this path by exponentiating score (thus making it positive) and normalizing it with respect to all possible paths. We define $\mathcal{P}$ as the set of possible tag paths in the constrained graph G, as shown in Figure 3. The log-probability of a sequence of tags $[t]_1^N$ given the lookup table representations $[u]_1^N$ is given by:

$$\log P([t]_1^N | [u]_1^N, \theta) = S([t]_1^N, [u]_1^N, \theta) \tag{3}$$
$$- \operatorname*{logadd}_{\forall [t']_1^N \in \mathcal{P}} S([t']_1^N, [u]_1^N, \theta))$$

where we adopt the notation $\operatorname{logadd}_{z_n} = \log\left(\sum_i e^{z_i}\right)$, as in [23].

Computing the log-likelihood efficiently is not straightforward, as the number of terms in the logadd grows exponentially with the length of the sentence. Fortunately, it can be computed in linear time with the Forward algorithm, which derives a recursion similar to the Viterbi algorithm (see [28]). The complete architecture is trained by simply backpropagating through this recursion, up to the lookup layers (for further details, see [14]). Note that the likelihood (3) corresponds to a standard CRF model for sequences. The only difference here is that the underlying model is non-linear, while CRFs are often considered as linear models.

4 Experiments

4.1 Corpus

Experiments were performed using the standard English Penn Treebank data set (Marcus et al., 1993). We used the classical parsing setup, with sections 02-21 used to train our model, section 22 used as validation for choosing all our hyper-parameters, and section 23 used for testing. We applied only a small subset of the typical pre-processing set over the data: (1) functional labels, traces were removed, (2) the PRT label was replaced as ADVP [1].

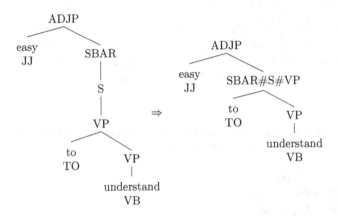

Fig. 5. Training corpus pre-processing. Original Penn Treebank trees containing non-terminal nodes with only one non-terminal node (left), and after concatenating those nodes (right).

The Penn Treebank data set contains non-terminal tree nodes which only have one non-terminal child, as shown in Figure 5. To avoid possible looping issues in our parsing algorithm (e.g. a node being repetitively tagged with two different tags in our iterative process), we transformed the *training* corpus so that non-terminal nodes having only one non-terminal child were merged together, and take as tag the concatenation of all merged node tags (see Figure 5). This way, the system learns that a node must contain at least two ancestors. The iterative process is thus guaranteed to converge. We kept only concatenated labels which occurred at least 30 times (corresponding to the lowest number of occurrences of the less common original parsing tag), leading to 11 additional parsing tags. Added to the original 26 parsing tags, this resulted in 161 tags produced

Table 2. Influence of different features. Results are given in terms of F1-score. POS = part-of-speech, hw = head-word, wi = word initialization from [27].

FEATURE	F1
WORD + POS	85.1
WORD + POS + HW	86.9
WORD + POS + WI	86.2
WORD + POS + HW + WI	88.3

Table 3. Results in terms of Precision (P), Recall (R), and F1 score. The reported time is the time to parse the full WSJ test corpus.

	MODEL	(R)	(P)	F1	(R)	(P)	F1	TIME
GENERATIVE	MAGERMAN (1995)	84.6	84.9	84.8				
	COLLINS (1999)	88.5	88.7	88.6	88.1	88.3	88.2	1247
	CHARNIAK (2000)	90.1	90.1	90.1	89.6	89.5	89.6	
GENERATIVE WITH RE-RANKING	HENDERSON (2004)				89.8	90.4	90.1	
	CHARNIAK AND JOHNSON (2005)			92.0			91.1	
	SOCHER ET AL (2013)			91.1			92.1	390
	MCCLOSKY ET AL (2006)						92.1	
PURELY DISCRIMINATIVE	PETROV AND KLEIN (2008)			90.0			89.4	
	CARRERAS ET AL. (2008)				90.7	91.4	91.1	
	OUR MODEL	88.4	89.0	88.7	88.0	88.6	88.3	110
	OUR MODEL (VOTING)	90.0	90.1	90.1	89.6	89.7	89.6	

by our parser. At test time, the inverse operation is performed: concatenated tag nodes are simply expanded into their original form.

$$WHADVP \rightarrow When$$
$$NP \rightarrow the\ little\ guy$$
$$ADJP \rightarrow frightened$$
$$NP \rightarrow the\ big\ guys$$
$$ADVP \rightarrow badly$$

4.2 Features

We consider the following features to train our architecture:

- Words and headwords:
  - For terminal nodes, the word itself, in low caps[1]. As in [2], words occurring 5 times or less were mapped to an "UNKNOWN" word.
  - For non-terminal nodes: headwords, following the procedure described in [2].

---

[1] Adding a capital feature had no impact on the performance of our parser. Note that POS tags were generated with the original caps in the sentence.

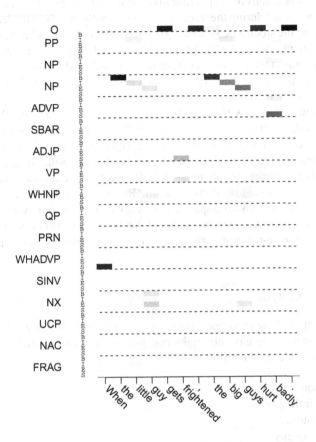

Fig. 6. Normalized scores from the network classifier (black means high score) for the sentence "When the little guy gets frightened, the big guys hurt badly.". Each tag is in BIOES form (y axis). Each ancestor in the input is on the x axis.

- POS tags (for terminals) or parsing tags of the node's ancestors (for non-terminals). POS tags were produced with SENNA [23].
- POS tags of headwords.

4.3   Results

We train the network using stochastic gradient descent over the available training data, until convergence on the validation set. We chose the following hyper-parameters according to the validation. Lookup-table sizes for the words and tags (part-of-speech and parsing) are 100 and 20, respectively. The window size for the tagger is $K = 7$ (3 neighbors from each side). The size of the tagger's hidden layer is $H = 500$. We used the word embeddings obtained from [27] to initialize the word lookup-table. These embeddings were then fine-tuned during the training process. Finally, we fixed the learning rate to $\lambda = 0.025$ during the stochastic gradient procedure. The only "trick" used during training was to divide the learning rate by the input size of each linear layer [29].

Table 2 shows the importance of the different features we used. Even though the training procedure is non-convex, the variance of the F1 score over 20 different runs (for the architecture Word + POS + hw + wi) was only 0.01.

Since our architecture performs the decoding very quickly, we additionally performed a voting procedure using several models learned from different random initializations. We averaged all neural network classifiers (ignoring their own respective CRF decoding part) and trained a new CRF on top of it (without fine-tuning any of the neural network classifiers). The scores obtained with 10 classifiers are shown in Table 3.

Results in Table 3 are reported in terms of recall (R), precision (P) and F1 score. Scores were obtained using the Evalb implementation[2]. We compare our system with several other parsers. We chose to report the scores of the three main generative parsers, as well as those of known re-ranking parsers. We also considered two major purely discriminative parsers.

4.4 Rule Prediction Analysis

Figure 6 shows the output of the classifier (applied on every possible window of size 7) for the sentence "When the little guy gets frightened, the big guys hurt badly.". For this sentence, the expected rules are the following:

$$
\begin{aligned}
\text{WHADVP} &\rightarrow \text{When} \\
\text{NP} &\rightarrow \text{the little guy} \\
\text{ADJP} &\rightarrow \text{frightened} \\
\text{NP} &\rightarrow \text{the big guys} \\
\text{ADVP} &\rightarrow \text{badly}
\end{aligned}
$$

It is interesting to see that the network alone is able to predict all the rules of the sentence. The CRF is however essential to produce a consistent output, by aggregating BIES prefixed chunks.

5   Conclusion

We presented a very simple model that is able to learn syntactic grammar rules surprisingly well, considering the simple features employed. This parser achieves performance very close to state-of-the-art re-ranking systems and is almost the best among the purely generative parsers. Due to its simplicity, there are many possibilities for further improvement. In particular, the head-word procedure from Collins could be revisited, e.g. by learning a higher-level chunk representation in the same spirit as [15]. We could also investigate re-ranking approaches, as well as the use of unlabeled corpora.

---

[2] Available at http://nlp.cs.nyu.edu/evalb/

Acknowledgments. This work was supported by NEC Laboratories America. We would like to thank Leonidas Lefakis and Pedro Oliveira Pinheiro for proofreading this paper and Dimitri Palaz for his contribution on figures 2 and 3.

References

1. Magerman, D.M.: Statistical decision-tree models for parsing. In: Proceedings of the 33rd Annual Meeting of the Association for Computational Linguistics (1995)
2. Collins, M.: Head-driven statistical models for natural language parsing. Comput. Linguist. (2003)
3. Charniak, E.: A maximum-entropy-inspired parser. In: Proceedings of the 1st North American Chapter of the Association for Computational Linguistics Conference (2000)
4. Ratnaparkhi, A.: Learning to parse natural language with maximum entropy models. Mach. Learn. (February 1999)
5. Costa, F., Frasconi, P., Lombardo, V., Soda, G.: Towards incremental parsing of natural language using recursive neural networks (2002)
6. Taskar, B., Klein, D., Collins, M., Koller, D., Manning, C.D.: Max-margin parsing. In: Proceedings of EMNLP (2004)
7. Henderson, J.: Discriminative training of a neural network statistical parser. In: Proceedings of the 42nd Annual Meeting on Association for Computational Linguistics (2004)
8. Charniak, E., Johnson, M.: Coarse-to-fine N-best parsing and MaxEnt discriminative reranking. In: Proceedings of the 43rd Annual Meeting on Association for Computational Linguistics (2005)
9. Turian, J., Melamed, I.D.: Advances in discriminative parsing. In: Proceedings of the Joint International Conference on Computational Linguistics and Association of Computational Linguistics, COLING/ACL (2006)
10. McClosky, D., Charniak, E., Johnson, M.: Effective self-training for parsing. In: Proceedings of the Main Conference on Human Language Technology Conference of the North American Chapter of the Association of Computational Linguistics (2006)
11. Finkel, J.R., Kleeman, A., Manning, C.D.: Efficient, feature-based, conditional random field parsing. In: Proc. ACL/HLT (2008)
12. Petrov, S., Klein, D.: Sparse multi-scale grammars for discriminative latent variable parsing. In: Proceedings of the Conference on Empirical Methods in Natural Language Processing (2008)
13. Carreras, X., Collins, M., Koo, T.: TAG, dynamic programming, and the perceptron for efficient, feature-rich parsing. In: Proceedings of the Twelfth Conference on Computational Natural Language Learning (2008)
14. Collobert, R.: Deep learning for efficient discriminative parsing. In: AISTATS (2011)
15. Socher, R., Bauer, J., Manning, C., Ng, A.: Parsing With Compositional Vector Grammars. In: ACL (2013)
16. Klein, D., Manning, C.: Accurate unlexicalized parsing. In: Proceedings of the 41st Annual Meeting on Association for Computational Linguistics, vol. 1 (2003)
17. Matsuzaki, T., Miyao, Y., Tsujii, J.: Probabilistic CFG with latent annotations. In: Proceedings of the 43rd Annual Meeting on Association for Computational Linguistics (2005)
18. Cohen, S.B., Collins, M.: Tensor decomposition for fast parsing with latent-variable PCFGs. In: Proceedings of NIPS (2012)
19. Collobert, R., Weston, J.: A unified architecture for natural language processing: Deep neural networks with multitask learning. In: International Conference on Machine Learning, ICML (2008)

20. Dhillon, P.S., Foster, D., Ungar, L.: Multi-view learning of word embeddings via cca. In: Advances in Neural Information Processing Systems, NIPS (2011)
21. Socher, R., Lin, C., Ng, A.Y., Manning, C.D.: Parsing natural scenes and natural language with recursive neural networks. In: ICML (2011)
22. Bengio, Y., Ducharme, R.: A neural probabilistic language model. In: NIPS 13 (2001)
23. Collobert, R., Weston, J., Bottou, L., Karlen, M., Kavukcuoglu, K., Kuksa, P.: Natural language processing (almost) from scratch. Journal of Machine Learning Research (2011)
24. Denker, J.S., Burges, C.J.C.: Image segmentation and recognition. In: The Mathematics of Induction (1995)
25. Bottou, L., LeCun, Y., Bengio, Y.: Global training of document processing systems using graph transformer networks. In: Proc. of Computer Vision and Pattern Recognition (1997)
26. Lafferty, J., McCallum, A., Pereira, F.: Conditional random fields: Probabilistic models for segmenting and labeling sequence data. In: Eighteenth International Conference on Machine Learning, ICML (2001)
27. Lebret, R., Collobert, R.: Word embeddings through hellinger PCA. In: Proceedings of the 14th Conference of the European Chapter of the Association for Computational Linguistics (2014)
28. Rabiner, L.R.: A tutorial on hidden markov models and selected applications in speech recognition. Proceedings of the IEEE, 257–286 (1989)
29. Plaut, D.C., Hinton, G.E.: Learning sets of filters using back-propagation. Computer Speech and Language (1987)

FILTA: Better View Discovery from Collections of Clusterings via Filtering

Yang Lei, Nguyen Xuan Vinh, Jeffrey Chan, and James Bailey

Department of Computing and Information Systems
University of Melbourne, Australia
yalei@student.unimelb.edu.au
{vinh.nguyen,jeffrey.chan,baileyj}@unimelb.edu.au

Abstract. Meta-clustering is a popular approach to find multiple clusterings in the datasest, which takes a large number of base clusterings as input for further user navigation and refinement. However, the effectiveness of meta-clustering is highly dependent on the distribution of the base clusterings and open challenges exist with regard to its stability and noise tolerance. In this paper we propose a simple and effective filtering algorithm (FILTA) that can be flexibly used in conjunction with any meta-clustering method. Given a (raw) set of base clusterings, FILTA employs information theoretic criteria to remove those having poor quality or high redundancy. Then this filtered set of clusterings is highly suitable for further exploration, particularly the use of visualization for determining the dominant views in the dataset. We evaluate FILTA on both synthetic and real world datasets, and see how its use can enhance view discovery for complex scenarios.

Keywords: Clustering, Meta-Clustering, Multiple Clusterings, Clustering Visualization.

1   Introduction

Clustering is one of the most important unsupervised techniques for discovering dataset structure. Many clustering methods focus on obtaining one single 'best' solution by optimizing a pre-defined criterion [11]. There are two limitations with this: firstly, data can be multi-faceted in nature. Particularly when the datasets are large and complex, there may be several useful clusterings that exist, not only one. Secondly, users may be seeking different perspectives on the same dataset, requiring multiple clustering solutions. This has stimulated considerable recent research on the topic of *multiple clustering analysis* [2].

Multiple clustering analysis aims to discover a set of reasonable and distinctive clustering solutions from the same dataset. Many methods have been proposed on this topic and one very popular technique is meta-clustering [3],[20]. Meta-clustering generates a large number of base clusterings using different procedures: running different clustering algorithms, running a specific algorithm several times with different initializations, or using random feature weighting in

T. Calders et al. (Eds.): ECML PKDD 2014, Part II, LNCS 8725, pp. 145–160, 2014.
© Springer-Verlag Berlin Heidelberg 2014

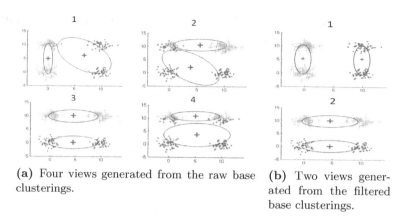

(a) Four views generated from the raw base clusterings.

(b) Two views generated from the filtered base clusterings.

Fig. 1. Two sets of views found using as input a) raw set of *unfiltered* base clusterings, and b) set of *filtered* base clusterings. Colours indicate clusters.

the distance function. These base clusterings may then be meta-clustered into groups. Further, clusterings within the same group can be combined using consensus (ensemble) clustering to generate a consensus view of that group. This results in one or more distinctive clusterings (views) of the dataset, each offering a different perspective or explanation.

A major drawback and challenge with the use of meta-clustering is that its effectiveness is highly dependent on the quality and diversity of the generated base clusterings. Specifically, if the base clusterings have high similarity, then further processing may generate multiple similar views. If the base clusterings are of low quality, then naive visualization or analysis will produce low quality views. Users may be misled by these similar or poor quality views.

We illustrate this problem with an example in Figure 1a, where the dataset consists of four Gaussian clusters. We generate a set of raw base clusterings via k-means (with $k = 2$ clusters) and random feature weighting (these base clusterings are not shown in the figure). These base clusterings are then meta-clustered into groups, and for each group a consensus view is extracted via consensus clustering. The views generated on the raw base clusterings are presented in Figure 1a. Observe that among these four views, some are rather similar (view2, view3 and view4) and some have poor quality (view1, view2 and view4). This stimulates the following question, which is the basis for our paper - *Can we apply a filtering step to the base clusterings and thus avoid discovering poor quality or redundant views?* Figure 1b provides intuition about the benefits of filtering. It shows the views generated using a filtered set of base clusterings as input. These are more natural views of the dataset, being both of high quality and non-redundant.

In more detail, we propose filtered meta-clustering (FILTA), aiming at detecting multiple high quality and distinctive views by filtering and then analyzing a given set of base clusterings. Algorithmically, we propose an information theoretic criterion to perform the filtering. In addition, we show how to employ a

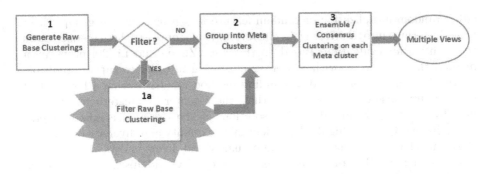

Fig. 2. The meta-clustering framework with proposed filtering step highlighted

visual method to automatically determine the dominant meta clusters within the filtered base clusterings. Finally, we perform consensus clustering on each meta cluster to identify the views. Figure 2 shows the whole process. The novelty of our approach lies in the addition of a filtering step to the existing meta-clustering framework, which is highlighted as step 1a in Figure 2. Our focus is on investigating how to filter the given raw base clusterings to generate a set of better views, in terms of quality and diversity, compared to the unfiltered meta-clustering. We assume that we are given a set of base clusterings and the generation of appropriate base clusterings (step 1) is outside the scope of this paper and is left for future work. An important advantage of our method is that the filtering step is independent of the other steps in this framework and thus may be easily integrated with them.

Our contributions can be summarized as follows:

- We identify limitations with the current pipeline for meta-clustering. In particular, its reliance on the quality and diversity of a set of (raw) base clusterings for generating high quality and diverse views.
- We propose a novel *filtering* based meta-clustering approach for discovering multiple high quality and diverse views from a given set of base clusterings. Our filtering step can enhance any existing meta-clustering method.
- We propose a mutual information based filtering criterion which considers the quality and diversity of base clusterings simultaneously. We provide a parameter that allows users to flexibly control the balance between less number of views but of higher quality or more of them but of relatively lower quality.

2   Related Work

Our research is related to several topics: meta-clustering, alternative clustering and cluster ensemble or consensus clustering.

Meta-Clustering aims to find multiple views by generating and evaluating a large set of base clusterings. In work [3], it first generates these base clusterings by either random initialization or random feature weighting. Then, it groups these base clusterings into multiple meta clusters and then presents these meta clusters to the users for evaluation. Based on this idea, Zhang and Li [20] proposed a method that extend [3] with consensus clustering in order to capture multiple views. Work in [14] proposed a sampling method for discovering a large set of good quality base clusterings. After that, the k-center [9] clustering method is used to select the k most dissimilar solutions as the views. The existing meta-clustering methods are highly dependent on the quality and diversity of the base clusterings for generating multiple high quality and diverse views.

Alternative Clustering discovers high quality and dissimilar views via searching in the clustering space guided by criteria about what constitutes an alternative. One may discover alternatives either iteratively or simultaneously. See [2] for a review. Compared with meta-clustering, alternative clustering is more efficient for discovering alternative views. However, it restricts the definition of an alternative to certain objective functions, which may cause the search process to miss some interesting views, due to mismatches between the objective function and the underlying view structure. It can be difficult to define an objective function characterizing what is an alternative, especially in the initial period of data analysis, when there is little information about the data available.

Cluster Ensemble or Consensus Clustering combines a collection of partitions of data into a single solution which aims to improve the quality and stability of individual clusterings [16]. However, instead of combining all the available clusterings into one single solution, it has been demonstrated that a better clustering can often be achieved by combining only a part of all the available solutions [8], that is the **cluster ensemble selection problem**. It has been shown that quality and diversity are two important factors which will influence the performance of cluster ensemble [8]. Cluster ensemble and the cluster ensemble selection methods typically focus on discovering a single high quality solution from a collection of clusterings, rather than multiple solutions.

Our proposed framework in Figure 2 combines all of the above clustering paradigms. The critical difference between our work compared to the others is that we place each clustering paradigm into its most relevant place. In particular, we employ alternative clustering as one of the mechanisms for generating the base clusterings. Alternative clustering employs objective functions to guide the search process, thus it may discover alternative views faster when compared to meta-clustering which employs a random clustering generation scheme (such as random initialization or random feature weighting). On the other hand, meta-clustering can cover the space of clusterings more comprehensively compared to alternative clustering, by flexibly employing different means of generation. Finally, we propose to group the clusterings and generate the consensus view for each group via consensus clustering. This is a more flexible approach than generating a single consensus view for the whole set of base clusterings, as the base clusterings may reflect very different structures of the data and thus may not be reasonably combined to produce a single consensus view.

3   FILTA: An Algorithm for Filtering Base Clusterings

Let us first introduce the notations used and a formal problem definition. Let $X = \{x_1, \ldots, x_n\}$ be a set of n objects, where $x_i \in \mathbb{R}^d$. These objects can be grouped into clusters (sets of objects). A clustering C is a hard partition of X, denoted by $C = \{c_1, \ldots, c_k\}$, where c_i is a cluster and $c_i \cap c_j = \emptyset, \bigcup c_i = X$. We denote the space of possible clusterings on X as $\mathcal{P}_{\mathcal{X}}$. We use $\mathcal{C}$ to denote a set of (base) clusterings, i.e., $\mathcal{C} = \{C_1, \ldots, C_l\}$. Let a set of views be denoted by $\mathcal{V} = \{V_1, \ldots, V_R\}$, where a view V_i is a clustering on X, $V_i \in \mathcal{P}_{\mathcal{X}}$. Even though a view is just a clustering, we use the view nomenclature to distinguish between the initial base clusterings and the final, selected clusterings (the set of views) at the end of the meta-clustering process. The quality of a clustering C is measured by a function $Q(C): \mathcal{P}_{\mathcal{X}} \to \mathbb{R}^+$, and the diversity between two clusterings can be computed according to a similarity measure $Sim(C_i, C_j): \mathcal{P}_{\mathcal{X}} \times \mathcal{P}_{\mathcal{X}} \to \mathbb{R}^+$. Our problem can be formalized as follows.

Problem Definition 1 *Given a set of raw base clusterings $\mathcal{C} = \{C_1, \ldots, C_l\}$, we seek a set of views $\mathcal{V} = \{V_1, \ldots, V_R\}$ generated from $\mathcal{C}$, such that, $\sum_{V_i \in \mathcal{V}} Q(V_i)$ is maximized and $\sum_{V_i, V_j \in \mathcal{V}, i \neq j} Sim(V_i, V_j)$ is simultaneously minimized.*

We solve this problem by selecting a subset of clusterings $\mathcal{C}'$, which are of high quality and diversity, from the given raw base clusterings $\mathcal{C}$. Next we discuss the quality and diversity criteria for clusterings.

3.1   Clustering Quality and Diversity Measures

We employ an information theoretic criterion, namely the mutual information for measuring both clustering quality and diversity. As a clustering quality measure, mutual information is a well-known criterion for clustering discovery, which can discover both linear and non-linear clusterings [5]. For measuring similarity between clusterings, mutual information can detect any kind of linear or non-linear relationship between random variables [17]. More specifically, the quality of a clustering C is measured by the amount of shared information with the data X, i.e., $I(X; C)$. Intuitively, the more information that is shared, the better that a clustering models the data. The mutual information between two clusterings $I(C_i; C_j)$ quantifies their similarity. Thus, the less mutual information shared, the more dissimilar they are. The average quality of the selected set of base clusterings can be optimized as:

$$\max_{\mathcal{C}'} \left\{ \frac{1}{|\mathcal{C}'|} \sum_{C_i \in \mathcal{C}'} I(X; C_i) \right\} \equiv \min_{\mathcal{C}'} \left\{ \frac{1}{|\mathcal{C}'|} \sum_{C_i \in \mathcal{C}'} H(X|C_i) \right\} \tag{1}$$

where the right hand side results from $I(X; C) = H(X) - H(X|C)$ and $H(X)$ is a constant (where $H(\cdot)$ is the Shannon entropy function). The diversity can be optimized by minimizing the average similarity between clusterings, as:

$$\min_{\mathcal{C}'} \left\{ \frac{1}{|\mathcal{C}'|^2} \sum_{C_i, C_j \in \mathcal{C}'} I(C_i; C_j) \right\}$$

Computation of the mutual information $I(X;C)$ requires the joint density function, $p(X,C)$, which is difficult to estimate for high dimensional data. Instead of directly estimating the joint densities, we may use the meanNN differential entropy estimator for computing the conditional entropy $H(X|C)$ [7], due to its desirable properties of efficiently estimating density functions in high dimensional data and being parameterless. The mutual information between two clusterings C_i and C_j is computed directly from their contingency table:

$$I(C_i; C_j) = \sum_{u \in C_i} \sum_{v \in C_j} p(u, v) \log \frac{p(u, v)}{p(u)p(v)} \tag{2}$$

where $p(v)$ is the fraction of data points in cluster v, and $p(u, v)$ is the fraction of points belonging to cluster u in C_i and v in C_j.

3.2   Filtering Criterion and Incremental Selection Strategy

We wish to select a subset of base clusterings, $\mathcal{C}'$, to achieve high quality and diversity simultaneously. Inspired by the mutual information based feature selection literature [13] which maximizes feature relevancy while minimizing feature redundancy, we propose a clustering selection criterion which combines the quality and diversity of clusterings:

$$\min_{\mathcal{C}' \subset \mathcal{C}, |\mathcal{C}'| = L} \left\{ \frac{1}{|\mathcal{C}'|} \sum_{C_i \in \mathcal{C}'} H(X|C_i) + \frac{\beta \beta_0}{|\mathcal{C}'|^2} \sum_{C_i, C_j \in \mathcal{C}', i \neq j} I(C_i; C_j) \right\} \tag{3}$$

where L is a user defined parameter specifying the number of base clusterings $\mathcal{C}'$ to be selected, and $\beta \in [0, \infty)$ is a trade-off parameter that balances the quality and diversity during selection. To make sure the second term is on the same scale as the first term, we set $\beta_0 = \max H(X|C_i) / \max I(C_i; C_j)$. Thus, our selection method aims to select L base clusterings $\mathcal{C}'$ from the given raw base clusterings $\mathcal{C}$, optimizing the dual-objective criterion in Equation (3).

A simple incremental search strategy can be used to select a good subset $\mathcal{C}'$ for the criterion (3) as follows. Initially, we select the clustering solution with the highest quality among the given clusterings $\mathcal{C}$. Then, we incrementally select the next solution from the set $\mathcal{C} \setminus \mathcal{C}'$ as:

$$\arg \min_{C_i \in \mathcal{C} \setminus \mathcal{C}'} \left\{ H(X|C_i) + \frac{\beta \beta_0}{|\mathcal{C}'|} \sum_{C_j \in \mathcal{C}'} I(C_i; C_j) \right\} \tag{4}$$

with the aim of selecting the next clustering with high quality and small average similarity with the selected ones in $\mathcal{C}'$. This process repeats until we reach the L desired number of base clusterings. The overall computational complexity of the filtering step costs $O(|\mathcal{C}| \cdot n^2 d)$, where n is the number of data observations and d is the number of data features.

4   Discovering the Clustering Views

We have obtained a filtered set of base clusterings after performing the filtering process. Next we group them into clusters at the meta level (step 2 in Figure 2) and then perform ensemble clustering on each meta cluster for view generation (step 3). We first explain the measure used to compute the similarity between the base clusterings, then explain a visualization technique called VAT for determining the potential number of meta clusters. We then introduce a method that combines with VAT to automatically determine the appropriate number of meta clusters and performs the grouping, and finally describe how to obtain the views from the meta clusters.

Measuring the Similarity between Clusterings: In order to divide the selected clusterings into groups, we need a similarity measure for pairwise clustering comparison. Several measures of clustering similarity have been proposed in the literature [11]. Here, we utilize the Adjusted Mutual Information(AMI) [18], which is an adjusted-for-chance version of the normalized mutual information [16]. The AMI between two clusterings C_i and C_j is defined as:

$$AMI(C_i; C_j) = \frac{I(C_i; C_j) - E\{I(C_i; C_j)\}}{\max\{H(C_i), H(C_j)\} - E\{I(C_i; C_j)\}} \qquad (5)$$

where the $E\{\cdot\}$ is the expected value of mutual information $I(C_i; C_j)$, and $H(C_i)$ is the entropy of the clustering C_i. The AMI is 1 when the two clusterings are identical and 0 when any commonality between the clusterings is due to chance. The distance between two clusterings is then $1 - AMI(C_i; C_j)$.

Grouping the Base Clusterings into Meta Clusters: After filtering the base clusterings to obtain $\mathcal{C}'$, we compute the pairwise dissimilarity matrix between all members of $\mathcal{C}'$ as a prelude to grouping them into meta clusters. There are two challenges for this grouping step: a) determining the number of relevant meta clusters; and b) partitioning the clusterings into meta clusters. Next, we will describe a visualization technique for assessing the number of meta clusters in a set of base clusterings. Then, an automatic method for determining the number of meta clusters and partitioning the clusterings into meta clusters will be presented.

The VAT method [19] is a visualization tool for cluster tendency assessment. By reordering a pairwise dissimilarity matrix of a set of data objects, it can reveal the hidden clustering structure of the data by visualizing the intensity image of the reordered dissimilarity matrix. The number of clusters in a set of data objects can be visually identified by the number of "dark blocks" displayed along the diagonal of the VAT image. In our work, each clustering can be taken as a data object, and we utilize the VAT method to visualize the number of potential meta clusters.

For grouping the set of clusterings, existing research uses hierarchical clustering [3],[20]. Our FILTA algorithm is not restricted to any particular grouping method. Here, we employ an automatic clustering method-CLODD which automatically extracts the number of clusters and produces a hard partition of the data objects based on a reordered dissimilarity matrix [10]. We obtain the reordered dissimilarity matrix by applying the VAT method to the dissimilarity matrix of clusterings. As mentioned above, there will be dense blocks along the diagonal of this ordered dissimilarity matrix if clusters exist in this set of clusterings. The CLODD algorithm discovers the number of meta clusters and produces a hard partition of these clusterings by optimizing an objective function which assesses the dense diagonal block structures of the reordered dissimilarity matrix.

Discovering the Views via Ensemble Clustering: In this final step, we use the MCLA ensemble clustering algorithm [16] to find a consensus view for each meta cluster. At the end of this step, we have a set of high quality and diverse views of the data.

5   Experimental Results

In this section, we use a synthetic and two real world datasets to compare the performance of our FILTA method against the existing meta-clustering methods, i.e., we compare the views generated from the filtered base clusterings against the views discovered from the raw base clusterings.

Experimental Setup: We generate 400 base clusterings for each dataset. Then FILTA and the proposed steps in Section 4 are applied. The base clusterings are generated using a combination of the following six clustering methods, some of which have been used previously in other meta-clustering algorithms:

- k-means with random initializations.
- random feature weighting method where feature weights are drawn from the zipf distribution[3].
- random sampling that selects $\{50\%, 60\%, 70\%, 80\%, 90\%, 100\%\}$ of objects and features, and then applying k-means on the sampled objects and features. Then the objects not initially sampled are assigned to the nearest clusters by the k-nearest neighbour method.
- spectral clustering method [15] using the similarity measure $S = exp(-\|x_i - x_j\|^2/\sigma^2)$ with the shape parameter $\sigma = \frac{\max\{\|x_i-x_j\|\}}{2^{k/8}}$, where k is randomly chosen from $k = 0, \ldots, 64$.
- EM-based mixture model clustering method with different initializations.
- an alternative clustering method, minCEntropy [17], with different reference clusterings generated by one of the above methods.

5.1   Evaluation of the Resulting Views

In our experiments, we use two measures for evaluating the discovered views. The Dunn Index is a popular internal clustering quality measure [6] and is defined as: $DI(C) = \frac{\min_{i \neq j}\{\delta(c_i, c_j)\}}{\max_{1 \leq w \leq k}\{\Delta(c_w)\}}$, where δ is the cluster to cluster distance and Δ is the cluster diameter. A larger DI is better. When we seek to compare against the ground truth labels, we use the adjusted mutual information (AMI).

Inspired by Mean Average Precision(MAP) [12], a popular measure for evaluating ranked retrieval of documents in information retrieval, we propose a mean best matching (MBM) score to test: i) (diversity) how many ground truth labels can be recovered by the top k views? and ii) (quality) how well do the top k views match the multiple sets of ground truth labels? Here, we select the top k views according to their quality (measured by the DI) and then we assess the matching between these views and the ground truth labels using AMI. In more detail, given multiple ground truth views $\mathcal{G} = \{G_1, \ldots, G_H\}$ and a set of ranked views $\mathcal{V}_r = \{V_{r_1}, \ldots, V_{r_m}\}$, the mean best matching score for the top k views $\mathcal{V}_{r_k} = \{V_{r_1}, \ldots, V_{r_k}\}$, where $k \leq m$, is defined as:

$$MBM(\mathcal{V}_{r_k}) = \sum_{i=1}^{H} \max_{V_j \in \mathcal{V}_{r_k}} AMI(G_i, V_j)/H \qquad (6)$$

5.2   Synthetic Dataset

In this section, we use a synthetic dataset to test whether our FILTA method is able to discover high quality and diverse views by filtering out poor quality and similar base clusterings. Our synthetic dataset consists of four Gaussian clusters. Each of the generated 400 raw base clusterings consists of two clusters. There are two high quality and dissimilar views within these base clusterings and we aim to recover these.

Using the unfiltered set of base clusterings, the CLODD method produced four meta clusters, highlighted by the green dashed line surrounding the blocks in Figure 3a. After performing the ensemble clustering on each meta cluster, four views are generated (see Figure 3b), with their numbers corresponding to the numbered blocks in the VAT diagram. We can see that views 2, 3 and 4 are similar and redundant, while views 1, 2 and 4 are of poor quality (the consensus clusters are spread out) and only view 3 is of good quality. Note that views 1 and 3 correspond to the two ground truth views included in the raw base clusterings. However, view 1 is of low quality, due to poor quality base clusterings included in its meta cluster. This experiment demonstrates that the meta-clustering methods may generate poor quality and similar views since it uses all the base clusterings, whether they are of high quality or not.

Next, we apply our FILTA algorithm on the same set of 400 base clusterings. We filter out 300 of the low quality and similar base clusterings setting $L = 100$ and $\beta = 0.1$ and the results are shown in Figure 4. The corresponding VAT diagram is presented in Figure 4a. Observe that there are two clearly separated blocks, indicating that there are two groups of clusterings that exist in the filtered base clusterings. The views generated based on the discovered two groups are presented in Figure 4b. We can see that these two views are of high quality, dissimilar and correspond to the two ground truth views. In addition, we can

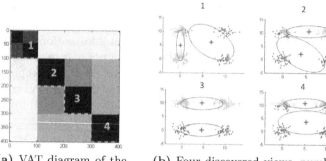

(a) VAT diagram of the 400 base clusterings.

(b) Four discovered views, numbered according to the diagonal blocks of Figure 3a.

Fig. 3. Four views discovered from the 400 raw base clusterings of the synthetic dataset. Each numbered block in Figure 3a represents a view.

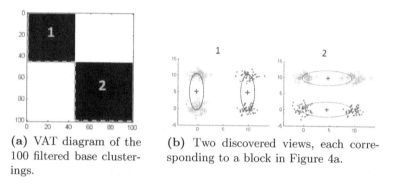

(a) VAT diagram of the 100 filtered base clusterings.

(b) Two discovered views, each corresponding to a block in Figure 4a.

Fig. 4. Two views discovered from the 100 filtered base clusterings of the synthetic dataset. Each numbered block in Figure 4a represents a view.

also observe that view 1 generated by FILTA has better quality compared to the view 1 generated by the unfiltered method (Figure 3b). It is because our filtering method filtered out the poor quality base clusterings in this group.

For a quantitative comparison of both approaches, we plot the MBM scores for the top 4 views in Figure 5a. Recall that FILTA only produced 2 views, hence it only has 2 scores in Figure 5a. Observe that for the top 1 view, both methods achieve the same MBM score, which means that the first view for each of these methods is of the same quality. For the top 2 views, the MBM score for the unfiltered method is lower than the FILTA method, because the view generated by the unfiltered method has a relative lower quality than the one generated by FILTA method. As we can see, with the increasing of value k, the MBM score for the unfiltered method does not change. It is because the third and fourth views of the unfiltered method do not perform better than the top 2 views meaning that they are either similar views to the others (redundant) or of poor quality.

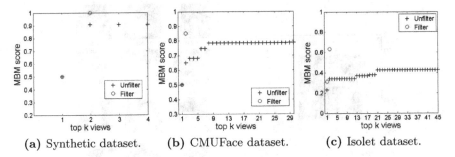

(a) Synthetic dataset.   (b) CMUFace dataset.   (c) Isolet dataset.

Fig. 5. The mean best matching (MBM) scores for the top k views of three datasets. One set of views is generated from the raw base clusterings (results represented by blue crosses), while the other set is from filtered ones (red circles).

5.3   CMUFace Dataset

We next show how FILTA performs for two real datasets. The CMUFace dataset from the UCI Machine Learning Repository [1] is a commonly used dataset for the discovery of alternative clusterings [4]. It contains 624 32 × 30 images of 20 persons, along with different features of these persons, e.g., pose (straight, left, right, up). Two dominant views exist in this dataset - identity and pose. In our experiment, we randomly select the images of three people and have 93 images in total. Again we generated 400 base clusterings and FILTA selected $L = 100$ with $\beta = 0.03$.

The results generated by the unfiltered method are shown in Figure 6. From the VAT diagram in Figure 6a, we can observe that there are a larger number of diagonal blocks. The CLODD algorithm produces 30 meta clusters and hence we have 30 views overall. Due to the limitation of space, we just show the top 4 views as measured by DI in Figure 6b. Each row in the figure is a view of three clusters, and each cluster of images is illustrated by its mean image. The number above each image is the purity score, which is the percentage of images, of a cluster, with the majority ground truth label (this can be labels from the identity or pose views). Higher purities are desirable. Consider Figure 6b. The view displayed in the first row corresponds to the person ground truth view, and the view in the second row corresponds to the pose one. However, the third and fourth row views are a combination of the other two. Their clusters mix poses and identities. From this experiment, we can see that the existing unfiltered meta-clustering methods can generate many poor quality and redundant views.

The results generated by our FILTA method are shown in Figure 7. As we can observe, the VAT diagram (Figure 7a) is less fuzzy compared with the one generated from the raw base clusterings (Figure 6a), has higher purity and includes two relatively well separated blocks. Again our filtering method has filtered out the poor quality and redundant base clusterings. Two views are generated according to the discovered two groups shown in the VAT diagram, and are shown in Figure 7b. They are the desired person and pose views. Compared with the

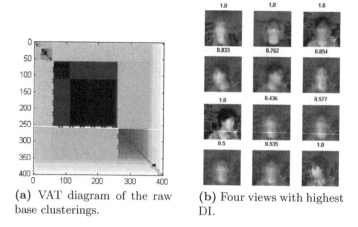

(a) VAT diagram of the raw base clusterings.

(b) Four views with highest DI.

Fig. 6. Views generated from the 400 raw base clusterings of the CMUFace data. The number above each image is its purity score.

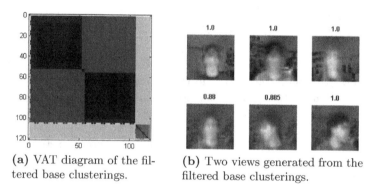

(a) VAT diagram of the filtered base clusterings.

(b) Two views generated from the filtered base clusterings.

Fig. 7. Views generated from the 100 filtered base clusterings of the CMUFace data. The number above each image in Figure 7b is its purity score.

pose view generated by the unfiltered method (Figure 6b), we get better quality in terms of the purity score shown above the image.

The MBM scores for these two sets of views are shown in Figure 5b. As we can see, the best MBM score for the unfiltered method is reached at the 8th view, implying that noisy results are present in the top 7 views. This result further demonstrates the influence of the quality and diversity of the base clusterings on the performance of the unfiltered meta-clustering methods.

5.4   Isolet Dataset

The isolet dataset from UCI machine learning repository [1] contains 7797 records with 617 features, which come from 150 subjects speaking the name of each letter of the alphabet twice. There are two views (speaker and letters) in this dataset.

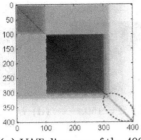

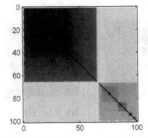

(a) VAT diagram of the 400 raw base clusterings.

(b) VAT diagram of the 100 filtered base clusterings.

Fig. 8. VAT diagrams for the raw and filtered sets of base clusterings of the Isolet dataset

In our experiment, we randomly selected 10 persons along with 10 letters, resulting in a 200 records dataset. We generate 400 base clusterings that contains the speaker and letter views, and select 100 base clusterings using FILTA ($\beta = 0.5$).

The results are shown in Figure 8. From the VAT diagram on the raw base clusterings (Figure 8a), we can observe that there are many small, dissimilar blocks in the right bottom corner of the VAT diagram, highlighted by the red dashed circle (dissimilarity indicated by the light shading of the area between the blocks). Each of them is taken as a view which results in 45 views overall. After applying our filtering method on the raw base clusterings, we obtain the VAT diagram in Figure 8b. As we can see, there are two explicit big blocks without those dissimilar individual meta clusters, which have been filtered out due to their poor quality.

The MBM scores for these two sets of views are shown in Figure 5c. It can be observed that the top 1 view generated by our FILTA method has higher quality than the one generated by the unfiltered method. In addition, the two views of FILTA capture the two ground truth views well. In contrast, the existing unfiltered method generated almost 45 views and the quality of the best matching views for the two ground truth views among the 45 views are not comparable with FILTA's. This result further shows that the input base clusterings, including low quality and redundant solutions, will lead to similar and poor quality views.

5.5 Impact of the Number of Selected Base Clusterings

The number of selected clusterings L does not have high impact on the quality of view generation by our method. We take the CMUFace dataset as the example to show the impact of L. In Figure 9, we show the VAT diagram constructed for $L = 50$ to 400 (filtered) base clusterings (recall that there are 400 raw base clusterings). We see that the VAT diagrams are mostly stable from $L = 50$ to 300, meaning that FILTA is quite robust to noise and relatively insensitive to the choice of L. From our experiments we found $L = 25\% \times l$ (l is the number of raw base clusterings) works well.

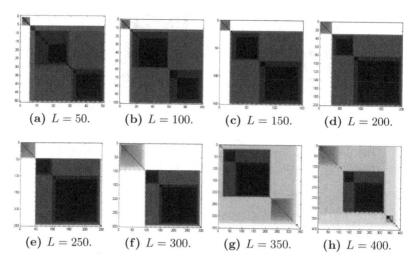

(a) $L = 50$.   (b) $L = 100$.   (c) $L = 150$.   (d) $L = 200$.

(e) $L = 250$.   (f) $L = 300$.   (g) $L = 350$.   (h) $L = 400$.

Fig. 9. VAT diagrams for different number of filtered base clusterings with $\beta = 0.03$ on the CMUFace data

5.6   Impact of the Regularization Parameter

The regularization parameter $\beta \in [0, \infty)$ balances the quality and diversity during the clustering filtering procedure. We have found that $\beta \in [0, 1]$ works fairly well. Essentially within this range, we place more emphasis on clustering quality. For example, when $\beta = 0.5$, it means we treat quality as twice as important as diversity. When $\beta \to 1$, the filtering process places equal emphasis on diversity, which generally increases the number of potential views but at the risk of including more poor quality solutions. In contrast, when $\beta \to 0$, the filtering procedure focuses on the quality, which will result in high quality views but some relevant views may be filtered out. Thus, users can tune this parameter according to their specific needs for view detection. Given that we usually do not have the cluster labels, the VAT diagrams can be used as one of the ways to help users for investigation. In particular, we propose to 'slide' β within the $[0, 1]$ range and inspect the VAT reordered matrix and the consensus views that emerge. We demonstrate the effect of β on the CMUFace dataset. Figure 10 shows how the MBM score changes as we vary β. As it can be seen, a $\beta = 0.03$ to 0.05 gives the best matching scores. To further confirm these are effective β values for this dataset, we illustrate a number of VAT diagrams (Figures 11) constructed from different β values and $L = 100$. The diagrams show that $\beta = 0.03$ (Figure 11b) discovers two relatively sharp dark blocks which are turned out to correspond to the two true views. Also, we can see that as β increases, the VAT diagram becomes more fuzzy, which means that the selected base clusterings are more diverse but their quality is decreasing. In this respect, our proposed framework is a useful tool to assist the discovery of novel views from the data.

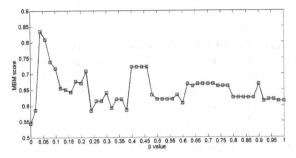

Fig. 10. The mean best matching scores(MBM) with different β on 100 filtered clusterings generated from CMUFace dataset

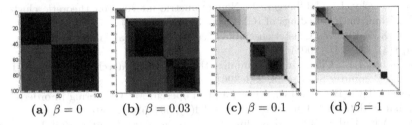

(a) $\beta = 0$ (b) $\beta = 0.03$ (c) $\beta = 0.1$ (d) $\beta = 1$

Fig. 11. VAT diagrams generated from 100 filtered base clusterings and different β values, for the CMUFace data

6   Conclusions

Meta-clustering is an important tool for discovering multiple views from data by analyzing a large set of raw base clusterings. It does not require any prior knowledge nor pose any assumption on the data, which especially suits exploratory data analysis. However, the generation of a large set of high-quality base clusterings is a challenging problem. There may exist poor quality and similar solutions which will affect the generation of high quality and diverse views.

In this paper we have introduced a clustering selection method for filtering out the poor quality and redundant clusterings from a set of raw base clusterings. This has the effect of lifting the quality of views generated by the meta-clustering methods applied to this set of filtered clusterings. In particular, we proposed a mutual information based filtering criterion which considers the quality and diversity of clusterings simultaneously. By optimizing this objective function via a simple incremental procedure, we can select a subset of good and diverse base clusterings. Meta-clustering on this filtered set of base clusterings can then yield multiple good and diverse views. We believe FILTA is a simple and useful tool to incorporate within the area of multiple clustering exploration and analysis.

References

1. Bache, K., Lichman, M.: UCI machine learning repository (2013)
2. Bailey, J.: Alternative clustering analysis: A review. In: Aggarwal, C., Reddy, C. (eds.) Data Clustering: Algorithms and Applications. CRC Press (2013)
3. Caruana, R., Elhaway, M., Nguyen, N., Smith, C.: Meta Clustering. In: Proceedings of ICDM, pp. 107–118 (2006)
4. Cui, Y., Fern, X.Z., Dy, J.G.: Multi-view clustering via orthogonalization. In: Proceedings of ICDM, pp. 133–142 (2007)
5. Dang, X.H., Bailey, J.: A hierarchical information theoretic technique for the discovery of non linear alternative clusterings. In: Proc. of KDD, pp. 573–582 (2010)
6. Davidson, I., Qi, Z.: Finding alternative clusterings using constraints. In: Proceedings of ICDM, pp. 773–778 (2008)
7. Faivishevsky, L., Goldberger, J.: Nonparametric information theoretic clustering algorithm. In: Proceedings of ICML, pp. 351–358 (2010)
8. Fern, X.Z., Lin, W.: Cluster ensemble selection. Statistical Analysis and Data Mining 1(3), 128–141 (2008)
9. Gonzalez, T.F.: Clustering to minimize the maximum intercluster distance. Theoretical Computer Science 38, 293–306 (1985)
10. Havens, T.C., Bezdek, J.C., Keller, J.M., Popescu, M.: Clustering in ordered dissimilarity data. Int. Journal of Int. Sys. 24(5), 504–528 (2009)
11. Jain, A.K., Dubes, R.C.: Algorithms for clustering data. Prentice-Hall, Inc. (1988)
12. Manning, C.D., Raghavan, P., Schütze, H.: Introduction to information retrieval, vol. 1. Cambridge university press, Cambridge (2008)
13. Peng, H., Long, F., Ding, C.: Feature selection based on mutual information criteria of max-dependency, max-relevance, and min-redundancy. IEEE Transactions on Pattern Analysis and Machine Intelligence 27(8), 1226–1238 (2005)
14. Phillips, J.M., Raman, P., Venkatasubramanian, S.: Generating a diverse set of high-quality clusterings. arXiv, 1108.0017 (2011)
15. Shi, J., Malik, J.: Normalized cuts and image segmentation. IEEE Transactions on Pattern Analysis and Machine Intelligence 22(8), 888–905 (2000)
16. Strehl, A., Ghosh, J.: Cluster ensembles—a knowledge reuse framework for combining multiple partitions. The Journal of Mach. Learn. Res. 3, 583–617 (2003)
17. Vinh, N.X., Epps, J.: minCEntropy: A novel information theoretic approach for the generation of alternative clusterings. In: Proc. of ICDM, pp. 521–530 (2010)
18. Vinh, N.X., Epps, J., Bailey, J.: Information theoretic measures for clusterings comparison: is a correction for chance necessary? In: Proceedings of ICML, pp. 1073–1080 (2009)
19. Wang, L., Nguyen, U.T.V., Bezdek, J.C., Leckie, C.A., Ramamohanarao, K.: iVAT and aVAT: Enhanced visual analysis for cluster tendency assessment. In: Zaki, M.J., Yu, J.X., Ravindran, B., Pudi, V. (eds.) PAKDD 2010, Part I. LNCS, vol. 6118, pp. 16–27. Springer, Heidelberg (2010)
20. Zhang, Y., Li, T.: Extending consensus clustering to explore multiple clustering views. In: Proceedings of SDM, pp. 920–931 (2011)

Nonparametric Markovian Learning of Triggering Kernels for Mutually Exciting and Mutually Inhibiting Multivariate Hawkes Processes

Remi Lemonnier[1,2] and Nicolas Vayatis[1]

[1] CMLA - ENS Cachan and CNRS, France
[2] 1000 Mercis, Paris, France

Abstract. In this paper, we address the problem of fitting multivariate Hawkes processes to potentially large-scale data in a setting where series of events are not only mutually-exciting but can also exhibit inhibitive patterns. We focus on nonparametric learning and propose a novel algorithm called MEMIP (Markovian Estimation of Mutually Interacting Processes) that makes use of polynomial approximation theory and self-concordant analysis in order to learn both triggering kernels and base intensities of events. Moreover, considering that N historical observations are available, the algorithm performs log-likelihood maximization in $O(N)$ operations, while the complexity of non-Markovian methods is in $O(N^2)$. Numerical experiments on simulated data, as well as real-world data, show that our method enjoys improved prediction performance when compared to state-of-the art methods like MMEL and exponential kernels.

1 Introduction

Multivariate Hawkes processes are a class of multivariate point processes which are often used to model counting processes where physicals events rate of occurrence usually depend on past occurences of many other events. This is typically the case for earthquakes aftershocks [1] and financial trade orders on marketplace [2,3,4,5], but also in other fields such as crime prediction [6], genome analysis [7] and more recently for modeling social interactions [8]. Multivariate Hawkes processes are fairly well-known from a probabilistic point of view : their Poisson cluster representation was outlined by the seminal paper of Hawkes and Oakes [9], stability conditions and sample path large deviations principles were derived in a sequence of papers by Bremaud and Massoulie (see e.g [10]). In the unidimensional case, Ogata [11] showed that the log-likelihood estimator enjoys usual convergence properties under mild regularity conditions. However, in practical applications, estimation of the triggering kernels g_{uv} has always been a difficult task. First, because Hawkes log-likelihood contains the logarithm of the weighted sum of triggering kernels, most of the aforementioned papers made the choice of fixing triggering kernels up to a normalization factor in order to

T. Calders et al. (Eds.): ECML PKDD 2014, Part II, LNCS 8725, pp. 161–176, 2014.
© Springer-Verlag Berlin Heidelberg 2014

ensure concavity, that is $g_{uv} = c_{uv} \cdot g$. Secondly, when computational efficiency is an issue, the dependency of the stochastic rate at a given time on all the past occurences implies quadratic complexity in the number of occurences for tasks like log-likelihood computation. This issue has often been tackled by choosing memoryless exponential triggering kernels, but the actual dynamics of kernels strongly depends on the field of application: price impacts of a given trade [12] and process of views of Youtube videos [13] were shown to be better described by slowly decaying power-law kernels whereas for DNA sequence modelization [7] kernels are known to have bounded support. Thus, it is highly desirable to estimate triggering kernels in a data-driven way instead of assuming a given parametric form. Nonparametric estimation has been successfully addressed for unidimensional [7,14] and symmetric bidimensional [12] Hawkes processes. In the case where triggering kernels are known to sparsely decompose over a dictionary of basis functions of bounded support (e.g for neuron spikes interactions), a LASSO-based algorithm with provable guarantees was derived in [15].

Recently, combining majorization-minimization techniques with resolution of a Euler-Lagrange equation, Zhou, Zha and Song [8] proposed what is to our knowledge the first nonparametric learning algorithm for general multivariate Hawkes processes. But although this work constitutes a significant improvement over existing parametric methods, it still relies on several assumptions. First, interactions between events are assumed to be "mutually-exciting", i.e $g_{uu'}$ are non-negative for all u, u'. We nevertheless argue that in real-world settings, there is no reason to think that interactions beween events are only mutually-exciting. Secondly, the background rates μ_u are assumed to be constant. While this is a common assumption for multivariate Hawkes processes, it was shown by [16] that estimating $\mu_u(t)$ from the data could lead to significant improvement. To address these different issues, we construct a novel algorithm MEMIP (Markovian Estimation of Mutually Interacting Processes) based on polynomial approximation of a mapping of the triggering kernels to $[0, 1]$. Our method does not assume non-negativity on triggering kernels and is able to estimate time-dependent background rate on a data-driven way. Moreover, by constructing a markovian and linear estimator, we carry the more appealing properties of the most widely used parametric setting, where triggering kernels are fixed to exponentials up to a normalization factor : concavity of the log-likelihood that ensures global convergence of the estimator, and $O(N)$ log-likelihood calculation in a single pass through the data. While giving a concave formulation of the exact log-likelihood that can be maximized by multiple optimization techniques, we propose an algorithm based on maximisation of a self-concordant approximation that is shown to outperform state-of-the-art methods on both simulated and real-world data sets.

The paper is organized as follows. In Section 2, we formally define multivariate Hawkes processes as well as the associated log-likelihood maximization problem. In section 3, we decompose the log-likelihood on a basis of memoryless triggering kernels. Through Section 4, we develop two novel algorithms for exact as well as fast approximate maximization of the log-likelihood, analyze their complexity and show numerical convergence results based on the properties of self-concordant functions. In section 5, we show that MEMIP significantly improves over state of the art on both synthetic and real world data sets for the tasks of predicting future events as well as estimating underlying dynamics of the Hawkes process.

2 Setup and Notations

2.1 Model Description and Notation

We consider a multivariate Hawkes process, that is a d-dimensional counting process $N(t) = \{N^u(t) : u = 1, \ldots, d\}$ for which the rate of occurence of each component $N^u(t)$ is defined by:

$$\lambda_u(t) = \left(\mu_u(t) + \sum_{v \in [1...d]} \sum_{t_v < t} g_{uv}(t - t_v) \right)_+ , \quad \forall u = 1, \ldots, d \qquad (1)$$

where $\mu_u(t)$ is the natural rate of occurence of events along dimension u. Note that the occurence of a given event affects stochastic rates of occurence of every dimension. With an empty history, events of type u will occur as if they were drawn from a non-homogeneous Poisson process of rate $\mu_u(t)$. The kernel function evaluation $g_{uv}(t - t_v)$ quantifies the change in the rate of occurence of event u at time t caused by the realization of event v at time t_v. Following the intuition, we can characterize three situations depending on the values taken by the kernel function at a given time lapse s:

- *Excitation* corresponds to the case where we have $g_{uv}(s) > 0$, *i.e.* an event of type v is more likely to occur if an event of type u has occured at a time distance of s.
- *Independence* is observed when $g_{uv}(s) = 0$, meaning that the realization of an event of type u has no effect on the rate of occurence of an event of type v at time distance s.
- *Inhibition* takes place when $g_{uv}(s) < 0$, *i.e.* an event of type v is less likely to occur if an event of type u occured at time distance s.

Such processes can be seen as a generalization over the common definition of multivariate Hawkes process where the kernels g_{uv} are non-negative and the componentwise background rate μ_u is often taken constant.

2.2 Log-Likelihood of Multivariate Hawkes Processes

Input Observations. We define a *realization* h of a multivariate point process by the triplet $T_h^-, T_h^+, (t_i^h, u_i^h)_{i \in [1...n_h]}$, where T_h^- and T_h^+ are respectively the beginning and the end of the observation period, and (t_i^h, u_i^h), for $i \in [1...n_h]$, is the sequence of the n_h events occuring during this period. In the rest of the paper, we will assume we are given n i.i.d realizations of a multivariate Hawkes process. Without loss of generality, we will assume $\min_h(T_h^-) = 0$ and take $T = \max_h(T_h^+)$.

Expression of the Log-Likelihood. We first set $\Lambda = \{\lambda_u \; : \; u = 1, \ldots, d\}$. For a general multivariate point process, the log-likelihood of the whole dataset $\mathcal{H}$ is given by (e.g. [17]):

$$\mathcal{L}(\Lambda, \mathcal{H}) = \sum_{u=1}^{d} \sum_{h \in \mathcal{H}} \int_{T_h^-}^{T_h^+} \ln(\lambda_u(s)) dN_h^u(s) - \sum_{u=1}^{d} \sum_{h \in \mathcal{H}} \int_{T_h^-}^{T_h^+} \lambda_u(s) ds \qquad (2)$$

where $\int f(s)) dN_h^u(s) = \sum_{i=1}^{n_h} f(t_i^h) 1 \{u_i^h = u\}$. In the case of a linear Hawkes process (1), we introduce $\Lambda = (M, G)$ where $M = \{\mu_u \; : \; u = 1, \ldots, d\}$ and $G = \{g_{u,v} \; : \; u, v = 1, \ldots, d\}$ and the log-likelihood can be rewritten as:

$$\mathcal{L}(M, G, \mathcal{H}) = \sum_{h \in \mathcal{H}} \sum_{i=1}^{n_h} \ln \left(\mu_{u_i^h}(t_i^h) + \sum_{j \; : \; t_j^h < t_i^h} g_{u_j^h, u_i^h}(t_i^h - t_j^h) \right)$$

$$- \sum_{u=1}^{d} \sum_{h \in \mathcal{H}} \int_{T_h^-}^{T_h^+} \left(\mu_u(s) + \sum_{j=1}^{n_h} 1 \{u_j^h = u\} \, g_{u,u_j}(s - t_j) \right)_+ ds \qquad (3)$$

Depending on the parametrization of triggering kernels g_{uv}, this log-likelihood may or may not be concave. For instance, in the widely used setting where the background rates μ_u are constant and the kernels g_{uv} are non-negative and fixed up to the normalization factor ν_{uv}, the log-likelihood is concave and can be relatively easily maximized. However, even for the simple case of nonnegative exponential kernels $g_{uv}(t) = \nu_{uv} \exp(-\alpha_j t)$ where $\nu_{uv} \geq 0$ the product term $\nu_{uv} \exp(-\alpha_v t)$ makes the log-likelihood not concave with respect to α_v. Therefore, global convergence of maximization methods is not guaranteed anymore.

3   Approximations of Multivariate Hawkes Processes on a Basis of Exponential Triggering Kernels

3.1   A K-approximation of the Multivariate Hawkes Process

For a given multivariate Hawkes process $\Lambda = (M, G)$, we consider finite approximations of the components of the rates of occurence μ_u and g_{uv}. We first introduce the following functions:

$$\forall y \in [-\ln(T)/\alpha, 1], \quad \nu_u(y) = \mu_u(-\ln(y)/\alpha) \quad \text{and} \quad f_{uv}(y) = g_{uv}(-\ln(y)/\alpha)$$

and we use Bernstein-type polynomial approximations of order K for ν_u and f_{uv}: there exist coefficients $X_{uv,k}^K$ such that

$$\forall y \in [-\ln(T)/\alpha, 1], \quad \widehat{\nu}^K(y) = \sum_{k=0}^{K} X_{u0,k}^K y^k \quad \text{and} \quad \widehat{f}_{uv}^K(y) = \sum_{k=0}^{K} X_{uv,k}^K y^k \; .$$

These polynomial approximations are known to converge with a polynomial rate for smooth functions (with first r derivatives continuously differentiable) and geometric rate for analytic functions (see below). The K-aproximation considered in this paper relies on a simple change of variable in the Bernstein approximations by setting: $y = \exp(-\alpha t)$. We can now introduce the linear approximation of a multivariate Hawkes process with exponential kernels:

$$\forall t \in [0,T], \quad \widehat{\mu}^K(t) = \sum_{k=0}^{K} X_{u0,k}^K \exp(-k\alpha t) \quad \text{and} \quad \widehat{g}_{uv}^K(t) = \sum_{k=0}^{K} X_{uv,k}^K \exp(-k\alpha t) .$$

Classical arguments from approximation theory [18] lead to the following proposition.

Proposition 1. *For any function Ψ defined over $[0,T]$, we consider the supremum norm $\|\Psi\|_{T,\infty} = \sup_{t \in [0,T]} |\Psi(t)|$. The K-approximations $(\widehat{\mu}_u^K)_{K \geq 1}$ and $(\widehat{g}_{uv}^K)_{K \geq 1}$ converge in supremum norm towards true functions μ_u and g_{uv} at the following rates:*

1. *if μ_u is C^r, $\left\|\mu_u(t) - \widehat{\mu}_u^K(t)\right\|_\infty^T = O(1/K^r)$*
2. *if μ_u is analytic, $\left\|\mu_u(t) - \widehat{\mu}_u^K(t)\right\|_\infty^T = O(\exp(-K))$*
3. *if g_{uv} is C^r, $\left\|g_{uv}(t) - \widehat{g}_{uv}^K(t)\right\|_\infty^T = O(1/K^r)$*
4. *if g_{uv} is analytic, $\left\|g_{uv}(t) - \widehat{g}_{uv}^K(t)\right\|_\infty^T = O(\exp(-K))$.*

Another property of the approximated multivariate Hawkes process is the Markov property of the counting process. We set $\widehat{N}^K(t)$ the d-dimensional Hawkes process uniquely defined by $\widehat{\lambda}^K = (\widehat{\mu}_u^K, \widehat{g}_{uv}^K)_{u,v}$.

Proposition 2. *Assume that the empirical estimate $\widehat{N}^K(t)$ of the multivariate Hawkes process is obtained after i.i.d. realizations of $N(t)$ over the time interval $[0,T]$. There exists $(\widehat{\ell}^0, \widehat{\ell}^1, \ldots, \widehat{\ell}^K)$ such that:*

$$\forall u \in \{1, \ldots, d\} , \quad \widehat{\lambda}^K(t) = \sum_{k=0}^{K} \left(\widehat{\ell}^k(t)\right)_+$$

and $(\widehat{N}^K(t), \widehat{\ell}^0(t), \widehat{\ell}^1(t), \ldots, \widehat{\ell}^K(t))$ is a Markov Process on $\mathbb{N}^d \times \mathbb{R}^{d(K+1)}$.

The proof results from the following decomposition of each occurrence rate in the approximation: $\forall u \geq 1$,

$$\widehat{\lambda}_u^K(t) = \left(X_{u0,0}^K + \sum_{k=1}^{K} \left(X_{u0,k}^K \exp(-k\alpha t) + \sum_{v \, : \, t_v < t} X_{uv,(k-1)}^K \exp(-k\alpha(t-t_v)) \right) \right.$$

$$\left. + \sum_{v \, : \, t_v < t} X_{uv,K}^K \exp(-(K+1)\alpha(t-t_v)) \right)_+$$

Markov property is then a direct consequence of the dynamics of the functions $\widehat{\ell}_u^k(t)$: they decay at rate $\exp(-k\alpha t)$ and jump by $X_{uv,(k-1)}^K$ whenever an event of type v occurs. As they entirely determine the stochastic rate which determines the conditional probability distribution of $\widehat{N}^K(t)$, the conditional probability distribution of future states of the process $(\widehat{N}^K(t), \widehat{\ell}^0(t), \widehat{\ell}^1(t), \ldots \widehat{\ell}^K(t))$ is uniquely determined by the present state.

3.2   A New Decomposition of the Log-Likelihood

The algorithms proposed in this paper rely on a novel expression of the log-likelihood over a basis of triggering kernels. We use exponential excitation functions to account for nonlinearity but our algorithms benefit from the properties of linear approximations. Based on the expression of the log-likelihood for general linear multivariate Hawkes process (3), we introduce the following notation to discover the specific expression for the K-approximation based on exponential triggering functions: $\forall u, v = 1, \ldots, d, \forall k = 1, \ldots, K, \forall h \in \mathcal{H}, \forall i = 1, \ldots, n_h,$

$$A_{uv,k}^{K,h,i} = \sum_{j\,:\,t_j^h < t_i^h} 1\left\{u_i^h = v, u_j^h = u\right\} \exp\left(-(k + 1\left\{u > 0\right\})\alpha(t_i^h - t_j^h)\right) \quad (4)$$

$$B_{0v,k}^{K,h}(s) = \exp(-k\alpha s) \quad (5)$$

$$B_{uv,k}^{K,h}(s) = \sum_{j\,:\,t_j^h < s} 1\left\{u_j^h = v\right\} \exp(-(k+1)\alpha(s - t_j^h)) \quad (6)$$

The key expression of the approximate log-likelihood can then be derived by plugging-in the previous notations and replacing the intrinsic parameters (M, G) by the linear coefficients X^K:

$$\mathcal{L}^K(X^K, \mathcal{H}) = \sum_{h\in\mathcal{H}}\sum_{i=1}^{n_h} \ln(A^{K,h,i}X^K) - \sum_{h\in\mathcal{H}}\int_0^{T_h}\left(\sum_{i=1}^{n_h} B^{K,h}(s)X^K\right)_+ ds \quad (7)$$

Note that the dependance of $\mathcal{L}^K$ on the history $\mathcal{H}$ is entirely expressed by vectors $(A^{K,h,i})_{h\in\mathcal{H},i\in[1...n_h]}$ and $(B^{K,h}(s))_{h\in\mathcal{H},s\in[0,T]}$. An important feature of the approximate log-likelihood expressed in the parameter space defined by linear decompositions onto bases of exponential triggering kernels is given in the following proposition.

Proposition 3. *The function $X \to \mathcal{L}^K(X, \mathcal{H})$ is concave.*

From there, we have a complete roadmap for the design of algorithms estimating the parameters of multidimensional Hawkes processes: the last propostion indicates that a proxy of the log-likelihood (3) can be globally maximized with tools of convex analysis. Moreover, thanks to the approximation rates of convergence (Proposition 1), triggering kernels can be accurately estimated for large K through maximization of the new objective (7). Finally, the Markov property is an important feature that will allow us to construct the vectors $(A^{K,h,i})$ and $(B^{K,h})$ with linear complexity.

4   Markovian Algorithms for the Estimation of Triggering Kernels

Computational tractability of algorithms on large data sets depends on the algorithmic complexity in the dominating dimensions of the problem. For realizations of multivariate Hawkes processes, dominating dimensions are almost always the total number of events $N = \sum_{h \in \mathcal{H}} n_h$ and the time of observation T. Indeed, it would be unrealistic to try to learn d^2 nonparametric functions in an infinite dimensional space with only N observations without the condition $N \gg d^2$. In the rest of the paper, we will therefore focus on constructing two algorithms with no more than linear complexity in N and T.

4.1   Exact Maximization of the Approximated Log-Likelihood

Vectors $(A^{K,h,i})_{h \in \mathcal{H}, i \in [1...n_h]}$ and $(B^{K,h}(s))_{h \in \mathcal{H}, s \in [0,T]}$ can be constructed in a single pass through the data by **Algorithm 1**.

Algorithm 1. Algorithm for construction of vectors $(A^{K,h,i})$ and $(B^{K,h}(s))$

Initialize $i = 0$ and fix a time step dt
for all h **do**
    Initialize $(C_{uv}^k = 0)_{u \geq 1, v \geq 1}$; $t = T_h^-$; $(D_{uv}^k(T_h^-) = 1_{\{u=0\}})_{u \geq 0, v \geq 1}$
    while $t < T_h^+$ **do**
        $t \leftarrow t + \delta t = \min(t + dt, t_i)$
        for all k,u,v **do**
            $C_{uv}^k \leftarrow C_{uv}^k \exp(-(k+1\,\{u>0\}\,\alpha\delta t)$, $D_{uv}^k \leftarrow D_{uv}^k \exp(-(k+1\,\{u>0\}\,\alpha\delta t)$
            $B_{uv,k}^{K,h}(t) \leftarrow D_{uv}^k$
        end for
        if $t = t_i$ **then**
            for all k,u **do**
                $A_{uv,k}^{K,h,i} \leftarrow C_{uu_i}^k$
            end for
            for all k,v **do**
                $C_{u_i v}^k \leftarrow C_{u_i v}^k + 1$, $D_{u_i v}^k \leftarrow D_{u_i v}^k + 1$
            end for
            $i \leftarrow i + 1$
        end if
    end while
end for

Complexity of Algorithm 1. With $M = T/dt$ the number of discretizations steps, construction of vectors $(A^{K,h,i})$ and $(B^{K,h}(s))$ has thus a complexity of $O(N+M)$. As each log-likelihood evaluation (7) requires $2N+M$ scalar products computations, various optimization techniques can be used to find the global maximum of $X \rightarrow \mathcal{L}^K(X, \mathcal{H})$ in $O(N + M)$ operations. On the contrary, a nonmarkovian estimator, even linear, would need at each time t to compute the values of triggering kernels between current time and all preceding occurence times, thus leading to a $O(\sum_h n_h^2)$ complexity. This construction is thus very often the bottleneck of the whole maximization procedure.

4.2 Relaxed Version of the Log-Likelihood

While the previous paragraph exposes a fully tractable method to estimate the triggering kernels for potentially large data sets, we now develop an approximate algorithm called MEMIP, for Markovian Estimation of Mutually Interacting Processes, that leads to a substantial speed-up, as well as theoretical guarantees in terms of efficiency. For this purpose, we approximate the log-likelihood $\mathcal{L}^K(M, G, \mathcal{H})$ by dropping the positive part in log-likelihood (3), *i.e.*

$$
\tilde{\mathcal{L}}^K(M, G, \mathcal{H}) = \sum_{h \in \mathcal{H}} \left(\sum_{i=1}^{n_h} \ln \left(\mu_{u_i^h}(t_i^h) + \sum_{j \,:\, t_j^h < t_i^h} g_{u_j^h, u_i^h}(t_i^h - t_j^h) \right) \right.
$$
$$
\left. - \sum_{u=1}^d \int_{T_h^-}^{T_h^+} \left(\mu_u(s) + \sum_{j=1}^{n_h} \mathbb{1}\left\{u_j^h = u\right\} g_{u, u_j}(s - t_j) \right) ds \right) \tag{8}
$$

which can be rewritten:

$$
\hat{\mathcal{L}}^K(X^K, \mathcal{H}) = \sum_{h \in \mathcal{H}} \left(\sum_{i=1}^{n_h} \ln(A^{K,h,i} X^K) \right) - \hat{B}^K X^K \tag{9}
$$

where $\hat{B}^K_{uv,k} = \sum_{h \in \mathcal{H}} \sum_{j=1}^{n_h} \mathbb{1}\left\{u_j^h = v\right\} \int_{T_h^-}^{T_h^+} \exp(-k\alpha(s - t_j^h))$.

Although $\hat{\mathcal{L}}^K(X, \mathcal{H})$ is an upper bound of the actual log-likelihood and it is not clear at first sight why its maximization should lead to large values of $\mathcal{L}^K(X, \mathcal{H})$, we point out that the difference $\hat{\mathcal{L}}^K(X, \mathcal{H}) - \mathcal{L}^K(X, \mathcal{H})$ is only caused by intervals where there exists $u \in [1...d]$ such that $\hat{\lambda}^K_u(t) = 0$. But maximizers of $\hat{\mathcal{L}}^K(X, \mathcal{H})$ are very unlikely to exhibit wide range of negative values in their triggering kernels because any single event realization with a predicted nonpositive stochastic rate yields $\hat{\mathcal{L}}^K(X, \mathcal{H}) = -\infty$. Therefore, we assume we can rely on this approximation in order to construct fast algorithms.

4.3 MEMIP: a Learning Algorithm for Fast Log-Likelihood Estimation

Since the gradient and the hessian matrix of $X \mapsto \hat{\mathcal{L}}^K(X, \mathcal{H})$ can be computed analytically and their size does not depend on N, we derive the proposed algorithm MEMIP on the base of successive Newton optimizations. In the following, we denote by $\text{NewtonArgMax}(f, x_0)$ the result of a Newton maximization of function f with starting point x_0 using a classical backtracking linesearch method. The main idea is to construct recursively a sequence $(\widehat{X^1}...\widehat{X^K})$ of maximizers of functions $(\hat{\mathcal{L}}^k)_{k \in [1...K]}$ by using $\text{NewtonArgMax}(\hat{\mathcal{L}}^{k-1}, \widehat{W}^{k-1})$ as the starting point $\widehat{W}^k$ of maximization of $\hat{\mathcal{L}}^k$. From the estimated sequence $(\widehat{X}^1...\widehat{X}^K)$, the best value of k can be estimated by cross-validation or various other model selection techniques. Interestingly, $A^{k,h,i} = (A^{K,h,i}_{\bullet,j})_{j \in [1...k]}$ and $B^k = (B^K_{\bullet,j})_{j \in [1...k]}$ such that only $(A^{K,h,i})_{h \in \mathcal{H}, i \in [1...n_h]}$ and B^K need to be computed.

Algorithm 2. Algorithm (MEMIP) for learning background rates and triggering kernels of a multivariate Hawkes process

---

input Mapping parameter $\alpha > 0$, maximal polynomial degree K, starting point $\widehat{W}^1 \in \mathbb{R}^{d(d+1)}$

  Construct $(A^{K,h,i})$ and B^K according to $O(N)$ modified version of **Algorithm 1**

  $\widehat{X}^1 \leftarrow \text{NewtonArgMax}(\widehat{\mathcal{L}}^1, \widehat{W}^1)$

  for $k \in [2...K]$ **do**

    $\widehat{W}^k = 0$

    for $j \in [1...k-1], u \in [1...d], v \in [0...d]$ **do**

      $\widehat{W}^k_{uv,j} = \widehat{X}^{k-1}_{uv,j}$

    end for

    $\widehat{X}^k \leftarrow \text{NewtonArgMax}(\widehat{\mathcal{L}}^k, \widehat{W}^k)$

  end for

Complexity of Algorithm 2. We obtain two substantial computational speed-ups compared to exact log-likelihood maximization. First, time discretization is no longer needed for the construction of B^K. Thus, vectors $(A^{K,h,i})$ and B^K can be constructed with the same procedure than **Algorithm 1** except that updates are made only on time occurence of events. Therefore, construction complexity is $O(N)$. Similarily, approximate log-likelihood evaluations are also of complexity $O(N)$. Secondly, the approximate log-likelihood is separable by type of event u : $\widehat{\mathcal{L}}^K = \sum_{u=1}^d \widehat{\mathcal{L}}^K_u$ where $\widehat{\mathcal{L}}^K_u$ only depends on background rate μ_u and triggering kernels $(g_{uv})_{v \in [1...d]}$. Maximization can thus be parallelized across the different dimensions. Note that because of the Hessian inversion at each Newton step, complexity in d of maximization of $\widehat{\mathcal{L}}^K_u$ is $O(d^3)$ for any u, which yields a $O(d^4)$ overall complexity. In cases where $N \gg d^2$ but $d^4 > N$, the use of quasi-Newton method might therefore be preferable. For instance, BFGS method enjoys superlinear convergence [19], and would lead to a $O(d^3)$ overall complexity, the maximization of each $\widehat{\mathcal{L}}^K_u$ requiring $O(d^2)$ operations.

4.4 Self-concordance Property and Numerical Convergence of MEMIP

Problem (9) can be solved by various optimisation techniques. **Algorithm 2** is actually based on the concept of *self-concordance* [20] that we apply to function $X \mapsto -\widehat{\mathcal{L}}^k(X, \mathcal{H})$. Self-concordant functions are, along with strongly-convex functions with Lipschitz-continuous Hessian matrices, a very important class of functions for which nonasymptotic upper bounds of the number of Newton steps necessary to reach precision ϵ is known. More specifically, the following property holds:

Proposition 4. *Starting from a d(d+1)-dimensional vector $\widehat{W}^1$, MEMIP constructs a sequence of K estimates $(\widehat{X}^1...\widehat{X}^K)$ verifying for any $k \in [1...K]$, $|\widehat{\mathcal{L}}^k(\widehat{X}^k, \mathcal{H}) - \sup_X(\widehat{\mathcal{L}}_k(X, \mathcal{H}))| \leq \epsilon$ in at most $C\left(\sup_X(\widehat{\mathcal{L}}_K(X, \mathcal{H})) - \widehat{\mathcal{L}}_1(\widehat{W}^1, \mathcal{H})\right) + K(\log_2 \log_2(1/\epsilon) + C\epsilon)$ Newton iterations.*

Lemma 1. *Using Newton method with backtracking line search from a starting point $x_0 \in \mathbf{R}^d$, there exists $C > 0$ depending only on the line search parameters such that the total number of Newton iterations needed to minimize a self-concordant function f up to a precision ϵ is upper bounded by $C(\sup(f) - f(x_0)) + \log_2 \log_2(\frac{1}{\epsilon})$.*

Proof of Proposition 4. Self-concordance of functions $(-\widehat{\mathcal{L}}_k)_{k \in [1...K]}$ is a direct consequence of self-concordance on $\mathbf{R}_+^*$ of $f : x \mapsto -\ln(x)$ and affine invariance properties of self-concordant functions. By applying the aforementioned lemma to function $-\widehat{\mathcal{L}}_k$ and starting point $\widehat{W}^k$ at each Newton optimization, we get the bound

$$C \sum_k \left(\sup_X(\widehat{\mathcal{L}}_k(X, \mathcal{H})) - \widehat{\mathcal{L}}^k(\widehat{W}^k, \mathcal{H}) \right) + K \log_2 \log_2(1/\epsilon) \qquad (10)$$

By construction of MEMIP iterates, we also have $\widehat{\mathcal{L}}^k(\widehat{W}^k, \mathcal{H}) = \widehat{\mathcal{L}}^{(k-1)}(\widehat{W}^k, \mathcal{H}) = \widehat{\mathcal{L}}^{(k-1)}(\widehat{X}^{k-1}, \mathcal{H})$ where the first equality holds because for any $u, v, \widehat{W}^k_{uv,k} = 0$ and the second because for any $u, v, j \leq k-1, \widehat{W}^{k-1}_{uv,j} = \widehat{X}^{k-1}_{uv,j}$. But for any $k \geq 2, \widehat{\mathcal{L}}^{k-1}(\widehat{X}^{k-1}, \mathcal{H}) \geq \sup_X(\widehat{\mathcal{L}}_{k-1}(X, \mathcal{H})) - \epsilon$. Therefore the bound reformulates as

$$C \sum_{k=1}^K \left(\sup_X(\widehat{\mathcal{L}}_k(X, \mathcal{H})) - \sup_X(\widehat{\mathcal{L}}_{k-1}(X, \mathcal{H})) \right) + K(\log_2 \log_2(1/\epsilon) + C\epsilon) \quad (11)$$

which proves Proposition 4, using the notation $\sup_X(\widehat{\mathcal{L}}_0(X, \mathcal{H})) = \widehat{\mathcal{L}}_1(\widehat{W}^1, \mathcal{H})$.

□

Remark. The previous proposition emphasizes the key role played by the starting point $\widehat{W}^1$ in the speed of convergence of Newton-like methods. In our case, a good choice is for instance to select it by classical non-negative maximization techniques for objectives of type (9) (see *e.g* [21]). Because these methods are quite fast, they can also be used for steps $k \in [2...K]$ in order to provide an alternative starting point $\widehat{W}^k_+$. The update $\widehat{X}^k$ is then given by either NewtonArgMax($\widehat{\mathcal{L}}^k, \widehat{W}^k$) or NewtonArgMax($\widehat{\mathcal{L}}^k, \widehat{W}^k_+$) depending on the most succesful maximization.

5   Experimental Results

We first evaluate MEMIP on realistic synthetic data sets. We compare it to MMEL [8] and fixed exponential kernels and show that MEMIP performs significantly better in terms of prediction and triggering kernels recovery.

5.1   Synthetic Data Sets: Experiment Setup and Results

Data Generation We simulate multivariate Hawkes processes by *Ogata modified thinning algorithm* (see e.g. [22]). Since each occurence can potentially increase stochastic rates of all events, special attention has to be paid to avoid *explosion, i.e* the occurence of an infinite number of events on a finite time window. In order to avoid such behavior, our simulated data sets verify the sufficient non-explosion condition $\rho(\Gamma) < 1$ where $\rho(\Gamma)$ denotes the spectral radius of the matrix $\Gamma = (\int_0^\infty |g_{uv}(t)dt|)_{uv}$ (see *e.g* [17]). We perform experiments on three different simulated data sets where triggering kernels are taken as

$$g_{uv}(t) = \nu_{uv} \frac{\sin\left(\frac{2\pi t}{\omega_{uv}} + \frac{\pi}{2}((u+v) \bmod 2)\right) + 2}{3(t+1)^2} \tag{12}$$

We sample the periods ω_{uv} from an uniform distribution over $[1,10]$. Absolute values of normalization factors ν_{uv} are sampled uniformally from $[0,1/d[$ and their sign is sampled from a Bernoulli law of parameter p. Except for the toy data set, background rates μ_v are taken constant and sampled in $[0,0.001]$. An important feature of this choice of triggering kernels and parameters is that resulting Hawkes processes respect the aforementioned sufficient non-explosion condition. For quantitative evaluation, we simulate two quite large data sets (1) $d = 300, p = 1$ (2) $d = 300, p = 0.9$. Thus, data set (1) contains realizations of purely mutually-exciting processes whereas data set (2) has 10% of inhibitive kernels. For each data set, we sample 10 sets of parameters $(\omega_{uv}, \nu_{uv})_{u\geq 1, v\geq 1}, (\mu_v)_{v\geq 1}$ and simulate 400,000 i.i.d realizations of the resulting Hawkes process over $[0,20]$. The first 200,000 are taken as training set and the remaining 200,000 as test set.

Evaluation Metrics We evaluate the different algorithms by two metrics: (a) *Diff* a normalized L^2 distance between the true and estimated triggering kernels, defined by

$$\text{Diff} = \frac{1}{d^2} \sum_{u=1}^d \sum_{v=1}^d \frac{\int (\widehat{g}_{uv} - g_{uv})^2}{\int \widehat{g}_{uv}^2 + \int g_{uv}^2} \tag{13}$$

, (b) *Pred* a prediction score on the test data set defined as follows. For each dimension $u \in [1...d]$ and occurence i in the test set, probability for that occurence to be of type u is given by $P_i^{true}(u) = \frac{\lambda_u(t_i)}{\sum_{v=1}^d \lambda_v(t_i)}$. Thus, defining $AUC(d,P)$ the area under ROC curve for binary task of predicting $(1_{\{u_i=u\}})_i$ with scores $(P_i^{true}(d))_i$ and $(P_i^{model}(d))_i$ the probabilities estimated by the evaluated model, we set

$$\text{Pred} = \frac{\sum_{u=1}^d (AUC(d, P^{model}) - 0.5)}{\sum_{u=1}^d (AUC(d, P^{true}) - 0.5)} \tag{14}$$

Baselines We compare MEMIP to (a) **MMEL** for which we try various sets of number of base kernels, total number of iterations and smoothing hyperparameter, (b) **Exp** the widely used setting where $g_{uv}(t) = \nu_{uv} \exp(-\alpha t)$ and only ν_{uv} are estimated from the data. In order to give this baseline more flexibility and prediction power, we allow negative values of ν_{uv}. We train three different versions with $\alpha \in \{0.1, 1.0, 10.0\}$.

Results Part 1: Visualization on a Toy Dataset In order to demonstrate the ability of MEMIP to discover the underlying dynamics of Hawkes processes even in presence of inhibition and varying background rates, we construct the following toy bidimensional data set. Amongst the four triggering kernels, g_{11} is taken negative and background rates are defined by $\mu_0 = \frac{cos(\frac{2\pi t}{\omega_0})+2}{1+t}$ and $\mu_1 = \frac{sin(\frac{2\pi t}{\omega_1})+2}{1+t}$ with parameters ω_0 and ω_1 sampled in $[5, 15]$. We sample a set of parameters $(\omega_{uv}, \nu_{uv})_{u\geq 1, v\geq 1}, (\mu_v)_{v\geq 1}$ and simulate 200,000 i.i.d realizations of the resulting Hawkes process. From Fig. 1, we observe that both compared methods MEMIP and MMEL accurately recover nonnegative triggering kernels g_{00}, g_{01} and g_{10}. However, MEMIP is also able to estimate the inhibitive g_{11} whereas MMEL predicts $g_{11} = 0$. Varying background rates μ_0 and μ_1 are also well estimated by MEMIP, whereas by construction MMEL and Exp only return constant values $\bar{\mu}_0$ and $\bar{\mu}_1$.

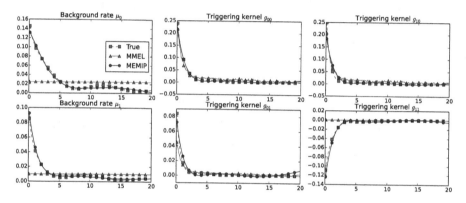

Fig. 1. Triggering kernels and background rates for toy data set estimated by MEMIP and MMEL algorithms vs true triggering kernels and background rate

Results Part 2: Prediction Score In order to evaluate **Pred** score of the competing methods on the generated data sets, we remove for each model the best and worst perfomance over the ten simulated processes, and average **Pred** over the eight remaining one. Empirical 10% confidence intervals are also indicated to assess significativity of the experimental results. From Table 1, we observe that MEMIP significantly outperforms the competing baselines for both data sets. Prediction rates are quite low for all methods which indicates a rather difficult prediction problem, as 90,000 nonparametric functions are indeed to be estimated from the data. In Fig. 2 , we study the sensitivity of **Pred** score to α and K for simulated data sets (1)(above) and (2)(below). Left plots show MEMIP and Exp **Pred** score with respect to α, as well as best MMEL average score across a broad range of hyperparameters. Empirical 10% confidence intervals are also plotted in dashed line. We see that MEMIP gives good results in a wide range of values of α, and outperforms the exponential baseline for all values of α. Right plots show MEMIP **Pred** score with respect to K for $\alpha = 0.1$, as well as best Exp and MMEL average score. We see that MEMIP achieves good prediction results for low values of K, and that taking $K > 10$ is not necessary. For very large values of α, we also note that MEMIP and Exp baseline are the same, because the optimal choice of K for MEMIP is $K = 1$.

Table 1. Pred score for prediction of the type of next event on simulated data sets

Dataset	MEMIP	MMEL	Exp
(1) d=300,p=1	$0.288 \in [0.258, 0.310]$	$0.261 \in [0.250, 0.281]$	$0.255 \in [0.236; 0.278]$
(2) d=300,p=0.9	$0.287 \in [0.266, 0.312]$	$0.261 \in [0.241, 0.280]$	$0.256 \in [0.242, 0.280]$

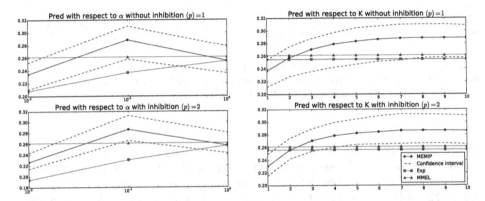

Fig. 2. Sensitivity to hyperparameters α (left) and K(right) for **Pred** score of MEMIP algorithm, compared to Exp and MMEL baselines on non-inhibitive simulated data set (above) and simulated data set with 10 % inhibitive kernels (below)

Results Part 3: Accuracy of Kernel Estimation Besides having a greater prediction power, we observe in Table **2** that MEMIP is also able to estimate the true values of triggering kernels more accurately on both data sets. In Fig. **3**, we study the sensitivity of **Diff** score to α and K for data sets (1)(above) and (2)(below). We see that the variance of **Diff** score is very low for MEMIP, and its fitting error is significatively lower than those of the baselines at level 10%.

Comparison to Related Work The closest work to ours is the algorithm MMEL derived in [8] by Zhou, Zha and Song. MMEL decomposes the triggering kernels on a low-rank set of basis functions, and makes use of EM-like methods in order to maximize the log-likelihood. Compared to MMEL, the proposed algorithm MEMIP enjoy three main improvements: 1) $O(N)$ complexity instead of $O(N^2)$, 2) global convergence of log-likelihood maximization, 3) the ability to learn negative projection coefficients $X_{uv,k}$ as well as varying background rates. Experimental results also suggest that MEMIP may outperform MMEL significantly even for non-inhibitive data set. Actually, even in purely mutually-exciting

Table 2. Diff score for triggering kernels recovery on simulated data sets

Dataset	MEMIP	MMEL	Exp
(1) d=300,p=1	$0.759 \in [0.755, 0.768]$	$0.807 \in [0.803, 0.814]$	$0.791 \in [0.788, 0.800]$
(2) d=300,p=0.9	$0.803 \in [0.793, 0.810]$	$0.839 \in [0.833, 0.844]$	$0.830 \in [0.818, 0.836]$

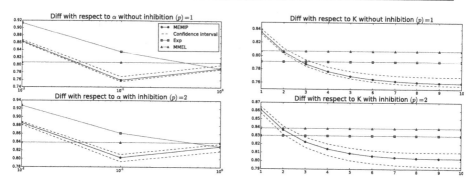

Fig. 3. Sensitivity to hyperparameters α (left) and K(right) for **Diff** score of MEMIP algorithm, compared to Exp and MMEL baselines on non-inhibitive simulated data set (above) and simulated data set with 10 % inhibitive kernels (below)

settings, these two algorithms can exhibit quite different behaviors due to their smoothing strategies. Indeed, because the log-likelihood (1) can be made arbitrarily high by the sequence of functions $(g_{uv}^n)_{n \in N}$ defined by $g_{uv}^n(t) = n1_{\{t \in T_{uv}\}}$ where $T_{uv} = \{t_v - t_u \mid (t_u < t_v \wedge (\exists h \in \mathcal{H} \mid (t_v, v) \in h \wedge (t_u, u) \in h))\}$, smoothing is mandatory when learning triggering kernels by means of log-likelihood maximization. Using a L^2 roughness norm penalization $\alpha \int_0^T g'^2$, MMEL can face difficult dilemmas when fitting power-laws fastly decaying around 0 : either under-estimating the rate when it is at its peak or lowering the smoothness parameter and being vulnerable to overfitting. On the contrary, MEMIP would face difficulties to perfectly fit periodic functions with a very small period, as the derivative of its order K estimates can only vanish $K - 1$ times.

5.2   Experiment on the MemeTracker Data Set

In order to show that the ability to estimate inhibitive triggering kenels and varying background rates yields better accuracy on real-world data sets, we compare the proposed method MEMIP to different baselines on the MemeTracker data set, following the experience plan exposed in [8]. MemeTracker contains links creation between some of the most popular websites between August 2008 and April 2009. We extract link creations between the top 100 popular websites and define the occurence of an event for the ith website as a link creation on this website to one the 99 other websites. We then use half of the data set as training data and the other half at test data on which each baseline is evaluated by average area under ROC curve for predicting future events. From Fig. **4**, we observe that the proposed method MEMIP achieves a better prediction score than both baselines. Left plot shows MEMIP and Exp prediction score with respect to α, as well as best MMEL score across a broad range of hyperparameters. We see that MEMIP gives good results in a very broad range of values of α, and significantly outperforms the exponential baseline for all values of α. Right plot shows MEMIP prediction score with respect to K for $\alpha = 0.01$, as well as best Exp and MMEL score. For $K = 10$, MEMIP achieves a prediction score of 0.8021 whereas best MMEL and Exp score are respectively 0.6928 and 0.7716. We note that, even for K as low as 3, MEMIP performs the prediction task quite accurately.

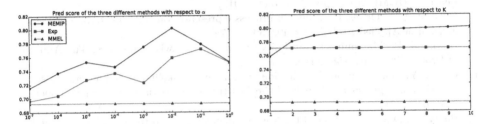

Fig. 4. Sensitivity to hyperparameters α (left) and K(right) for prediction score of MEMIP algorithm, compared to Exp and MMEL baselines on MemeTracker data set

6   Conclusions

In this paper, we propose MEMIP, which is to our knowledge the first method to learn nonparametrically triggering kernels of multivariate Hawkes processes in presence of inhibition and varying background rates. By relying on results of approximation theory, the triggering kernels are decomposed on a basis on memoryless exponential kernels. This maximization of the log-likelihood is then shown to reformulate as a concave maximization problem, that can be solved in linear complexity thanks to the Markov property verified by the proposed estimates. Experimental results on both synthetic and real-world data sets show that the proposed model is able to learn more accurately the underlying dynamics of Hawkes processes and therefore has a greater prediction power.

References

1. Ogata, Y.: Statistical models for earthquake occurrences and residual analysis for point processes. Journal of the American Statistical Association 83(401), 9–27 (1988)
2. Errais, E., Giesecke, K., Goldberg, L.R.: Pricing credit from the top down with affine point processes. In: Numerical Methods for Finance, pp. 195–201 (2007)
3. Bauwens, L., Hautsch, N.: Modelling financial high frequency data using point processes. Springer (2009)

4. Bacry, E., Delattre, S., Hoffmann, M., Muzy, J.F.: Modelling microstructure noise with mutually exciting point processes. Quantitative Finance 13(1), 65–77 (2013)
5. Alfonsi, A., Blanc, P.: Dynamic optimal execution in a mixed-market-impact Hawkes price model. ArXiv preprint ArXiv:1404.0648 (2014)
6. Mohler, G.O., Short, M.B., Brantingham, P.J., Schoenberg, F.P., Tita, G.E.: Self-exciting point process modeling of crime. Journal of the American Statistical Association 106(493), 100–108 (2011)
7. Reynaud-Bouret, P., Schbath, S.: Adaptive estimation for Hawkes processes; application to genome analysis. The Annals of Statistics 38(5), 2781–2822 (2010)
8. Zhou, K., Zha, H., Song, L.: Learning triggering kernels for multi-dimensional Hawkes processes. In: Proceedings of the 30th International Conference on Machine Learning (ICML 2013), pp. 1301–1309 (2013)
9. Hawkes, A.G., Oakes, D.: A cluster process representation of a self-exciting process. Journal of Applied Probability, 493–503 (1974)
10. Brémaud, P., Massoulié, L.: Stability of nonlinear Hawkes processes. The Annals of Probability, 1563–1588 (1996)
11. Ogata, Y.: The asymptotic behaviour of maximum likelihood estimators for stationary point processes. Annals of the Institute of Statistical Mathematics 30(1), 243–261 (1978)
12. Bacry, E., Dayri, K., Muzy, J.F.: Non-parametric kernel estimation for symmetric Hawkes processes. Application to high frequency financial data. The European Physical Journal B 85(5), 1–12 (2012)
13. Crane, R., Sornette, D.: Robust dynamic classes revealed by measuring the response function of a social system. Proceedings of the National Academy of Sciences 105(41), 15649–15653 (2008)
14. Lewis, E., Mohler, G.: A nonparametric EM algorithm for multiscale Hawkes processes. In: Joint Statisticals Meetings 2011 (2011)
15. Hansen, N.R., Reynaud-Bouret, P., Rivoirard, V.: Lasso and probabilistic inequalities for multivariate point processes. ArXiv preprint ArXiv:1208.0570 (2012)
16. Lewis, E., Mohler, G., Brantingham, P.J., Bertozzi, A.L.: Self-exciting point process models of civilian deaths in Iraq. Security Journal 25(3), 244–264 (2011)
17. Daley, D.J., Vere-Jones, D.: An introduction to the theory of point processes. Springer (2007)
18. Cheney, E.W., Cheney, E.W.: Introduction to approximation theory, vol. 3. McGraw-Hill, New York (1966)
19. Powell, M.: Some global convergence properties of a variable metric algorithm for minimization without exact line searches. Nonlinear Programming 9, 53–72 (1976)
20. Nesterov, Y., Nemirovskii, A.S., Ye, Y.: Interior-point polynomial algorithms in convex programming, vol. 13. SIAM (1994)
21. Seung, D., Lee, L.: Algorithms for non-negative matrix factorization. In: Advances in Neural Information Processing Systems, vol. 13, pp. 556–562 (2001)
22. Liniger, T.J.: Multivariate Hawkes processes. PhD thesis, Diss., Eidgenössische Technische Hochschule ETH Zürich, Nr. 18403 (2009)

Learning Binary Codes with Bagging PCA

Cong Leng[1], Jian Cheng[1,*], Ting Yuan[1], Xiao Bai[2], and Hanqing Lu[1]

[1] National Laboratory of Pattern Recognition, Institute of Automation,
Chinese Academy of Sciences, Beijing, China
{cong.leng,jcheng,tyuan,luhq}@nlpr.ia.ac.cn
[2] School of Computer Science and Engineering, Beihang University, Beijing, China
baixiao@buaa.edu.cn

Abstract. For the eigendecomposition based hashing approaches, the information caught in different dimensions is unbalanced and most of them is typically contained in the top eigenvectors. This often leads to an unexpected phenomenon that longer code does not necessarily yield better performance. This paper attempts to leverage the bootstrap sampling idea and integrate it with PCA, resulting in a new projection method called Bagging PCA, in order to learn effective binary codes. Specifically, a small fraction of the training data is randomly sampled to learn the PCA directions each time and only the top eigenvectors are kept to generate one piece of short code. This process is repeated several times and the obtained short codes are concatenated into one piece of long code. By considering each piece of short code as a "super-bit", the whole process is closely connected with the core idea of LSH. Both theoretical and experimental analyses demonstrate the effectiveness of the proposed method.

Keywords: Bootstrap, random, bagging, PCA, binary codes, Hamming ranking.

1 Introduction

Hashing based approximate nearest neighbor (ANN) search is crucial for many large scale machine learning and computer vision tasks, such as image retrieval [24], object detection [5] and 3D reconstruction [23]. In these tasks, one is often required to find the nearest neighbor for one point in a huge database. Nearest neighbor (NN) search is unfeasible in these scenarios because its high time complexity. In the hashing based approaches, binary codes will be generated for points in the database and similar points will have close codes. Searching will be very fast because the Hamming distance between binary codes can be efficiently calculated with XOR instruction in modern CPU. Furthermore, the binary code can be very compact for memory storage.

The most well-known hashing method is Locality Sensitive Hashing (LSH) [13,4,3,20], which generates a batch of random projections to embed the data into Hamming space. Owing to the inner randomness, LSH based methods are *data-independent* and have nice theoretical properties. Indyk *et al.* [13] proved that two similar samples would be embedded to close codes with high probability, as long as the hash functions are of locality-sensitive function family. Moreover, they proved that the collision

---

* Corresponding author.

T. Calders et al. (Eds.): ECML PKDD 2014, Part II, LNCS 8725, pp. 177–192, 2014.
© Springer-Verlag Berlin Heidelberg 2014

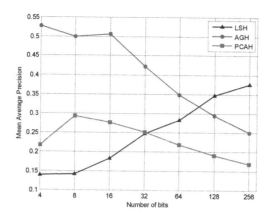

Fig. 1. Mean Average Precision of LSH, AGH and PCAH with various bits on toy data

probability would be higher as the code size increases [13]. However, in practice LSH typically needs very long codes and multiple tables to guarantee reasonable recall rate, which would degrade the search efficiency. By comparison, many recent hashing approaches attempt to learn data-aware hash functions by utilizing machine learning techniques. Several methods such as Spectral Hashing (SH) [26], Anchor Graph Hashing (AGH) [18], Iterative Quantization (ITQ) [7], Spherical Hashing (SpH) [11] and Kernel Supervised Hashing (KSH) [17] have been developed. These *data-dependent* methods aim to learn a set of projections, which are usually demonstrated to be more effective than the data-independent LSH.

From the mathematical perspective, lots of existing data-dependent hashing methods are based on eigendecomposition of matrix. Motivated by spectral clustering, SH [26] calculates the bits by thresholding a subset of eigenvectors of a Laplacian matrix of the similarity graph. AGH [18] follows the same idea of SH but utilizes anchor graph to overcome the computation problem in graph construction. Self-taught Hashing (STH) [29] first binarizes the eigenvectors of a normalized Laplacian and then learns SVMs as hash functions to overcome the out-of-sample extension problem. Other representative works include PCA Hashing (PCAH) [7], Semi-supervised Hashing (SSH) [24], Optimized Kernel Hashing (OKH) [8], etc.

For eigendecomposition based hashing algorithms, the information caught by different eigenvectors is unbalanced, that is, most of the information is typically contained in the top eigenvectors while the remainders are usually less informative or even noisy. Since each eigenvector is encoded into the same number of bits, the bits generated with the noisy eigenvectors will be noisy too, which will degrade the performance. As demonstrated in Fig.1, increasing number of bits leads to poorer mean average precision (MAP) performance on the toy data[1] with both PCAH and AGH. This is because more noise will be introduced while more eigenvectors are used. However, by intu-

---

[1] The used toy data is MNIST, which is available at http://yann.lecun.com/exdb/mnist/. 1,000 images are randomly selected as queries. The ground truth is defined as semantic neighbors based on digit labels.

ition, longer codes should catch more information than the shorter ones and give better retrieval performance, like that in LSH. On the other hand, although the short code generated with the top eigenvectors achieves decent performance, the data representation capability of short code is limited, and the most recently proposed hashing methods such as SpH [11] often use relatively longer code to get much better performance. Therefore, for eigendecomposition based methods, there exists a dilemma between the code length and performance.

Since most of information is contained in the top eigenvectors, one natural idea is to just use the top eigenvectors to generate a piece of short but strong code and repeat this process several times, and then concatenate these pieces of short codes into one piece of long code. However, it is obvious that if these pieces of short codes are identical, the obtained long code won't catch any more information. Thus, to get stronger long code, the short codes should be strong but also diverse. Inspired by the work of bagging strategy [2], in this paper, we adopt the widely used bootstrap technique [6] to generate diverse short codes. A new projection method, named Bagging PCA, is proposed and applied to learn effective binary codes. More specifically, each time we randomly sample a small subset of the training data to learn PCA directions, and only the top eigenvectors are kept to generate one piece of short code for all the data. This process will be repeated several times and afterwards the obtained many pieces of short codes will be concatenated into one piece of long code. The proposed hashing method, dubbed Bagging PCA Hashing (BPCAH), enjoys the following three appealing advantages:

- Since only the top eigenvectors are used every time, we can expect that the obtained binary codes would be more effective. Extensive experiments on three large scale datasets demonstrate that the proposed method outperforms several state-of-the-art hashing methods.
- It can be theoretically guaranteed that the longer codes tend to yield better results than the shorter ones under such a strategy. In addition, due to the randomness introduced by bagging, our method shares some common attributes with the well-known LSH, which is of nice theoretical properties.
- Because only PCA is used in the whole process, the proposed method is very suitable for large scale dataset. An important benefit of bagging scheme is that it is inherently favorable to parallel computing. Therefore, although the learning process has to be repeated several times, it can be completed in parallel with independent computation units.

2   Related Work

In this section, we give some backgrounds about hashing and introduce some related works which attempt to handle the unbalance problem in eigendecomposition based hashing methods. First of all, some notations are defined. Let $\mathcal{X} = \{x_1, x_2, \cdots, x_n\}$ denote a set of n data points, where $x_i \in \mathbb{R}^d$ is the ith data point. We denote $X = [x_1, x_2, \cdots, x_n] \in \mathbb{R}^{d \times n}$ as the data matrix. The binary code matrix of these points is $H = [h_1, h_2, \cdots, h_r] \in \{-1, 1\}^{n \times r}$, where r is the code length. Hashing code for one point is a row of H and denoted as $code(x_i)$.

As one of the most popular hashing methods, LSH randomly samples hash functions from a locality sensitive function family [1]. SimHash [4,13] and MinHash [3,20] are two widely adopted LSH schemes. MinHash is a technique for quickly calculating the Jaccard coefficient of two sets by estimating the resemblance similarity defined over binary vercotrs [3]. In contrast, SimHash is an LSH for the similarities (e.g., cosine similarity) which work on general real-valued data. As indicated in [21], when the data are high-dimensional and binary, MinHash tends to work better than SimHash. On the other hand, SimHash achieves better performance than MinHash on real-valued data. Specifically, to approximate the cosine similarity, Charikar [4] defined a hash function h as:

$$h(q) = \begin{cases} 1, & \text{if } w \cdot q > 0 \\ 0, & \text{if } w \cdot q < 0 \end{cases} \tag{1}$$

where w is a random vector from the d-dimensional Gaussian distribution $N(0, I_d)$. Although with abundant nice theoretical properties, these random projection based data-independent hashing methods are less discriminative over data and typically need very long codes to achieve satisfactory search performance.

Recently, many data-dependent hashing methods [26,18,15,11,24,29,16,7,9] have been proposed to learn data aware hash functions. As we have mentioned, many of them [26,18,24,29,8,7] are based on eigendecomposition of a matrix (e.g. Laplacian matrix). This brings the unbalance problem because the information caught in different eigenvectors is unbalanced. A few recent works have been proposed to address this problem.

In [25], instead of learning all the eigenvectors at once, Wang *et al.* proposed a sequential learning framework (USPLH) to learn hash function which tends to minimize the errors made by the previous one. Inspired by multiclass spectral clustering [28], in Iterative Quantization (ITQ) [7], Gong *et al.* proposed an alternating minimization scheme to learn an orthogonal transformation to the PCA projected data so as to minimize the quantization error of mapping the data to their corresponding binary codes (the vertices of binary hypercube). In Isotropic Hashing (IsoH) [15], Kong *et al.* proposed to learn projection functions which can produce projected dimensions with equal variances. Same as in ITQ, they tried to learn an orthogonal transformation to the PCA projected data by iteratively minimizing the reconstruction error of the covariance matrix and a diagonal matrix. Similar idea was adopted in [27], but in which the PCA projection was replaced with locality preserving projection (LPP) [10]. In these methods, longer codes often catch much more information and thus give better experimental results than the shorter ones. However, to the best of our knowledge, there still lack enough theoretical guarantee that the performance will be better as the size of codes increases, like in LSH.

The differences between our method and the previous works are obvious. Instead of minimizing the quantization error or toughly requiring each dimension to have equal variance, we leverage the bootstrap sampling scheme and integrate it with PCA. Every time only the informative top eigenvectors are used to learn short binary code. Owing to the sophisticated theories established in ensemble learning, our method enjoys several advantages which are lacking from previous works.

3   The Proposed Approach

Assuming the data X is zero-centered, for a projection $W = (w_1, w_2, \cdots, w_r) \in \mathbb{R}^{d \times r}$, code matrix can be written as $H = sgn(X^T W)$, where $sgn(\cdot)$ is the sign function. In general, for a code to be efficient, two requirements should be satisfied [26]: (1) each bit has a 50% chance of being +1 or -1, i.e. $\sum_i h_k(x_i) = 0, k = 1, 2, \cdots, r$; (2) different bits are independent of each other.

For the first requirement, Wang *et al.* [24] have proved that constraint $\sum_i h_k(x_i) = 0$ is equivalent to maximizing the variance for the k-th bit. The second "independent" requirement is often relaxed into the pairwise decorrelation of bits [26], i.e. $\frac{1}{n} H^T H = I$. In [24], it was further relaxed into the orthogonality constraints on the projection directions, i.e. $W^T W = I$. By dropping the non-differentiable $sgn(\cdot)$ function and trivially using $tr(AB) = tr(BA)$, overall, the objective can be formally written as [24,7,15]:

$$\max_{W \in \mathbb{R}^{d \times r}} \frac{1}{n} tr(W^T X X^T W)$$

$$s.t. \quad W^T W = I_r \tag{2}$$

This objective function is exactly the same as that of Principal Component Analysis (PCA). The optimized projection W can be obtained by solving the top r eigenvectors corresponding to the r biggest eigenvalues of the data covariance matrix $X X^T$.

3.1   Hashing with Bagging PCA

For the eigendecomposition based hashing method, e.g. PCAH, on the one hand, the amount of information caught in different dimensions differs significantly, but each dimension is encoded into the same bits of code. This brings the unbalance problem in the obtained binary codes. On the other hand, the most discriminative information is typically contained in the top eigenvectors so that the short code generated with only the top eigenvectors often yield better retrieval performance than longer ones in these methods. In spite of this, the data representation capability of short codes is limited, and the most recently proposed hashing methods often use relatively longer code to achieve better performance.

We notice that the bagging strategy can enhance advantages and avoid disadvantages of the eigendecomposition based method. Bagging [2] is a classical and efficient combining technique for improving weak classifiers, and is extensively applied to classification problems [22]. Bagging is based on bootstrap [6] and aggregating concepts, and incorporates the benefits of both approaches. Bootstrap is based on random sampling in all the training data with replacement. Taking a bootstrap replicate of p samples $X^{(i)} = [x_1^{(i)}, x_2^{(i)}, \cdots, x_p^{(i)}]$ from the whole training set $\mathcal{X} = \{x_1, x_2, \cdots, x_n\}$, one can sometimes avoid or get less misleading training samples in the bootstrap training set. Aggregating actually means combining the base weak learners, and the combined learner typically gives better results than individual base learner. In general, bagging is helpful to build a better learner on training sets with misleading samples.

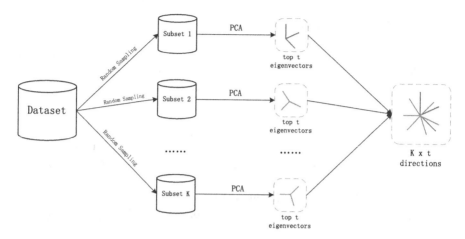

Fig. 2. The flowchart of Bagging PCA. At first, K bootstrap training set are randomly sampled from the whole training set. For each bootstrap, PCA is applied and only the top t eigenvectors (directions) are kept. Afterwards, the K diverse blocks of PCA directions are combined.

The motivation is clear: since most of the information is caught in the top eigenvectors, why not repeat using them to generate short codes and then concatenate them into long code. However, it is obvious that if the many pieces of short codes are identical, the obtained long code will not get any more information. Based on this idea, we propose a new projection method, named as Bagging PCA, and apply it to learn discriminative binary codes.

In the first step, a small subset of p samples $X^{(i)} = [x_1^{(i)}, x_2^{(i)}, \cdots, x_p^{(i)}]$ is randomly sampled from the training set X. With this bootstrap training set, we can learn the corresponding PCA projections and only the top t eigenvectors are kept. In other words, we optimize the following objective and get the optimized $W^{(i)} \in \mathbb{R}^{d \times t}$.

$$\max_{W^{(i)} \in \mathbb{R}^{d \times t}} \frac{1}{p} tr(W^{(i)^T} X^{(i)} X^{(i)^T} W^{(i)})$$
$$s.t. \quad W^{(i)^T} W^{(i)} = I_t \tag{3}$$

With the sampling and learning process repeated K times, we can obtain K diverse blocks of PCA directions, i.e. $\{W^{(i)}\}_{i=1}^{K}$. The final projection matrix is generated by combining the K blocks:

$$W = \left[W^{(1)}, W^{(2)}, \cdots, W^{(K)}\right] \in \mathbb{R}^{d \times Kt} \tag{4}$$

The binary code for each data x can be written as $code(x) = sgn(x^T W)$. Note that this equals to concatenate many pieces of short codes into one piece of long code because $sgn(x^T W) = [sgn(x^T W^{(1)}), sgn(x^T W^{(2)}), \cdots, sgn(x^T W^{(K)})]$.

In some extreme cases, there exists heavy unbalance even between the top eigenvectors. In these circumstances, a random rotation can be applied to the learned $W^{(i)}$.

Algorithm 1. Hashing with Bagging PCA

Input: Training set $\mathcal{X} = \{x_1, x_2, \cdots, x_n\}$, the number of piece of short codes K, the number of samples in each bootstrap replicate p, the code size of short code t.
Output: Hashing codes of $K \times t$ bits as well as $K \times t$ hash functions.

1: Generate K bootstrap replicates $\{X^{(i)}\}_{i=1}^K$. Each replicate $X^{(i)}$ contains p training samples randomly selected from the whole training set $\mathcal{X}$ with n samples.
2: **for** $i = 1, \cdots, K$ **do**
3:    Calculate the covariance matrix $C^{(i)} = X^{(i)} X^{(i)T}$.
4:    Get the t top eigenvectors $\{e_k^{(i)}\}_{k=1}^t$ of $C^{(i)}$, denoted as $W^{(i)} = [e_1^{(i)}, e_2^{(i)}, \cdots, e_t^{(i)}]$.
5:    Optional: Apply a random rotation to $W^{(i)}$: $W^{(i)} \longleftarrow W^{(i)} R$.
6:    **Coding:** for j = 1, $\cdots$, n, do
7:       $code^{(i)}(x_j) \longleftarrow sgn(x_j^T W^{(i)})$
8: **end for**
9: Concatenate the K pieces of short codes $\{code^{(i)}\}_{i=1}^K$ of each sample into one piece of $(K \times t)$ bits binary code $[code^{(1)}, code^{(2)}, \cdots, code^{(K)}]$.

Obviously, if $W^{(i)}$ is an optimal solution of Eq.(3), then so is $W^{(i)} R$ for any $t \times t$ orthogonal matrix R. As indicated in [14], a random rotation is helpful for balancing the information in top eigenvectors. The flowchart of Bagging PCA is shown in Fig.2, and the proposed strategy for binary codes learning can be summarized as in Algorithm 1.

3.2   Theoretical Analysis

In the last decades, mature theory frame has been established in ensemble learning to guarantee correctness of the algorithms such as bagging, random forest and boosting. In this subsection, we generalize the discussion to our bagging PCA hashing. In specific, we intend to theoretically guarantee that longer hashing codes tend to give better retrieval performance in our method.

Let $f : \mathcal{X} \times \mathcal{X} \rightarrow [0, 1]$ be a ground truth similarity function over a set of objects $\mathcal{X}$, where we can interpret $f(x, y)$ to mean that x and y are "similar" or "dissimilar". The essence of hashing is constructing codes such that the Hamming distance between $code(x)$ and $code(y)$ corresponds to the ground truth similarity $f(x, y)$, i.e. two similar items have a small Hamming distance, while two dissimilar items have a large Hamming distance. Although discussing the Hamming distance, as indicated in [17], the Hamming distance $d(x, y)$ between $code(x)$ and $code(y)$ can be directly converted to a similarity measure $S(x, y)$ in Hamming space:

$$S(x, y) = \frac{r - d(x, y)}{r} \tag{5}$$

where r is the code length. Obviously, $d(x, y) \in [0, r], S(x, y) \in [0, 1]$, and smaller Hamming distance corresponds to larger similarity. From this perspective, we can consider that the essence of hashing is constructing codes for the data whose similarities match the ground truth as better as possible.

Lemma 1: *Concatenating K pieces of short codes of t bits into one piece of $K \times t$ bits long code, then the similarity between two samples evaluated with the long code is the mean of those evaluated with the K pieces of short codes.*

Proof: Denote the Hamming distance and similarity between two samples evaluated with the short codes as $d_i(x, y)$ and $S_i(x, y)$, where $i = 1, 2, \cdots, K$. Similarly, denote that evaluated with the long code (obtained by concatenating short ones) as $d_l(x, y)$ and $S_l(x, y)$, respectively. It is obvious that $d_l(x, y) = \sum_{i=1}^{K} d_i(x, y)$. Then we have

$$
\begin{aligned}
S_l(x, y) &= \frac{K \times t - d_l(x, y)}{K \times t} \\
&= \frac{1}{K} \left(\frac{K \times t - \sum_{i=1}^{K} d_i(x, y)}{t} \right) \\
&= \frac{1}{K} \left(\sum_{i=1}^{K} \left(\frac{t - d_i(x, y)}{t} \right) \right) \\
&= \frac{1}{K} \sum_{i=1}^{K} S_i(x, y)
\end{aligned}
$$

Theorem 1: *Under the bootstrap sampling framework, the similarity between two samples evaluated with the long code (obtained by concatenating short codes) tend to be closer to the ground truth than that evaluated with the short code.*

Proof: As shown in Lemma 1, the aggregated similarity evaluated with long code is:

$$S_l(x, y) = \mathbf{E}[S_i(x, y)] \tag{6}$$

It is easy to find that:

$$
\begin{aligned}
\mathbf{E}[(f(x, y) - S_i(x, y))^2] = f^2(x, y) - 2f(x, y)\mathbf{E}[S_i(x, y)] \\
+ \mathbf{E}[S_i^2(x, y)]
\end{aligned} \tag{7}
$$

Since $E[Z^2] \geq (E[Z])^2$, we have

$$\mathbf{E}[S_i^2(x, y)] \geq (\mathbf{E}[S_i(x, y)])^2 \tag{8}$$

Eq.(7) can derive

$$\mathbf{E}[(f(x, y) - S_i(x, y))^2] \geq (f(x, y) - \mathbf{E}[S_i(x, y)])^2 \tag{9}$$

Plugging in Eq.(6), we have

$$(f(x, y) - S_l(x, y))^2 \leq \mathbf{E}[(f(x, y) - S_i(x, y))^2] \tag{10}$$

From Eq.(10) we can get some insights on how the longer code improve the ranking performance. The deviation of $S_l(x, y)$ to the true similarity $f(x, y)$ is smaller than that of $S_i(x, y)$ averaged over the bootstrap sampling distribution. As we have mentioned, the essence of hashing is constructing codes whose similarity can match the ground truth as better as possible. Therefore, we can draw the conclusion that longer codes tend to give better retrieval performance under such a strategy.

How much improvement we can get depends on how unequal the Eq.(8) is. The effect of diversity is clear. If $S_i(x, y)$ does not change too much with different i the two sides will be nearly equal, and bagging will not help. As an extreme example, if every time all the samples are used to train the model, $S_i(x, y)$ will be identical for different i, so the left side and right side of Eq.8 (so Eq.10) will be the same. But S_l is always not inferior to S_i in theory.

Bagging strategy has been proved to be an efficient way to reduce generalization error by combining results from multiple base classifiers. According to Hoeffding inequality [12], when the base classifiers are mutually independent, the generalization error of the ensemble reduces exponentially to the ensemble size, and ultimately approaches to zero as the ensemble size approaches to infinity. Similar theory can be applied in our approach, but here the generalization error is the deviation to the true similarity and the ensemble size is the length of code.

3.3 Connection with LSH

For LSH, Indyk *et al.* [13] have proved that two similar samples will be embedded into close codes with high probability and this probability will increase as the code size increases [13]. Actually, as pointed out in [7], LSH is guaranteed to yield exact Euclidean neighbors in the limit of infinitely many bits. As we have proved in Theorem 1, the obtained long code by concatenating short codes will result in smaller deviation to the true similarity, which is the same as in LSH.

On the other hand, if we treat one piece of short code in our method as a "super-bit", our method can be seen as a special case of LSH. The difference is that in LSH the hash functions are randomly generated but in our method we introduce randomness via randomly sampling the training data and every "super-bit" here is learned with consideration of the data. Our method enjoys the benefits of both data-independent and data-dependent methods.

3.4 Computation Complexity Analysis

There exists a straightforward question for Bagging PCA hashing, i.e. whether the bagging strategy will increase the computational complexity. The time complexity of PCA is $O(nd^2 + d^3)$, which is linear to the size of dataset. For the proposed BPCAH, each time the size of training set is p, which is much smaller than n. In addition, an important benefit of bagging scheme for hashing is that it is inherently favorable to parallel computing. With this benefit, although we have to repeat the learning process K times, it can be completed in parallel with K computation units. This characteristic is important for large scale dataset in real applications.

4 Experiments

To be consistent with the motivation and theoretical analysis given above, what we want to verify in experiments is threefold: (1) The proposed solution addressing the unbalance problem of eigendecomposition based methods outperforms other existing solutions. (2) Longer codes, as indicated in the previous section, do give better performance

Table 1. Description of Datasets

	CIFAR10	Tiny100K	GIST1M
Dimensionality	512	384	960
Size	60,000	100,000	1,000,000

than the shorter ones in the proposed method. (3) The proposed method outperforms other state-of-the-art hashing methods.

4.1 Experimental Setting

Datasets: Three large scale datasets are employed to evaluate the proposed method: CIFAR10, Tiny100K and GIST1M. CIFAR10[2] consists of 60K 32×32 color images in 10 classes, with 6000 images per class. We extract 512 dimensional GIST descriptor [19] to represent each image. Tiny100K consists of 100K tiny images randomly sampled from the original 80 million tiny images[3]. Each image is represented by 384 dimensional GIST descriptor. GIST1M contains one million 960 dimensional GIST descriptors extracted from random images. It is directly downloaded from the website[4]. We summarize the size and dimensionality of the datasets in Table 1. For each dataset, we randomly select 1,000 data points as queries and use the rest as gallery database and training set. Following [24,25], the top 2 percentile nearest neighbors in Euclidean space are taken as ground truth.

Compared Methods: We compare the proposed BPCAH with several state-of-the-arts hashing methods, including Locality Sensitive Hashing (LSH) [4], PCA Hashing (PCAH) [24], Anchor Graph Hashing (AGH) [18], Unsupervised Sequential Projection Learning Hashing (USPLH) [25], Iterative Quantization (ITQ) [7], Isotropic Hashing (IsoH) [15] and Spherical Hashing (SpH) [11]. As we have pointed out above, there are many versions of LSH. Here we adopt the algorithm proposed in [4] because the experimental data is dense and real-valued here. AGH is a popular eigendecomposition based hashing method. USPLH, ITQ, IsoH and our BPCAH are all based on PCA and aim to handle the unbalance problem in PCAH. SpH is a recently proposed hashing method which achieves promising retrieval performance on many datasets. We implement LSH and our method by ourselves, and use the source codes provided by the authors for all the other methods. In our method, there are two parameters to be set, the size of each bootstrap training set p and code length of each short code t. We set $p = 30\% \times n$ and $t = 16$ for all the comparisons.

Evaluation Criterions: To perform fair evaluation, we adopt the Hamming Ranking search commonly used in the literature. All points in the database are ranked according

---

[2] http://www.cs.toronto.edu/~kriz/cifar.html
[3] http://groups.csail.mit.edu/vision/TinyImages/
[4] http://corpus-texmex.irisa.fr/

to their Hamming distance to the query and the top K samples will be returned. The retrieval performance is measured with three widely used metrics: mean average precision (MAP), precision of the top K returned examples (Precision@K) and precision-recall curves. The MAP score is calculated by

$$MAP = \frac{1}{|Q|} \sum_{i=1}^{|Q|} \frac{1}{n_i} \sum_{k=1}^{n_i} precision(R_{ik})$$

where $q_i \in Q$ is a query, n_i is the number of points relevant to q_i in the dataset. Suppose the relevant points are ordered as $\{r_1, r_2, \cdots, r_{n_i}\}$, then R_{ik} is the set of ranked retrieval results from the top result until you get to point r_k.

4.2 Experimental Results and Analysis

MAP Scores: MAP is one of the most comprehensive criterions to evaluate the retrieval performance in the literature [25,7,15,11]. Table 2-4 show the MAP scores for all the algorithms on the three datasets. We observe that BPCAH achieves the highest MAP scores with different code lengths on all the datasets. Comparing the data dependent methods with the data-independent LSH, it can be observed that the data dependent methods like ITQ and SpH are generally better than LSH, especially with small code size. However, LSH results in higher MAP score as the code size increasing, for example, from 0.0946 (32 bits) to 0.2997 (256 bits) on CIFAR10. This behavior is due to the theoretical convergence guarantee of LSH that when enough bits are assigned, two similar samples will be embedded into close codes with high probability. By comparison, as the code size increases, the MAP scores of PCAH and AGH decrease. When the code size exceeds 64, the MAP score of AGH is even lower than LSH on all the datasets.

By comparing the MAP scores of our BPCAH with those of three other methods which also aim to address the unbalance problem of eigendecomposition based method, i.e. USPLH, ITQ and IsoH, we find that our method outperforms them with a large margin. This improvement is mainly due to that we only use the informative top eigenvectors and drop the noisy eigenvectors in learning each piece of short code in our method. These results imply that the proposed strategy of concatenating short codes generated with only the top eigenvectors into long code is very effective to handle the unbalance problem.

Considering the MAP scores of BPCAH with different code sizes, it is easy to find that our method yields to higher MAP score as the code size increases on all the datasets. The improvement from 32 bits to 64 bits is prominent, and becomes stable when the code size exceeds 128. This phenomenon is natural and easy to understand since in our method as the code size increases more useful information is integrated. From this perspective, our method is very similar to LSH. This verifies the claims made in the previous section, and it is an important characteristic of our method. Furthermore, our BPCAH consistently performs better than SpH, although the advantage is not so significant when the code size exceeds 256. SpH achieves promising performance on these datasets in the literature [11] and is the best competitor under most settings in our experiments. Given a MAP of 0.32 in Table 2, BPCAH needs to use about 48 bits to encode

Table 2. Mean Average Precision (MAP) scores for CIFAR10 dataset

Methods	Mean Average Precision					
	32-bits	48-bits	64-bits	96-bits	128-bits	256-bits
LSH	0.0946	0.1349	0.1591	0.2046	0.2344	0.2997
PCAH	0.0593	0.0596	0.0587	0.0567	0.0539	0.0464
AGH	0.1488	0.1528	0.1544	0.1511	0.1461	0.1290
USPLH	0.1048	0.1215	0.1213	0.1199	0.1196	0.1192
ITQ	0.2345	0.2604	0.2781	0.2999	0.3131	0.3450
IsoH	0.2138	0.2426	0.2612	0.2876	0.3058	0.3414
SpH	0.1745	0.2131	0.2422	0.2853	0.3198	0.3823
BPCAH	**0.2614**	**0.3170**	**0.3469**	**0.3679**	**0.3819**	**0.3926**

Table 3. Mean Average Precision (MAP) scores for Tiny100K dataset

Methods	Mean Average Precision					
	32-bits	48-bits	64-bits	96-bits	128-bits	256-bits
LSH	0.1280	0.1467	0.1746	0.2100	0.2401	0.2825
PCAH	0.0822	0.0770	0.0712	0.0635	0.0582	0.0460
AGH	0.1582	0.1621	0.1617	0.1547	0.1473	0.1292
USPLH	0.1240	0.1230	0.1229	0.1225	0.1222	0.1218
ITQ	0.2076	0.2294	0.2426	0.2593	0.2627	0.2821
IsoH	0.2008	0.2307	0.2428	0.2627	0.2726	0.2941
SpH	0.1872	0.2256	0.2463	0.2805	0.3066	0.3501
BPCAH	**0.2239**	**0.3040**	**0.3413**	**0.3669**	**0.3810**	**0.3930**

Table 4. Mean Average Precision (MAP) scores for GIST1M dataset

Methods	Mean Average Precision					
	32-bits	48-bits	64-bits	96-bits	128-bits	256-bits
LSH	0.0994	0.1242	0.1381	0.1697	0.1952	0.2453
PCAH	0.1088	0.0998	0.0914	0.0791	0.0718	0.0523
AGH	0.1346	0.1415	0.1455	0.1464	0.1460	0.1655
USPLH	0.1014	0.1006	0.1005	0.0999	0.0995	0.0990
ITQ	0.1878	0.2070	0.2188	0.2307	0.2383	0.2514
IsoH	0.1821	0.1999	0.2180	0.2334	0.2406	0.2592
SpH	0.1544	0.1946	0.2137	0.2447	0.2635	0.3082
BPCAH	**0.1961**	**0.2341**	**0.2517**	**0.2788**	**0.2951**	**0.3187**

each image while the best competitor SpH needs about 128 bits. In consequence, our method typically provides about three times more compact binary representation than other methods. Similar trends can be found in Table 3 and Table 4.

Precision@K: In some applications, what we really concern is the precision of the top K returned samples. For example, in real image retrieval system, most of the users only care about the returned images in the first page. Fig.3(a)-(c) show the precision of

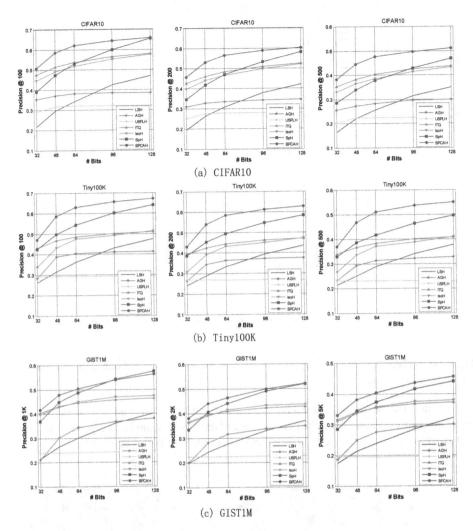

Fig. 3. Precision of top K returned of different methods on three datasets. (a) Precision@100,200,500 with various code sizes on CIFAR10. (b) Precision@100,200,500 with various code sizes on Tiny100K. (c) Precision@1K,2K,5K with various code sizes on GIST1M.

top K returned with different bits on the three datasets. For CIFAR10 and Tiny100K, precision on top 100, 200 and 500 is reported. For GIST1M, since the relevant samples for each query are more (top 2 percentile nearest neighbors are defined to be relevant), we report the precision on the top 1K, 2K and 5K. The performance of PCAH is not even comparable with other competitors as the code size exceeds 32. To avoid clutter, these curves and the subsequent results reported in this section omit the baseline method PCAH.

Our BPCAH consistently outperforms its competitors almost on all the cases. Once again, we gain remarkable improvement over USPLH, ITQ and IsoH. For instance, when the top 100 samples are returned in CIFAR10, our method achieves a precision

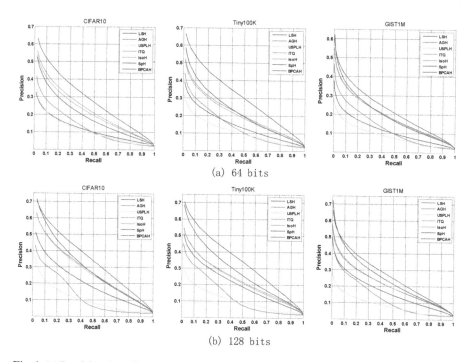

Fig. 4. (a) Precision-Recall curves with 64 bits on three datasets. (b) Precision-Recall curves with 64 bits on three datasets.

of 62% and the best competitor ITQ only arrives at 54%. We can also make some interesting observations about the performance of the other methods. In Fig.3(a)(c), ITQ and IsoH work relatively well for small code size and are better than SpH. However,

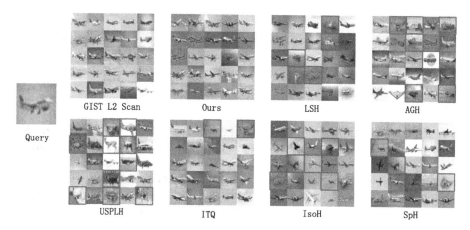

Fig. 5. Retrieval results on CIFAR10 using original gist descriptor, and binary codes build with different hashing methods. Top 25 returned are shown. We used 64-bit hashing codes, and show the false positives in red rectangle.

as the code size increases to 64, the performance of SpH rises rapidly and outperforms ITQ and IsoH (even ours in the left subfig of Fig.3(c)). As indicated in [11], this may be because the closed regions created by the hyperspheres is tighter than those created by the hyperplanes when multiple hyperspheres are used.

Precision-Recall Curves: Fig.4(a)(b) show the complete precision-recall curves on the three datasets with 64 bits and 128 bits. These detailed results are consistent with the trends that we discovered in the Table 2-4. Actually, MAP score is the area under the precision-recall curve. Fig.4 clearly shows the superiority of our BPCAH over other hashing methods.

In order to give an intuitive understanding about how these hashing methods work, Fig.5 shows an example with CIFAR10. An input query image on the left with 25 nearest neighbors using the original gist descriptor and binary codes built with different hashing methods are shown.

5   Conclusion and Future Work

In this paper, we proposed a new projection method named Bagging PCA for binary codes learning. The key idea is concatenating many pieces of diverse short codes into one piece of long code. In order to obtain diverse short codes, we adopted the bootstrap technique to learn PCA directions. Theoretical analysis and the connection with LSH were given. Extensive experiments on three large scale datasets demonstrated that our approach can outperform other state-of-the-arts hashing methods. Future work will explore the effectiveness of the proposed BPCAH on high-dimensional binary data such as text.

Acknowledgements. This work was supported in part by 973 Program (Grant No. 2010CB327905), National Natural Science Foundation of China (Grant No. 61170127, 61332016).

References

1. Andoni, A., Indyk, P.: Near-optimal hashing algorithms for approximate nearest neighbor in high dimensions. In: Proceeding of the Annual IEEE Symposium on Foundations of Computer Science (2006)
2. Breiman, L.: Bagging predictors. Machine Learning 24(2), 123–140 (1996)
3. Broder, A.Z., Charikar, M., Frieze, A.M., Mitzenmacher, M.: Min-wise independent permutations. In: Proceedings of the Annual ACM Symposium on Theory of Computing (1998)
4. Charikar, M.: Similarity estimation techniques from rounding algorithm. In: ACM Symposium on Theory of Computing, pp. 380–388 (2002)
5. Dean, T., Ruzon, M.A., Segal, M., Shlens, J., Vijayanarasimhan, S., Yagnik, J.: Fast, accurate detection of 100,000 object classes on a single machine. In: IEEE Conference on Computer Vision and Pattern Recognition (2013)
6. Efron, B., Tibshirani, R.: An introduction to the bootstrap, vol. 57. CRC press (1993)
7. Gong, Y., Lazebnik, S.: Iterative quantization: A procrustean approach to learning binary codes. In: IEEE Conference on Computer Vision and Pattern Recognition (2011)

8. He, J., Liu, W., Chang, S.F.: Scalable similarity search with optimized kernel hashing. In: Proceedings of the ACM SIGKDD Conference (2010)
9. He, K., Wen, F., Sun, J.: K-means hashing: an affinity-preserving quantization method for learning binary compact codes. In: IEEE Conference on Computer Vision and Pattern Recognition (2013)
10. He, X., Niyogi, P.: Locality preserving projections. In: Advances in Neural Information Processing Systems (2003)
11. Heo, J., Lee, Y., He, J., Chang, S., Yoon, S.: Spherical hashing. In: IEEE Conference on Computer Vision and Pattern Recognition (2012)
12. Hoeffding, W.: Probability inequalities for sums of bounded random variables. Journal of the American Statistical Association 58(301), 13–30 (1963)
13. Indyk, P., Motwani, R.: Approximate nearest neighbors: towards removing the curse of dimensionality. In: Proceedings of ACM Symposium on Theory of Computing (1998)
14. Jégou, H., Douze, M., Schmid, C., Pérez, P.: Aggregating local descriptors into a compact image representation. In: IEEE Conference on Computer Vision and Pattern Recognition, pp. 3304–3311. IEEE (2010)
15. Kong, W., Li, W.: Isotropic hashing. In: Advances in Neural Information Processing Systems (2012)
16. Leng, C., Cheng, J., Lu, H.: Random subspace for binary codes learning in large scale image retrieval. In: Proceedings of ACM SIGIR Conference, SIGIR (2014)
17. Liu, W., Wang, J., Ji, R., Jiang, Y., Chang, S.: Supervised hashing with kernels. In: IEEE Conference on Computer Vision and Pattern Recognition (2012)
18. Liu, W., Wang, J., Kumar, S., Chang, S.: Hashing with graphs. In: Proceedings of the International Conference on Machine Learning (2011)
19. Oliva, A., Torralba, A.: Modeling the shape of the scene: A holistic representation of the spatial envelope. International Journal of Computer Vision 42(3), 145–175 (2001)
20. Shrivastava, A., Li, P.: Fast near neighbor search in high-dimensional binary data. In: Flach, P.A., De Bie, T., Cristianini, N. (eds.) ECML PKDD 2012, Part I. LNCS, vol. 7523, pp. 474–489. Springer, Heidelberg (2012)
21. Shrivastava, A., Li, P.: In defense of minhash over simhash. In: Proceedings of International Conference on Artificial Intelligence and Statistics (2014)
22. Skurichina, M., Duin, R.P.: Bagging, boosting and the random subspace method for linear classifiers. Pattern Analysis & Applications 5(2), 121–135 (2002)
23. Strecha, C., Bronstein, A.M., Bronstein, M.M., Fua, P.: Ldahash: Improved matching with smaller descriptors. IEEE Transactions on Pattern Analysis and Machine Intelligence 34(1), 66–78 (2012)
24. Wang, J., Kumar, S., Chang, S.F.: Semi-supervised hashing for scalable image retrieval. In: IEEE Conference on Computer Vision and Pattern Recognition (2010)
25. Wang, J., Kumar, S., Chang, S.F.: Sequential projection learning for hashing with compact codes. In: Proceedings of International Conference on Machine Learning, pp. 1127–1134 (2010)
26. Weiss, Y., Torralba, A., Fergus, R.: Spectral hashing. In: Advances in Neural Information Processing Systems (2008)
27. Xu, B., Bu, J., Lin, Y., Chen, C., He, X., Cai, D.: Harmonious hashing. In: Proceedings of International Joint Conference on Artificial Intelligence (2013)
28. Yu, S.X., Shi, J.: Multiclass spectral clustering. In: Proceedings of the International Conference on Computer Vision (2003)
29. Zhang, D., Wang, J., Cai, D., Lu, J.: Self-taught hashing for fast similarity search. In: Proceedings of International ACM SIGIR Conference (2010)

Conic Multi-task Classification

Cong Li[1], Michael Georgiopoulos[1],
and Georgios C. Anagnostopoulos[2]

[1] University of Central Florida, Department of Electrical Engineering and Computer Science, 4000 Central Florida Blvd, Orlando, Florida, 32816, USA
congli@eecs.ucf.edu, michaelg@ucf.edu
[2] Florida Institute of Technology, Department of Electrical and Computer Engineering, 150 W University Blvd, Melbourne, FL 32901, USA
georgio@fit.edu

Abstract. Traditionally, Multi-task Learning (MTL) models optimize the average of task-related objective functions, which is an intuitive approach and which we will be referring to as Average MTL. However, a more general framework, referred to as Conic MTL, can be formulated by considering conic combinations of the objective functions instead; in this framework, Average MTL arises as a special case, when all combination coefficients equal 1. Although the advantage of Conic MTL over Average MTL has been shown experimentally in previous works, no theoretical justification has been provided to date. In this paper, we derive a generalization bound for the Conic MTL method, and demonstrate that the tightest bound is not necessarily achieved, when all combination coefficients equal 1; hence, Average MTL may not always be the optimal choice, and it is important to consider Conic MTL. As a byproduct of the generalization bound, it also theoretically explains the good experimental results of previous relevant works. Finally, we propose a new Conic MTL model, whose conic combination coefficients minimize the generalization bound, instead of choosing them heuristically as has been done in previous methods. The rationale and advantage of our model is demonstrated and verified via a series of experiments by comparing with several other methods.

Keywords: Multi-task Learning, Kernel Methods, Generalization Bound, Support Vector Machines.

1 Introduction

Multi-Task Learning (MTL) has been an active research field for over a decade, since its inception in [1]. By training multiple tasks simultaneously with shared information, it is expected that the generalization performance of each task can be improved, compared to training each task separately. Previously, various MTL schemes have been considered, many of which model the t-th task by a linear function with weight $\boldsymbol{w}_t, t = 1, \cdots T$, and assume a certain, underlying relationship between tasks. For example, the authors in [2] assumed all $\boldsymbol{w}_t$'s to

T. Calders et al. (Eds.): ECML PKDD 2014, Part II, LNCS 8725, pp. 193–208, 2014.
© Springer-Verlag Berlin Heidelberg 2014

be part of a cluster centered at $\bar{w}$, the latter one being learned jointly with w_t. This assumption was further extended to the case, where the weights w_t's can be grouped into different clusters instead of a single global cluster [3,4]. Furthermore, a widely held MTL assumption is that tasks share a common, potentially sparse, feature representation, as done in [5,6,7,8,9,10,11], to name a few. It is worth mentioning that many of these works allow features to be shared among only a subset of tasks, which are considered "similar" or "related" to each other, where the relevance between tasks is discovered during training. This approach reduces and, sometimes, completely avoids the effect of "negative transfer", *i.e.*, knowledge transferred between irrelevant tasks, which leads to degraded generalization performance. Several other recent works that focused on the discovery of task relatedness include [12,13,14,15]. Additionally, some kernel-based MTL models assume that the data from all tasks are pre-processed by a (partially) common feature mapping, thus (partially) sharing the same kernel function; see [16,17,18], again, to name a few.

Most of these previous MTL formulations consider the following classic setting: A set of training data $\{x_t^i, y_t^i\} \in \mathcal{X} \times \mathcal{Y}, i = 1, \cdots, N_t$ is provided for the t-th task ($t = 1, \cdots, T$), where $\mathcal{X}, \mathcal{Y}$ are the input and output spaces correspondingly. Each datum from the t-th task is assumed to be drawn from an underlying probability distribution $P_t(X_t, Y_t)$, where X_t and Y_t are random variables in the input and output space respectively. Then, a MTL problem is formulated as follows

$$\min_{w \in \Omega(w)} \sum_{t=1}^{T} f(w_t, x_t, y_t) \tag{1}$$

where $w \triangleq (w_1, \cdots, w_T)$ is the collection of all w_t's, and, similarly, $x_t \triangleq (x_t^1, \cdots, x_t^{N_t})$, $y_t \triangleq (y_t^1, \cdots, y_t^{N_t})$. f is a function common to all tasks. It is important to observe that, without the constraint $w \in \Omega(w)$, Problem (1) degrades to T independent learning problems. Therefore, in most scenarios, the set $\Omega(w)$ is designed to capture the inter-task relationships. For example, in [16], the model combines MTL with Multiple Kernel Learning (MKL), which is formulated as follows

$$f(w_t, x_t, y_t) \triangleq \frac{1}{2}\|w_t\|^2 + C \sum_{i=1}^{N_t} l(w_t, \phi_t(x_t^i), y_t^i) \tag{2}$$

$$\Omega(w) \triangleq \{w = (w_1, \cdots, w_T) : w_t \in \mathcal{H}_{\theta, \gamma_t}, \theta \in \Omega(\theta), \gamma \in \Omega(\gamma)\}$$

Here, l is a specified loss function, $\phi_t : \mathcal{X} \to \mathcal{H}_{\theta, \gamma_t}$ is the feature mapping for the t-th task, $\mathcal{H}_{\theta, \gamma_t}$ is the Reproducing Kernel Hilbert Space (RKHS) with reproducing kernel function $k_t \triangleq \sum_{m=1}^{M} (\theta_m + \gamma_t^m) k_m$, where $k_m : \mathcal{X} \times \mathcal{X} \to \mathbb{R}, m = 1, \cdots, M$ are pre-selected kernel functions. $\|w_t\| \triangleq \sqrt{\langle w_t, w_t \rangle}$ is the norm defined in $\mathcal{H}_{\theta, \gamma_t}$. Also, $\Omega(\theta)$ is the feasible set of $\theta \triangleq (\theta_1, \cdots, \theta_M)$, and, similarly, $\Omega(\gamma)$ is the feasible set of $\gamma \triangleq (\gamma_1, \cdots, \gamma_T)$. It is not hard to see that, in this setting, $\Omega(w)$ is designed such that all tasks partially share the same kernel function in a MKL manner, parameterized by the common coefficient $\boldsymbol{\theta}$ and task-specific coefficient $\boldsymbol{\gamma}_t, t = 1, \cdots, T$.

Another example, Sparse MTL [17], has the following formulation:

$$f(\boldsymbol{w}_t, \boldsymbol{x}_t, \boldsymbol{y}_t) \triangleq \sum_{i=1}^{N_t} l(\boldsymbol{w}_t, \phi_t(\boldsymbol{x}_t^i), y_t^i)$$

$$\Omega(\boldsymbol{w}) \triangleq \{\boldsymbol{w} = (\boldsymbol{w}_1, \cdots, \boldsymbol{w}_T) : \boldsymbol{w}_t \triangleq (\boldsymbol{w}_t^1, \cdots, \boldsymbol{w}_t^M), \sum_{m=1}^{M} (\sum_{t=1}^{T} \|\boldsymbol{w}_t^m\|^q)^{p/q} \leq R\}$$

(3)

where $\boldsymbol{w}_t^m \in \mathcal{H}_m, \forall m = 1, \cdots, M, t = 1, \cdots, T, \boldsymbol{w}_t \in \mathcal{H}_1 \times \cdots \times \mathcal{H}_M, 0 < p \leq 1,$ $1 \leq q \leq 2$. Note that although the original Sparse MTL is formulated as follows

$$\min_{\boldsymbol{w}} \sum_{m=1}^{M} (\sum_{t=1}^{T} \|\boldsymbol{w}_t^m\|^q)^{p/q} + C \sum_{t=1}^{T} \sum_{i=1}^{N_t} l(\boldsymbol{w}_t, \phi_t(\boldsymbol{x}_t^i), y_t^i) \tag{4}$$

due to the first part of Proposition 12 in [19], which we restate as Proposition 1 below[1], it is obvious that, for any $C > 0$, there exists a $R > 0$, such that Problem (1) and Problem (4) are equivalent.

Proposition 1. *Let $\mathcal{D} \subseteq \mathcal{X}$, and let $f, g : \mathcal{D} \mapsto \mathbb{R}$ be two functions. For any $\sigma > 0$, there must exist a $\tau > 0$, such that the following two problems are equivalent*

$$\min_{x \in \mathcal{D}} f(x) + \sigma g(x) \tag{5}$$

$$\min_{x \in \mathcal{D}, g(x) \leq \tau} f(x) \tag{6}$$

The formulation given in Problem (1), which we refer to as *Average MTL*, is intuitively appealing: It is reasonable to expect the average generalization performance of the T tasks to be improved, by optimizing the average of the T objective functions. However, as argued in [20], solving Problem (1) yields only a particular solution on the Pareto Front of the following Multi-Objective Optimization (MOO) problem

$$\min_{\boldsymbol{w} \in \Omega(\boldsymbol{w})} \boldsymbol{f}(\boldsymbol{w}, \boldsymbol{x}, \boldsymbol{y}) \tag{7}$$

where $\boldsymbol{f}(\boldsymbol{w}, \boldsymbol{x}, \boldsymbol{y}) \triangleq [f(\boldsymbol{w}_1, \boldsymbol{x}_1, \boldsymbol{y}_1), \cdots, f(\boldsymbol{w}_T, \boldsymbol{x}_T, \boldsymbol{y}_T)]'$. This is true, because scalarizing a MOO problem by optimizing different conic combinations of the objective functions, leads to the discovery of solutions that correspond to points on the convex part of the problem's Pareto Front [21, p. 178]. In other words, by

---

[1] Note that the difference between Proposition 1 here and Proposition 12 in [19] is that, Proposition 1 does not require convexity of f, g and $\mathcal{D}$; these are requirements necessary for the second part of Proposition 12 in [19], which we do not utilize here.

conically scalarizing Problem (7) using different $\boldsymbol{\lambda} \triangleq [\lambda_1, \cdots, \lambda_T]'$, $\lambda_t > 0, \forall t = 1, \cdots, T$, the optimization problem

$$\min_{\boldsymbol{w} \in \Omega(\boldsymbol{w})} \sum_{t=1}^{T} \lambda_t f(\boldsymbol{w}_t, \boldsymbol{x}_t, \boldsymbol{y}_t) \tag{8}$$

yields different points on the Pareto Front of Problem (7). Therefore, there is little reason to believe that the solution of Problem (8) for the special case of $\lambda_t = 1, \forall t = 1, \cdots, T$, i.e., the Average MTL's solution, is the best achievable. In fact, there might be other points on the Pareto Front that result in better generalization performance for each task, hence, yielding better average performance of the T tasks. Therefore, instead of solving Problem (1), one can accomplish this by optimizing Problem (8).

A previous work along these lines was performed in [20]. The authors considered the following MTL formulation, named *Pareto-Path MTL*

$$\min_{\boldsymbol{w} \in \Omega(\boldsymbol{w})} \left[\sum_{t=1}^{T} (f(\boldsymbol{w}_t, \boldsymbol{x}_t, \boldsymbol{y}_t))^p \right]^{1/p} \tag{9}$$

which, assuming all objective functions are positive, minimizes the L_p-norm of the objectives when $p \geq 1$, and the L_p-pseudo-norm when $0 < p < 1$. It was proven that, for any $p > 0$, Problem (9) is equivalent to Problem (8) with

$$\lambda_t = \begin{cases} \frac{f(\boldsymbol{w}_t, \boldsymbol{x}_t, \boldsymbol{y}_t)^{p-1}}{\sum_{t=1}^{T} (f(\boldsymbol{w}_t, \boldsymbol{x}_t, \boldsymbol{y}_t))^p} & \text{if } p > 1 \\ 1 & \text{if } p = 1 \\ \frac{\sum_{t=1}^{T} (f(\boldsymbol{w}_t, \boldsymbol{x}_t, \boldsymbol{y}_t))^{\frac{1-p}{p}}}{f(\boldsymbol{w}_t, \boldsymbol{x}_t, \boldsymbol{y}_t)^{1-p}} & \text{if } 0 < p < 1 \end{cases} , \forall t = 1, \cdots, T \tag{10}$$

Thus by varying $p > 0$, the solutions of Problem (9) trace a path on the Pareto Front of Problem (7). While Average MTL is equivalent to Problem (9), when $p = 1$, it was demonstrated that the experimental results are usually better when $p < 1$, compared to $p = 1$, in a Support Vector Machine (SVM)-based MKL setting. Regardless of the close correlation of the superior obtained results to our previous argument, the authors did not provide a rigorous basis of the advantage of considering an objective function other than the average of the T task objectives. Therefore, use of the L_p-(pseudo-)norm in the paper's objective function remains so far largely a heuristic element of their approach.

In light of the just-mentioned potential drawbacks of Average MTL and the lack of supporting theory in the case of Pareto-Path MTL, in this paper, we analytically justify why it is worth considering Problem (8), which we refer to as *Conic MTL*, and why it is advantageous. Specifically, a major contribution of this paper is the derivation of a generalization bound for Conic MTL, which illustrates that, indeed, the tightest bound is not necessarily achieved, when all λ_t's equal to 1. Therefore, it answers the previous question, and justifies the importance of considering Conic MTL. Also, as a byproduct of the generalization bound, in Section 2, we theoretically show the benefit of Pareto-Path MTL: the generalization bound of Problem (9) is usually tighter when $p < 1$, compared to the case, when $p = 1$. Therefore, it explains Pareto-Path MTL's superiority over Average MTL.

Regarding Conic MTL, a natural question is how to choose the coefficients λ_t's. Instead of setting them heuristically, such as what Pareto-Path MTL does, we propose a new Conic MTL model that learns the λ_t's by minimizing the generalization bound. It ensures that our new model achieves the tightest generalization bound compared to any other settings of the λ_t values and, potentially, leads to superior performance. The new model is described in Section 3 and experimentally evaluated in Section 4. The experimental results verified our theoretical conclusions: Conic MTL can indeed outperform Average MTL and Pareto-Path MTL in many scenarios and, therefore, learning the coefficients λ_t's by minimizing the generalization bound is reasonable and advantageous. Finally, we summarize our work in Section 5.

In the sequel, we'll be using the following notational conventions: vector and matrices are denoted in boldface. Vectors are assumed to be columns vectors. If v is a vector, then v' denotes the transposition of v. Vectors $\mathbf{0}$ and $\mathbf{1}$ are the all-zero and all-one vectors respectively. Also, $\succeq, \succ, \preceq$ and $\prec$ between vectors will stand for the component-wise $\geq, >, \leq$ and $<$ relations respectively. Similarly, for any v, v^p represents the component-wise exponentiation of v.

2   Generalization Bound

Similar to previous theoretical analyses of MTL methods [22,23,24,25,26,27], in this section, we derive the Rademacher complexity-based generalization bound for Conic MTL, *i.e.*, Problem (8). Specifically, we assume the following form of f and $\Omega(w)$ for classification problems:

$$f(w_t, x_t, y_t) \triangleq \frac{1}{2}\|w_t\|^2 + C\sum_{i=1}^{N} l(y_t^i \langle w_t, \phi(x_t^i) \rangle) \tag{11}$$

$$\Omega(w) \triangleq \{w = (w_1, \cdots, w_T) : w_t \in \mathcal{H}_\theta, \theta \in \Omega(\theta)\}$$

where l is the margin loss:

$$l(x) = \begin{cases} 0 & \text{if } \rho \leq x \\ 1 - x/\rho & \text{if } 0 \leq x \leq \rho \\ 1 & \text{if } x \leq 0 \end{cases} \tag{12}$$

$\phi : \mathcal{X} \to \mathcal{H}_\theta$ is the common feature mapping for all tasks. $\mathcal{H}_\theta$ is the RKHS defined by the kernel function $k \triangleq \sum_{m=1}^{M} \theta_m k_m$, where $k_m : \mathcal{X} \times \mathcal{X} \to \mathbb{R}, m = 1, \cdots, M$ are the pre-selected kernel functions. Furthermore, we assume the training data $\{x_t^i, y_t^i\} \in \mathcal{X} \times \mathcal{Y}, t = 1, \cdots, T, i = 1, \cdots, N$ are drawn from the probability distribution $P_t(X_t, Y_t)$, where X_t and Y_t are random variables in the input and output space respectively. Note that, here, we assumed all tasks have equal number of training data and share a common kernel function.

These two assumptions were made to simplify notation and exposition, and they do not affect extending our results to a more general case, where an arbitrary number of training samples is available for each task and partially shared kernel functions are used; in the latter case, only relevant tasks may share the common kernel function, hence, reducing the effect of "negative transfer".

Substituting (11) into Problem (8) and based on Proposition 1, it is not hard to see that for any C in Eq. (11), there exist a $R > 0$ such that Problem (8) is equivalent to the following problem

$$\min_{\boldsymbol{w} \in \Omega(\boldsymbol{w})} \sum_{t=1}^{T} \sum_{i=1}^{N_t} \lambda_t l(y_t^i \langle \boldsymbol{w}_t, \phi(\boldsymbol{x}_t^i) \rangle)$$
$$s.t. \sum_{t=1}^{T} \lambda_t \|\boldsymbol{w}_t\|^2 \leq R \tag{13}$$

Obviously, solving Problem (13) is the process of choosing the $\boldsymbol{w}$ in the hypothesis space $\mathcal{F}_\lambda$, such that the empirical loss, *i.e.*, the objective function of Problem (13), is minimized. The relevant hypothesis space is defined below:

$$\mathcal{F}_\lambda \triangleq \{\boldsymbol{w} = (\boldsymbol{w}_1, \cdots, \boldsymbol{w}_T) : \sum_{t=1}^{T} \lambda_t \|\boldsymbol{w}_t\|^2 \leq R, \boldsymbol{w}_t \in \mathcal{H}_\theta, \boldsymbol{\theta} \in \Omega(\boldsymbol{\theta})\} \tag{14}$$

By defining the Conic MTL expected error $er(\boldsymbol{w})$ and empirical loss $\hat{er}_\lambda(\boldsymbol{w})$ as follows

$$er(\boldsymbol{w}) = \frac{1}{T} \sum_{t=1}^{T} E[\mathbf{1}_{(-\infty,0]}(Y_t \langle \boldsymbol{w}_t, \phi(X_t) \rangle)] \tag{15}$$

$$\hat{er}_\lambda(\boldsymbol{w}) = \frac{1}{TN} \sum_{t=1}^{T} \sum_{i=1}^{N} \lambda_t l(y_t^i \langle \boldsymbol{w}_t, \phi(\boldsymbol{x}_t^i) \rangle) \tag{16}$$

one of our major contribution is the following theorem, which gives the generalization bound of Problem (13) in the context of MKL-based Conic MTL for any $\lambda_t \in (1, r_\lambda), \forall t = 1, \cdots, T$, where r_λ is a pre-specified upper-bound for the λ_t's.

Theorem 1. *For fixed $\rho > 0$, $r_\lambda \in \mathbb{N}$ with $r_\lambda > 1$, and for any $\boldsymbol{\lambda} = [\lambda_1, \cdots, \lambda_T]'$, $\lambda_t \in (1, r_\lambda), \forall t = 1, \cdots, T$, $\boldsymbol{w} \in \mathcal{F}_\lambda$, $0 < \delta < 1$, the following generalization bound holds with probability at least $1 - \delta$:*

$$er(\boldsymbol{w}) \leq \hat{er}_\lambda(\boldsymbol{w}) + \frac{\sqrt{2}r_\lambda}{\rho} R(\mathcal{F}_\lambda) + \sqrt{\frac{9}{TN} \ln \left(\frac{2r_\lambda}{T} \sum_{t=1}^{T} \frac{1}{\lambda_t}\right)} + \sqrt{\frac{9 \ln \frac{1}{\delta}}{2TN}} \tag{17}$$

where $R(\mathcal{F}_\lambda)$ is the empirical Rademacher complexity of the hypothesis space $\mathcal{F}_\lambda$, which is defined as

$$R(\mathcal{F}_\lambda) \triangleq \frac{2}{TN} E[\sup_{w \in \mathcal{F}_\lambda} \sum_{t=1}^{T} \sum_{i=1}^{N} \sigma_t^i \langle \boldsymbol{w}_t, \phi(\boldsymbol{x}_t^i) \rangle] \tag{18}$$

and the σ_t^i's are i.i.d. Rademacher-distributed (i.e., Bernoulli(1/2)-distributed random variables with sample space $\{-1, +1\}$).

Based on Theorem 1, one is motivated to choose λ that minimizes the generalization bound, instead of heuristically selecting λ as in Eq. (10), which was suggested in [20]. Indeed, doing so does not guarantee obtaining the tightest generalization bound.

However, prior to proposing our new Conic MTL model that minimizes the generalization bound, it is still of interest to theoretically analyze why Pareto-Path MTL, *i.e.*, Problem (9), usually enjoys better generalization performance when $0 < p < 1$, rather than when $p = 1$, as described in Section 1. While the analysis is not given in [20], fortunately, we can provide some insights of the good performance of the model, when $0 < p < 1$, by utilizing Theorem 1 and with the help of the following two theorems.

Theorem 2. *For $\lambda \succ 0$, the empirical Rademacher complexity $R(\mathcal{F}_\lambda)$ is monotonically decreasing with respect to each $\lambda_t, t = 1, \cdots, T$.*

Theorem 3. *Assume $f(\boldsymbol{w}_t, \boldsymbol{x}_t, \boldsymbol{y}_t) > 0, \forall t = 1, \cdots, T$. For λ that is defined in Eq. (10), when $0 < p < 1$, we have $\lambda_t > 1$ and λ_t is monotonically decreasing with respect to $p, \forall t = 1, \cdots, T$.*

Based on Eq. (10), if $f(\boldsymbol{w}_t, \boldsymbol{x}_t, \boldsymbol{y}_t) > 0, \forall t = 1, \cdots, T$, there must exist a fixed $r_\lambda > 0$, such that $\lambda_t \in (1, r_\lambda), \forall t = 1, \cdots, T$. Therefore we can analyze the generalization bound of Pareto-Path MTL based on Theorem 1, when $0 < p < 1$. Although Theorem 1 is not suitable for the case when $p = 1$, we can approximate its bound by letting p to be infinitely close to 1.

The above two theorems indicate that the empirical Rademacher complexity for the hypothesis space of Pareto-Path MTL monotonically increases with respect to p, when $0 < p < 1$. Therefore, the second term in the generalization bound decreases as p decreases. This is also true for the third term in the bound, based on Theorem 3. Thus, it is not a surprise that the generalization performance is usually better when $0 < p < 1$ than when $p = 1$, and it is reasonable to expect the performance to get improved when p decreases. In fact, such a monotonicity is reported in the experiments of [20]: the classification accuracy is usually monotonically increasing, when p decreases. It is worth mentioning that, although rarely observed, we may not have such monotonicity in performance, if the first term in the generalization bound, *i.e.*, the empirical loss, grows quickly as p decreases. However, the monotonic behavior of the generalization bound (except the empirical loss) is still sufficient for explaining the experimental results of Problem (9), which justifies the rationale of employing an arbitrarily weighted conic combination of objective functions instead of using the average of these functions.

Finally, we provide two theorems that not only are used in the proof of Theorem 1, but also may be of interest on their own accord. Subsequently, in the next section, we describe our new MTL model.

Theorem 4. *Given* $\gamma \triangleq [\gamma_1, \cdots, \gamma_T]'$ *with* $\gamma \succ 0$, *define*

$$R(\mathcal{F}_\lambda, \gamma) = \frac{2}{TN} E[\sup_{w \in \mathcal{F}_\lambda} \sum_{t=1}^{T} \sum_{i=1}^{N} \gamma_t \sigma_t^i \langle w_t, \phi(x_t^i) \rangle] \tag{19}$$

For fixed $\lambda \succ 0$, $R(\mathcal{F}_\lambda, \gamma)$ *is monotonically increasing with respect to each* γ_t.

Theorem 5. *For fixed* $r_\lambda \geq 1$, $\rho > 0$, $\lambda = [\lambda_1, \cdots, \lambda_T]', \lambda_t \in [1, r_\lambda], \forall t = 1, \cdots, T$, *and for any* $w \in \mathcal{F}_\lambda$, $0 < \delta < 1$, *the following generalization bound holds with probability at least* $1 - \delta$:

$$er(w) \leq \hat{er}_\lambda(w) + \frac{r_\lambda}{\rho} R(\mathcal{F}_\lambda) + \sqrt{\frac{9 \ln \frac{1}{\delta}}{2TN}} \tag{20}$$

Note that the difference between Theorem 5 and Theorem 1 is that, Theorem 1 is valid for *any* $\lambda_t \in (1, r_\lambda)$, while Theorem 5 is only valid for *fixed* $\lambda_t \in [1, r_\lambda]$. While the bound given in Theorem 1 is more general, it is looser due to the additional third term in (17) and due to the factor $\sqrt{2}$ multiplying the empirical Rademacher complexity.

3   A New MTL Model

In this section, we propose our new MTL model. Motivated by the generalization bound in Theorem 1, our model is formulated to select w and λ by minimizing the bound

$$\hat{er}_\lambda(w) + \frac{\sqrt{2}r_\lambda}{\rho} R(\mathcal{F}_\lambda) + \sqrt{\frac{9}{TN} \ln \left(\frac{2r_\lambda}{T} \sum_{t=1}^{T} \frac{1}{\lambda_t} \right)} + \sqrt{\frac{9 \ln \frac{1}{\delta}}{2TN}} \tag{21}$$

instead of choosing the coefficients λ heuristically, such as via Eq. (10) in [20]. Note that the bound's last term does not depend on any model parameters, while the third term has only a minor effect on the bound, when $\lambda_t \in (1, r_\lambda)$. Therefore, we omit these two terms, and propose the following model:

$$\min_{w, \lambda} \hat{er}_\lambda(w) + \frac{\sqrt{2}r_\lambda}{\rho} R(\mathcal{F}_\lambda) \tag{22}$$

$$s.t. \ w \in \mathcal{F}_\lambda, 1 \prec \lambda \prec r_\lambda 1.$$

Furthermore, due to the complicated nature of $R(\mathcal{F}_\lambda)$, it is difficult to optimize Problem (22) directly. Therefore, in the following theorem, we prove an upper bound for $R(\mathcal{F}_\lambda)$, which yields a simpler expression. We remind the readers that the hypothesis space $\mathcal{F}_\lambda$ is defined as

$$\mathcal{F}_\lambda \triangleq \{w = (w_1, \cdots, w_T) : \sum_{t=1}^{T} \lambda_t \|w_t\|^2 \leq R, w_t \in \mathcal{H}_\theta, \theta \in \Omega(\theta)\} \quad (23)$$

where $\mathcal{H}_\theta$ is the RKHS defined by the kernel function $k \triangleq \sum_{m=1}^{M} \theta_m k_m$.

Theorem 6. *Given the hypothesis space $\mathcal{F}_\lambda$, the empirical Rademacher complexity can be upper-bounded as follows:*

$$R(\mathcal{F}_\lambda) \leq \frac{2}{TN} \sqrt{\sum_{t=1}^{T} \frac{1}{\lambda_t} E\left[\sqrt{\sup_{w \in \mathcal{F}_1} \sum_{t=1}^{T} \left(\sum_{i=1}^{N} \sigma_t^i \langle w_t, \phi(x_t^i)\rangle\right)^2}\right]} \quad (24)$$

where the feasible region of w, i.e., $\mathcal{F}_1$, is the same as $\mathcal{F}_\lambda$ but with $\lambda = 1$.

Note that, for a given $\Omega(\theta)$, the expectation term in (24) is a constant. If we define

$$s \triangleq E\left[\sqrt{\sup_{w \in \mathcal{F}_1} \sum_{t=1}^{T} \left(\sum_{i=1}^{N} \sigma_t^i \langle w_t, \phi(x_t^i)\rangle\right)^2}\right] \quad (25)$$

we arrive at our proposed MTL model:

$$\min_{w, \lambda} \sum_{t=1}^{T} \sum_{i=1}^{N} \lambda_t l(y_t^i \langle w_t, \phi(x_t^i)\rangle) + \frac{2\sqrt{2}sr_\lambda}{\rho} \sqrt{\sum_{t=1}^{T} \frac{1}{\lambda_t}}$$

$$\text{s.t. } w_t \in \mathcal{H}_\theta, \forall t = 1, \cdots, T \quad (26)$$

$$\theta \in \Omega(\theta), \sum_{t=1}^{T} \lambda_t \|w_t\|^2 \leq R, 1 \prec \lambda \prec r_\lambda 1.$$

The next proposition provides an equivalent optimization problem, which is easier to solve.

Proposition 2. *For any fixed $C > 0$, $s > 0$ and $r_\lambda > 0$, there exist $R > 0$ and $a > 0$ such that Problem (26) and the following optimization problem are equivalent*

$$\min_{w, \lambda, \theta} \sum_{t=1}^{T} \lambda_t \left(\sum_{m=1}^{M} \frac{\|w_t^m\|^2}{2\theta_m} + C \sum_{i=1}^{N} \sum_{m=1}^{M} l(y_t^i \langle w_t^m, \phi_m(x_t^i)\rangle)\right)$$

$$\text{s.t. } w_t^m \in \mathcal{H}_m, \forall t = 1, \cdots, T, m = 1, \cdots, M, \quad (27)$$

$$\theta \in \Omega(\theta), \sum_{t=1}^{T} \frac{1}{\lambda_t} \leq a, 1 \prec \lambda \prec r_\lambda 1.$$

where $\mathcal{H}_m$ is the RKHS defined by the kernel function k_m, and $\phi_m : \mathcal{X} \to \mathcal{H}_m$.

It is worth pointing out that, Problem (27) minimizes the generalization bound (21) for *any* $\Omega(\boldsymbol{\theta})$. A typical setting is to adapt the L_p-norm MKL method by letting $\Omega(\boldsymbol{\theta}) \triangleq \{\boldsymbol{\theta} = [\theta_1, \cdots, \theta_M]' : \boldsymbol{\theta} \succeq \mathbf{0}, \|\boldsymbol{\theta}\|_p \leq 1\}$, where $p \geq 1$. Alternatively, one may want to employ the *optimal neighborhood kernel* method [28] by letting $\Omega(\boldsymbol{\theta}) \triangleq \{\boldsymbol{\theta} = [\theta_1, \cdots, \theta_M]' : \sum_{t=1}^{T} \|\boldsymbol{K}_t - \hat{\boldsymbol{K}}_t\|_F \leq R_k, \boldsymbol{K}_t \triangleq \sum_{m=1}^{M} \theta_m \boldsymbol{K}_t^m\}$, where $\boldsymbol{K}_t^m \in \mathbb{R}^{N \times N}$ is the kernel matrix whose (i,j)-th element is calculated as $k_m(\boldsymbol{x}_t^i, \boldsymbol{x}_t^j)$, and $\hat{\boldsymbol{K}}_t$'s are the kernel matrices evaluated by a pre-defined kernel function on the training data of the t-th task.

By assuming $\Omega(\boldsymbol{\theta})$ to be a convex set and electing the loss function l to be convex in the model parameters (such as the hinge loss function), Problem (27) is jointly convex with respect to both $\boldsymbol{w}$ and $\boldsymbol{\theta}$. Also, it is separately convex with respect to $\boldsymbol{\lambda}$. Therefore, it is straightforward to employ a block-coordinate descent method to optimize Problem (27). Finally, it is worth mentioning that, by choosing to employ the hinge loss function, the generalization bound in Theorem 1 still holds, since the hinge loss upper-bounds the margin loss for $\rho = 1$. Therefore, our model still minimizes the generalization bound.

3.1   Incorporating L_p-Norm MKL

In this paper, we specifically consider endowing our MTL model with L_p-norm MKL, since it can be better analyzed theoretically, is usually easy to optimize and, often, yields good performance outcomes.

Although the upper bound in Theorem 6 is suitable for any $\Omega(\boldsymbol{\theta})$, it might be loose due to its generality. Another issue is that the expectation present in the bound is still hard to calculate. Therefore, as we consider L_p-norm MKL, it is of interest to derive a bound specifically for it, which is easier to calculate and is potentially tighter.

Theorem 7. *Let* $\Omega(\boldsymbol{\theta}) \triangleq \{\boldsymbol{\theta} = [\theta_1, \cdots, \theta_M]' : \boldsymbol{\theta} \succeq \mathbf{0}, \|\boldsymbol{\theta}\|_p \leq 1\}$, $p \geq 1$, *and* $\boldsymbol{K}_t^m \in \mathbb{R}^{N \times N}, t = 1, \cdots, T, m = 1, \cdots, M$ *be the kernel matrix, whose (i,j)-th element is defined as* $k_m(\boldsymbol{x}_t^i, \boldsymbol{x}_t^j)$. *Also, define* $\boldsymbol{v}_t \triangleq [tr(\boldsymbol{K}_t^1), \cdots, tr(\boldsymbol{K}_t^M)]' \in \mathbb{R}^M$. *Then, we have*

$$R(\mathcal{F}_{\boldsymbol{\lambda}}) \leq \frac{2\sqrt{2R p^*}}{TN} \sqrt{\sum_{t=1}^{T} \frac{1}{\lambda_t} \|\boldsymbol{v}_t\|_{p^*}} \qquad (28)$$

where $p^* \triangleq \frac{p}{p-1}$.

Following a similar procedure to formulating our general model Problem (27), we arrive at the following L_p-norm MKL-based MTL problem

$$\min_{w,\lambda,\theta} \sum_{t=1}^{T} \lambda_t \Big(\sum_{m=1}^{M} \frac{\|w_t^m\|^2}{2\theta_m} + C \sum_{i=1}^{N} \sum_{m=1}^{M} l(y_t^i \langle w_t^m, \phi(x_t^i) \rangle) \Big)$$

$$s.t. \ w_t^m \in \mathcal{H}_m, \forall t = 1, \cdots, T, m = 1, \cdots, M,$$

$$\theta \succeq 0, \|\theta\|_p \leq 1, \tag{29}$$

$$\sum_{t=1}^{T} \frac{\|v_t\|_{p^*}}{\lambda_t} \leq a, 1 \prec \lambda \prec r_\lambda 1.$$

which, based on (21) and (28), minimizes the generalization bound. Note that, due to the bound that is specifically derived for L_p-norm MKL, the constraint $\sum_{t=1}^{T} \frac{1}{\lambda_t} \leq a$ in Problem (27) is changed to $\sum_{t=1}^{T} \frac{\|v_t\|_{p^*}}{\lambda_t} \leq a$ in the previous problem. However, when all kernel matrices K_t^m's have the same trace (as is the case, when all kernel functions are normalized, such that $k_m(x, x) = 1, \forall m = 1, \cdots, M, x \in \mathcal{X}$), for a given $p \geq 1$, $\|v_t\|_{p^*}$ has the same value for all $t = 1, \cdots, T$. In this case, Problem (29) is equivalent to Problem (27).

4   Experiments

In this section, we conduct a series of experiments with several data sets, in order to show the merit of our proposed MTL model by comparing it to a few other related methods.

4.1   Experimental Settings

In our experiments, we specifically evaluate the L_p-norm MKL-based MTL model, *i.e.*, Problem (29), on classification problems using the hinge loss function. To solve Problem (29), we employed a block-coordinate descent algorithm, which optimizes each of the three variables w, λ and θ in succession by holding the remaining two variables fixed. Specifically, in each iteration, three optimization problems are solved. First, for fixed λ and θ, the optimization with respect to w can be split into T independent SVM problems, which are solved via LIBSVM [29]. Next, for fixed w and θ, the optimization with respect to λ is convex and is solved using CVX [30][31]. Finally, minimizing with respect to θ, while w and λ are held fixed, has a closed-form solution:

$$\theta^* = \left(\frac{v}{\|v\|_{\frac{p}{p+1}}} \right)^{\frac{1}{p+1}} \tag{30}$$

where $v \triangleq [v_1, \cdots, v_M]'$ and $v_m \triangleq \sum_{t=1}^{T} \|w_t^m\|, \forall m = 1, \cdots, M$. Although more efficient algorithms may exist, we opted to use this simple and easy-to-implement algorithm, since the optimization strategy is not the focus of our paper[2].

---

[2] Our MATLAB implementation is located at http://github.com/congliucf/ECML2014

For all experiments, 11 kernels were selected for use: a Linear kernel, a 2nd-order Polynomial kernel and Gaussian kernels with spread parameter values $\{2^{-7}, 2^{-5}, 2^{-3}, 2^{-1}, 2^0, 2^1, 2^3, 2^5, 2^7\}$. Parameters C, p and a were selected via cross-validation. Our model is evaluated on 6 data sets: 2 real-world data sets from the UCI repository [32], 2 handwritten digits data sets, and 2 multi-task data sets, which we detail below.

The Wall-Following Robot Navigation (*Robot*) and Vehicle Silhouettes (*Vehicle*) data sets were obtained from the UCI repository. The *Robot* data, consisting of 4 features per sample, describe the position of the robot, while it navigates through a room following the wall in a clockwise direction. Each sample is to be classified according to one of the following four classes: "Move-Forward", "Slight-Right-Turn", "Sharp-Right-Turn" and "Slight-Left-Turn". On the other hand, the *Vehicle* data set is a collection of 18-dimensional feature vectors extracted from images. Each datum should be classified into one of four classes: "4 Opel", "SAAB", "Bus" and "Van".

The two handwritten digit data sets, namely $MNIST^3$ and $USPS^4$, consist of grayscale images of handwritten digits from 0 to 9 with 784 and 256 features respectively. Each datum is labeled as one of ten classes, each of which represents a single digit. For these four multi-class data sets, an equal number of samples from each class were chosen for training. Also, we approached these multi-class problems as MTL problems using a one-vs.-one strategy and the averaged classification accuracy is calculated for each data set.

The last two data sets, namely $Letter^5$ and $Landmine^6$, correspond to pure multi-task problems. Specifically, the *Letter* data set involves 8 tasks: "C" vs. "E", "G" vs. "Y", "M" vs. "N", "A" vs. "G", "I" vs. "J", "A" vs. "O", "F" vs. "T" and "H" vs. "N". Each letter is represented by a 8×16 pixel image, which forms a 128-dimensional feature vector. The goal for this problem is to correctly recognize the letters in each task. On the other hand, the *Landmine* data set consists of 29 binary classification tasks. Each datum is a 9-dimensional feature vector extracted from radar images that capture a single region of landmine fields. The goal for each task is to detect landmines in specific regions. For the experiments involving these two data sets, we re-sampled the data such that, for each task, the two classes contain equal number of samples.

In all our experiments, we considered training set sizes of 10%, 20% and 50% of the original data set. As an exception, for the *Landmine* data set, we did not use the 10% of the original set for training due to its small size; instead, we used 20%, 30% and 50%.

We compared our method with five different Multi-Task MKL (MT-MKL) methods. The first one is Pareto-Path MTL, *i.e.*, Problem (9), which was originally proposed in [20]. One can expect our new method to outperform it in most cases, since our method selects λ by minimizing the generalization bound, while

---

[3] Available at: http://yann.lecun.com/exdb/mnist/

[4] Available at: http://www.cs.nyu.edu/~roweis/data.html

[5] Available at: http://multitask.cs.berkeley.edu/

[6] Available at: http://people.ee.duke.edu/~lcarin/LandmineData.zip

Pareto-Path MTL selects its value heuristically via Eq. (10). The second method we compared with is the L_p-norm MKL-based Average MTL, which is the same as our method for $\lambda = 1$. As we argued earlier in the introduction, minimizing the averaged objective does not necessarily guarantee the best generalization performance. By comparing with Average MTL, we expect to verify our claim experimentally. Moreover, we compared with two other popular MT-MKL methods, namely Tang's Method [16] and Sparse MTL [17]. These two methods were outlined in Section 1. Finally, we considered the baseline approach, which trains each task individually via a traditional single-task L_p-norm MKL strategy.

4.2 Experimental Results

Table 1 provides the obtained experimental results based on the settings that were described in the previous sub-section. More specifically, in Table 1, we

Table 1. Comparison of Multi-task Classification Accuracy between Our Method and Five Other Methods. Averaged performances of 20 runs over randomly sampled training set are reported.

Robot	Our Method	Pareto	Average	Tang	Sparse	Baseline
10%	**95.83**	95.07	95.16	93.93	94.69	95.54
20%	**97.11**	96.11	95.90	96.36	96.56	95.75
50%	**98.41**	96.80	96.59	97.21	98.09	96.31

Vehicle	Our Method	Pareto	Average	Tang	Sparse	Baseline
10%	**80.10**	80.05	79.77	78.47	79.28	78.01
20%	84.69	**85.33**	85.22	83.98	84.44	84.37
50%	**89.90**	88.04	87.93	88.13	88.57	87.64

Letter	Our Method	Pareto	Average	Tang	Sparse	Baseline
10%	83.00	**83.95**	81.45	80.86	83.00	81.33
20%	87.13	**87.51**	86.42	82.95	87.09	86.39
50%	90.47	**90.61**	90.01	84.87	90.65	89.80

Landmine	Our Method	Pareto	Average	Tang	Sparse	Baseline
20%	**70.18**	69.59	67.24	66.60	58.89	66.64
30%	**74.52**	74.15	71.62	70.89	65.83	71.14
50%	**78.26**	77.42	76.96	76.08	75.82	76.29

MNIST	Our Method	Pareto	Average	Tang	Sparse	Baseline
10%	**93.59**	89.30	88.81	92.37	93.48	88.71
20%	**96.08**	95.02	94.95	95.94	95.96	94.81
50%	97.44	96.92	96.98	97.47	**97.53**	97.04

USPS	Our Method	Pareto	Average	Tang	Sparse	Baseline
10%	**94.61**	90.22	90.11	93.20	94.52	89.02
20%	**97.44**	96.26	96.25	97.37	97.53	96.17
50%	**98.98**	98.51	98.59	98.96	**98.98**	98.49

report the average classification accuracy of 20 runs over a randomly sampled training set. Moreover, the best performance among the 6 competing methods is highlighted in boldface. To test the statistical significance of the differences between our method and the 5 other methods, we employed a t-test to compare mean accuracies using a significance level of $\alpha = 0.05$. In the table, underlined numbers indicate the results that are statistically significantly worse than the ones produced by our method.

When analyzing the results in Table 1, first of all, we observe that the optimal result is almost always achieved by the two Conic MTL methods, namely our method and Pareto-Path MTL. This result not only shows the advantage of Conic MTL over Average MTL, but also demonstrates the benefit compared to other MTL methods, such as Tang's MTL and Sparse MTL. Secondly, it is obvious that our method can usually achieve better result than Pareto-Path MTL; as a matter of fact, in many cases the advantage is statistically significant. This observation validates the underlying rationale of our method, which chooses the coefficient λ by minimizing the generalization bound instead of using Eq. (10). Finally, when comparing our method against the five alternative methods, our results are statistically better most of the time, which further emphasizes the benefit of our method.

5   Conclusions

In this paper, we considered the MTL problem that minimizes the conic combination of objectives with coefficients λ, which we refer to as Conic MTL. The traditional MTL method, which minimizes the average of the task objectives (Average MTL), is only a special case of Conic MTL with $\lambda = 1$. Intuitively, such a specific choice of λ should not necessarily lead to optimal generalization performance.

This intuition motivated the derivation of a Rademacher complexity-based generalization bound for Conic MTL in a MKL-based classification setting. The properties of the bound, as we have shown in Section 2, indicate that the optimal choice of λ is indeed not necessarily equal to 1. Therefore, it is important to consider different values for λ for Conic MTL, which may yield tighter generalization bounds and, hence, better performance. As a byproduct, our analysis also explains the reported superiority of Pareto-Path MTL [20] over Average MTL.

Moreover, we proposed a new Conic MTL model, which aims to directly minimize the derived generalization bound. Via a series of experiments on six widely utilized data sets, our new model demonstrated a statistically significant advantage over Pareto-Path MTL, Average MTL, and two other popular MT-MKL methods.

Acknowledgments. Cong Li acknowledges support from National Science Foundation (NSF) grants No. 0806931 and No. 0963146. Furthermore, Michael Georgiopoulos acknowledges support from NSF grants No. 0963146, No. 1200566, and

No. 1161228. Also, Georgios C. Anagnostopoulos acknowledges partial support from NSF grant No. 1263011. Note that any opinions, findings, and conclusions or recommendations expressed in this material are those of the authors and do not necessarily reflect the views of the NSF. Finally, the authors would like to thank the three anonymous reviewers, that reviewed this manuscript, for their constructive comments.

References

1. Caruana, R.: Multitask learning. Machine Learning 28, 41–75 (1997)
2. Evgeniou, T., Pontil, M.: Regularized multi–task learning. In: Proceedings of the Tenth ACM SIGKDD International Conference on Knowledge Discovery and Data Mining, pp. 109–117. ACM (2004)
3. Zhong, L.W., Kwok, J.T.: Convex multitask learning with flexible task clusters. In: Proceedings of the 29th International Conference on Machine Learning, ICML 2012 (2012)
4. Zhou, J., Chen, J., Ye, J.: Clustered multi-task learning via alternating structure optimization. In: Advances in Neural Information Processing Systems, pp. 702–710 (2011)
5. Obozinski, G., Taskar, B., Jordan, M.I.: Joint covariate selection and joint subspace selection for multiple classification problems. Statistics and Computing 20(2), 231–252 (2010)
6. Jalali, A., Sanghavi, S., Ruan, C., Ravikumar, P.K.: A dirty model for multi-task learning. In: Advances in Neural Information Processing Systems, pp. 964–972 (2010)
7. Gong, P., Ye, J., Zhang, C.: Multi-stage multi-task feature learning. The Journal of Machine Learning Research 14(1), 2979–3010 (2013)
8. Liu, J., Ji, S., Ye, J.: Multi-task feature learning via efficient l 2, 1-norm minimization. In: Proceedings of the Twenty-Fifth Conference on Uncertainty in Artificial Intelligence, pp. 339–348. AUAI Press (2009)
9. Fei, H., Huan, J.: Structured feature selection and task relationship inference for multi-task learning. In: 2011 IEEE 11th International Conference on Data Mining (ICDM), pp. 171–180 (2011)
10. Argyriou, A., Evgeniou, T., Pontil, M.: Convex multi-task feature learning. Machine Learning 73, 243–272 (2008)
11. Kang, Z., Grauman, K., Sha, F.: Learning with whom to share in multi-task feature learning. In: Proceedings of the 28th International Conference on Machine Learning, ICML 2011 (2011)
12. Zhang, Y., Yeung, D.Y.: A convex formulation for learning task relationships in multi-task learning. ArXiv e-prints (2012)
13. Zhang, Y.: Heterogeneous-neighborhood-based multi-task local learning algorithms. In: Advances in Neural Information Processing Systems, pp. 1896–1904 (2013)
14. Romera-Paredes, B., Argyriou, A., Berthouze, N., Pontil, M.: Exploiting unrelated tasks in multi-task learning. In: International Conference on Artificial Intelligence and Statistics, pp. 951–959 (2012)
15. Pu, J., Jiang, Y.G., Wang, J., Xue, X.: Multiple task learning using iteratively reweighted least square. In: Proceedings of the 23rd International Joint Conference on Artificial Intelligence, pp. 1607–1613 (2013)

16. Tang, L., Chen, J., Ye, J.: On multiple kernel learning with multiple labels. In: Proceedings of the 21st International Joint Conference on Artifical Intelligence, pp. 1255–1260 (2009)
17. Rakotomamonjy, A., Flamary, R., Gasso, G., Canu, S.: $l_p - l_q$ penalty for sparse linear and sparse multiple kernel multitask learning. IEEE Transactions on Neural Networks 22, 1307–1320 (2011)
18. Samek, W., Binder, A., Kawanabe, M.: Multi-task learning via non-sparse multiple kernel learning. In: Real, P., Diaz-Pernil, D., Molina-Abril, H., Berciano, A., Kropatsch, W. (eds.) CAIP 2011, Part I. LNCS, vol. 6854, pp. 335–342. Springer, Heidelberg (2011)
19. Kloft, M., Brefeld, U., Sonnenburg, S., Zien, A.: l_p-norm multiple kernel learning. Journal of Machine Learning Research 12, 953–997 (2011)
20. Li, C., Georgiopoulos, M., Anagnostopoulos, G.C.: Pareto-Path Multi-Task Multiple Kernel Learning. ArXiv e-prints (April 2014)
21. Boyd, S., Vandenberghe, L.: Convex Optimization. Cambridge University Press (2004)
22. Ando, R.K., Zhang, T.: A framework for learning predictive structures from multiple tasks and unlabeled data. Journal of Machine Learning Research 6 (2005)
23. Maurer, A.: The rademacher complexity of linear transformation classes. In: Lugosi, G., Simon, H.U. (eds.) COLT 2006. LNCS (LNAI), vol. 4005, pp. 65–78. Springer, Heidelberg (2006)
24. Maurer, A.: Bounds for linear multi-task learning. Journal of Machine Learning Research 7, 117–139 (2006)
25. Kakade, S.M., Shalev-Shwartz, S., Tewari, A.: Regularization techniques for learning with matrices. Journal of Machine Learning Research 13, 1865–1890 (2012)
26. Maurer, A., Pontil, M.: Structured sparsity and generalization. Journal of Machine Learning Research 13, 671–690 (2012)
27. Pontil, M., Maurer, A.: Excess risk bounds for multitask learning with trace norm regularization. In: Conference on Learning Theory, pp. 55–76 (2013)
28. Liu, J., Chen, J., Chen, S., Ye, J.: Learning the optimal neighborhood kernel for classification. In: Proceedings of the 21st International Joint Conference on Artifical Intelligence, pp. 1144–1149 (2009)
29. Chang, C.C., Lin, C.J.: LIBSVM: A library for support vector machines. ACM Transactions on Intelligent Systems and Technology 2, 27:1–27:27 (2011), Software available at http://www.csie.ntu.edu.tw/~cjlin/libsvm
30. Grant, M.C., Boyd, S.P.: Graph implementations for nonsmooth convex programs. In: Blondel, V.D., Boyd, S.P., Kimura, H. (eds.) Recent Advances in Learning and Control. LNCIS, vol. 371, pp. 95–110. Springer, Heidelberg (2008), http://stanford.edu/~boyd/graph_dcp.html
31. Grant, M., Boyd, S.: CVX: Matlab software for disciplined convex programming, version 1.21 (April 2011)
32. Frank, A., Asuncion, A.: UCI machine learning repository (2010)

Bi-directional Representation Learning
for Multi-label Classification

Xin Li and Yuhong Guo

Department of Computer and Information Sciences
Temple University
Philadelphia, PA 19122, USA
{xinli,yuhong}@temple.edu

Abstract. Multi-label classification is a central problem in many application domains. In this paper, we present a novel supervised bi-directional model that learns a low-dimensional mid-level representation for multi-label classification. Unlike traditional multi-label learning methods which identify intermediate representations from either the input space or the output space but not both, the mid-level representation in our model has two complementary parts that capture intrinsic information of the input data and the output labels respectively under the autoencoder principle while augmenting each other for the target output label prediction. The resulting optimization problem can be solved efficiently using an iterative procedure with alternating steps, while closed-form solutions exist for one major step. Our experiments conducted on a variety of multi-label data sets demonstrate the efficacy of the proposed bi-directional representation learning model for multi-label classification.

1 Introduction

Multi-label classification is a central problem in many areas of data analysis, where each data instance can simultaneously have multiple class labels. For example, in image labelling [3,13], an image can contain multiple objects of interest and thus have multiple annotation tags; in text categorization [20], a webpage can be assigned into multiple related topic categories; similarly, in gene or protein function prediction [4], a gene or protein can exhibit multiple functions. Moreover, in these multi-label classification problems, strong label co-occurrences and label dependencies usually exist. For example, an object *"computer"* often appears together with the object *"desk"*, but is rarely seen together with the object *"cooking pan"*. Hence different from the standard multi-class problems where each instance is mapped to a single class label, multi-label classification needs to map each instance to typically a few interdependent class labels in a relatively large output space.

One straightforward approach for multi-label classification is to decompose the multi-label learning problem into a set of independent binary classification problems [16], which however has the obvious drawback of ignoring the interdependencies among multiple binary prediction tasks. Exploiting label prediction

T. Calders et al. (Eds.): ECML PKDD 2014, Part II, LNCS 8725, pp. 209–224, 2014.
© Springer-Verlag Berlin Heidelberg 2014

dependence is critical for multi-label learning, especially when the label information is sparse. A group of methods in the literature explore label prediction dependencies or correlations by identifying mid-level low-dimensional representations shared across labels from the input data [8,21,26,22,14,23]. Many other methods exploit the label dependency information directly in the output space [4,6,12,24,25]. Moreover, a number of recent works perform label space reduction to produce a low-dimensional intermediate label representation to facilitate multi-label classification with many labels [1,11,19,25]. They demonstrate that even simple label space reduction can capture intrinsic information in the output space for multi-label classification tasks while reducing the computational cost.

In this paper, we propose a novel bi-directional model for multi-label classification, which introduces a compact mid-level representation layer between the input features and the output labels to capture the common prediction representations shared across multiple labels. The mid-level representation layer is constructed from both input and output spaces, and it has two complementary parts, one of which captures the *predictive* low-dimensional semantic representation of the input features and the other captures the *predictable* low-dimensional intrinsic representation of the output labels. These two parts augment each other to integrate information encoded in both the feature and label spaces and enhance the overall multi-label classification.

This bi-directional model exploits the autoencoder principle [9] to generate the mid-level representation from two directions, while extending this principle by promoting the discriminability of the mid-level representation for predicting the target output labels. We formalize this model as a joint optimization problem over all the encoding/decoding/prediction parameters. We show that this optimization problem can be solved using an iterative optimization algorithm with alternating steps, in which one major step has efficient closed-form solution. We conduct experiments on a variety of multi-label classification data sets. The results show the proposed model outperforms its one-directional component models, and a number of multi-label classification methods.

The remainder of the paper is organized as follows. We review related works in Section 2 and present the proposed model in Section 3. The experiments are reported in Section 4. We finally conclude the paper in Section 5.

2   Related Work

Multi-label classification has received significant attention from machine learning community in recent years. There is a rich body of work on multi-label learning in the literature. In this section, we present a brief review over existing methods that are closely related to the proposed work.

One direction of multi-label learning research exploits bottom-up learning schemes that induce intermediate layers from the inputs to bridge the gap between the input features and output labels. For example, [10] trained multiple base models on randomly selected subsets of the input features and then combined the multiple models. This method however ignores label dependencies.

To amend this drawback, [26] proposed a multi-label dimensionality reduction method, which induces a low-dimensional feature space by maximizing the dependence between the original feature description and the class labels. [5] first augmented the original input features with the output of base binary classifiers and then performed multi-label learning on the augmented representation. A few other methods perform feature selection for multi-label learning [23,15]. [23] used a combination of PCA and genetic algorithms to search for the best feature subset. [15] proposed a feature selection algorithm that takes feature-label correlation into account based on a symmetrical uncertainty measure. A more advanced method [8] performs sparse feature learning with sparsity inducing norms to capture common predictive model structures across labels. The methods in [21,22,14] explore common subspaces shared among multiple labels.

Another set of methods exploit top-down learning schemes and induce alternative label representations from the original label space to facilitate multi-label learning. [24] applied canonical correlation analysis to extract error-correcting code words as intermediate prediction outputs between the input features and the original output labels. The work in [25] further enhanced this output coding learning scheme with a maximum margin output coding (MMOC) method, which formulates the output coding problem in a max-margin form to capture both discriminability and predictability of the output codes. [7] extended the kernel techniques widely used in the input feature space into the output label space to induce kernelized outputs in a large margin multi-label learning framework.

Moreover, a number of works pursue label space dimensionality reduction to produce intermediate low-dimensional label representations. An early work in [11] employs random label projection to address multi-label classification with a large number of labels. It first projects the high-dimensional label vectors to a low-dimensional space using a random transformation matrix, and then learns a multi-dimension regression model with the transformed labels. For a test instance, the estimated label vector from the regression models is then projected from the low-dimensional space back to the original high-dimensional label space. Following this work, a number of improvements have been proposed. [19] proposed a principal label space transformation method, which employs the principal component analysis to reduce the original label matrix to a low-dimensional space. Unlike random projections, the PCA dimensionality reduction minimizes the encoding error between the projected label matrix and the original label matrix. Subsequently, [1] proposed a conditional principal label space transformation method. It is a feature-aware method, which simultaneously minimizes both the label encoding error and the least squares linear regression error in the reduced label space. [27] proposed a Gaussian random projection method for label space transformation. Though these label space transformation methods have demonstrated that the intrinsic information of output labels can be captured in a low-dimensional output space, they mainly focus on reducing the computational cost without much loss of the multi-label classification performance.

Different from these existing methods, the proposed bi-directional model in this paper integrates the strengths of both bottom-up and top-down intermediate

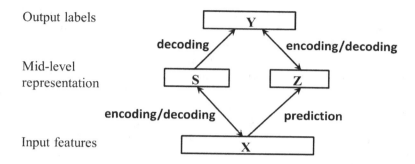

Fig. 1. The proposed bi-directional model. **X** denotes the input features and **Y** denotes the output labels. The mid-level latent layer has two parts, **S** and **Z**. **S** is the low-dimensional representation of the input features, and **Z** is the low-dimensional representation of the output labels. Both **S** and **Z** contribute to the decoding of the target output labels.

representation learning schemes in a complementary structure, aiming to improve multi-label classification performance. Our learning framework is not about producing any ad-hoc latent representations from the inputs and outputs, but aims to augment each other from two directions. Our model extends the generative autoencoder principle [9] in a discriminative way, sharing some similarity with the multi-class zero-shot learning approach in [18]. The work in [18] nevertheless is still a one-directional method that learns low-dimensional semantic representations from the inputs.

3   A Bi-directional Model for Multi-label Classification

Traditional multi-label models typically learn a mapping function from the input space to the output space directly. Recent studies show that an intermediate representation can be very useful for bridging the original inputs and outputs, as we discussed in Section 2. Nevertheless, all these previous works have focused on one-directional representation learning using either bottom-up or top-down schemes. In this section, we present a novel bi-directional representation learning model for multi-label classification, which has a hybrid mid-level representation layer that captures both feature-sourced and label-sourced intrinsic information of the data in a complementary way, aiming to boost the learning performance.

Figure 1 shows the proposed bi-directional model. In this model, **X** and **Y** denote the input features and the output labels respectively. The latent mid-level layer has two parts, **S** and **Z**, which encode information from two directions in a low-dimensional space. The low-dimensional representation code **S** is constructed from the input **X** using the autoencoder principle such that **S** can be produced from **X** with an encoding function and **X** can be reconstructed from **S** with a decoding function. This mechanism ensures that **S** captures the intrinsic information stored in the input features. The latent representation code **Z** is produced from the output $\mathbf{Y}$ with an encoding function, while its predictability from the input $\mathbf{X}$ is simultaneously promoted with a prediction function. To ensure the informativeness of the latent layer for the target output prediction, both $\mathbf{S}$ and $\mathbf{Z}$ are used to predict the output $\mathbf{Y}$ with a joint decoding function. With such a learning structure, we expect $\mathbf{S}$ and $\mathbf{Z}$ can contribute complementary information for accurate identification of $\mathbf{Y}$. Moreover, in the test phase, multi-label classification can be naturally achieved by first following the double line information flow from $\mathbf{X}$ to $\mathbf{S}$ and $\mathbf{Z}$, and then decoding $\mathbf{Y}$ from the concatenation of $\mathbf{S}$ and $\mathbf{Z}$.

Below we will introduce the components of the proposed model and the training algorithm in detail. The following notations will be used. We assume a set of t training instances, (X, Y), is given, where $X \in \mathbb{R}^{t \times d}$ is the input data matrix and $Y \in \{0, 1\}^{t \times k}$ is the label indicator matrix. The low-dimensional representation matrix of X is denoted as $S \in \mathbb{R}^{t \times m}$ for $m < d$, and the low-dimensional representation matrix of Y is denoted as $Z \in \mathbb{R}^{t \times n}$ for $n < k$. We use $\mathbf{1}$ to denote any column vector with all 1 values, assuming its length can be determined from the context; use I_t to denote an identity matrix with size t; and use "$\circ$" to denote the Hadamard product operator between two matrices.

3.1   Framework: Encoding, Prediction and Decoding

The learning framework on the proposed model involves three major components: encoding, prediction and decoding, which follows but extends the standard autoencoder models. We will introduce each of them below.

Encoding. We propose to use typical sigmoid based functions to perform nonlinear encoding over the input data matrix X and the output label matrix Y, which map X to the low-dimensional latent representation matrix S and map Y to the low-dimensional latent representation matrix Z respectively. The two encoder functions are compositions of the standard sigmoid function and linear functions:

$$S = \sigma(XW_x + \mathbf{1}\mathbf{b}_x^\top), \tag{1}$$
$$Z = \sigma(YW_y + \mathbf{1}\mathbf{b}_y^\top) \tag{2}$$

where $(W_x \in \mathbb{R}^{d \times m}, \mathbf{b}_x \in \mathbb{R}^m)$ and $(W_y \in \mathbb{R}^{k \times n}, \mathbf{b}_y \in \mathbb{R}^n)$ are the linear model parameters for the two encoder functions respectively; $\sigma(x) = 1/(1 + \exp(-x))$ is a sigmoid function that encodes entry-wise nonlinear transformation of the input. Moreover the sigmoid functions in the two encoder functions also put the values of S and Z in a comparable range.

A linear version of the encoder functions can be obtained by simply dropping the outer sigmoid functions.

Prediction. The encoder function over Y produces a low-dimensional mid-level prediction target Z for the input data. To ensure the information flow from the input data to the output labels, we consider a prediction function, $f : \mathcal{X} \to \mathcal{Z}$, that maps X to the low-dimensional representation code Z. In particular, we consider the following linear regression function:

$$\hat{Z} = f(X) = X\Omega + \mathbf{1}\mathbf{q}^{\top} \tag{3}$$

where $\Omega \in \mathbb{R}^{d \times n}$ and $\mathbf{q} \in \mathbb{R}^n$ are the prediction model parameters, and $\hat{Z}$ denotes the prediction value matrix. This prediction function will be learned by minimizing a prediction loss $\ell_p(Z, \hat{Z})$ between the low-dimensional matrix Z and the prediction value matrix $\hat{Z}$ produced by the predictor over the input data. It thus enforces the predictability of Z from the inputs. This component is also necessary for exploiting the latent representation part $\mathbf{Z}$ in the test phase. In the test phase, where Y is unknown, one can produce Z from the input data using this prediction function and then use Z and S to reconstruct Y.

Decoding. There are two decoder functions, $g(\cdot)$ and $h(\cdot)$, in our model to reconstruct the observed data X and Y from the latent low-dimensional code matrices S and Z. The decoder function $g : \mathcal{S} \to \mathcal{X}$ reconstructs the input data X in its original space from its low-dimensional representation S. We consider a linear decoder function:

$$\hat{X} = g(S) = SU + \mathbf{1}\mathbf{d}^{\top} \tag{4}$$

where $U \in \mathbb{R}^{m \times d}$ and $\mathbf{d} \in \mathbb{R}^d$ are decoding parameters, and $\hat{X}$ is the reconstructed data matrix in the original input space. This decoder function and the encoder function from X to S in Eq. (1) together form an *autoencoder* model. An *autoencoder* in general is a two-layered construction, in which a first layer encodes the input data into the latent low-dimensional representation and a second layer decodes this representation back into the original data space. The *autoencoder* is trained to minimize the reconstruction error in the original data space. In our model, the two-layered constructions of the autoencoder correspond to the encoding from X to S and the decoding from S to $\hat{X}$ respectively. We will minimize a decoding loss $\ell_d(X, \hat{X})$ between the original input data X and the reconstructed data $\hat{X}$ in the training process.

The reconstruction of Y however is not a standard decoding problem in an autoencoder, since it is the key step for the overall multi-label classification and information from the input data should be taken into account. Hence instead of reconstructing it from its low-dimensional representation Z, we consider a decoder function $h : \mathcal{S} \times \mathcal{Z} \to \mathcal{Y}$ that reconstructs Y from a concatenated low-dimensional space of the latent representation codes S and Z. Specifically, we use the following linear decoding function:

$$\hat{Y} = h(S, Z) = SA_s + ZA_z + \mathbf{1}\mathbf{a}^{\top} \tag{5}$$

where $A_s \in \mathbb{R}^{m \times k}$, $A_z \in \mathbb{R}^{n \times k}$ and $\mathbf{a} \in \mathbb{R}^k$ are the decoder parameters, and $\hat{Y}$ is the reconstructed data matrix in the original output label space. We will minimize a decoding loss $\ell_d(Y, \hat{Y})$ between the original Y and the reconstructed $\hat{Y}$ in the training process. Since the latent representation matrices, S and Z, will be simultaneously induced in the training process, we expect the latent S and Z identified will complement each other in this final decoder function to accurately reconstruct the output label matrix Y. The encoding and decoding process between the output layer and the mid-level representation layer can be viewed as an augmented autoencoder.

3.2   Optimization Formulation

Given the framework introduced above, we formulate the bi-directional model training as a joint optimization problem over all the model parameters that minimizes a regularized linear combination of the prediction loss and the two decoding losses:

$$\min_{W_x, b_x, W_y, b_y, A_s, A_z, a, U, d, \Omega, q} \mathcal{L}(X, Y) \tag{6}$$

such that

$$\mathcal{L}(X, Y) = \ell_p(Z, \hat{Z}) + \eta \ell_d(Y, \hat{Y}) + \rho \ell_d(X, \hat{X}) + \tag{7}$$
$$\alpha_x \mathcal{R}(W_x) + \alpha_y \mathcal{R}(W_y) + \alpha_u \mathcal{R}(U) + \alpha_s \mathcal{R}(A_s) + \alpha_z \mathcal{R}(A_z) + \alpha_o \mathcal{R}(\Omega)$$

where the trade-off parameters η and ρ are used to adjust the relative degrees of focus on the three different loss terms; all the other α_* trade-off parameters adjust the degrees of regularization over the model parameter matrices; $\mathcal{R}(\cdot)$ denotes the regularization function. Note with the encoding parameters, (W_x, b_x) and (W_y, b_y), the latent representation matrices S and Z are directly available through the nonlinear functions in Eq. (1) and Eq. (2) respectively. This objective function expresses the following properties we expect from the latent representations: 1) S and Z should be low-dimensional (enforced by the encoding model parameters); 2) S should preserve as much information as possible from X (enforced by $\ell_d(X, \hat{X})$); 3) Z should preserve information from Y that is complementary to S for the reconstruction of Y (enforced by $\ell_d(Y, \hat{Y})$), while being predictable from X (enforced by $\ell_p(Z, \hat{Z})$); and 4) the concatenation of S and Z should be discriminative for the target output label matrix Y (enforced by $\ell_d(Y, \hat{Y})$).

To produce a concrete training problem, we use least squares loss functions for both the prediction loss and the two decoding losses, such that

$$\ell_p(Z, \hat{Z}) = \|Z - \hat{Z}\|_F^2 = \|Z - f(X)\|_F^2 \tag{8}$$
$$\ell_d(X, \hat{X}) = \|X - \hat{X}\|_F^2 = \|X - g(S)\|_F^2 \tag{9}$$
$$\ell_d(Y, \hat{Y}) = \|Y - \hat{Y}\|_F^2 = \|Y - h(S, Z)\|_F^2 \tag{10}$$

where $\| \cdot \|_F$ denotes the Frobenius norm of a matrix. We use the square of Frobenius norm as the regularization function $\mathcal{R}$ over the parameter matrices, such that $\mathcal{R}(\cdot) = \| \cdot \|_F^2$.

Moreover, among the three loss terms in the objective function (7), the decoding loss $\ell_d(X, \hat{X})$ in (9) is used to ensure the input data can be reconstructed from its low-dimensional representation S with a small error by using the decoder function $g(\cdot)$. The decoder function $g(\cdot)$ is not directly involved in the overall target label prediction process, since S can be produced from the original input matrix X with the encoding function. The necessity of having this decoding component in the framework, hence the decoding loss $\ell_d(X, \hat{X})$ in the optimization objective, can be questioned. In our empirical study later, we investigated this issue by dropping $g(\cdot)$, hence $\ell_d(X, \hat{X})$ and $\mathcal{R}(U)$, from the optimization problem. Our results suggest the decoder component $g(\cdot)$ is useful and the decoding loss $\ell_d(X, \hat{X})$ should be included in the learning process.

3.3 Optimization Algorithm

The minimization problem in (6) involves two sets of parameters, the encoder parameters, $\{W_x, \mathbf{b}_x, W_y, \mathbf{b}_y\}$, and the decoder and prediction model parameters, $\{A_s, A_z, \mathbf{a}, U, \mathbf{d}, \Omega, \mathbf{q}\}$. We develop an iterative optimization algorithm that conducts optimization over these two groups of model parameters in an alternating way in each iteration.

We first randomly initialize the model parameters. Then in each iteration, we perform the following two steps.

Step I. In this step, given the current encoder parameters, $\{W_x, \mathbf{b}_x, W_y, \mathbf{b}_y\}$, to be fixed, we optimize the decoder and prediction model parameters to minimize the objective function in (7). We first compute the latent matrices S and Z according to Eq. (1) and Eq. (2) respectively. Given S and Z, the joint minimization problem in (6) can be decomposed into the following sub-optimization problems over the decoder and prediction model parameters:

$$\min_{A_s, A_z, \mathbf{a}} \quad \eta \|Y - h(S, Z)\|_F^2 + \alpha_s \|A_s\|_F^2 + \alpha_z \|A_z\|_F^2 \tag{11}$$

$$\min_{\Omega, \mathbf{q}} \quad \|Z - f(X)\|_F^2 + \alpha_o \|\Omega\|_F^2 \tag{12}$$

$$\min_{U, \mathbf{d}} \quad \rho \|X - g(S)\|_F^2 + \alpha_u \|U\|_F^2 \tag{13}$$

which have the following three sets of closed-form solutions respectively:

$$\left.\begin{aligned}
A_s &= (S^\top H S + \tfrac{\alpha_s}{\eta} I_m)^{-1} S^\top H(Y - Z A_z) \\
A_z &= (Z^\top H Z + \tfrac{\alpha_z}{\eta} I_n)^{-1} Z^\top H(Y - S A_s) \\
\mathbf{a} &= \tfrac{1}{t}(Y - S A_s - Z A_z)^\top \mathbf{1}
\end{aligned}\right\} \tag{14}$$

$$\left.\begin{aligned}
\Omega &= (X^\top H X + \alpha_o I_d)^{-1} X^\top H Z \\
\mathbf{q} &= \tfrac{1}{t}(Z - X \Omega)^\top \mathbf{1}
\end{aligned}\right\} \tag{15}$$

$$\left.\begin{aligned}
U &= (S^\top H S + \tfrac{\alpha_u}{\rho} I_m)^{-1} S^\top H X \\
\mathbf{d} &= \tfrac{1}{t}(X - S U)^\top \mathbf{1}
\end{aligned}\right\} \tag{16}$$

where $H = I_t - \tfrac{1}{t}\mathbf{1}\mathbf{1}^\top$ is a centering matrix of size t.

Algorithm 1. The Bi-Directional Learning Algorithm

1: **Initialize** all the model parameters.
2: **repeat**
3:     **Step I**: given current encoder parameters $\{W_x, \mathbf{b}_x, W_y, \mathbf{b}_y\}$, update the decoder and prediction model parameters, $\{A_s, A_z, \mathbf{a}, U, \mathbf{d}, \Omega, \mathbf{q}\}$, with closed-form solutions in Eq. (14)–(16).
4:     **Step II**: given current $\{A_s, A_z, \mathbf{a}, U, \mathbf{d}, \Omega, \mathbf{q}\}$, perform optimization over encoder parameters $\{W_x, \mathbf{b}_x, W_y, \mathbf{b}_y\}$ to minimize the objective $\mathcal{J}$ in (18) using gradient descent with line search.
5: **until** Convergence or maximum number of iterations is reached

---

Step II. In this step, given the current decoder and prediction model parameters, $\{A_s, A_z, \mathbf{a}, U, \mathbf{d}, \Omega, \mathbf{q}\}$, to be fixed, we optimize the encoder parameters, $\{W_x, \mathbf{b}_x, W_y, \mathbf{b}_y\}$, to minimize the objective function in (7). This leads to the following minimization problem:

$$\min_{W_x, \mathbf{b}_x, W_y, \mathbf{b}_y} \quad \ell_p(Z, \hat{Z}) + \eta \ell_d(Y, h(S, Z)) + \rho \ell_d(X, g(S)) + \alpha_x \mathcal{R}(W_x) + \alpha_y \mathcal{R}(W_y)$$

(17)

where $\hat{Z}$ is pre-computed with the fixed parameters via Eq. (3). By expressing S and Z in terms of the encoder functions in Eq. (1) and Eq. (2), the objective function in (17) can be written as

$$\mathcal{J} = \|\hat{Z} - \sigma(YW_y + \mathbf{1b}_y^\top)\|_F^2$$
$$+ \eta \|\sigma(XW_x + \mathbf{1b}_x^\top)A_s + \sigma(YW_y + \mathbf{1b}_y^\top)A_z + \tilde{Y}\|_F^2$$
$$+ \rho \|\sigma(XW_x + \mathbf{1b}_x^\top)U + \tilde{X}\|_F^2 + \alpha_x \|W_x\|_F^2 + \alpha_y \|W_y\|_F^2 \quad (18)$$

where $\tilde{Y}$ and $\tilde{X}$ are defined as

$$\tilde{Y} = \mathbf{1a}^\top - Y, \qquad \tilde{X} = \mathbf{1d}^\top - X. \tag{19}$$

We use a gradient descent algorithm with line search [17] to solve the minimization problem in (17), which requires computing the gradients of the objective function $\mathcal{J}$ regarding the decoder parameters $\{W_x, \mathbf{b}_x, W_y, \mathbf{b}_y\}$.

The overall optimization algorithm for training the bi-directional model is summarized in Algorithm 1.

Test Phase: In the test phase, given a new instance $\mathbf{x}$, we first produce the latent representations, $\mathbf{s} = \sigma(\mathbf{x}W_x + \mathbf{1b}_x^\top)$ and $\mathbf{z} = f(\mathbf{x})$. Then the final output can be computed by $\mathbf{y} = h(\mathbf{s}, \mathbf{z})$. The labels of $\mathbf{x}$ can be determined by simply rounding the entries of $\mathbf{y}$ to 0s or 1s.

4   Experimental Results

Data Sets. To evaluate the proposed bi-directional model for multi-label classification, we conducted experiments on 5 different types of real-world data sets:

Table 1. The statistic information of the data sets used in the experiments

Data set	corel5k	delicious	yeast	genbase	mirflickr
Num. of Instances	4999	5000	2417	662	5000
Num. of Features	512	500	103	1186	512
Num. of Labels	209	918	14	15	38
Label Cardinality	3.32	18.72	4.24	1.16	4.71

two image data sets (*corel5k* [3] and *mirflickr* [13]), one text set (*delicious* [20]), and two biology data sets (*yeast* [4] and *genbase* [2]). For each big data set with more than $10k$ instances, we randomly sampled a 5000-instance subset to use. We conducted experiments with 5-fold cross-validation and dropped the labels that do not appear in at least one of the five fold partitions. The statistic information of all 5 data sets used is summarized in Table 1, where label cardinality denotes the average number of labels assigned to each instance.

Methods. We compared the proposed bi-directional multi-label learning method with the following multi-label learning methods:

- *Binary relevance* (BR). This baseline method decomposes multi-label classification into a set of independent binary classification problems via the one-vs-all scheme. We used linear SVM classifiers for the binary problems.

- *Multi-label Output Codes using CCA* (MOC-CCA) [24]. This method performs error-correcting coding for the labels based on canonical correlation analysis (CCA).

- *Multi-Label Learning using Local Correlation* (ML-LOC) [12]. Instead of assuming global label correlations, this method separates instances into different groups and allows label correlations to be exploited locally.

- *Calibrated Separation Ranking Loss* (CSRL) [6]. This method performs large margin multi-label learning based on a novel loss function.

Experimental Setting. In each iteration of the 5-fold cross-validation, the training set is further randomly divided into two parts for parameter selection: 80% for model training and 20% for parameter evaluation. For the proposed method, there are a number of parameters need to be determined. We fixed the regularization parameters as relatively small values, such as $\alpha_x = \alpha_y = \alpha_o = 0.05$, $\alpha_s = \alpha_z = 0.05\eta$, and $\alpha_u = 0.05\rho$. The trade-off parameters, ρ and η, and the dimensions of latent representations, m and n, are automatically selected in the learning phase. The values of ρ and η are both selected from $\{0.1, 1, 10, 100\}$. The candidate values for m and n however vary across data sets since the feature and label dimensions are different for different data sets. We used $m \in \{20, 50, 100\}$ and $n \in \{20, 60, 80\}$ on *corel5k*; used $m, n \in \{20, 50, 100\}$ on *delicious*; used $m \in \{20, 50\}$ and $n \in \{5, 10\}$ on *yeast*; used $m \in \{20, 50, 100\}$ and $n \in \{5, 10\}$ on *genbase*; and used $m \in \{20, 50, 100\}$ and $n \in \{5, 15, 25\}$ on *mirflickr*. For the comparison methods, we performed parameter selection using the same scheme. For *BR*, the trade-off parameter C is selected from

Table 2. The average and standard deviation results in terms of Hamming Loss(%). Lower values indicate better performance.

Data set	BR	MOC-CCA	ML-LOC	CSRL	Proposed
corel5k	0.9 ± 0.03	0.6 ± 0.02	0.6 ± 0.01	$\mathbf{0.4 \pm 0.02}$	$\mathbf{0.4 \pm 0.02}$
delicious	7.2 ± 0.04	10.3 ± 0.06	17.5 ± 0.07	4.5 ± 0.03	$\mathbf{3.9 \pm 0.03}$
yeast	8.6 ± 0.03	4.1 ± 0.03	14.2 ± 0.05	6.7 ± 0.04	$\mathbf{4.1 \pm 0.04}$
genbase	0.6 ± 0.01	0.5 ± 0.02	1.2 ± 0.01	0.5 ± 0.03	$\mathbf{0.3 \pm 0.01}$
mirflickr	3.2 ± 0.02	3.1 ± 0.04	8.1 ± 0.05	$\mathbf{2.3 \pm 0.03}$	2.5 ± 0.02

Table 3. The average and standard deviation results in terms of Macro-F1 measure(%). Larger values indicate better performance.

Data set	BR	MOC-CCA	ML-LOC	CSRL	Proposed
corel5k	2.1 ± 0.02	5.6 ± 0.01	3.9 ± 0.02	5.3 ± 0.05	$\mathbf{8.0 \pm 0.02}$
delicious	6.5 ± 0.04	13.9 ± 0.04	10.0 ± 0.09	14.1 ± 0.03	$\mathbf{15.3 \pm 0.02}$
yeast	30.1 ± 0.14	38.1 ± 0.17	39.7 ± 0.18	39.3 ± 0.12	$\mathbf{40.6 \pm 0.15}$
genbase	46.4 ± 0.08	61.3 ± 0.05	55.1 ± 0.07	61.5 ± 0.12	$\mathbf{63.8 \pm 0.05}$
mirflickr	18.1 ± 0.14	22.5 ± 0.19	21.2 ± 0.16	24.9 ± 0.20	$\mathbf{25.6 \pm 0.13}$

$\{0.1, 1, 10, 50, 100\}$; for *MOC-CCA*, the number of canonical components d is selected from $[1, \min(\#\text{features}, \#\text{labels})]$ and the trade-off parameter λ is selected from $\{0.25, 1, 4\}$; for *ML-LOC*, the parameters are selected as $\lambda_1 \in \{0.1, 1, 10\}$, $\lambda_2 \in \{1, 10, 100\}$, and $m \in \{10, 15, 20\}$; for CSRL, the trade-off parameter C is selected from $\{0.1, 1, 10\}$.

4.1 Multi-label Classification Results

We evaluated the performance of each comparison method with three criteria: *Hamming Loss*, *Macro-F1*, and *Micro-F1*. These three criteria measure the multi-label classification performance from different aspects. The 5-fold cross validation results for all comparison methods over the five data sets are reported in Table 2 – Table 4 in terms of the three evaluation criteria respectively. Both the average result values and their standard deviations are reported.

From the results in Table 2, we can see that the multi-label comparison method *MOC-CCA* does not show consistent advantages over the baseline *BR* in terms of hamming loss, *ML-LOC* even has inferior performance on most data sets comparing to *BR*, while *CSRL* and the proposed approach outperform *BR* across all data sets. Moreover, the proposed approach produces the best results among all the comparison methods on four out of the total five data sets. Hamming loss however may prefer extreme prediction results without balancing the prediction recall and precision. Table 3 presents the comparison results in terms of Macro-F1 score which takes both prediction recall and precision into account. In terms of Macro-F1, the proposed approach consistently outperforms all the other methods across all the five data sets. In particular, on *corel5k*, the proposed

Table 4. The average and standard deviation results in terms of Micro-F1 measure(%). Larger values indicate better performance.

Data set	BR	MOC-CCA	ML-LOC	CSRL	Proposed
corel5k	7.6 ± 0.09	**11.1 ± 0.10**	9.3 ± 0.14	10.6 ± 0.10	**11.1 ± 0.08**
delicious	16.5 ± 0.07	**23.8 ± 0.07**	16.1 ± 0.08	22.9 ± 0.03	23.2 ± 0.03
yeast	52.4 ± 0.22	64.8 ± 0.20	67.7 ± 0.25	60.1 ± 0.16	**68.2 ± 0.17**
genbase	54.8 ± 0.12	**71.4 ± 0.17**	71.1 ± 0.21	65.3 ± 0.17	70.9 ± 0.18
mirflickr	24.1 ± 0.45	32.1 ± 0.49	36.7 ± 0.69	31.5 ± 0.32	**36.8 ± 0.25**

method produces incredible improvement over the other comparison methods. It improves the performance of *MOC-CCA* by more than 40%, improves the performance of *CSRL* by more than 50%, and improves the performance of *ML-LOC* by more than 100%. Moreover, all the four multi-label learning methods greatly outperform the baseline *BR* across all the data sets in terms of Macro-F1. *Table 4* presents the comparison results in terms of Micro-F1 score. We can see that the four multi-label learning methods again outperform the baseline *BR* across all the data sets, except on *delicious* where *ML-LOC* has slightly inferior result. Among the multi-label learning methods, the proposed approach produces the best results on three data sets, while presenting results very close to the best ones on the remaining two data sets. It demonstrates more consistent good performance across different types of data sets. All these results suggest our proposed bi-directional model is effective for multi-label classification by capturing complementary information from both inputs and outputs.

4.2   Study of the Bi-directional Model

To gain a deeper understanding over the novel bi-directional learning scheme, we have also conducted experiments to investigate the influence of different components in the proposed bi-directional model.

First, we investigated the capacity of the bi-directional model by comparing the full model to its two essential one-directional components, the bottom-up component and the top-down component. The mid-level representation of the bi-directional full model has two complementary parts, S and Z. The bottom-up component and the top-down component consider solely the feature-sourced mid-level representation S and the label-sourced mid-level representation Z respectively, by deactivating the other component. We denote the full model with $S + Z$ and denote the two component models with S and Z respectively. Parameter selection for the two component models is conducted using the same procedure introduced before. The comparison results for these three models are reported in Table 5. We can see that the two one-directional component models have strengths on different data sets. The label-sourced one-directional model Z performs better than the feature-sourced one-directional model S on *corel5k* and *delicious* where the label space is large, while model S performs better on the other three data sets. Nevertheless, the proposed bi-directional model greatly

Table 5. Comparison results over the bi-directional model and its two one-directional components. $S+Z$ denotes the proposed bi-directional model, S denotes the bottom-up one-directional model and Z denotes the top-down one-directional model.

Measure	Method	corel5k	delicious	yeast	genbase	mirflickr
Hamming Loss	$S+Z$	0.4 ± 0.02	3.9 ± 0.03	4.1 ± 0.04	0.3 ± 0.01	2.5 ± 0.02
	S	0.7 ± 0.03	5.0 ± 0.05	5.8 ± 0.03	0.5 ± 0.04	2.9 ± 0.05
	Z	0.6 ± 0.05	4.4 ± 0.02	7.2 ± 0.02	2.4 ± 0.05	2.6 ± 0.05
Macro-F1	$S+Z$	8.0 ± 0.02	15.3 ± 0.02	40.6 ± 0.15	63.8 ± 0.05	25.6 ± 0.13
	S	4.1 ± 0.05	8.8 ± 0.04	34.2 ± 0.12	48.4 ± 0.17	18.2 ± 0.18
	Z	6.2 ± 0.05	11.8 ± 0.03	26.6 ± 0.12	44.1 ± 0.10	19.1 ± 0.13
Micro-F1	$S+Z$	11.1 ± 0.08	23.2 ± 0.03	68.2 ± 0.17	70.9 ± 0.18	36.8 ± 0.25
	S	8.5 ± 0.12	17.1 ± 0.05	60.2 ± 0.11	58.3 ± 0.14	28.7 ± 0.21
	Z	9.8 ± 0.10	21.8 ± 0.06	55.6 ± 0.14	50.5 ± 0.17	28.4 ± 0.44

outperforms both one-directional component models across all the five data sets in terms of all the three measures. This suggests that the proposed bi-directional model can successfully integrate the strengths of its one-directional components in a complementary way and has capacity of capturing useful information from both the input and output spaces.

Next, we have also compared the proposed approach with two of its alternative versions: one drops the decoder component $g(\cdot)$ and is denoted as "*Proposed w/o g*"; the other removes the nonlinear $\sigma(\cdot)$ function from encoding and uses linear encoder functions, which is denoted as "*Proposed w/o σ*". We conducted the comparison experiment using varying latent dimension sizes, m and n, on the *corel5k* data set. We first set $n = 20$ and studied the performance of the three methods by varying the m value within the set $\{20, 50, 100, 300\}$. Then we set $m = 50$ and vary the n value within the set $\{20, 60, 100, 140\}$. The experimental results are presented in Figure 2, in terms of the logarithm of the three evaluation criteria. We can see that the proposed full model clearly outperforms the other two variants across all learning scenarios, which suggests that both the decoder component $g(\cdot)$ and the nonlinear encoders are important in our bi-directional model. From Figure 2 (a)-(c), we can see that with fixed n value, the performance of all the three methods becomes stable when the m value reaches 50. This suggests that with even a reasonably small m value such as $m = 50$, our model can already capture the intrinsic information from the input data. On the other hand, from Figure 2 (d)-(f), we can see that the performance of all the three methods deteriorates when the n value becomes larger than 60. This suggests that if the latent representation size of the output labels is too large, noise may be introduced in augmenting the latent component produced from the input data and hence harm the performance.

In summary, all these experimental results demonstrated the compactness and effectiveness of the novel bi-directional model for multi-label classification.

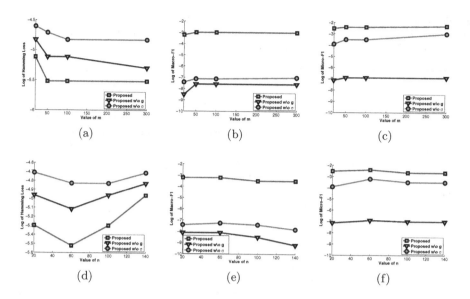

Fig. 2. Impact of the latent dimension sizes m and n over the model performance on *corel5k*. In the top row, n is fixed to 20, and m varies from 20 to 300. In the bottom row, m is fixed to 50, and n varies from 20 to 140. The first column shows the results in terms of log hamming loss, the middle column shows the results in terms of log Macro-F1, and the right column shows the results in terms of log Micro-F1.

5   Conclusion

In this paper, we proposed a novel bi-directional representation learning model for multi-label classification, which has a two-part latent representation layer constructed from both the input data and the output labels. The two latent parts augment each other by integrating information from both the feature and label spaces to enhance the overall multi-label classification. We formulated multi-label learning over this model as a joint minimization problem over all parameters of the component functions, and developed an iterative optimization algorithm with alternating steps to perform optimization. We conducted experiments on five real world multi-label data sets by comparing the proposed method to a number of previously developed multi-label classification methods and a few variant models. Our experimental results suggest that our proposed model is compact and showed that it outperformed all the other comparison methods.

References

1. Chen, Y., Lin, H.: Feature-aware label space dimension reduction for multi-label classification. In: Proceedings of NIPS (2012)
2. Diplaris, S., Tsoumakas, G., Mitkas, P.A., Vlahavas, I.P.: Protein classification with multiple algorithms. In: Bozanis, P., Houstis, E.N. (eds.) PCI 2005. LNCS, vol. 3746, pp. 448–456. Springer, Heidelberg (2005)

3. Duygulu, P., Barnard, K., de Freitas, J.F.G., Forsyth, D.: Object recognition as machine translation: Learning a lexicon for a fixed image vocabulary. In: Heyden, A., Sparr, G., Nielsen, M., Johansen, P. (eds.) ECCV 2002, Part IV. LNCS, vol. 2353, pp. 97–112. Springer, Heidelberg (2002)
4. Elisseeff, A., Weston, J.: A kernel method for multi-labelled classification. In: Proceedings of NIPS (2001)
5. Godbole, S., Sarawagi, S.: Discriminative methods for multi-labeled classification. In: Dai, H., Srikant, R., Zhang, C. (eds.) PAKDD 2004. LNCS (LNAI), vol. 3056, pp. 22–30. Springer, Heidelberg (2004)
6. Guo, Y., Schuurmans, D.: Adaptive large margin training for multilabel classification. In: Proceedings of AAAI (2011)
7. Guo, Y., Schuurmans, D.: Multi-label classification with output kernels. In: Blockeel, H., Kersting, K., Nijssen, S., Železný, F. (eds.) ECML PKDD 2013, Part II. LNCS, vol. 8189, pp. 417–432. Springer, Heidelberg (2013)
8. Guo, Y., Xue, W.: Probablistic mult-label classification with sparse feature learning. In: Proceedings of IJCAI (2013)
9. Hinton, G., Salakhutdinov, R.: Reducing the dimensionality of data with neural networks. Science 313, 504–507 (2006)
10. Ho, T.: The random subspace method for constructing decision forests. IEEE TPAMI 20(8) (August 1998)
11. Hsu, D., Kakade, S., Langford, J., Zhang, T.: Multi-label prediction via compressed sensing. In: Proceedings of NIPS (2009)
12. Huang, S., Zhou, Z.: Multi-label learning by exploiting label correlations locally. In: Proceedings of AAAI (2012)
13. Huiskes, M., Lew, M.: The MIR Flickr retrieval evaluation. In: Proceedings of ACM MIR (2008)
14. Ji, S., Tang, L., Yu, S., Ye, J.: Extracting shared subspace for multi-label classification. In: Proceedings of KDD (2008)
15. Lastra, G., Luaces, O., Quevedo, J.R., Bahamonde, A.: Graphical feature selection for multilabel classification tasks. In: Gama, J., Bradley, E., Hollmén, J. (eds.) IDA 2011. LNCS, vol. 7014, pp. 246–257. Springer, Heidelberg (2011)
16. Lewis, D., Yang, Y., Rose, T., Li, F.: RCV1: A new benchmark collection for text categorization research. JMLR 5, 361–397 (2004)
17. Nocedal, J., Wright, S.J.: Numerical Optimization. Springer, New York (2006)
18. Sharmanska, V., Quadrianto, N., Lampert, C.H.: Augmented attribute representations. In: Fitzgibbon, A., Lazebnik, S., Perona, P., Sato, Y., Schmid, C. (eds.) ECCV 2012, Part V. LNCS, vol. 7576, pp. 242–255. Springer, Heidelberg (2012)
19. Tai, F., Lin, H.: Multilabel classification with principal label space transformation. In: Proceedings of Inter. Workshop on Learning from Multi-Label Data (2010)
20. Tsoumakas, G., Katakis, I., Vlahavas, I.: Effective and efficient multilabel classification in domains with large number of labels. In: ECML/PKDD Workshop on Mining Multidimensional Data (2008)
21. Yan, R., Tesic, J., Smith, J.: Model-shared subspace boosting for multi-label classification. In: Proceedings of KDD (2007)
22. Yu, K., Yu, S., Tresp, V.: Multi-label informed latent semantic indexing. In: Proceedings of the Annual ACM SIGIR Conference (2005)
23. Zhang, M., Peña, J., Robles, V.: Feature selection for multi-label naive bayes classification. Inf. Sci. 179(19) (September 2009)
24. Zhang, Y., Schneider, J.: Multi-label output codes using canonical correlation analysis. In: Proceedings of AISTATS (2011)

25. Zhang, Y., Schneider, J.: Maximum margin output coding. In: Proceedings of ICML (2012)
26. Zhang, Y., Zhou, Z.: Multilabel dimensionality reduction via dependence maximization. In: Proceedings of AAAI (2008)
27. Zhou, T., Tao, D., Wu, X.: Compressed labeling on distilled lablsets for multi-label learning. Machine Learning 88, 69–126 (2012)

Optimal Thresholding of Classifiers
to Maximize F1 Measure

Zachary C. Lipton, Charles Elkan, and Balakrishnan Naryanaswamy

University of California, San Diego
La Jolla, California 92093-0404, USA
{zlipton,celkan,muralib}@cs.ucsd.edu

Abstract. This paper provides new insight into maximizing F1 measures in the context of binary classification and also in the context of multilabel classification. The harmonic mean of precision and recall, the F1 measure is widely used to evaluate the success of a binary classifier when one class is rare. Micro average, macro average, and per instance average F1 measures are used in multilabel classification. For any classifier that produces a real-valued output, we derive the relationship between the best achievable F1 value and the decision-making threshold that achieves this optimum. As a special case, if the classifier outputs are well-calibrated conditional probabilities, then the optimal threshold is half the optimal F1 value. As another special case, if the classifier is completely uninformative, then the optimal behavior is to classify all examples as positive. When the actual prevalence of positive examples is low, this behavior can be undesirable. As a case study, we discuss the results, which can be surprising, of maximizing F1 when predicting 26,853 labels for Medline documents.

Keywords: supervised learning · text classification · evaluation methodology · F score · F1 measure · multilabel learning · binary classification.

1 Introduction

Performance measures are useful for comparing the quality of predictions across systems. Some commonly used measures for binary classification are accuracy, precision, recall, F1 measure, and Jaccard index [15]. Multilabel classification is an extension of binary classification that is currently an area of active research in supervised machine learning [18]. Micro averaging, macro averaging, and per instance averaging are three commonly used variations of F1 measure used in the multilabel setting. In general, macro averaging increases the impact on final score of performance on rare labels, while per instance averaging increases the importance of performing well on each example [17]. In this paper, we present theoretical and experimental results on the properties of the F1 measure. For concreteness, the results are given specifically for the F1 measure and its multilabel variants. However, the results can be generalized to $F\beta$ measures for $\beta \neq 1$.

T. Calders et al. (Eds.): ECML PKDD 2014, Part II, LNCS 8725, pp. 225–239, 2014.
© Springer-Verlag Berlin Heidelberg 2014

	Actual Positive	Actual Negative
Predicted Positive	tp	fp
Predicted Negative	fn	tn

Fig. 1. Confusion Matrix

Two approaches exist for optimizing performance on the F1 measure. Structured loss minimization incorporates the performance measure into the loss function and then optimizes during training. In contrast, plug-in rules convert the numerical outputs of classifiers into optimal predictions [5]. In this paper, we highlight the latter scenario, and we differentiate between the beliefs of a system and predictions selected to optimize alternative measures. In the multilabel case, we show that the same beliefs can produce markedly dissimilar optimally thresholded predictions depending upon the choice of averaging method.

It is well-known that F1 is asymmetric in the positive and negative class. Given complemented predictions and complemented true labels, the F1 measure is in general different. It also generally known that micro F1 is affected less by performance on rare labels, while macro F1 weighs the F1 achieved on each label equally [11]. In this paper, we show how these properties are manifest in the optimal threshold for making decisions, and we present results that characterize that threshold. Additionally, we demonstrate that given an uninformative classifier, optimal thresholding to maximize F1 predicts all instances positive regardless of the base rate.

While F1 measures are widely used, some of their properties are not widely recognized. In particular, when choosing predictions to maximize the expected F1 measure for a set of examples, each prediction depends not only on the conditional probability that the label applies to that example, but also on the distribution of these probabilities for all other examples in the set. We quantify this dependence in Theorem 1, where we derive an expression for optimal thresholds. The dependence makes it difficult to relate predictions that are optimally thresholded for F1 to a system's predicted conditional probabilities.

We show that the difference in F1 measure between perfect predictions and optimally thresholded random guesses depends strongly on the base rate. As a consequence, macro average F1 can be argued not to treat labels equally, but to give greater emphasis to performance on rare labels. In a case study, we consider learning to tag articles in the biomedical literature with MeSH terms, a controlled vocabulary of 26,853 labels. These labels have heterogeneously distributed base rates. Our results imply that if the predictive features for rare labels are lost (because of feature selection or from another cause) then the optimal thresholds to maximize macro F1 lead to predicting these rare labels frequently. For the case study application, and likely for similar ones, this behavior is undesirable.

2   Definitions of Performance Measures

Consider binary class prediction in the single or multilabel setting. Given training data of the form $\{\langle x_1, y_1 \rangle, \ldots, \langle x_n, y_n \rangle\}$ where each x_i is a feature vector of

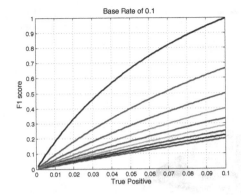 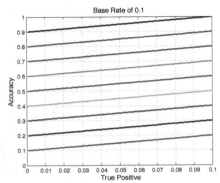

Fig. 2. Holding base rate and fp constant, F1 is concave in tp. Each line is a different value of fp.

Fig. 3. Unlike F1, accuracy offers linearly increasing returns. Each line is a fixed value of fp.

dimension d and each $\boldsymbol{y}_i$ is a binary vector of true labels of dimension m, a probabilistic classifier outputs a model that specifies the conditional probability of each label applying to each instance given the feature vector. For a batch of data of dimension $n \times d$, the model outputs an $n \times m$ matrix C of probabilities. In the single-label setting, $m = 1$ and C is an $n \times 1$ matrix, i.e. a column vector.

A decision rule $D(C) : \mathbb{R}^{n \times m} \to \{0,1\}^{n \times m}$ converts a matrix of probabilities C to binary predictions P. The gold standard $G \in \{0,1\}^{n \times m}$ represents the true values of all labels for all instances in a given batch. A performance measure M assigns a score to a prediction given a gold standard:

$$M(P,G) : \{0,1\}^{n \times m} \times \{0,1\}^{n \times m} \to \mathbb{R} \in [0,1].$$

The counts of true positives tp, false positives fp, false negatives fn, and true negatives tn are represented via a confusion matrix (Figure 1).

Precision $p = tp/(tp + fp)$ is the fraction of all positive predictions that are actual positives, while recall $r = tp/(tp + fn)$ is the fraction of all actual positives that are predicted to be positive. By definition, the F1 measure is the harmonic mean of precision and recall: $F1 = 2/(1/r + 1/p)$. By substitution, F1 can be expressed as a function of counts of true positives, false positives and false negatives:

$$F1 = \frac{2tp}{2tp + fp + fn}. \tag{1}$$

The harmonic mean expression for F1 is undefined when $tp = 0$, but the alternative expression is undefined only when $tn = n$. This difference does not impact the results below.

Before explaining optimal thresholding to maximize F1, we first discuss some properties of F1. For any fixed number of actual positives in the gold standard, only two of the four entries in the confusion matrix (Figure 1) vary independently. This is because the number of actual positives is equal to the sum $tp + fn$ while

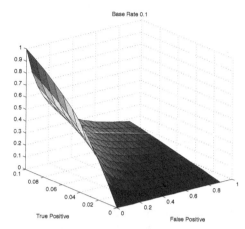

Fig. 4. Given a fixed base rate, the F1 measure is a nonlinear function with two degrees of freedom

the number of actual negatives is equal to the sum $tn + fp$. A second basic property of F1 is that it is nonlinear in its inputs. Specifically, fixing the number fp, F1 is concave as a function of tp (Figure 2). By contrast, accuracy is a linear function of tp and tn (Figure 3).

As mentioned in the introduction, F1 is asymmetric. By this, we mean that the score assigned to a prediction P given gold standard G can be arbitrarily different from the score assigned to a complementary prediction P^c given complementary gold standard G^c. This can be seen by comparing Figure 2 with Figure 5. The asymmetry is problematic when both false positives and false negatives are costly. For example, F1 has been used to evaluate the classification of tumors as benign or malignant [1], a domain where both false positives and false negatives have considerable costs.

While F1 was developed for single-label information retrieval, as mentioned there are variants of F1 for the multilabel setting. Micro F1 treats all predictions on all labels as one vector and then calculates the F1 measure. Specifically,

$$tp = 2 \sum_{i=1}^{n} \sum_{j=1}^{m} \mathbb{1}(P_{ij} = 1)\mathbb{1}(G_{ij} = 1).$$

We define fp and fn analogously and calculate the final score using (1). Macro F1, which can also be called per label F1, calculates the F1 for each of the m labels and averages them:

$$F1_M(P, G) = \frac{1}{m} \sum_{j=1}^{m} F1(P_{:j}, G_{:j}).$$

The per instance F1 measure is similar, but averages F1 over all n examples:

$$F1_I(P, G) = \frac{1}{n} \sum_{i=1}^{n} F1(P_{i:}, G_{i:}).$$

Accuracy is the fraction of all instances that are predicted correctly:

$$A = \frac{tp + tn}{tp + tn + fp + fn}.$$

Accuracy is adapted to the multilabel setting by summing tp and tn for all labels and then dividing by the total number of predictions:

$$A(P, G) = \frac{1}{nm} \sum_{i=1}^{n} \sum_{j=1}^{m} \mathbb{1}(P_{ij} = G_{ij}).$$

The Jaccard index, a monotonically increasing function of F1, is the cardinality of the intersection of the predicted positive set and the actual positive set divided by the cardinality of their union:

$$J = \frac{tp}{tp + fn + fp}.$$

3   Prior Work

Motivated by the widespread use of the F1 measure in information retrieval and in single and multilabel binary classification, researchers have published extensively on its optimization. The paper [8] proposes an outer-inner maximization technique for F1 maximization, while [4] studies extensions to the multilabel setting, showing that simple threshold search strategies are sufficient when individual probabilistic classifiers are independent. Finally, the paper [6] describes how the method of [8] can be extended to efficiently label data points even when classifier outputs are dependent. More recent work in this direction can be found in [19]. However, none of this work directly identifies the relationship of the optimal threshold to the maximum achievable F1 measure over all thresholds, as we do here.

While there has been work on applying general constrained optimization techniques to related measures [13], research often focuses on specific classification methods. In particular, the paper [16] studies F1 optimization for conditional random fields and [14] discusses similar optimization for SVMs. In our work, we study the consequences of maximizing F1 for the general case of any classifier that outputs real-valued scores.

A result similar to a special case below, Corollary 1, was recently derived in [20]. However, the derivation there is complex and does not prove the more general Theorem 1, which describes the optimal decision-making threshold even when the scores output by a classifier are not probabilities.

The batch observation is related to the note in [9] that given a fixed classifier, a specific example may or may not cross the decision threshold, depending on the other examples present in the test data. However, the previous paper does not characterize what this threshold is, nor does it explain the differences between predictions made to optimize micro and macro average F1.

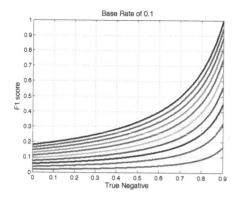

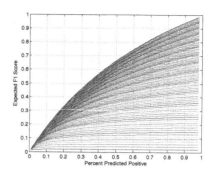

Fig. 5. F1 measure for fixed base rate and number fn of false negatives. F1 offers increasing marginal returns as a function of tn. Each line is a fixed value of fn.

Fig. 6. The expected F1 measure of an optimally thresholded random guess is highly dependent on the base rate

4   Optimal Decision Rule for F1 Maximization

In this section, we provide a characterization of the decision rule that maximizes the F1 measure, and, for a special case, we present a relationship between the optimal threshold and the maximum achievable F1 value.

We assume that the classifier outputs real-valued scores s and that there exist two distributions $p(s|t = 1)$ and $p(s|t = 0)$ that are the conditional probability of seeing the score s when the true label t is 1 or 0, respectively. We assume that these distributions are known in this section; the next section discusses an empirical version of the result. Note also that in this section tp etc. are fractions that sum to one, not counts.

Given $p(s|t = 1)$ and $p(s|t = 0)$, we seek a decision rule $D : s \to \{0, 1\}$ mapping scores to class labels such that the resulting classifier maximizes F1. We start with a lemma that is valid for any D.

Lemma 1. *The true positive rate* $tp = b \int_{s:D(s)=1} p(s|t = 1)ds$ *where the base rate is* $b = p(t = 1)$.

Proof. Clearly $tp = \int_{s:D(s)=1} p(t = 1|s)p(s)ds$. Bayes rule says that $p(t = 1|s) = p(s|t = 1)p(t = 1)/p(s)$. Hence $tp = b \int_{s:D(s)=1} p(s|t = 1)ds$.

Using three similar lemmas, the entries of the confusion matrix are

$$tp = b \int_{s:D(s)=1} p(s|t = 1)ds$$

$$fn = b \int_{s:D(s)=0} p(s|t = 1)ds$$

$$fp = (1 - b) \int_{s:D(s)=1} p(s|t = 0)ds$$

$$tn = (1 - b) \int_{s:D(s)=0} p(s|t = 0)ds.$$

The following theorem describes the optimal decision rule that maximizes F1.

Theorem 1. *An example with score s is assigned to the positive class, that is $D(s) = 1$, by a classifier that maximizes F1 if and only if*

$$\frac{b \cdot p(s|t = 1)}{(1 - b) \cdot p(s|t = 0)} \geq J \tag{2}$$

where $J = tp/(fn + tp + fp)$ is the Jaccard index of the optimal classifier, with ambiguity given equality in (2).

Before we provide the proof of this theorem, we note the difference between the rule in (2) and conventional cost-sensitive decision making [7] or Neyman-Pearson detection. In both the latter approaches, the right hand side J is replaced by a constant λ that depends only on the costs of $0 - 1$ and $1 - 0$ classification errors, and not on the performance of the classifier on the entire batch. We will later describe how the relationship can lead to undesirable thresholding behavior for applications in the multilabel setting.

Proof. Divide the domain of s into regions of fixed size. Suppose that the decision rule $D(\cdot)$ has been fixed for all regions except a particular region denoted Δ around a point s. Write $P_1(\Delta) = \int_\Delta p(s|t = 1)ds$ and define $P_0(\Delta)$ similarly.

Suppose that the F1 achieved with decision rule D for all scores besides those in Δ is $F1 = 2tp/(2tp + fn + fp)$. Now, if we add Δ to the positive region of the decision rule, $D(\Delta) = 1$, then the new F1 measure is

$$F1' = \frac{2tp + 2bP_1(\Delta)}{2tp + 2bP_1(\Delta) + fn + fp + (1 - b)P_0(\Delta)}.$$

On the other hand, if we add Δ to the negative region of the decision rule, $D(\Delta) = 0$, then the new F1 measure is

$$F1'' = \frac{2tp}{2tp + fn + bP_1(\Delta) + fp}.$$

We add Δ to the positive region only if $F1' \geq F1''$. With some algebraic simplification, this condition becomes

$$\frac{bP_1(\Delta)}{(1 - b)P_0(\Delta)} \geq \frac{tp}{tp + fn + fp}.$$

Taking the limit $|\Delta| \to 0$ gives the claimed result.

If, as a special case, the model outputs calibrated probabilities, that is $p(t = 1|s) = s$ and $p(t = 0|s) = 1 - s$, then we have the following corollary.

Corollary 1. *An instance with predicted probability s is assigned to the positive class by the decision rule that maximizes F1 if and only if $s \geq F/2$ where the F1 measure achieved by this optimal decision rule is $F = 2tp/(2tp + fn + fp)$.*

Proof. Using the definition of calibration and then Bayes rule, for the optimal decision surface for assigning a score s to the positive class

$$\frac{p(t = 1|s)}{p(t = 0|s)} = \frac{s}{1 - s} = \frac{p(s|t = 1)b}{p(s|t = 0)(1 - b)}. \tag{3}$$

Incorporating (3) in Theorem 1 gives

$$\frac{s}{1 - s} \geq \frac{tp}{fn + tp + fp}$$

and simplifying results in

$$s \geq \frac{tp}{2tp + fn + fp} = F/2.$$

Thus, the optimal threshold in the calibrated case is half the maximum F1 value.

5   Consequences of the Optimal Decision Rule

We demonstrate two consequences of designing classifiers that maximize the F1 measure, which we call the batch observation and the uninformative classifier observation. We will later show with a case study that these can combine to produce surprising and potentially undesirable predictions when macro F1 is optimized in practice.

The batch observation is that a label may or may not be predicted for an instance depending on the distribution of conditional probabilities (or scores) for other instances in the same batch. Earlier, we observed a relationship between the optimal threshold and the maximum achievable F1 value, and demonstrated that this maximum depends on the distribution of conditional probabilities for all instances. Therefore, depending upon the set in which an instance is placed, its conditional probability may or may not exceed the optimal threshold. Note that because an F1 value cannot exceed 1, the optimal threshold cannot exceed 0.5.

Consider for example an instance with conditional probability 0.1. It will be classified as positive if it has the highest probability of all instances in a batch. However, in a different batch, where the probabilities predicted for all other instances are 0.5 and n is large, the maximum achievable F1 measure is close to $2/3$. According to the results above, we will then classify this last instance as negative because it has a conditional probability less than $1/3$.

An uninformative classifier is one that predicts the same score for all examples. If these scores are calibrated conditional probabilities, then the base rate is predicted for every example.

Theorem 2. *Given an uninformative classifier for a label, optimal thresholding to maximize expected F1 results in classifying all examples as positive.*

Proof. Given an uninformative classifier, we seek the threshold that maximizes $\mathbb{E}(F1)$. The only choice is how many labels to predict. By symmetry between the instances, it does not matter which instances are labeled positive.

Let $a = tp + fn$ be the number of actual positives and let $c = tp + fp$ be a fixed number of positive predictions. The denominator of the expression for F1 in Equation (1), that is $2tp + fp + fn = a + c$, is then constant. The number of true positives, however, is a random variable. Its expected value is equal to the sum of the probabilities that each example predicted positive actually is positive:

$$\mathbb{E}(F1) = \frac{2\sum_{i=1}^{c} b}{a+c} = \frac{2c \cdot b}{a+c}$$

where $b = a/n$ is the base rate. To maximize this expectation as a function of c, we calculate the partial derivative with respect to c, applying the product rule:

$$\frac{\partial}{\partial c}\mathbb{E}(F1) = \frac{\partial}{\partial c}\frac{2c \cdot b}{a+c} = \frac{2b}{a+c} - \frac{2c \cdot b}{(a+c)^2}.$$

Both terms in the difference are always positive, so we can show that the derivative is always positive by showing that

$$\frac{2b}{a+c} > \frac{2c \cdot b}{(a+c)^2}.$$

Simplification gives the condition $1 > c/(a+c)$. As this condition always holds, the derivative is always positive. Therefore, whenever the frequency of actual positives in the test set is nonzero, and the predictive model is uninformative, then expected F1 is maximized by predicting that all examples are positive.

Figure 6 shows that for small base rates, an optimally thresholded uninformative classifier achieves $\mathbb{E}(F1)$ close to 0, while for high base rates $\mathbb{E}(F1)$ is close to 1. We revisit this point in the next section in the context of maximizing macro F1.

6   Multilabel Setting

Different measures are used to measure different aspects of a system's performance. However, changing the measure that is optimized can change the optimal predictions. We relate the batch observation to discrepancies between predictions that are optimal for micro versus macro averaged F1. We show that while performance on rare labels is unimportant for micro F1, macro F1 is dominated by performance on these labels. Additionally, we show that macro averaging F1 can conceal the occurrence of uninformative classifier thresholding.

Consider the equation for micro averaged F1, for m labels with base rates b_i. Suppose that tp, fp, and fn are fixed for the first $m-1$ labels, and suppose that

b_m is small compared to the other b_i. Consider (i) a perfect classifier for label m, (ii) a trivial classifier that never predicts label m, and (iii) a trivial classifier that predicts label m for every example. The perfect classifier increases tp by a small amount $b_m \cdot n$, the number of actual positives for the rare label m, while contributing nothing to the counts fp and fn:

$$F1' = \frac{2(tp + b_m \cdot n)}{2(tp + b_m \cdot n) + fp + fn}.$$

The trivial classifier that never predicts label m increases fn by the same small amount:

$$F1'' = \frac{2tp}{2tp + fp + (fn + b_m \cdot n)}.$$

Finally, the trivial classifier that predicts label m for every example increases fp by a large amount $n(1 - b_m)$. Clearly this last classifier leads to micro average F1 that is much worse than that of the perfect classifier for label m. However, $F1'$ and $F1''$ both tend to the same value, namely $2tp/(2tp + fp + fn)$, as b_m tends to zero. Hence, for a label with very small base rate, a perfect classifier does not improve micro F1 noticeably compared to a trivial all-negative classifier. It is fair to say that performance on rare labels is unimportant for micro F1.

Now consider the context of macro F1, where separately calculated F1 measures over all labels are averaged. Consider the two label case where one label has a base rate of 0.5 and the other has a base rate of 0.1. The corresponding F1 measures for trivial all-positive classifiers are 0.67 and 0.18 respectively. Thus the macro F1 for trivial classifiers is 0.42. An improvement to perfect predictions on the rare label increases macro F1 to 0.83, while the same improvement on the common label only increases macro F1 of 0.59. Hence it is fair to say that macro F1 emphasizes performance on rare labels, even though it weights performance on every label equally.

For a rare label with an uninformative predictive model, micro F1 is optimized by classifying all examples as negative, while macro F1 is optimized by classifying all examples as positive. Optimizing micro F1 as compared to macro F1 can be thought of as choosing optimal thresholds given very different batches. If the base rates and distributions of conditional probabilities predicted for instances vary from label to label, so will the optimal binary predictions. Generally, labels with small base rates and less informative classifiers will be over-predicted when maximizing macro F1, and under-predicted when maximizing micro F1. We present empirical evidence of this phenomenon in the following case study.

7   Case Study

This section discusses a case study that demonstrates how in practice, thresholding to maximize macro F1 can produce undesirable predictions. To our knowledge, a similar real-world case of pathological behavior has not been previously described in the literature, even though macro averaging F1 is a common approach.

MeSH term	count	maximum F1	threshold
Humans	2346	0.9160	0.458
Male	1472	0.8055	0.403
Female	1439	0.8131	0.407
Phosphinic Acids	**1401**	$1.544 \cdot 10^{-4}$	$7.71 \cdot 10^{-5}$
Penicillanic Acid	**1064**	$8.534 \cdot 10^{-4}$	$4.27 \cdot 10^{-4}$
Adult	1063	0.7004	0.350
Middle Aged	1028	0.7513	0.376
Platypus	**980**	$4.676 \cdot 10^{-4}$	$2.34 \cdot 10^{-4}$

Fig. 7. Selected frequently predicted MeSH terms. Columns show the term, the number of times it is predicted for a given test set, the empirical maximum achieved F1 measure, and the empirical threshold that achieves this maximum. When F1 is optimized separately for each term, low thresholds are chosen for rare labels (bold) with uninformative classifiers.

We consider the task of assigning tags from a controlled vocabulary of 26,853 MeSH terms to articles in the biomedical literature based on their titles and abstracts. We represent each abstract as a sparse bag-of-words vector over a vocabulary of 188,923 words. The training data consists of a matrix A with n rows and d columns, where n is the number of abstracts and d is the number of features in the bag of words representation. We apply a tf-idf text preprocessing step to the bag of words representation to account for word burstiness [10] and to elevate the impact of rare words.

Because linear regression models can be trained for multiple labels efficiently, we choose linear regression as a predictive model. Note that square loss is a proper loss function and yields calibrated probabilistic predictions [12]. Further, to increase the speed of training and prevent overfitting, we approximate the training matrix A by a rank restricted A_k using singular value decomposition. One potential consequence of this rank restriction is that the signal of extremely rare words can be lost. This can be problematic when rare terms are the only features of predictive value for a label.

Given the probabilistic output of each classifier and the results relating optimal thresholds to maximum attainable F1, we designed three different plug-in rules to maximize micro, macro and per instance average F1. Inspection of the predictions to maximize micro F1 revealed no irregularities. However, inspecting the predictions thresholded to maximize performance on macro F1 showed that several terms with very low base rates were predicted for more than a third of all test documents. Among these terms were "Platypus", "Penicillanic Acids" and "Phosphinic Acids" (Figure 7).

In multilabel classification, a label can have low base rate and an uninformative classifier. In this case, optimal thresholding requires the system to predict all examples positive for this label. In the single-label case, such a system would achieve a low F1 and not be used. But in the macro averaging multilabel case, the extreme thresholding behavior can take place on a subset of labels, while the system manages to perform well overall.

8   A Winner's Curse

In practice, decision rules that maximize F1 are often set empirically, rather than analytically. That is, given a set of validation examples with predicted scores and true labels, rules for mapping scores to labels are selected that maximize F1 on the validation set. In such situations, the optimal threshold can be subject to a winner's curse [2] where a sub-optimal threshold is chosen because of sampling effects or limited training data. As a result, the future performance of a classifier using this threshold is worse than the anticipated performance. We show that threshold optimization for F1 is particularly susceptible to this phenomenon.

In particular, different thresholds have different rates of convergence of empirical F1 with number of samples n. As a result, for a given n, comparing the empirical performance of low and high thresholds can result in suboptimal performance. This is because, for a fixed number of samples, some thresholds converge to their true error rates fast, while others have higher variance and may be set erroneously. We demonstrate these ideas for a scenario with an uninformative model, though they hold more generally.

Consider an uninformative model, for a label with base rate b. The model is uninformative in the sense that output scores are $s_i = b + n_i$ for examples i, where $n_i = \mathcal{N}(0, \sigma^2)$. Thus, scores are uncorrelated with and independent of the true labels. The empirical accuracy for a threshold t is

$$A(t) = \frac{1}{n}\sum_{i \in +} \mathbf{1}[s_i \geq t] + \frac{1}{n}\sum_{i \in -} \mathbf{1}[s_i \leq t] \tag{4}$$

where $+$ and $-$ index the positive and negative class respectively. Each term in Equation (4) is the sum of $O(n)$ i.i.d random variables and has exponential (in n) rate of convergence to the mean irrespective of the base rate b and the threshold t. Thus, for a fixed number T of threshold choices, the probability of choosing the wrong threshold is less than $T2^{-\epsilon n}$ where ϵ depends on the distance between the optimal and next nearest threshold. Even if errors occur, the most likely errors are thresholds close to the true optimal threshold, a consequence of Sanov's theorem [3].

Consider how to select an F1-maximizing threshold empirically, given a validation set of ground truth labels and scores from an uninformative classifier. The scores s_i can be sorted in decreasing order (w.l.o.g.) since they are independent of the true labels for an uninformative classifier. Based on the sorted scores, we empirically select the threshold that maximizes the F1 measure F on the validation set. The optimal empirical threshold will lie between two scores that include the value $F/2$, when the scores are calibrated, in accordance with Theorem 1.

The threshold s that classifies all examples positive (and maximizes F1 analytically by Theorem 2) has an empirical F1 value close to its expectation of $\frac{2b}{1+b} = \frac{2}{1+1/b}$ since tp, fp and fn are all estimated from the entire data. Consider a threshold s that classifies only the first example positive and all others negative. With probability b, this threshold has F1 value $2/(2 + b \cdot n)$, which is worse than that of the optimal threshold only when

$$b \geq \frac{\sqrt{1 + 8/n} - 1}{2}.$$

Despite the threshold s being far from optimal, it has a constant probability of having a better F1 value on validation data, a probability that does not decrease with n, for $n < (1 - b)/b^2$. Therefore, optimizing F1 has a sharp threshold behavior, where for $n < (1 - b)/b^2$ the algorithm incorrectly selects large thresholds with constant probability, whereas for larger n it correctly selects small thresholds. Note that identifying optimal thresholds for F1 is still problematic since it then leads to issues identified in the previous section. While these issues are distinct, they both arise from the nonlinearity of the F1 measure and its asymmetric treatment of positive and negative labels.

Figure 8 shows the result of simulating this phenomenon, executing 10,000 runs for each setting of the base rate, with $n = 10^6$ samples for each run used to set the threshold. Scores are chosen using variance $\sigma^2 = 1$. True labels are assigned at the base rate, independent of the scores. The threshold that maximizes F1 on the validation set is selected. We plot a histogram of the fraction predicted positive as a function of the empirically chosen threshold. There is a shift from predicting almost all positives to almost all negatives as the base rate is decreased. In particular, for low base rate, even with a large number of samples, a small fraction of examples are predicted positive. The analytically derived optimal decision in all cases is to predict all positive, i.e. to use a threshold of 0.

9   Discussion

In this paper, we present theoretical and empirical results describing properties of the F1 performance measure for binary and multilabel classification. We relate the best achievable F1 measure to the optimal decision-making threshold and show that when a classifier is uninformative, classifying all instances as positive maximizes F1. In the multilabel setting, this behavior is problematic when the measure to maximize is macro F1 and for some labels their predictive model is uninformative. In contrast, we demonstrate that given the same scenario, micro F1 is maximized by predicting those labels for all examples to be negative. This knowledge can be useful as such scenarios are likely to occur in settings with a large number of labels. We also demonstrate that micro F1 has the potentially undesirable property of washing out performance on rare labels.

No single performance measure can capture every desirable property. For example, for a single binary label, separately reporting precision and recall is more informative than reporting F1 alone. Sometimes, however, it is practically necessary to define a single performance measure to optimize. Evaluating competing systems and objectively choosing a winner presents such a scenario. In these cases, it is important to understand that a change of performance measure can have the consequence of dramatically altering optimal thresholding behavior.

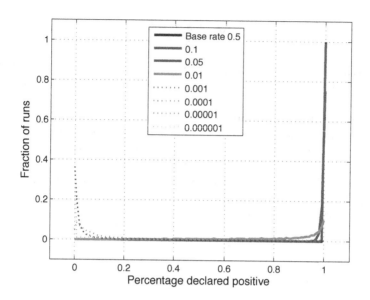

Fig. 8. The distribution of experimentally chosen thresholds changes with varying base rate b. For small b, a small fraction of examples are predicted positive even though the optimal thresholding is to predict all positive.

References

1. Akay, M.F.: Support vector machines combined with feature selection for breast cancer diagnosis. Expert Systems with Applications 36(2), 3240–3247 (2009)
2. Capen, E.C., Clapp, R.V., Campbell, W.M.: Competitive bidding in high-risk situations. Journal of Petroleum Technology 23(6), 641–653 (1971)
3. Cover, T.M., Thomas, J.A.: Elements of information theory. John Wiley & Sons (2012)
4. del Coz, J.J., Diez, J., Bahamonde, A.: Learning nondeterministic classifiers. Journal of Machine Learning Research 10, 2273–2293 (2009)
5. Dembczynski, K., Kotłowski, W., Jachnik, A., Waegeman, W., Hüllermeier, E.: Optimizing the F-measure in multi-label classification: Plug-in rule approach versus structured loss minimization. In: ICML (2013)
6. Dembczyński, K., Waegeman, W., Cheng, W., Hüllermeier, E.: An exact algorithm for F-measure maximization. In: Neural Information Processing Systems (2011)
7. Elkan, C.: The foundations of cost-sensitive learning. In: International Joint Conference on Artificial Intelligence, pp. 973–978 (2001)
8. Jansche, M.: A maximum expected utility framework for binary sequence labeling. In: Annual Meeting of the Association for Computational Linguistics, p. 736 (2007)
9. Lewis, D.D.: Evaluating and optimizing autonomous text classification systems. In: Proceedings of the 18th Annual International ACM SIGIR Conference on Research and Development in Information Retrieval, pp. 246–254. ACM (1995)

10. Madsen, R., Kauchak, D., Elkan, C.: Modeling word burstiness using the Dirichlet distribution. In: Proceedings of the International Conference on Machine Learning (ICML), pp. 545–552 (August 2005)
11. Manning, C., Raghavan, P., Schütze, H.: Introduction to information retrieval, vol. 1. Cambridge University Press (2008)
12. Menon, A., Jiang, X., Vembu, S., Elkan, C., Ohno-Machado, L.: Predicting accurate probabilities with a ranking loss. In: Proceedings of the International Conference on Machine Learning (ICML) (June 2012)
13. Mozer, M.C., Dodier, R.H., Colagrosso, M.D., Guerra-Salcedo, C., Wolniewicz, R.H.: Prodding the ROC curve: Constrained optimization of classifier performance. In: NIPS, pp. 1409–1415 (2001)
14. Musicant, D.R., Kumar, V., Ozgur, A., et al.: Optimizing F-measure with support vector machines. In: FLAIRS Conference, pp. 356–360 (2003)
15. Sokolova, M., Lapalme, G.: A systematic analysis of performance measures for classification tasks. Information Processing and Management 45, 427–437 (2009)
16. Suzuki, J., McDermott, E., Isozaki, H.: Training conditional random fields with multivariate evaluation measures. In: Proceedings of the 21st International Conference on Computational Linguistics and the 44th Annual Meeting of the Association for Computational Linguistics, pp. 217–224. Association for Computational Linguistics (2006)
17. Tan, S.: Neighbor-weighted k-nearest neighbor for unbalanced text corpus. Expert Systems with Applications 28, 667–671 (2005)
18. Tsoumakas, G., Katakis, I.: Multi-label classification: An overview. International Journal of Data Warehousing and Mining 3(3), 1–13 (2007)
19. Ye, N., Chai, K.M., Lee, W.S., Chieu, H.L.: Optimizing F-measures: A tale of two approaches. In: Proceedings of the International Conference on Machine Learning (2012)
20. Zhao, M.J., Edakunni, N., Pocock, A., Brown, G.: Beyond Fano's inequality: Bounds on the optimal F-score, BER, and cost-sensitive risk and their implications. Journal of Machine Learning Research 14(1), 1033–1090 (2013)

Randomized Operating Point Selection in Adversarial Classification

Viliam Lisý, Robert Kessl, and Tomáš Pevný

Agent Technology Center, Department of Computer Science
Faculty of Electrical Engineering, Czech Technical University in Prague,
Prague, Czech Republic
{viliam.lisy,robert.kessl,tomas.pevny}@agents.fel.cvut.cz

Abstract. Security systems for email spam filtering, network intrusion detection, steganalysis, and watermarking, frequently use classifiers to separate malicious behavior from legitimate. Typically, they use a fixed operating point minimizing the expected cost / error. This allows a rational attacker to deliver invisible attacks just below the detection threshold. We model this situation as a non-zero sum normal form game capturing attacker's expected payoffs for detected and undetected attacks, and detector's costs for false positives and false negatives computed based on the Receiver Operating Characteristic (ROC) curve of the classifier. The analysis of Nash and Stackelberg equilibria reveals that using a randomized strategy over multiple operating points forces the rational attacker to design less efficient attacks and substantially lowers the expected cost of the detector. We present the equilibrium strategies for sample ROC curves from network intrusion detection system and evaluate the corresponding benefits.

Keywords: Game theory, operating point selection, receiver operating characteristic, adversarial machine learning, misclassification cost.

1 Introduction

Receiver operating characteristics (ROC) graph is a curve showing dependency of true positive rate (y-axis) and false positive rate (x-axis) of a classifier. The most attractive property of ROC curves is their insensitivity to changes in class distributions and costs of wrong decisions on different classes. Both these properties are almost certainly user specific and often non-stationary in security applications. For example, in spam detection the proportion of spam volume changes from month to month, and the cost of receiving spam can be different for different users. It is therefore important to determine operating points of classifiers (e.g., thresholds) systematically based on the costs in a specific deployment. A well-established result [20] shows that the threshold minimizing the (Bayesian) cost corresponds to the tangent of the ROC curve of slope defined by the ratio of mis-classification costs weighted by class probabilities. This method for selecting thresholds is routinely used in many domains, including steganalysis[11], watermarking [13,6], and fraud detection[17].

T. Calders et al. (Eds.): ECML PKDD 2014, Part II, LNCS 8725, pp. 240–255, 2014.
© Springer-Verlag Berlin Heidelberg 2014

Similarly to [3], we argue that the method is optimal only in non-adversarial setting without a rational attacker actively avoiding the detection. If knowledgeable attackers are present, such as in network intrusion detection, spam filtering, steganalysis, watermarking, etc., this standard operating point is sub-optimal.

Our approach is to model the canonical machine learning problem of setting the optimal operating point based on ROC in scenarios with a rational attacker as a two-player normal-form game. The goal of the defender is to detect a presence of an attack, for which she uses a publicly known classifier. The goal of the attacker is to design data samples (an attack) maximizing his benefit yet having a good chance of being undetected. The set of thresholds (or other parameters of the classifier) is the set of strategies for both players, as it is assumed the attacker can design data-samples not detected at a given threshold [4,1].

We present, compute, and analyze two different solution concepts in this game. The first is the Nash equilibrium, which is the most standard solution concept for situations where players interact only once and decide about their strategies simultaneously. The second is the Stackelberg equilibrium, which has been recently very popular in security domains [21]. The latter assumes that one player, typically defender, computes its strategy and discloses it to the other player before the game is played. The other player (attacker) can then play optimally with respect to this strategy. This better describes the situation, when the classifier (detection system) is publicly known, as the attacker may even run his own copy of the classifier to verify undetectability of his attacks.

The main results of our analysis are that in adversarial setting, the defender can substantially reduce its expected cost by randomizing over a larger set of thresholds. We formally prove that in some games, no finite number of thresholds is sufficient for the optimal randomization, but a reasonably sparse discretization is often sufficient to guarantee strategies with performance close to the optimum.

Throughout the paper, we use a simple running example from the network security domain: The attacker tries to gain remote access to a server using a brute force password attack. The attacker knows that the defender deploys an intrusion detection system (IDS) and if she detects the attack, she will block attacker's IP address. The detector needs to decide how many passwords per second he will try. The defender has to decide how many login attempts per second is enough to manually inspect the incident. If she sets the threshold too high, the attacker has a good chance of succeeding in the attack. On the other hand, too low thresholds force her to inspect many false alarms caused by users who forgot their password.

2   Related Work

Only a few papers addressed operating point selection in game theoretic framework. Cavusoglu et al. [3] is one of the first papers advocating the importance of game theoretic models in configuration of detectors of malicious behavior. As in multiple similar papers, e.g.,[2], their models assume that the players can possibly make randomized decisions about whether to attack or whether to manually check a specific incident, but still allow choosing only a single fixed operating point as the optimal configuration based on an ROC curve. The main difference of our approach is that we propose randomization over multiple operating points, which determines strategies with respect to whole ROC curve resulting into lower defender's costs.

A game theoretic model of randomized threshold selection is presented in [8]. The rational attacker tries to hide its preferences by distributing his attacks between the preferred and nonpreferred target, while the defender sets a threshold for the number of attack attempts on the more valuable target. In contrast to our paper, this model is not connected to machine learning theory and general classifier characteristics, such as ROC. Furthermore, is requires a discrete set of thresholds and it analyses only Nash equilibria.

The use of Stackelberg equilibrium in our work has been inspired mainly by the recent progress in research and practical applications of resource allocation security games [21]. While there are several parallels, the class of games we study here is substantially different. The resource allocation games assume a specific utility structure, which causes the Nash and Stackelberg equilibria to prescribe the same strategies [12]. As we show in experiments, this is not always the case in our model. Also, there is no connection between these models and machine learning and all the studied models of resource allocation games are fundamentally discrete.

Recent works [4,1,10] from different domains show that for a fixed detector an attacker can devise an invisible attack just below the detection threshold. This paper uses the following generalization: against every detector from a set of detectors, an attacker can plant an invisible attack, providing he knows the detector. Since every detector has certain false positive and false negative rate, the set can be described by ROC curve, which can be parameterized by a single parameter – threshold (more on this in the next section). Hence, thresholds used in discussions of operating point selection serve here as an abstraction linking a single parameter to a particular classifiers from a possibly rich set. With respect to cited works on evasion attack, this simplification does not decrease generality of the presented approach.

3   Background

A two player *normal form game* is defined by a set of players $\mathcal{I}$; set of actions for each player $\mathcal{A}_i, i \in \mathcal{I}$; and utility functions $u_i : \mathcal{A} \to \mathbf{R}$ for each player and *action profile* from $\mathcal{A} = \Pi_{i \in \mathcal{I}} \mathcal{A}_i$. A *(mixed) strategy* of a player $\sigma_i \in \Sigma_i$ is a probability distribution over her actions and a *pure strategy* is a strategy playing only one of the actions. The utility functions can be extended to mixed strategies by taking expectation over players' randomization. For a strategy profile $\sigma \in \Sigma = \Pi_{i \in \mathcal{I}} \Sigma_i$, we denote σ_i the strategy of player i and σ_{-i} the strategy of the other player. A strategy profile σ^* is an ϵ-*Nash Equilibrium*

$$\text{if } u_i(\sigma_i, \sigma_{-i}^*) - u_i(\sigma^*) \leq \epsilon \ \forall i \in \mathcal{I}, \sigma_i \in \Sigma_i.$$

A strategy profile is an exact Nash equilibrium (NE) if $\epsilon = 0$.

A *Stackelberg Equilibrium* (SE) assumes that one of the players is a leader who commits to a strategy and discloses it to the other player (termed follower). The other player then plays the action that maximizes her utility. For a two player game with leader i is (σ_i^*, a_{-i}^*) a SE if

$$u_i(\sigma_i^*, a_{-i}^*) \geq u_i(\sigma_i, a_{-i}^*) \; \forall \sigma_i \in \Sigma_i$$
$$\& \; u_{-i}(\sigma_i^*, a_{-i}^*) \geq u_{-i}(\sigma_i^*, a_{-i}) \; \forall a_{-i} \in \mathcal{A}_{-i}.$$

The first line says that the leader does not have an incentive to change the strategy and the second line says the follower plays the best response to the leader's strategy. The *value* of the equilibrium for player i is $v_i = u_i(\sigma_i^*, a_{-i}^*)$. If the follower breaks ties in favor of the leader, it is a *Strong* Stackelberg Equilibrium (SSE). Breaking ties in favor of the leader is generally not a restrictive assumption, because minimal perturbation of any SE can ensure this choice is the only optimal for the follower [22].

Receiver Operating Characteristic (ROC) of a classifier is a parametric curve describing dependency of true positive rate (rate of successfully detected attacks) on the false positive rate (rate of benign events flagged as alarms). Each value of the parameter corresponds to a single point on the curve with specific true and false positive rates, which is also called operating point. Without loss of generality we assume the curve to be parameterized by a detection threshold $t \in$ **T**, but other parameterizations such as different penalties on error on different classes during training of the classifier are indeed possible. ROC curve is non-decreasing in the false positives rate, but we do not assume it to be necessarily concave.

In reality the operating point of a classifier can be controlled by more than one parameter, for example by varying costs of errors on different classes during training. Nevertheless, for every false positive rate the rational defender always chooses a classifier with the highest detection accuracy. Consequently a particular false positive rate is linked to a particular classifier which is in this paper abstracted by a threshold. By similar reasoning it can be assumed that the ROC curve is non-decreasing, because in the defender does not have any incentive to use classifier with higher false positive rate and smaller detection accuracy.

With respect to the above arguments it is assumed that there is a bijective decreasing mapping between false positive rate and the threshold, which means that the higher threshold implies smaller false positive rate. $R_{FP} : \mathbf{T} \to \langle 0, 1 \rangle$ maps thresholds to false positive rate.

4   Game Model

We formalize the operating point selection in presence of adversary as a two-player non-zero-sum normal form game with continuous strategy spaces.

Players: The two players in the game are the defender (denoted d) and the attacker (denoted a). In reality the game will be played between one defender and many different attackers, but since at this point we assume all attackers to share the same costs and penalties, they can be represented as a single player. We plan to generalize the model to the Bayesian game setting [18] in future work.

Actions: The action sets of the players are identical. Each player selects a threshold from a set $\mathbf{T}$, which can be mapped to $\langle 0, 1 \rangle$ without loss of generality. If the defender selects a threshold $t_d \in \mathbf{T}$, all attacks stronger than this threshold are detected. If the attacker selects a threshold $t_a \in \mathbf{T}$, he plays an attack of maximal intensity undetected by the detector with threshold t_a. The attacker is detected if $t_a > t_d$.

Utility functions: The utility functions of the players depend on ROC curves, defender's costs for processing false positives and cost of missed detection, and attackers reward for successful attack and penalty for the attack being detected. Formal definitions of all quantities are following: $ROC : \langle 0, 1 \rangle \to \langle 0, 1 \rangle$ is the receiver operation characteristic of the classifier; $C^{FP} \in \mathbf{R}_0^+$ is the defender's cost of processing a false positive and $C^{FN} : \mathbf{T} \to \mathbf{R}$ is a non-decreasing defender's cost of missing an attack of certain intensity; $r_a : \mathbf{T} \to \mathbf{R}_0^+$ is the non-decreasing attacker's reward for performing an undetected attack and $p_a \in \mathbf{R}_0^+$ is the attacker's penalty for being detected while performing an attack. We allow the attacker to choose not to attack for zero reward and penalty. We further assume not all attackers being rational, as there is $A_r \in \mathbf{R}_0^+$ times more rational attackers than non-rational, who attack with the same intensity regardless of the classifier's setting. Strategies of irrational attackers are reflected in the true positives of the ROC.

In our running example, the attacker's reward $r_a(t)$ can be the number of passwords he tries per second without being detected; $C^{FN}(t)$ can be $c \cdot r_a(t)$ for some scaling factor c and p_a being the penalty the attacker suffers if his attack IP address is blocked. The notion of attack intensity in other domains could represent the entropy of attack sources in a DDoS attack, the amount of information injected to a media file in steganalysis, or the negative of distortion caused to the media file in watermarking.

Based on the inputs above, we define the defender's background cost for irrational attackers as the standard classification cost used in non-adversarial setting:

$$c_d^b(t) = R_{FP}(t) \cdot C^{FP} + (1 - ROC(t)) \cdot C^{FN}(t) \qquad (1)$$

For rational attackers playing a threshold t the defender suffers an additional penalty for the undetected attacks:

$$c_d^r(t) = A_r \cdot C^{FN}(t). \qquad (2)$$

The utility function of the defender is the negative of the sum of the background cost and the cost for rational attacks if undetected:

$$u_d(t_d, t_a) = \begin{cases} -c_d^b(t_d) - c_d^r(t_a) \text{ if } t_d \geq t_a \\ -c_d^b(t_d) \text{ otherwise.} \end{cases}$$

The utility of the attacker is his reward in case of being undetected and the negative penalty when he is detected:

$$u_a(t_d, t_a) = \begin{cases} r_a(t_a) \text{ if } t_d \geq t_a \\ -p_a \text{ otherwise.} \end{cases}$$

5 Game Model Properties

The ROC curves from real problems are usually estimated from data samples without clear analytical formulations. For this reason we base our study on a discretized version of the game, which means that optimal strategies are only approximated. We therefore first derive approximation bounds of Nash (NE) and Stackelberg equilibria (SSE) between discretized and continuous version of the problem. Then we show that even if we can get arbitrarily good approximations with finite sets of thresholds, creating the exact optimal randomized strategy may require using infinitely many thresholds. Finally, we show that some subsets of thresholds will never be used by a rational defender and can be disregarded in the strategy computation.

Proposition 1. *Let v_d be the value of SSE of the continous game for the defender; $(t_i) = t_0 < t_1 < \cdots < t_n$; $t_i \in \mathbf{T}$; $t_0 = \min(\mathbf{T})$; $t_n = \max(\mathbf{T})$ be a discretization of the set of applicable thresholds and*

$$\Delta = \max_{i \in \{0,\ldots,n-1\}} \{\max\{c_d^r(t_{i+1}) - c_d^r(t_i), \max_{t \in (t_i, t_{i+1})} c_d^b(t) - \min_{t \in (t_i, t_{i+1})} c_d^b(t)\}\},$$

be the maximal difference between the highest and the lowest point in the defender's cost functions within one interval. Then there is a mixed strategy selecting only the thresholds from (t_i), such that its expected value for the defender is at least $v_d - 2\Delta$.

Proof. Assume (D, t_a) are the cumulative distribution function[1] (CDF) of the defender's strategy and threshold selection of the attacker in a SSE of the continuous game. Let $t_j \in (t_i)$ be a threshold in the discretization, such that $t_a \in (t_j, t_{j+1})$. We construct a CDF D' lower than D in the interval (t_j, t_{j+1}) and higher then D outside, so that the attacker still plays in (t_j, t_{j+1}) and the cost of the defender is not increased substantially.

$$D'(t) = \begin{cases} D(t_{i+1}) \ \forall t \in (t_i, t_{i+1}) \ i \neq j \\ D(t_j) \ \forall t \in (t_j, t_{j+1}) \end{cases} \tag{3}$$

The expected utility of the attacker for playing threshold t in response to distribution D is

$$u_a(D, t) = (1 - D(t))r_a(t) - D(t)p_a \tag{4}$$

---

[1] Probability that randomly selected threshold is below the input parameter, i.e., $D(t_d) = P(t \leq t_d)$.

While CDFs and r_a are non-decreasing, the attacker's expected utility with D' cannot increase outside (t_j, t_{j+1}) and cannot decrease in the interval. Hence, he will keep playing to interval (t_j, t_{j+1}) and even if he modifies his strategy within this interval, it will not increase the costs of the defender by more than Δ in the c_d^r component of her utility by definition of Δ. Furthermore, any time the defender would play $t \in (t_i, t_{i+1})$ with distribution D, she plays one of the bounds instead with D'. For each of these bounds, she has the cost c_d^b at most Δ more than with t. In the worst case, the defender will suffer the increased cost in both components. □

Proposition 2. *Let (D, A) be CDFs of strategies in NE of the operating point selection game discretized to $(t_i) \subseteq \mathbf{T}$ and Δ defined as in Proposition 1, then (D, A) is a 2Δ-NE of the continuous game.*

Proof. Attacker: The attacker's expected utility cannot be increased by playing thresholds not included in the discretization. For any $t \in (t_i, t_{i+1})$ the attacker might consider, he can only improve his payoff by playing t_{i+1} instead. Recall that playing the same threshold as the defender results to an undetected attack. The probability of detection by D is the same on the whole interval and $r_a(t)$ is non-decreasing.

Defender: The best response to any mixed strategy can always be found in pure strategies. Assume that $t_d \in (t_i, t_{i+1})$ is the best response of the defender to the attacker's strategy A. From definition of Δ, $u_d(t_d, A) \leq u_d(t_i, A) + 2\Delta$, because it can differ by Δ in each component of the utility function. The defender has no incentive to deviate to t_i from a discrete NE strategy, because this threshold was considered in its computation; hence, $u_d(t_i, A) \leq u_d(D, A)$. Combining the two inequalities gives us $u_d(t_d, A) \leq u_d(D, A) + 2\Delta$. □

It is important to realize that Δ is not a parameter of the problem, but rather a guide for creating a suitable discretization. The goal is to select a discretization, such that Δ is small. Δ can even be selected in advance and then the algorithm to compute a matching discretization could just swipe through the interval of possible thresholds and add a new threshold to the discretization always when one of the relevant functions changes its value by more than Δ. For a function with range $[0,1]$ and $\Delta = \frac{1}{n}$, a monotonic function will require at most n thresholds; convex/concave function at most $2n$ thresholds. If we want to guarantee less than 5% error from the optimum in the worst case, we can always choose 40 thresholds for the monotonous c_d^r function to keep $\Delta = 2.5\%$ for this component of its definition. If the c_d^b function is convex (which seems to be the case in the real world examples presented in our experiments), we will need at most additional 80 thresholds to guarantee even this component of definition of Δ to be 2.5%. Moreover, as we explain later, we can remove some portions of the thresholds completely form consideration.

The above discussion shows that the error in the quality of the produced solutions caused by discretization is bounded and we can always choose a relatively small discretization of set $\mathbf{T}$ that guarantees a low error. We further show that there are instances of the game, in which the optimal solution of the discretized version of the game is always worse than the optimal solution of the continuous game.

Proposition 3. *There are continuous operating point selection games, in which the optimal strategy requires the defender to use infinitely many thresholds.*

Proof. Let the mapping from thresholds $\langle 0, 1 \rangle$ to false positive rate be $R_{FP}(t) = (1-t)$; $ROC(t) = \min(2(1-t), 1)$; the misclassification costs $C^{FN}(t) = C^{FP} = 1$; and twice as much rational as background attackers $(A_r = 2)$. Then

$$c_d^b(t) = 1 - t \text{ on } \langle 0, \tfrac{1}{2} \rangle \text{ and } t \text{ on } \langle \tfrac{1}{2}, 1 \rangle \tag{5}$$

$$\text{and } c_d^r(t) = 2 \text{ for } t \in (0, 1). \tag{6}$$

In this case, the rational defender prefers to prevent the rational attacker from attacking at all, even if it meant setting detection threshold to 0. $c_d^r(t_a)$ is always larger than $c_d^b(t_d)$ for any $t_a > 0$.

Assume the rational attacker's penalty $p_a = 1$ and reward $r_a(t) = 1 + t$. The attacker will not attack if $u_a(D, t) \leq 0$ for all $t \in \mathbf{T}$, because it can always get zero utility by not attacking. If we assume this is an equality, we can derive

$$D(t) = \frac{t+1}{t+2} \Rightarrow D(0) = \tfrac{1}{2}, D(\tfrac{1}{2}) = \tfrac{3}{5}. \tag{7}$$

If the rational attacker does not attack at all, the defender prefers to play $t = \tfrac{1}{2}$, as it minimizes c_d^b. Therefore, the optimal continuous strategy for this situation is to play D on $\langle 0, \tfrac{1}{2} \rangle$ and set $D(t) = 1$ for all larger thresholds. D is strictly increasing and c_d^b strictly decreasing on $\langle 0, \tfrac{1}{2} \rangle$. If the defender wants to prevent the rational attack with a discrete distribution, her CDF has to be larger or equal to D for each threshold. If it is lower, the attacker has positive utility for attacking. Hence, she plays $D'(t_j) = D(t_{j+1}); \forall t_j \in (t_i)$; i.e., threshold t_j with probability $\pi(t_j) = D(t_{j+1}) - D(t_j)$. Her cost $u_d(D', 0) = \sum_i \pi(t_i) c_d^b(t_i)$ can always be decreased by adding any new threshold in $\langle 0, \tfrac{1}{2} \rangle$. The monotonicity of the involved functions implies that for any new threshold $t_m \in (t_i, t_{i+1})$ holds

$$(D(t_{j+1}) - D(t_j)) \, c_d^b(t_j) > (D(t_{j+1}) - D(t_m)) \, c_d^b(t_j) + (D(t_m) - D(t_j)) \, c_d^b(t_m).$$

$\square$

Besides proving that it might not be possible to play optimally with a finite number of thresholds, the previous proposition also demonstrates how the model motivates the attacker to perform weaker attacks. The mechanism is the same even if it is beneficial only to reduce the attack strength and not to prevent it completely.

Based on the previous propositions, we can choose a discretization of set $\mathbf{T}$ that guarantees a low error. Below we show that considering only a subset of $\mathbf{T}$ for discretization is sufficient.

Proposition 4. *If $t_d^* = \arg\min\{c_d^b(t) + c_d^r(t)\}$, then a rational defender will never play a threshold t for which*

$$c_d^b(t) > c_d^b(t_d^*) + c_d^r(t_d^*)$$

Proof. Threshold t_d^* is the best pure strategy for the Stackelberg setting and the maximin strategy for the defender. The defender can always guarantee payoff at least $c_d^b(t_d^*) + c_d^r(t_d^*)$ for any strategy of the attacker. Hence, she will not play a threshold that certainly induces a higher cost.   □

5.1   Concavity of ROC Curves

The machine learning literature often assumes that the ROC curves are concave [9]. If an ROC curve is not concave then there is a false positive rate b so that

$$ROC(b) < \frac{b-a}{c-a}ROC(c) + \frac{c-b}{c-a}ROC(a) \tag{8}$$

for some $a < b < c$. If we use, instead of the threshold corresponding to b (t_b), the threshold for c (t_c) with probability $\frac{b-a}{c-a}$ and the threshold for a (t_a) with probability $\frac{c-b}{c-a}$, then the expected false positive rate is still b, but the true positive rate is the right hand side in equation 8. This randomization creates a classifier with strictly better expected performance than the classifier described by the original ROC; therefore, all ROC curves are assumed to be concave.

We argue that this well-known procedure is correct only for the traditional settings without rational attackers. In their presence, playing the probability distribution on t_a and t_c is not strategically equivalent to playing t_b. Recall that playing t_b motivates the rational attacker to play t_b as well; however, playing the randomization over t_a and t_c will motivate the attacker to play either t_a or t_c (depending on his costs), but generally not to play t_b. The attacker playing one of t_a or t_c may induce a substantially different cost to the defender compared to playing t_b. Consequently the widely adopted procedure to make ROC curves convex is not applicable in presence of rational attackers, as it results into misrepresenting the actual costs for the defender. We are not aware of any existing work presenting a similar observation.

6   Experimental Evaluation

This section experimentally demonstrates that the proposed game-theoretic approach randomizing among multiple detector's operating points forces rational attackers to attack with lower intensity than with a single threshold optimized against non-rational attackers. Consequently, the expected cost of the classifier is reduced. The use-case is an intrusion detection system (IDS) where results of this paper can be readily applied. We show three kinds of experiments: (i) experiments with multiple thresholds showing the strategies computed for specific ROC curves and the cost reduction they provide; (ii) experiments with only two thresholds which provide better insight into the rationale behind the model; and (iii) an experiments varying attacker's penalty showing the effect of this parameter on computed strategies.

In our experiments, we use ROC curves of a real world IDS from [19]. Each ROC curve has 100 points representing thresholds used by the detector. As explained in the previous section, we do not assume ROC curves to be concave, because the stochastic concave envelope changes the solution space. We have chosen the cost of false positives to be $C^{FP} = 15$, to represent that typical IDS faces much more benign traffic than actual attacks; and the amount of rational attackers to be the same as the background attackers $A_r = 1$. We set the attacker's penalty $p_a = 2$, as the cost for having the IP address blocked. The attacker's reward is a linear function in terms of attack intensity, i.e., ith point on the ROC curve, in the order of increasing false positive rate, is assigned the reward $\frac{(100-i)}{100} \cdot 10$. Choosing lower attack strength is equivalent to choosing a lower threshold to attack. Note that if the defender lowers the detection threshold, then it increases its false positive rate. The graphs in this section show false positive rate increasing on x-axis from left to right to have the ROC curves in their standard form. In all these graphs, the threshold grows on the same axes from right to left.

6.1 Computing the Equilibria and Scalability

Computing a NE may be computationally expensive as it belongs to the PPAD complexity class [7]. However, we do not reach the scalability limits of the standard equilibrium computation tools with our model. In our experiments, we use the Gambit [16] implementation of an algorithm for computing all Nash equilibria [15]. On standard Intel i5 2.3GHz laptop computer, the computation takes up to 1 minute for 12 thresholds, up to 5 minutes for 13 and up to an hour for 14 thresholds. Even though the algorithm computes all NE, it always found only a single NE in our experiments. It indicates that these games generally have a unique NE. We intend to formally study this property in our future research. In practical application, a more scalable algorithm computing only one NE can be used [14]. The gambit implementation of this algorithm is able to compute the strategy for 100 thresholds in less than one minute. Since the calculation of NE can be done off-line on a computer cluster, we do not consider the computational complexity here to be an issue.

Computing the SSE is a polynomial problem and we used the multiple linear programs (LP) method described in [5] with IBM CPLEX 12.4 as the LP solver. Computation of SEE even for 100 thresholds takes up to 10 seconds.

6.2 Multiple Thresholds

Figure 1 presents the results of the games defined based on ROCs of detectors of Secure Shell (SSH) password cracking and Skype supernodes. The games are discretized to contain 11 thresholds. Ten thresholds are chosen to be equidistant on the range of false positive values $[0, 1.0]$ and the last one is the optimal fixed threshold t_d^* defined in Proposition 4. This threshold is marked by the vertical lines in the graphs. White / black bars in Figures 1(a,b,d,e) correspond to the defender's / attacker's probability of selecting the threshold at their position. The curve in these figures is the ROC. In Figures 1(b,e) we can see that the attacker is forced to pick the lowest threshold played by the defender in the case of Stackelberg equilibria (SSE). Although the defender plays most often the high

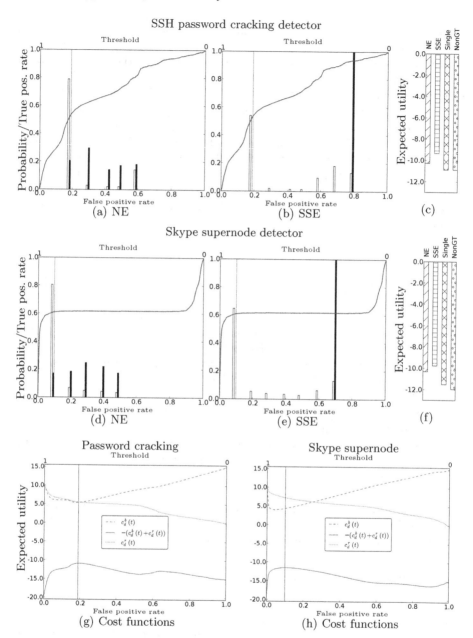

Fig. 1. The top two rows show the main results for ROC curves corresponding to detection of SSH pasword cracking and Skype supernodes. Figures (a-b) and (d-e) show the probability distributions of using thresholds for the defender (white bars) and the attacker (black bars) in the Nash and Stackelberg equilibria of the game. Figures (c) and (f) show the expected utility of these strategies in comparison to the optimal fixed thresholds selection considering the rational opponent (Single) and the standard Bayesian threshold disregarding the rational opponent (NonGT). Figures (g) and (h) show the relevant cost functions and the optimal fixed thresholds for reference.

threshold optimizing her background cost (54% of time), she plays the lower thresholds sufficiently often to force rational attackers to use weaker attacks. The threshold corresponding to the false positive rate just below 0.8 is played 13.5% of time. In these cases, the rational attacker uses less than half the attack intensity it would use without the randomization, i.e., the threshold marked by the vertical line. In the Nash equilibria presented in Figures 1(a,d), the attacker uses all the thresholds played by the defender with almost uniform probability. If the defender does not commit to a strategy in advance, the attacker also needs to randomize to prevent exploitation of his strategy by the defender. This is the main source of the defender's higher cost with NE compared to SSE.

Figures 1(c,f) present the expected costs of the Nash (NE) and Stackelberg (SSE) equilibrium strategies compared to the single threshold maximizing the utility defined as t_d^* in Proposition 4 (Single), and the standard Bayesian cost minimizing threshold disregarding the rational attackers (NonGT). The value of the SSE is better than the value of the NE. In both graphs, it is more than 10% better than the fixed operation point selection (t_d^*). The difference between the expected utility for fixed threshold selection considering and disregarding the rational attacker is quite low. Figures 1(g,h) present the utility function components computed based on the ROCs. The vertical line marks the optimal fixed game theoretic threshold (t_d^*) and the optimal threshold disregarding the rational opponents would be the minimum of $c_d^b(t)$.

Besides the results presented in this figure, we computed the expected utility values for 34 other ROC curves from [19]. The improvements of using multiple thresholds against the optimal fixed threshold (t_d^*) is between 5% and 20% (average 15%) for the Stackelberg equilibria and between 0.5% and 14% (average 9%) for the Nash equilibria.

6.3   Two Thresholds

Figure 2(a) shows the ROC curve of the horizontal scan detector we used for experiments with two thresholds. The first threshold is fixed in the optimal static thresholds selection t_d^* from Proposition 4 and the second varies over the x-axis of the graphs. Figure 2(b) presents the expected value of both equilibria and the probability of playing the threshold other than t_d^* if we optimally randomize only among these two thresholds. The vertical lines denote the position of the optimal fixed threshold and the horizontal line denotes its utility. The graph shows that substantial reduction of cost is possible already with two thresholds (top), and that even though values differ, the different equilibria suggest playing the same threshold with the same probability on large portion of possible thresholds (bottom). Furthermore, the probability of playing a low threshold (high false positive rate) quickly drops to zero at a point when it would no longer increase the defender's utility. Recall that the NE and SSE strategies overlap completely in well studied resource allocation security games [12]. Better understanding of when this happens in our model could enable reuse of many interesting results, such as efficient computation of strategies for Bayesian games with different player types.

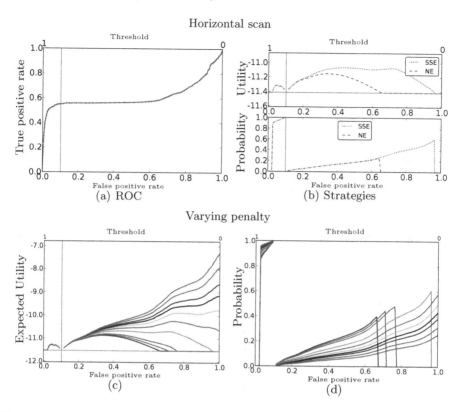

Fig. 2. Randomizing over two thresholds: first fixed at the optimal static threshold and the second varying on the x-axis. The graphs show (a) ROC curve; (b) upper: defender solution values, lower: second threshold probability; (c) the defender's SSE values for penalty for detected attack varying in the range [0.1, 20] – curves from bottom to top; and (d) are the probabilities corresponding to (c) – curves from top to bottom.

6.4   Varying Penalty

Figures 2(c,d) present the effect of attacker's penalty set to 0.1, 0.5, 1, 3, 5, 7, 9, 11, 15, 20 on the defender's payoff and probabilities in the scenario with two thresholds. We use the same setting as in the previous subsection, i.e., using one fixed thresholds t_d^* and changing the other threshold. Figure 2(c) shows that increasing the penalty increases the defender's payoff: the lines (from bottom up) represents the defender's utility with increasing penalty. Figure 2(d) with the probabilities of the alternative threshold selection shows that increasing the penalty decreases the probability of playing the second threshold: the lines (from top down) correspond to increasing attacker's penalty. At $p_a = 5$, detecting the IP address by the defender has so high penalty for the attacker that if the defender chooses the lowest threshold sufficiently often, the attacker stops attacking at all.

7   Conclusions

We analyze the problem of classifier operation point selection in presence of rational adversaries, applicable in various real world domains, such as network intrusion detection, spam filtering, steganalysis, or watermarking. We formalize it in game theoretic framework and focus on two well-known solution concepts: the more standard Nash equilibrium and the Stackelberg equilibrium commonly used in security domains. While it is not clear how to find these solutions exactly for the exact (continuous) games, we formally prove that we can create a discretized version of the game, which is solvable by standard techniques and its solution is a good approximation of the optimal solution of the original game.

We have experimentally evaluated the benefits of the model on a set of ROC curves originating from a real-world intrusion detection system. Using game theoretic randomization over multiple thresholds improves the defender's expected cost by up to 20% for some types of attacks, compared to using just single optimal threshold. This cost reduction is caused by the rational attacker selecting more than two times smaller attack strength in response to the randomization. While randomizing among larger number of thresholds is generally better, we show that substantial improvements can be achieved also by using only two different thresholds. We analyze this simplified case showing the main mechanisms by which the randomized strategies operate. Furthermore, we show that the Nash and Strong Stackelberg equilibrium strategies overlap on some subsets of threshold selections as in the resource allocation security games, but it is not true in general. This motivates more detailed study of the relation of these two models.

The future work on the proposed model may include generalization of the model to allow optimizing the thresholds against multiple different types of adversaries with different reward and cost functions. We would also like to generalize the model to allow multiple classifiers with correlated outputs and further analyze the relation to resource allocation games and other formal properties of the proposed model.

Acknowledgments. This work was supported by Ministry of the Interior of the Czech Republic, grant number VG20122014079. The work of T. Pevný on this paper was supported by the Grant Agency of Czech Republic under the project P103/12/P514.

References

1. Biggio, B., Corona, I., Maiorca, D., Nelson, B., Šrndić, N., Laskov, P., Giacinto, G., Roli, F.: Evasion attacks against machine learning at test time. In: Blockeel, H., Kersting, K., Nijssen, S., Železný, F. (eds.) ECML PKDD 2013, Part III. LNCS (LNAI), vol. 8190, pp. 387–402. Springer, Heidelberg (2013)

2. Cárdenas, A.A., Baras, J.S., Seamon, K.: A framework for the evaluation of intrusion detection systems. In: 2006 IEEE Symposium on Security and Privacy, pp. 15–77. IEEE (2006)
3. Cavusoglu, H., Raghunathan, S.: Configuration of detection software: A comparison of decision and game theory approaches. Decision Analysis 1(3), 131–148 (2004)
4. Comesana, P., Pérez-Freire, L., Pérez-González, F.: Blind newton sensitivity attack. In: IEE Proceedings of the Information Security, vol. 153, pp. 115–125. IET (2006)
5. Conitzer, V., Sandholm, T.: Computing the optimal strategy to commit to. In: Proceedings of the 7th ACM Conference on Electronic Commerce, pp. 82–90. ACM (2006)
6. Cox, I., Miller, M., Bloom, J., Fridrich, J., Kalker, T.: Digital Watermarking and Steganography. Cambridge University Press (2008)
7. Daskalakis, C., Goldberg, P.W., Papadimitriou, C.H.: The complexity of computing a nash equilibrium. SIAM Journal on Computing 39(1), 195–259 (2009)
8. Dritsoula, L., Loiseau, P., Musacchio, J.: Computing the nash equilibria of intruder classification games. In: Grossklags, J., Walrand, J. (eds.) GameSec 2012. LNCS, vol. 7638, pp. 78–97. Springer, Heidelberg (2012)
9. Flach, P.A., Wu, S.: Repairing concavities in roc curves. In: Proceedings of the 19th International Joint Conference on Artificial Intelligence, IJCAI 2005, pp. 702–707. Morgan Kaufmann Publishers Inc., San Francisco (2005)
10. Fogla, P., Lee, W.: Evading network anomaly detection systems: Formal reasoning and practical techniques. In: Proceedings of the 13th ACM Conference on Computer and Communications Security, CCS 2006, pp. 59–68. ACM, New York (2006)
11. Fridrich, J.: Steganography in Digital Media: Principles, Algorithms, and Applications. Cambridge University Press (2009)
12. Korzhyk, D., Yin, Z., Kiekintveld, C., Conitzer, V., Tambe, M.: Stackelberg vs. nash in security games: An extended investigation of interchangeability, equivalence, and uniqueness. Journal of Artificial Intelligence Research 41(2), 297–327 (2011)
13. Kutter, M., Petitcolas, F.A.: Fair benchmark for image watermarking systems. In: Electronic Imaging 1999, pp. 226–239. International Society for Optics and Photonics (1999)
14. Lemke, C.E., Howson Jr, J.T.: Equilibrium points of bimatrix games. Journal of the Society for Industrial & Applied Mathematics 12(2), 413–423 (1964)
15. Mangasarian, O.L.: Equilibrium points of bimatrix games. Journal of the Society for Industrial & Applied Mathematics 12(4), 778–780 (1964)
16. McKelvey, R.D., McLennan, A.M., Turocy, T.L.: Gambit: Software tools for game theory, version 13.1.1 (2013), http://www.gambit-project.org
17. Ogwueleka, F.N.: Data mining application in credit-card fraud detection system. Journal of Engineering Science and Technology 6(3), 311–322 (2011)
18. Paruchuri, P., Pearce, J.P., Marecki, J., Tambe, M., Ordóñez, F., Kraus, S.: Efficient algorithms to solve bayesian stackelberg games for security applications. In: AAAI, pp. 1559–1562 (2008)
19. Pevny, T., Rehak, M., Grill, M.: Detecting anomalous network hosts by means of pca. In: 2012 IEEE International Workshop on Information Forensics and Security (WIFS), pp. 103–108 (December 2012)
20. Provost, F., Fawcett, T.: Robust classification for imprecise environments. Mach. Learn. 42(3), 203–231 (2001)

21. Tambe, M.: Security and Game Theory: Algorithms, Deployed Systems, Lessons Learned. Cambridge University Press (2011)
22. Von Stengel, B., Zamir, S.: Leadership with commitment to mixed strategies. Tech. Rep. LSE-CDAM-2004-01, Centre for Discrete and Applicable Mathematics, London School of Economics and Political Science (2004)

Hierarchical Latent Tree Analysis for Topic Detection

Tengfei Liu, Nevin L. Zhang, and Peixian Chen

Department of Computer Science and Engineering
The Hong Kong University of Science and Technology
{liutf,lzhang,pchenac}@cse.ust.hk

Abstract. In the LDA approach to topic detection, a topic is determined by identifying the words that are used with high frequency when writing about the topic. However, high frequency words in one topic may be also used with high frequency in other topics. Thus they may not be the best words to characterize the topic. In this paper, we propose a new method for topic detection, where a topic is determined by identifying words that appear with high frequency in the topic and low frequency in other topics. We model patterns of word co-occurrence and co-occurrences of those patterns using a hierarchy of discrete latent variables. The states of the latent variables represent clusters of documents and they are interpreted as topics. The words that best distinguish a cluster from other clusters are selected to characterize the topic. Empirical results show that the new method yields topics with clearer thematic characterizations than the alternative approaches.

1 Introduction

Topic models have been the focus of much research in the past decade. The predominant methods are latent Dirichlet allocation (LDA) [3] and its variants [2,11]. These methods assume a generating process for the documents. For example, to generate one document, LDA first samples a multinomial distribution over topics, then it repeatedly samples a topic according to this distribution and samples a word from the topic. In this setting, a topic is defined as a multinomial distribution over the entire vocabulary. Each document is viewed as a probabilistic mixture of all the topics. The topics and the topic composition of each document are inferred by inverting the generating process using statistical techniques such as variational inference [3] and Gibbs sampling [2].

The topic definition in LDA or its variants models how frequent an author would use each word in the vocabulary when writing about a topic. A few words with high frequency are usually selected to interpret the topic [3]. However, this does not consider the differences in word usage between the documents about the topic and the documents not about the topic. High frequency words in one topic might also appear with high frequency in other topics. They may be common words for multiple topics and contain little content information to the specific topic. Thus the high frequency words are not necessarily the best words to describe the topic. To better characterize a topic, it would be advisable to consider the words that appear with high probability in the documents about the topic, while appear with low probability in the documents not on the topic. We call such words the characteristic words of the topic.

T. Calders et al. (Eds.): ECML PKDD 2014, Part II, LNCS 8725, pp. 256–272, 2014.
© Springer-Verlag Berlin Heidelberg 2014

When writing an article on a topic, an author is likely to use the characteristic words along with other non-characteristic words. When describing a topic, however, it would be better to focus on the characteristic words. For example, if we try to describe the topic *military*, we may use only a few words such as *troop*, *army*, *soldier*, *weapon*, *gun*, *bomb*, *tank*, *missile* and so on. To write an article on military, on the other hand, we might use many other words. The characteristic words of the topic consist of only a small fraction of all the words in an article on the topic.

In this paper, we propose a new method for topic detection that determines topics by identifying their characteristic words. The key idea is to model patterns of word co-occurrence and co-occurrence of those patterns using a hierarchy of discrete latent variables. Each latent variable represents a soft partition of the documents based on some word co-occurrence patterns. The states of the latent variable correspond to document clusters in the partition. They are interpreted as topics. Each document may belong to multiple clusters in different partitions. In other words, a document might belong to two or more topics 100% simultaneously. For each topic, the words that best distinguish the topic from other topics are selected to describe the topic. These words are usually the words appear with high probability in the documents belonging to the topic, while appear with low probability in the documents belonging to other topics.

This paper builds upon previous work on latent tree models (LTMs) by Zhang [19], which are tree-structured probabilistic graphical models where leaf nodes represent observed variables and internal nodes represent latent variables. When applied to text data, LTMs are effective in systematically discovering patterns of word co-occurrence [13,12]. In this work, we introduce semantically higher level latent variables to model co-occurrence of those patterns, resulting in hierarchical latent tree models (HLTMs). The latent variables at higher levels of the hierarchy correspond to more general topics, while the latent variables at lower levels correspond to more specific topics. The proposed method for topic detection is therefore called *hierarchical latent tree analysis (HLTA)*.

The remainder of this paper is organized as follows. In Section 2 we briefly introduce the concepts of LTMs. In Section 3 we present the HLTA algorithm. We will explain how to find the patterns of word co-occurrence and then how to aggregate these patterns for better topic detection. In Section 4, we present empirical results and compare HLTA with alternative methods. Finally, conclusions are drawn in Section 6.

2 Basics of Latent Tree Models

A *latent tree model (LTM)* is a Markov random field over an undirected tree where leaf nodes represent observed variables and internal nodes represent latent variables. LTMs were originally called hierarchical latent class models [19] to underline the fact that they are a generalization of *latent class model (LCM)* [1]. Figure 1 shows an example LTM that was learned from a collection of documents. The words at the bottom are binary variables that indicate the presence or absence of the words in a document. The Z_i's are the latent variables. They are discrete and their *cardinalities*, i.e., the number of states, are given in parentheses.

For technical convenience, we often root an LTM at one of its latent nodes and regard it as a directed graphical model, i.e., a *Bayesian network* [16]. Then all the

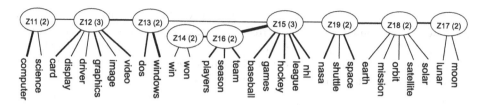

Fig. 1. A latent tree model for a toy text data set. The edge widths visually show the strength of correlation between variables. They are computed from the probability distributions of the model.

edges are directed away from the root. The numerical information of the model includes a marginal distribution for the root and one conditional distribution for each edge. For example, for edge $Z15 \rightarrow hockey$, it is associated with probability $P(hockey \mid Z15)$. The conditional distribution associated with each edge characterizes the probabilistic dependence between the two nodes that the edge connects. The product of all those distributions defines a joint distribution over all the latent and observed variables.

In general, suppose there are n observed variables $\mathbf{X} = \{X_1, \ldots, X_n\}$ and m latent variables $\mathbf{Z} = \{Z_1, \ldots, Z_m\}$. Denote the parent of a variable Y as $pa(Y)$ and let $pa(Y)$ be a empty set when Y is the root. Then the LTM defines a joint distribution over all observed and latent variables as follows:

$$P(X_1, \ldots, X_n, Z_1, \ldots, Z_m) = \prod_{Y \in \mathbf{X} \cup \mathbf{Z}} P(Y \mid pa(Y)) \tag{1}$$

3   Topic Detection with Hierarchical Latent Tree Models

In this section, we describe the new method HLTA. Conceptually, the method has three steps: (1) Discover patterns of word co-occurrences; (2) Build a hierarchy by recursively discovering co-occurrence patterns of patterns; and (3) Extract topics from the resulting hierarchy. The pseudo code of the algorithm is given in Algorithm 1. We describe the steps in details in the following subsections.

3.1   Discovering Co-occurrence of Words

In HLTA, words are regarded as binary variables which indicate the presence or absence of the words in the documents. Documents are represented as binary vectors. The first step of HLTA is to identify the patterns of word co-occurrence. This is done by partitioning the word variables into clusters such that variables in each cluster are closely correlated and the correlations among variables in each cluster can be properly modeled with only one latent variable. The Bridged-Islands algorithm (BI) [12] is used for this purpose (Line 3).

Figure 1 shows the latent tree model that BI obtained on a toy data set. We see that the word variables are partitioned into 9 clusters. One example cluster is {*baseball, games, hockey, league, nhl*}. The five word variables are connected to the same latent variable $Z15$ and are hence called *siblings*. The cluster is then called a sibling cluster. It

---

Algorithm 1. HLTA(D_{in}, δ, k)

Inputs: D_{in}: input dataset; δ: Bayes factor threshold, k: maximum level of latent variables.
Output: An HLTM and the topics.

1: $D \leftarrow D_{in}, Level \leftarrow 0, M_{whole} \leftarrow \emptyset$.
2: **while** $Level < k$ or D contains more than two variables **do**
3:     $M \leftarrow$ Bridged-Islands(D, δ)
4:     **if** $M_{whole} = \emptyset$ **then**
5:        $M_{whole} \leftarrow M$
6:     **else**
7:        $M_{whole} \leftarrow$ MergeModel(M_{whole}, M)
8:     **end if**
9:     $D' \leftarrow$ ProjectData(M_{whole}, D_{in})
10:    $D \leftarrow D', Level \leftarrow Level + 1$
11: **end while**
12: Output topics in different levels and return M_{whole}.

---

is apparent that the words in each sibling cluster are semantically correlated and tend to co-occur. The correlations among the word variables in each sibling cluster are modeled by a latent variable. In fact, every latent variable in the model is connected to at least one word variable. Because of this, the model is called a *flat latent tree model*.

In the following, we briefly describe how BI works. The reader is referred to [12] for the details. In general, BI is a greedy algorithm that aims at finding the flat latent tree model with the highest Bayesian Information Criterion (BIC) score [17]. It proceeds in four steps: (1) partition the set of variables into sibling clusters; (2) introduce a latent variable for each sibling cluster; (3) connect the latent variables to form a tree; (4) refine the model based on global considerations.

To identify potential siblings, BI considers how closely correlated each pair of variables are in terms of mutual information. The mutual information (MI) $I(X;Y)$ [8] between the two variables X and Y is defined as follows:

$$I(X;Y) = \sum_{X,Y} P(X,Y) \log \frac{P(X,Y)}{P(X)P(Y)}, \tag{2}$$

where the summation is taken over all possible states of X and Y. The distributions $P(X,Y)$, $P(X)$ and $P(Y)$ are estimated from data.

To determine the first sibling cluster, BI maintains a working set S of variables that initially consists of the pair of variables with the highest MI. Other variables are added to the set one by one. At each step, BI chooses to add the variable X that maximizes the quantity $\max_{Z \in S} I(X;Z)$. After each step of expansion, BI performs a Bayesian statistical test to determine whether correlations among the variables in S can be properly modeled using one single latent variable. The test is called *uni-dimensionality test* or simply *UD-test*. The expansion stops when the UD-test fails.

To perform the UD-test, BI first projects the original data set D onto the working set S to get a smaller data set D'. Then it obtains from D' the best LTMs m_1 and m_2 that contains only 1 latent variable or no more than 2 latent variables respectively. BI concludes that the *UD-test* passes if and only if one of the two conditions is satisfied :

(1) m_2 contains only one latent variable, or (2) m_2 contains two latent variables and

$$BIC(m_2 \mid D') - BIC(m_1 \mid D') \le \delta, \tag{3}$$

where δ is a threshold parameter. The left hand side of this inequation is an approximation to the natural logarithm of the Bayes factor [10] for comparing model m_2 with model m_1.

To illustrate the process, suppose that the working set $S=\{X_1, X_2\}$ initially. Then X_3 and X_4 were subsequently added to S and the UD-test passed on both cases. Now consider adding X_5. Assume the models m_1 and m_2 for the set $S=\{X_1, X_2, X_3, X_4, X_5\}$ are as shown in Figure 2, and suppose the BIC score of m_2 exceeds that of m_1 by threshold δ. Then UD-test fails and BI stops growing the set S.

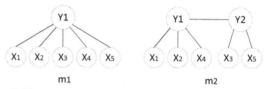

Fig. 2. The two models m_1 and m_2 considered in the UD-test

When the UD-test fails, model m_2 gives us two potential sibling clusters. If one of the two potential sibling clusters contains both the two initial variables, it is picked. Otherwise, BI picks the one with more variables and breaks ties arbitrarily. In the example, the two potential sibling clusters are $\{X_1, X_2, X_4\}$ and $\{X_3, X_5\}$. BI picks $\{X_1, X_2, X_4\}$ because it contains both the two initial variables X_1 and X_2.

After the first sibling cluster is determined, BI removes the variables in the cluster from the data set, and repeats the process to find other sibling clusters. This continues until all variables are grouped into sibling clusters.

After the sibling clusters are determined, BI introduces a latent variable for each sibling cluster. The cardinality of latent variable is automatically determined during the learning process. All the latent variables are further connected to form a tree structure by using Chow-Liu's algorithm [7]. At the end, BI carries out adjustments to the structure based on global considerations.

3.2 Discovering Co-occurrence of Patterns

As illustrated in Figure 1, BI yields flat latent tree models where the latent variables capture patterns of word co-occurrences. The patterns themselves may co-occur. Such higher level co-occurrence patterns can be discovered by recursively applying BI. This is what the HLTA algorithm is designed to do.

For reasons that will become clear later, we call the latent variables for word co-occurrence patterns level-1 latent variables. To discover higher level co-occurrence patterns, we first project the data onto the latent space spanned by the level-1 latent variables (Line 9). This is done by carrying out inference in the current model M_{whole}. For each data case d_i and each level-1 latent variable $Z1j$, we compute the posterior probability $P(Z1j|d_i, M_{whole})$, and assign the data case to the state with the maximum probability. In other words, we set the value of $Z1j$ to the state with highest

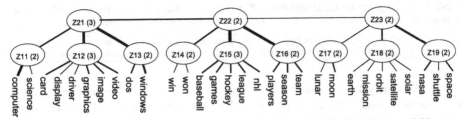

Fig. 3. The hierarchical latent tree model obtained by HLTA. The level-1 latent variables are found by running BI on the original data (*cf.* Figure 1) , and the level-2 latent variables are found by running BI on the data projected to the space spanned by the level-1 latent variables.

posterior probability. The values of the level-1 latent variables make up the projected data set D' (Line 9).

Next, we run BI algorithm on the projected data (Line 3). The resulting model is denoted as M in Algorithm 1. In this model, the level-1 latent variables are partitioned into sibling clusters and a level-2 latent variable is introduced for each cluster. The level-1 latent variables capture patterns of word co-occurrence, while level-2 variables capture the co-occurrence patterns of those patterns.

Now we have two flat LTMs, M_{whole} and M. M_{whole} consists of word variables and level-1 latent variables, and M consists of level-1 latent variables (viewed as observed variables) and level-2 latent variables. At the third step (Line 7), the two models are merged into a new LTM that consists of two levels of latent variables. Let us illustrate this using Figure 1. Assume the level-1 latent variables are partitioned into three clusters {Z11, Z12, Z13}, {Z14, Z15, Z16} and {Z17, Z18, Z19}, and suppose the corresponding level-2 latent variables are Z21, Z22 and Z23 respectively. Then, after merging M_{whole} and M at Line 7, we get the model shown in Figure 3. Basically, we stack M on top of M_{whole}, and then remove the connections among the level-1 latent variables. In the merged model M_{whole}, the parameters that involve the top level latent variables are estimated using Expectation-Maximization (EM) algorithm [9], and the values for other parameters are copied from the previous M_{whole} model.

After the level-2 latent variables are added to the model, we repeat the process to add more levels of latent variables until a predetermined number of levels k is reached, or when there are no more than two latent variables at the top level.

3.3   Topic Extraction

At the end of the while-loop (Line 11), HLTA has built a model with multiple levels of latent variables. The top level consists of either a single latent variable or multiple latent variables connected up in a tree structure. The other levels consist of multiple latent variables, each of which is connected to one latent variable at the level above and several variables at the level below. The bottom level consists of the word variables. We call the model a hierarchical latent tree model (HLTM). Each latent variable in the model represents a soft partition of the documents and its states can be interpreted as topics. In the last step (Line 12), HLTA computes descriptions of topics.

To see how the topics should be described, first consider the level-1 latent variable $Z15$ in Figure 3. It is directly connected to *baseball, games, hockey, league,* and *nhl*.

Table 1. Topics given by latent variable $Z22$ in Figure 3. The font sizes for probability values indicate their magnitude, while the font sizes of words indicate the discerning index.

$Z22$									
S0 (87%)	team 0.01	players 0.01	baseball 0	season 0	hockey 0	games 0.02	league 0	nhl 0
S1 (13%)	team 0.42	players 0.29	baseball 0.28	season 0.26	hockey 0.26	games 0.31	league 0.22	nhl 0.18

During data analysis, $Z15$ was introduced to model the correlations among those five word variables. Hence those words form the base for interpreting $Z15$. For level-2 latent variable $Z22$, it was introduced during data analysis to model the correlations among the level-1 latent variables $Z14$, $Z15$ and $Z16$. Hence all the words in the subtree rooted at $Z22$ form the base for its interpretation. We call the collection of those words the semantic base of $Z22$.

For a latent variable at a high level of the hierarchy, the semantic base can be large. To deal with the issue, we introduce the concept of *effective semantic base*. Sort all the word variables in the semantic base of a latent variable Z as X_1, X_2, $\cdots$, X_m in descending order of their mutual information $I(Z; X_i)$ with Z. Consider the mutual information $I(Z; X_1 \cdots X_i)$ between Z and the first i word variables. It monotonically increases with i, and reaches the maximum when $i = m$. The ratio $I(Z; X_1 \cdots X_i)/I(Z; X_1 \cdots X_m)$ is called the *information coverage* of the first i variables [4]. We define the *effective semantic base* of Z to be the collection of first i word variables for which the information coverage exceeds 0.95. Intuitively, $I(Z; X_1 \cdots X_m)$ is the amount of information about Z that is contained in its semantic base. The effective semantic base covers 95% of that information, hence is sufficient to determine the meaning of Z. The effective semantic bases for the level-2 latent variables in Figure 3 are shown in the following.

$Z21$	windows dos computer card graphics video image
$Z22$	team baseball players hockey season games league nhl
$Z23$	space nasa orbit shuttle mission earth moon solar

The meaning of a state of a latent variable is determined by the conditional distributions of the word variables from the effective semantic base. For example, the latent variable $Z22$ in Figure 3 has two states S0 and S1. The conditional probabilities (i.e., P(word=1|$Z22$=Si)) are given in Table 1. We see that in the cluster $Z22$=S1, the words *team, players, baseball* etc. occur with relatively high probabilities. It can be interpreted as the topic *sports*. On the other hand, the words seldom occur in cluster $Z22$=S0 which can be considered as a background topic. Those two topics consist of 13% (P($Z22$=S1)=0.13) and 87% (P($Z22$=S0)=0.87) of the documents respectively.

To highlight the importance of words in characterizing a topic, we introduce the concept of *discerning index*. Let Z be a latent variable that has two or more states, and W be a word variable. For a given state s of Z, let Z_s be another variable that takes two possible values 0 and 1, with Z_s=1 meaning Z=s and Z_s=0 meaning $Z \neq s$. The discerning index of W for Z=s is the mutual information $I(W, Z_s)$ between W and Z_s. The higher the index, the more important W is for distinguishing the cluster Z=s from other clusters in the partition given by Z. Usually, the words that occur with high probabilities in Z=s and low probabilities when $Z \neq s$ have high discerning index values. In Table 1, the order and font sizes of the words are determined according to the discerning index. The font sizes for the probability values are determined by their own magnitude. This visualization scheme is proposed so that the thematic meaning of each topic is readily visible to the reader.

At line 12, HLTA computes the effective semantic base of each latent variable and, for each topic, the discerning index and occurrence probability of each word from the base. The probability of each topic in the entire corpus and the probability of each document belonging to each topic are also computed. All the computations need to perform inference in the resulting model, which can be done in HLTM by applying standard algorithms, e.g. clique tree propagation [18].

4 Empirical Results

In this section, we demonstrate the characteristics of HLTA topics and the topic hierarchy. We also compare the predictive performance of HLTA on several data sets which include: (1) NIPS[1] data: 1,740 NIPS articles published from 1988 to 1999; (2) JACM[2] data: 536 abstracts from the Journal of the ACM; (3) Newsgroup[3] data: about 20,000 newsgroup documents. For JACM data, all 1,809 words were used in the experiments. For Newsgroup and NIPS data, the vocabulary was restricted to 1,000. The stop words and words appear in less than ten papers were removed. Then we computed the TF-IDF value of each word in each document. The top 1,000 words with highest average TF-IDF value were selected. The code and data sets used in the experiment are available at: `http://www.cse.ust.hk/~lzhang/ltm/index.htm`.

4.1 Results on the NIPS Data

We first show the results of HLTA on NIPS data.

Model Structure. The analysis resulted in a hierarchical LTM with 382 latent variables arranged in 5 levels. There are 279, 72, 21, 8 and 2 latent variables on levels 1, 2, 3, 4 and 5 respectively. Table 2 shows part of the hierarchical structure. Table representation is used instead of the tree structure to save space. We see that, for example, the word variables *bayesian, posterior, prior, bayes, priors, framework, gamma* and *normal* are connected to the level-1 latent variable Z106; the level-1 latent variables Z106-Z112 are connected to the level-2 latent variable Z203; the level-2 latent variables Z201-Z205 are connected to the level-3 latent variable Z301; the level-3 latent variables Z301-Z302 are connected to a latent variable at level 4 (which is not shown); and so on.

The words are displayed in different font sizes to indicate their mutual information with the level-1 latent variables to which they are connected. For example, Z106 is more strongly correlated with *bayesian* than *normal*. The names of the latent variables are also displayed in different font sizes to indicate their mutual information with the parent latent variables at the next higher level.

---

[1] `http://www.cs.nyu.edu/~roweis/data.html`
[2] `http://www.cs.princeton.edu/~blei/downloads/`
[3] `http://qwone.com/~jason/20Newsgroups/`

Table 2. Part of the hierarchical latent tree model obtained by HLTA on the NIPS data

Z301(2)	Z201(2)	Z101(2): likelihood conditional log em maximum ix derived mi ; Z102(2): statistical statistics ; z103(2): density densities
	Z202(2)	Z104(2): entropy divergence mutual ; Z105(2): variables variable
	Z203(3)	Z106(3): bayesian posterior prior priors bayes framework gamma normal ; Z107(2): probabilistic distributions probabilities ; z108(2): inference gibbs sampling generative uncertainty ; Z109(2): mackay independent averaging ensemble uniform ; Z110(2): belief graphical variational ; z111(2): monte carlo ; z112(2): uk ac generalisation
	Z204(3)	Z113(2): mixture mixtures latent fit ; Z114(3): multiple hierarchical individual sparse missing multi significant index represent hme ; Z115(2): experts expert gating ; z116(2): weighted sum weighting ; z117(2): scale scales scaling
	Z205(3)	Z118(3): estimate estimation estimated estimates estimating measure deviation ; Z119(2): estimator true unknown ; Z120(2): sample samples ; z121(2): assumption assume assumptions assumed ; z122(2): observations observation observed ; Z123(2): computed compute
Z302(3)	Z206(4)	Z124(2): gaussian covariance variance gaussians program provided ; Z125(2): subspace dimensionality orthogonal reduction ; Z126(3): component components principal pca decomposition ; Z127(2): dimension dimensional dimensions vectors ; Z128(2): matrix matrices diagonal ; Z129(2): exp cr exponential ; Z130(2): noise noisy robust ; Z131(2): projection projections pursuit operator ; Z132(2): radial basis rbf ; Z133(2): column row ; Z134(2): eq fig proc
	Z207(2)	Z135(2): eigenvalues eigenvalue eigenvectors identical ; Z136(2): ij product wij bi ; Z137(2): modes mode
	Z208(2)	Z138(2): mixing coefficients inverse joint smooth smoothing ; Z139(2): blind ica separation sejnowski natural concept ; Z140(2): sources source
Z303(2)	Z209(2)	Z141(2): classification classifier classifiers nn ; Z142(2): class classes
	Z210(2)	Z143(2): discriminant discrimination fisher ; Z144(2): labels label labeled
	Z211(2)	Z145(2): handwritten digit digits le ; Z146(2): character characters handwriting
Z304(3)	Z212(3)	Z147(2): regression regularization generalization risk ; Z148(3): vapnik svm margin support vc dual fraction ; Z149(2): kernel kernels ; Z150(2): empirical drawn theoretical ; z151(2): xi yi xj zi xl gi
	Z213(2)	Z152(2): validation cross bias ; Z153(2): stopping pruning criterion obs ; Z154(2): prediction predictions predict predicted predictive
	Z214(2)	Z155(2): machines machine boltzmann ; Z156(2): boosting adaboost weak

We can see from Table 2 that many level-1 latent variables represent thematically meaningful patterns. Examples include Z101 (*likelihood conditional log etc.*), Z106 (*Bayesian posterior prior etc.*), Z108 (*inference gibbs sampling etc.*), Z111 (*monte carlo*), Z124 (*gaussian covariance variance etc.*), Z125 (*subspace dimensionality orthogonal reduction*), Z139 (*blind ica separation*), Z145 (*handwritten digit*), Z148 (*vapnik svm margin support etc.*), Z155 (*machines Boltzmann*), Z156 (*boosting adaboost*).

For latent variables at level 2 and level 3, their effective semantic bases are given in Table 3. For latent variables at different levels, a higher level latent variable represents a partition of documents based on a wider selection of words than its children. It is usually about a general concept that has several aspects. We see in Table 3 that Z301 is about probabilistic method, while its children cover likelihood (Z201), entropy (Z202), Bayesian (Z203), mixture (Z204) and estimate (Z205). Z302 is about the use of Gaussian covariance matrix, while its children cover PCA (Z206), eigenvalue/vector (Z207), and blind source separation (Z208); Z303 is about classification, while its

Table 3. Effective semantic bases of level-2 and level-3 latent variables

---

Z301 likelihood bayesian statistical conditional posterior probabilistic density log mixture prior bayes distributions estimate priors

   Z201 likelihood statistical conditional density log em statistics

   Z202 entropy variables variable divergence

   Z203 bayesian posterior probabilistic prior bayes distributions priors inference monte carlo probabilities

   Z204 mixture mixtures experts hierarchical latent expert weighted sparse

   Z205 estimate estimation estimated estimates estimating estimator sample true samples observations

---

Z302 gaussian covariance matrix variance eigenvalues eigenvalue exp gaussians pca principal matrices eigenvectors component noise

   Z206 gaussian matrix covariance pca variance principal subspace dimensionality projection exp gaussians

   Z207 eigenvalues eigenvalue eigenvectors

   Z208 blind mixing ica coefficients inverse separation sources joint

---

Z303 classification classifier classifiers class classes handwritten discriminant digit

   Z209 classification classifier classifiers class

   Z210 discriminant label labels discrimination

   Z211 handwritten digit character digits characters

Z304 regression validation vapnik svm machines regularization margin generalization boosting kernel kernels risk empirical

   Z212 regression vapnik svm margin kernel regularization generalization kernels support xi risk

   Z213 validation cross stopping pruning prediction predictions

   Z214 machines boosting machine boltzmann

---

children cover discriminant and handwritten digit/character recognition. Z304 is about regression, while its children cover SVM, cross validation, and boosting.

Topics. Each latent variable in the model represents a soft partition of the documents. Each state of the latent variables corresponds to a cluster in the partition and can be interpreted as a topic. As discussed in Section 3.3, we characterize the topic using words from its effective semantic base with the highest discerning indices, that is, the words that best distinguish the documents in the cluster from documents not in the cluster. For non-background topics, those usually are the words that appear with high probability in the cluster and low probability in other clusters.

Take Z301 as an example. It has two states and hence partitions the documents into two clusters. The characterizations of the two clusters are given in Table 4. We see that, for Z301=S1, the words *likelihood*, *Bayesian* and *statistical* are placed at the beginning of the list. They have the highest discerning indices. Their probabilities of occurring in Z301=S1 are 0.58, 0.45 and 0.73 respectively, which are significantly larger than those for Z301=S0, which are 0.06, 0.03, and 0.24 respectively. On the other hand, the word *estimate* also occurs with high probability (0.64) in Z301=S1. However, it has low discerning index for Z301=S1 because its probability in the other cluster Z301=S0 is also relatively high (0.25). It is clear from the characterization that the topic Z301=S1 is about general probabilistic method, while Z301=S0 is a background topic. They consist of 34% and 66% of the documents respectively.

Table 4 shows characterizations of topics given by ten latent variables. The three level-3 latent variables (i.e., Z301, Z303 and Z304) are chosen because they are about the basic topics covered in a typical machine learning course, namely probabilistic methods, classification and regression. The others are selected level-2 and level-1

Table 4. Example topics found by HLTA on NIPS data. For each topic, only the words in the effective semantic base are shown. The order and font sizes of words in each topic are determined by discerning index. The font sizes of word occurrence probabilities simply reflect their magnitude.

Z301	S0 (66%)	likelihood 0.06 bayesian 0.03 statistical 0.24 conditional 0.05 posterior 0.04 density 0.13 probabilistic 0.06 log 0.18 bayes 0.02 mixture 0.06 prior 0.15 estimate 0.25 distributions 0.15 priors 0.01
	S1 (34%)	likelihood 0.58 bayesian 0.45 statistical 0.73 conditional 0.4 posterior 0.37 density 0.52 probabilistic 0.42 log 0.58 bayes 0.29 mixture 0.4 prior 0.54 estimate 0.64 distributions 0.52 priors 0.22
Z203	S1 (19%)	probabilistic 0.48 distributions 0.58 probabilities 0.49 bayesian 0.4 prior 0.52 posterior 0.32 bayes 0.26 priors 0.17 inference 0.22 carlo 0.05 monte 0.05
	S2 (11%)	bayesian 0.75 monte 0.54 carlo 0.53 posterior 0.64 inference 0.58 prior 0.79 priors 0.41 bayes 0.47 probabilistic 0.54 distributions 0.63 probabilities 0.54
Z113 S1 (19%)		mixture 0.76 mixtures 0.53 latent 0.17 fit 0.34
Z303 S1 (30%)		classification 0.81 classifier 0.48 classifiers 0.38 class 0.69 classes 0.53 handwritten 0.22 discriminant 0.15 digit 0.2
Z211 S1 (13%)		handwritten 0.58 digit 0.52 character 0.54 digits 0.42 characters 0.31
Z145 S1 (12%)		handwritten 0.72 digit 0.64 digits 0.52
Z146 S1 (9%)		character 0.84 characters 0.49 handwriting 0.24
Z304	S1 (25%)	regression 0.36 validation 0.33 regularization 0.21 generalization 0.5 risk 0.15 empirical 0.31 svm 0 boosting 0 machines 0.1 vapnik 0.09 margin 0.04 kernels 0.07 kernel 0.11
	S2 (7%)	machines 0.76 svm 0.42 vapnik 0.53 margin 0.44 boosting 0.3 kernel 0.55 kernels 0.4 regression 0.49 validation 0.42 generalization 0.62 regularization 0.29 empirical 0.44 risk 0.21
Z213 S1 (19%)		validation 0.57 cross 0.55 stopping 0.24 pruning 0.18 prediction 0.48 predictions 0.35
Z156 S1 (3%)		boosting 0.9 adaboost 0.35 weak 0.39

latent variables under the level-3 latent variables. The background topics are not shown except Z301=S0. These topics show clear thematic meaning. We can see that Z301=S1 is about probabilistic method in general, while its subtopics Z203=S2 is about Bayesian-monte-carlo, Z203=S1 is about probabilistic method not involving monte-carlo, Z113=S1 is about mixture models. Topic Z303=S1 is about classification, while its subtopics Z211=S1 is about digit/character classification, Z145=S1 is about handwritten digit classification, Z146=S1 is about handwritten character classification. Z304=S1 is about regression in general, while its subtopics Z304=S2 is about SVM, Z213=S1 is about cross validation and Z156=S1 is about boosting.

Comparisons with LDA. To better appreciate the topics found by HLTA, it is necessary to compare them with those detected by the LDA approach [3]. In this section, we run LDA on the NIPS data to find 150 topics. The documents are represented as bags-of-words in LDA, while as binary vectors in HLTA.

Table 5 shows the LDA topics that are the closest in meaning to the HLTA topics shown in Table 4. They are selected using the top three words of the HLTA topics. The LDA topic that best matches the HLTA topic is selected manually. The LDA approach produces flat topics. It does not organize the topics in a hierarchical structure as in HLTA. Thus in this section, we focus on the topics. Compared the HLTA topics with the LDA topics, we can find that they differ in two fundamental ways. First, an HLTA topic corresponds to a collection of documents. As such, we can talk about the size of a topic, which is the fraction of documents belonging to the topic among all documents.

Table 5. LDA topics that correspond to the HLTA topics in Table 4. Only the top eight words are shown for each topic.

HLTA topic	LDA topic
Z301=S1	T-25: bayesian .16 prior .13 posterior .10 evidence .04 bayes .04 priors .03 log .03 likelihood .03
Z203=S2	T-97: gaussian .23 monte .07 carlo .07 covariance .05 variance .05 processes .04 williams .03 exp .02
Z113=S1	T-78: mixture .13 em .12 likelihood .11 gaussian .05 log .04 maximum .04 mixtures .04 latent .03
Z303=S1	T-139: classification .20 class .19 classifier .15 classifiers .09 classes .09 decision .03 bayes .02 labels .01
Z211=S1	T-120: recognition .18 character .09 digit .06 characters .06 digits .05 handwritten .05 segmentation .04 le .02
Z304=S1	T-84: regression .15 risk .08 variance .08 bias .08 confidence .04 empirical .03 smoothing .03 squared .03
Z304=S2	T-141: kernel .16 support .09 svm .05 kernels .05 machines .04 margin .03 vapnik .02 feature .02
Z213=S1	T-146: cross .21 validation .19 stopping .07 generalization .05 selection .04 early .04 fit .02 statistics .02
Z156=S1	T-45: margin .09 hypothesis .08 weak .07 boosting .06 generalization .05 adaboost .04 algorithms .03 base .02

In Table 4, the numbers shown in parenthesis indicate the size of each HLTA topic. On the other hand, LDA treats each document as a mixture of topics. It is possible to aggregate the topic proportions of all documents. However, the aggregated quantity would be the fraction of words belonging to that topic, not the fraction of documents.

A more important difference lies in the way topics are characterized. When picking words to characterize a topic, HLTA uses discerning index which considers two factors: (1) the word occurrence probability in documents belonging to the topic, and (2) the word occurrence probability in documents not on the topic. This results in clear and clean topic characterizations. The consideration of the second factor implies that HLTA is unlikely to pick polysemous words when characterizing a topic, because such words are used in multiple topics. This should not affect topic identification as long as there are words peculiar to each topic. In contrast, LDA considers only the first factor, and the resulting topic characterizations are sometimes not clear.

For example, we first look at Z113 which has two states. According to Table 4, the words *mixture*, *mixtures* and *latent* have the highest discerning indices in Z113=S1, which indicate they occur with high probability in Z113=S1 and low probability in background topic (i.e., Z113=S0). Z113=S1 is clearly about mixture models. The closest LDA topic is Topic T-78. As shown in Table 5, the leadings words in the topic are *mixture*, *em* and *likelihood*. The words *em* and *likelihood* are not characteristic of mixture models because they are used more often in other situations such as the handling of missing data. The HLTA characterization seems cleaner.

For Z156 in Table 4, the words with highest discerning indices in Z156=S1 are *boosting*, *adaboost* and *weak*, which occur with high probability in Z156=S1 and low probability in background topic (i.e., Z156=S0). Z156=S1 is clearly about boosting. The closest LDA topic is Topic T-45. As shown in Table 5, the leadings words in the topic are *margin*, *hypothesis*, *weak*, *generalization* and *boosting*. It is not clear to us at the first sight what this topic is about.

Comparisons with HLDA. The HLDA [2] approach, which is an extension of LDA, can also learn topic hierarchies from data. To compare HLTA with HLDA, we trained a three-level HLDA[4] on NIPS data. The hyperparameters of HLDA are set according to

---

[4] The code of HLDA is obtained from: http://www.cs.princeton.edu/~blei/topicmodeling.html

Table 6. Part of the topic hierarchy produced by HLDA on NIPS data. Only the top ten words are shown for each topic.

[Topic L3] units hidden layer unit weight test noise inputs trained patterns
[Topic L2-0] gaussian likelihood density log mixture em prior posterior estimate estimation
[Topic L1-0] kernel xi pca kernels feature regression matrix support svm principal
[Topic L1-1] evidence bayesian gaussian posterior prior approximation field mackay variational exp
[Topic L1-2] validation cross generalization stopping variance examples early prediction estimator penalty
[Topic L1-3] sampling carlo bayesian monte prior predictive posterior inputs priors loss
[Topic L1-4] class bayes matrix coding learned max nearest classes classifier classifiers
[Topic L1-5] propagation belief inference jordan nodes tree variational variables product graphical
[Topic L2-1] recognition image feature block le images address features handwritten digit
[Topic L1-6] characters character recognition net field segmentation fields word digits window
[Topic L1-7] image images digits digit transformation convex pixel generative object control
[Topic L2-2] regression image classification representation mixture prediction capacity selection weight classifier
[Topic L1-8] adaboost cost boosting margin potential algorithms ct hypothesis base weak
[Topic L1-9] transform pca dimension coding reduction mixture image images reconstruction grid

the settings used in [2]. Table 6 shows part of the topic hierarchy. Only the root topic (i.e., Topic-L3) and the topics that match the HLTA topics in Table 4 are presented.

A comparison of Tables 4 and 6 suggests that the thematic meaning of the topic hierarchy given by HLDA is not as clear as that given by HLTA. For example, we can first look at Topic L2-0 in Table 6. The top words of Topic L2-0 are *gaussian, likelihood, density* and *log*. It can be interpreted as a topic about probabilistic methods. Most subtopics of Topic L2-0 are about probabilistic methods, e.g., Topic L1-1 (evidence-bayesian-gaussian-posterior) and Topic L1-3 (sampling-carlo-bayesian -monte). However, there are also subtopics that are not about probabilistic methods. In particular, Topic L1-2 (validation-cross-generalization-stopping) is about cross validation. It can hardly be viewed as a subtopic of probabilistic methods. In contrast, all the subtopics of the HLTA topic Z301=S1 are about probabilistic methods.

As another example, consider the HLDA topics Topic L1-6 (characters-character -recognition-net) and Topic L1-7 (image-images-digits-digits). They can be interpreted as "character recognition" and "digit recognition" respectively. However, the meaning of their parent topic, i.e., Topic L2-1 (recognition-image-feature-block), is not clear to us. The HLTA topics Z145=S1 (handwritten-digit-digits) and Z146=S1 (character -characters-handwriting), as shown in Table 4, are about the same topics as Topic L1-7 and Topic L1-6. They seem to give better characterizations of the topics. More importantly, they are subtopics of Z303=S1 (classification-classifier-classifiers-class) and Z211=S1 (handwritten-digit-character-digits), which is clearly reasonable.

In summary, in HLTA, the topics at higher level are more general topics since they are defined on a larger semantic base. A subtopic, on the other hand, is defined on a subset of the semantic base of its parent topic. Thus the subtopics in HLTA are more specific topics. They are semantically close to their parent topics since they share part of the semantic base. Compared to HLDA topics in Table 6, the parent topic and child topics in HLTA, as shown in Table 4, show higher semantic closeness.

Topic Semantic Coherence. To quantitatively compare the quality of topics found by LDA, HLDA and HLTA on NIPS data, we compute their *topic coherence scores* [14]. The topic coherence score for topic t is defined as

$$C(t, W^{(t)}) = \sum_{m=2}^{M} \sum_{l=1}^{m-1} \log \frac{D(w_m^{(t)}, w_l^{(t)}) + 1}{D(w_l^{(t)})}, \tag{4}$$

where $W^{(t)} = \{w_1^{(t)}, ..., w_M^{(t)}\}$ are the top M words used to characterize topic t, $D(w_i)$ is the document frequency of word w_i and $D(w_i, w_j)$ is the co-document frequency of words w_i and w_j. Document frequency is the number of documents containing the words. Given two collections of topics, the one with higher average topic coherence is regarded as better. For comparability, all topics from the two collections should be of the same length, that is, characterized by the same number (i.e., M) of words.

For HLTA, it produced 140 non-background topics by latent variables from levels 2, 3 and 4. We consider two scenarios in terms of the number of words we use to characterize the topics. In the first scenario, we set M=10. The level-2 topics are excluded in this scenario because the semantic bases of some level-2 latent variables consist of fewer than 10 words. As a result, there are only 47 topics. In the second scenario, we set M=4. Here all the 140 topics are included. LDA was instructed to find 47 and 140 topics for the two scenarios respectively. HLDA produced 179 topics. For comparability, 47 and 140 topics were sampled for the two scenarios respectively.

The average coherence of the topics produced by the three methods are given in Table 7. we see that the value for HLTA is significantly higher than those for LDA and HLDA in both scenarios. The statistics suggest that HLTA has found, on average, thematically more coherent topics than LDA and HLDA.

Table 7. Average topic coherence for topics found by LDA, HLDA and HLTA on NIPS data

	M	NUMBER OF TOPICS	AVG. COHERENCE
HLTA(L3-L4)	10	47	-47.26
LDA	10	47	-55.38
HLDA-S	10	47	-62.83
HLDA	10	179	-63.32
HLTA(L2-L3-L4)	4	140	-5.89
LDA	4	140	-7.81
HLDA-S	4	140	-7.97
HLDA	4	179	-7.98

4.2 Likelihood Comparison

Having compared HLTA, LDA and HLDA in terms of the topics they produce, we next compare them as methods for text modeling. The comparison is in terms of per-document held-out log likelihood. For compatibility, LDA was run on both the count data and the binary version data. The results on the count data are denoted as LDA-C-100. For the binary data, several possibilities were tried for the number of topics, namely 20, 40, 60 and 80. For HLDA, the hyperparameters are set according to the settings used in [2]. For HLTA, three possibilities were tried for the UD-test threshold δ, namely 1, 3 and 5, which are suggested by [10].

Table 8. Per-document held-out log likelihood for HLTA, HLDA and LDA on test data. Results are averaged over five-fold cross validation.

	JACM	NIPS	NEWSGROUP
HLTA-1	-226±6	-394± 3	-113±1
HLTA-3	-226±6	-394± 3	-113±1
HLTA-5	-226±6	-394± 3	-113±1
HLDA	-220±39	-529±23	-98±8
LDA-20	-489±20	-1229±18	-197±2
LDA-40	-498±20	-1240±17	-199±2
LDA-60	-505±20	-1250±18	-199±2
LDA-80	-510±21	-1257±18	-199±2
LDA-C-100	-819±37	-3413± 35	-289±6

The results are given in Table 8. They show that HTLA and HLDA performed much better than LDA on all the data sets in the sense that the models they produced are much better in predicting unseen data than those obtained by LDA. The differences become larger when LDA was run on count data (last row of Table 8) rather than binary data. HLTA is better than HLDA on the NIPS data. However, HLDA is slightly better than HLTA on the other two data sets. In terms of running time, HLDA and HLTA are significantly slower than LDA. For example, for the binary version NIPS data, LDA took about 3.5 hours, while HLTA and HLDA took about 17 hours and 68 hours respectively.

5   Related Work

There are some other methods for learning latent tree models. We refer the readers to [15] for a detailed survey. Most of the methods are designed for density estimation [6], latent structure discovery [5] and multi-dimensional clustering [4]. None of these methods are designed for topic detection. More importantly, these methods do not provide a principled way to extract the topics from the model when they are applied on text data.

HLTM also resembles hierarchical clustering. However, there are fundamental differences between the two models. First, an HLTM is a probabilistic graphical model which allows inference among variables, while the structure given by traditional hierarchical clustering is not. Second, an HLTM can be also seen as a clustering tool which clusters the data points and variables simultaneously, while hierarchical clustering can only be used to cluster either data points or variables.

6   Conclusions and Future Directions

We propose a new method called HLTA for topic detection. The idea is to model patterns of word co-occurrence and co-occurrence of those patterns using a hierarchical latent tree model. Each latent variable in HLTM represents a soft partition of documents. The document clusters in each partition are interpreted as topics. Each topic is characterized using the words that occur with high probability in documents belonging to the topic and occur with low probability in documents not belonging to the topic. Empirical results indicate that HLTA can identify rich and thematically meaningful topics of various generality. In addition, HLTA can determine the number of topics automatically and organize topics into a hierarchy. Currently, HLTA treats words as binary variables. One future direction is to extend it so that it can handle count data. A second direction is to develop faster algorithms for learning hierarchical latent tree models.

Acknowledgments. We thank the anonymous reviewers for their valuable comments. Research on this paper was supported by Guangzhou HKUST Fok Ying Tung Research Institute.

References

1. Bartholomew, D., Knott, M., Moustaki, I.: Latent Variable Models and Factor Analysis. A Unified Approach. John Wiley & Sons (2011)
2. Blei, D., Griffiths, T., Jordan, M.: The nested chinese restaurant process and bayesian nonparametric inference of topic hierarchies. J. ACM 57(2), 7:1–7:30 (2010)
3. Blei, D., Ng, A., Jordan, M.: Latent dirichlet allocation. the Journal of Machine Learning Research 3, 993–1022 (2003)
4. Chen, T., Zhang, N.L., Liu, T.F., Poon, K.M., Wang, Y.: Model-based multidimensional clustering of categorical data. Artif. Intell. 176(1), 2246–2269 (2012)
5. Chen, T., Zhang, N.L., Wang, Y.: Efficient model evaluation in the search-based approach to latent structure discovery. In: 4th European Workshop on Probabilistic Graphical Models, pp. 57–64 (2008)
6. Choi, N.J., Tan, V.Y.F., Anandkumar, A., Willsky, A.: Learning latent tree graphical models. Journal of Machine Learning Research 12, 1771–1812 (2011)
7. Chow, C.K., Liu, C.N.: Approximating discrete probability distributions with dependence trees. IEEE Transactions on Information Theory 14(3), 462–467 (1968)
8. Cover, T., Thomas, J.: Elements of Information Theory. Wiley-Interscience (2006)
9. Dempster, A., Laird, N., Rubin, D.: Maximum likelihood from incomplete data via the em algorithm. J. Royal Statistical Society, Series B 39(1), 1–38 (1977)
10. Kass, R., Raftery, A.: Bayes factor. Journal of American Statistical Association 90(430), 773–795 (1995)
11. Lafferty, J., Blei, D.: Correlated topic models. In: NIPS, pp. 147–155 (2005)
12. Liu, T.F., Zhang, N.L., Chen, P., Liu, H., Poon, K.M., Wang, Y.: Greedy learning of latent tree models for multidimensional clustering. Machine Learning (2013), doi: 10.1007/s10994-013-5393-0
13. Liu, T.F., Zhang, N.L., Poon, K.M., Liu, H., Wang, Y.: A novel ltm-based method for multi-partition clustering. In: 6th European Workshop on Probabilistic Graphical Models, pp. 203–210 (2012)
14. Mimno, D., Wallach, H.M., Talley, E., Leenders, M., McCallum, A.: Optimizing semantic coherence in topic models. In: EMNLP, pp. 262–272 (2011)
15. Mourad, R., Sinoquet, C., Zhang, N.L., Liu, T.F., Leray, P.: A survey on latent tree models and applications. J. Artif. Intell. Res. (JAIR) 47, 157–203 (2013)
16. Pearl, J.: Probabilistic Reasoning in Intelligent Systems: Networks of Plausible Inference. Morgan Kaufmann Publishers Inc. (1988)

17. Schwarz, G.: Estimating the dimension of a model. The Annals of Statistics 6, 461–464 (1978)
18. Shafer, G., Shenoy, P.: Probability propagation. Annals of Mathematics and Artificial Intelligence 2(1-4), 327–351 (1990)
19. Zhang, N.L.: Hierarchical latent class models for cluster analysis. Journal of Machine Learning Research 5(6), 697–723 (2004)

Bayesian Models for Structured Sparse Estimation via Set Cover Prior

Xianghang Liu[1,2], Xinhua Zhang[1,3], and Tibério Caetano[1,2,3]

[1] Machine Learning Research Group, National ICT Australia, Sydney and Canberra, Australia
[2] School of Computer Science and Engineering, The University of New South Wales, Australia
[3] Research School of Computer Science, The Australian National University, Australia
{xianghang.liu,xinhua.zhang,tiberio.caetano}@nicta.com.au

Abstract. A number of priors have been recently developed for Bayesian estimation of sparse models. In many applications the variables are simultaneously relevant or irrelevant in groups, and appropriately modeling this correlation is important for improved sample efficiency. Although group sparse priors are also available, most of them are either limited to disjoint groups, or do not infer sparsity at group level, or fail to induce appropriate patterns of support in the posterior. In this paper we tackle this problem by proposing a new framework of prior for overlapped group sparsity. It follows a hierarchical generation from group to variable, allowing group-driven shrinkage and relevance inference. It is also connected with set cover complexity in its maximum a posterior. Analysis on shrinkage profile and conditional dependency unravels favorable statistical behavior compared with existing priors. Experimental results also demonstrate its superior performance in sparse recovery and compressive sensing.

1 Introduction

Sparsity is an important concept in high-dimensional statistics [1] and signal processing [2] that has led to important application successes. It reduces model complexity and improves interpretability of the result, which is critical when the number of explanatory variables p in the problem is much higher than the number of training instances n.

From a Bayesian perspective, the sparsity of a variable β_i is generally achieved by shrinkage priors, which often take the form of scale mixture of Gaussians: $\beta_i | z_i \sim \mathcal{N}(0, z_i)$. z_i indicates the relevance of β_i, and a broad range of priors on z_i has been proposed. For example, the spike and slab prior [3, 4] uses a Bernoulli variable for z_i, which allows β_i to be exactly zero with a positive probability. Absolutely continuous alternatives also abound [5], *e.g.*, the horseshoe prior [6, 7] which uses half-Cauchy on z_i and offers robust shrinkage in the posterior. Interestingly, the maximum a posterior (MAP) inference often corresponds to deterministic models based on sparsity-inducing regularizers, *e.g.* Lasso [8] when z_i has a gamma distribution [9, 10]. In general, the log-posterior can be non-concave [11, 12].

However, many applications often exhibit additional structures (correlations) in variables rather than being independent. Groups may exist such that variables of each group are known a priori to be jointly relevant or irrelevant for data generation. Encoding this

T. Calders et al. (Eds.): ECML PKDD 2014, Part II, LNCS 8725, pp. 273–289, 2014.
© Springer-Verlag Berlin Heidelberg 2014

knowledge in the prior proves crucial for improved accuracy of estimation. The simplest case is when the groups are disjoint, and they form a partition of the variable set. This allows the relevance indicator z_i of all variables in each group to be tied, forming a group indicator which is endowed with a zero-centered prior as above [13, 14]. In particular, a gamma prior now yields the Bayesian group Lasso [15], and its MAP is the group Lasso [16] which allows group information to *provably* improve sample efficiency [17]. More refined modeling on the sparsity within each group has also been explored [18, 19]. We overview the related background in Section 2.

However, groups do overlap in many practical applications, *e.g.* gene regulatory network in gene expression data [20], and spatial consistency in images [21]. Techniques that deal with this scenario start to diverge. A commonly used class of method employs a Markov random field (MRF) to enforce smoothness over the relevance indicator of all variables within each group [22–24]. However, this approach does not infer relevance at the group level, and does not induce group-driven shrinkage.

Another popular method is to directly use the Bayesian group Lasso, despite the loss of hierarchical generative interpretation due to the overlap. Its MAP inference has also led to a rich variety of regularizers that promote structured sparsity [21, 25], although statistical justification for the benefit of using groups is no longer rich and solid. Moreover, Bayesian group Lasso tends to shrink a whole group based on a complexity score computed from its constituent variables. So the support of the posterior β tends to be the complement of the union of groups, rather than the union of groups as preferred by many applications.

To address these issues, we propose in Section 3 a hierarchical model by placing relevance priors on groups only, while the variable relevance is derived (probabilistically) from the set of groups that involve it. This allows direct inference of group relevance, and is amenable to the further incorporation of hierarchies among groups. All previously studied sparsity-inducing priors on relevance variables can also be adopted naturally, leading to a rich family of structured sparse prior. The MAP of our model turns out exactly the set cover complexity, which provably reduces sample complexity for *overlapped* groups [26].

Although in appearance our model simply reverses the implication of relevance in Bayesian group Lasso, it amounts to considerably more desirable shrinkage profile [7]. In Section 4, detailed analysis based on horseshoe prior reveals that set cover priors retain the horseshoe property in its posterior, shrinking reasonably for small response and diminishing when response grows. Surprisingly, these properties are not preserved by the other structured alternatives. Also observed in set cover prior is the favorable conditional dependency between relevance variables, which allows them to "explain-away" each other through the overlap of two groups they each belong to. Experimental results in Section 5 confirm that compared with state-of-the-art structured priors, the proposed set cover prior outperforms in sparse recovery and compressive sensing on both synthetic data and real image processing datasets.

Note different from [27] and [28], we do not introduce regression variables that account for interactions between features, *i.e.* β_{ij} for $x_i x_j$.

2 Preliminaries on Sparse Priors

In a typical setting of machine learning, we are given n training examples $\{\mathbf{x}_i, y_i\}_{i=1}^n$, where $\mathbf{x}_i \in \mathbb{R}^p$ represents a vector of p features/variables, and y_i is the response that takes value in $\mathbb{R}$ for regression, and in $\{-1, 1\}$ for classification. Our goal is to learn a linear model $\boldsymbol{\beta} \in \mathbb{R}^p$, or a distribution of $\boldsymbol{\beta}$, such that $\mathbf{x}_i'\boldsymbol{\beta}$ agrees with y_i. This problem is usually ill-posed, especially when $p \gg n$ as considered in this work. Therefore prior assumptions are required and here we consider a popular prior that presumes $\boldsymbol{\beta}$ is sparse. In Bayesian methods, the compatibility between y_i and $\mathbf{x}_i'\boldsymbol{\beta}$ is enforced by a likelihood function, which is typically normal for regression (*i.e.*, $y|\mathbf{x}, \boldsymbol{\beta} \sim \mathcal{N}(\mathbf{x}'\boldsymbol{\beta}, \sigma^2)$), and Bernoulli for classification. σ is a pre-specified constant.

The simplest form of sparsity is enforced on each element of $\boldsymbol{\beta}$ independently through priors on β_i. Most existing models use a scalar mixture of normals that correspond to the graphical model $z_i \to \beta_i$ [27, 29, 30]:

$$\pi(\beta_i) = \int \mathcal{N}(\beta_i; 0, \sigma_0^2 z_i) f(z_i) \mathrm{d}z_i. \tag{1}$$

Here σ_0^2 can be a constant, or endowed with a prior. Key to the model is the latent conditional variance z_i, which is often interpreted as *relevance* of the variable β_i. Larger z_i allows β_i to take larger absolute value, and by varying the mixing distribution f of z_i we obtain a range of priors on $\boldsymbol{\beta}$, differing in shrinkage profile and tail behavior. For example, the spike and slab sparse prior [3, 4] adopts

$$f_{\mathrm{SS}}(z_i) = p_0 \delta(z_i - 1) + (1 - p_0) \delta(z_i), \tag{2}$$

where δ is the Dirac impulse function and p_0 is the prior probability that β_i is included. Absolutely continuous distributions of z_i are also commonly used. An inverse gamma distribution on z_i leads to the Student-t prior, and automatic relevance determination [ARD, 9] employs $f(z_i) \propto z_i^{-1}$. Indeed, a number of sparsity-inducing priors can be unified using the generalized beta mixture [5, 31]:

$$z_i | \lambda_i \sim \mathrm{Ga}(a, \lambda_i), \quad \text{and} \quad \lambda_i \sim \mathrm{Ga}(b, d). \tag{3}$$

Here Ga stands for the gamma distribution with shape and rate (*inverse* scale) parameters. In fact, z_i follows the generalized beta distribution of the second kind:

$$\mathrm{GB2}(z_i; 1, d, a, b) = z_i^{a-1}(1 + z_i/d)^{-a-b} d^{-a} / B(a, b), \tag{4}$$

where $B(a, b)$ is the beta function. When $a = b = \frac{1}{2}$, it yields the horseshoe prior on $\boldsymbol{\beta}$ [6]. The normal-exponential-gamma prior and normal-gamma prior [32] can be recovered by setting $a = 1$ and $b = d \to \infty$ respectively. In the intersection of these two settings is the Bayesian Lasso: $\pi(\boldsymbol{\beta}) \sim \exp(-\|\boldsymbol{\beta}\|_1)$ [10], where $\|\boldsymbol{\beta}\|_p := (\sum_i |\beta_i|^p)^{\frac{1}{p}}$ for $p \geq 1$.

To lighten notation, in the case of spike and slab we will also use z_i to represent Bernoulli variables valued in $\{0, 1\}$. So integrating over $z_i \geq 0$ with respect to the density in (2) can be interpreted as weighted sum over $z_i \in \{0, 1\}$.

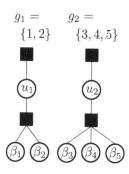

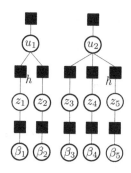

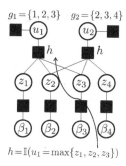

Fig. 1. Group spike and slab

Fig. 2. Nested spike and slab

Fig. 3. Group counting prior for spike and slab

2.1 Disjoint Groups

In many applications, prior knowledge is available that the variables can be partitioned into *disjoint* groups $g_i \subseteq [p] := \{1, \ldots, p\}$, and all variables in a group tend to be positively correlated, *i.e.* relevant or irrelevant simultaneously. Denote $\mathcal{G} = \{g_1, g_2, \ldots\}$. [13] generalized the spike and slab prior to this scenario by introducing a scalar parameter of relevance for each group: $u_g \sim f_{SS}$, and extending (1) into a scalar mixture of *multivariate* normal

$$\pi(\boldsymbol{\beta}_g) = \int \mathcal{N}(\boldsymbol{\beta}_g; \mathbf{0}, \boldsymbol{\Lambda}_g u_g) f(u_g) \mathrm{d}u_g \ \forall\, g \in \mathcal{G}. \tag{5}$$

Here $\boldsymbol{\beta}_g$ encompasses all variables in the group g, and $\boldsymbol{\Lambda}_g$ is a diagonal matrix of variance. See Figure 1 for the factor graph representation that will facilitate a unified treatment of other models below. As a result, *correlation* is introduced among all variables in each group. Using exactly the same density f as above (but on u_g here), one may recover the group horseshoe, group ARD [14], and Bayesian group Lasso [15]:

$$\pi(\boldsymbol{\beta}) \propto \exp(-\|\boldsymbol{\beta}\|_{\mathcal{G}}), \text{ where } \|\boldsymbol{\beta}\|_{\mathcal{G}} = \sum_g \|\boldsymbol{\beta}_g\|_p. \tag{6}$$

Common choices of p are 2 and ∞. To further model the sparsity of different variables within a group, [18] proposed a nested spike and slab model as shown in Figure 2. The key idea is to employ *both* Bernoulli variables z_i and u_g that encode the relevance of variables and groups respectively, and to define the spike and slab distribution of β_i conditional on $u_g = 1$. In particular, z_i must be 0 if $u_g = 0$, *i.e.* group g is excluded. This relation is encoded by a factor between z_i and u_g:

$$h(z_i, u_g) = \begin{cases} p_0^{z_i}(1-p_0)^{1-z_i} & \text{if } u_g = 1 \\ \mathbb{I}(z_i = 0) & \text{if } u_g = 0 \end{cases}, \ \forall i \in g. \tag{7}$$

Here $\mathbb{I}(\cdot) = 1$ if $\cdot$ is true, and 0 otherwise.

3   Structured Prior with Overlapped Groups

In many applications, groups may overlap and fully Bayesian treatments for this setting have become diverse.

Group Counting Prior (GCP). A straightforward approach is to ignore the fact of overlapping, and simply use the group Lasso prior in (6). This idea is also used in deterministic overlapped group Lasso [16]. When $p = \infty$, the norm in (6) is the Lovász extension of the group counting penalty [33] which, in the case of spike and slab prior on β_i, can be written in terms of the binary relevance indicator $\mathbf{z} := \{z_i\} \in \{0,1\}^p$

$$\Omega(\mathbf{z}) = \prod_{g \in \mathcal{G}} p_0^{u_g}(1 - p_0)^{1-u_g}, \quad \text{where } u_g = \max_{i:i \in g} z_i. \tag{8}$$

So a group is deemed as relevant ($u_g = 1$) if, and only if, any variable in the group is relevant ($z_i = 1$). The factor graph is given in Figure 3, with a Bernoulli potential on u_g. However, since this prior promotes u_g to be 0 (*i.e.* zero out all variables in the group g), the support of β in the posterior tends to be the complement of a union of groups. Although this may be appropriate for some applications, the support is often more likely to be the union of groups.

MRF Prior. Instead of excluding groups based on its norm, the MRF prior still places sparsity-inducing priors on each variable β_i, but further enforces *consistency* of relevance within each group via z_i. For example, assuming the variables are connected via an undirected graph where each edge $(i, j) \in E$ constitutes a group, [22, 34] extended the spike and slab prior by incorporating a pairwise MRF over the relevance indicators z_i: $\exp(- \sum_{(i,j) \in E} R_{ij} \mathbb{I}(z_i \neq z_j))$.

As a key drawback of the above two priors, they do not admit a generative hierarchy and perform no inference at the group level. To address these issues, we next construct a hierarchical generative model which explicitly characterizes the relevance of both groups and variables, as well as their conditional correlations.

3.1   Set Cover Prior (SCP)

To better clarify the idea, we first focus on spike and slab prior where sparsity can be easily modeled by Bernoulli variables z_i and u_g. Recall the nested model in Figure 2, where each group has a Bernoulli prior, and each variable z_i depends on the unique group that it belongs to. Now since multiple groups may be associated with each node, it will be natural to change the dependency into some arithmetics of these group indicators. In Figure 4, we show an example with[1]

$$h(z_i, \{u_g : i \in g\}) = \mathbb{I}(z_i \leq \max\{u_g : i \in g\}). \tag{9}$$

This means a variable can be relevant only if any group including it is also relevant. Although this appears simply reversing the implication relations between group and variable in the group counting prior, it does lead to a hierarchical model and enjoys much more desirable statistical properties as will be shown in Section 4.

---

[1] This defines a potential in an MRF; there is no explicit prior on z_i.

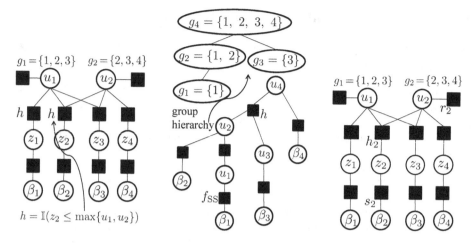

Fig. 4. Set cover prior for spike and slab

Fig. 5. Set cover prior for spike and slab with tree hierarchy. u_j corresponds to g_j. $h = \mathbb{I}(u_4 \geq \max\{u_2, u_3\})$.

Fig. 6. Set cover prior using horseshoe. $r_2 = \mathrm{Ga}(u_2; \frac{1}{2}, \frac{1}{2})$. $h_2 = \mathrm{Ga}(z_2; \frac{1}{2}, \max\{u_1, u_2\})$.

By endowing a Bernoulli prior on all u_g with $\Pr(u_g = 1) = p_0 < 0.5$ (*i.e.* favoring sparsity), we complete a generative prior of β in a spike and slab fashion. Given an assignment of $\mathbf{z}$, it is interesting to study the mode of $\{u_g\}$, which is the solution to

$$\min_{\{u_g\}} \sum_g u_g, \quad \text{s.t. } u_g \in \{0, 1\}, \quad \sum_{g:i\in g} u_g \geq z_i, \ \forall\, i. \tag{10}$$

This turns out to have exactly the same complexity as set cover [26]. It seeks the smallest number of groups such that their union covers the set of variables. Hence we will call this prior as "set cover prior". This optimization problem is NP-hard in general, and some benefit in sample complexity is established by [26].

A number of extensions follow directly. Additional priors (*e.g.* MRF) can be placed on variables z_i. The max in (9) can be replaced by min, meaning that a variable can be selected only if *all* groups involving it are selected. Other restrictions such as limiting the number of selected variables in each (selected) group can also be easily incorporated [35]. Moreover, groups can assume a hierarchical structure such as tree, *i.e.* $g \cap g' \in \{g, g', \emptyset\}$ for all g and g'. Here the assumption is that if a node g' is included, then all its ancestors $g \supset g'$ must be included as well [21, 36]. This can be effectively enforced by adding a factor h that involves each group g and its children $\mathrm{ch}(g)$ (see Figure 5):

$$h(g, \mathrm{ch}(g)) = \mathbb{I}(u_g \geq \max_{g'\in\mathrm{ch}(g)} u_{g'}). \tag{11}$$

When the groups are disjoint, both set cover and group counting priors are equivalent to group spike and slab.

3.2   Extension to Generalized Beta Mixture

The whole framework is readily extensible to the continuous sparse priors such as horseshoe and ARD. Using the interpretation of z_i and u_g as relevance measures, we could simply replace the function $\mathbb{I}$ that tests equality by the Dirac impulse function δ, and apply various types of continuous valued priors on z_i and u_g. This is indeed feasible for GCP, *e.g.* encode the continuous variant of (8) using the generalized beta mixture in (3)

$$h(u_g, \{z_i : i \in g\}) = \delta(u_g - \max\{z_i : i \in g\}), \quad h(u_g) = \mathrm{GB2}(u_g; 1, d, a, b). \quad (12)$$

Here more flexibility is available when z_i is continuous valued, because the max can be replaced by multiplication or summation, which promotes or suppresses sparsity respectively [27, Theorem 1, 2].

However problems arise in SCP if we directly use

$$z_i = \max_{g:i \in g} u_g \quad \text{or} \quad \min_{g:i \in g} u_g, \quad \text{where } u_g \sim \mathrm{GB2}(u_g; 1, d, a, b), \quad (13)$$

because it leads to singularities in the prior distribution on $\mathbf{z}$. To smooth the prior, we resort to arithmetic combinations of the *intermediate* variables in the generative process of the prior on u_g. Note that in (3), d is a scale parameter, while a and b control the behavior of the distribution of z_i close to zero and on the tail, respectively. A smaller value of λ_i places more probability around 0 in z_i, encouraging a sparser β_i. So a natural way to combine the group prior is:

$$z_i | \{u_g\} \sim \mathrm{Ga}(a, \max_{g:i \in g} u_g), \quad \text{and} \quad u_g \sim \mathrm{Ga}(b, d), \quad (14)$$

where max allows z_i to pick up the most sparse tendency encoded in all u_g of the associated groups[2]. Changing it to min leads to adopting the least sparse one. The resulting graphical model is given in Figure 6. Here u_g has a gamma distribution, playing the same role of relevance measure as in the normal-gamma prior on β_i [32]. The SCP constructed in (14) is no longer equivalent to the group priors in Section 2.1, even when the groups are disjoint.

In fact, the arithmetics that combine multiple groups can be carried out at an even higher level of the generative hierarchy. For example, in the horseshoe prior where $a = b = 1/2$, one may introduce an additional layer of mixing over the scale parameter d, making it an arithmetic combination of u_g of the associated groups. We leave this possibility for future exploration.

Notice [38] used a partial least squares approach based on an MRF of binary selectors of groups and variables. However their method is confined to spike and slab, because these two groups of indicators are *not* coupled by the potential function, but by imposing external restrictions on the admissible joint assignment that is valued in $\{0, 1\}$. It also brings much challenge in MCMC inference.

4   Analysis of Structured Sparse Prior

Although the above three types of priors for structured sparsity appear plausible, their statistical properties differ significantly as we study in this section. Here in addition to the robust shrinkage profile studied by [6], we also compare the conditional correlation among variables when the groups overlap.

---

[2] See more detailed discussions in Appendix A of the full paper [37] on how a greater value of the second argument (rate, *i.e.* inverse scale) of a Gamma distribution induces higher sparsity in β_i.

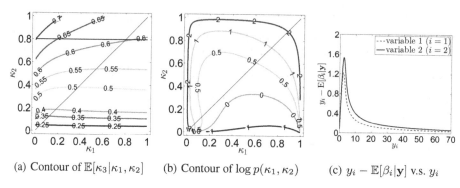

(a) Contour of $\mathbb{E}[\kappa_3|\kappa_1,\kappa_2]$ ǃǃ (b) Contour of $\log p(\kappa_1,\kappa_2)$ ǃǃ (c) $y_i - \mathbb{E}[\beta_i|\mathbf{y}]$ v.s. y_i

Fig. 7. Set cover prior. The contour levels in panel (b) are: $-1, 0, 0.5, 1, 2$.

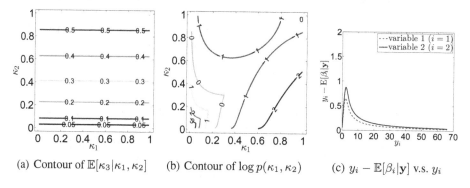

(a) Contour of $\mathbb{E}[\kappa_3|\kappa_1,\kappa_2]$ ǃǃ (b) Contour of $\log p(\kappa_1,\kappa_2)$ ǃǃ (c) $y_i - \mathbb{E}[\beta_i|\mathbf{y}]$ v.s. y_i

Fig. 8. Group counting prior. The contour levels in panel (b) are: $-2, -1, 0, 1, 2, 3$.

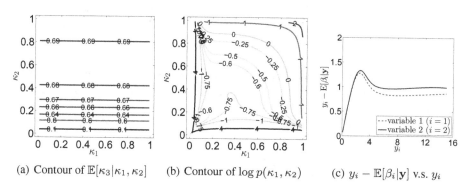

(a) Contour of $\mathbb{E}[\kappa_3|\kappa_1,\kappa_2]$ ǃǃ (b) Contour of $\log p(\kappa_1,\kappa_2)$ ǃǃ (c) $y_i - \mathbb{E}[\beta_i|\mathbf{y}]$ v.s. y_i

Fig. 9. MRF prior ($\alpha = 0.01$). The contour levels in panel (b) are $-4, -1, -0.75, \ldots, 0, 1, 2$.

Consider $p = 3$ variables, and there are two groups $\{1, 2\}$ and $\{2, 3\}$ which overlap on variable 2. The design matrix X is the 3×3 identity matrix I ($n = 3$), and the observation $\mathbf{y}|\beta \sim \mathcal{N}(X\beta, I) = \mathcal{N}(\beta, I)$. Let $\sigma_0 = 1$. Then the expected posterior value of β given $\mathbf{z}$ has a closed form $\mathbb{E}[\beta_i|z_i, y_i] = (1-\kappa_i)y_i$ where $\kappa_i = 1/(1+z_i)$ is a random shrinkage coefficient. The distribution of κ_i is determined entirely by the prior on z_i, and a larger value of κ_i means a greater amount of shrinkage towards the origin.

As a concrete example, we study the horseshoe prior with $a = b = d = 1/2$. GCP and SCP use the formulae (12) and (14), respectively. The MRF prior attaches a horseshoe potential on each β_i, and in addition employs a smooth MRF $\exp(-\alpha(z_1 - z_2)^2 - \alpha(z_2 - z_3)^2)$ with $\alpha = 0.01$. We use a Gaussian MRF because there is no need of shrinking the difference.

4.1 Conditional Dependency (Explain-Away Effect)

We first consider the conditional distribution of κ_3 given κ_1 and κ_2. Since it is hard to visualize a function of three variables, we show in panel (a) of Figures 7 to 9 the mean $\mathbb{E}[\kappa_3|\kappa_2, \kappa_1]$ under the three priors. Clearly the mean of κ_3 does not change with κ_1 in GCP and MRF prior, because z_3 is simply independent of z_1 given z_2. The mean of κ_3 grows monotonically with κ_2, as MRF favors small difference between z_2 and z_3 (hence between κ_2 and κ_3), and in SCP smaller κ_2 clamps a larger value of $\max\{z_2, z_3\}$, shifting more probability mass towards greater z_3 which results in a lower mean of κ_3.

Interestingly when κ_2 is large, the SCP allows the mean of κ_3 to decrease when κ_1 grows. See, e.g., the horizonal line at $\kappa_2 = 0.8$ in Figure 7a. To interpret this "explain-away" effect, first note a greater value of κ_2 means z_2 has a higher inverse scale. Due to the max in (14), it implies that either u_1 or u_2 is large, which means κ_1 or κ_3 is large since variables 1 and 3 belong to a single group only. Thus when κ_1 is small, κ_3 receives more incentive to be large, while this pressure is mitigated when κ_1 itself increases. On the other hand when κ_2 is small, κ_1 and κ_3 must be both small, leading to the flat contour lines.

4.2 Joint Shrinkage

Next we study the joint density of κ_1 and κ_2 plotted in panel (b). SCP exhibits a 2-D horseshoe shaped joint density in Figure 7b, which is desirable as it prefers either large shrinkage or little shrinkage. In GCP, however, the joint density of (κ_1, κ_2) concentrates around the origin in Figure 8b. Indeed, this issue arises even when there are only two variables, making a single group. Fixing κ_2 and hence z_2, $\max\{z_1, z_2\}$ does not approach 0 even when z_1 approaches 0 (i.e. κ_1 approaches 1). So it is unable to exploit the sparsity-inducing property of horseshoe prior which places a sharply growing density towards the origin. The joint density of MRF is low when κ_1 and κ_2 are both around the origin, although the marginal density of each of them seems still high around zero.

4.3 Robust Marginal Shrinkage

Finally we investigate the shrinkage profile via the posterior mean $\mathbb{E}[\boldsymbol{\beta}|\mathbf{y}]$, with $\mathbf{z}$ integrated out. Let $q(\mathbf{z})$ be proportional to the prior density on $\mathbf{z}$ (note the group counting and MRF priors need a normalizer). Then $\mathbb{E}[\beta_i|\mathbf{y}] = \gamma_i^{(1)}/\gamma_i^{(0)}$, where for $k \in \{0, 1\}$

$$\gamma_i^{(k)} = \int \beta_i^k q(\mathbf{z}) \prod_j \mathcal{N}(\beta_j; 0, z_j)\mathcal{N}(y_j; \beta_j, 1)\mathrm{d}\beta_j \mathrm{d}\mathbf{z}, \tag{15}$$

and $\int \beta_j^k \mathcal{N}(\beta_j; 0, z_j)\mathcal{N}(y_j; \beta_j, 1)\mathrm{d}\beta_j = \sqrt{\dfrac{1}{8\pi}} \dfrac{z_j^k}{(1 + z_j)^{k+\frac{1}{2}}} \exp\left(\dfrac{-y_j^2}{2 + 2z_j}\right).$ (16)

Panel (c) of Figures 7 to 9 shows $y_i - \mathbb{E}[\beta_i|y_i]$ (the amount of shrinkage) as a function of y_i, for variables $i \in \{1, 2\}$. All y_j $(j \neq i)$ are fixed to 1. In Figure 7c, Both SCP and GCP provide valuable robust shrinkage, with reasonable shrinkage when y_i is small in magnitude, and diminishes as y_i grows. And as expected, variable 2 shrinks more than variable 1. In SCP, variable 2 takes the sparser state between variables 1 and 3 via the max in (14), while in GCP variable 2 contributes to both sparsity-inducing priors of u_1 and u_2 in (12). Notice that for small y_1, GCP is not able to yield as much shrinkage as SCP. This is because for small y_1, z_1 is believed to be small, and hence the value of $\max\{z_1, z_2\}$ is dominated by the belief of z_2 (which is larger). This prevents z_1 from utilizing the horseshoe prior around zero. The case for y_2 is similar.

In fact, we can theoretically establish the robust shrinkage of SCP for any group structure under the current likelihood $\mathbf{y}|\boldsymbol{\beta} \sim \mathcal{N}(\boldsymbol{\beta}, I)$.

Theorem 1. *Suppose SCP uses horseshoe prior in (14) with $a = b = 1/2$. Then for any group structure,* $\lim_{y_i \to \infty}(y_i - \mathbb{E}[\beta_i|\mathbf{y}]) = 0$ *with fixed values of $\{y_j : j \neq i\}$.*

Proof. (sketch) The key observation based on (15) and (16) is that $\mathbb{E}[\beta_i|\mathbf{y}] - y_i = \frac{\partial}{\partial y_i} \log F(\mathbf{y})$ where $F(\mathbf{y})$ is given by

$$\int_{\mathbf{z}} \prod_j (1 + z_j)^{\frac{-1}{2}} \exp\left(\frac{-y_j^2}{2 + 2z_j}\right) q(\mathbf{z})\mathrm{d}\mathbf{z} \tag{17}$$

$$= \int_{\mathbf{u}} \prod_j \left(\int_{z_j} (1 + z_j)^{\frac{-1}{2}} \exp\left(\frac{-y_j^2}{2 + 2z_j}\right) \mathrm{Ga}(z_j; a, \max_{g:j \in g} u_g)\mathrm{d}z_j\right) \prod_g \mathrm{Ga}(u_g; b, d)\mathrm{d}\mathbf{u}$$

The rest of the proof is analogous to [6, Theorem 3]. The detailed proof is provided in Appendix B of the longer version of the paper [37].   □

By contrast, MRF is unable to drive down the amount of shrinkage when the response y_i is large. To see the reason (*e.g.* for variable 1), note we fix y_2 to 1. Since MRF enforces smoothness between z_1 and z_2, the fixed value of y_2 (hence its associated belief of z_2) will prevent z_1 to follow the increment of y_1, disallowing z_1 to utilize the heavy tail of horseshoe prior. The amount of shrinkage gets larger when α increases.

To summarize, among the three priors only the set cover prior enjoys all the three desirable properties namely conditional dependency, significant shrinkage for small observation, and vanishing shrinkage for large observations.

5   Experimental Results

We next study the empirical performance of SCP, compared with GCP, and MRF priors [34]. Since the MRF prior therein is restricted to spike and slab, to simplify comparison

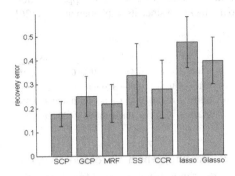

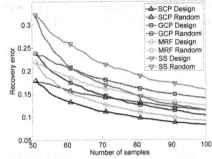

Fig. 10. Recovery rate for sparse signal   **Fig. 11.** Sequential experimental design for sparse recovery

we also base SCP and GCP on spike and slab. This allows convenient application of expectation propagation for posterior inference [EP, 39, 40], where all discrete factors are approximated by Bernoulli messages [34]. At each iteration, messages are passed from top to the bottom in Figure 4, and back up. Other inference algorithms are also possible, such as MCMC [*e.g.*, 38], and variational Bayes [41]. Since the Bayesian models used here are typically multi-modal and the mean of the posterior is generally more important, we choose to use EP in our experiment, although it will also be interesting to try other methods.

Empirically EP always converged within 10 iterations, with change of message fallen below 1e-4. The loops make it hard to analyze the local or global optimality of EP result. But in practice, we did observe that with different initializations, EP always converged to the same result on all experiments, being highly reliable. To give an example of computational efficiency, in image denoising (Section 5.5, $p = n = 4096$), it took only 0.5 seconds per image and per iteration to compute messages related to the prior, while common techniques for Gaussian likelihood allowed its related messages to be computed in 1-2 seconds.

As a baseline, we also tried spike and slab prior with non-overlapping groups (GSS) if reasonable non-overlapping group approximation is available, or even without groups (SS). Furthermore we consider three state-of-the-art frequentist methods, including Lasso, group Lasso (GLasso), and coding complexity regularization [CCR, 26]. Groups are assumed available as prior knowledge.

5.1 Sparse Signal Recovery

We first consider a synthetic dataset for sparse signal reconstruction with $p = 82$ variables [42]. $\{\beta_i\}$ was covered by 10 groups of 10 variables, with an overlap of two variables between two successive groups: $\{1, \ldots, 10\}, \{9, \ldots, 18\}, \ldots, \{73, \ldots, 82\}$. The support of β was chosen to be the union of group 4 and 5, with the non-zero entries generated from *i.i.d.* Gaussian $\mathcal{N}(0, 1)$. We used $n = 50$ samples, with the elements of the design matrix $X \in \mathbb{R}^{n \times p}$ and the noisy measurements $\mathbf{y}$ drawn by

$$X_{ij} \overset{i.i.d.}{\sim} \mathcal{N}(0, 1), \quad \mathbf{y} = X\beta + \epsilon, \quad \epsilon_i \overset{i.i.d.}{\sim} \mathcal{N}(0, 1). \tag{18}$$

Table 1. Recovery error for network sparsity. GSS is not included as disjoint group approximation is not clear for general graph structure. CCR is also not included since its implementation in [26] does not allow flexible specification of groups.

	SCP	GCP	MRF	SS	Lasso	GLasso
Jazz	**0.264** ± 0.083	0.312 ± 0.068	0.338 ± 0.149	0.398 ± 0.188	0.489 ± 0.101	0.456 ± 0.107
NetScience	**0.067** ± 0.005	0.093 ± 0.058	0.167 ± 0.110	0.188 ± 0.113	0.394 ± 0.045	0.383 ± 0.048
Email	**0.106** ± 0.025	**0.104** ± 0.054	0.243 ± 0.105	0.310 ± 0.130	0.432 ± 0.049	0.420 ± 0.057
C.elegans	**0.158** ± 0.034	0.163 ± 0.025	0.184 ± 0.057	0.225 ± 0.101	0.408 ± 0.068	0.394 ± 0.068

We used recovery error as the performance measure, which is defined as $||\hat{\beta} - \beta||_2/||\beta||_2$ for the posterior mean $\hat{\beta}$. X and β were randomly generated for 100 times, and we report the mean and standard deviation of recovery error. An extra 10 runs were taken to allow all models to select the hyper-parameters that optimize the performance on the 10 runs. This scheme is also used in subsequent experiments.

In Figure 10, SCP clearly achieves significantly lower recovery error than all other methods. MRF is the second best, followed by GCP. This suggests that when β is generated over the union of some groups, SCP is indeed most effective in harnessing this knowledge. Bayesian models for structured sparse estimation also outperform vanilla Bayesian models for independent variables (SS), as well as frequentist methods (CCR, Lasso, GLasso),

5.2   Sequential Experimental Design

A key advantage of Bayesian model is the availability of uncertainty estimation which facilitates efficient sequential experimental design [13]. We randomly generated a data pool of 10,000 examples based on (18), and initialized the training set with $n = 50$ randomly selected examples (*i.e.* revealing their response y_i). Then we gradually increased the size of training set up to $n = 100$. At each iteration, one example was selected and its response y_i was revealed for training. In the random setting examples were selected uniformly at random, while in sequential experimental design, typically the example with the highest uncertainty was selected. For each candidate example $\mathbf{x}$, we used $\mathbf{x}'V\mathbf{x}$ as the uncertainty measure, where V is the current approximated posterior covariance matrix. The whole experiment was again repeated for 100 times, and the average recovery error is shown.

In Figure 11, for all models sequential experimental design is significantly more efficient in reducing the recovery error compared with random design. In particular, SCP achieves the steepest descent in error with respect to the number of measurements. This again confirms the superiority of SCP in modeling group structured sparsity in comparison to GCP and MRF. SS performs worst as it completely ignores the structure.

5.3   Network Sparsity

Following [34] and [43], we next investigate the network sparsity where each node is a variable and each edge constitutes a group (*i.e.* all groups have size 2). We tried on four network structures: Email ($p = 1,133$, #edge=5,451), C.elegans (453, 2,015),

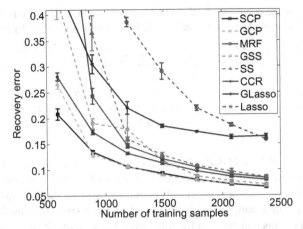

Fig. 12. Recovery error for background subtraction

Jazz (198, 2,742), NetScience (1,589, 2,742).[3] See network properties in Table 1. We picked a subset of edges uniformly at random, and added their two incident nodes to the support of β. By adjusting the number of selected edges, the size of the support of β is $0.25p$, and the nonzero elements in β were sampled from $\mathcal{N}(0,1)$. The design matrix X and the response y were drawn from (18). We used $n = \lfloor p/2 \rfloor$ examples as in [43].

The average recovery error of 100 runs is shown in Table 1. Again SCP yields significantly lower error than all other algorithms, except for a tie with GCP on Email. GCP outperforms MRF, which in turn defeats all other methods that do not faithfully model the group structure.

5.4 Background Subtraction

We next consider real-world applications in compressive sensing with overlapped group sparsity. Here the data generating process is beyond our control. In video surveillance, the typical configuration of the images are the sparse foreground objects on the static backgrounds. Our task here is to recover the sparse background subtracted images via compressive sensing.

Our experimental setting follows [26][4]. The spatial consistency is an important prior knowledge on 2D image signals which has been successfully leveraged in various applications. Specifically pixels in a spatial neighborhood are likely to be background or foreground at the same time. Edges connecting pixels to its four neighbors are used in the MRF prior to encourage the consistency between adjacent pixels. For GSS which requires no overlap between groups, we simply defined the groups as non-overlapped 3×3 patches. For the rest structured priors, we defined groups as the overlapped 3×3 patches. Singleton groups were also added to deal with isolated foreground pixels. Each image is sized 80×80 ($p = 6,400$). We varied the number of image (n) from 600 to 2400.

---

[3] Downloaded from http://www-personal.umich.edu/~mejn/netdata

[4] Video from http://homepages.inf.ed.ac.uk/rbf/CAVIARDATA1

Table 2. PSNR in image denoising. MRF and GSS are not included because in the case of hierarchical structure, it is not clear how to enforce MRF, or approximate a tree by disjoint groups.

	SCP	GCP	SS	Lasso	GLasso	CCR
House	**28.77** ± 0.04	28.13 ± 0.06	27.72 ± 0.06	27.22 ± 0.02	27.24 ± 0.03	27.79 ± 0.04
Lenna	**28.27** ± 0.03	27.65 ± 0.02	27.28 ± 0.02	26.95 ± 0.03	27.15 ± 0.01	27.11 ± 0.02
pepperr	**26.57** ± 0.03	25.87 ± 0.01	25.75 ± 0.03	25.06 ± 0.05	25.39 ± 0.06	25.51 ± 0.04
Boat	**26.80** ± 0.01	26.24 ± 0.01	26.09 ± 0.01	25.65 ± 0.01	26.05 ± 0.02	25.65 ± 0.01
Barbara	**24.93** ± 0.02	24.56 ± 0.02	24.43 ± 0.02	24.23 ± 0.01	24.77 ± 0.02	24.34 ± 0.01

Figure 12 shows SCP and GCP achieve significantly lower recovery error than other methods on any number of measurement. The prior of spatial consistency does help improve the recovery accuracy, especially when the size of the training set is small. With sufficient training samples, both structured and non-structured methods can have accurate recovery. This can be seen by comparing Lasso with GLasso, as well as SCP with GCP, GSS, and SS. The superiority of SCP and GCP over GSS corroborates the importance of accommodating more flexible group definitions.

5.5 Image Denoising with Tree-Structured Wavelets

Our last set of experiment examines the effectiveness of structured sparse priors for modeling hierarchical sparsity. The task is to restore 2D images which are contaminated by noise via compressive sensing on 2D wavelet basis. The setting is similar to [26] and [21]. 2D wavelet basis at different resolution levels is used as dictionary to get sparse representation of images. There is a natural hierarchical structure in the wavelet coefficients: a basis **b** can be defined as the parent of all such basis at finer resolution and whose support is covered by the support of **b**. Such tree-structured dependency corresponds to the nature of multi-resolution wavelet analysis and have been proven empirically effective in sparse representation of signals.

We choose the orthogonal Haar wavelet basis and the classical quad-tree structure on the 2D wavelet coefficients. We use $\text{PSNR} := \log_{10}(\frac{255^2}{\text{MSE}})$ to measure the quality of recovery. The benchmark set consists of five standard testing images: house, Lenna, boat, Barbara and pepper. We added Gaussian white noise $\mathcal{N}(0, 25^2)$ to the original images. The PSNR of the resulting noisy image is around 20. The images were divided into non-overlapped patches sized 64×64. Each patch is recovered independently with six levels of 2D Haar wavelet basis. For each method, we selected the parameters with the highest PSNR.

The recovery result is shown in Table 2. SCP delivers the highest PSNR in denoising on all test images, demonstrating the power of hierarchical structure prior to improve the recovery accuracy. Figure 15 in Appendix C of [37] shows a visual comparison of the denoising results, and it can be observed that SCP outperforms other methods in removing noise and preserving details in the image.

6   Conclusion and Discussion

We proposed a framework of set cover prior for modeling structured sparsity with overlapped groups. Its behavior is analyzed and empirically it outperforms existing competent structured priors. For future work, it will be interesting to further model sparsity within each group [18, 44]. Extension to other learning tasks is also useful, *e.g.* multi-task learning [45, 46].

Acknowledgements. NICTA is funded by the Australian Government as represented by the Department of Broadband, Communications and the Digital Economy and the Australian Research Council through the ICT Centre of Excellence program.

References

[1] Bühlmann, P., van de Geer, S.: Statistics for High-Dimensional Data. Springer (2011)

[2] Eldar, Y., Kutyniok, G. (eds.): Compressed Sensing: Theory and Applications, Cambridge (2012)

[3] George, E., McCulloch, R.: Variable selection via Gibbs sampling. Journal of the American Statistical Association 88, 881–889 (1993)

[4] Mitchell, T.J., Beauchamp, J.J.: Bayesian variable selection in linear regression. Journal of the American Statistical Association 83(404), 1023–1032 (1988)

[5] Armagan, A., Dunson, D., Clyde, M.: Generalized Beta mixtures of Gaussians. In: NIPS (2011)

[6] Carvalho, C., Polson, N., Scott, J.: The horseshoe estimator for sparse signals. Biometrika 97, 465–480 (2010)

[7] Carvalho, C., Polson, N., Scott, J.: Handling sparsity via the horseshoe. In: AI-STATS (2009)

[8] Tibshirani, R.: Regression shrinkage and selection via the lasso. Journal of the Royal Statistical Society, Series B 58, 267–288 (1996)

[9] Tipping, M.: Sparse Bayesian learning and the relevance vector machine. Journal of Machine Learning Research 1, 211–244 (2001)

[10] Park, T., Casella, G.: The Bayesian lasso. Journal of the American Statistical Association 103(482), 618–686 (2008)

[11] Griffin, J., Brown, P.: Bayesian adaptive lassos with non-convex penalization. Australian & New Zealand Journal of Statistics 53(4), 423–442 (2011)

[12] Fan, J., Li, R.: Variable selection via nonconcave penalized likelihood and its oracle properties. Journal of the American Statistical Association 96, 1348–1360 (2001)

[13] Hernández-Lobato, D., Hernández-Lobato, J.M., Dupont, P.: Generalized spike and slab priors for Bayesian group feature selection using expectation propagation. Journal of Machine Learning Research 16, 1891–1945 (2013)

[14] Ji, S., Dunson, D., Carin, L.: Multitask compressive sensing. IEEE Trans. Signal Processing 57(1), 92–106 (2009)

[15] Raman, S., Fuchs, T., Wild, P., Dahl, E., Roth, V.: The Bayesian group-lasso for analyzing contingency tables. In: ICML (2009)

[16] Yuan, M., Lin, Y.: Model selection and estimation in regression with grouped variables. Journal of the Royal Statistical Society, Series B 68(1), 49–67 (2006)

[17] Huang, J., Zhang, T.: The benefit of group sparsity. Annals of Stat. 38, 1978–2004 (2010)

[18] Yen, T.-J., Yen, Y.-M.: Grouped variable selection via nested spike and slab priors. ArXiv 1106.5837 (2011)

[19] Suo, Y., Dao, M., Tran, T., Srinivas, U., Monga, V.: Hierarchical sparse modeling using spike and slab priors. In: ICASSP (2013)

[20] Li, C., Li, H.: Network-constrained regularization and variable selection for analysis of genomic data. Biometrics 24(9), 1175–1182 (2008)

[21] Jenatton, R., Mairal, J., Obozinski, G., Bach, F.: Proximal methods for hierarchical sparse coding. Journal of Machine Learning Research 12, 2297–2334 (2011)

[22] Li, F., Zhang, N.: Bayesian variable selection in structured high-dimensional covariate spaces with applications in genomics. Journal of the American Statistical Association 105(491), 1201–1214 (2010)

[23] Pan, W., Xie, B., Shen, X.: Incorporating predictor network in penalized regression with application to microarray data. Biometrics 66(2), 474–484 (2010)

[24] Stingo, F., Vannucci, M.: Variable selection for discriminant analysis with Markov random field priors for the analysis of microarray data. Bioinformatics 27(4), 495–501 (2011)

[25] Zhao, P., Rocha, G., Yu, B.: Grouped and hierarchical model selection through composite absolute penalties. Annals of Stat. 37(6A), 3468–3497 (2009)

[26] Huang, J., Zhang, T., Metaxas, D.: Learning with structured sparsity. Journal of Machine Learning Research 12, 3371–3412 (2011)

[27] Griffin, J., Brown, P.: Hierarchical sparsity priors for regression models. Arxiv:1307.5231 (2013)

[28] Yuan, M., Joseph, V.R., Zou, H.: Structured variable selection and estimation. Annals of Applied Statistics 3, 1738–1757 (2009)

[29] Griffin, J., Brown, P.: Some priors for sparse regression modelling. Bayesian Analysis 8(3), 691–702 (2013)

[30] Palmer, J.A., Wipf, D.P., Kreutz-Delgado, K., Rao, B.D.: Variational EM algorithms for non-Gaussian latent variable models. In: NIPS (2005)

[31] Bernardo, J., Smith, A.: Bayesian Theory. Wiley (1994)

[32] Griffin, J., Brown, P.: Inference with normal-gamma prior distributions in regression problems. Bayesian Analysis 5(1), 171–188 (2010)

[33] Obozinski, G., Bach, F.: Convex relaxation for combinatorial penalties. Technical Report HAL 00694765 (2012)

[34] Hernández-Lobato, J.M., Hernández-Lobato, D., Suárez, A.: Network-based sparse Bayesian classification. Pattern Recognition 44(4), 886–900 (2011)

[35] Argyriou, A., Foygel, R., Srebro, N.: Sparse prediction with the k-support norm. In: NIPS (2012)

[36] He, L., Carin, L.: Exploiting structure in wavelet-based Bayesian compressive sensing. IEEE Trans. Signal Processing 57(9), 3488–3497 (2009)

[37] Liu, X., Zhang, X., Caetano, T.: Bayesian models for structured sparse estimation via set cover prior. Technical report (2014), http://users.cecs.anu.edu.au/~xzhang/papers/LiuZhaCae14_long.pdf

[38] Stingo, F., Chen, Y., Tadesse, M., Vannucci, M.: Incorporating biological information into linear models: A Bayesian approach to the selection of pathways and genes. Annals of Applied Statistics 5(3), 1978–2002 (2011)

[39] Minka, T.: A Family of Algorithms for Approximate Bayesian Inference. PhD thesis, MIT (2001)

[40] Seeger, M.: Bayesian inference and optimal design for the sparse linear model. Journal of Machine Learning Research 9, 759–813 (2008)

[41] Carbonetto, P., Stephens, M.: Scalable variational inference for bayesian variable selection in regression, and its accuracy in genetic association studies. Bayesian Analysis 7(1), 73–108 (2012)

[42] Jacob, L., Obozinski, G., Vert, J.-P.: Group lasso with overlap and graph lasso. In: ICML (2009)

[43] Mairal, J., Yu, B.: Supervised feature selection in graphs with path coding penalties and network flows. Journal of Machine Learning Research 14, 2449–2485 (2013)

[44] Rockova, V., Lesaffre, E.: Incorporating grouping information in Bayesian variable selection with applications in genomics. Bayesian Analysis 9(1), 221–258 (2014)

[45] Hernández-Lobato, D., Hernández-Lobato, J.M.: Learning feature selection dependencies in multi-task learning. In: NIPS (2013)

[46] Hernández-Lobato, D., Hernández-Lobato, J.M., Helle-putte, T., Dupont, P.: Expectation propagation for Bayesian multi-task feature selection. In: ECML (2010)

Preventing Over-Fitting of Cross-Validation with Kernel Stability

Yong Liu and Shizhong Liao*

School of Computer Science and Technology
Tianjin University, Tianjin, China
szliao@tju.edu.cn

Abstract. Kernel selection is critical to kernel methods. Cross-validation (CV) is a widely accepted kernel selection method. However, the CV based estimates generally exhibit a relatively high variance and are therefore prone to over-fitting. In order to prevent the high variance, we first propose a novel version of stability, called kernel stability. This stability quantifies the perturbation of the kernel matrix with respect to the changes in the training set. Then we establish the connection between the kernel stability and variance of CV. By restricting the derived upper bound of the variance, we present a kernel selection criterion, which can prevent the high variance of CV and hence guarantee good generalization performance. Furthermore, we derive a closed form for the estimate of the kernel stability, making the criterion based on the kernel stability computationally efficient. Theoretical analysis and experimental results demonstrate that our criterion is sound and effective.

1 Introduction

Kernel methods, such as support vector machine (SVM) [36], kernel ridge regression (KRR) [32] and least squares support vector machine (LSSVM) [35], have been widely used in machine learning and data mining. The performance of these algorithms greatly depends on the choice of kernel function, hence kernel selection becomes one of the key issues both in recent research and application of kernel methods [9].

It is common to select the kernel selection for kernel methods based on the generalization error of learning algorithms. However, the generalization error is not directly computable, as the probability distribution generating the data is unknown. Therefore, it is necessary to resort to estimates of the generalization error, either via testing on some data unused for learning (hold-out testing or cross-validation techniques) or via a bound given by theoretical analysis. To derive the theoretical upper bounds of the generalization error, some measures are introduced: such as VC dimension [36], Rademacher complexity [2], regularized risk [33], radius-margin bound [36], compression coefficient [26], Bayesian regularisation [7], influence function [14], local Rademacher complexity [11], and eigenvalues perturbation [23], etc.

---

* Corresponding author.

T. Calders et al. (Eds.): ECML PKDD 2014, Part II, LNCS 8725, pp. 290–305, 2014.
© Springer-Verlag Berlin Heidelberg 2014

While there have been many interesting attempts to use the theoretical bounds of generalization error or other techniques to select kernel functions, the most commonly used and widely accepted kernel selection method is still cross-validation. However, the cross-validation based estimates of performance generally exhibit a relatively high variance and are therefore prone to over-fitting [19,27,7,8]. To overcome this limitation, we introduce a notion of kernel stability, which quantifies the perturbation of the kernel matrix when removing an arbitrary example from the training set. We illuminate that the variance of cross-validation for KRR, LSSVM and SVM can be bounded based on the kernel stability. To prevent the high variance of cross-validation, we propose a novel kernel selection criterion by restricting the derived upper bound of the variance. Therefore, the kernel chosen by this criterion can avoid over-fitting of cross-validation. Furthermore, the closed form of the estimate of the kernel stability is derived, making the kernel stability computationally efficient. Experimental results demonstrate that our criterion based on kernel stability is a good choice for kernel selection. To our knowledge, this is the first attempt to use the notion of stability to entirely quantify the variance of cross-validation for kernel selection.

The rest of the paper is organized as follows. Related work and preliminaries are respectively introduced in Section 2 and Section 3. In Section 4, we present the notion of kernel stability, and use this stability to derive the upper bounds of the variance of cross-validation for KRR, LSSVM and SVM. In Section 5, we propose a kernel selection criterion by restricting these bounds. In Section 6, we analyze the performance of our proposed criterion compared with other state-of-the-art kernel selection criteria. Finally, we conclude in the last section.

2   Related Work

Cross-validation has been studied [27,19,3,15] and used in practice for many years. However, analyzing the variance of cross-validation is tricky. Bengio and Grandvalet [3] asserted that there exists no universal unbiased estimator of the variance of cross-validation. Blum et al. [4] showed that the variance of the cross-validation estimate is never larger than that of a single holdout estimate. Kumar et al. [20] generalized the result of [4] considerably, quantifying the variance reduction as a function of the algorithm's stability. Unlike the above work which considers the link between the variance of the cross-validation estimate and that of the single holdout estimate, in this paper we consider bounding the variance of cross-validation for some kernel methods, such as KRR, LSSVM and SVM, based on an appropriately defined notion of stability for kernel selection.

The notion of stability has been studied in various contexts over the past years. Rogers and Wagner [31] presented the definition of weak hypothesis stability. Kearns and Ron [16] defined the weak-error stability in the context of proving sanity check bounds. Kutin and Niyogi [21] defined the uniform stability notion; see also the work of Bousquet and Elisseeff [5]. The notions of mean square stability and the loss stability were introduced by Kumar et al. [20], which are closely related to the leave-one-out cross-validation. Unfortunately, for most of these notions of stability, proposed to derive the theoretical generalization error bounds, it is difficult to compute their specific values [28]. Thus, these notions of stability are hard to be used in practical kernel selection. To address this issue, we propose a new version of stability, which is defined on a kernel function, are computationally efficient and practical for kernel selection.

3   Preliminaries and Notations

Let $S = \{z_i = (\boldsymbol{x}_i, y_i)\}_{i=1}^n$ be a sample set of size n drawn i.i.d from a fixed, but unknown probability distribution P on $\mathcal{Z} = \mathcal{X} \times \mathcal{Y}$, where $\mathcal{X}$ is the input space and $\mathcal{Y}$ is the output space. Let $K : \mathcal{X} \times \mathcal{X} \to \mathbb{R}$ be a kernel. The reproducing kernel Hilbert space (RKHS) $\mathcal{H}_K$ associated with K is defined to be the completion of the linear span of the set of functions $\{K(\boldsymbol{x}, \cdot) : \boldsymbol{x} \in \mathcal{X}\}$ with the inner product denoted as $\langle \cdot, \cdot \rangle_K$ satisfying

$$\left\langle \sum_{i=1}^n \alpha_i K(\boldsymbol{x}_i, \cdot), \sum_{i=1}^n \beta_i K(\boldsymbol{x}_i', \cdot) \right\rangle_K = \sum_{i,j=1}^n \alpha_i \beta_j K(\boldsymbol{x}_i, \boldsymbol{x}_j').$$

We assume that $|y| \leq M$ for all $y \in \mathcal{Y}$ and $K(x,x) \leq \kappa$ for all $x \in \mathcal{X}$.

The learning algorithms we study here are the regularized algorithms:

$$f_S := \underset{f \in \mathcal{H}_K, b \in \mathbb{R}}{\arg\min} \left\{ \frac{1}{|S|} \sum_{z \in S} \ell(y_i, f(\boldsymbol{x}_i) + b) + \lambda \|f\|_K^2 \right\},$$

where $\ell(\cdot, \cdot)$ is a loss function, λ is the regularization parameter and $|S|$ is the size of S. KRR, LSSVM, and SVM are the special cases of the regularized algorithms. For KRR,

$$b = 0 \text{ and } \ell(f(\boldsymbol{x}), y) = (y - f(\boldsymbol{x}))^2,$$

for LSSVM

$$\ell(f(\boldsymbol{x}) + b, y) = (y - f(\boldsymbol{x}) - b)^2,$$

and for SVM

$$\ell(f(\boldsymbol{x}) + b, y) = \max(0, 1 - y(f(\boldsymbol{x}) + b)).$$

The (empirical) loss of the hypothesis f_S on a set Q is defined as

$$\ell_{f_S}(Q) = \frac{1}{|Q|} \sum_{z \in Q} \ell(f_S(\boldsymbol{x}), y).$$

Let $S_1, \ldots, S_k$ be a random equipartition of S into k parts, called folds, with $|S_i| = \lfloor \frac{n}{k} \rfloor$. We learn k different hypotheses with $f_{S \backslash S_i}$ being the hypothesis learned on all of the data except for the ith fold; Let $m = (k-1)k/n$ be the size of the training set for each of these k hypotheses. The k-fold cross-validation hypothesis, f_{kcv}, which picks one of the $\{f_{S\backslash S_i}\}_{i=1}^k$ uniformly at random. The (empirical) loss of f_{kcv} is defined as

$$\ell_{f_{kcv}}(S) = \frac{1}{k}\sum_{i=1}^k \frac{1}{|S_i|} \sum_{z \in S_i} \ell(f_{S\backslash S_i}(\boldsymbol{x}), y).$$

4   Variance Bounds of Cross-Validation

k-fold cross-validation (k-CV) is the most widely accepted method for kernel selection. However, it is known to exhibit a relatively high variance $\text{var}_S\left(\ell_{f_{kcv}}(S)\right)$,

$$\text{var}_S\left(\ell_{f_{kcv}}(S)\right) = \underset{S \sim \mathcal{Z}^n}{\mathbb{E}}\left[\ell_{f_{kcv}}(S) - \underset{S \sim \mathcal{Z}^n}{\mathbb{E}}\left[\ell_{f_{kcv}}(S)\right]\right]^2.$$

Therefore, k-CV is prone to over-fitting [19,27,7,8]. Obviously, $\text{var}_S\left(\ell_{f_{kcv}}(S)\right)$ is not directly computable, as the probability distribution is unknown. In the next subsection, we will define a new notion of stability to bound $\text{var}_S\left(\ell_{f_{kcv}}(S)\right)$.

4.1   Kernel Stability

The way of making the definition of kernel stability is to start from the goal: to get bounds on the variance of CV and want these bounds to be tight when the kernel function satisfies the kernel stability.

It is well known that the kernel matrix contains most of the information needed by kernel methods. Therefore, we introduce a new notion of stability to quantify the perturbation of the kernel matrix with respect to the changes in the training set for kernel selection.

To this end, let $T = \{\boldsymbol{x}_i\}_{i=1}^m$ and the ith removed set T^i be

$$T^i = \{\boldsymbol{x}_1, \ldots, \boldsymbol{x}_{i-1}, \boldsymbol{x}_{i+1}, \ldots, \boldsymbol{x}_m\}.$$

Denote the kernel matrix $\boldsymbol{K}$ as $[K(\boldsymbol{x}_i, \boldsymbol{x}_j)]_{i,j=1}^m$, and let $\boldsymbol{K}^i$ be the $m \times m$ ith removed kernel matrix with

$$\begin{cases} [\boldsymbol{K}^i]_{jk} = K(\boldsymbol{x}_j, \boldsymbol{x}_k) & \text{if } j \neq i \text{ and } k \neq i, \\ [\boldsymbol{K}^i]_{jk} = 0 & \text{if } j = i \text{ or } k = i. \end{cases}$$

One can see that $\boldsymbol{K}^i$ can be considered as the kernel matrix with respect to the removed set T^i.

Definition 1 (Kernel Stability). *A kernel function K is of β kernel stability if the following holds: $\forall x_i \in \mathcal{X}, i = 1, \ldots, m,$*

$$\forall i \in \{1, \ldots, m\}, \|\boldsymbol{K} - \boldsymbol{K}^i\|_2 \leq \beta,$$

where $\boldsymbol{K}$ and $\boldsymbol{K}^i$ are the kernel matrices with respect to T and T^i, respectively. $\|\boldsymbol{K} - \boldsymbol{K}^i\|_2$ is the 2-norm of $[\boldsymbol{K} - \boldsymbol{K}^i]$, that is, the largest eigenvalue of $[\boldsymbol{K} - \boldsymbol{K}^i]$.

According to the above definition, the kernel stability is used to quantify the perturbation of the kernel matrix when an arbitrary example is removed. Different from the existing notions of stability, see, e.g., [31,16,5,21,29,12,34] and the references therein, our proposed stability is defined on the kernel matrix. Therefore, we can estimate its value from empirical data, which makes this stability usable for kernel selection in practice.

4.2   Upper Bounds via Kernel Stability

We will show that the kernel stability can yield the upper bounds of the variance of CV for KRR, LSSVM and SVM.

Kernel Ridge Regression. KRR has been successfully applied to solve regression problems, which is a special case of the regularized algorithms when the loss function

$$b = 0 \text{ and } \ell(f(\boldsymbol{x}), y) = (f(\boldsymbol{x}) - y)^2.$$

Theorem 1. *If the kernel function K is of β kernel stability, then for KRR,*

$$\operatorname*{var}_{S}(\ell_{f_{kcv}}(S)) \leq C_1\beta^2,$$

where $C_1 = 8\left(\frac{\kappa^2 M^2}{\lambda^3 m} + \frac{\kappa M^2}{\lambda^2 m}\right)^2$.

Proof. The proof is given in Appendix A.

This theorem shows that small β can restrict the value of $\operatorname{var}_S(\ell_{f_{kcv}}(S))$. Thus, we can select the kernel which has small β to prevent the over-fitting of CV caused by the high variance.

Least Squares Support Vector Machine. LSSVM is a popular learning machine for solving classification problems, its loss function is the square loss

$$\ell(f(\boldsymbol{x}) + b, y) = (y - f(\boldsymbol{x}) - b)^2.$$

Theorem 2. *If the kernel function K is of β kernel stability, then for LSSVM,*

$$\operatorname*{var}_{S}(\ell_{f_{kcv}}(S)) \leq C_2\beta^2,$$

where $C_2 = \left(\frac{2(\kappa+1)^2}{\lambda^3 m} + \frac{2(\kappa+1)}{\lambda^2 m}\right)^2$.

Proof. The proof is given in Appendix B.

Similar with KRR, this theorem also show that we can choose the kernel function which has small β to prevent the high variance for LSSVM.

Support Vector Machine. The loss function of SVM is the hinge loss

$$\ell(f(\boldsymbol{x}) + b, y) = \max(0, 1 - y(f(\boldsymbol{x}) + b)).$$

Theorem 3. *If the kernel function K is of β kernel stability, then for SVM,*

$$\operatorname*{var}_{S}(\ell_{f_{kcv}}(S)) \leq C_3 \beta^{\frac{1}{2}} \left(1 + C_4 \beta^{\frac{1}{4}}\right)^2,$$

where $C_3 = 8\lambda^2 \kappa^{\frac{3}{2}}$ and $C_4 = \left[\frac{1}{4\kappa}\right]^{\frac{1}{4}}$.

Proof. The proof is given in Appendix C.

The bound we obtain for SVM is different from our bounds for KRR and LSSVM. This is mainly due to the difference between the hinge loss and the squared loss.

5   Kernel Selection Criterion

Theorem 1, 2 and 3 show that the variance of CV can be bounded via the kernel stability. Thus, to prevent over-fitting caused by the high variance, it is reasonable to use the following criterion for kernel selection:

$$\operatorname*{arg\,min}_{K \in \mathcal{K}} \ell_{f_{kcv}}(S) + \frac{\eta}{n}\beta,$$

where η is a trade-off parameter and $\mathcal{K}$ is an candidate set of kernel functions. However, by the definition of the kernel stability, we need to try all the possibilities of the training set to compute β, which is infeasible in practice. We should estimate it from the available empirical data. Therefore, we consider using the following kernel stability criterion in practice:

$$\operatorname*{arg\,min}_{K \in \mathcal{K}} k\text{-KS}(K) = \ell_{f_{kcv}}(S) + \frac{\eta}{n} \cdot \max_{i \in \{1,\dots,n\}} \|\boldsymbol{K} - \boldsymbol{K}^i\|_2.$$

This criterion consists of two parts: bias and variance. $\ell_{f_{kcv}}(S)$ can be considered as the bias, and $\max_{i \in \{1,\dots,n\}} \|\boldsymbol{K} - \boldsymbol{K}^i\|_2$ is the variance.

To apply this criterion, we should compute $\|\boldsymbol{K} - \boldsymbol{K}^i\|_2$, which requires the calculation of the eigenvalues of $[\boldsymbol{K} - \boldsymbol{K}^i]$, $i = 1, \dots, n$. It is computationally expensive. Fortunately, this problem can be effectively solved by using the closed form of $\|\boldsymbol{K} - \boldsymbol{K}^i\|_2$ given by the following theorem.

Theorem 4. $\forall\, S \in \mathcal{Z}^n$ *and* $i \in \{1, \dots, n\}$,

$$\|\boldsymbol{K} - \boldsymbol{K}^i\|_2 = \frac{K_{ii} + \sqrt{K_{ii}^2 + 4\sum_{j=1, j \neq i}^{n} K_{ji}^2}}{2}.$$

Proof. By the definitions of $\boldsymbol{K}$ and $\boldsymbol{K}^i$, it is easy to verify that the characteristic polynomial of $[\boldsymbol{K} - \boldsymbol{K}^i]$ is

$$\det(t\boldsymbol{I} - (\boldsymbol{K} - \boldsymbol{K}^i)) = t^{n-2}(t^2 - \boldsymbol{K}_{ii}t - \sum_{j=1,j\neq i}^{n} \boldsymbol{K}_{ji}^2).$$

Thus, the eigenvalues of $\boldsymbol{K} - \boldsymbol{K}^i$ is

$$\sigma(\boldsymbol{K} - \boldsymbol{K}^i) = \left\{ \frac{\boldsymbol{K}_{ii} \pm \sqrt{\boldsymbol{K}_{ii}^2 + 4\sum_{j=1,j\neq i}^{n} \boldsymbol{K}_{ji}^2}}{2}, \overset{n-2}{\overbrace{0,\ldots,0}} \right\}.$$

So, the largest eigenvalue is

$$\frac{\boldsymbol{K}_{ii} + \sqrt{\boldsymbol{K}_{ii}^2 + 4\sum_{j=1,j\neq i}^{n} \boldsymbol{K}_{ji}^2}}{2}.$$

Hence we complete the proof of Theorem 4.

This theorem shows that only $O(n^2)$ is needed to compute

$$\max_{i\in\{1,\ldots,n\}} \|\boldsymbol{K} - \boldsymbol{K}^i\|_2,$$

making the criterion based on kernel stability computationally efficient.

Remark 1. Instead of choosing a single kernel, several authors consider combining multiple kernels by some criteria, called multiple kernel learning (MKL), see, e.g., [22,1,30,18,25], etc. Our criterion k-KS(K) can also be applied to MKL:

$$\arg\min_{\boldsymbol{\mu}=(\mu_1,\ldots,\mu_k)} k\text{-KS}(K_{\boldsymbol{\mu}}), \text{s.t.} \|\boldsymbol{\mu}\|_p = 1, \boldsymbol{\mu} \geq 0,$$

where $K_{\boldsymbol{\mu}} = \sum_{i=1}^{k} \mu_i K_i$, which can be efficiently solved using gradient-based algorithms [17]. However, in this paper we mainly want to verify the effectiveness of our kernel stability criterion. Therefore, in our experiments, we focus on comparing our criterion with other popular kernel selection criteria.

5.1   Time Complexity Analysis

To compute our kernel stability criterion k-KS(K), we need kF to calculate $\ell_{f_{kcv}}(S)$, where F is the time complexity of training on the data set of size $(k-1)k/n$, n is the size of the training set. We also need $O(n^2)$ to compute

$$\max_{i\in\{1,\ldots,n\}} \|\boldsymbol{K} - \boldsymbol{K}^i\|_2.$$

Thus, the overall time complexity of k-KS(K) is

$$O(kF + n^2).$$

Remark 2. In our previous work [24], we presented a strategy for approximating the k-fold CV based on the Bouligand influence function [10]. This approximate method requires the solution of the algorithm only once, which can dramatically improve the efficiency. Thus, the time complexity of the approximate k-KS(K) can reduce to $O(F + n^2)$.

6   Experiments

In this section, we will compare our proposed kernel selection criteria (k-KS, $k = 5, 10$) with 5 popular kernel selection criteria: 5-fold cross-validation (5-CV), 10-fold cross-validation (10-CV), the efficient leave-one-out cross-validation (ELOO) [6], Bayesian regularisation (BR) [7], and the latest eigenvalues perturbation criterion (EP) [23]. The evaluation is made on 9 popular data sets from LIBSVM Data. All data sets are normalized to have zero-means and unit-variances on every attribute to avoid numerical problems. We use the popular Gaussian kernels

$$K(\boldsymbol{x}, \boldsymbol{x}') = \exp\left(-\frac{\|\boldsymbol{x} - \boldsymbol{x}'\|_2^2}{2\tau}\right)$$

as our candidate kernels,

$$\tau \in \{2^i, i = -10, -9, \ldots, 10\}.$$

For each data set, we have run all the methods 10 times with randomly selected 70% of all data for training and the other 30% for test. The learning machine we considered is LSSVM.

6.1   Accuracy

In this subsection, we will compare the performance of 5-KS (ours), 10-KS (ours), 5-CV, 10-CV, ELOO, BR and EP. In the first experiment, we set $\eta = 1$ (the parameter of the 5-KS and 10-KS criterion, we will explore the effect of this parameter in the next experiment). The average test errors are reported in Table 1. The elements in this table are obtained as follows: For each training set and each regularization parameter[1] λ, $\lambda \in \{10^i, i = -4, \ldots, -1\}$, we choose the kernel by each kernel selection criterion on the training set, and evaluate the test error for the chosen parameters on the test set. The results in Table 1 can be summarized as follows: (a) k-KS gives better results than k-CV on most data sets, $k = 5, 10$. In particular, for each λ, k-KS outperforms k-CV on 8 (or more) out of 9 sets, and also give results closed to results of k-CV on the remaining set. Thus, it indicates that using the kernel stability to restrict the high variance

---

[1] the value of λ we set seems too small at first sight, but in fact, the regularized algorithm we considered in this paper is $\frac{1}{n}\sum_{i=1}^{n} \ell(f(x_i), y_i) + \lambda\|f\|_K^2$, while other authors usually ignore $1/n$. Therefore, the value of λ in our paper is $1/n$ time of that of regularized algorithms other authors considered.

Table 1. The test errors (%) with standard deviations of 5-KS (ours), 10-KS (ours), 5-CV, 10-CV, ELOO, BR and EP. For each training set, each regularization parameter λ ($\lambda \in \{10^{-i}, i = -4, \ldots, -1\}$), we choose the kernel by each kernel selection criterion on the training set, and evaluate the test error for the chosen kernel on test set.

$\lambda = 0.0001$

Method	5-CV	5-KS	10-CV	10-KS	ELOO	BR	EP
australian	14.78 ± 2.4	**13.23** ± 2.0	15.46 ± 1.6	15.27 ± 1.4	14.30 ± 2.6	14.30 ± 2.6	14.15 ± 2.9
heart	19.01 ± 3.9	18.52 ± 3.1	18.27 ± 3.7	**18.17** ± 2.3	18.52 ± 3.4	18.52 ± 3.3	17.83 ± 6.1
ionosphere	4.76 ± 1.5	4.38 ± 1.5	4.67 ± 1.7	**4.35** ± 1.9	5.33 ± 2.2	5.13 ± 2.1	6.76 ± 3.9
breast	3.61 ± 0.8	**3.41** ± 0.7	3.51 ± 0.6	3.45 ± 0.6	3.41 ± 0.9	3.41 ± 0.9	5.27 ± 1.4
diabetes	23.65 ± 3.7	23.65 ± 3.7	23.83 ± 3.3	23.91 ± 2.9	23.04 ± 2.5	**22.96** ± 2.7	30.26 ± 2.3
german	24.60 ± 1.3	**24.17** ± 1.2	25.80 ± 1.1	24.67 ± 1.3	24.67 ± 1.4	24.60 ± 1.3	29.67 ± 3.1
liver	26.92 ± 2.5	27.12 ± 2.0	27.50 ± 2.6	**26.15** ± 1.0	26.55 ± 1.3	26.73 ± 2.0	30.08 ± 4.7
sonar	14.19 ± 4.0	13.23 ± 3.5	13.87 ± 2.9	**12.58** ± 2.4	12.90 ± 4.9	12.90 ± 4.9	17.74 ± 6.7
a2a	18.44 ± 1.0	17.14 ± 0.9	17.94 ± 1.0	**15.20** ± 0.6	17.91 ± 1.0	17.34 ± 1.0	18.92 ± 1.5

$\lambda = 0.001$

Method	5-CV	5-KS	10-CV	10-KS	ELOO	BR	EP
australian	14.30 ± 1.2	**12.19** ± 1.3	13.30 ± 0.7	12.30 ± 0.4	14.43 ± 0.9	13.43 ± 0.9	16.29 ± 3.4
heart	18.27 ± 6.7	15.80 ± 4.0	17.80 ± 4.3	15.31 ± 4.2	14.83 ± 4.8	**14.07** ± 4.9	20.77 ± 6.3
ionosphere	5.33 ± 1.9	**3.81** ± 1.9	5.33 ± 1.9	4.38 ± 1.6	6.48 ± 2.2	6.48 ± 2.2	5.38 ± 4.6
breast	3.61 ± 0.8	3.42 ± 0.8	3.51 ± 0.8	**3.22** ± 0.6	3.32 ± 0.8	3.32 ± 0.8	5.56 ± 1.1
diabetes	23.65 ± 2.4	23.48 ± 1.8	23.83 ± 2.3	23.45 ± 2.2	24.22 ± 1.9	**23.22** ± 1.9	26.52 ± 0.4
german	25.07 ± 2.4	**24.60** ± 2.4	23.93 ± 1.1	23.87 ± 0.9	24.60 ± 2.4	24.67 ± 2.4	25.13 ± 2.1
liver	29.04 ± 3.5	28.46 ± 3.3	27.12 ± 4.6	**26.54** ± 1.8	27.12 ± 2.7	26.82 ± 2.6	28.46 ± 3.3
sonar	14.84 ± 6.7	13.87 ± 4.5	11.61 ± 5.9	**11.55** ± 5.7	13.55 ± 6.7	13.83 ± 6.8	13.90 ± 7.3
a2a	17.23 ± 1.1	**15.51** ± 0.9	17.35 ± 1.0	16.11 ± 1.0	16.91 ± 0.9	16.94 ± 0.9	19.71 ± 1.1

$\lambda = 0.01$

Method	5-CV	5-KS	10-CV	10-KS	ELOO	BR	EP
australian	14.59 ± 2.0	13.82 ± 1.9	14.98 ± 2.0	**13.72** ± 2.0	14.01 ± 2.1	14.01 ± 2.1	16.54 ± 3.4
heart	18.27 ± 2.3	17.78 ± 2.2	18.27 ± 1.6	18.02 ± 0.6	**17.28** ± 1.5	17.78 ± 2.0	19.74 ± 6.6
ionosphere	4.95 ± 1.5	**4.38** ± 1.2	4.95 ± 1.5	5.14 ± 1.2	5.14 ± 2.1	5.14 ± 2.1	9.52 ± 3.3
breast	3.51 ± 0.7	3.46 ± 0.6	3.80 ± 0.6	3.80 ± 0.6	3.75 ± 1.3	**3.41** ± 1.0	7.02 ± 0.8
diabetes	24.00 ± 1.4	**22.30** ± 1.3	23.48 ± 1.4	23.83 ± 1.3	23.83 ± 1.7	23.83 ± 1.7	25.83 ± 1.8
german	26.40 ± 0.9	26.33 ± 0.7	26.47 ± 0.9	**24.93** ± 0.4	26.87 ± 1.1	26.25 ± 1.5	28.67 ± 1.3
liver	28.85 ± 1.8	**25.27** ± 1.2	30.00 ± 2.6	28.65 ± 2.4	28.46 ± 2.3	28.65 ± 2.3	29.42 ± 3.0
sonar	14.52 ± 5.4	13.55 ± 2.9	13.23 ± 4.4	12.23 ± 3.5	12.58 ± 3.8	**11.94** ± 4.0	14.74 ± 4.0
a2a	18.88 ± 2.0	**17.76** ± 1.1	18.97 ± 1.9	17.82 ± 1.7	18.76 ± 2.1	18.41 ± 2.5	20.15 ± 2.5

$\lambda = 0.1$

Method	5-CV	5-KS	10-CV	10-KS	ELOO	BR	EP
australian	14.30 ± 2.1	13.53 ± 1.8	14.30 ± 2.1	14.20 ± 0.7	13.91 ± 1.4	**13.41** ± 1.7	14.54 ± 2.9
heart	19.51 ± 3.2	19.26 ± 2.9	19.26 ± 3.4	**18.26** ± 2.9	19.51 ± 3.2	19.26 ± 3.4	22.65 ± 4.0
ionosphere	12.76 ± 10.5	8.38 ± 3.7	9.14 ± 4.2	**8.28** ± 3.5	9.33 ± 4.3	9.14 ± 4.6	12.95 ± 3.9
breast	3.92 ± 1.4	3.22 ± 1.3	3.62 ± 1.3	**3.12** ± 1.4	3.41 ± 1.1	3.21 ± 1.6	4.98 ± 1.0
diabetes	29.65 ± 1.8	29.83 ± 2.0	29.74 ± 2.1	**29.48** ± 1.8	29.57 ± 1.7	29.40 ± 1.6	31.40 ± 1.4
german	31.21 ± 1.7	27.40 ± 1.5	27.40 ± 1.4	26.51 ± 1.2	**25.40** ± 1.1	29.40 ± 1.6	31.40 ± 1.4
liver	33.46 ± 7.4	**31.08** ± 7.0	33.08 ± 8.0	32.42 ± 6.3	31.85 ± 8.6	38.65 ± 5.8	33.08 ± 8.0
sonar	27.81 ± 9.6	**26.06** ± 9.3	27.42 ± 9.3	27.10 ± 8.7	26.77 ± 9.3	26.77 ± 9.3	27.10 ± 5.0
a2a	24.68 ± 1.7	**22.18** ± 1.5	25.68 ± 1.9	22.68 ± 1.8	24.68 ± 1.7	24.31 ± 0.8	23.21 ± 1.7

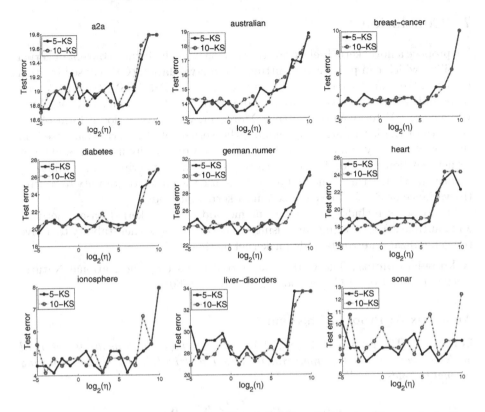

Fig. 1. The average test errors using 5-KS and 10-KS on different η. The regularization parameter λ is set as 0.001 (in Table 1, one can see that for most data sets, $\lambda = 0.001$ can achieve good results. Thus, we only consider setting $\lambda = 0.001$). For each training set, each η, we choose the kernel by 5-KS and 10-KS kernel selection criteria on the training set, and evaluate the test errors for the chosen parameters on test set.

of cross-validation can guarantee good generalization performance; (b) k-KS is better than BR on most data sets. In particular, for each λ, k-KS outperforms k-CV on 6 (or more) out of 9 sets; (c) BR is comparable or better than ELOO on most data sets; (d) The performances of the 5-KS and 10-KS are comparable.

6.2   Effect of the Parameter η

In this experiments, we will explore the effect of the η. The average test errors on different η are given in Figure 1. For each training set, each η, we choose the kernel by 5-KS and 10-KS kernel selection criteria on the training set, and evaluate the test errors for the chosen parameters on test set. It turns out that η is robust, and the test errors are not very sensitive w.r.t $\eta \in [2^{-2}, 2^5]$ on most data sets. Moreover, we find that $\eta \in [2^{-2}, 2^5]$ is a good choice for k-KS. Thus, we can select $\eta \in [2^{-2}, 2^5]$ in practice.

7 Conclusion

We propose a novel kernel selection criterion via a newly defined concept of kernel stability, which can prevent over-fitting of cross-validation (CV) caused by high variance. We illuminate that the variance of CV for KRR, LSSVM and SVM can be bounded with the kernel stability, so we can use this stability to control the variance of CV to avoid over-fitting. Moreover, we derive a closed form of the estimate of the kernel stability, making the kernel selection criterion based on the kernel stability computationally efficient and practically useful. Finally, our kernel selection criterion is theoretically justified and experimentally validated. To our knowledge, this is the first attempt to use the notion of stability to control the variance of CV for kernel selection in kernel methods.

Future work includes extending our method to other kernel based methods and multiple kernel learning, and using the notion of the kernel stability to derive the generalization error bounds for kernel methods.

Acknowledgments. The work is supported in part by the National Natural Science Foundation of China under grant No. 61170019.

Appendix A: Proof of Theorem 1

Lemma 1 (Proposition 1 in [13]). *Let h' denote the hypothesis returned by KRR when using the approximate kernel matrix $\boldsymbol{K}'$. Then, the following inequality holds for all $\boldsymbol{x} \in \mathcal{X}$:*

$$|h'(\boldsymbol{x}) - h(\boldsymbol{x})| \leq \frac{\kappa M}{\lambda^2 m} \|\boldsymbol{K}' - \boldsymbol{K}\|_2.$$

Definition 2 (Loss stability [20]). *The loss stability of a learning algorithm A trained on m examples and with respect to a loss ℓ is defined as*

$$\mathrm{ls}_{m,\ell}(A) = \mathbb{E}_{T:|T|=m, z', z} \left[\left(\ell'_{A(T)}(z) - \ell'_{A(T^{z'})}(z) \right)^2 \right],$$

where $T^{z'}$ denote the set of examples obtained by replacing an example chosen uniformly at random from T by z'. A learning algorithm A is γ-loss stable if $\mathrm{ls}_{m,\ell}(A) \leq \gamma$.

Lemma 2 (Theorem 1 in [20]). *Consider any learning algorithm A that is γ-loss stable with respect to ℓ. Then*

$$\operatorname*{var}_S(\ell_{f_{\mathrm{kcv}}}(S)) \leq \frac{1}{k} \operatorname*{var}_S(\ell_{f_{S \setminus S_1}}(S_1)) + \left(1 - \frac{1}{k}\right) \gamma.$$

Proof (of Theorem 1). Note that $f_{T^i}(\boldsymbol{x})$ is the hypothesis returned by KRR using $\boldsymbol{K}^i$. According to the definition of β kernel stability, we have $\|\boldsymbol{K} - \boldsymbol{K}^i\|_2 \leq \beta$. By Lemma 1,

$$|f_T(\boldsymbol{x}) - f_{T^i}(\boldsymbol{x})| \leq \frac{\kappa M}{\lambda^2 m} \|\boldsymbol{K}^i - \boldsymbol{K}\|_2 \leq \frac{\beta \kappa M}{\lambda^2 m}. \tag{1}$$

Since $f_T(x) = \sum_{i=1}^{m} \alpha_i K(x, x_i) = k_x \alpha$, where $\alpha = [K + m\lambda I]^{-1}y$ and $k_x = (K(x, x_1), \ldots, K(x, x_m))^{\mathrm{T}}$. Thus, we have

$$
\begin{aligned}
|f_T(x)| = |k_x^{\mathrm{T}} \alpha| &= |k_x [K + m\lambda I]^{-1} y| \\
&\leq \|k_x\| \|y\| \|[K + m\lambda I]^{-1}\|_2 \\
&\leq \frac{\kappa\sqrt{m} M \sqrt{m}}{m\lambda} \\
&= \frac{\kappa M}{\lambda}.
\end{aligned}
\tag{2}
$$

Thus, $\forall z \in \mathcal{Z}$, $\forall T \in \mathcal{Z}^m$ and $\forall i \in \{1, \ldots, m\}$,

$$
\begin{aligned}
&|\ell_{f_T}(z) - \ell_{f_{T^i}}(z)| \\
&= |(f_T(x) - y)^2 - (f_{T^i}(x) - y)^2| \\
&= |(f_T(x) - f_{T^i}(x))(f_T(x) + f_{T^i}(x) - 2y)| \\
&\leq \left(\frac{\beta \kappa M}{\lambda^2 m}\right) \cdot \left(\frac{2\kappa M}{\lambda} + 2M\right).
\end{aligned}
\tag{3}
$$

According to Lemma 2 in [20], we have

$$
\mathrm{ls}_{m,\ell}(A) \leq \mathbb{E}_{T, z', z}\left[\left(\ell_{A(T)}(z) - \ell_{A(T^{z'})}(z)\right)^2\right].
\tag{4}
$$

So, from (3), $\forall T, \forall z, \forall z'$, $\left(\ell_{f_T}(z) - \ell_{f_{T^{z'}}}(z)\right)^2 \leq$

$$
\begin{aligned}
&\leq \left(\left|\ell_{f_T}(z) - \ell_{f_{T^i}}(z)\right| + \left|\ell_{f_{T^i}}(z) - \ell_{A(T^{z'})}(z)\right|\right)^2 \\
&\leq \left(2\beta\left(\frac{2\kappa^2 M}{\lambda^3 m} + \frac{2\kappa M^2}{\lambda^2 m}\right)\right)^2 \\
&= C_1 \beta^2.
\end{aligned}
$$

Thus, according to (4), we have

$$
\mathrm{ls}_{m,\ell}(A) \leq C_1 \beta^2 = \gamma.
\tag{5}
$$

According to Lemma 5 in [20], we have

$$
\begin{aligned}
\mathrm{var}_S(\ell_{f_{S\backslash S_1}}(S_1)) &= \mathrm{cov}(\ell_{f_{S\backslash S_1}}(S_1), \ell_{f_{S\backslash S_1}}(S_1)) \\
&= \mathbb{E}_{S\backslash S_1, z_1', z_2}\left[\left(\ell'_{f_{S\backslash S_1}}(z_2) - \ell'_{f_{(S\backslash S_1)^{z_1'}}}(z_2)\right)^2\right] \\
&= \mathrm{ls}_{m,\ell}(A) \\
&\leq C_1 \beta^2 = \gamma \text{ (According to (5))}.
\end{aligned}
\tag{6}
$$

Substituting (5) and (6) into Lemma 2, we complete the proof of Theorem 1.

Appendix B: Proof of Theorem 2

Proof (of Theorem 2). For LSSVM,

$$f_T(\boldsymbol{x}) = \boldsymbol{k}_{\boldsymbol{x}}^{\mathrm{T}}\boldsymbol{\alpha} + b = \tilde{\boldsymbol{k}}_{\boldsymbol{x}} \boldsymbol{M}^{-1}\tilde{\boldsymbol{y}},$$

where

$$\tilde{\boldsymbol{k}}_{\boldsymbol{x}} = (K(\boldsymbol{x}, \boldsymbol{x}_1), \dots, K(\boldsymbol{x}, \boldsymbol{x}_m), 1), \tilde{\boldsymbol{y}} = [y_1, \dots, y_m, 0]^{\mathrm{T}}$$

and

$$\boldsymbol{M} = \begin{bmatrix} \boldsymbol{K} + m\lambda \boldsymbol{I} & \boldsymbol{1} \\ \boldsymbol{1}^{\mathrm{T}} & 0 \end{bmatrix},$$

Thus, it is easy to verify that

$$|f_T(\boldsymbol{x}) - f_{T^i}(\boldsymbol{x})| = |\tilde{\boldsymbol{k}}_{\boldsymbol{x}}(\boldsymbol{M}^{-1}\tilde{\boldsymbol{y}} - \boldsymbol{M}_i^{-1}\tilde{\boldsymbol{y}})|,$$

where

$$\boldsymbol{M}_i = \begin{bmatrix} \boldsymbol{K}^i + m\lambda \boldsymbol{I} & \boldsymbol{1} \\ \boldsymbol{1}^{\mathrm{T}} & 0 \end{bmatrix}.$$

Thus,

$$
\begin{aligned}
|f_T(\boldsymbol{x}) - f_{T^i}(\boldsymbol{x})| &\leq \|\tilde{\boldsymbol{k}}_{\boldsymbol{x}}\| \|\boldsymbol{M}^{-1} - \boldsymbol{M}_i^{-1}\|_2 \|\tilde{\boldsymbol{y}}\| \\
&\leq \sqrt{m\kappa^2 + 1} \|\boldsymbol{M}^{-1} - \boldsymbol{M}_i^{-1}\| \sqrt{m} \\
&\leq m(\kappa + 1) \|\boldsymbol{M}^{-1}(\boldsymbol{M} - \boldsymbol{M}_i)\boldsymbol{M}_i^{-1}\|_2 \\
&\leq m(\kappa + 1) \|\boldsymbol{M}^{-1}\|_2 \|\boldsymbol{M} - \boldsymbol{M}_i\|_2 \|\boldsymbol{M}_i^{-1}\|_2 \\
&\leq m(\kappa + 1) \frac{\|\boldsymbol{M} - \boldsymbol{M}_i\|_2}{m^2\lambda^2} \\
&\leq \frac{\kappa + 1}{m\lambda^2}\beta.
\end{aligned}
$$

Similar with the proof of Eq (2), we can obtain $f_T(\boldsymbol{x}) \leq \frac{\kappa+1}{\lambda}$. Thus, we have

$$
\begin{aligned}
|\ell_{f_T}(z) - \ell_{f_{T^i}}(z)| &= |(f_T(\boldsymbol{x}) - y)^2 - (f_{T^i}(\boldsymbol{x}) - y)^2| \\
&= |(f_T(\boldsymbol{x}) - f_{T^i}(\boldsymbol{x}))(f_T(\boldsymbol{x}) + f_{T^i}(\boldsymbol{x}) - 2y)| \\
&\leq \left(\frac{\kappa + 1}{m\lambda^2}\beta\right)\left(\frac{2\kappa + 2}{\lambda} + 2\right).
\end{aligned}
$$

Similar with the proof of (5) and (6), it is easy to verify that

$$\mathrm{ls}_{m,\ell}(A) \leq C_2\beta^2 = \gamma$$

and

$$\operatorname*{var}_{S}(\ell_{f_{S\setminus S_1}}(S_1)) \leq C_2\beta^2 = \gamma.$$

From Lemma 2, we prove Theorem 2.

Appendix C: Proof of Theorem3

Lemma 3 (Proposition 2 in [13]). *Let h' denote the hypothesis returned by SVMs when using the approximate kernel matrix $\boldsymbol{K}'$. Then, the following inequality holds for all $\boldsymbol{x} \in \mathcal{X}$:*

$$|h'(\boldsymbol{x}) - h(\boldsymbol{x})| \leq \sqrt{2}\lambda\kappa^{\frac{3}{4}}\|\boldsymbol{K}' - \boldsymbol{K}\|_2^{\frac{1}{4}}\left[1 + \left[\frac{\|\boldsymbol{K}' - \boldsymbol{K}\|_2}{4\kappa}\right]^{\frac{1}{4}}\right].$$

Proof (of Theorem 3). Note that $f_{T^i}(\boldsymbol{x})$ is the hypothesis returned by SVM using the ith removed kernel matrix $\boldsymbol{K}^i$. By Lemma 3 and the definition of β kernel stability,

$$|f_T(\boldsymbol{x}) - f_{T^i}(\boldsymbol{x})| \leq \sqrt{2}\lambda\kappa^{\frac{3}{4}}\beta^{\frac{1}{4}}\left[1 + \left[\frac{\beta}{4\kappa}\right]^{\frac{1}{4}}\right].$$

Since the hinge loss ℓ is 1-Lipschitz, so $\forall z, T, z'$

$$|\ell_{f_T}(z) - \ell_{f_{T^i}}(z)| \leq \sqrt{2}\lambda\kappa^{\frac{3}{4}}\beta^{\frac{1}{4}}\left[1 + \left[\frac{\beta}{4\kappa}\right]^{\frac{1}{4}}\right].$$

Similar with the proof of (5) and (6), we can obtain that

$$\mathrm{ls}_{m,\ell}(A) \leq C_3\beta^{\frac{1}{2}}\left(1 + C_3\beta^{\frac{1}{4}}\right)^2 = \gamma$$

and

$$\mathrm{var}_S(\ell_{f_{S\setminus S_1}}(S_1)) \leq C_3\beta^{\frac{1}{2}}\left(1 + C_3\beta^{\frac{1}{4}}\right)^2 = \gamma.$$

Thus, Theorem 3 follows from substituting the above two equations to Lemma 2.

References

1. Bach, F., Lanckriet, G., Jordan, M.: Multiple kernel learning, conic duality, and the SMO algorithm. In: Proceedings of the 21st International Conference on Machine Learning (ICML 2004), pp. 41–48 (2004)
2. Bartlett, P.L., Mendelson, S.: Rademacher and Gaussian complexities: Risk bounds and structural results. Journal of Machine Learning Research 3, 463–482 (2002)
3. Bengio, Y., Grandvalet, Y.: No unbiased estimator of the variance of k-fold cross-validation. Journal of Machine Learning Research 5, 1089–1105 (2004)
4. Blum, A., Kalai, A., Langford, J.: Beating the hold-out: Bounds for k-fold and progressive cross-validation. In: Proceedings of the 12nd Annual Conference on Computational Learning Theory (COLT 1999), pp. 203–208 (1999)
5. Bousquet, O., Elisseeff, A.: Stability and generalization. Journal of Machine Learning Research 2, 499–526 (2002)
6. Cawley, G.C.: Leave-one-out cross-validation based model selection criteria for weighted LS-SVMs. In: Proceeding of the International Joint Conference on Neural Networks (IJCNN 2006), pp. 1661–1668 (2006)

7. Cawley, G.C., Talbot, N.L.C.: Preventing over-fitting during model selection via Bayesian regularisation of the hyper-parameters. Journal of Machine Learning Research 8, 841–861 (2007)
8. Cawley, G.C., Talbot, N.L.C.: On over-fitting in model selection and subsequent selection bias in performance evaluation. Journal of Machine Learning Research 11, 2079–2107 (2010)
9. Chapelle, O., Vapnik, V., Bousquet, O., Mukherjee, S.: Choosing multiple parameters for support vector machines. Machine Learning 46(1-3), 131–159 (2002)
10. Christmann, A., Messem, A.V.: Bouligand derivatives and robustness of support vector machines for regression. Journal of Machine Learning Research 9, 915–936 (2008)
11. Cortes, C., Kloft, M., Mohri, M.: Learning kernels using local Rademacher complexity. In: Advances in Neural Information Processing Systems 25 (NIPS 2013), pp. 2760–2768. MIT Press (2013)
12. Cortes, C., Mohri, M., Pechyony, D., Rastogi, A.: Stability of transductive regression algorithms. In: Proceedings of the 25th International Conference on Machine Learning (ICML 2008), pp. 176–183 (2008)
13. Cortes, C., Mohri, M., Talwalkar, A.: On the impact of kernel approximation on learning accuracy. In: Proceedings of the 13rd International Conference on Artificial Intelligence and Statistics (AISTATS 2010), pp. 113–120 (2010)
14. Debruyne, M., Hubert, M., Suykens, J.A.: Model selection in kernel based regression using the influence function. Journal of Machine Learning Research 9, 2377–2400 (2008)
15. Geras, K.J., Sutton, C.: Multiple-source cross-validation. In: Proceedings of the 30th International Conference on Machine Learning (ICML 2013), pp. 1292–1300 (2013)
16. Kearns, M.J., Ron, D.: Algorithmic stability and sanity-check bounds for leave-one-out cross-validation. Neural Computation 11(6), 1427–1453 (1999)
17. Keerthi, S.S., Sindhwani, V., Chapelle, O.: An efficient method for gradient-based adaptation of hyperparameters in SVM models. In: Advances in Neural Information Processing Systems 19 (NIPS 2007), pp. 673–680. MIT Press (2007)
18. Kloft, M., Brefeld, U., Sonnenburg, S., Zien, A.: l_p-norm multiple kernel learning. Journal of Machine Learning Research 12, 953–997 (2011)
19. Kohavi, R.: A study of cross-validation and bootstrap for accuracy estimation and model selection. In: Proceedings of the 14th International Conference on Artificial Intelligence (IJCAI 1995), pp. 1137–1143 (1995)
20. Kumar, R., Lokshtanov, D., Vassilvitskii, S., Vattani, A.: Near-optimal bounds for cross-validation via loss stability. In: Proceedings of the 30th International Conference on Machine Learning (ICML 2013), pp. 27–35 (2013)
21. Kutin, S., Niyogi, P.: Almost-everywhere algorithmic stability and generalization error. In: Proceedings of the 18th Conference in Uncertainty in Artificial Intelligence (UAI 2002), pp. 275–282 (2002)
22. Lanckriet, G.R.G., Cristianini, N., Bartlett, P.L., Ghaoui, L.E., Jordan, M.I.: Learning the kernel matrix with semidefinite programming. Journal of Machine Learning Research 5, 27–72 (2004)
23. Liu, Y., Jiang, S., Liao, S.: Eigenvalues perturbation of integral operator for kernel selection. In: Proceedings of the 22nd ACM International Conference on Information and Knowledge Management (CIKM 2013), pp. 2189–2198 (2013)
24. Liu, Y., Jiang, S., Liao, S.: Efficient approximation of cross-validation for kernel methods using Bouligand influence function. In: Proceedings of the 31st International Conference on Machine Learning (ICML 2014 (1)), pp. 324–332 (2014)

25. Liu, Y., Liao, S., Hou, Y.: Learning kernels with upper bounds of leave-one-out error. In: Proceedings of the 20th ACM International Conference on Information and Knowledge Management (CIKM 2011), pp. 2205–2208 (2011)
26. Luxburg, U.V., Bousquet, O., Schölkopf, B.: A compression approach to support vector model selection. Journal of Machine Learning Research 5, 293–323 (2004)
27. Ng, A.Y.: Preventing "overfitting" of cross-validation data. In: Proceeding of the 14th International Conference on Machine Learning (ICML 1997), pp. 245–253 (1997)
28. Nguyen, C.H., Ho, T.B.: Kernel matrix evaluation. In: Proceedings of the 20th International Joint Conference on Artificial Intelligence (IJCAI 2007), pp. 987–992 (2007)
29. Poggio, T., Rifkin, R.M., Mukherjee, S., Niyogi, P.: General conditions for predictivity in learning theory. Nature 428(6981), 419–422 (2004)
30. Rakotomamonjy, A., Bach, F., Canu, S., Grandvalet, Y.: SimpleMKL. Journal of Machine Learning Research 9, 2491–2521 (2008)
31. Rogers, W.H., Wagner, T.J.: A finite sample distribution-free performance bound for local discrimination rules. The Annals of Statistics 6, 506–514 (1978)
32. Saunders, C., Gammerman, A., Vovk, V.: Ridge regression learning algorithm in dual variables. In: Proceedings of the 15th International Conference on Machine Learning (ICML 1998), pp. 515–521 (1998)
33. Schölkopf, B., Smola, A.J.: Learning with kernels. MIT Press, Cambridge (2002)
34. Shalev-Shwartz, S., Shamir, O., Srebro, N., Sridharan, K.: Learnability, stability and uniform convergence. Journal of Machine Learning Research 11, 2635–2670 (2010)
35. Suykens, J.A.K., Vandewalle, J.: Least squares support vector machine classifiers. Neural Processing Letters 9(3), 293–300 (1999)
36. Vapnik, V.: The nature of statistical learning theory. Springer (2000)

Experimental Design in Dynamical System Identification: A Bandit-Based Active Learning Approach

Artémis Llamosi[1,4], Adel Mezine[1], Florence d'Alché-Buc[1,3], Véronique Letort[2], and Michèle Sebag[3]

[1] Informatique Biologie Intégrative et Systèmes Complexes (IBISC),
Université d'Evry-Val d'Essonne, France
`artemis.llamosi@univ-paris-diderot.fr`
`{amezine,florence.dalche@ibisc.univ-evry.fr}`
[2] Ecole Centrale Paris, 92295 Châtenay-Malabry Cedex
`veronique.letort@ecp.fr`
[3] TAO, INRIA Saclay
Laboratoire de Recherche en Informatique (LRI), CNRS, Université Paris Sud, Orsay,
France
`Michele.Sebag@lri.fr`
[4] Laboratoire Matière et Systèmes Complexes, Université Paris Diderot & CNRS,
75013 Paris, France
INRIA Paris-Rocquencourt, Rocquencourt, 78153 Le Chesnay, France

Abstract. This study focuses on dynamical system identification, with the reverse modeling of a gene regulatory network as motivating application. An active learning approach is used to iteratively select the most informative experiments needed to improve the parameters and hidden variables estimates in a dynamical model given a budget for experiments. The design of experiments under these budgeted resources is formalized in terms of sequential optimization. A local optimization criterion (reward) is designed to assess each experiment in the sequence, and the global optimization of the sequence is tackled in a game-inspired setting, within the Upper Confidence Tree framework combining Monte-Carlo tree-search and multi-armed bandits.

The approach, called EDEN for Experimental Design for parameter Estimation in a Network, shows very good performances on several realistic simulated problems of gene regulatory network reverse-modeling, inspired from the international challenge DREAM7.

Keywords: Active learning, experimental design, parameter estimation, Monte-Carlo tree search, Upper Confidence Tree, ordinary differential equations, e-science, gene regulatory network.

1 Introduction

A rising application field of Machine Learning, e-science is concerned with modeling phenomena in e.g. biology, chemistry, physics or economics. The main goals

T. Calders et al. (Eds.): ECML PKDD 2014, Part II, LNCS 8725, pp. 306–321, 2014.
© Springer-Verlag Berlin Heidelberg 2014

of e-science include the prediction, the control and/or the better understanding of the phenomenon under study. While black-box models can achieve prediction and control goals, models consistent with the domain knowledge are most desirable in some cases, particularly so in domains where data is scarce and/or expensive.

This paper focuses on the identification of dynamical systems from data, with gene regulatory network reverse modeling as motivating application [26]. We chose the framework of parametric ordinary differential equations (ODE) [18,11] whose definition is based on the domain knowledge of the studied field. Our goal is restricted to *parametric identification*. Formally, it is assumed that the structure of the ODE model is known; the modeling task thus boils down to finding its m-dimensional parameter vector $\boldsymbol{\theta}$. Setting the ODE parameter values, also referred to as reverse-modeling, proceeds by solving an optimization problem on $\mathbb{R}^m$, with two interdependent subtasks. The first one is to define the target optimization criterion; the second one is to define the experimental setting, providing evidence involved in the optimization process.

Regarding the first subtask, it must be noticed that parametric ODE identification faces several difficulties: i) the behavior described by the ODE model is hardly available in closed form when the ODE is nonlinear and numerical integration is required to identify the parameters, ii) the experimental evidence is noisy, iii) the data is scarce due to the high costs of experiments, iv) in some cases the phenomenon is partially observed and therefore depends on hidden state variables. To overcome at least partially these difficulties, several estimation methods have been employed using either frequentist [11] or Bayesian inference [27]. In case of hidden variables, Expectation-Maximization approaches and filtering approaches with variants devoted to nonlinear systems such as the Unscented Kalman Filter (UKF) [31] and the Extended Kalman Filter [37] have been applied to ODE estimation.

Overall, the main bottleneck for parameter estimation in complex dynamical systems is the non-identifiability issue, when different parameter vectors θ might lead to the same response under some experimental stimuli[1]. The non-identifiability issue is even more critical when models involving a high-dimensional parameter vector $\boldsymbol{\theta}$ must be estimated using limited evidence, which is a very common situation. To mitigate the non-identifiability of parameters and hidden states in practice, the e-scientist runs complementary experiments and gets additional observations. Ideally, these observations show some new aspects of the dynamical behavior (e.g. the *knock-out* of a gene in an organism), thereby breaking the non-identifiability of parameters. The selection of such (expensive) complementary experiments is referred to as *design of experiments (DOE)*. The point is to define the optimal experiments in the sense of some utility function, usually measuring the uncertainties on $\boldsymbol{\theta}$ (including the non-identiabilities), and depending on the experiment, the data observed from it and the quality of estimates produced by some chosen estimation procedure. For instance, the utility

---

[1] See [32] for a presentation of non-identifiability issues, beyond the scope of this paper.

function can refer to the (trace of) the covariance matrix of the parameter estimate. DOE has been thoroughly studied from a statistical point of view for various parameter estimation problems (see [36,16,25]), within a frequentist or a Bayesian framework [10,28]. Usual definitions of utility include functions of Fisher information in the frequentist case [35] or of the variance of the estimated posterior distribution in the Bayesian case [5]. In this work, we focus on the case of sequential experimental design [33,6], which is the most realistic situation for experimentations in a wet laboratory.

The limitations of current standard sequential DOE is twofold. Firstly, it seldom accounts for the cost of the experiments and the limited budget constraint on the overall experiment campaign. Secondly and most importantly, it proceeds along a myopic strategy, iteratively selecting the most informative experiment until the budget is exhausted.

The contribution of the present paper is to address both above limitations, formulating DOE as an *active learning* problem. Active learning [12,13,14,3] allows the learner to ask for data that can be useful to improve its performance on the task at hand. In this work, we consider active learning as a *one-player game* similarly to the work of [34] devoted to supervised learning and propose a strategy to determine an optimal set of experiments complying with the limited budget constraint. The proposed approach is inspired by the Upper Confidence Tree (UCT) [24,34], combining Monte-Carlo tree search (MCTS) [7] and multi-armed bandits [9]. Formally, a reward function measuring the utility of a set of experiments is designed, and UCT is extended to yield the optimal set of experiments (in the sense of the defined reward function) aimed at the estimation of parameters and hidden variables in a multivariate dynamical system.

The approach is suitable for any problem of parameter estimation in ODE where various experiments can be defined: those experiments can correspond to the choice of the sampling time of observation, the initial condition in which the system is primary observed or some intervention on the system itself. In this work, the approach is illustrated considering the reverse modeling of gene regulatory networks (GRN) in systems biology [11,26]. GRN are dynamical systems able to adapt to various input signals (e.g. hormones, drugs, stress, damage to the cell). GRN identification is a key step toward biomarkers identification [4] and therapeutical targeting [23].

After an introduction of the problem formalization, the paper gives an overview of the proposed approach, based on an original reward function and extending the UCT approach. A proof of concept of the presented approach on three realistic reverse-modeling problems, inspired by the international DREAM7 [15] challenge, is then presented in the application section.

2   Problem Setup

We consider a dynamical system whose state at time t is the d-dimensional vector $\mathbf{x}(t)^T = [x_1(t) \ldots x_d(t)]$ and whose dynamics are modeled by the following first-order ODE:

$$\dot{\mathbf{x}}(t) = \mathbf{f}(\mathbf{x}(t), \mathbf{u}(t); \boldsymbol{\theta}) \ , \tag{1}$$

where $\dot{\mathbf{x}}(t) = \frac{d\mathbf{x}(t)}{dt}$ denotes the first order derivative of $\mathbf{x}(t)$ with respect to time, function $\mathbf{f}$ is a non linear mapping, $\boldsymbol{\theta}$ is the m-dimensional parameter vector, $\mathbf{u}(t)$ is an exogenous input to the system. Let us first assume that we partially observe its behavior given some initial condition $\mathbf{x}(0) = \mathbf{x}_0$ and with some neutral input $\mathbf{u}(t) = g_0(t)$, e.g. without any *intervention* (as defined below). Let $\mathbf{H}$ be the observation model, typically a projection of $\mathbb{R}^d$ in a lower dimensional space $\mathbb{R}^p$ $(p < d)$, $\mathbf{Y}_0 = (\mathbf{y}_{t_k}^0)_{k=0,\dots,n-1}$, a time series of n p-dimensional observations and $(\epsilon_{t_k})_{k=0,\dots,n-1}$, n i.i.d realizations of a p-dimensional noise. For sake of simplicity, $\mathbf{y}_{t_k}$ (resp. ϵ_{t_k}) will be noted $\mathbf{y}_k$ (resp. ϵ_k). Given these assumptions, the observations and the states of the system [31] can now be expressed as follows: given $k = 0,\dots,n-1$:

$$\mathbf{x}(0) = \mathbf{x}_0$$

$$\mathbf{x}(t_{k+1}) = \mathbf{x}(t_k) + \int_{t_k}^{t_{k+1}} f(\mathbf{x}(\tau), \mathbf{u}(\tau), \boldsymbol{\theta}) d\tau$$

$$\mathbf{y}_k = \mathbf{H}(\mathbf{x}(t_k), \mathbf{u}(t), \boldsymbol{\theta}) + \epsilon_k \ . \tag{2}$$

This model can be seen as a special state-space model where the hidden process is deterministic and computed using a numerical integration. Different tools such as nonlinear filtering approaches such as Unscented Kalman Filtering (UKF) [31] and extended Kalman Filtering (EKF) [37] can be applied. However it is well known that nonlinearity and limited amount of data can lead to practical non-identifiability of parameters. Namely, two different parameter solutions can provide the same likelihood value. A well known way to address this issue is to intervene on the dynamical system to perform additional experiments producing observations that exhibit different kinetics. It can consist either in perturbing the system, e.g. forcing the level of a state variable to be zero, or in changing the observation model by allowing to observe different state variables. To benefit from these new data during the estimation phase, the ODE model must account for all the available experiments defined by a finite set of size E: $\mathcal{E} = \mathcal{E}_0 = \{e_1,\dots,e_E\}$. This can be done by defining adequately the exogenous input $\mathbf{u}(t)$ among a set of intervention functions $\mathbf{g}_e(t), e \in \mathcal{E}$ as shown in the application section.

Choosing the appropriate interventions (experiments) to apply to the system in order to produce better estimates of parameter and hidden states is the purpose of this work. We are especially interested in an *active learning algorithm* that sequentially selects at each step ℓ, the next experiment e_ℓ^* among the candidate experiments of the set $\mathcal{E}_\ell = \mathcal{E}_{\ell-1} - \{e_{\ell-1}^*\}$, that will produce the most useful dataset for the estimation task. Contrary to the purely statistical approaches of experimental design, ours aims at offering the possibility to anticipate on the fact that one given experiment will be followed by others. The search for an optimal $e_\ell^* \in \mathcal{E}_\ell$ thus depends on the potential subsequent sequences of experiments, their total number being limited by a finite affordable *budget* to account for the cost of real experiments.

Algorithm 1. EDEN or *real game*

Initialization (section 3.2)
while (budget not exhausted) and (estimates not accurate) **do**
  Design a new experiment using Upper Confidence Tree (UCT) as in Algorithm 2
  (section 3.3)
  Perform the proposed experiment and re-estimate parameters with the augmented
  dataset (section 3.4)
  Evaluate the estimates (section 3.5)
end while

---

3   Game-Based Active Learning for DOE

Please note that in the following, to simplify the description of the approach, we will only talk about parameter estimates, implying hidden state and parameter estimates.

3.1   Complete Algorithm

Active learning of parameters and hidden states in differential equations is considered as a one-player game. The goal of the game is to provide the most accurate estimates of parameters and hidden states. Before the game begins, a first estimate of the hidden states and parameters is obtained using an initial dataset (here unperturbed, termed *wild type*). Then, at each turn, the player chooses and buys an experiment and receives the corresponding dataset. This new dataset is incorporated into the previous dataset and parameters are re-estimated. This procedure, described in Algorithm 1, is repeated until the quality of estimates is sufficiently high or the player has exhausted the budget.

3.2   Initialization

At the beginning of the game, the player is given:

- An initial dataset, here a time series $\mathbf{Y}_0 : \{\mathbf{y}_0^0, \ldots, \mathbf{y}_{n-1}^0\}$, corresponding to the partial observation of the wild type system measured at time $t_0, \ldots, t_{n-1}$ with given initial condition.
- A system of parametric ordinary differential equations, $\mathbf{f}$, of the form of Eq. (1) and an observation model $\mathbf{H}$.
- A set of experiments $\mathcal{E} = \mathcal{E}_0 = \{1, \ldots, E\}$ along with their cost (for simplicity and without loss of generality, the cost is assumed in this work to be equal to 1 for all experiments).
- A total budget: $B \in \mathbb{R}$ (here the total number of experiments we can conduct) and an optimizing horizon T which states how many experiments we optimize jointly at each iteration of Algorithm 1.
- A *version space*, Θ, which represents all the probable parametrization of our system, compatible with the observed initial set. More precisely, it consists

of a candidate set of *hypotheses* $\Theta(\mathbf{Y}_0) = \{\boldsymbol{\theta}_1^{*(0)}, \boldsymbol{\theta}_2^{*(0)}, \ldots, \boldsymbol{\theta}_m^{*(0)}\}$: a parameter vector can be considered as a hypothesis, i.e. included in the version space, if the simulated trajectories of the observed state variables it generates are consistent with the available dataset. The initial version space is built from the means of the posterior distributions of parameters estimated from the initial dataset $\mathbf{Y}_0$. Building up on previous works [31], we learn m Unscented Kalman Filter (UKF), as described in 3.4, starting with m different initializations and flat priors.

- A reward (or utility) function used in the design procedure, described in 3.3.

3.3   Design of Experiment Using Upper Confidence Tree

The ℓth move of the *real game*, i.e. the EDEN protocol, consists of running a Monte-Carlo Tree Search (MCTS) in order to find the best first experiment to perform given it is followed by $T - 1$ experiments (or less if we have a remaining budget that does not allow for $T - 1$ experiments).

The utility of a sequence of experiments is inherently a random variable because of the uncertainty on the true system (the true parameter vector $\boldsymbol{\theta}_{true}$ is not known), but also because of the particular realization of the measurement noise (note that for stochastic models, additional uncertainty would come from the process noise). In addition, the utility of a sequence of experiments is not additive in the single experiments' utilities. Therefore we optimize a tuple of experiments (with size of the horizon, *i.e.* according to the available budget), even though only the first experiment of the sequence will be performed at a given iteration of EDEN. This problem is addressed by seeing the sequence of experiments as arms in a *multi armed bandit* (MAB) problem.

Upper Confidence Tree (UCT). UCT, extending the multi-armed bandit setting to tree-structured search space [24], is one of the most popular algorithm in the MCTS family and was also proposed to solve the problem of active learning in a supervised framework by [34]. Its application to sequential design under budgeted resources is to our knowledge an original proposal. A sketch is given in Algorithm 2.

UCT simultaneously explores and builds a search tree, initially restricted to its root node, along N tree-walks. Each tree-walk involves several phases:

The **bandit phase** starts from the root node (where all available experiments are represented by accessible nodes) and iteratively select experiments until arriving at an unknown node or a leaf (distance T from the root). Experiment selection is handled as a MAB problem. The selected experiment $\tilde{e}_\ell$ in $\mathcal{E}_{\ell,known}$ maximizes the Upper Confidence Bound [1]:

$$\tilde{e}_\ell = \arg \max_{node_i \in \mathcal{E}_{\ell,known}} \left(\hat{\mu}_i + C \times \sqrt{\frac{log(\sum_j n_j)}{n_i}} \right). \tag{3}$$

where :

- $\mathcal{E}_{\ell,known}$ is the set of known nodes (already visited) which are accessible from the current position (ℓ^{th} experiment in the Path).
- $\hat{\mu}_i$ is the mean utility of node i
- n_i is the number of time node i has been visited before
- C is a tuning constant that favors exploration when high and exploitation when low. Its value is problem specific and must be compared to both the number of possible experiments and the overall mean utility of a sequence of experiments. In the illustration, we used $C = \sqrt{10}$.

The bandit phase stops upon arriving at an unknown node (or leaf). Then in the **tree building phase**, a new experiment is selected at random and added as a child node of this current leaf node. This is repeated until arriving in a terminal state as determined by the size of the horizon. Overall, we can summarize this procedure as going from root to a leaf following a path of length T. When children nodes are known, the UCB criterion is applied, when they are not known, a random choice is performed and the node is created. At this point the reward R of the whole sequence of experiments is computed and used to update the cumulative reward estimates in all nodes visited during the tree-walk.

One of the great features of the proposed method is to perform a *biased* Monte Carlo tree search thanks to the UCB criterion which preserves optimality asymptotically and ensures we build an UCT. After some pure random exploration of the tree, this criterion makes a rational trade-off between exploration (valuation of untested sequences of experiments) and exploitation (improving the estimation of mean utility for an already tested sequence).

When a sufficient number of tree walks has been performed, we select the next experiment to make among the nodes (experiments) directly connected to the root. This choice is based on the best mean score (but could have been selected by taking the most visited node: when the number of tree walks is sufficiently high, these two options give the same results).

Surrogate Hypothesis. A reward function is thus required that measures how informative a sequence of experiment is. The tricky issue is that the true parameter vector θ_{true} is not known and therefore cannot be used as a reference for evaluating the obtained estimates. As in [34], we proceed by associating to each tree-walk a surrogate hypothesis θ^*, drawn from the version space $\Theta_{\ell-1}$, that will represent the true parameter θ_{true} in the current tree walk. The reward R attached to this tree walk is computed by i) estimating the parameters $\hat{\theta}$ from the obtained dataset; ii) evaluating the estimate $\hat{\theta}$.

Here we present two different approaches to evaluate this estimate and thus to calculate the reward. The reward R_1 calculates a quantity related to the (log) empirical bias of the parameter estimate. The average reward associated to a node of the tree, i.e. to a sequence of experiments, thus estimates the expectation over $\Theta_{\ell-1}$ of the estimation error yielded by the choice of this sequence, e.g. the (log) bias of the parameter estimate. The reward R_2 calculates the empirical variance of the parameter estimate and thus does not use the current surrogate hypothesis θ^*.

Algorithm 2. UCT pseudo-code

```
 1: Input:
 2: Hypothesis Space: Θ
 3: Budget: B
 4: Max Horizon: T
 5: Maximal number of tree-walks: N
 6: Initialize :
 7: walk = 1
 8: while walk ≤ N do
 9:    current_node = root
10:    Sample a surrogate hypothesis: θ ∼ Θ
11:    Path = {current_node}
12:    Init virtual budget: b = min(B, T)
13:    while b ≥ min_{i∈ε}(cost(e_i)) do
14:       e = UCB(current_node)
15:       current_node = e
16:       Path = {Path ∪ current_node}
17:       b = b − cost(e)
18:    end while
19:    Reward = R(Path, θ*)
20:    Update path score: Update(Path, Reward)
21:    walk = walk + 1
22: end while
23: e* = MaxReward(root)
```

Reward 1. The concept of this utility function is to quantify how well the selected experiments allow the parameters' estimation to converge towards the true parameters. At each turn ℓ, the uncertainty on the true parameters of the system is captured by the distribution of likely parameter candidates $\boldsymbol{\theta}^* \in \Theta_{\ell-1}$. The utility function for $R1$ does not require any specific assumption on the model itself and only requires an estimation method and a way to value the quality of that estimation. It is computed using the following procedure: Let $\boldsymbol{\theta}^* \in \Theta$ be the current surrogate hypothesis, and Estimation : (prior $\pi, \tilde{\mathbf{Y}}_{1:k}(\boldsymbol{\theta}^*)) \mapsto \hat{\boldsymbol{\theta}}$ be an estimation procedure, here bayesian, where π is some prior distribution on $\boldsymbol{\theta}$, $\tilde{\mathbf{Y}}_{1:k}(\boldsymbol{\theta}^*)$ is the set of simulated data according to the observation model given in the problem setting and corresponding to a sequence of k experiments, $\tilde{e}_{1:k}$, when considering the surrogate hypothesis $\boldsymbol{\theta}^*$ as the true parameters. We can evaluate this estimation by comparing the estimated parameters, $\hat{\boldsymbol{\theta}} = E[\boldsymbol{\theta}|\tilde{\mathbf{Y}}_{1:k}(\boldsymbol{\theta}^*))]$ to the current $\boldsymbol{\theta}^*$. In this work we use the following metric to measure the quality of estimate $\hat{\boldsymbol{\theta}}$, based on the DREAM 7 challenge [30]:

$$d(\boldsymbol{\theta}^*, \hat{\boldsymbol{\theta}}) = \sum_{i=1}^{m} ln\left(\frac{\theta_i^*}{\hat{\theta}_i}\right)^2. \tag{4}$$

Where θ_i^* is the ith component of θ_i^* and we sum over all components. Overall, this defines a semi-metric (lacking triangular inequality) on the space of parameters. This semi-metric is proportional to the mean squared logarithmic ratio of the parameters, and so penalizes fold changes in parameters' values. This is especially relevant in estimating biological parameter values that can span several orders of magnitude and where observables may be very insensitive to some parameter values [17]. With all these notations, the utility function returned at each iteration of the MCTS is:

$$r1(\tilde{\mathbf{Y}}_{1:k}(\boldsymbol{\theta}^*), \pi, \boldsymbol{\theta}^*) = -d(\boldsymbol{\theta}^*, \hat{\boldsymbol{\theta}}) \ . \tag{5}$$

In this work, we chose the prior π as a Gaussian distribution whose mean is a randomly perturbed $\boldsymbol{\theta}^*$ (mean of $\pi = \boldsymbol{\theta}^*.\epsilon$) with $\epsilon \sim \mathcal{N}(1, 0.1)$. The prior covariance is set to the identity matrix. Because of the prior π, we only perform an estimation around the target value $\boldsymbol{\theta}^*$, which explains why this reward is called *local*. The rationale behind this being that, assuming our representation of the version space is fine enough, we will always find a sample not too far away from the true value (here, further than 10% on average for each dimension). In the end, since this function is called within a MCTS framework, the relevant utility for the selection of experiments is the average over different calls to the function. Given $Path = (\tilde{e}_{1:k})$ the sequence of chosen experiments, we have:

$$R_1(\tilde{e}_{1:k}, \boldsymbol{\theta}^*) = \mathbb{E}_\epsilon[r1(\tilde{\mathbf{Y}}_{1:k}(\boldsymbol{\theta}^*), \pi_\epsilon, \boldsymbol{\theta}^*)] \ . \tag{6}$$

Thus R_1 compares the expectation of the posterior probability defined from data $\tilde{\mathbf{Y}}_{1:k}(\boldsymbol{\theta}^*)$ produced by experiments $\tilde{e}_{1:k}$ to the parameter $\boldsymbol{\theta}^*$, coordinate by coordinate. The main interest of this reward function is that it can be straightforwardly applied to any estimation method and that its only significant assumption is that the version space is fine-grained enough. In this respect, it is said to be *agnostic*. On the other hand, its main drawback is that it is usually computationally expensive (depending on the estimation scheme used).

Reward 2. In the second reward, we also solve an estimation problem using a joint UKF starting from a Gaussian prior centered on the surrogate θ^* and with an identity covariance matrix. The reward is classically defined in relation to the evolution of the trace of the covariance of the posterior :

$$R_2(\tilde{e}_{1:k}, \boldsymbol{\theta}^*) = -\sum_{i=1}^m VAR[\theta_i|\tilde{\mathbf{Y}}_{1:k}(\boldsymbol{\theta}^*)] \ . \tag{7}$$

3.4   Performing Experiments and Re-estimation of Parameters and Hidden Variables

Having estimated an optimal sequence of experiments, we will perform (or simulate noisy data for the in-silico illustration) one experiment only. This allows us to subsequently choose the next experiment benefiting from the new information brought by the genuine acquired data.

An estimation procedure is required in the *virtual games* (MCTS iterations) each time a reward of a sequence of experiments has to be calculated, as well as in the *real game*, when the real data are acquired. In the case of the *virtual game*, each experiment $\tilde{e}$ in a sequence to be evaluated corresponds to the basal model perturbed with some specific exogenous input $\mathbf{u}(t) = \mathbf{g}_{\tilde{e}}(t)$. Therefore, the learning problem turns to the joint estimation of different models sharing some parameters (and some states). To achieve this joint learning task, we propose an original strategy that consists of aggregating the different models corresponding to different interventions into a single fused model. Then, we apply a Bayesian filtering approach devoted to nonlinear state-space models, the UKF, to the new system. Parameters are estimated together with hidden states using an augmented state approach. UKF provides an approximation of the posterior probability of $\boldsymbol{\theta}$ given the multiple time series corresponding to the multiple experiments, allowing to calculate the different rewards, the bias-like reward R_1 or the variance reward R_2.

In the case of the *real game*, at each turn, different models have to be jointly learnt from the previous datasets and the new one, just acquired after a purchase of an experiment. This can be performed exactly the same way using UKF on a single fused model.

3.5 Evaluation of the Quality of Estimates in the Real Game

During the real game, the quality of the estimate is measured by the trace of the covariance of the UKF estimate.

4 Application to Reverse-Modeling of Gene Regulatory Networks

4.1 Model Setting

Let us consider a simple gene regulatory network that implements the transcriptional regulatory mechanisms at work in the cell [26]. We denote by d the number of genes and assume, for the sake of simplicity, that one gene codes for one protein. In contrast, a gene can be regulated by several genes, including self-regulation, with interactions of several possible types, additive or multiplicative: a gene j is said to regulate a gene i if the level of expression of gene j influences the level of expression of gene i. The vector $\mathbf{r}(t) \in \mathbb{R}^d$ denotes the expression levels (mRNA concentration) of the d genes at time t and the vector $\mathbf{p}(t) \in \mathbb{R}^d$, the concentration of the encoded proteins. Similarly to one of the challenges [30,15] in DREAM6 (2011) and DREAM7 (2012), we consider a problem of parameter and hidden variable estimation in a Hill kinetics model. In the numerical simulations, we apply EDEN on 3 different reverse-modeling problems of GRN of increasing size (3, 5, 7) whose graphs are represented in Fig. 1.

In the following, we introduce the ODE system and the set of experiments on the second target dynamical system composed of 5 genes. The dynamics of this network can be represented by the following system of differential equations associated to the regulation graph represented in Fig. 1B:

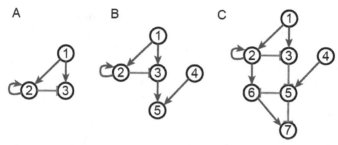

Fig. 1. Regulation graph of the 3 models. Blue arrows represent activations and red bars represent inhibitions.

$$\dot{r}_1(t) = \gamma_1 - k_1^r \cdot r_1(t)$$
$$\dot{r}_2(t) = \gamma_2 \cdot \left(h_{21}^+(t) + h_{22}^+(t)\right) - k_2^r \cdot r_2(t)$$
$$\dot{r}_3(t) = \gamma_3 \cdot h_{31}^+(t) \cdot h_{32}^-(t) - k_3^r \cdot r_3(t)$$
$$\dot{r}_4(t) = \gamma_4 - k_4^r \cdot r_4(t)$$
$$\dot{r}_5(t) = \gamma_5 \cdot \left(h_{53}^+(t) + h_{54}^+(t)\right) - k_5^r \cdot r_5(t)$$
$$\dot{p}_i(t) = \rho_i \cdot r_i(t) - k_i^p \cdot p_i(t), \quad \forall i = \{1, \ldots, 5\} \ . \tag{8}$$

where h_{ji}^+ is the Hill function for activation defined as: $h_{ji}^+(t) = \frac{p_j(t)^2}{K_{ji}^2 + p_j(t)^2}$ and h_{ij}^- is the Hill function for inhibition defined as: $h_{ji}^-(t) = \frac{K_{ji}^2}{K_{ji}^2 + p_j(t)^2}$. The parameters K_{ji} is called dissociation constant of the regulation of gene i by the protein p_j. The set of parameter to estimate is then $\boldsymbol{\theta} = [(\gamma_i)_{i=\{1,\ldots,5\}}, (K_{ji})_{\{(i,j),\ j\to i\}}]$ and the state vector: $\mathbf{x}(t)^T = [\mathbf{r}(t)\ \mathbf{p}(t)]^T$. As in the DREAM7 challenge, the initial conditions are chosen as: $\mathbf{x}_0 = [0.4\ 0.7\ 0.5\ 0.1\ 0.9\ 0.4\ 0.3\ 1.0\ 1.0\ 0.8]^T$ and are used to simulate as well the wild-type as the perturbation experiments. As for the observations, only one type of state variable, protein concentrations or mRNA concentrations, can be measured at a time, the other one being then considered as hidden state.

Two kinds of perturbations are considered: the knock-out (ko) that fully re-presses the expression of the targeted gene, and the over-expression (oe) that accelerates the translation of the targeted protein. In our problem, only one perturbation can be applied at a time. In order to simulate the behavior of the perturbed system, we introduce in the model two types of intervention functions, $\mathbf{g_{oe}}$ and $\mathbf{g_{ko}}$, for each gene. The wildtype system corresponds to the case of these two control variables being equal to 1. Taking $g_{ko}^i(t) = 0, \forall t \geq 0$, for gene i, simulates a knock-out on this gene by removing the production term of mRNA and protein. For instance, under a knock-out on gene 1, the equations for mRNA 1 write as:

$$\dot{r}_1(t) = g_{ko1}(t) \cdot \gamma_1$$

Taking $g_{oe}^i(t) = 2, \forall t \geq 0$, for protein i, simulates the corresponding over-expression since the production term of the protein concentration p_i is then doubled. Applying this perturbation on gene 1 gives:

$$\dot{p}_1(t) = g_{oe1}(t) \cdot \rho_1 \cdot r_1(t) - k_1^p \cdot p_1(t) \ . \tag{9}$$

Overall, 11 perturbations are considered including the wild-type, with two possible observation models (either protein or mRNA concentrations for each of them), giving in total 22 potential experiments to perform for the 5-genes network. For the 3- (resp. 7) genes network, 14 (resp. 30) experiments are considered.

4.2   Numerical Results

In this section, we describe the results we obtained on the systems described in the previous section. These simulations of an experimental design problem were performed using the two reward functions R_1 and R_2 described in (6) and (7) with an hypothesis space Θ represented by samples (1000 if not stated otherwise). A convergence criterion was used in order to limit the number of tree walks performed in the MCTS phase of the algorithm. This criterion allowed to stop tree walks as soon as the mean utility associated to all experiments did not change by more than 10% over the last 20 walks. For all details, you are encouraged to request our Matlab© code (based on the *pymaBandits* framework [20,8,19]).

Figure 4.2 shows that some well chosen experiments provide a significant (more than 100 folds) reduction of the uncertainty on parameters' value. But some of the additional experiments can lead to only marginally decreasing the quality of estimation. The experiments chosen with R_1 or R_2 are not the same. We also see in Figure 4.2 B that the number of samples forming the version space can change significantly the performance as the algorithm takes into account uncertainty on the system which is related to the number of samples. Although the results after 5 experiment purchases are similar, the same performance can be achieved with only 3 experiments if uncertainty is properly accounted for.

Figure 3 A reports the scores obtained by applying the reward R_1 for 3 sizes of network. All were using an horizon of $T = 3$ experiments and a version space represented by 1000 samples. Concerning the scores, the more complex problem leads nearly only to a reduction on the uncertainty but could not improve significantly the estimations. This is because increasing complexity implies usually more non-identifiability and requires a larger budget. These results also illustrate the complexity of experimental design: since less genes means less means to acquire data indirectly on a gene's parameters, the 3-gene network is not significantly better estimated than the 5 genes network with the same learning horizon. Concerning the computation scaling, networks of 3, 5 and 7 genes have respectively required 6, 14 and 24 hours on a quad-core Intel i7 processor at 4,5 GHz.

Current methods for sequential experimental design generally optimize one experiment at a time. Even if we only acquire one experiment per iteration of

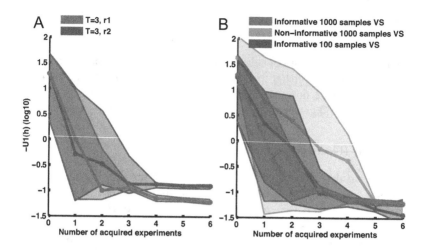

Fig. 2. A. Impact of the reward function: evolution of the (log10) scores for the 5-gene network, $d(\theta_{true}, \theta)$ for $\theta \in \Theta$ (1000 samples) using either R_1 or R_2. Are plotted the minimum, the maximum and the average over all Θ. Starting from a large Θ which only included information from the unperturbed (*wildtype*) observations, a well chosen sequence of experiments can lead to very significant improvement of the estimation of the parameters. B. Effect of the version space representation: We compare the performance on the 5- gene model, with an horizon T=3 for EDEN run starting from either the typical 1000 samples Θ which contains the information from the *wildtype* observation, the same but using 1000 uniformly distributed samples (non-informative) or an 100 sample version space (taking the best 100 samples in terms of mean squared deviation from the prior information) from the informative version space.

EDEN, we optimize sequences of experiments up to a given *learning horizon*, T. This obviously implies sampling in a much bigger space of possible designs ($O(|\mathcal{E}|^T)$) but allows at the same time to consider experimental strategies that mitigate the risk of individual experiments when the outcome is uncertain. As we can see on figure 3 B, a different learning horizon can change importantly the speed of reduction of uncertainty and estimation quality. Interestingly for that particular problem, an horizon of 3 lead the algorithm to take some risk (given the uncertainty on the system) that did not pay-off as a greedy version (T=1) performs better. But with a larger horizon (T=5), the risk mitigation is differently considered by the algorithm and an excellent performance is achieved in 2 experiments only.

5   Conclusion

We developed an active learning approach, EDEN, based on a one-player game paradigm to improve parameters and hidden states estimates of a dynamical system. This setting is identical with that of active learning [34] for supervised

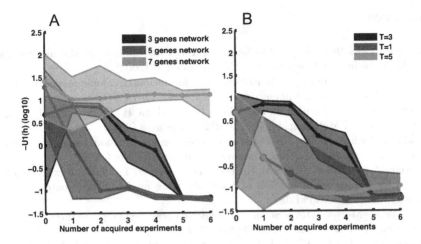

Fig. 3. A. Performance on problems of increasing complexity: evolution of the (log10) scores $d(\theta_{true}, \theta)$ for the 3,5 and 7-gene networks, for $\theta \in \Theta$ (1000 samples) using R_1. Are plotted the minimum, the maximum and the average over all Θ. B. Performance for various learning horizon T=1, 3 and 5 for the 3 genes model, using R_1 and a 1000 samples version space.

learning, where theoretical guarantees have been given along the following lines. The active learning (here experimental design) problem is equivalent to a reinforcement learning problem that can be expressed formally in terms of a Markov Decision Process; this problem is intractable but approximation with asymptotic guarantees are provided by the UCT algorithm [24,29,34]. Future work will focus on lightening these guarantees in the framework of dynamical system identification.

Furthermore, to our knowledge, this is the first application of UCT-based approaches to sequential experimental design for dynamical nonlinear systems, opening the door to a very large number of potential applications in scientific fields where experiments are expensive. The versatility of the proposed framework allows to extend it in various ways. Different dynamical models including stochastic ones can be in principle used with this strategy while other rewards can be designed. An interesting perspective is also to link the theoretical guarantee brought by UCT with the framework of Bayesian experimental design [28]. Finally, considering the scalability issue, we notice that the learning horizon (the number of experiments we jointly optimize) does not need to scale with the size of the model. In fact, the relevant horizon is the number of experiments allowing to eliminate the non-identifiability for the set of parameters of a given model. This means that the approach can be in principle extended to larger systems. Although automated experimental design approaches are still an exception in wet laboratories, some pioneering works on the robot scientist *Adam* [21,22] show the immense potential offered by realistic and practice-oriented active learning in biology and other experimental sciences.

References

1. Auer, P., Cesa-Bianchi, N., Fischer, P.: Finite-time analysis of the multiarmed bandit problem. Machine Learning 47(2-3), 235–256 (2002)
2. Asprey, S.P., Macchietto, S.: Statistical tools for optimal dynamic model building. Computers & Chemical Engineering 24:2017, 1261–1267 (2000)
3. Azimi, J., Fern, A., Fern, X.Z., Borradaile, G., Heeringa, B.: Batch Active Learning via Coordinated Matching. In: ICML 2012. Omnipress (2012)
4. Baldi, P., Hatfield, G.W.: DNA microarrays and gene expression: from experiments to data analysis and modeling. Cambridge University Press (2002)
5. Bandara, S., Schlöder, J.P., Eils, R., Bock, H.G., Meyer, T.: Optimal experimental design for parameter estimation of a cell signaling model. PLoS Computational Biology 5(11), e1000558 (2009)
6. Blot, W.J., Meeter, D.A.: Sequential experimental design procedures. Jour. of American Statistical Association 68(343), 343 (1973)
7. Browne, C.B., Powley, E., Whitehouse, D., Lucas, S.M., Cowling, P.I., Rohlfshagen, P., Colton, S.: A survey of Monte-Carlo tree search methods. Intelligence and AI 4(1), 1–49 (2012)
8. Cappé, O., Garivier, A., Maillard, O.A.: Kullback-Leibler upper confidence bounds for optimal sequential allocation. Annals of Statistics, 1–56 (2012)
9. Coquelin, P.A., Munos, R.: Bandit algorithms for tree search. In: Proc. of Int. Conf. on Uncertainty in Artificial Intelligence (2007)
10. Chaloner, K., Verdinelli, I.: Bayesian experimental design: a review. Statistical Science 10(3), 273–304 (1995)
11. Chou, I.C., Voit, E.O.: Recent developments in parameter estimation and structure identification of biochemical and genomic systems. Math. Biosci. 219(2), 57–83 (2009)
12. Cohn, D., Atlas, L., Ladner, R.: Improving generalization with active learning. Mach. Learn. 15(2), 201–221 (1994)
13. Dasgupta, S.: Analysis of a greedy active learning strategy. In: NIPS 17, pp. 337–344. MIT Press (2005)
14. Hanneke, S.: A bound on the label complexity of agnostic active learning. In: ICML 2007, pp. 353–360. ACM (2007)
15. The dream project website: http://www.the-dream-project.org/
16. Franceschini, G., Macchietto, S.: Model-based design of experiments for parameter precision. Chemical Engineering Science 63(19), 4846–4872 (2008)
17. Gutenkunst, R.N., Waterfall, J.J., Casey, F.P., Brown, K.S., Myers, C.R., Sethna, J.P.: Universally sloppy parameter sensitivities in systems biology models. PLoS Computational Biology (10), 1871–1878 (2007)
18. Hirsch, M.W., Smale, S.: Differential Equations, Dynamical Systems, and Linear Algebra. Academic press (1974)
19. Kaufmann, E., Cappé, O., Garivier, A.: On bayesian upper confidence bounds for bandit problems. In: Proc. AISTATS, JMLR W&CP, La Palma, Canary Islands, vol. 22, pp. 592–600 (2012)
20. Garivier, A., Cappé O.: The KL-UCB Algorithm for bounded stochastic bandits and beyond. In: COLT, Budapest, Hungary (2011)
21. King, R.D., Whelan, K.E., Jones, F.M., Reiser, P.G., Bryant, C.H., Muggleton, S.H., Kell, D.B., Oliver, S.G.: Functional genomic hypothesis generation and experimentation by a robot scientist. Nature 427(6971), 247–252 (2004)

22. King, R.D., Rowland, J., Oliver, S.G., Young, M., Aubrey, W., ..., Whelan, K.E., Clare, A.: The automation of science. Science 324(5923), 85–89 (2009)
23. Kitano, H.: Computational systems biology. Nature 420(6912), 206–210 (2002)
24. Kocsis, L., Szepesvári, C.: Bandit based Monte-Carlo planning. In: Fürnkranz, J., Scheffer, T., Spiliopoulou, M. (eds.) ECML 2006. LNCS (LNAI), vol. 4212, pp. 282–293. Springer, Heidelberg (2006)
25. Kreutz, C., Timmer, J.: Systems biology: experimental design. FEBS Journal 276, 923–942 (2009)
26. Lawrence, N., Girolami, M., Rattray, M., Sanguinetti, G. (eds.): Learning and inference in computational systems biology. MIT Press (2010)
27. Mazur, J., Ritter, D., Reinelt, G., Kaderali, L.: Reconstructing nonlinear dynamic models of gene regulation using stochastic sampling. BMC Bioinformatics 10(1), 448 (2009)
28. Mazur, J.: Bayesian Inference of Gene Regulatory Networks: From parameter estimation to experimental design. Ph.D. Dissertation, University of Heidelberg, Germany (2012)
29. Ortner, R.: Online regret bounds for markov decision processes with deterministic transitions. In: Freund, Y., Györfi, L., Turán, G., Zeugmann, T. (eds.) ALT 2008. LNCS (LNAI), vol. 5254, pp. 123–137. Springer, Heidelberg (2008)
30. Prill, R.J., Marbach, D., Saez-Rodriguez, J., Sorger, P.K., Alexopoulos, L.G., Xue, X., Clarke, N.D., Altan-Bonnet, G., Stolovitzky, G.: Towards a rigorous assessment of systems biology models: the DREAM3 challenges. PLoS One 5, e9202 (2010)
31. Quach, M., Brunel, N., d'Alche Buc, F.: Estimating parameters and hidden variables in non-linear state-space models based on odes for biological networks inference. Bioinformatics 23, 3209–3216 (2007)
32. Raue, A., Kreutz, C., Maiwald, T., Bachmann, J., Schilling, M., Klingmüller, U., Timmer, J.: Structural and practical identifiability analysis of partially observed dynamical models by exploiting the profile likelihood. Bioinformatics 25(15), 1923–1929 (2009)
33. Robbins, H.: Some aspects of the sequential design of experiments. Bulletin of the American Mathematical Society 58(5), 527–535 (1952)
34. Rolet, P., Sebag, M., Teytaud, O.: Boosting active learning to optimality: A tractable Monte-Carlo, billiard-based algorithm. In: Buntine, W., Grobelnik, M., Mladenić, D., Shawe-Taylor, J. (eds.) ECML PKDD 2009, Part II. LNCS, vol. 5782, pp. 302–317. Springer, Heidelberg (2009)
35. Ruess, J., Milias-Argeitis, A., Lygeros, J.: Designing experiments to understand the variability in biochemical reaction networks. Journal of the Royal Society Interface, 1742–5662 (2013)
36. Walter, E., Pronzato, L.: Identification of parametric models from experimental data. Communications and control engineering. Springer (1997)
37. Wang, Z., Liu, X., Liu, Y., Liang, J., Vinciotti, V.: An extended Kalman filtering approach to modeling nonlinear dynamic gene regulatory networks via short gene expression time series. IEEE/ACM TCBB 6(3), 410–419 (2009)

On the Null Distribution of the Precision and Recall Curve

Miguel Lopes and Gianluca Bontempi

Machine Learning Group, ULB
Interuniversity Institute of Bioinformatics in Brussels (IB)^2
Brussels, Belgium

Abstract. Precision recall curves (pr-curves) and the associated area under (AUPRC) are commonly used to assess the accuracy of information retrieval (IR) algorithms. An informative baseline is random selection. The associated probability distribution makes it possible to assess pr-curve significancy (as a p-value relative to the null of random). To our knowledge, no analytical expression of the null distribution of empirical pr-curves is available, and the only measure of significancy used in the literature relies on non-parametric Monte Carlo simulations. In this paper, we derive analytically the expected null pr-curve and AUPRC, for different interpolation strategies. The AUPRC variance is also derived, and we use it to propose a continuous approximation to the null AUPRC distribution, based on the beta distribution. Properties of the empirical pr-curve and common interpolation strategies are also discussed.

Keywords: Information retrieval, precision-recall curves, statistical significancy assessment.

1 Introduction

Information retrieval (IR) aims to identify relevant instances out of a larger pool of relevant and non relevant items [12]. Common applications include web search, fraud detection, text mining, identification and recommendation in multimedia, and network inference in bioinformatics [10]. The outcome of an IR algorithm is typically a rank of instances according to a relevance score, from which it is easy to derive a selection of relevant items by fixing a certain threshold. Such selection is equivalent to a binary classification, where the relevant and non relevant instances are termed positives and negatives. In the remainder of the paper, positive/negative instances are equivalent to relevant/non relevant instances.

Precision-recall curves (pr-curves) plot precision vs. recall values and are commonly used to assess the performance of information retrieval systems and binary classifiers. Instances are ranked according to a confidence score that they are of the positive class, are incrementally selected and at each selection step, values of precision and recall are computed.

Often a single number is preferred for comparative purposes, and the area under the pr-curve ($0 \leq \text{AUPRC} \leq 1$) is typically used (a higher area corresponding to a higher retrieval accuracy). Its maximum value of 1 corresponds to

T. Calders et al. (Eds.): ECML PKDD 2014, Part II, LNCS 8725, pp. 322–337, 2014.
© Springer-Verlag Berlin Heidelberg 2014

the optimal configuration where all the positive instances are ranked before all negative ones.

Alternative performance measures to pr-curves are F-scores and receiving operating characteristic (ROC) curves. The first combine a given recall and respective precision into a single value, and therefore are not as informative as pr-curves, as performance is only measured in one point in the curve. ROC curves plot the true positive rate vs the false positive rate. One advantage of ROC analysis is that the expected ROC curve, in the null hypothesis of random selection, is invariant with the positive/negative class distribution and takes the form of a diagonal from 0 to 1 (with a respective area under the curve 0.5). Regarding pr-curves, on the contrary, the expected null curve depends on the class distribution and on the number of positive and negative instances. Pr-curves can be considered to be a more informative indicator of performance than ROC curves, on class imbalanced problems [6]. This is due to the fact that ROC analysis is less sensitive to variations in the number of false positives in highly unbalanced situations where the positive class is minoritary (ie. "a large change in the number of false positives can lead to a small change in the false positive rate used in ROC analysis" [6]).

Due to the discrete nature of pr-curves, the plotting of the obtained empirical precision and recall values requires the adoption of an interpolation strategy. An alternative is the estimation of smooth pr-curves. In the literature, nonparametric (eg. boostrap-based [5]) and parametric approaches [4] can be found - the latter assuming an intrinsic continuous (eg. normal) probability distribution describing the class decision, for the two classes (of negatives and positives).

This paper only deals with the first case, of the empirical pr-curve. The analysis of the pr-curve from this perspective is useful as the empirical AUPRC is closely related to average precision measures, commonly used in IR. In this paper, the interpolation of discrete points is done for all possible points (ie. precision and recall is computed for each ranked element). The interpolation of points corresponding to distant values of recall is also considered in the literature [6, 8].

The significancy assessment of obtained curves is an important issue, and, contrary to ROC analysis, with established techniques to infer the standard error of the AUC [3], the significance analysis of pr-curves (and area under) is relatively less developed. A common approach is the estimation of confidence intervals for pr-curves and AUPRC. As with smoothing pr-curve estimates, these can rely on bootstrap or continuous probability distributions assumptions [2].

The null pr-curve is the curve resultant from random ranking, and can be a useful baseline in problems of low precision (eg. network inference in bioinformatics). The distribution of the null pr-curve (and of the respective AUPRC) allows to infer statistical significancy, and also makes it possible to compare AUPRC values independently of their class distribution (by comparing the associated p-values/z-scores). To our knowledge, the only used approach to estimate empirical AUPRC significancy is non-parametric [10, 13], based on the estimation of an empirical null distribution by Monte Carlo. This approach may require a high number of simulations for a good approximation, and the sampling error is also subject to uncertainty.

The main contribution of this paper is an analytical derivation of the expected null pr-curve and of the mean and variance of the respective AUPRC. On their own, these parameters can be used to assess the approximation error of Monte Carlo results. We also suggest a continuous approximation to the null AUPRC distribution, based on the beta distribution, which can be used as an alternative to possibly lengthy Monte Carlo simulations (particularly when the number of instances is high, as illustrated in section 5).

The outline of the paper is the following. Section 2 describes different strategies to interpolate pr-curves and compute the respective AUPRC. Section 3 contains the analytical derivation of the expected (maximum and minimum) precision for a given recall value, as they are sufficient for the expected null pr-curve. Section 4 uses the previous results to derive the mean and the variance of the AUPRC distribution, and suggests the beta distribution as a continuous approximation of the discrete AUPRC probability distribution.

Section 5 describes how to use the proposed method to assess AUPRC significancy, and presents an experimental comparison with Monte Carlo. The expected error of Monte Carlo simulations, regarding the mean and variance of the null AUPRC, is also investigated.

2 Interpolation of the Discrete Pr-curve

Consider an IR problem with a finite number of items, among which P are relevant. From the ranking of these items a set of precision and recall values is obtained (a precision and a recall to each ranked item). Assume the adoption of an empirical pr-curve (as discussed in the section 1). The interpolation of the points in the curve is not straightforward since several values of precision can be associated to the same recall (i.e. if a selected instance in the ranking is not positive, precision falls and recall remains constant). A pr-curve is then characterized by a saw-tooth shape, which is the more accentuated the lower is the value of P. There are different approaches to interpolate the points in the curve. A common approach (named *interpolated average precision*) assigns to each recall the value of the maximum precision at that recall, or at higher recall values [9]. An alternative consists in considering that the precision between two consecutive recall values is constant and equal to the maximal precision value associated to the higher recall (an approach named *average precision*). A more precise way to interpolate pr-curves is to connect the minimum precision at a given recall and the maximum precision at the subsequent recall [13]. Consider a scenario with P positive instances. At a point A, after a selection of N instances, there are TP true positives and FP false positives (precision is $\frac{TP}{TP+FP}$ and recall is $\frac{TP}{P}$). If the next selected instance is positive, the precision moves to $\frac{TP+1}{TP+FP}$ and the recall moves to $\frac{TP+1}{P}$. Let this be point B. Point A corresponds to the minimum precision at the recall $\frac{TP}{P}$, and point B corresponds to the maximum precision at the recall $\frac{TP+1}{P}$. In these two points, the precision p as

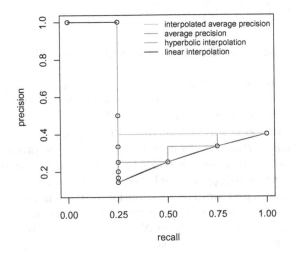

Fig. 1. Different ways to interpolate an empirical pr-curve, $P = 4$. The first selected instance is positive, the next six are negatives, and the last three are positives.

a function of the recall r takes the following hyperbolic form: $p = \frac{rP}{rP+FP}$. The simpler linear interpolation between points A and B is a close approximation of the hyperbolic function, particularly between close recall points. It returns area values necessarily lower, making this approximation a close lower bound. The estimation of the AUPRC is straightforward in all interpolation strategies (except in the hyperbolic interpolation), reducing to a sum of trapezoid areas. Regarding the hyperbolic interpolation, a way to estimate the AUPRC is to incrementally increase it every time there is a recall increase. If we integrate p from point A to point B, we have the area below p, between these points. The integration of p for values of r between $\frac{TP}{P}$ and $\frac{TP+1}{P}$ (the area between these points) is easily derived[1] [13]. Figure 1 illustrates the differences in the different interpolation approaches described, for a number of positive instances of 4 (and same number of different recall values). The first selected instance is positive, the next six are negatives, and the last are positives.

3   The Expected Null Pr-curve

This section derives analytically the expected maximum and minimum (and average) precision for a given recall in the case of random ranking. A value of recall can be associated with multiple values of precision, but only the maximum

---

[1] $\Delta A = \frac{1}{P}\left(1 - FP \ln\left(\frac{FP+TP+1}{FP+TP}\right)\right).$

and minimum determine the pr-curve (as discussed in 2). In this section we also investigate the difference between the maximum and minimum null precision, as a function of P and N.

3.1   Expected Maximum Precision for a Given Recall

Let N be the total number of instances, and P the number of positive instances. Suppose we have a random ranking where n denotes the position of the k-th positive instance. This implies that for the recall $r = \frac{k}{P}$ we obtain the precision, $p = \frac{k}{n}$. This precision p is the maximum precision that can be obtained for the recall r, since further selections will either cause the precision at recall r to go down (false positive) or the recall to increase to $\frac{k+1}{P}$ (true positive). The probability that the k-th positive selected instance is the n-th selected can be obtained with the hypergeometric distribution (returning the probability of selecting k positive instances in n draws, without replacement, on a population of size N, with P positive instances (and $N - P$ negative instances). The probability that the k-th positive instance is the n-th selected instance is equal to the probability of selecting $k - 1$ positive instances in $n - 1$ draws (without replacement), multiplied by the probability of selecting a positive instance in the next draw (which is $\frac{P-(k-1)}{N-(n-1)}$). The first multiplicand is returned by the hypergeometric distribution:

$$\mathcal{P}_h(k - 1, n - 1, N, P) = \frac{\binom{N-n+1}{P-k+1}\binom{n-1}{k-1}}{\binom{N}{P}} \tag{1}$$

We will denote the probability that the k-th positive selected instance is the n-th selected instance by $\mathcal{P}_{sel}(k, n, N, P)$. This probability is also known as the negative (or inverse) hypergeometric probability [7]. We define it as:

$$\mathcal{P}_{sel}(k, n, N, P) = \frac{\binom{N-n+1}{P-k+1}\binom{n-1}{k-1}}{\binom{N}{P}}\left(\frac{P - k + 1}{N - n + 1}\right) = \frac{\binom{N-n}{P-k}\binom{n-1}{k-1}}{\binom{N}{P}} \tag{2}$$

Note that the probability that the first randomly ranked instance is a positive one (i.e. $n = 1$ and $k = 1$) is $\frac{P}{N}$ while the maximum precision at the recall level $\frac{k}{P}$ is $p_{max}(k) = \frac{k}{n}$. Therefore, the expected maximum precision for a recall $\frac{k}{P}$ of a random selection, is:

$$\langle p_{max}\rangle(k) = \sum_{n=k}^{n=N} \frac{k}{n}\mathcal{P}_{sel}(k, n, N, P) \tag{3}$$

3.2   Expected Minimum Precision for a Given Recall

The probability that the k-th positive instance has the n-th position in the ranking equals the probability that the minimum precision at the recall $\frac{k-1}{P}$

is $\frac{k-1}{n-1}$. Therefore, the expected minimum precision at recall $\frac{k-1}{P}$, of a random ranking, is:

$$\langle p_{min}(k-1) \rangle = \sum_{n=k}^{n=N} \begin{cases} \frac{k-1}{n-1} \mathcal{P}_{sel}(k,n,N,P), & n > 1 \\ \mathcal{P}_{sel}(k,n,N,P) & n = 1 \end{cases} \tag{4}$$

Note that if $n = k = 1$ the first selected instance is a positive one and the value of precision is not defined when the recall is zero. In this case we set the value of such zero-recall precision equal to one (as in figure 1). Otherwise, the precision at recall zero is also zero. Equation (4) can be simplified as it follows. If $k = n = 1$, $\mathcal{P}_{sel}(k,n,N,P)$ is equal to $\frac{P}{N}$. If $k > 1$ (i.e. the recall is $\frac{k-1}{P}$), equation (4) becomes:

$$\langle p_{min}(k-1) \rangle = \sum_{n=k}^{n=N} \left(\frac{k-1}{n-1} \right) \mathcal{P}_{sel}(k,n,N,P) =$$

$$= \sum_{n=k}^{n=N} \left(\frac{k-1}{n-1} \right) \frac{\binom{N-n}{P-k}\binom{n-1}{k-1}}{\binom{N}{P}} = \frac{1}{\binom{N}{P}} \sum_{n=k}^{n=N} \binom{n-2}{k-2}\binom{N-n}{P-k} \tag{5}$$

Since according to the Chu-Vandermonde identity [1],

$$\sum_{a=0}^{c} \binom{a}{b}\binom{c-a}{d-b} = \binom{c+1}{d+1} \tag{6}$$

if we set $a = n - 2$, $b = k - 2$, $c = N - 2$ and $d = P - 2$, $\langle p_{min}(k-1) \rangle$ becomes:

$$\langle p_{min}(k-1) \rangle = \frac{1}{\binom{N}{P}} \binom{N-1}{P-1} = \frac{P}{N} \tag{7}$$

This implies that the expected minimum precision for any value of recall is constant and equal to $\frac{P}{N}$. An horizontal approximation of the null pr-curve of value $\frac{P}{N}$ necessarily underestimates the true null pr-curve (for any interpolation strategy that takes into account not only the minimum precision, but also the maximum precision). A discussion of this dissimilarity, and its dependence with P and N, is presented in 3.4.

3.3 Expected Average Precision for a Given Recall

For a comparative purpose we also derive the expected average precision for a given recall $\frac{k}{P}$, of a random selection. It is estimated as follows:

$$\langle p_{avg}(k) \rangle = \sum_{n=k}^{N} \left(\mathcal{P}_{sel}(n,k,N,P) \sum_{n^*=0}^{N-n} \mathcal{P}_{sel}^* P_{avg}^* \right) \tag{8}$$

where

$$\mathcal{P}_{sel}^* = \mathcal{P}_{sel}(1, n^* + 1, N - n, P - k) \tag{9}$$

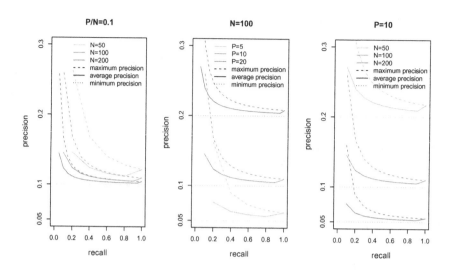

Fig. 2. Expected maximum, average and minimum precision for different values of recall, and for different combinations of P and N

and

$$p_{avg}^*(n^*, n) = \begin{cases} \frac{k}{n} & n^* = 0 \\ \frac{1}{n^*+1}\left(\frac{k}{n} + \sum_{n^{**}=1}^{n^*} \frac{k}{n^{**}}\right) & n^* > 0 \end{cases} \quad (10)$$

For a given recall $\frac{k}{P}$ we estimate the probability that the n-th selected instance is the k-th positive, and that the $(n+n^*+1)$-th selected instance is the $(k+1)$-th positive. This probability is multiplied by the average of the precision $\frac{k}{n^{**}}$ for all values of n^{**} between n and $n + n^*$ (for k positive selected instances). The sum of this product, for all possible values of n and n^* gives the expected average precision for a given recall.

3.4 Difference between Expected Maximum Precision and Expected Minimum Precision

On the basis of previous results we can bound the gap between the expected maximum precision and expected minimum precision for a given recall $\frac{k}{P}$. The expected minimum precision, for any recall, is $\frac{P}{N}$. Since $\langle p_{min}(k) \rangle = \langle p_{min}(k - 1) \rangle$, as $\langle p_{min}(k) \rangle$ is constant and does not depend on k, we obtain

$$\langle p_{max}(k) \rangle - \langle p_{min}(k) \rangle = \langle p_{max}(k) \rangle - \langle p_{min}(k - 1) \rangle =$$

$$= \sum_{n=k}^{n=N} \mathcal{P}_{sel}(k, n, N, P)\left(\frac{k}{n} - \frac{k-1}{n-1}\right) = \sum_{n=k}^{n=N} \frac{\binom{N-n}{P-k}\binom{n-1}{k-1}}{\binom{N}{P}}\left(\frac{n-k}{n(n-1)}\right) =$$

$$= \sum_{n=k}^{n=N} \frac{\binom{N-n}{P-k}\binom{n-2}{k-1}}{\binom{N}{P}} \left(\frac{1}{n}\right) \tag{11}$$

By replacing the term $\left(\frac{1}{n}\right)$ with $\left(\frac{1}{n-2}\right)$ we obtain an upper bound of the difference $\langle p_{max}(k)\rangle - \langle p_{min}(k)\rangle$. This upper bound is:

$$\sum_{n=k}^{n=N} \frac{\binom{N-n}{P-k}\binom{n-2}{k-1}}{\binom{N}{P}} \left(\frac{1}{n-2}\right) = \sum_{n=k}^{n=N} \frac{\binom{N-n}{P-k}\binom{n-3}{k-2}}{\binom{N}{P}(k-1)} = \tag{12}$$

(and using again the Chu-Vandermonde identity (6))

$$= \frac{\binom{N-2}{P-1}}{(k-1)\binom{N}{P}} = \frac{\binom{N}{P}\frac{(N-P)P}{(N-2)N}}{(k-1)\binom{N}{P}} = \frac{(N-P)P}{(k-1)(N-2)N} \tag{13}$$

Equation (13) represents the difference between the expected maximum precision at recall $\frac{k}{P}$ and the expected minimum precision $\frac{P}{N}$. Let us consider also the relative difference (divided by $\frac{P}{N}$). For a given recall, if P is fixed, this difference decreases with N (but the relative difference increases). If N is fixed, the difference (and relative) difference decreases with P. If $\frac{P}{N}$ is fixed, the difference increases with the number of instances.

This behavior is illustrated in the figure 2, which illustrates the expected maximum, average and minimum precision as a function of recall, for different combinations of P and N. Note that there is an uptick in the average precision curve when it reaches the last recall value. This is due to the fact that when the last positive instance is selected, the curve is completed and there are no more selections - at recall 1, the maximum and average precision are the same.

The fact that the null pr-curve tends to $\frac{P}{N}$ in the asymptotic case (i.e. $P \to \infty$ and finite P/N) becomes clear. If a value of recall $\frac{k}{P}$ is finite and if $P \to \infty$, then k must also tend to infinite (and equation (13), representing an upper bound between the expected maximum and minimum null precision, tends to 0).

4   The Null Distribution of the AUPRC

In this section, and using the previous results, we derive the expected value and variance of the AUPRC. In section 2) we presented the common interpolation strategies average precision, hyperbolic interpolation and linear interpolation. In what follows we will consider the AUPRC when these three strategies are used.

4.1   Expected Value of the AUPRC of a Random Selection

Let us denote by AUPRC^{pmax} the AUPRC returned by average precision (average of the maximum precision for all recall values), by AUPRC^{hyp} the one returned by hyperbolic interpolation and by AUPRC^{lin} the one computed with linear interpolation. From equations (4-7) we can estimate the average maximum precision and the average minimum precision of a random ranking for all recall values. The average maximum precision for non-zero recall is:

$$\langle p_{max} \rangle = \frac{1}{P} \sum_{k=1}^{P} \langle p_{max}(k) \rangle \tag{14}$$

and the average minimum precision is:

$$\langle p_{min} \rangle = \frac{1}{P} \sum_{k=0}^{P-1} \frac{P}{N} = \frac{P}{N} \tag{15}$$

It follows that the expected AUPRC^{pmax} and AUPRC^{lin} of a random ranking of instances are:

$$\langle \text{AUPRC}_{random}^{pmax} \rangle = \langle p_{max} \rangle \tag{16}$$

$$\langle \text{AUPRC}_{random}^{lin} \rangle = \frac{\langle p_{max} \rangle + \langle p_{min} \rangle}{2} \tag{17}$$

The expected $\text{AUPRC}_{random}^{hyp}$ is estimated as follows: each time there is an increase in recall, there is an increase in the AUPRC. This increase is equal to:

$$\Delta A(k,n) = \begin{cases} \frac{1}{P}\left(1 - (n-k)\ln(\frac{n}{n-1})\right) & n > 1 \\ \frac{1}{P} & n = 1 \end{cases} \tag{18}$$

where $k = TP + 1$ and $n = TP + FP + 1$ (see section 2). The value in (18) is the area that is added to the AUPRC when the k-th selected positive instance is the n-th selected. We can estimate the expected added area for any value of k:

$$\langle \Delta A(k) \rangle = \sum_{n=k}^{N} \Delta A(k,n) \mathcal{P}_{sel}(k,n,N,P) \tag{19}$$

The expected AUPRC of a random ranking, using the hyperbolic interpolation, is given by:

$$\langle \text{AUPRC}_{random}^{hyp} \rangle = \sum_{k=1}^{P} \langle \Delta A(k) \rangle \tag{20}$$

Finally, the expected AUPRC of a random ranking in the asymptotic case (i.e. assuming that the number of instances is infinite, and $\frac{P}{N}$ is finite, see section 3.4) is:

$$\langle \text{AUPRC}_{random}^{lim} \rangle = \frac{P}{N} \tag{21}$$

Figure 3 shows the expected AUPRC for the different interpolation methods, and for the asymptotic continuous case, as a function of $\frac{P}{N}$. In the right plots, P is fixed (equal to 10), and in the left plots N is fixed (equal to 100). The plots on the top show the absolute expected AUPRC, whereas the bottom plots show the relative difference, the difference between the expected AUPRC and

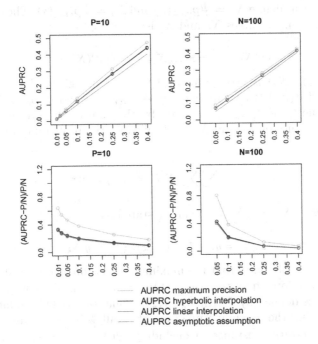

Fig. 3. estimated expected AUPRC of a random ranking as a function of $\frac{P}{N}$ when $P=10$, and $N=100$

the asymptotic AUPRC ($\frac{P}{N}$), divided by the latter. An increase in N leads to a decrease in the absolute difference between the AUPRC estimations (top-right plot), but leads to an increase in the relative difference (bottom-right plot). An increase in P leads to a decrease in both absolute and relative differences (in line with the results of section 3.4).

4.2 Variance of the AUPRC of a Random Selection

For the sake of simplicity we will consider here only the average precision approach, though the obtained results can be easily extended to the linear interpolation approach, and with some more difficulty, to the hyperbolic interpolation. The expected AUPRC of a random selection, using the average precision interpolation, is given in (16): the AUPRC is the average of the expected maximum precision for the different recall values. This is equivalent to the sum of the expected maximum precision for the different values of recall, divided by P. The variance of a sum of random variables is equal to the sum of the values of their covariance matrix. In our case, we have:

$$\text{Var}(\text{AUPRC}_{random}^{pmax}) = \sum_{k=1}^{P}\sum_{j=k}^{P}\text{Cov}\left(\frac{1}{P}p_{max}(k),\frac{1}{P}p_{max}(j)\right) \qquad (22)$$

For simplicity, let us define $X_k = \frac{1}{P}p_{max}(k)$ and $X_j = \frac{1}{P}p_{max}(j)$. The covariance between two random variables X_k and X_j is:

$$\text{Cov}(X_k, X_j) = \mathbb{E}(X_k X_j) - \mathbb{E}(X_k)\mathbb{E}(X_j) \tag{23}$$

The term $\mathbb{E}(X_k)\mathbb{E}(X_j)$ is straightforward to estimate, given that we already have $\mathbb{E}(X_k)$ and $\mathbb{E}(X_j)$ (they are given by $\frac{1}{P}\langle p_{max}(k)\rangle$ and $\frac{1}{P}\langle p_{max}(j)\rangle$, in the equation 3). The term $\mathbb{E}(X_k X_j)$ is given by the sum of the product between all the possible values of X_k and X_j, multiplied by the probability of observing them:

$$\mathbb{E}(X_k X_j) = \sum_x \sum_y P(X_k = x, X_j = y)xy \tag{24}$$

The probability $P(X_k = x, X_j = y)$ can be stated as:

$$P(X_k = x, X_j = y) = P(X_k = x)P(X_j = y | X_k = x) \tag{25}$$

$P(X_k = x)$ is the probability that the maximum precision at the recall $\frac{k}{P}$ is xP. $P(X_k = x)$ is therefore equal to $P(p_{max}(k) = xP)$. The probability mass function of $p_{max}(k)$ is defined by $P_{sel}(k, n_k, N, P)$ (equation 2). This equation gives the probability that the maximum precision at recall $\frac{k}{P}$ is $\frac{k}{n_k}$, assuming N total instances and P positive instances. Concluding, $P(X_k = x) = P_{sel}(k, n_k, N, P)$, given that $x = \frac{k}{Pn_k}$ and $X_k = \frac{1}{P}p_{max}(k)$.

The conditional probability $P(X_j = y | X_k = x)$ is the probability that the maximum precision at the recall $\frac{j}{P}$ is yP, given that the maximum precision at the recall $\frac{k}{P}$ is xP. If we define $yP = \frac{j}{n_j}$ and $xP = \frac{k}{n_k}$, and on the condition that $j > k$, $P(X_j = y | X_k = x)$ is the probability that the $(j - k)$-th positive selected instance is the $(n_j - n_k)$-th selected instance, in a population of $N - n_k$ instances, and $P - k$ positive instances. This probability is given by $P_{sel}(j - k, n_j - n_k, N - n_k, P - k)$ (2). Let's denote it simply by P_{sel}^k. We can now rewrite equation (24) as:

$$\mathbb{E}(X_k X_j) = \sum_{n_k=1}^{N} \sum_{n_j=n_k+j-k}^{N} P_{sel}(k, n_k, N, P)P_{sel}^k \frac{k}{n_k P}\frac{j}{n_j P} \tag{26}$$

If $j < k$, $\mathbb{E}(X_k X_j) = \mathbb{E}(X_j X_k)$. If $j = k$, $\mathbb{E}(X_k X_k)$ is given by equation (3): $\mathbb{E}(X_k) = \frac{1}{P}\langle p_{max}(k)\rangle$, and $P_{sel}(k, n, N, P)$ describes the probability mass function of $p_{max}(k)$. X_k takes values in $\frac{k}{Pn}$, $n = k, k+1, ...N$. Therefore, we have:

$$\mathbb{E}(X_k X_k) = \sum_{n=k}^{n=N} \left(\frac{k}{Pn}\right)^2 P_{sel}(k, n, N, P) \tag{27}$$

$\text{Cov}(X_k, X_j)$ can now be estimated as in (23), and the variance of the AUPRC of a random selection, if the pr-curve is interpolated using the average precision interpolation, is given in (22).

4.3   Distribution of the AUPRC

Knowing the mean and variance of the AUPRC, and the fact that it is contained in a finite interval (between a minimum and 1) suggests the beta distribution as a candidate for a parametric approximation to the AUPRC probability distribution. The beta is a continuous distribution with finite support, fully described with two parameters characterizing the mean and variance, and two parameters defining the support interval. Note that being a continuous distribution, it can only be an approximation of the true (discrete) distribution. The two shape parameters of the beta distribution, α and β, can be estimated through the methods of moments approach:

$$\alpha = x^* \left(\frac{x^*(1 - x^*)}{v^*} - 1 \right) \tag{28}$$

and

$$\beta = (1 - x^*) \left(\frac{x^*(1 - x^*)}{v^*} - 1 \right) \tag{29}$$

where x^* and v^* are the normalized mean and variance [11]:

$$x^* = \frac{\langle \text{AUPRC}_{random}^{pmax} \rangle - \min(\text{AUPRC}_{random}^{pmax})}{1 - \min(\text{AUPRC}_{random}^{pmax})} \tag{30}$$

and

$$v^* = \frac{\text{Var}(\text{AUPRC}_{random}^{pmax})}{(1 - \min(\text{AUPRC}_{random}^{pmax}))^2} \tag{31}$$

In the equations above, the value 1 in the denominator corresponds to the maximum $\text{AUPRC}_{random}^{pmax}$. The minimum is attained when the last P ranked instances are all positive. Since precision is non-zero only in the last k recall points the minimum is returned by:

$$\min(\text{AUPRC}_{random}^{pmax}) = \frac{1}{P} \sum_k^P \frac{k}{N - P + k} \tag{32}$$

The support of the resulting standard beta distribution should be re-transformed: multiplying by the range $(1 - \min(\text{AUPRC}_{random}^{pmax}))$, and addition of $\min(\text{AUPRC}_{random}^{pmax})$.

5   Empirical Assessment

This section illustrates the usefulness of the analytical derivation of the previous sections, by considering a common retrieval task in bioinformatics: the identification of pairwise gene interactions, among all the possible interactions between the genes composing a network. We assess the quality of the inference with $\text{AUPRC}_{random}^{pmax}$ (in the following, simply referred to as AUPRC), the area under the pr-curve returned by the average precision approach. Consider a network of 10 genes and 15 effective interactions. The number of total instances N

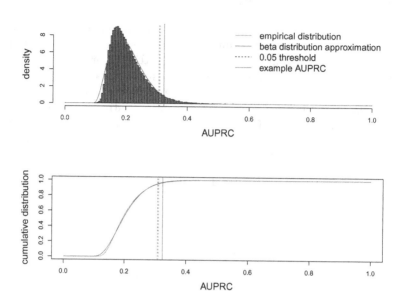

Fig. 4. AUPRC probability and cumulative distribution, empirical and estimated, when $N = 90$ and $P = 15$

is 90 (the number of different ordered pairs of different genes) and the number of positive instances P is 15.

The performance of a network inference algorithm is assessed by the AUPRC and an associated p-value. We compare our analytical result with a non-parametric Monte Carlo approach, based on an empirical probability distribution obtained by a large number of assessments of random selections (number of simulations equal to 100000).

The empirical mean and variance estimations as well as the analytical expected values computed with equations (16) and (22) are shown in table II. Both the expected value and variance are very close in both estimations (a relative difference of less than 0.001).

Table 1. Expected and empirical values for the AUPRC of a random ranking, with $N = 90$ and $P = 15$ ($R = 100000$ samples generated)

	mean	variance
expected	0.2049	0.002990
empirical	0.2051	0.002991

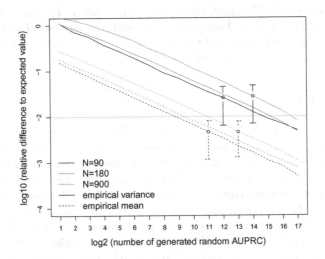

Fig. 5. Relative difference of the mean and variance of the empirical AUPRC distribution, relative to the expected values, as a function of the number of random AUPRC generations

Figure 4 illustrates both the empirical and the beta probability densities (top) and distributions (bottom). The parameters of the beta distribution are computed as discussed in Section 4.3. It can be seen that the beta distribution reasonably approximates the distribution of AUPRC. The AUPRC value corresponding to a 0.05 p-value is drawn vertically. The empirical and estimated values are 0.311 and 0.307. The green vertical line represents an illustrative AUPRC, obtained using a network inference algorithm. The obtained AUPRC is considered statistically significant (0.05 cutoff), using either Monte Carlo, or the beta distribution approximation. This approach can be extended to any IR problem.

A second experiment concerns the study of the impact of the number of simulations on the accuracy of the empirical AUPRC distribution. Figure 5 shows the relative difference between the empirical and expected AUPRC mean and variance (relative to the expected values), as a function of the number of simulated AUPRC. We performed this experiment for $N = 90, 180, 900$, keeping P fixed. The values shown are the average of 1000 simulations (the variance is also plotted for $N = 90$ and $N = 900$, for simplicity for a couple of simulation numbers only). The number of simulated AUPRC is shown in a logarithmic scale of base 2, and goes from 2 to 131072 (2^{17}). The relative difference is shown in a logarithmic scale of base 10. A threshold for the relative difference corresponding to 0.01 (10^{-2}) is drawn. As expected, the empirical mean and variance tend to the respective expected values, as the number of simulated AUPRC increases. When $N = 90$, the relative difference of both the mean and variance is below 0.01

when the number of simulations is higher than 32768 (2^{15}). When $N = 900$ this number increases to 131072 simulations (2^{17}). The number of needed simulations to approximate the true distribution increases with the number of instances. As described, the expected value and variance of the null AUPRC can be used to control the sampling error of empirical estimations.

6 Conclusion

The null pr-curve and AUPRC is a useful baseline in the assessment of challenging IR tasks, when outperforming random is not guaranteed. Monte Carlo methods are typically used to estimate the probability distribution of the null AUPRC, and to assess statistical significancy. However, these methods can be computationally intensive, requiring a high number of simulations to accurately approximate the true distribution of the null AUPRC, particularly when the number of instances is high. In this paper, we deduced analytically expressions for the expected null pr-curve, and AUPRC mean and variance. These can be used to assess the quality of empirical Monte Carlo null distributions. Complementary, and as an alternative to Monte Carlo, they can be used to estimate a parametric approximation to the null AUPRC probability distribution - a continuous approximation based on the beta distribution is suggested.

References

1. Askey, R.: Orthogonal polynomials and special functions (1975); Lectures given at the National Science Foundation regional conference held at Virginia Polytechnic Institute in June 1974
2. Boyd, K., Eng, K.H., David Page, C.: Area under the precision-recall curve: Point estimates and confidence intervals. In: Blockeel, H., Kersting, K., Nijssen, S., Železný, F. (eds.) ECML PKDD 2013, Part III. LNCS, vol. 8190, pp. 451–466. Springer, Heidelberg (2013),
   http://dblp.uni-trier.de/db/conf/pkdd/pkdd2013-3.html#BoydEP13
3. Bradley, A.P.: The use of the area under the roc curve in the evaluation of machine learning algorithms. Pattern Recognition 30, 1145–1159 (1997)
4. Brodersen, K.H., Ong, C.S., Stephan, K.E., Buhmann, J.M.: The binormal assumption on precision-recall curves. In: Proceedings of the 2010 20th International Conference on Pattern Recognition, ICPR 2010, pp. 4263–4266. IEEE Computer Society, Washington, DC (2010)
5. Clémençon, S., Vayatis, N.: Nonparametric estimation of the precision-recall curve. In: Proceedings of the 26th Annual International Conference on Machine Learning, ICML 2009, pp. 185–192. ACM, New York (2009)
6. Davis, J., Goadrich, M.: The relationship between precision-recall and roc curves. In: Proceedings of the 23rd International Conference on Machine Learning, ICML 2006, pp. 233–240. ACM, New York (2006)
7. Forbes, C., Evans, M., Hastings, N., Peacock, B.: Statistical Distributions. Wiley (2011)
8. Keilwagen, J., Grosse, I., Grau, J.: Area under precision-recall curves for weighted and unweighted data. PLoS One 9(3), e92209 (2014)

9. Manning, C.D., Raghavan, P., Schütze, H.: Introduction to Information Retrieval. Cambridge University Press, New York (2008)
10. Marbach, D., Costello, J.C., Küffner, R., Vega, N.M., Prill, R.J., Camacho, D.M., Allison, K.R., Kellis, M., Collins, J.J., Stolovitzky, G., the DREAM5 Consortium, Saeys, Y.: Wisdom of crowds for robust gene network inference. Nature Methods 9(8), 796–804 (2012)
11. Natrella, M.: NIST/SEMATECH e-Handbook of Statistical Methods. NIST/SEMATECH (July 2010)
12. Van Rijsbergen, C.J.: Information Retrieval, 2nd edn. Butterworth-Heinemann, Newton (1979)
13. Gustavo Stolovitzky, Robert J. Prill, and Andrea Califano

Linear State-Space Model
with Time-Varying Dynamics

Jaakko Luttinen, Tapani Raiko, and Alexander Ilin

Aalto University, Finland

Abstract. This paper introduces a linear state-space model with time-varying dynamics. The time dependency is obtained by forming the state dynamics matrix as a time-varying linear combination of a set of matrices. The time dependency of the weights in the linear combination is modelled by another linear Gaussian dynamical model allowing the model to learn how the dynamics of the process changes. Previous approaches have used switching models which have a small set of possible state dynamics matrices and the model selects one of those matrices at each time, thus jumping between them. Our model forms the dynamics as a linear combination and the changes can be smooth and more continuous. The model is motivated by physical processes which are described by linear partial differential equations whose parameters vary in time. An example of such a process could be a temperature field whose evolution is driven by a varying wind direction. The posterior inference is performed using variational Bayesian approximation. The experiments on stochastic advection-diffusion processes and real-world weather processes show that the model with time-varying dynamics can outperform previously introduced approaches.

1 Introduction

Linear state-space models (LSSM) are widely used in time-series analysis and modelling of dynamical systems [1,2]. They assume that the observations are generated linearly from hidden states with a linear dynamical model that does not change with time. The assumptions of linearity and constant dynamics make the model easy to analyze and efficient to learn.

Most real-world processes cannot be accurately described by linear Gaussian models, which motivates more complex nonlinear state-space models (see, e.g., [3,4]). However, in many cases processes behave approximately linearly in a local regime. For instance, an industrial process may have a set of regimes with very distinct but linear dynamics. Such processes can be modelled by switching linear state-space models [5,6] in which the transition between a set of linear dynamical models is described with hidden Markov models. Thus, these models have a small number of states defining their dynamics.

Instead of having a small number of possible states of the process dynamics, some processes may have linear dynamics that change continuously in time. For instance, physical processes may be characterized by linear stochastic partial

T. Calders et al. (Eds.): ECML PKDD 2014, Part II, LNCS 8725, pp. 338–353, 2014.
© Springer-Verlag Berlin Heidelberg 2014

differential equations but the parameters of the equations may vary in time. Simple climate models may use the advection-diffusion equation in which the diffusion and the velocity field parameters define how the modelled quantity mixes and moves in space and time. In a realistic scenario, these parameters are time-dependent because, for instance, the wind modelled by the velocity field changes with time.

This paper presents a Bayesian linear state-space model with time-varying dynamics. The dynamics at each time is formed as a linear combination of a set of state dynamics matrices, and the weights of the linear combination follow a linear Gaussian dynamical model. The main difference to switching LSSMs is that instead of having a small number of dynamical regimes, the proposed model allows for an infinite number of them with a smooth transition between them. Thus, the model can adapt to small changes in the system. This work is an extension of an abstract [7] which presented the basic idea without the Bayesian treatment. The model bears some similarity to relational feature learning in modelling sequential data [8].

Posterior inference for the model is performed using variational Bayesian (VB) approximation because the exact Bayesian inference is intractable [9]. In order for the VB learning algorithm to converge fast, the method uses a similar parameter expansion that was introduced in [10]. This parameter expansion is based on finding the optimal rotation in the latent subspace and it may improve the speed of the convergence by several orders of magnitude.

The experimental section shows that the proposed LSSM with time-varying dynamics is able to learn the varying dynamics of complex physical processes. The model predicts the processes better than the classical LSSM and the LSSM with switching dynamics. It finds latent processes that describe the changes in the dynamics and is thus able to learn the dynamics at each time point accurately. These experimental results are promising and suggest that the time-varying dynamics may be a useful tool for statistical modelling of complex dynamical and physical processes.

2   Model

Linear state-space models assume that a sequence of M-dimensional observations $(\mathbf{y}_1, \ldots, \mathbf{y}_N)$ is generated from latent D-dimensional states $(\mathbf{x}_1, \ldots, \mathbf{x}_N)$ following a first-order Gaussian Markov process:

$$\mathbf{y}_n = \mathbf{C}\mathbf{x}_n + \text{noise}, \tag{1}$$

$$\mathbf{x}_n = \mathbf{W}\mathbf{x}_{n-1} + \text{noise}, \tag{2}$$

where noise is Gaussian, $\mathbf{W}$ is the $D \times D$ state dynamics matrix and $\mathbf{C}$ is the $M \times D$ loading matrix. Usually, the latent space dimensionality D is assumed to be much smaller than the observation space dimensionality M in order to model the dependencies of high-dimensional observations efficiently. Because the state dynamics matrix is constant, the model can perform badly if the dynamics of the modelled process changes in time.

In order to model changing dynamics, the constant dynamics in (2) can be replaced with a state dynamics matrix $\mathbf{W}_n$ which is time-dependent. Thus, (2) is replaced with

$$\mathbf{x}_n = \mathbf{W}_n \mathbf{x}_{n-1} + \text{noise.} \tag{3}$$

However, modelling the unknown time dependency of $\mathbf{W}_n$ is a challenging task because for each $\mathbf{W}_n$ there is only one transition $\mathbf{x}_{n-1} \to \mathbf{x}_n$ which gives information about each $\mathbf{W}_n$.

Previous work modelled the time-dependency using switching state dynamics [6]. It means having a small set of matrices $\mathbf{B}_1, \dots, \mathbf{B}_K$ and using one of them at each time step:

$$\mathbf{W}_n = \mathbf{B}_{z_n}, \tag{4}$$

where $z_n \in \{1, \dots, K\}$ is a time-dependent index. The indices z_n then follow a first-order Markov chain with an unknown state-transition matrix. The model can be motivated by dynamical processes which have a few states with different dynamics and the process jumps between these states.

This paper presents an approach for continuously changing time-dependent dynamics. The state dynamics matrix is constructed as a linear combination of K matrices:

$$\mathbf{W}_n = \sum_{k=1}^{K} s_{kn} \mathbf{B}_k. \tag{5}$$

The mixing weight vector $\mathbf{s}_n = \begin{bmatrix} s_{1n} \dots s_{Kn} \end{bmatrix}^{\mathrm{T}}$ varies in time and follows a first-order Gaussian Markov process:

$$\mathbf{s}_n = \mathbf{A}\mathbf{s}_{n-1} + \text{noise,} \tag{6}$$

where $\mathbf{A}$ is the $K \times K$ state dynamics matrix of this latent mixing-weight process. The model with switching dynamics in (4) can be interpreted as a special case of (5) by restricting the weight vector $\mathbf{s}_n$ to be a binary vector with only one non-zero element. However, in the switching model, $\mathbf{s}_n$ would follow a first-order Markov chain, which is different from the first-order Gaussian Markov process used in the proposed model. Compared to models with switching dynamics, the model with time-varying dynamics allows the state dynamics matrix to change continuously and smoothly.

The model is motivated by physical processes which roughly follow stochastic partial differential equations but the parameters of the equations change in time. For instance, a temperature field may be modelled with a stochastic advection-diffusion equation but the direction of the wind may change in time, thus changing the velocity field parameter of the equation.

2.1 Prior Probability Distributions

We give the proposed model a Bayesian formulation by setting prior probability distributions for the variables. It roughly follows the linear state-space model formulation in [9,10] and the principal component analysis formulation in [11]. The likelihood function is

$$p(\mathbf{Y}|\mathbf{C}, \mathbf{X}, \tau) = \prod_{n=1}^{N} \mathcal{N}(\mathbf{y}_n | \mathbf{C}\mathbf{x}_n, \mathrm{diag}(\boldsymbol{\tau})^{-1}), \tag{7}$$

where $\mathcal{N}(y|m, v)$ is the Gaussian probability density function of y with mean m and covariance v, and $\mathrm{diag}(\boldsymbol{\tau})$ is a diagonal matrix with elements $\tau_1, \dots, \tau_M$ on the diagonal. For simplicity, we used isotropic noise ($\tau_m = \tau$) in our experiments.

The loading matrix $\mathbf{C}$ has the following prior, which is also known as an automatic relevance determination (ARD) prior [11]:

$$p(\mathbf{C}|\boldsymbol{\gamma}) = \prod_{d=1}^{D} \mathcal{N}(\mathbf{c}_d | \mathbf{0}, \mathrm{diag}(\boldsymbol{\gamma})^{-1}), \qquad p(\boldsymbol{\gamma}) = \prod_{d=1}^{D} \mathcal{G}(\gamma_d | a_\gamma, b_\gamma), \tag{8}$$

where $\mathbf{c}_d$ is the d-th row vector of $\mathbf{C}$, the vector $\boldsymbol{\gamma} = \begin{bmatrix} \gamma_1 \dots \gamma_D \end{bmatrix}^{\mathrm{T}}$ contains the ARD parameters, and $\mathcal{G}(\gamma|a, b)$ is the gamma probability density function of γ with shape a and rate b.

The latent states $\mathbf{X} = \begin{bmatrix} \mathbf{x}_0 \dots \mathbf{x}_N \end{bmatrix}$ follow a first-order Gaussian Markov process, which can be written as

$$p(\mathbf{X}|\mathbf{W}_n) = \mathcal{N}(\mathbf{x}_0 | \boldsymbol{\mu}_0^{(x)}, \boldsymbol{\Lambda}_0^{-1}) \prod_{n=1}^{N} \mathcal{N}(\mathbf{x}_n | \mathbf{W}_n \mathbf{x}_{n-1}, \mathbf{I}), \tag{9}$$

where $\boldsymbol{\mu}_0^{(x)}$ and $\boldsymbol{\Lambda}_0$ are the mean and precision of the auxiliary initial state $\mathbf{x}_0$. The process noise covariance matrix can be an identity matrix without loss of generality because any rotation can be compensated in $\mathbf{x}_n$ and $\mathbf{W}_n$. The initial state $\mathbf{x}_0$ can be given a broad prior by setting, for instance, $\boldsymbol{\mu}_0^{(x)} = \mathbf{0}$ and $\boldsymbol{\Lambda}_0 = 10^{-6} \cdot \mathbf{I}$.

The state dynamics matrices $\mathbf{W}_n$ are a linear combination of matrices $\mathbf{B}_k$ which have the following ARD prior:

$$p(\mathbf{B}_k|\boldsymbol{\beta}_k) = \prod_{c=1}^{D} \prod_{d=1}^{D} \mathcal{N}(b_{kcd} | 0, \beta_{kd}^{-1}), \ p(\beta_{dk}) = \mathcal{G}(\beta_{kd} | a_\beta, b_\beta), \ k = 1, \dots, K, \tag{10}$$

where $b_{kcd} = [\mathbf{B}_k]_{cd}$ is the element on the c-th row and d-th column of $\mathbf{B}_k$. The ARD parameter β_{kd} helps in pruning out irrelevant components in each matrix.

In order to keep the formulas less cluttered, we use the following notation: $\mathbf{B}$ is a $K \times D \times D$ tensor. When using subscripts, the first index corresponds to the index of the state dynamics matrix, the second index to the rows of the matrices and the third index to the columns of the matrices. A colon is used to denote that all elements along that axis are taken. Thus, for instance, $\mathbf{B}_{k::}$ is $\mathbf{B}_k$ and $\mathbf{B}_{:d:}$ is a $K \times D$ matrix obtained by stacking the d-th row vectors of $\mathbf{B}_k$ for each k.

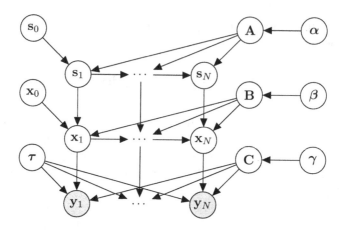

Fig. 1. The graphical model of the linear state-space model with time-varying dynamics

The mixing weights $\mathbf{S} = \begin{bmatrix} \mathbf{s}_0 \dots \mathbf{s}_N \end{bmatrix}$ have first-order Gaussian Markov process prior

$$p(\mathbf{S}|\mathbf{A}) = \mathcal{N}(\mathbf{s}_0|\boldsymbol{\mu}_0^{(s)}, \mathbf{V}_0^{-1}) \prod_{n=1}^{N} \mathcal{N}(\mathbf{s}_n|\mathbf{A}\mathbf{s}_{n-1}, \mathbf{I}), \qquad (11)$$

where, similarly to the prior of $\mathbf{X}$, the parameters $\boldsymbol{\mu}_0^{(s)}$ and $\mathbf{V}_0$ are the mean and precision of the auxiliary initial state $\mathbf{s}_0$, and the noise covariance can be an identity matrix without loss of generality. The initial state $\mathbf{s}_0$ can be given a broad prior by setting, for instance, $\boldsymbol{\mu}_0^{(s)} = \mathbf{0}$ and $\mathbf{V}_0 = 10^{-6} \cdot \mathbf{I}$.

The state dynamics matrix $\mathbf{A}$ of the latent mixing weights $\mathbf{s}_n$ is given an ARD prior

$$p(\mathbf{A}|\boldsymbol{\alpha}) = \prod_{k=1}^{K} \mathcal{N}\left(\mathbf{a}_k|\mathbf{0}, \operatorname{diag}(\boldsymbol{\alpha})^{-1}\right), \qquad p(\boldsymbol{\alpha}) = \prod_{k=1}^{K} \mathcal{G}(\alpha_k|a_\alpha, b_\alpha) \qquad (12)$$

where $\mathbf{a}_k$ is the k-th row of $\mathbf{A}$, and $\boldsymbol{\alpha} = \begin{bmatrix} \alpha_1 \dots \alpha_K \end{bmatrix}^{\mathrm{T}}$ contains the ARD parameters.

Finally, the noise parameter is given a gamma prior

$$p(\boldsymbol{\tau}) = \prod_{m=1}^{M} \mathcal{G}(\tau_m|a_\tau, b_\tau). \qquad (13)$$

The hyperparameters of the model can be set, for instance, as $a_\alpha = b_\alpha = a_\beta = b_\beta = a_\gamma = b_\gamma = a_\tau = b_\tau = 10^{-6}$ to obtain broad priors for the variables. Small values result in approximately non-informative priors which are usually a good choice in a wide range of problems.

For the experimental section, we constructed the LSSM with switching dynamics by using a hidden Markov model (HMM) for the state dynamics matrix

$\mathbf{W}_n$. The HMM had an unknown initial state and a state transition matrix with broad conjugate priors. We used similar prior probability distributions in the classical LSSM with constant dynamics, the proposed LSSM with time-varying dynamics, and the LSSM with switching dynamics for the similar parts of the models.

3   Variational Bayesian Inference

As the posterior distribution is analytically intractable, it is approximated using variational Bayesian (VB) framework, which scales well to large applications compared to Markov chain Monte Carlo (MCMC) methods [12]. The posterior approximation is assumed to factorize with respect to the variables:

$$p(\mathbf{X},\mathbf{C},\boldsymbol{\gamma},\mathbf{B},\boldsymbol{\beta},\mathbf{S},\mathbf{A},\boldsymbol{\alpha},\boldsymbol{\tau}|\mathbf{Y}) \approx q(\mathbf{X})q(\mathbf{C})q(\boldsymbol{\gamma})q(\mathbf{B})q(\boldsymbol{\beta})q(\mathbf{S})q(\mathbf{A})q(\boldsymbol{\alpha})q(\boldsymbol{\tau}). \quad (14)$$

This approximation is optimized by minimizing the Kullback-Leibler divergence from the true posterior by using the variational Bayesian expectation-maximization (VB-EM) algorithm [13]. In VB-EM, the posterior approximation is updated for the variables one at a time and iterated until convergence.

3.1   Update Equations

The approximate posterior distributions have the following forms:

$$q(\mathbf{X}) = \mathcal{N}([\mathbf{X}]_:|\boldsymbol{\mu}_x,\boldsymbol{\Sigma}_x), \qquad\qquad q(\boldsymbol{\tau}) = \prod_{m=1}^{M} \mathcal{G}(\tau_m|\bar{a}_\tau^{(m)},\bar{b}_\tau^{(m)}), \qquad (15)$$

$$q(\mathbf{C}) = \prod_{m=1}^{M} \mathcal{N}(\mathbf{c}_m|\boldsymbol{\mu}_c^{(m)},\boldsymbol{\Sigma}_c^{(m)}), \qquad q(\boldsymbol{\gamma}) = \prod_{d=1}^{D} \mathcal{G}(\gamma_d|\bar{a}_\gamma^{(d)},\bar{b}_\gamma^{(d)}), \qquad (16)$$

$$q(\mathbf{B}) = \prod_{d=1}^{D} \mathcal{N}([\mathbf{B}_{:d:}]_:|\boldsymbol{\mu}_b^{(d)},\boldsymbol{\Sigma}_b^{(d)}), \qquad q(\boldsymbol{\beta}) = \prod_{k=1}^{K}\prod_{d=1}^{D} \mathcal{G}(\beta_{kd}|\bar{a}_\beta^{(kd)},\bar{b}_\beta^{(kd)}), \quad (17)$$

$$q(\mathbf{S}) = \mathcal{N}([\mathbf{S}]_:|\boldsymbol{\mu}_s,\boldsymbol{\Sigma}_s), \qquad\qquad\qquad\qquad\qquad\qquad\qquad\qquad (18)$$

$$q(\mathbf{A}) = \prod_{k=1}^{K} \mathcal{N}(\mathbf{a}_k|\boldsymbol{\mu}_a^{(k)},\boldsymbol{\Sigma}_a^{(k)}), \qquad q(\boldsymbol{\alpha}) = \prod_{k=1}^{K} \mathcal{G}(\alpha_k|\bar{a}_\alpha^{(k)},\bar{b}_\alpha^{(k)}), \qquad (19)$$

where $[\mathbf{X}]_:$ is a vector obtained by stacking the column vectors $\mathbf{x}_n$. It is straightforward to derive the following update equations of the variational parameters:

$$\bar{a}_\tau^{(m)} = a_\tau + \frac{N_m}{2}, \qquad\qquad\qquad \bar{b}_\tau^{(m)} = b_\tau + \frac{1}{2}\sum_{n\in\mathcal{O}_{m:}} \xi_{mn}, \qquad (20)$$

$$\boldsymbol{\Sigma}_c^{(m)} = \left(\langle\mathrm{diag}(\boldsymbol{\gamma})\rangle + \sum_{n\in\mathcal{O}_{m:}} \langle\tau_m\rangle\langle\mathbf{x}_n\mathbf{x}_n^\mathrm{T}\rangle\right)^{-1}, \quad \boldsymbol{\mu}_c^{(m)} = \boldsymbol{\Sigma}_c^{(m)}\sum_{n\in\mathcal{O}_{m:}} y_{mn}\langle\tau_m\rangle\langle\mathbf{x}_n\rangle, \quad (21)$$

$$\bar{a}_\gamma^{(d)} = a_\gamma + \frac{M}{2}, \qquad\qquad\qquad \bar{b}_\gamma^{(d)} = b_\gamma + \frac{1}{2}\sum_{m=1}^{M} \langle c_{md}^2\rangle, \qquad (22)$$

$$\Sigma_b^{(d)} = \left(\langle \text{diag}(\boldsymbol{\beta}) \rangle + \sum_{n=1}^{N} \boldsymbol{\Omega}_n \right)^{-1}, \qquad \boldsymbol{\mu}_b^{(d)} = \Sigma_b^{(d)} \sum_{n=1}^{N} \left[\langle \mathbf{s}_n \rangle \langle x_{dn} \mathbf{x}_{n-1}^{\mathrm{T}} \rangle \right]_{:}, \quad (23)$$

$$\Sigma_a^{(k)} = \left(\langle \text{diag}(\boldsymbol{\alpha}) \rangle + \sum_{n=1}^{N} \langle \mathbf{s}_{n-1} \mathbf{s}_{n-1}^{\mathrm{T}} \rangle \right)^{-1}, \qquad \boldsymbol{\mu}_a^{(k)} = \Sigma_a^{(k)} \sum_{n=1}^{N} \langle s_{kn} \mathbf{s}_{n-1} \rangle, \qquad (24)$$

$$\bar{a}_\alpha^{(k)} = a_\alpha + \frac{K}{2}, \qquad\qquad \bar{b}_\alpha^{(k)} = b_\alpha + \frac{1}{2} \sum_{i=1}^{K} \langle a_{ik}^2 \rangle, \qquad (25)$$

where $\mathcal{O}_{m:}$ is the set of time instances n for which the observation y_{mn} is not missing, N_m is the size of the set $\mathcal{O}_{m:}$, $\xi_{mn} = \langle (y_{mn} - \mathbf{c}_m^{\mathrm{T}} \mathbf{x}_n)^2 \rangle$, $\boldsymbol{\Omega}_n = \langle \mathbf{x}_{n-1} \mathbf{x}_{n-1}^{\mathrm{T}} \rangle \otimes \langle \mathbf{s}_n \mathbf{s}_n^{\mathrm{T}} \rangle$, and $\otimes$ denotes the Kronecker product. The computation of the posterior distribution of $\mathbf{X}$ and $\mathbf{S}$ is more complicated and will be discussed next.

The time-series variables $\mathbf{X}$ and $\mathbf{S}$ can be updated using algorithms similar to the Kalman filter and the Rauch-Tung-Striebel smoother. The classical formulations of those algorithms do not work for VB learning because of the uncertainty in the dynamics matrix [9,14]. Thus, we used a modified version of these algorithms as presented for the classical LSSM in [10]. The algorithm performs a forward and a backward pass in order to find the required posterior expectations.

The explicit update equations for $q(\mathbf{X})$ can be written as:

$$\Sigma_x^{-1} = \begin{bmatrix} \boldsymbol{\Lambda}_0 + \langle \mathbf{W}_1^{\mathrm{T}} \mathbf{W}_1 \rangle & -\langle \mathbf{W}_1 \rangle^{\mathrm{T}} & & \\ -\langle \mathbf{W}_1 \rangle & \mathbf{I} + \langle \mathbf{W}_2^{\mathrm{T}} \mathbf{W}_2 \rangle + \boldsymbol{\Psi}_1 & \ddots & \\ & \ddots & \ddots & -\langle \mathbf{W}_N \rangle^{\mathrm{T}} \\ & & -\langle \mathbf{W}_N \rangle & \mathbf{I} + \boldsymbol{\Psi}_N \end{bmatrix}, \quad (26)$$

$$\boldsymbol{\mu}_x = \Sigma_x \begin{bmatrix} \boldsymbol{\Lambda}_0 \boldsymbol{\mu}_0^{(x)} \\ \sum_{m \in \mathcal{O}_{:1}} y_{m1} \langle \tau_m \rangle \langle \mathbf{c}_m \rangle \\ \vdots \\ \sum_{m \in \mathcal{O}_{:N}} y_{mN} \langle \tau_m \rangle \langle \mathbf{c}_m \rangle \end{bmatrix}, \qquad (27)$$

where $\mathcal{O}_{:n}$ is the set of indices m for which the observation y_{mn} is not missing, $\boldsymbol{\Psi}_n = \sum_{m \in \mathcal{O}_{:n}} \langle \tau_m \rangle \langle \mathbf{c}_m \mathbf{c}_m^{\mathrm{T}} \rangle$, $\langle \mathbf{W}_n \rangle = \sum_{k=1}^{K} \langle s_{kn} \rangle \langle \mathbf{B}_k \rangle$, and $\langle \mathbf{W}_n^{\mathrm{T}} \mathbf{W}_n \rangle = \sum_{k=1}^{K} \sum_{l=1}^{K} [\langle \mathbf{s}_n \mathbf{s}_n^{\mathrm{T}} \rangle]_{kl} \langle \mathbf{B}_k^{\mathrm{T}} \mathbf{B}_l \rangle$. Instead of using standard matrix inversion, one can utilize the block-banded structure of Σ_x^{-1} to compute the required expectations $\langle \mathbf{x}_n \rangle$, $\langle \mathbf{x}_n \mathbf{x}_n^{\mathrm{T}} \rangle$ and $\langle \mathbf{x}_n \mathbf{x}_{n-1}^{\mathrm{T}} \rangle$ efficiently. The algorithm for the computations is presented in [10].

Similarly for **S**, the explicit update equations are

$$\Sigma_s^{-1} = \begin{bmatrix} \mathbf{V}_0 + \langle \mathbf{A}^\mathrm{T}\mathbf{A} \rangle & -\langle \mathbf{A} \rangle^\mathrm{T} & & \\ -\langle \mathbf{A} \rangle & \mathbf{I} + \langle \mathbf{A}^\mathrm{T}\mathbf{A} \rangle + \Theta_1 & \ddots & \\ & \ddots & \ddots & -\langle \mathbf{A} \rangle^\mathrm{T} \\ & & -\langle \mathbf{A} \rangle & \mathbf{I} + \Theta_N \end{bmatrix}, \qquad (28)$$

$$\boldsymbol{\mu}_s = \Sigma_s \begin{bmatrix} \mathbf{V}_0 \boldsymbol{\mu}_0^{(s)} \\ \sum_{d=1}^D \langle \mathbf{B}_{:d:} \rangle \langle x_{d1} \mathbf{x}_0^\mathrm{T} \rangle \\ \vdots \\ \sum_{d=1}^D \langle \mathbf{B}_{:d:} \rangle \langle x_{dN} \mathbf{x}_{N-1}^\mathrm{T} \rangle \end{bmatrix}, \qquad (29)$$

where $\Theta_n = \sum_{i=1}^D \sum_{j=1}^D [\langle \mathbf{x}_{n-1} \mathbf{x}_{n-1}^\mathrm{T} \rangle]_{ij} \cdot \langle \mathbf{B}_{::i} \mathbf{B}_{::j}^\mathrm{T} \rangle$. The required expectations $\langle \mathbf{s}_n \rangle$, $\langle \mathbf{s}_n \mathbf{s}_n \rangle$ and $\langle \mathbf{s}_n \mathbf{s}_{n-1} \rangle$ can be computed efficiently by using the same algorithm as for **X** [10].

The VB learning of the LSSM with switching dynamics is quite similar to the equations presented above. The main difference is that the posterior distribution of the discrete state variable z_n is computed by using alpha-beta recursion [12]. The update equations for the state transition probability matrix and the initial state probabilities are straightforward because of the conjugacy. The expectations $\langle \mathbf{W}_n \rangle$ and $\langle \mathbf{W}_n^\mathrm{T} \mathbf{W}_n \rangle$ are computed by averaging $\langle \mathbf{B}_k \rangle$ and $\langle \mathbf{B}_k^\mathrm{T} \mathbf{B}_k \rangle$ over the state probabilities $\mathrm{E}[z_n = k]$.

3.2   Practical Issues

The main practical issue with the proposed model is that the VB learning algorithm may converge to bad local minima. As a solution, we found two ways of improving the robustness of the method. The first improvement is related to the updating of the posterior approximation and the second improvement is related to the initialization of the approximate posterior distributions.

The first practical tip is that one may want to run the VB updates for the lower layers of the model hierarchy first for a few times before starting to update the upper layers. Otherwise, the hyperparameters may learn very bad values because the child variables have not yet been well estimated. Thus, we updated **X**, **C**, **B** and τ 5–10 times before updating the hyperparameters and the upper layers. However, this procedure requires a reasonable initialization.

We initialized **X** and **C** randomly but for **S** and **B** we used a bit more complicated approach. One goal of the initialization was that the model would be close to a model with constant dynamics. Thus, the first component in **S** was set to a constant value and the corresponding matrix $\mathbf{B}_k$ was initialized as an identity matrix. The other components in **S** and **B** were random but their scale was a bit smaller so that the time variation in the resulting state dynamics matrix $\mathbf{W}_n$ was small initially. Obviously, this initialization leads to a bias towards a constant component in $\mathbf{S}$ but this is often realistic as the system probably has some average dynamics and deviations from it.

3.3   Rotations for Faster Convergence

One issue with the VB-EM algorithm for state-space models is that the algorithm may converge extremely slowly. This happens if the variables are strongly correlated because they are updated only one at a time causing zigzagging and small updates. This effect can be reduced by the parameter expansion approach, which finds a suitable auxiliary parameter connecting several variables and then optimizes this auxiliary parameter [15,16]. This corresponds to a parameterized joint optimization of several variables.

A suitable parameter expansion for state-space models is related to the rotation of the latent sub-space [17,10]. It can be motivated by noting that the latent variable $\mathbf{X}$ can be rotated arbitrarily by compensating it in $\mathbf{C}$:

$$\mathbf{y}_n = \mathbf{C}\mathbf{x}_n = \mathbf{C}\mathbf{R}^{-1}\mathbf{R}\mathbf{x}_n = \left(\mathbf{C}\mathbf{R}^{-1}\right)\left(\mathbf{R}\mathbf{x}_n\right) \quad \text{for all non-singular } \mathbf{R}. \quad (30)$$

The rotation of $\mathbf{X}$ must also be compensated in the dynamics $\mathbf{W}_n$ as

$$\mathbf{R}\mathbf{x}_n = \mathbf{R}\mathbf{W}_n\mathbf{R}^{-1}\mathbf{R}\mathbf{x}_{n-1} = \left(\mathbf{R}\mathbf{W}_n\mathbf{R}^{-1}\right)\left(\mathbf{R}\mathbf{x}_{n-1}\right). \quad (31)$$

The rotation $\mathbf{R}$ can be used to parameterize the posterior distributions $q(\mathbf{X})$, $q(\mathbf{C})$ and $q(\mathbf{B})$. Optionally, the distributions of the hyperparameters $q(\boldsymbol{\gamma})$ and $q(\boldsymbol{\beta})$ can also be parameterized. Optimizing the posterior approximation with respect to $\mathbf{R}$ is efficient and leads to significant improvement in the speed of the VB learning. Details for the procedure in the context of the classical LSSM can be found in [10].

Similarly to $\mathbf{X}$, the latent mixing weights $\mathbf{S}$ can also be rotated as

$$[\mathbf{W}_n]_{d:} = \mathbf{B}_{:d:}^{\mathrm{T}}\mathbf{s}_n = \mathbf{B}_{:d:}^{\mathrm{T}}\mathbf{R}^{-1}\mathbf{R}\mathbf{s}_n = \left(\mathbf{B}_{:d:}^{\mathrm{T}}\mathbf{R}^{-1}\right)\left(\mathbf{R}\mathbf{s}_n\right), \quad (32)$$

where $[\mathbf{W}_n]_{d:}$ is the d-th row vector of $\mathbf{W}_n$. The rotation must also be compensated in the dynamics of $\mathbf{S}$ as

$$\mathbf{R}\mathbf{s}_n = \mathbf{R}\mathbf{A}\mathbf{R}^{-1}\mathbf{R}\mathbf{s}_{n-1} = \left(\mathbf{R}\mathbf{A}\mathbf{R}^{-1}\right)\left(\mathbf{R}\mathbf{s}_{n-1}\right). \quad (33)$$

Thus, the rotation corresponds to a parameterized joint optimization of $q(\mathbf{S})$, $q(\mathbf{A})$, $q(\mathbf{B})$, and optionally also $q(\boldsymbol{\alpha})$ and $q(\boldsymbol{\beta})$. Note that the optimal rotation of $\mathbf{S}$ can be computed separately from the optimal rotation for $\mathbf{X}$.

The extra computational cost by the rotation speed up is small compared to the computational cost of one VB update of all variables. Thus, the rotation can be computed at each iteration after the variables have been updated. If for some reason the computation of the optimal rotation is slow, one can use the rotations less frequently, for instance, after every ten updates of all variables, and still gain similar performance improvements. However, as was shown in [10], the rotation transformation is essential even for the classical LSSM, thus ignoring it may lead to extremely slow convergence and poor results. Thus, we used the rotation transformation for all methods in the next section.

4 Experiments

We compare the proposed linear state-space model with time-varying dynamics (LSSM-TVD) to the classical linear-state space model (LSSM) and the linear state-space model with switching dynamics (LSSM-SD) using three datasets: a one-dimensional signal with changing frequency, a simulated physical process with time-varying parameters, and real-world daily temperature measurements in Europe. The methods are evaluated by their ability to predict missing values and gaps in the observed processes.

4.1 Signal with Changing Frequency

We demonstrate the LSSM with time-varying dynamics using an artificial signal with changing frequency. The signal is defined as

$$f(n) = \sin(a \cdot (n + c\sin(b \cdot 2\pi)) \cdot 2\pi), \quad n = 0, \ldots, 999 \qquad (34)$$

where $a = 0.1$, $b = 0.01$ and $c = 8$. The resulting signal is shown in Fig. 2(a). The signal was corrupted with Gaussian noise having zero mean and standard deviation 0.1 to simulate noisy observations. In order to see how well the different methods can learn the dynamics, we created seven gaps in the signal by removing 15 consecutive observations to produce each gap. In addition, 20% of the remaining observations were randomly removed. Each method (LSSM, LSSM-SD and LSSM-TVD) used $D = 5$ dimensions for the latent states $\mathbf{x}_n$. The LSSM-SD and LSSM-TVD used $K = 4$ state dynamics matrices $\mathbf{B}_k$.

Figures 2(b)-(d) show the posterior distribution of the latent noiseless function f for each method. The classical LSSM is not able to capture the dynamics and the reconstructions over the gaps are bad and have high variance. The LSSM-SD learns two different states for the dynamics corresponding to a lower and a higher frequency. The reconstructions over the gaps are better than with the LSSM, but it still has quite a large variance and the fifth gap is reconstructed using a wrong frequency. The gap reconstructions have large variance because the two state dynamics matrices learned by the model do not fit the process very well so the model assumes a larger innovation noise in the latent process $\mathbf{X}$. In contrast to that, the LSSM-TVD learns the dynamics practically perfectly and even learns the dynamics of the process which changes the frequency. Thus, the LSSM-TVD is able to make nearly perfect predictions over the gaps and the variance is small. It also prunes out one state dynamics matrix, thus using effectively only three dimensions for the latent mixing-weight process.

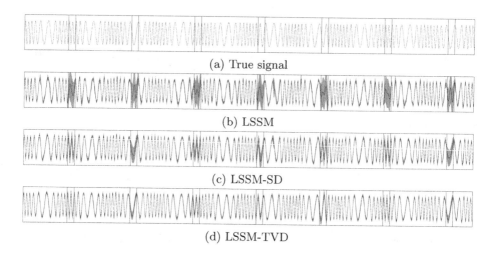

(a) True signal

(b) LSSM

(c) LSSM-SD

(d) LSSM-TVD

Fig. 2. Results for the signal with changing frequency: (a) the true signal, (b) the classical LSSM, (c) the LSSM with switching dynamics, (d) the LSSM with time-varying dynamics. In (b)-(d), the posterior mean is shown as solid black line, two standard deviations are shown as a gray area, and the true signal is shown in red for comparison. Vertical lines mark the seven gaps that contain no observations.

4.2   Stochastic Advection-Diffusion Process

We simulated a physical process with time-dependent parameters in order to compare the considered approaches. The physical process is a stochastic advection-diffusion process, which is defined by the following partial differential equation:

$$\frac{\partial f}{\partial t} = \delta \nabla^2 f - \mathbf{v} \cdot \nabla f + R, \tag{35}$$

where f is the variable of interest, δ is the diffusivity, $\mathbf{v}$ is the velocity field and R is a stochastic source. We have assumed that the diffusivity is a constant and the velocity field describes an incompressible flow. The velocity field $\mathbf{v}$ changes in time. This equation could describe, for instance, air temperature and the velocity field corresponds to winds with changing directions. The spatial domain was a torus, a two-dimensional manifold with periodic boundary conditions.

The partial differential equation (35) is discretized using the finite difference method. This is used to generate a discretized realization of the stochastic process by iterating over the time domain. The stochastic source R is a realization from a spatial Gaussian process at each time step. The two velocity field components are modelled as follows:

$$\mathbf{v}(t+1) = \sqrt{\rho} \cdot \mathbf{v}(t) + \sqrt{1-\rho} \cdot \boldsymbol{\xi}(t+1), \tag{36}$$

where $\rho \in (0,1)$ controls how fast the velocity changes and $\boldsymbol{\xi}(t+1)$ is Gaussian noise with zero mean and variance which was chosen appropriately. Thus, there

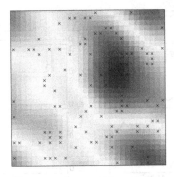

Fig. 3. One of the simulated processes at one time instance. Crosses denote the locations that were used to collect the observations. Note that the domain is a torus, that is, a 2-dimensional manifold with periodic boundaries.

are actually two sources of randomness in the stochastic process: the random source R and the randomly changing velocity field $\mathbf{v}$.

The data were generated from the simulated process as follows: Every 20-th sample was kept in the time domain, which resulted in $N = 2000$ time instances. From the spatial discretization grid, $M = 100$ locations were selected randomly as the measurement locations (corresponding to weather stations). The simulated values were corrupted with Gaussian noise to obtain noisy observations.

We used four methods in this comparison: LSSM, LSSM-SD and LSSM-TVD with $D = 30$ dimensions for the latent states $\mathbf{x}_n$, and LSSM with $D = 60$ to see if adding more dimensions improves the performance of the classical LSSM. Both the LSSM-SD and LSSM-TVD used $K = 5$ state dynamics matrices $\mathbf{B}_k$.

For measuring the performance of the methods, we generated two test sets. First, we created 18 gaps of 15 consecutive time points, that is, the observations from all the spatial locations were removed over the gaps and the corresponding values of the noiseless process f formed the first test set. Second, we randomly removed 20% of the remaining observations and used the corresponding values of the process f as the second test set. The tests were performed for five simulated processes. Figure 3 shows one process at one time instance as an example.[1]

Table 1 shows the root-mean-square errors (RMSE) of the mean reconstructions for both the generated gaps and the randomly removed values. The results for each of the five process realizations are shown separately. It appears that using $D = 60$ components does not significantly change the performance of the LSSM compared to using $D = 30$ components. Also, the LSSM-SD performs practically identically to the LSSM. The LSSM-SD used effectively two or three state dynamics matrices. However, this does not seem to help in modelling the variations in the dynamics and the learned model performs similarly to the LSSM. In contrast to that, the proposed LSSM-TVD has the best performance

---

[1] http://users.ics.aalto.fi/jluttine/ecml2014/ contains a video visualization of each of the simulated processes.

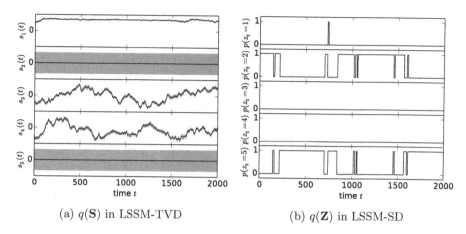

(a) $q(\mathbf{S})$ in LSSM-TVD   (b) $q(\mathbf{Z})$ in LSSM-SD

Fig. 4. Results for the advection-diffusion experiments. (a) The posterior mean and two standard deviations of the latent mixing weights by the LSSM-TVD. (b) The posterior probability of each state transition matrix as a function of time in the LSSM-SD.

in each experiment. For the test set of random values, the difference is not large because the reconstruction is mainly based on the correlations between the locations rather than the dynamics. However, in order to accurately reconstruct the gaps, the model needs to learn the changes in the dynamics. The LSSM-TVD reconstructs the gaps more accurately than the other methods, because it finds latent mixing weights $\mathbf{s}_n$ which model the changes in the dynamics.

Figure 4(a) shows the the posterior distribution of the $K = 5$ mixing-weight signals $\mathbf{S}$ in one experiment: the first signal is practically constant corresponding to the average dynamics, the third and the fourth signals correspond to the changes in the two-dimensional velocity field, and the second and the fifth signals have been pruned out as they are not needed. Thus, the method was able to learn the effective dimensionality of the latent mixing-weight process, which suggests that the method is not very sensitive to the choice of K as long as it is large enough. The results look similar in all the experiments with the LSSM-TVD and in every experiment the LSSM-TVD found one constant and two varying components. Thus, the posterior distribution of $\mathbf{S}$ might give insight on some latent processes that affect the dynamics of the observed process.

Table 1. Results for five stochastic advection-diffusion experiments. The root-mean-square errors (RMSE) have been multiplied by a factor of 1000 for clarity.

Method	RMSE for gaps					RMSE for random				
	1	2	3	4	5	1	2	3	4	5
LSSM $D = 30$	104	107	102	94	104	34	38	39	34	34
LSSM $D = 60$	105	107	110	98	108	35	39	40	35	35
LSSM-SD $D = 30$	106	117	113	94	102	35	37	39	34	34
LSSM-TVD $D = 30$	**73**	**81**	**75**	**67**	**82**	**30**	**34**	**35**	**31**	**30**

Table 2. GSOD reconstruction errors of the test sets in degrees Celsius for five runs

Method	RMSE for gaps					RMSE for randomly missing				
	1	2	3	4	5	1	2	3	4	5
LSSM	1.748	1.753	1.758	1.744	1.751	0.935	0.937	0.935	0.933	0.934
LSSM-SD	1.800	1.801	1.796	1.777	1.788	0.936	0.938	0.936	0.934	0.935
LSSM-TVD	**1.661**	**1.650**	**1.659**	**1.653**	**1.660**	0.935	0.937	0.935	0.932	0.934

This experiment showed that the LSSM-SD is not good at modelling linear combinations of the state dynamics matrices. Interestingly, although the VB update formulas average the state dynamics matrices by their probabilities resulting in a convex combination, this mixing is not very prominent in the approximate posterior distribution as seen in Fig. 4(b). Most of the time, only one state dynamics matrix is active with probability one. This happens because the prior penalizes switching between the matrices and one of the matrices is usually much better than the others on average over several time steps.

4.3   Daily Mean Temperature

The third experiment used real-world temperature measurements in Europe. The data were taken from the global surface summary of day product produced by the National Climatic Data Center (NCDC) [18]. We studied daily mean temperature measurements roughly from the European area[2] in 2000–2009. Stations that had more than 20% of the measurements missing were discarded. This resulted in $N = 3653$ time instances and $M = 1669$ stations for the analysis.

The three models were learned from the data. They used $D = 80$ dimensions for the latent states. The LSSM-SD and the LSSM-TVD used $K = 6$ state dynamics matrices. We formed two test sets similarly to the previous experiment. First, we generated randomly 300 2-day gaps in the data, which means that measurements from all the stations were removed during those periods of time. Second, 20% of the remaining data was used randomly to form another test set.

Table 2 shows the results for five experiments using different test sets. The methods reconstructed the randomly formed test sets equally well suggesting that learning more complex dynamics did not help and the learned correlations between the stations was sufficient. However, the reconstruction of gaps is more interesting because it measures how well the method learns the dynamical structure. This reconstruction shows consistent performance differences between the methods: The LSSM-SD is slightly worse than the LSSM, and the LSSM-TVD outperforms the other two. Because climate is a chaotic process, the modelling is extremely challenging and predictions tend to be far from perfect. However, these results suggest that the time-varying dynamics might offer a promising improvement to the classical LSSM in statistical modelling of physical processes.

---

[2] The longitude of the studied region was in range $(-13, 33)$ and the latitude in range $(35, 72)$.

5   Conclusions

This paper introduced a linear state-space model with time-varying dynamics. It forms the state dynamics matrix as a time-varying linear combination of a set of matrices. It uses another linear state-space model for the mixing weights in the linear combination. This is different from previous time-dependent LSSMs which use switching models to jump between a small set of states defining the model dynamics.

Both the LSSM with switching dynamics and the proposed LSSM are useful but they are suitable for slightly different problems. The switching dynamics is realistic for processes which have a few possible states that can be quite different from each other but each of them has approximately linear dynamics. The proposed model, on the other hand, is realistic when the dynamics vary more freely and continuously. It was largely motivated by physical processes based on stochastic partial differential equations with time-varying parameters.

The experiments showed that the proposed LSSM with time-varying dynamics can capture changes in the underlying dynamics of complex processes and significantly improve over the classical LSSM. If these changes are continuous rather than discrete jumps between a few states, it may achieve better modelling performance than the LSSM with switching dynamics. The experiment on a stochastic advection-diffusion process showed how the proposed model adapts to the current dynamics at each time and finds the current velocity field which defines the dynamics.

The proposed model could be further improved for challenging real-world spatio-temporal modelling problems. First, the spatial structure could be taken into account in the prior of the loading matrix using, for instance, Gaussian processes [19]. Second, outliers and badly corrupted measurements could be modelled by replacing the Gaussian observation noise distribution with a more heavy-tailed distribution, such as the Student-t distribution [20].

The method was implemented in Python as a module for an open-source variational Bayesian package called BayesPy [21]. It is distributed under an open license, thus making it easy for others to apply the method. In addition, the scripts for reproducing all the experimental results are also available.[3]

Acknowledgments. We would like to thank Harri Valpola, Natalia Korsakova, and Erkki Oja for useful discussions. This work has been supported by the Academy of Finland (project number 134935).

References

1. Bar-Shalom, Y., Li, X.R., Kirubarajan, T.: Estimation with Applications to Tracking and Navigation. Wiley-Interscience (2001)
2. Shumway, R.H., Stoffer, D.S.: Time Series Analysis and Its Applications. Springer (2000)

---

[3] Experiment scripts available at `http://users.ics.aalto.fi/jluttine/ecml2014/`

3. Ghahramani, Z., Roweis, S.T.: Learning nonlinear dynamical systems using an EM algorithm. In: Advances in Neural Information Processing Systems, pp. 431–437 (1999)
4. Valpola, H., Karhunen, J.: An unsupervised ensemble learning method for nonlinear dynamic state-space models. Neural Computation 14(11), 2647–2692 (2002)
5. Ghahramani, Z., Hinton, G.E.: Variational learning for switching state-space models. Neural Computation 12, 963–996 (1998)
6. Pavlovic, V., Rehg, J.M., MacCormick, J.: Learning switching linear models of human motion. In: Advances in Neural Information Processing Systems 13, pp. 981–987. MIT Press (2001)
7. Raiko, T., Ilin, A., Korsakova, N., Oja, E., Valpola, H.: Drifting linear dynamics (abstract). In: International Conference on Artificial Intelligence and Statistics, AISTATS 2010 (2010)
8. Michalski, V., Memisevic, R., Konda, K.: Modeling sequential data using higher-order relational features and predictive training. ArXiv preprint ArXiv:1402.2333 (2014)
9. Beal, M.J.: Variational algorithms for approximate Bayesian inference. PhD thesis, Gatsby Computational Neuroscience Unit, University College London (2003)
10. Luttinen, J.: Fast variational Bayesian linear state-space model. In: Blockeel, H., Kersting, K., Nijssen, S., Železný, F. (eds.) ECML PKDD 2013, Part I. LNCS, vol. 8188, pp. 305–320. Springer, Heidelberg (2013)
11. Bishop, C.M.: Variational principal components. In: Proceedings of the 9th International Conference on Artificial Neural Networks (ICANN 1999), pp. 509–514 (1999)
12. Bishop, C.M.: Pattern Recognition and Machine Learning, 2nd edn. Information Science and Statistics. Springer, New York (2006)
13. Beal, M.J., Ghahramani, Z.: The variational Bayesian EM algorithm for incomplete data: with application to scoring graphical model structures. Bayesian Statistics 7, 453–464 (2003)
14. Barber, D., Chiappa, S.: Unified inference for variational Bayesian linear Gaussian state-space models. In: Advances in Neural Information Processing Systems 19. MIT Press (2007)
15. Liu, C., Rubin, D.B., Wu, Y.N.: Parameter expansion to accelerate EM: the PX-EM algorithm. Biometrika 85, 755–770 (1998)
16. Qi, Y.A., Jaakkola, T.S.: Parameter expanded variational Bayesian methods. In: Advances in Neural Information Processing Systems 19, pp. 1097–1104. MIT Press (2007)
17. Luttinen, J., Ilin, A.: Transformations in variational Bayesian factor analysis to speed up learning. Neurocomputing 73, 1093–1102 (2010)
18. NCDC: Global surface summary of day product, http://www.ncdc.noaa.gov/cgi-bin/res40.pl (Online; accessed April 16, 2014)
19. Luttinen, J., Ilin, A.: Variational Gaussian-process factor analysis for modeling spatio-temporal data. In: Advances in Neural Information Processing Systems 22. MIT Press (2009)
20. Luttinen, J., Ilin, A., Karhunen, J.: Bayesian robust PCA of incomplete data. Neural Processing Letters 36(2), 189–202 (2012)
21. Luttinen, J.: BayesPy – Bayesian Python, http://www.bayespy.org

An Online Policy Gradient Algorithm for Markov Decision Processes with Continuous States and Actions

Yao Ma[1], Tingting Zhao[1], Kohei Hatano[2], and Masashi Sugiyama[1]

[1] Tokyo Institute of Technology,
2-12-1 O-okayama, Meguro, Tokyo 152-8552, Japan
{yao@sg.,tingting@sg.,sg@}cs.titech.ac.jp
[2] Kyushu University,
744 Motooka, Nishi, Fukuoka, 819-0395, Japan
hatano@inf.kyushu-u.ac.jp

Abstract. We consider the learning problem under an online Markov decision process (MDP), which is aimed at learning the time-dependent decision-making policy of an agent that minimizes the regret — the difference from the best fixed policy. The difficulty of online MDP learning is that the reward function changes over time. In this paper, we show that a simple online policy gradient algorithm achieves regret $O(\sqrt{T})$ for T steps under a certain concavity assumption and $O(\log T)$ under a strong concavity assumption. To the best of our knowledge, this is the first work to give an online MDP algorithm that can handle continuous state, action, and parameter spaces with guarantee. We also illustrate the behavior of the online policy gradient method through experiments.

Keywords: Markov decision process, Online learning.

1 Introduction

The *Markov decision process* (MDP) is a popular framework of reinforcement learning for sequential decision making [6], where an agent takes an action depending on the current state, moves to the next state, receives a reward based on the last transition, and this process is repeated T times. The goal is to find an optimal decision-making policy (i.e., a conditional probability density of action given state) that maximizes the expected sum of rewards over T steps.

In the standard MDP formulation, the reward function is fixed over iterations. On the other hand, in this paper, we consider an online MDP scenario where the reward function changes over time — it can be altered even adversarially. The goal is to find the best *time-dependent* policy that minimizes the *regret*, the difference from the best fixed policy. We expect the regret to be $o(T)$, by which the difference from the best fixed policy vanishes as T goes to infinity.

The *MDP expert* algorithm (MDP-E), which chooses the current best action at each state, was shown to achieve regret $O(\sqrt{T \log |A|})$ [1,2], where $|A|$

T. Calders et al. (Eds.): ECML PKDD 2014, Part II, LNCS 8725, pp. 354–369, 2014.
© Springer-Verlag Berlin Heidelberg 2014

denotes the cardinality of the action space. Although this bound does not explicitly depend on the cardinality of the state space, the algorithm itself needs an expert algorithm for each state. Another algorithm called the *lazy follow-the-perturbed-leader* (lazy-FPL) divides the time steps into short periods and policies are updated only at the end of each period using the average reward function [8]. This lazy-FPL algorithm was shown to have regret $O(T^{3/4+\epsilon} \log T(|S| + |A|)|A|^2)$ for $\epsilon \in (0, 1/3)$. The online MDP algorithm called the *online relative entropy policy search* is considered in [9], which was shown to have regret $O(L^2 \sqrt{T \log(|S||A|/L)})$ for state space with L-layered structure. However, the regret bounds of these algorithms explicitly depend on $|S|$ and $|A|$, and the algorithms cannot be directly implemented for problems with continuous state and action spaces. The *online algorithm for Markov decision processes* was shown to have regret $O(\sqrt{T \log |\Pi|} + \log |\Pi|)$ with changing transition probability distributions, where $|\Pi|$ it the cardinality of the policy set [11]. Although sub-linear bounds still hold for continuous policy spaces, the algorithm cannot be used with infinite policy candidates directly.

In this paper, we propose a simple *online policy gradient* (OPG) algorithm that can be implemented in a straightforward manner for problems with continuous state and action spaces[1]. Under the assumption that the expected average reward function is concave, we prove that the regret of our OPG algorithm is $O(\sqrt{T}(F^2 + N))$, which is independent of the cardinality of the state and action spaces, but is dependent on the diameter F and dimension N of the parameter space. Furthermore, regret $O(N^2 \log T)$ is also proved under a strongly concavity assumption on the expected average reward function. We numerically illustrate the superior behavior of the proposed OPG in continuous problems over MDP-E with different discretization schemes.

2   Online Markov Decision Process

In this section, we formulate the problem of online MDP learning.

An online MDP is specified by

- State space $s \in S$, which could be either continuous or discrete.
- Action space $a \in A$, which could be either continuous or discrete.
- Transition density $p(s'|s, a)$, which represents the conditional probability density of next state s' given current state s and action a to be taken.
- Reward function sequence $r_1, r_2, \ldots, r_T$, which are fixed in advance and will not change no matter what action is taken.

An online MDP algorithm produces a stochastic policy $\pi(a|s, t)$[2], which is a conditional probability density of action a to be taken given current state s at

---

[1] Our OPG algorithm can also be seen as an extension of the *online gradient descent* algorithm [10] to online MDPs problems, by decomposing the objective function.

[2] The stochastic policy incorporates exploratory actions, and exploration is usually required for getting a better policy in the learning process.

time step t. In other words, an online MDP algorithm $\mathcal{A}$ outputs parameter $\boldsymbol{\theta} = [\theta^{(1)}, \ldots, \theta^{(N)}]^\top \in \Theta \subset \mathbb{R}^N$ of stochastic policy $\pi(\boldsymbol{a}|\boldsymbol{s}; \boldsymbol{\theta})$.

Thus, algorithm $\mathcal{A}$ gives a sequence of policies:

$$\pi(\boldsymbol{a}|\boldsymbol{s}; \boldsymbol{\theta}_1), \pi(\boldsymbol{a}|\boldsymbol{s}; \boldsymbol{\theta}_2), \ldots, \pi(\boldsymbol{a}|\boldsymbol{s}; \boldsymbol{\theta}_T).$$

We denote the expected cumulative rewards over T time steps of algorithm $\mathcal{A}$ by

$$R_{\mathcal{A}}(T) = \mathbb{E}\left[\sum_{t=1}^{T} r_t(\boldsymbol{s}_t, \boldsymbol{a}_t)\Big|\mathcal{A}\right].$$

Suppose that there exists $\boldsymbol{\theta}^*$ such that policy $\pi(\boldsymbol{a}|\boldsymbol{s}; \boldsymbol{\theta}^*)$ maximizes the expected cumulative rewards:

$$R_{\boldsymbol{\theta}^*}(T) = \mathbb{E}\left[\sum_{t=1}^{T} r_t(\boldsymbol{s}_t, \boldsymbol{a}_t)\Big|\boldsymbol{\theta}^*\right] = \sup_{\boldsymbol{\theta}\in\Theta} \mathbb{E}\left[\sum_{t=1}^{T} r_t(\boldsymbol{s}_t, \boldsymbol{a}_t)\Big|\boldsymbol{\theta}\right],$$

where $\mathbb{E}$ denotes the expectation. Our goal is to design algorithm $\mathcal{A}$ that minimizes the *regret* against the best offline policy defined by

$$L_{\mathcal{A}}(T) = R_{\boldsymbol{\theta}^*}(T) - R_{\mathcal{A}}(T).$$

If the regret is bounded by a sub-linear function with respect to T, the algorithm $\mathcal{A}$ is shown to be asymptotically as powerful as the best offline policy.

3   Online Policy Gradient (OPG) Algorithm

In this section, we introduce an online policy gradient algorithm for solving the online MDP problem.

Different from the previous works, we do not use the expert algorithm in our method, because it is not suitable to handling continuous state and action problems. Instead, we consider a gradient-based algorithm which updates the parameter of policy $\boldsymbol{\theta}$ along the gradient direction of the expected average reward function at time step t.

More specifically, we assume that the target MDP $\{S, A, p, \pi, r\}$ is *ergodic*. Then it has the unique stationary state distribution $d_{\boldsymbol{\theta}}(\boldsymbol{s})$:

$$d_{\boldsymbol{\theta}}(\boldsymbol{s}) = \lim_{T\to\infty} p(\boldsymbol{s}_T = \boldsymbol{s}|\boldsymbol{\theta}).$$

Note that the stationary state distribution satisfies

$$d_{\boldsymbol{\theta}}(\boldsymbol{s}') = \int_{\boldsymbol{s}\in S} d_{\boldsymbol{\theta}}(\boldsymbol{s}) \int_{\boldsymbol{a}\in A} \pi(\boldsymbol{a}|\boldsymbol{s}; \boldsymbol{\theta}) p(\boldsymbol{s}'|\boldsymbol{s}, \boldsymbol{a}) \mathrm{d}\boldsymbol{a}\mathrm{d}\boldsymbol{s}.$$

Let $\rho_t(\boldsymbol{\theta})$ be the expected average reward function of policy $\pi(\boldsymbol{a}|\boldsymbol{s}; \boldsymbol{\theta})$ at time step t:

$$\rho_t(\boldsymbol{\theta}) = \int_{\boldsymbol{s}\in S} d_{\boldsymbol{\theta}}(\boldsymbol{s}) \int_{\boldsymbol{a}\in A} r_t(\boldsymbol{s}, \boldsymbol{a})\pi(\boldsymbol{a}|\boldsymbol{s}; \boldsymbol{\theta})\mathrm{d}\boldsymbol{a}\mathrm{d}\boldsymbol{s}. \tag{1}$$

Then our *online policy gradient (OPG) algorithm* is given as follows:

– Initialize policy parameter $\boldsymbol{\theta}_1$.
– for $t = 1$ to ∞

   1. Observe current state $\boldsymbol{s}_t = \boldsymbol{s}$.
   2. Take action $\boldsymbol{a}_t = \boldsymbol{a}$ according to current policy $\pi(\boldsymbol{a}|\boldsymbol{s};\boldsymbol{\theta}_t)$.
   3. Observe reward r_t from the environment.
   4. Move to next state $\boldsymbol{s}_{t+1}$.
   5. Update the policy parameter as

$$\boldsymbol{\theta}_{t+1} = P\left(\boldsymbol{\theta}_t + \eta_t \nabla_{\boldsymbol{\theta}} \rho_t(\boldsymbol{\theta}_t)\right), \tag{2}$$

where $P(\boldsymbol{\vartheta}) = \arg\min_{\boldsymbol{\theta} \in \Theta} \|\boldsymbol{\vartheta} - \boldsymbol{\theta}\|$ is the projection function, $\eta_t = \frac{1}{\sqrt{t}}$ is the step size, and $\nabla_{\boldsymbol{\theta}} \rho_t(\boldsymbol{\theta})$ is the gradient of $\rho_t(\boldsymbol{\theta})$:

$$\nabla_{\boldsymbol{\theta}} \rho_t(\boldsymbol{\theta}) \equiv \left[\frac{\partial \rho_t(\boldsymbol{\theta})}{\partial \theta^{(1)}}, \cdots, \frac{\partial \rho_t(\boldsymbol{\theta})}{\partial \theta^{(N)}}\right]^{\top}$$

$$= \int_{\boldsymbol{s} \in S} \int_{\boldsymbol{a} \in A} d_{\boldsymbol{\theta}}(\boldsymbol{s}) \pi(\boldsymbol{a}|\boldsymbol{s};\boldsymbol{\theta})(\nabla_{\boldsymbol{\theta}} \ln d_{\boldsymbol{\theta}}(\boldsymbol{s}) + \nabla_{\boldsymbol{\theta}} \ln \pi(\boldsymbol{a}|\boldsymbol{s};\boldsymbol{\theta}))$$

$$\times\, r_t(\boldsymbol{s},\boldsymbol{a}) d\boldsymbol{a} d\boldsymbol{s}.$$

If it is time-consuming to obtain the exact stationary state distribution, gradients estimated by a reinforcement learning algorithm may be used instead in practice.

When the reward function does not changed over time, the OPG algorithm is reduced to the ordinary policy gradient algorithm [7], which is an efficient and natural algorithm for continuous state and action MDPs. The OPG algorithm can also be regarded as an extension of the *online gradient descend* algorithm [10], which maximizes $\sum_{t=1}^{T} \rho_t(\boldsymbol{\theta}_t)$, not $\mathbb{E}\left[\sum_{t=1}^{T} r_t(\boldsymbol{s}_t, \boldsymbol{a}_t)|\mathcal{A}\right]$. As we will prove in Section 4, the regret bound of the OPG algorithm is $O(\sqrt{T})$ under a certain concavity assumption and $O(\log T)$ under a strong concavity assumption. Unlike previous works, this bound does not depend on the cardinality of state and action spaces. Therefore, the OPG algorithm would be suitable to handling continuous state and action online MDPs.

4   Regret Analysis under Concavity

In this section, we provide a regret bound for the OPG algorithm.

4.1   Assumptions

First, we introduce the assumptions required in the proofs. Some assumptions have already been used in related works for discrete state and action MDPs, and we extend them to continuous state and action MDPs.

Assumption 1. *For two arbitrary distributions* d *and* d' *over* S *and for every policy parameter* $\boldsymbol{\theta}$, *there exists a positive number* τ *such that*

$$\int_{\boldsymbol{s}\in S}\int_{\boldsymbol{s}'\in S}|d(\boldsymbol{s})-d'(\boldsymbol{s})|p(\boldsymbol{s}'|\boldsymbol{s};\boldsymbol{\theta})\mathrm{d}\boldsymbol{s}'\mathrm{d}\boldsymbol{s}\leq e^{-1/\tau}\int_{\boldsymbol{s}\in S}|d(\boldsymbol{s})-d'(\boldsymbol{s})|\mathrm{d}\boldsymbol{s},$$

where

$$p(\boldsymbol{s}'|\boldsymbol{s};\boldsymbol{\theta})=\int_{\boldsymbol{a}\in A}\pi(\boldsymbol{a}|\boldsymbol{s};\boldsymbol{\theta})p(\boldsymbol{s}'|\boldsymbol{s},\boldsymbol{a})\mathrm{d}\boldsymbol{a},$$

and τ *is called the* mixing time *[1,2].*

Assumption 2. *For two arbitrary policy parameters* $\boldsymbol{\theta}$ *and* $\boldsymbol{\theta}'$ *and for every* $\boldsymbol{s}\in S$, *there exists a constant* $C_1>0$ *depending on the specific policy model* π *such that*

$$\int_{\boldsymbol{a}\in A}|\pi(\boldsymbol{a}|\boldsymbol{s};\boldsymbol{\theta})-\pi(\boldsymbol{a}|\boldsymbol{s};\boldsymbol{\theta}')|\mathrm{d}\boldsymbol{a}\leq C_1\|\boldsymbol{\theta}-\boldsymbol{\theta}'\|_1.$$

The Gaussian policy is a common choice in continuous state and action MDPs. Below, we consider the Gaussian policy with mean $\mu(\boldsymbol{s})=\boldsymbol{\theta}^{\top}\phi(\boldsymbol{s})$ and standard deviation σ, where $\boldsymbol{\theta}$ is the policy parameter and $\phi(\boldsymbol{s}):S\to\mathbb{R}^N$ is the basis function. The KL-divergence between these two policies is

$$D(p(\cdot|s;\theta)\|p(\cdot|s;\theta'))=\int_{a\in A}\mathcal{N}_{\theta,\sigma}(a)\left\{\log\mathcal{N}_{\theta,\sigma}(a)-\log\mathcal{N}_{\theta',\sigma}(a)\right\}\mathrm{d}a$$

$$=\int_{a\in A}\mathcal{N}_{\theta,\sigma}(a)\left\{\frac{1}{2\sigma^2}\left(-(a-\theta)^2+(a-\theta')^2\right)\right\}\mathrm{d}a$$

$$=\frac{\|\phi(\boldsymbol{s})\|_\infty}{2\sigma}\|\boldsymbol{\theta}-\boldsymbol{\theta}'\|^2.$$

By Pinsker's inequality, the following inequality holds:

$$\|p(\cdot|s,\theta)-p(\cdot|s,\theta')\|_1\leq\frac{\|\phi(\boldsymbol{s})\|_\infty}{\sigma}\|\boldsymbol{\theta}-\boldsymbol{\theta}'\|_1. \tag{3}$$

This implies that the Gaussian policy model satisfies Assumption 2 with $C_1=\frac{\|\phi(\boldsymbol{s})\|_\infty}{\sigma}$. Note that we do not specify any policy model in the analysis, and therefore other stochastic policy models could also be used in our algorithm.

Assumption 3. *All the reward functions in online MDPs are bounded. For simplicity, we assume that the reward functions satisfy*

$$r_t(\boldsymbol{s},\boldsymbol{a})\in[0,1],\forall\boldsymbol{s}\in S,\forall\boldsymbol{a}\in A,\forall t=1,\dots,T.$$

Assumption 4. *For all* $t=1,\dots,T$, *the second derivative of the expected average reward function satisfies*

$$\nabla_\theta^2\rho_t(\boldsymbol{\theta})\leq 0. \tag{4}$$

This assumption means that the expected average reward function is concave, which is currently our sufficient condition to guarantee the $O(\sqrt{T})$-regret bound for the OPG algorithm.

4.2   Regret Bound

We have the following theorem.

Theorem 1. *The regret against the best offline policy of the OPG algorithm is bounded as*

$$L_A(T) \le \sqrt{T}\frac{F^2}{2} + \sqrt{T}C_2N + 2\sqrt{T}\tau^2 C_1 C_2 N + 4\tau,$$

where F is the diameter of Θ and $C_2 = \frac{2C_1 - C_1 e^{-1/\tau}}{1 - e^{-1/\tau}}$.

To prove the above theorem, we decompose the regret in the same way as the previous work [1,2,3,4]:

$$
\begin{aligned}
L_A(T) =& R_{\theta^*}(T) - R_A(T) \\
\le& \left(R_{\theta^*}(T) - \sum_{t=1}^{T} \rho_t(\theta^*) \right) + \left(\sum_{t=1}^{T} \rho_t(\theta^*) - \sum_{t=1}^{T} \rho_t(\theta_t) \right) \\
& + \left(\sum_{t=1}^{T} \rho_t(\theta_t) - R_A(T) \right).
\end{aligned}
\tag{5}
$$

In the OPG method, $\rho_t(\theta)$ is used for optimization, and the expected average reward is calculated by the stationary state distribution $d_\theta(s)$ of the policy parameterized by θ. However, the expected reward at time step t is calculated by $d_{\theta,t}$, which is the state distribution at time step t following policy $\pi(a|s;\theta)$. This difference affects the first and third terms of the decomposed regret (5).

Below, we bound each of the three terms in Lemma 1, Lemma 2, and Lemma 3, which are proved later.

Lemma 1.

$$\left| R_{\theta^*}(T) - \sum_{t=1}^{T} \rho_t(\theta^*) \right| \le 2\tau.$$

The first term has already been analyzed for discrete state and action online MDPs in [1,2], and we extended it to continuous state and action spaces in Lemma 1.

Lemma 2. *The expected average reward function satisfies*

$$\left| \sum_{t=1}^{T} (\rho_t(\theta^*) - \rho_t(\theta_t)) \right| \le \sqrt{T}\frac{F^2}{2} + \sqrt{T}C_2N.$$

Lemma 2 is obtained by using the result of [10].

Lemma 3.

$$\left| R_A(T) - \sum_{t=1}^{T} \rho_t(\theta_t) \right| \le 2\tau^2 C_1 C_2 N \sqrt{T} + 2\tau.$$

Lemma 3 is similar to Lemma 5.2 in [2], but our bound does not depend on the cardinality of state and action spaces.

Combining Lemma 1, Lemma 2, and Lemma 3, we can immediately obtain Theorem 1.

If the reward function is strongly concave for all $t = 1, \ldots, T$, the bound of the OPG algorithm is $O(\log T)$ which is proved in Section 5.

4.3   Proof of Lemma 1

The following proposition holds, which can be obtained by recursively using Assumption 1:

Proposition 1. *For any policy parameter* $\boldsymbol{\theta}$, *the state distribution* $d_{\boldsymbol{\theta},t}$ *at time* t *and stationary state distribution* $d_{\boldsymbol{\theta}}$ *satisfy*

$$\int_{\boldsymbol{s} \in S} |d_{\boldsymbol{\theta},t}(\boldsymbol{s}) - d_{\boldsymbol{\theta}}(\boldsymbol{s})| \mathrm{d}\boldsymbol{s} \leq 2e^{-t/\tau}.$$

The first part of the regret bound in Theorem 1 is caused by the difference between the state distribution at time t and the stationary state distribution following the best offline policy parameter $\boldsymbol{\theta}^*$.

$$\left| R_{\boldsymbol{\theta}^*}(T) - \sum_{t=1}^{T} \rho_t(\boldsymbol{\theta}^*) \right| = \left| \sum_{t=1}^{T} \left[\int_{\boldsymbol{s} \in S} d_{\boldsymbol{\theta}^*,t}(\boldsymbol{s}) \int_{\boldsymbol{a} \in A} r_t(\boldsymbol{s}, \boldsymbol{a}) \pi(\boldsymbol{a}|\boldsymbol{s}; \boldsymbol{\theta}^*) \mathrm{d}\boldsymbol{s}\mathrm{d}\boldsymbol{a} \right. \right.$$
$$\left. \left. - \int_{\boldsymbol{s} \in S} d_{\boldsymbol{\theta}^*}(\boldsymbol{s}) \int_{\boldsymbol{a} \in A} r_t(\boldsymbol{s}, \boldsymbol{a}) \pi(\boldsymbol{a}|\boldsymbol{s}; \boldsymbol{\theta}^*) \mathrm{d}\boldsymbol{s}\mathrm{d}\boldsymbol{a} \right] \right|$$
$$\leq \sum_{t=1}^{T} \int_{\boldsymbol{s} \in S} |d_{\boldsymbol{\theta}^*,t}(\boldsymbol{s}) - d_{\boldsymbol{\theta}^*}(\boldsymbol{s})| \, \mathrm{d}\boldsymbol{s}$$
$$\leq 2 \sum_{t=1}^{T} e^{-t/\tau}$$
$$\leq 2\tau,$$

which concludes the proof.

4.4   Proof of Lemma 2

The following proposition is a continuous extension of Lemma 6.3 in [2]:

Proposition 2. *For two policies with different parameters* $\boldsymbol{\theta}$ *and* $\boldsymbol{\theta}'$, *an arbitrary distribution* d *over* S, *and the constant* $C_1 > 0$ *given in Assumption 2, it holds that*

$$\int_{\boldsymbol{s} \in S} d(\boldsymbol{s}) \int_{\boldsymbol{s}' \in S} |p(\boldsymbol{s}'|\boldsymbol{s}; \boldsymbol{\theta}) - p(\boldsymbol{s}'|\boldsymbol{s}; \boldsymbol{\theta}')| \mathrm{d}\boldsymbol{s}'\mathrm{d}\boldsymbol{s} \leq C_1 \|\boldsymbol{\theta} - \boldsymbol{\theta}'\|_1,$$

where

$$p(s'|s; \boldsymbol{\theta}) = \int_{a \in A} \pi(a|s; \boldsymbol{\theta}) p(s'|s, a) \mathrm{d}a.$$

Then we have the following proposition, which is proved in Section 4.6:

Proposition 3. *For all $t = 1, \ldots, T$, the expected average reward function $\rho_t(\boldsymbol{\theta})$ for two different parameters $\boldsymbol{\theta}$ and $\boldsymbol{\theta}'$ satisfies*

$$|\rho_t(\boldsymbol{\theta}) - \rho_t(\boldsymbol{\theta}')| \leq C_2 \|\boldsymbol{\theta} - \boldsymbol{\theta}'\|_1.$$

From Proposition 3, we have the following proposition:

Proposition 4. *Let*

$$\boldsymbol{\theta} = [\theta^{(1)}, \ldots, \theta^{(i)}, \ldots, \theta^{(N)}],$$
$$\boldsymbol{\theta}' = [\theta^{(1)}, \ldots, \theta^{(i)'}, \ldots, \theta^{(N)}],$$

and suppose that the expected average reward $\rho_t(\boldsymbol{\theta})$ for all $t = 1, \ldots, T$ is Lipschitz continuous with respect to each dimension $\theta^{(i)}$. Then we have

$$|\rho_t(\boldsymbol{\theta}) - \rho_t(\boldsymbol{\theta}')| \leq C_2 |\theta^{(i)} - \theta^{(i)'}|, \forall i = 1, \ldots, N.$$

Form Proposition 4, we have the following proposition:

Proposition 5. *For all $t = 1, \ldots, T$, the partial derivative of expected average reward function $\rho_t(\boldsymbol{\theta})$ with respect to $\theta^{(i)}$ is bounded as*

$$\left| \frac{\partial \rho_t(\boldsymbol{\theta})}{\partial \theta^{(i)}} \right| \leq C_2, \forall i = 1, \ldots, N,$$

and $\|\nabla_{\boldsymbol{\theta}} \rho_t(\boldsymbol{\theta})\|_1 \leq N C_2$.

From Proposition 5, the result of online convex optimization [10] is applicable to the current setup. More specifically we have

$$\sum_{t=1}^{T} (\rho_t(\boldsymbol{\theta}^*) - \rho_t(\boldsymbol{\theta}_t)) \leq \frac{F^2}{2} \sqrt{T} + C_2 N \sqrt{T},$$

which concludes the proof.

4.5 Proof of Lemma 3

The following proposition holds, which can be obtained from Assumption 2 and

$$\|\boldsymbol{\theta}_t - \boldsymbol{\theta}_{t+1}\|_1 \leq \eta_t \|\nabla_{\boldsymbol{\theta}} \rho_t(\boldsymbol{\theta}_t)\|_1 \leq C_2 N \eta_t.$$

Proposition 6. *Consecutive policy parameters* $\boldsymbol{\theta}_t$ *and* $\boldsymbol{\theta}_{t+1}$ *given by the OPG algorithm satisfy*

$$\int_{\boldsymbol{a} \in A} |\pi(\boldsymbol{a}|\boldsymbol{s}; \boldsymbol{\theta}_t) - \pi(\boldsymbol{a}|\boldsymbol{s}; \boldsymbol{\theta}_{t+1})| \mathrm{d}\boldsymbol{a} \le C_1 C_2 N \eta_t.$$

From Proposition 2 and Proposition 6, we have the following proposition:

Proposition 7. *For consecutive policy parameters* $\boldsymbol{\theta}_t$ *and* $\boldsymbol{\theta}_{t+1}$ *given by the OPG algorithm and arbitrary transition probability density* $p(\boldsymbol{s}'|\boldsymbol{s}, \boldsymbol{a})$, *it holds that*

$$\int_{\boldsymbol{s} \in S} d(\boldsymbol{s}) \int_{\boldsymbol{s}' \in S} \int_{\boldsymbol{a} \in A} p(\boldsymbol{s}'|\boldsymbol{s}, \boldsymbol{a})$$
$$\times |\pi(\boldsymbol{a}|\boldsymbol{s}; \boldsymbol{\theta}_t) - \pi(\boldsymbol{a}|\boldsymbol{s}; \boldsymbol{\theta}_{t+1})| \mathrm{d}\boldsymbol{a} \mathrm{d}\boldsymbol{s}' \mathrm{d}\boldsymbol{s} \le C_1 C_2 N \eta_t.$$

Then the following proposition holds, which is proved in Section 4.6 following the same line as Lemma 5.1 in [2]:

Proposition 8. *The state distribution* $d_{\mathcal{A},t}$ *given by algorithm* $\mathcal{A}$ *and the stationary state distribution* $d_{\boldsymbol{\theta}_t}$ *of policy* $\pi(\boldsymbol{a}|\boldsymbol{s}; \boldsymbol{\theta}_t)$ *satisfy*

$$\int_{\boldsymbol{s} \in S} |d_{\boldsymbol{\theta}_t}(\boldsymbol{s}) - d_{\mathcal{A},t}(\boldsymbol{s})| \mathrm{d}\boldsymbol{s} \le 2\tau^2 \eta_{t-1} C_1 C_2 N + 2e^{-t/\tau}.$$

Although the original bound given in [1,2] depends on the cardinality of the action space, it is not the case in the current setup.

Then the third term of the decomposed regret (5) is expressed as

$$\left| R_{\mathcal{A}}(T) - \sum_{t=1}^{T} \rho_t(\boldsymbol{\theta}_t) \right| = \left| \sum_{t=1}^{T} \int_{\boldsymbol{s} \in S} d_{\mathcal{A},t}(\boldsymbol{s}) \int_{\boldsymbol{a} \in A} r_t(\boldsymbol{s}, \boldsymbol{a}) \pi(\boldsymbol{a}|\boldsymbol{s}; \boldsymbol{\theta}_t) \mathrm{d}\boldsymbol{a} \mathrm{d}\boldsymbol{s} \right.$$
$$\left. - \sum_{t=1}^{T} \int_{\boldsymbol{s} \in S} d_{\boldsymbol{\theta}_t}(\boldsymbol{s}) \int_{\boldsymbol{a} \in A} r_t(\boldsymbol{s}, \boldsymbol{a}) \pi(\boldsymbol{a}|\boldsymbol{s}; \boldsymbol{\theta}_t) \mathrm{d}\boldsymbol{a} \mathrm{d}\boldsymbol{s} \right|$$
$$\le \sum_{t=1}^{T} \int_{\boldsymbol{s} \in S} |d_{\mathcal{A},t}(\boldsymbol{s}) - d_{\pi_t}(\boldsymbol{s})| \mathrm{d}\boldsymbol{s}$$
$$\le 2\tau^2 C_1 C_2 N \sum_{t=1}^{T} \eta_t + 2 \sum_{t=1}^{T} e^{-t/\tau}$$
$$\le 2\tau^2 C_1 C_2 N \sqrt{T} + 2\tau,$$

which concludes the proof.

4.6   Proof of Proposition 3

For two different parameters θ and θ', we have

$$
\begin{aligned}
|\rho_t(\theta) - \rho_t(\theta')| &= \left| \int_{s \in S} d_\theta(s) \int_{a \in A} \pi(a|s; \theta) r_t(s, a) \mathrm{d}a \mathrm{d}s \right. \\
&\quad \left. - \int_{s \in S} d_{\theta'}(s) \int_{a \in A} \pi(a|s; \theta') r_t(s, a) \mathrm{d}a \mathrm{d}s \right| \\
&\leq \int_{s \in S} |d_\theta(s) - d_{\theta'}(s)| \int_{a \in A} \pi(a|s; \theta) r_t(s, a) \mathrm{d}a \mathrm{d}s \\
&\quad + \int_{s \in S} d_{\theta'}(s) \int_{a \in A} |\pi(a|s; \theta) - \pi(a|s; \theta')| \, r_t(s, a) \mathrm{d}a \mathrm{d}s.
\end{aligned}
\tag{6}
$$

The first equation comes from Eq.(1), and the second inequality is obtained from the triangle inequality. Since Assumption 2 and Assumption 3 imply

$$
\int_{s \in S} d_{\theta'}(s) \int_{a \in A} |\pi(a|s; \theta) - \pi(a|s; \theta')| r_t(s, a) \mathrm{d}a \mathrm{d}s \leq C_1 \|\theta - \theta'\|_1,
$$

and also

$$
\int_{a \in A} \pi(a|s; \theta) r_t(s, a) \mathrm{d}a \leq 1,
$$

Eq.(6) can be written as

$$
\begin{aligned}
|\rho_t(\theta) - \rho_t(\theta')| &\leq \int_{s \in S} |d_\theta(s) - d_{\theta'}(s)| \mathrm{d}s + C_1 \|\theta - \theta'\|_1 \\
&= \int_{s \in S} \int_{s' \in S} |d_\theta(s') p(s|s'; \theta) - d_{\theta'}(s') p(s|s'; \theta')| \mathrm{d}s' \mathrm{d}s \\
&\quad + C_1 \|\theta - \theta'\|_1 \\
&\leq \int_{s \in S} \int_{s' \in S} |d_\theta(s') p(s|s'; \theta) - d_{\theta'}(s') p(s|s'; \theta)| \mathrm{d}s' \mathrm{d}s \\
&\quad + \int_{s \in S} \int_{s' \in S} d_{\theta'}(s') |p(s|s'; \theta) - p(s|s'; \theta')| \mathrm{d}s' \mathrm{d}s \\
&\quad + C_1 \|\theta - \theta'\|_1 \\
&\leq e^{-1/\tau} \int_{s \in S} |d_\theta(s) - d_{\theta'}(s)| \mathrm{d}s + 2C_1 \|\theta - \theta'\|_1.
\end{aligned}
$$

The second equality comes from the definition of the stationary state distribution, and the third inequality can be obtained from the triangle inequality. The last inequality follows from Assumption 1 and Proposition 2. Thus, we have

$$
|\rho_t(\theta) - \rho_t(\theta')| \leq \frac{2C_1 - C_1 e^{-1/\tau}}{1 - e^{-1/\tau}} \|\theta - \theta'\|_1,
$$

which concludes the proof.

4.7 Proof of Proposition 8

This proof is following the same line as Lemma 5.1 in [2].

$$\int_{s \in S} |d_{\mathcal{A},k}(s) - d_{\theta_t}(s)| ds$$

$$= \int_{s \in S} \int_{s' \in S} |d_{\mathcal{A},k-1}(s')p(s|s';\theta_k) - d_{\theta_t}(s')p(s|s';\theta_t)| ds' ds$$

$$\leq \int_{s \in S} \int_{s' \in S} |d_{\mathcal{A},k-1}(s')p(s|s';\theta_t) - d_{\theta_t}(s')p(s|s';\theta_t)| ds' ds$$

$$+ \int_{s \in S} \int_{s' \in S} |d_{\mathcal{A},k-1}(s')p(s|s';\theta_k) - d_{\mathcal{A},k-1}(s')p(s|s';\theta_t)| ds' ds$$

$$\leq e^{-1/\tau} \int_{s \in S} |d_{\mathcal{A},k-1}(s) - d_{\theta_t}(s)| ds + 2(t-k)C_1 C_2 N \eta_{t-1}. \tag{7}$$

The first equation comes from the definition of the stationary state distribution, and the second inequality can be obtained by the triangle inequality. The third inequality holds from Assumption 1 and

$$\int_{s \in S} \int_{s' \in S} |d_{\mathcal{A},k-1}(s')p(s|s';\theta_k) - d_{\mathcal{A},k-1}(s')p(s|s';\theta_t)| ds$$

$$\leq C_1 \|\theta_t - \theta_k\|_1$$

$$\leq C_1 \sum_{i=k}^{t-1} \eta_i \|\nabla_\theta \rho_i(\theta_i)\|_1$$

$$\leq 2(t-k)C_1 C_2 N \eta_{t-1}.$$

Recursively using Eq.(7), we have

$$\int_{s \in S} |d_{\mathcal{A},t}(s) - d_{\pi_t}(s)| ds \leq 2 \sum_{k=2}^{t} e^{-(t-k)/\tau}(t-k)C_1 C_2 N \eta_{t-1} + 2e^{-t/\tau}$$

$$\leq 2\tau^2 C_1 C_2 N \eta_{t-1} + 2e^{-t/\tau},$$

which concludes the proof.

5   Regret Analysis under Strong Concavity

In this section, we derive a shaper regret bound for the OPG algorithm under a strong concavity assumption.

Theorem 1 shows the theoretical guarantee of the OPG algorithm with the concave assumption. If the expected reward function is strongly concave, i.e.,

$$\nabla_\theta^2 \rho_t \leq -H I_N,$$

where H is a positive constant and I_N is the $N \times N$ identity matrix, we have following theorem.

Theorem 2. *The regret against the best offline policy of the OPG algorithm is bounded as*

$$L_\mathcal{A}(T) \leq \frac{C_2^2 N^2}{2H}(1 + \log T) + \frac{2\tau^2 C_1 C_2 N}{H} \log T + 4\tau,$$

with step size $\eta_t = \frac{1}{Ht}$.

We again consider the same decomposition as Eq.(5), the first term of the regret bound is exactly the same as Lemma 1. The second and third parts are given by the following propositions.

Given the strongly concavity assumption and step size $\eta_t = \frac{1}{Ht}$, the following proposition holds:

Proposition 9.

$$\sum_{t=1}^{T}(\rho_t(\boldsymbol{\theta}^*) - \rho_t(\boldsymbol{\theta}_t)) \leq \frac{C_2^2 N^2}{2H}(1 + \log T).$$

The proof is following the same line as [12], i.e., by the Taylor approximation, the expected average reward function can be decomposed as

$$
\begin{aligned}
\rho_t&(\boldsymbol{\theta}^*) - \rho_t(\boldsymbol{\theta}_t) \\
&= \nabla_\theta \rho_t(\boldsymbol{\theta}_t)^\top (\boldsymbol{\theta}^* - \boldsymbol{\theta}_t) + \frac{1}{2}(\boldsymbol{\theta}_t)^\top (\boldsymbol{\theta}^* - \boldsymbol{\theta}_t)^\top \nabla_\theta^2 \rho_t(\boldsymbol{\xi}_t)(\boldsymbol{\theta}_t)^\top (\boldsymbol{\theta}^* - \boldsymbol{\theta}_t) \\
&\leq \nabla_\theta \rho_t(\boldsymbol{\theta}_t)^\top (\boldsymbol{\theta}^* - \boldsymbol{\theta}_t) - \frac{H}{2}\|\boldsymbol{\theta}^* - \boldsymbol{\theta}_t\|^2.
\end{aligned}
\tag{8}
$$

Given the parameter updating rule,

$$\nabla_\theta \rho_t(\boldsymbol{\theta}^* - \boldsymbol{\theta}_t) = \frac{1}{2\eta_t}\left((\boldsymbol{\theta}^* - \boldsymbol{\theta}_t)^2 - (\boldsymbol{\theta}^* - \boldsymbol{\theta}_{t+1})^2\right) + \eta_t\|\nabla_\theta \rho_t(\boldsymbol{\theta}_t)\|^2,$$

summing up all T terms of (8) and setting $\eta_t = \frac{1}{Ht}$ yield

$$
\begin{aligned}
\sum_{t=1}^{T}(\rho_t(\boldsymbol{\theta}^* - \boldsymbol{\theta}_t)) &\leq \sum_{t=1}^{T}\left(\frac{1}{\eta_{t+1}} - \frac{1}{\eta_t} - H\right)\|\boldsymbol{\theta}^* - \boldsymbol{\theta}_t\|^2 + \|\nabla_t \rho_t(\boldsymbol{\theta}_t)\|^2 \sum_{t=1}^{T}\eta_t \\
&\leq \frac{C_2^2 N^2}{2H}(1 + \log T).
\end{aligned}
$$

From the proof of Lemma 3, the bound of the third part with the strongly concavity assumption is given by following proposition.

Proposition 10.

$$\sum_{t=1}^{T}\rho_t(\boldsymbol{\theta}_t) - R_\mathcal{A}(T) \leq \frac{2\tau^2 C_1 C_2 N}{H}\log T + 2\tau. \tag{9}$$

The result of Proposition 10 is obtained by following the same line as the proof of Lemma 3 with different step sizes. Combining Lemma 1, Proposition 9, and Proposition 10, we can obtain Theorem 2.

6   Experiments

In this section, we illustrate the behavior of the OPG algorithm.

6.1   Target Tracking

The task is to let an agent track an abruptly moving target located in one-dimensional real space $S = \mathbb{R}$. The action space is also one-dimensional real space $A = \mathbb{R}$, and we can change the position of the agent as $s' = s + a$. The reward function is given by evaluating the distance between the agent and target as

$$r_t(s, a) = e^{-|s+a-\text{tar}(t)|},$$

where $\text{tar}(t)$ denotes the position of the target at time step t. Because the target is moving abruptly, the reward function is also changing abruptly. As a baseline method for comparison, we consider the MDP-E algorithm [1,2], where the exponential weighted average algorithm is used as the best expert. Since MDP-E can handle only discrete states and actions, we discretize the state and action space. More specifically, the state space is discretized as

$$(-\infty, -6], (-6, -6+c], (-6+c, -6+2c], \ldots, (6, +\infty),$$

and the action space is discretized as

$$-6, -6+c, -6+2c, \ldots, 6.$$

We consider the following 5 setups for c:

$$c = 6, 2, 1, 0.5, 0.1.$$

In the experiment, the stationary state distribution and the gradient are estimated by policy gradient theorem estimator[5]. $I = 20$ independent experiments are run with $T = 100$ time steps, and the average return $J(T)$ is used for evaluating the performance:

$$J(T) = \frac{1}{I} \sum_{i=1}^{I} \left[\sum_{t=1}^{T} r_t(s_t, a_t) \right].$$

The results are plotted in Figure 1, showing that the OPG algorithm works better than the MDP-E algorithm with the best discretization resolution. This illustrates the advantage of directly handling continuous state and action spaces without discretization.

6.2   Linear-Quadratic Regulator

The *linear-quadratic regulator* (LQR) is a simple system, where the transition dynamics is linear and the reward function is quadratic. A notable advantage of LQR is that we can compute the best offline parameter [5]. Here, an online LQR system is simulated to illustrate the parameter update trajectory of the OPG algorithm.

Let state and action spaces be one-dimensional real: $S = \mathbb{R}$ and $A = \mathbb{R}$. Transition is deterministically performed as

$$s' = s + a.$$

The reward function is defined as

$$r_t(s, a) = -\frac{1}{2}Q_t s^2 - \frac{1}{2}R_t a^2,$$

where $Q_t \in \mathbb{R}$ and $R_t \in \mathbb{R}$ are chosen from $\{1, \dots, 10\}$ uniformly for each t. Thus, the reward function is changing abruptly.

We use the Gaussian policy with mean parameter $\mu \cdot s$ and standard deviation parameter $\sigma = 0.1$, i.e., $\theta = \mu$. The best offline parameter is given by $\theta^* = -0.98$, and the initial parameter for the OPG algorithm is set at $\theta_1 = -0.5$.

In the top graph of Figure 2, a parameter update trajectory of OPG in an online LQR problem is plotted by the red line, and the best offline parameter is denoted by the black line. This shows that the OPG solution quickly approaches the best offline parameter.

Next, we also include the Gaussian standard deviation σ in the policy parameter, i.e., $\theta = (\mu, \sigma)^\top$. When σ takes a value less than 0.001 during gradient update iterations, we project it back to 0.001. A parameter update trajectory is plotted in the bottom graph of Figure 2, showing again that the OPG solution smoothly approaches the best offline parameter along μ.

7   Conclusion

In this paper, we proposed an online policy gradient method for continuous state and action online MDPs, and showed that the regret of the proposed method is $O(\sqrt{T})$ under a certain concavity assumption. A notable fact is that the regret bound does not depend on the cardinality of state and action spaces, which makes the proposed algorithm suitable in handling continuous states and actions. Furthermore, we also established the $O(\log T)$ regret bound under a strongly concavity assumption. Through experiments, we illustrated that directly handling continuous state and action spaces by the proposed method is more advantageous than discretizing them.

Our future work will extend the current theoretical analysis to non-concave expected average reward functions, where gradient-based algorithms suffer from the local optimal problem. Another important challenge is to develop an effective method to estimate the stationary state distribution which is required in our algorithm.

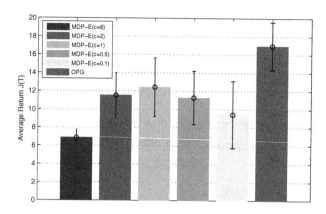

Fig. 1. Average returns of the OPG algorithm and the MDP-E algorithm with different discretization resolution c

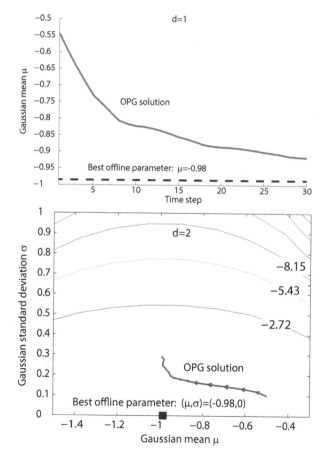

Fig. 2. Trajectory of the OPG solutions and the best offline parameter

Acknowledgments. YM was supported by the MEXT scholarship and the CREST program. KH was supported by MEXT KAKENHI 25330261 and 24106010. MS was supported by KAKENHI 23120004.

References

1. Even-Dar, E., Kakade, S.M., Mansour, Y.: Experts in a Markov Decision Process. In: Advances in Neural Information Processing System 17, pp. 401–408. MIT Press, Cambridge (2005)
2. Even-Dar, E., Karade, S.M., Mansour, Y.: Online Markov Decision Processes. Mathematics of Operations Research 34(3), 726–736 (2009)
3. Neu, G., György, A., Szepesvári, C., Antos, A.: Online Markov Decision Processes under Bandit Feedback. In: Advances in Neural Information Processing Systems 23, pp. 1804–1812 (2010)
4. Neu, G., György, A., Szepesvári, C.: The Online Loop-free Stochastic Shortest-path Problem. In: Conference on Learning Theory, pp. 231–243 (2010)
5. Peter, J., Schaal, S.: Policy Gradient Methods for Robotics. In: Proceedings of the IEEE/RSJ International Conference on Intelligent Robots and Systems (2006)
6. Sutton, R.S., Barto, A.G.: Reinforcement Learning: An Introduction. MIT Press (1998)
7. Williams, R.J.: Simple Statistical Gradient-following Algorithms for Connetionist Reinforcement Learning. Machine Learning 8(3-4), 229–256 (1992)
8. Yu, J.Y., Mannor, S., Shimkin, N.: Markov Decision Processes with Arbitrary Reward Processes. Mathematics of Operations Research 34(3), 737–757 (2009)
9. Zimin, A., Neu, G.: Online Learning in Episodic Markovian Decision Processes by Relative Entropy Policy Search. In: Advances in Neural Information Processing Systems 26, pp. 1583–1591 (2013)
10. Zinkevich, M.: Online Convex Programming and Generalized Infinitesimal Gradient Ascent. In: International Conference on Machine Learning, pp. 928–936. AAAI Press (2003)
11. Abbasi-Yadkori, Y., Bartlett, P., Kanade, V., Seldin, Y., Szepesvári, C.: Online Learning in Markov Decision Processes with Adversarially Chosen Transition Probability Distributions. In: Advances in Neural Information Processing Systems 26 (2013)
12. Hazan, E., Agarwal, A., Kale, S.: Logarithmic Regret Algorithms for Online Convex Optimization. Machine Learning 69, 169–192 (2007)

GMRF Estimation under Topological and Spectral Constraints

Victorin Martin[1], Cyril Furtlehner[2], Yufei Han[2,3], and Jean-Marc Lasgouttes[3]

[1] Mines ParisTech, Centre for Robotics, France
[2] Inria Saclay–Île-de-France, TAO team, France
[3] Inria Paris–Rocquencourt, RITS team, France

Abstract. We investigate the problem of Gaussian Markov random field selection under a non-analytic constraint: the estimated models must be compatible with a fast inference algorithm, namely the Gaussian belief propagation algorithm. To address this question, we introduce the $\star$-IPS framework, based on iterative proportional scaling, which incrementally selects candidate links in a greedy manner. Besides its intrinsic sparsity-inducing ability, this algorithm is flexible enough to incorporate various spectral constraints, like e.g. walk summability, and topological constraints, like short loops avoidance. Experimental tests on various datasets, including traffic data from San Francisco Bay Area, indicate that this approach can deliver, with reasonable computational cost, a broad range of efficient inference models, which are not accessible through penalization with traditional sparsity-inducing norms.

Keywords: Iterative proportional scaling, Gaussian belief propagation, walk-summability, Gaussian Markov Random Field.

1 Introduction

The Gaussian belief propagation algorithm [2] (GaBP) is an efficient distributed inference algorithm, well adapted to online inference on large scale Gaussian Markov random fields (GMRF). However, since it may encounter convergence problems, especially with non-sparse structures, it can be of practical interest to construct off-line a GMRF which is compatible with GaBP. When selecting such a model from observations, we potentially face a difficult constrained problem. In the present work, we propose to solve it in an approximate but satisfactory manner, with good accuracy and limited computational cost. To achieve this, we combine various methods, which have been discussed in the context of sparse inverse covariance matrix estimation [1,7,15].

The GMRF distribution is naturally characterized by a mean vector $\boldsymbol{\mu} \in \mathbb{R}$ and a positive definite precision (or concentration) matrix $\mathbf{A}$, which is simply the inverse of the covariance matrix $\mathbf{C}$. Zero entries in the precision matrix $\mathbf{A}$ indicate conditionally independent pairs of variables. This gives a graphical representation of dependencies: two random variables are conditionally independent if, and only if, there is no direct edge between them. Observations are

T. Calders et al. (Eds.): ECML PKDD 2014, Part II, LNCS 8725, pp. 370–385, 2014.
© Springer-Verlag Berlin Heidelberg 2014

summarized in an empirical covariance matrix $\hat{\mathbf{C}} \in \mathbb{R}^{N \times N}$ of a random vector $\mathbf{X} = (X_i)_{i \in \{1, \ldots, N\}}$, and we look for a GMRF model with sparse precision matrix $\mathbf{A}$. The model estimation problem can be expressed as the maximization of the log-likelihood:

$$\mathbf{A} = \operatorname*{argmax}_{\mathbf{M} \in \mathcal{S}^{BP}_{++}} \mathcal{L}(\mathbf{M}), \quad \mathcal{L}(\mathbf{M}) \stackrel{\text{def}}{=} \log \det(\mathbf{M}) - \operatorname{Tr}(\mathbf{M}\hat{\mathbf{C}}),$$

where $\mathcal{S}^{BP}_{++}$ formally represents the set of positive definite matrices corresponding to some GaBP-compatible GMRF.

Without any constraint on $\mathbf{M}$, the maximum likelihood estimate is trivially $\mathbf{A} = \hat{\mathbf{C}}^{-1}$. However, enforcing sparsity with simple thresholding of small magnitude entries may easily ruin the positive definiteness of the estimated precision matrix. In the context of structure learning, where meaningful interactions have to be determined, for instance among genes in genetic networks, the maximization is classically performed on the set of positive definite matrices, after adding to the log-likelihood a continuous penalty function P that imitates the L_0 norm. The Lasso penalty, a convex relaxation of the problem, uses the L_1 norm, measuring the amplitudes of off-diagonal entries in $\mathbf{A}$ [7,9]. Various optimization schemes have been proposed to solve it efficiently [1,7]. However, the L_1 norm penalty suffers from a modeling bias, due to excessive penalization of truly large magnitudes entries of $\mathbf{A}$. To overcome this issue, concave functions, that perform constant penalization to the large magnitudes, have been proposed. Experimental results indicate promising improvements compared to Lasso penalty by reducing bias, while conserving the sparsity-introducing capability [6,11].

In our context, where compatibility with GaBP has to be imposed, sparsity is a desirable feature, albeit without much guarantee: specific topological properties, like the presence of short loops, are likely to damage the GaBP compatibility, even on a sparse graph. Some spectral properties, e.g. walk-summability [12], which guarantee the compatibility with GaBP based inference, might be relevant too. In order to incorporate these explicitly, we propose an efficient constrained model selection framework called $\star$-IPS, where $\star$ stands for the imposed constraints. Approaches based on the iterative proportional scaling (IPS) procedure [17] have already been discussed for tackling the original sparse inverse covariance matrix problem [10,15]. A first contribution of this paper is to improve its performance by combining it with block update techniques used in [1,7], along with providing some precision guarantee based on duality. Our second and main contribution is to exploit the incremental nature of the method to impose, for a reasonable cost, both topological and/or spectral constraints, to generate GMRF models compatible with GaBP, achieving a very good trade-off between computational cost and precision in inference tasks, as shown experimentally.

The paper is organized as follows. The principles of IPS are described in Section 2. Our method includes a likelihood maximization step at fixed graph structure, for which we give a stopping criterion based on duality. In Section 3, we propose several constraints improving GaBP compatibility of the estimated models, and show how to introduce them in our framework. In Section 4, we describe $\star$-IPS as a whole, discuss its complexity and provide some implementation details. Finally Section 5 is devoted to numerical experiments, both on synthetic data and on real traffic data coming from $\approx 10^3$ fixed sensors in the San Francisco Bay Area, to illustrate the use of the method for traffic applications.

2   IPS-Based GMRF Selection

Iterative proportional scaling has been proposed for contingency table estimation [4] and extended further to MRF maximum likelihood estimation [17]. Assuming the structure of the graph is known, it appears to be less efficient than other gradient based methods [13]. Conversely, local changes based on single row-column updates have been shown to be very efficient, even in the first order setting [7]. In our work, we combine the benefits of the incremental characteristics of IPS to identify links (Section 2.1), with the efficiency of row-column update to optimize their parameters at fixed structure (Section 2.2).

2.1   Optimal 1-Link Perturbation

Suppose that we are given a set of single and pairwise empirical marginals $\hat{p}_i$ and $\hat{p}_{ij}$ from a real-valued random vector $\mathbf{X} = (X_i)_{i \in \{1...N\}}$, and a candidate distribution $\mathcal{P}^{(n)}$, based on the dependency graph $\mathcal{G}^{(n)}$. Let us first describe optimal link addition in terms of likelihood. Let $\mathcal{P}^{(n)}$ be the reference distribution, to which we want to add one factor ψ_{ij} to produce the distribution

$$\mathcal{P}^{(n+1)}(\mathbf{x}) = \mathcal{P}^{(n)}(\mathbf{x}) \times \psi_{ij}(x_i, x_j).$$

This is a special case of IPS and the optimal perturbation is

$$\psi_{ij}(x_i, x_j) = \frac{\hat{p}_{ij}(x_i, x_j)}{p_{ij}^{(n)}(x_i, x_j)}, \tag{1}$$

where $p_{ij}^{(n)}$ is the (i, j) pairwise marginal of $\mathcal{P}^{(n)}$. The correction to the log-likelihood can then be written as a Kullback-Leibler divergence:

$$\Delta\mathcal{L} = D_{KL}(\hat{p}_{ij} \| p_{ij}^{(n)}) = \iint \hat{p}_{ij}(u, v) \log \frac{\hat{p}_{ij}(u, v)}{p_{ij}^{(n)}(u, v)} du dv.$$

Sorting all the links w.r.t. this quantity yields the optimal 1-link correction to be made. Hence, the best candidate is the one for which the current model yields the joint marginal $p_{ij}^{(n)}$ that is most divergent from $\hat{p}_{ij}$. Note that the update mechanism can in fact also be applied if the link is already present.

In the general case, computing the pairwise marginals $\{p_{ij}, (ij) \notin \dot{\mathcal{G}}^{(n)}\}$ is expensive. However, in the GMRF family, these marginals depend only on the covariance matrix associated to $\mathcal{P}^{(n)}$. The correction factor (1) reads in that case

$$\psi_{ij}(x_i, x_j) = \exp\left[-\frac{1}{2}(x_i, x_j)^T (\hat{\mathbf{C}}_{\{ij\}}^{-1} - \mathbf{C}_{\{ij\}}^{-1})(x_i, x_j)\right],$$

where $\mathbf{C}_{\{ij\}}$ (resp. $\hat{\mathbf{C}}_{\{ij\}}$) represents the restricted 2×2 covariance matrix corresponding to the pair (X_i, X_j) of the current model $\mathcal{P}^{(n)}$ (resp. of the empirical distribution $\hat{\mathcal{P}}$) specified by precision matrix $\mathbf{A} = \mathbf{C}^{-1}$ (resp. $\hat{\mathbf{A}} = \hat{\mathbf{C}}^{-1}$). Let $[\mathbf{C}_{\{ij\}}]$ denote the $N \times N$ matrix formed by completing $\mathbf{C}_{\{ij\}}$ with zeros. The new model obtained after adding or changing link (i, j) reads

$$\mathbf{A}' = \mathbf{A} + [\hat{\mathbf{C}}_{\{ij\}}^{-1}] - [\mathbf{C}_{\{ij\}}^{-1}] \stackrel{\text{def}}{=} \mathbf{A} + [\mathbf{V}], \tag{2}$$

with a log-likelihood variation given by:

$$\Delta\mathcal{L} = \frac{C_{ii}\hat{C}_{jj} + C_{jj}\hat{C}_{ii} - 2C_{ij}\hat{C}_{ij}}{\det(\mathbf{C}_{\{ij\}})} - 2 - \log\frac{\det(\hat{\mathbf{C}}_{\{ij\}})}{\det(\mathbf{C}_{\{ij\}})}. \tag{3}$$

For a 2×2 perturbation matrix $\mathbf{V} = \mathbf{V}_{\{ij\}}$, the Sherman–Morrison–Woodbury (SMW) formula allows us to efficiently compute the new covariance matrix as

$$\mathbf{C}' = \mathbf{A}'^{-1} = \mathbf{A}^{-1} - \mathbf{A}^{-1}[\mathbf{C}_{\{ij\}}^{-1}](\mathbf{I} - [\hat{\mathbf{C}}_{\{ij\}}][\mathbf{C}_{\{ij\}}^{-1}])\mathbf{A}^{-1}. \tag{4}$$

The number of operations needed to maintain the covariance matrix – and to keep track of all pairwise marginals – after each addition is therefore $\mathcal{O}(N^2)$. This technical point is determinant to the usefulness of our approach. The identity $\det(A') = \det(A) \times \det(\hat{\mathbf{C}}_{\{ij\}}) / \det(\mathbf{C}_{\{ij\}})$ ensures that the new precision matrix remains definite positive when both $\mathbf{C}_{\{ij\}}$ and $\hat{\mathbf{C}}_{\{ij\}}$ are non-degenerate.

It is also possible to remove links, so that, with help of a penalty coefficient per link, the model can be optimized with a desired connectivity level. For a GMRF with precision matrix $\mathbf{A}$, removing the link (i, j) amounts to setting the entry A_{ij} to zero, and thus $\psi_{ij}(x_i, x_j) = \exp(A_{ij}x_ix_j)$. The corresponding change of log-likelihood is then

$$\Delta\mathcal{L} = \log(1 - 2A_{ij}C_{ij} - A_{ij}^2 \det(\mathbf{C}_{\{ij\}})) + 2A_{ij}\hat{C}_{ij},$$

and, using again the SMW formula, we get the new covariance matrix

$$\mathbf{C}' = \mathbf{C} - \frac{A_{ij}}{1 - 2A_{ij}C_{ij} - A_{ij}^2 \det(\mathbf{C}_{\{ij\}})} \mathbf{C}[\mathbf{B}_{\{ij\}}]\mathbf{C},$$

with

$$\mathbf{B}_{\{ij\}} \stackrel{\text{def}}{=} \begin{bmatrix} A_{ij}C_{jj} & 1 - A_{ij}C_{ij} \\ 1 - A_{ij}C_{ij} & A_{ij}C_{ii} \end{bmatrix}.$$

In this case, the positive-definiteness of $\mathbf{A}'$ needs to be checked and we have

$$\det(\mathbf{A}') = \det(\mathbf{A})\left[1 - \alpha(C_{ij} - \sqrt{C_{ii}C_{jj}})\right]\left[1 - \alpha(C_{ij} + \sqrt{C_{ii}C_{jj}})\right],$$

so that $\mathbf{A}'$ is definite positive if the following condition is verified:

$$\frac{1}{C_{ij} - \sqrt{C_{ii}C_{jj}}} < A_{ij} < \frac{1}{\sqrt{C_{ii}C_{jj}} + C_{ij}}.$$

2.2 Block Updates

When a new link is added, existing links become detuned by a slight amount. As pointed out, the optimal update given in Section 2.1 is actually indifferent to whether the considered link exists or not. This means that, after a while, detuned links may be automatically updated if the likelihood gain exceeds the one obtained by adding a new link. We observe in practice that, when many links have been added, all the existing links are slightly detuned, which eventually causes suboptimal or bad decisions for the next links, resulting in a significant departure of the learning curve from the optimal one (see Fig. 2-left in Section 5). However, correcting existing links can become very time consuming, the update of one single link having the same computational cost $\mathcal{O}(N^2)$ as the addition of one link. There are various options to address this problem. To keep the algorithm fast, robust and simple, we choose to stay with the logic of coordinate descent, by remarking that local updates are still possible via a single row-column update of the precision matrix, as originally proposed in [1] and refined in [7]. In our context, the method is based on the following expression of the log determinant of the precision matrix $\mathbf{A}$:

$$\log\det(\mathbf{A}) = \log\det(\mathbf{A}_{\backslash i\backslash i}) + \log(A_{ii} - \mathbf{A}_i^T \mathbf{A}_{\backslash i\backslash i}^{-1} \mathbf{A}_i),$$

where $\mathbf{A}_{\backslash i\backslash i}$ is the block matrix obtained after taking aside the ith row and column and $\mathbf{A}_i$ is the ith column vector of $\mathbf{A}$ without A_{ii}. The direct optimization of the log-likelihood w.r.t. $\mathbf{A}_i$ and A_{ii} yields the following updated values:

$$A'_{ii} = \frac{1}{\hat{C}_{ii}} + \mathbf{A}_i^T \mathbf{A}_{\backslash i\backslash i}^{-1} \mathbf{A}_i \quad \text{and} \quad \mathbf{A}'_i = \left[\mathbf{I}_{V(i)} \mathbf{A}_{\backslash i\backslash i}^{-1} \mathbf{I}_{V(i)}\right]^{-1} \mathbf{I}_{V(i)} \frac{\hat{\mathbf{C}}_i}{\hat{C}_{ii}}, \quad (5)$$

where $\hat{\mathbf{C}}_i$ represents the ith column vector of $\hat{\mathbf{C}}$, $V(i)$ the set of neighbors of i in the current graph, and $\mathbf{I}_{V(i)}$ the identity restricted to entries $j \in \{i\} \cup V(i)$. Note that this solution involve the inverse $\mathbf{A}_{\backslash i\backslash i}^{-1}$ of a matrix of size $N-1$. It is related to $\mathbf{C} = \mathbf{A}^{-1}$ as follows:

$$\mathbf{A}_{\backslash i\backslash i}^{-1} = \mathbf{C}_{\backslash i\backslash i} - \frac{\mathbf{C}_i \mathbf{C}_i^T}{C_{ii}}.$$

The overall cost for updating column (and row) i is thus $\mathcal{O}(|V(i)|^3)$ for the inversion of $\left[\mathbf{I}_{V(i)} \mathbf{A}_{\backslash i\backslash i} \mathbf{I}_{V(i)}\right]$ and $\mathcal{O}(N^2)$ to update the covariance matrix $\mathbf{C}$ after this change. The log-likelihood gain $\Delta\mathcal{L}$ reads

$$-\log\left[\hat{C}_{ii}(A_{ii} - \mathbf{A}_i^T \mathbf{A}_{\backslash i\backslash i}^{-1} \mathbf{A}_i)\right] - 2(\mathbf{A}'_i - \mathbf{A}_i)^T \hat{\mathbf{C}}_i - \left[\frac{1}{\hat{C}_{ii}} + \mathbf{A}'^T_i \mathbf{A}_{\backslash i\backslash i}^{-1} \mathbf{A}'_i - A_{ii}\right]\hat{C}_{ii}.$$

2.3 Stopping Criterion

If the set of links to be optimized is given by some graph $\mathcal{G}$, the likelihood optimization is a convex problem. Let us investigate its dual properties. Let $\mathbf{A}$

denote the precision matrix, and $\mathbf{\Pi}$ a Lagrange matrix multiplier, that imposes the structure given by $\mathcal{G}$. The support of $\mathbf{\Pi}$ is the complementary graph of $\mathcal{G}$: $\forall(i,j) \in \mathcal{G}, \Pi_{ij} = 0, \forall i, \Pi_{ii} = 0$ and $\mathbf{\Pi}$ is symmetric. Then, given $\mathbf{\Pi}$, we want to optimize

$$\mathbf{A}_{\mathbf{\Pi}} = \underset{\mathbf{M}}{\text{argmin}} \ \text{Tr}(\mathbf{M}\mathbf{\Pi}) + f(\mathbf{M}),$$

with $f(\mathbf{M}) \overset{\text{def}}{=} \text{Tr}(\mathbf{M}\hat{\mathbf{C}}) - \log\det(\mathbf{M})$ being convex for given support $\mathcal{G}$. The explicit solution is

$$\mathbf{A}_{\mathbf{\Pi}} = (\hat{\mathbf{C}} + \mathbf{\Pi})^{-1}. \tag{6}$$

We assume that $\mathbf{\Pi}$ is such that $\hat{\mathbf{C}} + \mathbf{\Pi}$ is positive definite, so the dual optimization problem reads $\mathbf{Y} = \text{argmax}_{\mathbf{\Pi}} \ g(\mathbf{\Pi})$, with $g(\mathbf{\Pi}) \overset{\text{def}}{=} N + \log\det(\hat{\mathbf{C}} + \mathbf{\Pi})$. The problem is now concave and, because of the barrier resulting from the log term, we are certain to have a positive definite solution. Thus, for any matrix $\mathbf{\Pi}$, such that $\hat{\mathbf{C}} + \mathbf{\Pi}$ is definite positive, $g(\mathbf{\Pi})$ is a lower bound of the log-likelihood. The support of $\mathbf{\Pi}$ represents the set of links to be removed from the precision matrix. Once optimality is reached for $\mathbf{\Pi}$, all non-zero entries Π_{ij} correspond to vanishing coefficients A_{ij} in (6). We may proceed as before, by computing the potential log-likelihood gain $\Delta\mathcal{L}$ for such local transformations of the covariance matrix. Local moves in the dual formulation deal with the covariance matrix instead of the precision matrix in the primal one. Let $\mathbf{C}$ and $\mathbf{C}'$ be two covariance matrices differing by a single modification on $\mathbf{\Pi}$ with $\mathbf{A} = \mathbf{C}^{-1}$ and $\mathbf{A}' = \mathbf{C}'^{-1}$. We have

$$\det(\mathbf{C}') = \det(\mathbf{C})\big(1 + 2\Pi_{ij}A_{ij} - \Pi_{ij}^2 \det(\mathbf{A}_{\{ij\}})\big),$$

with $\det(\mathbf{A}_{\{ij\}}) > 0$, since $\mathbf{A}$ is definite positive. Maximizing the log-likelihood variation yields the optimal values

$$\Pi_{ij} = \frac{A_{ij}}{\det(\mathbf{A}_{\{ij\}})} \quad \text{and} \quad \Delta\mathcal{L} = \log\left(1 + \frac{A_{ij}^2}{\det(\mathbf{A}_{\{ij\}})}\right).$$

In practice, we will not use this backward scheme: its computational cost is prohibitive, since the complementary graph, composed of links to be removed, is dense. However, this dual formulation will help us to build a confidence interval. During the greedy procedure, we always have to maintain $\mathbf{C} = \mathbf{A}^{-1}$ but $\mathbf{C}$ cannot be used directly to get a dual cost because, except at convergence, it does not fulfill the dual constraints $C_{ij} = \hat{C}_{ij}, \forall(i,j) \in \mathcal{G}$.

Let $\mathbf{\Pi}^{\|}$ be the correction matrix with coefficients $\Pi_{ij}^{\|} \overset{\text{def}}{=} (\hat{C}_{ij} - C_{ij})\mathbb{1}_{\{(i,j)\in G\}}$. Provided that $\tilde{\mathbf{C}} \overset{\text{def}}{=} \mathbf{A}^{-1} + \mathbf{\Pi}^{\|}$ is definite positive, which happens when $\mathbf{A}$ is close enough to the optimum $\mathbf{A}^\star$, it satisfies the dual constraints yielding the confidence bound

$$\log\det(\tilde{\mathbf{C}}) + N \leq -\mathcal{L}(\mathbf{A}^\star) \leq \text{Tr}(\mathbf{A}\hat{\mathbf{C}}) - \log\det(\mathbf{A}).$$

We have

$$\log\det(\tilde{\mathbf{C}}) = -\log\det(\mathbf{A}) + \log\det\left(\mathbf{I} + \mathbf{A}\mathbf{\Pi}^{\|}\right),$$

with both $\mathbf{A}$ and $\boldsymbol{\Pi}^\|$ sparse matrices, so the determinant can be estimated in $\mathcal{O}(N^2K)$ operations by expanding the logarithm at order 2 in $\mathbf{A}\boldsymbol{\Pi}^\|$. It leads to the following bound

$$\Delta\mathcal{L} \leq \frac{1}{2}\,\mathrm{Tr}\big(\mathbf{A}\boldsymbol{\Pi}^\|\mathbf{A}\boldsymbol{\Pi}^\|\big),$$

which will be used in practice as a stopping criterion for the link updates.

3   Introducing Constraints for GaBP Compatibility

Usually, GMRF estimation intends to describe a dependency structure. We pursue here another aim: finding a model suitable for fast inference. While inference in GMRF models can always be performed exactly in $\mathcal{O}(N^3)$ through matrix inversion, this may not be fast enough for some "real-time" applications on large networks. The GaBP algorithm [2] is a fast alternative to matrix inversion for sparse GMRF, which uses message passing along links in the graph $\mathcal{G}$, and thus, assuming it converges, will perform the inference in $\mathcal{O}(mKN)$, where K is the mean connectivity of $\mathcal{G}$ and m the maximum number of iterations before convergence. An important property of the GaBP algorithm is that, whenever it converges, it provides the exact mean values for all variables [18]. Variances are however generally incorrect [12]. Having a sparse GMRF gives no guarantee about its compatibility with GaBP, so we need to impose more precise constraints on the precision matrix and to the graph structure. In this section, we make such constraints explicit and show how to impose them in the framework of Section 2.

3.1   Spectral Constraints

The most precise condition known for convergence of GaBP is walk-summability (WS) [12]. Let $\mathbf{R}(\mathbf{A}) \stackrel{\text{def}}{=} \mathbf{A} - \mathbf{diag}(\mathbf{A})$ contain the off-diagonal terms of $\mathbf{A}$, and let $\rho(\cdot)$ denote the spectral radius of a matrix, that is, the maximal modulus of its eigenvalues. The two equivalent necessary and sufficient conditions for WS that we will use are:

(i) The matrix $\mathbf{W}(\mathbf{A}) \stackrel{\text{def}}{=} \mathbf{diag}(\mathbf{A}) - |\mathbf{R}(\mathbf{A})|$ is definite positive;
(ii) $\rho(|\mathbf{R}'(\mathbf{A})|) < 1$, with $\mathbf{R}'(\mathbf{A})_{ij} \stackrel{\text{def}}{=} \frac{R(\mathbf{A})_{ij}}{\sqrt{A_{ii}A_{jj}}}$.

Let us consider a GMRF with WS precision matrix $\mathbf{A}$ and investigate under which conditions the model remains WS after a perturbation of a link (i,j). The following proposition gives a sufficient condition:

Proposition 1. *Let $\mathbf{A}$ be a WS precision matrix and denote $\mathbf{W} \stackrel{\text{def}}{=} \mathbf{W}(\mathbf{A})$. The matrix $\mathbf{A}' = \mathbf{A} + [\mathbf{V}_{\{ij\}}]$ is WS if*

$$\Theta(\alpha) > 0, \forall \alpha \in [0,1], \tag{7}$$

where Θ is the following function

$$\Theta(\alpha) \stackrel{\text{def}}{=} \det(W_{\{ij\}}^{-1}) \left(\alpha^2 V_{ii} V_{jj} - (|A_{ij}| - |\alpha V_{ij} + A_{ij}|)^2\right)$$
$$+ \alpha \left(W_{ii}^{-1} V_{ii} + W_{jj}^{-1} V_{jj}\right) + 2\left(|A_{ij}| - |\alpha V_{ij} + A_{ij}|\right) W_{ij}. \quad (8)$$

In order to check condition (7), it is necessary to solve two quadratic equations. Note however that knowledge of the matrix $\mathbf{W}(\mathbf{A})^{-1}$ is also mandatory. We will discuss this point at the end of this section.

Proof. The sufficient condition is obtained as follows: for $\alpha \in [0,1]$ we have

$$\mathbf{W}(\mathbf{A} + \alpha \mathbf{V}) = \mathbf{W}(\mathbf{A}) + [\phi(\alpha \mathbf{V}, \mathbf{A})],$$

with

$$\phi(\mathbf{V}, \mathbf{A}) \stackrel{\text{def}}{=} \begin{bmatrix} V_{ii} & |A_{ij}| - |V_{ij} + A_{ij}| \\ |A_{ij}| - |V_{ji} + A_{ji}| & V_{jj} \end{bmatrix}.$$

$\mathbf{W}(\mathbf{A})$ being invertible, the determinant of $\mathbf{W}(\mathbf{A} + \alpha \mathbf{V})$ is expressed as

$$\det(\mathbf{W}(\mathbf{A} + \alpha \mathbf{V})) = \det\left(\mathbf{W}(\mathbf{A})\right) \det\left(\mathbf{I} + \mathbf{W}(\mathbf{A})^{-1}[\phi(\alpha \mathbf{V}, \mathbf{A})]\right),$$

and we can check that $\Theta(\alpha) = \det(\mathbf{I} + \mathbf{W}(\mathbf{A})^{-1}[\phi(\alpha \mathbf{V}, \mathbf{A})])$, with Θ defined in (8). The spectrum of $\mathbf{W}(\mathbf{A} + \alpha \mathbf{V})$ being a continuous function of α, as the roots of a polynomial, the condition (7) follows.

Note that the special case of removing one link of the graph always preserves the WS property. Indeed, it will change the matrix $\mathbf{A}$ in $\mathbf{A}'$ such as $|\mathbf{R}'(\mathbf{A}')| \leq |\mathbf{R}'(\mathbf{A})|$ where $\leq$ denotes element-wise comparison. Then, using elementary results on positive matrices [16, p. 22], $\rho(|\mathbf{R}'(\mathbf{A}')|) \leq \rho(|\mathbf{R}'(\mathbf{A})|)$ and thus $\mathbf{A}'$ is WS whenever $\mathbf{A}$ is WS.

As we shall see in the numerical experiments, imposing WS is generally too restrictive. It is easy to find non WS models which are still GaBP compatible. The above principle allows us however to impose a weaker spectral constraint: imposing that the matrix $\mathbf{diag}(\mathbf{A}) - \mathbf{R}(\mathbf{A})$ remains definite positive. This is equivalent to constrain the spectral radius $\rho(\mathbf{R}'(\mathbf{A}))$ to be strictly lower than 1 and it is a necessary condition for GaBP convergence [12]. We call that condition "weak walk-summability" (WWS) as a relaxation of the WS condition. We obtain the following condition

Proposition 2. *Let $\mathbf{A}$ be a WWS precision matrix, i.e. such as $\rho(\mathbf{R}'(\mathbf{A})) < 1$, and $\mathbf{S}(\mathbf{A}) \stackrel{\text{def}}{=} \mathbf{diag}(\mathbf{A}) - \mathbf{R}(\mathbf{A})$. The matrix $\mathbf{A}' = \mathbf{A} + [\mathbf{V}_{ij}]$ is WWS if*

$$\Gamma(\alpha) > 0, \forall \alpha \in [0,1], \quad (9)$$

with Γ the following degree 2 polynomial and $\mathbf{S} \stackrel{\text{def}}{=} \mathbf{S}(\mathbf{A})$

$$\Gamma(\alpha) \stackrel{\text{def}}{=} \alpha^2 \det(\mathbf{V} S_{\{i,j\}}^{-1}) + \alpha(V_{ii} S_{ii}^{-1} + V_{jj} S_{jj}^{-1} - 2V_{ij} S_{ij}^{-1}) + 1.$$

In order to check condition (9), we have to solve a quadratic equation. As in Proposition 1, we need to keep track of an inverse matrix, in this case $\mathbf{S(A)}^{-1}$.

Proof. Mimicking the proof of Proposition 2, we define, for $\alpha \in [0, 1]$,

$$\mathbf{M}(\alpha) \overset{\text{def}}{=} \mathbf{diag}(\mathbf{A} + \alpha[\mathbf{V}]) - \mathbf{R}(\mathbf{A} + \alpha[\mathbf{V}]) = \mathbf{S(A)} + \alpha \ \mathbf{diag}([\mathbf{V}]) - \mathbf{R}([\mathbf{V}]),$$

and we have

$$\det(\mathbf{M}(\alpha)) = \det(\mathbf{S(A)}) \det \left(\mathbf{I} + \alpha \mathbf{S}^{-1} \left[\mathbf{diag}(\mathbf{V}) - \mathbf{R}(\mathbf{V}) \right] \right) = \det(\mathbf{S(A)}) \Gamma(\alpha).$$

The spectrum of $\mathbf{M}(\alpha)$ being a continuous function of α, condition (9) follows.

Both (7) and (9) are only sufficient conditions for spectral constraint conservation after a pairwise perturbation. However, there are only a few cases where they lead to rejection of a valid perturbation. Indeed, it means that at least one eigenvalue goes to zero for some $\alpha \in]0, 1[$ and is positive again for $\alpha = 1$.

We have pointed out that checking sufficient condition (7) (resp. (9)) imposes to keep track of the inverse matrix $\mathbf{W(A)}^{-1}$ (resp. $\mathbf{S(A)}^{-1}$). This will not impact the overall complexity of the algorithm since, using the SMW formula, it can be done in $\mathcal{O}(N^2)$ operations, like for keeping track of the covariance matrix.

Note that, if we want to maintain these spectral constraints, we will not be able in practice to use the column updates described in Section 2.2. Indeed, computing the optimal perturbation is in this case costly, and we have no easy way to check whether it leads to a new admissible model, since our method would involve higher order characteristic polynomials.

3.2   Topological Constraints

We present in this section another approach, based mainly on empirical knowledge about the belief propagation algorithm. Belief propagation has been designed as an exact procedure on trees [14] and short loops are usually believed to cause convergence troubles. In the extreme case, where we forbid the addition of any loop, the best precision matrix estimate based on likelihood is known to be the max-spanning tree w.r.t. mutual information [3]. Since this is usually not enough, we propose here that the estimated precision matrix contains no loops of size smaller than ℓ. This is quite easy to impose: when adding a link (i, j), we have to search if i is in the neighborhood of j of depth $\ell - 1$. The computational cost is $\mathcal{O}(K^\ell)$, with K the connectivity of $\mathcal{G}$.

We can impose a more precise condition using the fact that, in the absence of frustrated loops, the GaBP algorithm is always convergent [12]. A frustrated loop is a loop along which the product of partial correlations is negative. Preventing the formation of frustrated loops is very similar to the previous loop constraint; the search cost is the same, the only difference is that we will avoid only this kind of loops. This last constraint cannot be imposed with guarantees during the block updates since the sign of partial correlations along edges may change. Prohibiting frustrated loops would require to store all the loops in the graph, which is by far too costly. However, experimental results show that a change of sign usually corresponds to small partial correlations, which are less likely to cause convergence issues.

4   Algorithm Description and Complexity

In this section, we give an overview of $\star$-IPS, leaving aside the backtracking option. A formal implementation[1] is given in Algorithm 1. Note that we suppose that the initial point of the algorithm corresponds to an empty graph. We may as well start from any precision matrix $\mathbf{A}$, provided that we have computed $\mathbf{C} = \mathbf{A}^{-1}$ and $\mathbf{W}(\mathbf{A})^{-1}$ or $\mathbf{S}(\mathbf{A})^{-1}$ if we want to impose spectral constraints.

Algorithm 1. $\star$-constrained Iterative Proportional Scaling ($\star$-IPS). The function *Check_constraint* returns true if the perturbation of $\widehat{ij}$ leads to a model compatible with the given constraint $\star$.

Inputs: $\star$: constraint type ($\varnothing$, WS, WWS, ℓ-LOOP, ℓ-FLOOP);
   $\hat{\mathbf{C}}$: empirical covariance matrix.
Parameters: ϵ and ϵ_u: stopping criteria on the log-likelihood and for the update step;
   δK: connectivity increment after which an update step is performed.

1: $\mathbf{A} = \mathbf{C} = \mathbf{W}^{-1} = \mathbf{S}^{-1} = \mathrm{diag}(1)$.
2: **while** $\Delta\mathcal{L}_{\max} > \epsilon$ and $n_{\mathrm{iter}} < n_{\max}$ **do**
3:   $\Delta\mathcal{L}_{\max} \leftarrow 0$, $n_{\mathrm{iter}} \leftarrow n_{\mathrm{iter}} + 1$
4:   **for** all pairs of nodes $\widehat{ij}$ **do**
5:     compute $\Delta\mathcal{L}^{ij}$ using (3)
6:     **if** $\Delta\mathcal{L}^{ij} > \Delta\mathcal{L}_{\max}$ **then**
7:       **if** Check_constraint$(\star, \widehat{ij}, \mathbf{A})$ **then**
8:         $\Delta\mathcal{L}_{\max} \leftarrow \Delta\mathcal{L}^{ij}$ and $\mathbf{V} \leftarrow [\mathbf{V}_{\{i,j\}}]$ defined in (2)
9:   $\mathbf{A} \leftarrow \mathbf{A} + \mathbf{V}$, update $\mathbf{C}$ using (4).
10:   **if** $\star =$ WS or WWS **then**
11:     update $\mathbf{W}^{-1}$ or $\mathbf{S}^{-1}$ using the SMW formula.
12:   **if** connectivity has increased by δK **then**
13:     **while** $\mathrm{Tr}\left(\mathbf{A}\mathbf{\Pi}^{\|}\mathbf{A}\mathbf{\Pi}^{\|}\right) > \epsilon_u$ **do**
14:       **for** $\widehat{ij} \mid A_{ij} \neq 0$ **do**
15:         $\Delta\mathcal{L}_{\max} \leftarrow 0$, compute $\Delta\mathcal{L}^{ij}$ using (3)
16:         **if** $\Delta\mathcal{L}^{ij} > \Delta\mathcal{L}_{\max}$ **then**
17:           **if** Check_constraint$(\star, \widehat{ij}, \mathbf{A})$ **then**
18:             $\Delta\mathcal{L}_{\max} \leftarrow \Delta\mathcal{L}^{ij}$ and $\mathbf{V} \leftarrow [\mathbf{V}_{\{i,j\}}]$ defined in (2)
19:       **if** $\star =$ WS or WWS **then**
20:         $\mathbf{A} \leftarrow \mathbf{A} + \mathbf{V}$, update $\mathbf{C}$ and $\mathbf{W}^{-1}$ or $\mathbf{S}^{-1}$ using the SMW formula.
21:       **else**
22:         Block updates (5) of i and j corresponding to $\Delta\mathcal{L}_{\max}$. Update $\mathbf{C}$ using the SMW formula.

Let us clarify the complexity of this algorithm. Each link addition or update has a cost $\mathcal{O}(N^2)$ to update the covariance matrix. If spectral constraints are

---

[1] The source code is available at http://www.rocq.inria.fr/~lasgoutt/star-ips/

imposed, it is necessary to keep track of another inverse matrix, which requires as well $\mathcal{O}(N^2)$ operations and does not change the complexity. Adding M links will therefore require at least $\mathcal{O}(MN^2)$ operations. This means that our algorithm complexity is in $\mathcal{O}(N^3)$ in the sparse regions, whereas it becomes $\mathcal{O}(N^4)$ in the dense ones. Note that this complexity does not take into account the time spent in link updates. As pointed out in Section 2.2, this update step is useful to avoid departing from the optimal learning curve. For a given bound ϵ_u, we observe numerically that the number of updates is $\mathcal{O}(N)$, regardless of the mean connectivity in the sparse regime, so this adds up another $\mathcal{O}(N^3)$ computational cost. Note that the critical parts of the algorithm, which are the update of the covariance matrix and the search for the perturbation that maximizes likelihood increase, can easily be parallelized.

Let us anticipate on the application to emphasize the usefulness of our algorithm, which complexity is comparable to a direct covariance matrix inversion. Suppose that the workflow is:

1. select off-line a GMRF model based on an empirical covariance matrix;
2. use the above model to perform inference for a "real-time" application (which here means at most a few minutes).

We may allow the first task to take a few hours, and thus matrix inversion is acceptable for quite large networks. However, the resultant model will not be suited to GaBP in the inference task and we will have to resort to exact inference, through a matrix inversion of complexity $\mathcal{O}((N - n_o)^3)$, where n_o is the number of observed variables. Using our sparse GaBP-compatible model instead, with a mean connectivity K, the approximate inference complexity $\mathcal{O}(mK(N - n_o))$ will then typically scale down from a few hours to a few seconds or a few minutes depending on the needed precision.

5   Experimental Results

To have some elements of comparison, let us first quickly describe traditional ways to tackle the maximum likelihood estimation problem with penalty-induced sparsity constraints. It involves a maximization of the form

$$\mathbf{A} = \underset{\mathbf{M} \in \mathcal{S}_{++}}{\operatorname{argmax}}\ \log \det(\mathbf{M}) - \operatorname{Tr}(\mathbf{M}\hat{\mathbf{C}}) + \lambda P(\mathbf{M}),$$

where $\mathcal{S}_{++}$ is the set of positive definite matrices. A classical penalization function P is a continuous approximation to the discrete L_0 norm like the "seamless L_0 penalty" (SELO) proposed in [11]

$$P(x) = \log \left(\frac{2|x| + \tau}{|x| + \tau} \right).$$

In the following, we set $\tau = 5.10^{-3}$, which is empirically good enough. We propose to use the Doubly Augmented Lagrange method [5] to solve this penalized log-determinant programming.

The second method used for comparison is QUIC [9], which uses the L_1 norm as penalty. This is a second order optimization method, leading to superlinear convergence rates. We perform it directly on the empirical covariance matrix with different values for the regularization coefficient λ. Once the structure has been found, it is necessary to maximize the likelihood. According to our experiments, L_1 norm penalty leads to poor precision matrices in terms of likelihood, even if it may be very efficient to find an existing sparse graph structure.

We can now compare the performance of $\star$-IPS and sparsity penalized likelihood optimization. For generating one single GMRF with a given designed sparsity level, both methods are comparable in terms of computational cost, while $\star$-IPS is faster in very sparse regime. Due to its incremental nature, it also has the advantage of generating a full Pareto set of approximate solutions. To assess the quality of the $\star$-IPS model selection, we first look into data fitting accuracy through log-likelihood, and then investigate its compatibility with GaBP inference.

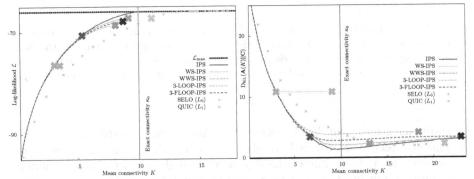

Fig. 1. Left: Log-likelihood $\mathcal{L}$ as a function of mean connectivity K for $\star$-IPS with different constraints, SELO and QUIC, all computed from the exact covariance matrix **C**. Right: Kullback-Leibler divergence to the actual distribution as a function of mean connectivity (estimations based on an empirical covariance matrix $\hat{\mathbf{C}}$, generated with 1000 samples). The end of GaBP compatibility for each algorithm is indicated by ×'s.

Likelihood and GaBP Compatibility Trade-off. The first test is performed on a randomly generated GMRF of 100 variables. The structure of its precision matrix is an Erdős-Rényi random graph, where each link is assigned a value with random sign and magnitude (between 0.1 and 0.8). A diagonal term is added to make it definite positive. The results of the different algorithms are shown in Fig. 1. Both SELO and IPS algorithm are able to find the true graph with the exact covariance matrix. As expected, WS is a very strict constraint and yields low connectivity models. Using WWS instead yields better GaBP compatibility, but provides no guarantee about it. However, this constraint can be enough to get a GaBP-compatible model with (almost) maximal likelihood (Fig. 1-right). For both models, QUIC is clearly sub-optimal regarding sparsity/likelihood trade-off.

Fig. 1-right illustrates ⋆-IPS performance in terms of overfitting. This overfitting starts after the Kullback-Leibler divergence reaches a minimum. This can happen as well as before as after the end of GaBP compatibility. Detecting this point is a classical but difficult statistical problem and further investigations are needed to find the best criterion adapted to our case. The second test (Fig. 2) is performed

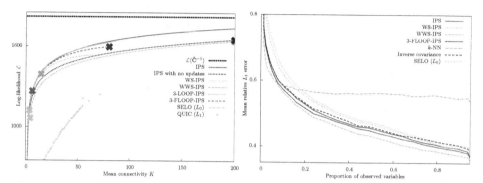

Fig. 2. Left: log-likelihood $\mathcal{L}$ as a function of mean connectivity K. Right: mean relative L_1 reconstruction error as a function of the fraction of observed stations on the San Francisco Bay Area network for various methods; results are averaged over 100 sample test experiments and normalized by the score obtained with daytime moving average predictor.

on traffic data from the San Francisco Bay Area[2]. Each sample data is a N-dimensional vector of observed speeds $\{\hat{V}_i, i = 1 \ldots N\}$, giving a snapshot of the network at a given time of the day, as measured by a set of fixed sensors. After filtering out inactive sensors, we finally kept 1020 variables, for which we had data from January to June 2013. The travel time distribution at each link, being bounded with heavy tail, is far from being Gaussian. In order to work with normal variables, we make the following transformation

$$Y_i = \varPhi^{-1}\left(\frac{1 + \hat{F}_i(V_i)}{2}\right), \quad \forall i = 1 \ldots N, \tag{10}$$

which maps the speed V_i to the positive domain of a standard Gaussian variable Y_i, where $\hat{F}_i$ and $\varPhi$ are respectively the cumulative distribution functions of the speeds and of a normal distribution. The input of the algorithms we are comparing is the covariance matrix of the vector $\mathbf{Y}$. This mapping is important to use the selected GMRF for the inference tasks in the next section.

Fig. 2-left compares the performance of some ⋆-IPS variants with methods based on penalized norms. IPS with update but no constraint is comparable to SELO optimization, albeit much faster to generate the Pareto set, while QUIC is by far the weakest contender. In fact, QUIC is not performing well because, in this case, there is no true underlying sparse dependency graph. Both IPS

---

[2] Available at the Caltrans PeMS website: http://pems.dot.ca.gov/

and SELO loose the compatibility with GaBP at low likelihood and mean connectivity (respectively at $K < 6.5$ and $K < 4$). By contrast, imposing the "no frustrated loops of size 3" constraint (3-FLOOP-IPS) yields a nearly optimal $\mathcal{L}(K)$ path, up to some flat regime, which endpoint is still GaBP compatible. This is the best trade-off which can be found among all constraints that we have tested. While the WS constraint is again too restrictive, we notice that WWS yields models which are all the way compatible, but with a suboptimal $\mathcal{L}(K)$ path. This is partly due to the absence of block updates, replaced by less efficient local updates. Actually, we see that the WWS $\mathcal{L}(K)$ roughly follows the one obtained with ∅-IPS (no update at all), which also delivers only GaBP-compatible models. Our interpretation is that updating the links has the effect of reducing some uncorrelated noise otherwise present in the approximate model. At some point, it may spoil the GaBP compatibility because of stronger correlations being taken into account.

Inverse Models for GaBP Inference. Our original motivation for this work is to provide models for travel time inference for large scale traffic network in real time [8]. In this application, from an historical data set, we have to build a GMRF reflecting the mutual information between traffic levels among different segments of the traffic network. Then, in real time, GaBP runs on this GMRF to propagate the information given by observed segments to the other ones.

In our experiment with PeMS data, both the historical and test data sets contain samples of a N-dimensional vector of 5-minutes averaged speeds, obtained from fixed sensors. Given a sample, for which a proportion ρ of the variables is observed, we want to infer the states of the $(1 - \rho)N$ unobserved variables. In practice, we proceed gradually on each test sample, by revealing the variables in a random order, and plot the relative L_1 error $|\hat{v} - v|/v$ made by the inference model on the unobserved variables as a function of ρ, aggregated over 100 different test samples. The error is measured on the speed, after inference has been done in the space defined by (10).

In principle, ⋆-IPS does not require complete samples and even knowledge of the whole covariance matrix is actually not mandatory. But, for this highway dataset, the samples have no missing values, which allows us to use a brute force k-NN predictor for the sake of comparison. The setting for k-NN is as follows: k samples out of the whole training set are selected according to their mean L_1 distance on the observed variables. Then, for each unobserved variable, the median value is extracted from the k selected samples as a predictor. In the experiments, the value $k = 70$ has been determined to yield the best k-NN predictions.

Fig. 2-right compares the prediction made by ⋆-IPS with the respective results of SELO, full inverse covariance matrix model and k-NN. GMRF models and k-NN behave very differently. k-NN seems to capture rapidly ($\rho \leq 0.1$) the global network behavior, but remains flat after that, despite the additional information. By contrast, GMRF models performance always improves with new information, because of their local nature. Moreover, while constraints applied to IPS offer us more precise models, the role of $\mathcal{L}$ as proxy is not completely respected.

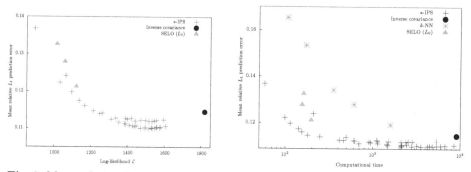

Fig. 3. Mean relative L_1 reconstruction error as a function of log-likelihood (left) and of computational time cumulated over 100 experiments (right). The different results for k-NN are obtained with historical datasets of sizes 10^3, 2.10^3, 5.10^3, 10^4 and 3.10^4.

This is due to overfitting problems: as we can see, the full inverse covariance matrix model behaves worse than the $K = 6.5$ model obtained by simple IPS. This appears clearly in Fig. 3-left, where the prediction error reaches a minimum before increasing again with $\mathcal{L}$. Finally, Fig. 3-right shows various trade-offs between precision and efficiency of the models. We clearly see that $\star$-IPS extracts more precise and less time-consuming models for traffic states reconstruction. For instance, the highest precision is obtained with WWS-IPS for $K \approx 50$, leading to a 10-fold time reduction of the computational time w.r.t. the full inverse covariance model, with a gain of 5% in precision.

6   Conclusion

In this paper, we revisit IPS and show that it provides an efficient framework to find GMRF models with constraints more specific than sparsity. Comparisons show the merits of the proposed $\star$-IPS in terms of flexibility, likelihood values reached and diversity of solutions, since a Pareto set can be delivered for the computational cost of one estimation.

In terms of trade-off between sparsity and likelihood, $\star$-IPS is comparable to the SELO approach, with less computational cost. In contrast, L_1 based methods do not provide satisfying results in our problem setting.

In addition, the flexibility of $\star$-IPS allows one to embed additional and rather exotic but useful constraints for GaBP compatibility, which is not simple to do with traditional penalized likelihood approaches. Experiments show that both topological and spectral constraints are useful. While the walk-summability constraint seems too strict to be useful in practice, relaxing it to weak walk-summability leads to very good models. At the same time, avoiding only frustrated triangles give very satisfactory results in our experiments.

Still in this context, the overfitting problem seems completely open to us. Classical information-theoretic criteria failed in our tests to locate properly where to stop in the incremental link addition process. In fact, we observe that both the link-addition and the link-update procedure can lead to overfitting, and the design of a specific criterion able to avoid this deserves further investigation.

References

1. Banerjee, O., El Ghaoui, L., d'Aspremont, A.: Model selection through sparse maximum likelihood estimation for multivariate Gaussian or binary data. JMLR 9, 485–516 (2008)
2. Bickson, D.: Gaussian Belief Propagation: Theory and Application. Ph.D. thesis, Hebrew University of Jerusalem (2008)
3. Chow, C., Liu, C.: Approximating discrete probability distributions with dependence trees. IEEE Transactions on Information Theory (1968)
4. Darroch, J., Ratcliff, D.: Generalized iterative scaling for log-linear models. Ann. Math. Statistics 43, 1470–1480 (1972)
5. Dong, B., Zhang, Y.: An efficient algorithm for l_0 minimization in wavelet frame based image restoration. Journal of Scientific Computing 54(2-3) (2012)
6. Fan, J., Li, R.: Variable selection via nonconcave penalized likelihood and its oracle properties. Journal of American Statistical Association (2001)
7. Friedman, J., Hastie, T., Tibshirani, R.: Sparse inverse covariance estimation with the graphical lasso. Biostatistics 9(3), 432–441 (2008)
8. Furtlehner, C., Han, Y., Lasgouttes, J.M., Martin, V., Marchal, F., Moutarde, F.: Spatial and temporal analysis of traffic states on large scale networks. In: ITSC, pp. 1215–1220 (2010)
9. Hsieh, C., Sustik, M.A., Dhillon, I.S., Ravikumar, K.: Sparse inverse covariance matrix estimation using quadratic approximation. In: NIPS (2011)
10. Jalali, A., Johnson, C.C., Ravikumar, P.D.: On learning discrete graphical models using greedy methods. In: NIPS, pp. 1935–1943 (2011)
11. Lee Dicker, B.H., Lin, X.: Variable selection and estimation with the seamless-L_0 penalty. Statistica Sinica 23(2), 929–962 (2012)
12. Malioutov, D., Johnson, J., Willsky, A.: Walk-sums and Belief Propagation in Gaussian graphical models. JMLR 7, 2031–2064 (2006)
13. Malouf, R.: A comparison of algorithms for maximum entropy parameter estimation. In: COLING, pp. 49–55 (2002)
14. Pearl, J.: Probabilistic Reasoning in Intelligent Systems: Network of Plausible Inference. Morgan Kaufmann (1988)
15. Scheinberg, K., Rish, I.: Learning sparse gaussian markov networks using a greedy coordinate ascent approach. In: Balcázar, J.L., Bonchi, F., Gionis, A., Sebag, M. (eds.) ECML PKDD 2010, Part III. LNCS, vol. 6323, pp. 196–212. Springer, Heidelberg (2010)
16. Seneta, E.: Non-negative matrices and Markov chains. Springer (2006)
17. Speed, T., Kiiveri, H.: Gaussian Markov distributions over finite graphs. The Annals of Statistics 14(1), 138–150 (1986)
18. Weiss, Y., Freeman, W.T.: Correctness of belief propagation in Gaussian graphical models of arbitrary topology. Neural Computation (2001)

Rate-Constrained Ranking and the Rate-Weighted AUC

Louise A.C. Millard[1,2,3], Peter A. Flach[1,3], and Julian P.T. Higgins[2,4]

[1] Intelligent Systems Laboratory, University of Bristol, United Kingdom
[2] School of Social and Community Medicine, University of Bristol, United Kingdom
[3] MRC Integrative Epidemiology Unit, University of Bristol, United Kingdom
[4] Centre for Reviews and Dissemination, University of York, York, United Kingdom
{Louise.Millard,Peter.Flach,Julian.Higgins}@bristol.ac.uk

Abstract. Ranking tasks, where instances are ranked by a predicted score, are common in machine learning. Often only a proportion of the instances in the ranking can be processed, and this quantity, the predicted positive rate (PPR), may not be known precisely. In this situation, the evaluation of a model's performance needs to account for these imprecise constraints on the PPR, but existing metrics such as the area under the ROC curve (AUC) and early retrieval metrics such as normalised discounted cumulative gain (NDCG) cannot do this. In this paper we introduce a novel metric, the rate-weighted AUC (rAUC), to evaluate ranking models when constraints across the PPR exist, and provide an efficient algorithm to estimate the rAUC using an empirical ROC curve. Our experiments show that rAUC, AUC and NDCG often select different models. We demonstrate the usefulness of rAUC on a practical application: ranking articles for rapid reviews in epidemiology.

1 Introduction and Motivation

The work reported in this paper was motivated by the task of undertaking rapid reviews of clinical trials. A rapid review should follow the broad principles of a systematic review, where a medical research question is asked (such as the effect of a drug on a disease) and the evidence from all relevant research articles is compiled to give a better estimate of the drugs effect than each individual study provides. However, a rapid review needs to be performed under strict time and resource constraints, so it may not be possible to review all relevant articles. Currently, a rapid review is performed by human reviewers who search online medical research databases for articles reporting clinical trials of a particular research question [7]. In order to retrieve a set of articles that can be reviewed in the allotted time, the reviewer may iteratively refine the search query until the number of articles is deemed manageable.

An important consideration when performing a rapid review is the quality of each study. Low-quality studies are more likely to give a biased estimate of the research question and may need to be excluded from the review or considered with caution [9]. Therefore, the aim of a rapid review can be described as maximising the number of high-quality articles assessed, given the particular time

T. Calders et al. (Eds.): ECML PKDD 2014, Part II, LNCS 8725, pp. 386–403, 2014.
© Springer-Verlag Berlin Heidelberg 2014

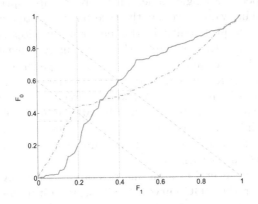

Fig. 1. Two hypothetical ROC curves (x-axis: false positive rate, y-axis: true positive rate), example PPR isometrics (diagonal lines with $PPR = 0.5$ (top) and $PPR = 0.3$ (bottom); the slope of -1 indicates a uniform class distribution) and example partial AUC bounds (vertical lines)

constraints of the review. The iterative search method described above is a rather crude approach that does not consider article quality, and can be thought of as a classification of articles as *included* or *excluded* from the review. We suggest that this can be greatly improved by instead learning a model for estimating the article's study quality, and using the model's scores to rank the studies under review, such that the most reliable research is assessed first. The reviewers can then simply review the articles in decreasing order of estimated quality until they run out of time. There is no need to specify a classification threshold.

This approach suggests that a good model is one that exhibits good ranking behaviour with respect to study quality, with particular emphasis on the proportion of articles that can reasonably be processed. The total amount of time available for a review and the number of articles returned from the initial search query is typically known. Given an estimate of the time it will take a reviewer to assess a single article, the proportion of articles in the search results that is expected to be processed can be inferred. In terms of binary classification this proportion is the predicted positive rate (PPR). If the PPR is known precisely, finding the best model is straight-forward. Figure 1 illustrates this with two hypothetical ROC curves where neither curve dominates the other. The two dashed lines show two example PPR values that could be inferred for a rapid review. We can see that the PPR value affects which model is chosen. The (solid) green model is chosen when $PPR = 0.5$ and the (dashed) blue model is chosen when $PPR = 0.3$, as these models have the highest recall at these respective points on the ROC curves.

However, the PPR inferred depends on the time needed to review a single article, and this is not known precisely. Articles vary in length and difficulty and hence it is only possible to estimate a probability distribution across the

PPR, rather than specify a single value. Therefore, an appropriate measure of rate-constrained ranking would average the true positive rate across each value of PPR, weighted by its probability. In this paper we develop such a measure. In addition to our motivating example, there are many other tasks that are rate constrained, with uncertainty across the rates. In general, these tasks are restricted to a fixed budget of a resource such as time or money, where the exact expenditure for each instance is not known precisely. Another example is telephone sales, which is restricted by the allocated number of person hours, such that when ranking a database of customers to determine those most likely to show interest, it is not known exactly how many customers will be contacted as the time per phone call is variable.

There are, of course, several existing metrics often used to evaluate ranking tasks. The area under the ROC curve (AUC), which estimates the probability that a random positive is ranked higher than a random negative, measures ranking performance across the entire ROC curve, treating all regions as equally important [5]. An alternative to the AUC, the partial AUC (pAUC), has previously been suggested [4]. This metric also weights uniformly, but restricts to a range of false positive rate (or true positive rate) values. For instance, the two solid vertical lines of Figure 1 show example pAUC bounds, constrained to false positive rates between 0.2 and 0.4. We require a metric that weights the area under the ROC curve with respect to the PPR, but the AUC weights the area uniformly and the pAUC can only weight across true positive or false positive rates, components of the PPR, and not the PPR itself.

Early retrieval tasks are those where examples near the top of the ranking are more important, as these examples are more likely to be processed. Several metrics in several fields have been proposed to address this problem, such as normalised discounted cumulative gain (NDCG) [10]. However, as we demonstrate later this metric and related ones assume that the likelihood of stopping at a particular position in the ranking is always higher nearer the top which is not necessarily the case when rates are constrained.

A key contribution of this paper is the derivation of a new metric, the rate-weighted AUC (rAUC), to evaluate models for rate-constrained ranking tasks (Section 3). We prove that the rAUC and rate-weighted expected recall are linearly related given a fixed class distribution. Furthermore, we provide an efficient algorithm to estimate the rAUC using an empirical ROC curve (Section 4). Finally, we demonstrate that given rate constraints the rAUC chooses the optimal model while the AUC and NDCG metrics often choose a suboptimal model (Section 5).

2   Notation and Basic Definitions

We follow the notation of [8]. We assume a two-class classification problem with instance space $\mathcal{X}$. The positive and negative classes are denoted by 0 and 1, respectively. The learner outputs a score $s(x) \in [0, 1]$ for each instance $x \in \mathcal{X}$, such that higher scores express a stronger belief that x belongs to class 1.

The score densities and cumulative distributions are denoted by f_k and F_k for class $k \in \{0,1\}$. Given a threshold at score t the true positive rate (also called sensitivity or positive recall) is $P(s(x) \leqslant t|k = 0) = F_0(t)$ and the false positive rate is $P(s(x) \leqslant t|k = 1) = F_1(t)$. The true negative rate, also called specificity or negative recall, is $1 - F_1(t)$.

The proportions of positives and negatives are denoted by π_0 and π_1 respectively. Accuracy acc at threshold t is a weighted average of positive and negative recall:

$$acc(t) = \pi_0 F_0(t) + \pi_1 (1 - F_1(t)) \tag{1}$$

Similarly, the proportion of positive predictions at threshold t (the predicted positive rate) is a weighted average of the true and false positive rates:

$$r(t) = \pi_0 F_0(t) + \pi_1 F_1(t) \tag{2}$$

This is the predicted positive rate, which we abbreviate to the rate.

A ROC curve is a plot of true positive rate on the y-axis against false positive rate on the x-axis. The area under the ROC curve (AUC) is the true positive rate averaged over all false positive rates:

$$AUC = \int_0^1 F_0 \, dF_1 = \int_{-\infty}^{+\infty} F_0(t) f_1(t) dt \tag{3}$$

Alternative parameterisations are possible; in this paper we are particularly interested in a parametrisation by rate.

Metrics such as predicted positive rate can be depicted in ROC space using isometrics – points on ROC space that have the same value for a given metric [6]. For instance, several combinations of false and true positive values result in the same rate (Equation 2), and this can be shown as a straight line drawn in ROC space.

3   The Rate-Weighted AUC

The aim of a rate-constrained ranking task is to maximise the expected true positive rate given a probability distribution across the rates. Common formulations of the AUC are given as an expectation of the true positive rate across all false positive rates or the thresholds (Equation 3). It is not possible to apply a weight across rates using these formulations, because they are given in terms of expectations over F_1 and t, rather than the rate. The following section derives the AUC as an expectation across rates, such that the derived formula can be altered to weight the AUC with respect to the rate.

Accuracy isometrics in ROC space are lines of constant accuracy with slope π_1/π_0 [6]. Similarly, rate isometrics are lines of constant rate with slope $-\pi_1/\pi_0$. Examples are shown in Figures 2a and Figure 2c for uniform and non-uniform class distributions, respectively.

Definition 1. Rate-accuracy space is a plot of rate on the x-axis and accuracy on the y-axis. Rate-recall space is a plot of rate on the x-axis and recall on the y-axis. Where positive recall is used, rate-recall space is denoted rate-$F_0(r)$ space. Where negative recall is used, rate-recall space is denoted rate-$(1 - F_1(r))$ space.

We translate the ROC curve to rate-accuracy and rate-recall spaces using a linear transformation, such that the AUC can be calculated in this space instead. The ROC curve of Figure 2a is transformed into the rate-accuracy curve shown in Figure 2b, and the rate-recall curves shown in Figures 2e and 2f, for positive and negative recall respectively. We can see that the transformations into rate-accuracy and rate-recall spaces result in unreachable areas. The upper bounds of the rate-accuracy and rate-recall curves correspond to the ROC curve of a perfect classifier, and the lower bounds to that of a pessimal classifier.

Definition 2. The lower bounds in x-y space are given by a function $f_{min}(x)$ specifying the minimum possible value of y at each value of x. The upper bounds in x-y space are given by a function $f_{max}(x)$ specifying the maximum possible value of y at each value of x.

We now focus on rate-accuracy space, but a similar derivation can be given for rate-recall space (given in Theorem 5). In rate-accuracy space, the lower and upper bounds of accuracy at rate r are given by:

$$acc_{min}(r) = |\pi_1 - r| \qquad acc_{max}(r) = 1 - |\pi_0 - r| \qquad (4)$$

These are derived from Equation 1 and the fact that acc_{min} corresponds to points with $F_0 = 0$ when $r \leqslant \pi_1$ and points with $F_1 = 1$ when $r \geqslant \pi_1$, and acc_{max} corresponds to points with $F_1 = 0$ when $r \leqslant \pi_0$ and points with $F_0 = 1$ when $r \geqslant \pi_0$.

Clearly, a ROC curve can only cross each rate isometric at a single point, which allows us to reformulate the AUC in terms of accuracy and rates in order to apply a weight across rates. Accuracy difference acc_{dif} is the difference in the accuracy value of the ROC curve with the minimum possible accuracy value for a given rate:

$$acc_{dif}(r) = acc(r) - acc_{min}(r) \qquad (5)$$

Theorem 3. *The AUC is equal to the normalised accuracy difference across all rates $r \in [0, 1]$:*

$$AUC = \frac{1}{K_{acc}} \int_0^1 acc_{dif}(r)dr \qquad (6)$$

where K_{acc} is constant for a fixed class distribution:

$$K_{acc} = \int_0^1 \left(acc_{max}(r) - acc_{min}(r) \right) dr \qquad (7)$$

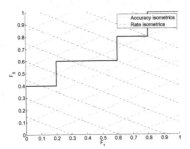

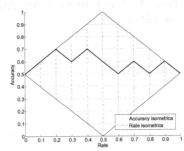

(a) Example ROC curve with rate and accuracy isometrics for $\pi_0 = \pi_1 = \frac{1}{2}$.

(b) Rate-accuracy curve corresponding to ROC curve shown in Figure 2a.

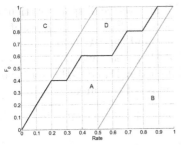

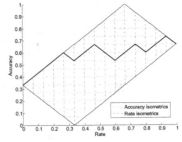

(c) Example ROC curve with rate and accuracy isometrics for $\pi_0 = \frac{2}{3}$, $\pi_1 = \frac{1}{3}$.

(d) Rate-accuracy curve corresponding to ROC curve shown in Figure 2c.

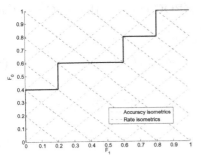

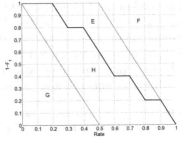

(e) Rate-recall curve for the positive class of ROC curve shown in Figure 2a.

(f) Rate-recall curve for the negative class, of ROC curve shown in Figure 2a.

Fig. 2. Example ROC curves, rate-accuracy curves and rate-recall curves

Theorem 3 holds as transforming a ROC curve from ROC to rate-accuracy space requires only linear transformations such that the relative areas under and above the curve within the transformed bounds of the original ROC space remains the same. This reformulation of AUC in terms of rates allows us to introduce a rate-constrained generalisation.

Definition 4. The rate-weighted AUC of a ROC curve is the AUC weighted across the rates:

$$rAUC = \frac{1}{K_{acc,w(r)}} \int_0^1 w(r) acc_{dif}(r) dr \qquad (8)$$

where $w(r)$ is a density over the rate and $K_{acc,w(r)}$ is given by:

$$K_{acc,w(r)} = \int_0^1 w(r) \left(acc_{max}(r) - acc_{min}(r)\right) dr \qquad (9)$$

Theorem 5. *The rAUC is equal to the normalised F_0 difference weighted across all rates. With a slight abuse of notation we use $F_k(r)$ to mean $F_k(F^{-1}(r))$.*

$$rAUC = \frac{1}{K_{F_0,w(r)}} \int_0^1 w(r) \left(F_0(r) - F_{0,min}(r)\right) dr \qquad (10)$$

where

$$K_{F_0,w(r)} = \int_0^1 w(r) \left(F_{0,max}(r) - F_{0,min}(r)\right) dr \qquad (11)$$

and $F_{0,min}(r) = \max\left(0, \frac{r-\pi_1}{\pi_0}\right)$, $F_{0,max}(r) = \min\left(1, \frac{r}{\pi_0}\right)$.

Clearly, we can derive an analogous result using negative recall $(1 - F_1(r))$ instead of positive recall $(F_0(r))$. The area under the rate-recall curve is the expected recall (positive or negative) given a uniform distribution across the rates. This makes the formulation of the rAUC in rate-recall space particularly interesting, as we can infer the relationship between $\mathbb{E}[F_0]$ – the quantity we intend to maximise in rate-constrained ranking – and the rAUC.

Rate-recall space, as shown in Figures 2e and 2f can be divided into 4 distinct regions, for both positive and negative recall (labelled A-D and E-H respectively). We use A both to label the area A and as the mass of this area.

Theorem 6. *The rate-weighted expected true positive rate is related to the rAUC, given a distribution over the rates, by:*

$$\mathbb{E}[F_0] = (1 - B - C)rAUC + B \qquad (12)$$

where $C = \int_0^{\pi_0} w(r) \left[\frac{\pi_0 - r}{\pi_0}\right] dr$ and $B = \int_{\pi_1}^1 w(r) \frac{r - \pi_1}{\pi_0} dr$.

Proof. Rate-F_0 space is bounded by $r = 0$, $r = 1$, $F_0 = 0$ and $F_0 = 1$, such that the total weighted mass of this area $\int_0^1 w(r) dr = 1$, hence $A + B + C + D = 1$. As $rAUC = \frac{A}{A+D}$, it follows that:

$$\mathbb{E}[F_0] = \frac{A + B}{A + B + C + D} = A + B = rAUC(A + D) + B = (1 - B - C)rAUC + B \qquad (13)$$

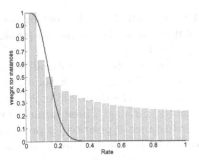

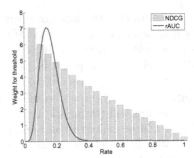

(a) Weights across instances, representing the likelihood an instance will be processed.

(b) Weights across thresholds, representing the likelihood a rate will be the threshold position (the instance at this rate will be the last to be processed).

Fig. 3. NDCG discrete weights (using log base 2) assuming 20 instances and rAUC continuous weights using beta distribution ($\alpha = 6.23$, $\beta = 32.80$). Weights across instances in left figure are equivalent to weights across thresholds in right figure, respectively.

Area C is the triangular region bounded by the lines $r = 0$, $F_0 = 1$ and $F_0 = \frac{r}{\pi_0}$. The weighted mass of C is given by:

$$C = \int_0^{\pi_0} w(r) \frac{\pi_0 - r}{\pi_0} dr \qquad (14)$$

Area B is the triangular region bounded by the lines $r = 1$, $F_0 = 1$ and $\frac{r - \pi_1}{\pi_0}$. The weighted mass of B is given by:

$$B = \int_{\pi_1}^{1} w(r) \frac{r - \pi_1}{\pi_0} dr \qquad (15)$$

$\square$

This completes the proof.

B and C depend only on the class and weight distributions, which implies that the relationship between $\mathbb{E}[F_0]$ and rAUC depends only on these and not the shape of the ROC curve. Therefore, maximising $\mathbb{E}[F_0]$ is equivalent to maximising $\mathbb{E}[rAUC]$, which means that rAUC is a suitable metric to evaluate models for rate-constrained ranking.

3.1   Comparing the Weights of NDCG and rAUC

Normalised discounted cumulative gain (NDCG) is given by:

$$NDCG = \frac{1}{K} \cdot \sum_{i=1}^{n} \frac{1}{log_b(i + 1)} rel_i \qquad (16)$$

where $rel_i \in [0, 1]$ is the label of example at rank i, which can be continuous or binary and denotes the relevance of the example. K is the maximum possible DCG for a ranking of size n: $K = \sum_{i=1}^{n} 1/log_b(i + 1)$.

NDCG weights each point in the ranking according to the probability that this instance will be processed. In contrast, the rAUC weights each point in the ranking according to the probability this point will be the threshold index, such that processing will terminate at this point in the ranking. These formulations are closely related, since the probability that an instance at position i is processed is the probability that an instance at a position after i is the threshold index. For example, if a person is processing 20 articles, the probability they will review the article at rank position 10 equals the probability they will stop processing articles at a position between articles 10 and 20. Hence, the relationship between the two weighting methods is given by:

$$w_{instance}(i) = 1 - CDF_{w_{threshold}}(i) \qquad (17)$$

where i is the position in the ranking and CDF denotes the cumulative distribution function.

The instance weights of NDCG are shown in Figure 3a, and the equivalent threshold weights are shown in Figure 3b. Here we use $rel_i = 1 - k_i$, where $k_i \in \{0, 1\}$. We can see that the weight of each threshold index decreases as we move further down the ranking. This is a key restriction of the NDCG (and related metrics), as it is not always the case that a ranking is more likely to be processed up to the rank positions nearer the top, as in our motivating example. Figure 3b also shows an example density across thresholds, using a beta distribution, where processing is most likely to stop at a rate of 14%. The corresponding instance weights are shown in Figure 3a. Note that by shifting the beta distribution to the right we will create a situation where a number of top-ranked instances will receive the highest weight, something which is not possible with NDCG.

4   Algorithm to Calculate the rAUC of an Empirical ROC Curve

We now use rate-accuracy space to compute the rAUC. A similar algorithm could be implemented in rate-recall space (of either positive or negative recall). Algorithm 1 estimates the rAUC from an empirical ROC curve, where the number of positive N^+ and negative N^- instances is known ($N = N^- + N^+$). This algorithm is similar to the standard AUC $O(N)$ algorithm [5] where the ROC space is processed one vertical (or horizontal) slice at a time. As can be seen in Figure 4, the area under the ROC curve in rate-accuracy space is composed of a series of vertical slices of width $\frac{1}{N}$, each corresponding to an instance. Ties triangles may also exist, each of which corresponds to a set of instances with the same score. The rAUC is calculated as a summation of the weighted mass of all vertical slices and ties triangles, normalised by the weighted mass of the whole rate-accuracy space. The algorithm we propose has four functions: $rAUC$,

$SAUC$, $VAUC$ and $TAUC$. The $rAUC$ function is the main function that iterates through the ranking of instances counting the number of positive and negative instances with the same score (which we call a ties section), and calling the $SAUC$ function when a new score is reached.

The $SAUC$, $VAUC$ and $TAUC$ functions calculate the mass of each ties section (which may consist of only one instance if it has a unique score). The $SAUC$ function simply calls the VAUC and TAUC functions. The negative instances are processed before the positives as when there is a ties triangle the shape of the area under this triangle in rate-accuracy space is given by the area of the negative instances, followed by the positive instances in this ties section. The $VAUC$ function computes the mass of a vertical slice of the area under the curve, using two equations depending whether the current instance is positive or negative. The accuracy difference equation is used, which is computed in terms of r and either F_0 or F_1 depending if the instance is negative or positive respectively (as for instance, if the instance is positive the value of F_1 stays constant). The $TAUC$ function computes the mass of the ties triangle (which is not shown

---

Algorithm 1. The rAUC algorithm. *scores*: list of scores of instances, in decreasing magnitude. *x*: list of class labels corresponding to the instances of *score*. N^+: number of positive instances. N^-: number of negative instances.

---

procedure RAUC($scores, x, N^+, N^-$)

   $\pi_0 \leftarrow N^+/(N^+ + N^-)$; $\pi_1 \leftarrow N^-/(N^+ + N^-)$; $N \leftarrow N^+ + N^-$

   $a_u \leftarrow 0$; $TP \leftarrow 0$; $FP \leftarrow 0$

   $N_{ties}^+ \leftarrow 0$; $N_{ties}^- \leftarrow 0$; $score_{ties} \leftarrow -1$

   for $i = 1$ to N **do**

      if $score_{ties} = scores(i)$ **then**

         if x_i is POSITIVE **then**

            $N_{ties}^+ \leftarrow N_{ties}^+ + 1$

         else

            $N_{ties}^- \leftarrow N_{ties}^- + 1$

         end if

      else

         $[FP, TP, a_u] \leftarrow SAUC(a_u, N_{ties}^-, N_{ties}^+, FP, TP, N^-, N^+)$

         if x_i is POSITIVE **then**

            $N_{ties}^+ \leftarrow 1$; $N_{ties}^- \leftarrow 0$

         else

            $N_{ties}^+ \leftarrow 0$; $N_{ties}^- \leftarrow 1$

         end if

         $score_{ties} \leftarrow score(i)$

      end if

   end for

   $[FP, TP, a_u] \leftarrow SAUC(a_u, N_{ties}^-, N_{ties}^+, FP, TP, N^-, N^+)$

   $a \leftarrow K(w, \pi_0, \pi_1)$

   $rAUC \leftarrow \frac{a_u}{a}$

   Return $rAUC$

end procedure

Algorithm 1. (continued)

$\quad$ **procedure** SAUC(a_u, N_{ties}^-, N_{ties}^+, FP,TP, N^-, N^+)

$\quad$ **if** $N_{ties}^- \geq 1$ **then**

$\qquad$ $FP_{prev} \leftarrow FP$; $FP \leftarrow FP + N_{ties}^-$

$\qquad$ $a_u \leftarrow a_u + VAUC(FP_{prev}, TP, FP, TP, N, 0)$

$\quad$ **end if**

$\quad$ **if** $N_{ties}^+ \geq 1$ **then**

$\qquad$ $TP_{prev} \leftarrow TP$; $TP \leftarrow TP + N_{ties}^+$

$\qquad$ $a_u \leftarrow a_u + VAUC(FP, TP_{prev}, FP, TP, N, 1)$

$\quad$ **end if**

$\quad$ **if** $N_{ties}^+ \geq 1$ & $N_{ties}^- \geq 1$ **then**

$\qquad$ $a_u \leftarrow a_u + TAUC(FP, TP, N_{ties}^-, N_{ties}^+, N^-, N^+)$

$\quad$ **end if**

$\quad$ **Return** $[FP, TP, a_u]$

$\quad$ **end procedure**

$\quad$ **procedure** VAUC($FP_{prev}, TP_{prev}, FP, TP, N, label$)

$\qquad$ $start \leftarrow FP_{prev} + TP_{prev}$, $end \leftarrow FP + TP$

$\qquad$ $f_1 \leftarrow \frac{FP_{prev}}{nTotalMinus}$; $f_0 \leftarrow \frac{TP_{prev}}{nTotalPlus}$

$\qquad$ **for** $i = start$ to end **do**

$\qquad$ **if** $label = 0$ **then**

$\qquad\quad$ $FP_{prev} \leftarrow FP_{prev} + 1$; $f_1 \leftarrow \frac{FP_{prev}}{nTotalMinus}$

$\qquad\quad$ $a_u \leftarrow a_u + \int_{\frac{i}{nTotal}}^{\frac{i+1}{nTotal}} (w(r)(2\pi_1 f_0 + \pi_1 - r - |r - \pi_1|)dr$

$\qquad$ **else**

$\qquad\quad$ $TP_{prev}, \leftarrow TP_{prev}, +1$; $f_0 \leftarrow \frac{TP_{prev},}{nTotalPlus}$

$\qquad\quad$ $a_u \leftarrow a_u + \int_{\frac{i}{nTotal}}^{\frac{i+1}{nTotal}} (w(r) (2 (r - \pi_1 f_1) + \pi_1 - r - | r - \pi_1 |) dr$

$\qquad$ **end if**

$\qquad$ **end for**

$\qquad$ **Return** a_u

$\quad$ **end procedure**

---

in Algorithm 1 due to space constraints). A ties triangle T is composed of 2 sub-triangles T_A and T_B where $T = T_A \cup T_B$. T_A and T_B adjoin on line H, where H is fixed along the rate isometric that passes through the right angled corner of T (see Figure 4).

We calculate the mass of a ties triangle by first finding the length H and the rate at each corner of T, labelled P_1, P_2 and P_3 in Figure 4. The weighted mass of the ties triangle is then the summation of T_A and T_B, which are given by:

$$T_A = \int_{r_1}^{r_2} w(r) \cdot H \cdot \frac{r - r_1}{r_2 - r_1} dr \qquad T_B = \int_{r_2}^{r_3} w(r) \cdot H \cdot \frac{1 - (r - r_2)}{r_3 - r_2} dr \qquad (18)$$

where r_1, r_2 and r_3 are the rates at P_1, P_2 and P_3 respectively. The rAUC of a ROC curve is computed in $O(N)$ time. Algorithm 4 appears more lengthy compared to the standard AUC algorithm that is calculated in ROC space because each step across rate-accuracy space corresponds to a negative or positive

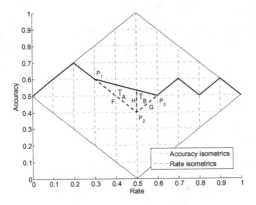

Fig. 4. Example rate-accuracy curve with a ties section (set of instances with the same score). Lengths and angles used to calculate rAUC are labelled.

instance and the height of the curve changes within this step. The standard AUC algorithm makes a step only when the instance is (for example) positive and (given this instance is not tied with another) the height of the ROC curve is constant within this step. The change in height at each step in rate-accuracy space means that the mass of the positive and negative vertical sections (and ties triangle) can only be calculated after the ties section has ended, hence the $SAUC$ function is needed to do this.

5   Experimental Evaluation

We used 5 UCI datasets (vote, autos, credit-g, breast-w and colic) to generate a set of models using 3 learning algorithms (naive Bayes, decision trees and one-rule). We chose a binary variable for each dataset as the label, and learnt 10 models with each dataset/model pair using bootstrap samples of 54% of the data, resulting in 150 generated models. We computed the AUC and NDCG metrics, and rAUC for each of these models, for 5 beta distributions with alpha and beta (α, β) values: $(3,19)$, $(7,15)$, $(11,11)$, $(15,7)$, and $(19,3)$, shown in Figure 5. We use NDCG with log base 10.

Figure 6 shows the AUC and NDCG values, compared with the rAUC values, for each model. Each model is shown by 5 points with a single AUC / NDCG value and variable rAUC value (for each of the 5 rate distributions of Figure 5). The variance of the rAUC for each ROC curve across the 5 beta distributions ranges from 0 to 0.260 for these datasets. Spearman's rank correlations between the model rankings using each rate distribution are given in Table 1. The correlation of the rAUC with the AUC varied between 0.872 and 0.975, depending on the rate distribution. We should note that a proportion of the generated models have very high AUC values, and therefore very high rAUC values for most rate distributions (see Figure 6a). To correct for this inflation of the correlation values, Table 1 also shows reduced correlations when restricted to models

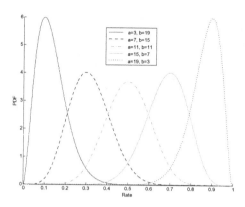

Fig. 5. Beta distributions across the rate

with $AUC \leqslant 0.95$. Correlations between the NDCG and rAUC metrics decrease dramatically when the mode of the beta distribution increases as expected.

The correlation between rAUC metrics using rate distributions that weight different portions of ROC space is low in general. For instance, the rAUC values using rate distributions with $\alpha = 3$, $\beta = 19$ and $\alpha = 19$, $\beta = 3$ have a Spearman's rank correlation of 0.610. This highlights the importance of using a rate distribution with an appropriate degree of uncertainty, as if it is incongruous with the true probability distribution a suboptimal model may be chosen.

5.1   Application to Screening for Rapid Reviews

We demonstrate the rAUC using our motivating example described in the introduction: ranking research articles for rapid reviews in epidemiology. We formulate this task in terms of a rate-constrained ranking problem. To reiterate,

Table 1. Spearman's rank correlations comparing the rankings of the 150 models, ranked using the rAUC (with rate distributions of Figure 5), NDCG and AUC

	$\alpha = 3$ $\beta = 19$	$\alpha = 7$ $\beta = 15$	$\alpha = 11$ $\beta = 11$	$\alpha = 15$ $\beta = 7$	$\alpha = 19$ $\beta = 3$
NDCG	0.565	0.438	0.235	0.092	0.018
AUC	0.872	0.951	0.975	0.927	0.886
AUC $\leqslant 0.95$	0.725	0.902	0.961	0.829	0.703
$\alpha = 3$ $\beta = 19$		0.923	0.791	0.676	0.610
$\alpha = 7$ $\beta = 15$			0.931	0.823	0.764
$\alpha = 11$ $\beta = 11$				0.964	0.925
$\alpha = 15$ $\beta = 7$					0.982

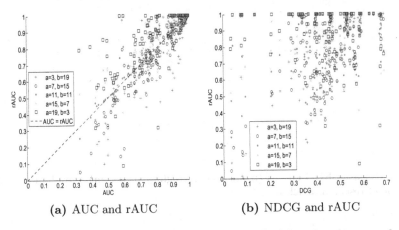

(a) AUC and rAUC   (b) NDCG and rAUC

Fig. 6. Comparison of metrics for 150 models generated with various learners, datasets and distributions over rates

the articles are ranked by estimated study quality, and the objective is to maximise the number of high-quality articles the reviewer assesses given the rate constraints. In this setting, the rate is the proportion of articles that the team reviews, which is not known precisely. The search will return M articles and the reviewers are allotted T hours to complete the review. We use elicitation to determine appropriate parameters for the rate distribution, a method commonly used in epidemiology to establish feasible parameters for a distribution where there is no data from which to infer this. For simplicity, we consider the case of only one reviewer, who estimated the minimum (t_0) and maximum (t_1) time per article, t, the number of minutes they will on average expect to take to assess a single article.

We model t as a inverse beta distribution (with bounds $[\frac{T}{M}, \infty]$), having 0.95 probability of being in the range $[t_0, t_1]$. The rate (the proportion of articles that are reviewed) is given by: $r = \frac{T}{M \cdot t}$. This relationship with t infers a beta distribution across the rates.

We suppose a hypothetical and realistic rapid review where the search returns $M = 2,500$ articles and a reviewer is given 120 person hours ($T = 7,200$ minutes) in which to perform the review. We imagine that the reviewer states they will take between 10 and 45 minutes to assess a single article, which we use to specify two quantiles of t ($0.025 = CDF_t(0, 10)$ and $0.975 = CDF_t(0, 45)$) which we convert to equivalent quantiles of r ($0.975 = CDF_r(0, 0.288)$ and $0.025 = CDF_r(0, 0.064)$). We use the *beta.select* function of the *LearnBayes* R package [1] to find the α and β parameters with these quantiles, giving $\alpha = 6.23$ and $\beta = 32.80$ (shown in Figure 3 (right)).

We use a dataset consisting of 315 full-text articles reporting the results from randomised controlled trials, each labelled with a binary value denoting whether blinding – an indicator of study quality – has been adequately carried out (as described in the article). There were an approximately equal number of articles

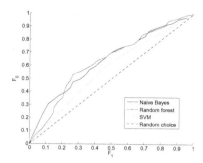

Fig. 7. Consensus ROC curves (using rate-averaging) predicting the blinding risk of bias value of research articles

of each class. We created a set of preliminary models using a bag of words representation, and evaluate these using 10 fold cross validation.

We generated consensus ROC curves for 3 learning algorithms: naive Bayes, decision tree and support vector machine (SVM), shown in Figure 7. A consensus curve represents an average across the ROC curves of all folds [13]. We used *rate-averaging* to generate our consensus curves, previously referred to as pooling [2], where the average of the true and false positive rates at each rate are calculated and then used to generate a single curve. This is appropriate for our rate-constrained task as the points of the consensus curves are the average performance given a particular rate constraint.

The random forest, naive Bayes and SVM models gave a mean rAUC (AUC) of 0.689 (0.636), 0.781 (0.639), and 0.639 (0.570), respectively, across the 10 folds. A two-tailed paired t-test of the AUC values of each model across the 10 cross validation folds, found no difference between the random forest and naive Bayes models (p = 0.884). A t-test using the rAUC values found the naive Bayes model is better than the random forest model for this rate distribution ($p = 0.021$). The random forest and naive Bayes models clearly dominate the SVM model such that the SVM model would be inferior for any rate distribution. However, we have shown that while the random forest and naive Bayes models are similar in terms of ranking performance across the entire ranking, the naive Bayes model is much better than the random forest when considering which rate values are more likely for this particular rapid review.

We thus clearly see that the weight distribution for rate-weighted AUC can be derived directly from the parameters of the rapid review task, in a way that could not be achieved with metrics such as the pAUC.

6   Related Work

The AUC is a popular choice to assess the performance of ranking models, estimating the probability that a randomly chosen positive instance is ranked higher than a randomly chosen negative instance, thus representing ranking performance across the entire dataset. Historically, the AUC has often been used as a measure of ranking performance without consideration for the particular task at hand. However, when the performance of a learner in particular regions of ROC space has more importance than other areas for a particular task, the AUC is not an appropriate choice.

Alternatives to the AUC have previously been suggested to allow differential importance across true positive or true negative rates, for empirical [4, 12] and analytical [14] ROC curves. As mentioned in the introduction, [4] propose a partial AUC metric (pAUC) to restrict the evaluation of the AUC to a range of false positive or true positive rate values. The pAUC measure is appropriate when it is required that either the true positive or false positive rates fall in a particular range. This metric could be generalised using weights rather than bounds (as we have used for the rAUC), which may be more appropriate where there is a non-uniform probability distribution across either the true or false positive rate. Furthermore, a recent variant of the AUC called the half-AUC was proposed by [3], and evaluates the AUC in only half of the ROC space, either where true positive rate is less than true negative rate or true positive rate is greater than true negative rate, giving two distinct regions that can be assessed.

Several metrics have been suggested for early retrieval tasks, where evaluation focuses on the top of the rankings. Precision@k gives the precision at the top k results of a ranking, thus weighting each example uniformly within this section of the ranking. NDCG [10,11], is one of several metrics that give decreasing weights to examples along the ranking, as discussed in Section 3.1. Others include; robust initial enhancement (RIE) [15], the Boltzmann-enhanced Discrimination of ROC (BEDROC) [17], concentrated ROC (CROC) [16] and sum of the log ranks (SLR) [18]. The instance weights used by these approaches all share the characteristic that they translate into monotonically decreasing rate weights, which as demonstrated before is inappropriate for rate-constrained ranking tasks.

7   Conclusions

In this paper we have introduced a new ranking measure, the rate-weighted AUC (rAUC), to better reflect model performance when the task is constrained by a probability distribution across the predicted positive rate, which we refer to as the rate. The AUC is equivalent to the rAUC given a uniform distribution across the rates. Furthermore, if the rate is fixed then models can be compared by simply comparing the recall at the point on the ROC curve with this rate. We have derived the rAUC from both rate-recall and rate-accuracy space, and introduced rate-recall space as a visualisation of model performance. Furthermore, the rAUC is a linear transformation of rate-weighted expected recall (both the positive and negative respectively), given fixed class and rate distributions. We have described an $O(N)$ algorithm to calculate an estimate of the true rAUC using a data sample.

Our experiments have shown large variability of the rAUC as the rate distribution varies. A comparison with NDCG found low correlations indicating that when the likelihood that the processing will stop at a particular position in the ranking is lower nearer the top of the ranking than elsewhere, NDCG may be inappropriate. Furthermore, a comparison with the AUC shows that often the rAUC prefers different models. Finally, we have also demonstrated how this approach can be usefully applied to real world tasks, using the example of ranking research articles for rapid reviews in epidemiology.

Acknowledgments. LACM is funded by a UK Medical Research Council studentship. This work was also supported by Medical Research Council grant MC_UU_12013/1-9.

References

1. Albert, J.: Learnbayes: Functions for learning Bayesian inference. R package version 2.12 (2008)
2. Bradley, A.P.: The use of the area under the ROC curve in the evaluation of machine learning algorithms. Pattern Recognition 30(7), 1145–1159 (1997)
3. Bradley, A.P.: Half-AUC for the evaluation of sensitive or specific classifiers. Pattern Recognition Letters 38, 93–98 (2014)
4. Dodd, L.E., Pepe, M.S.: Partial AUC estimation and regression. Biometrics 59(3), 614–623 (2003)
5. Fawcett, T.: An introduction to ROC analysis. Pattern Recognition Letters 27(8), 861–874 (2006)
6. Flach, P.A.: The geometry of ROC space: Understanding machine learning metrics through ROC isometrics. In: Proceedings of the 20th International Conference on Machine Learning, ICML 2003, pp. 194–201 (2003)
7. Ganann, R., Ciliska, D., Thomas, H.: Expediting systematic reviews: Methods and implications of rapid reviews. Implementation Science 5(1), 56 (2010)
8. Hand, D.J.: Measuring classifier performance: A coherent alternative to the area under the ROC curve. Machine Learning 77(1), 103–123 (2009)
9. Higgins, J., Altman, D.G.: Assessing risk of bias in included studies. In: Cochrane Handbook for Systematic Reviews of Interventions. Cochrane Book Series, pp. 187–241 (2008)
10. Järvelin, K., Kekäläinen, J.: IR evaluation methods for retrieving highly relevant documents. In: Proceedings of the 23rd Annual International ACM SIGIR Conference on Research and Development in Information Retrieval, pp. 41–48. ACM (2000)
11. Jarvelin, K., Kekalainen, J.: Cumulated gain-based evaluation of IR techniques. ACM Transactions on Information Systems (TOIS) 20(4), 422–446 (2002)
12. Jiang, Y., Metz, C.E., Nishikawa, R.M.: A receiver operating characteristic partial area index for highly sensitive diagnostic tests. Radiology 201(3), 745–750 (1996)
13. Macskassy, S.A., Provost, F., Rosset, S.: ROC confidence bands: An empirical evaluation. In: Proceedings of the 22nd International Conference on Machine Learning, ICML 2005, pp. 537–544. ACM (2005)
14. McClish, D.K.: Analyzing a portion of the ROC curve. Medical Decision Making 9(3), 190–195 (1989)
15. Sheridan, R.P., Singh, S.B., Fluder, E.M., Kearsley, S.K.: Protocols for bridging the peptide to nonpeptide gap in topological similarity searches. Journal of Chemical Information and Computer Sciences 41(5), 1395–1406 (2001)

16. Swamidass, J., Azencott, C.-A., Daily, K., Baldi, P.: A CROC stronger than ROC: measuring, visualizing and optimizing early retrieval. Bioinformatics 26(10), 1348–1356 (2010)
17. Truchon, J.-F., Bayly, C.I.: Evaluating virtual screening methods: good and bad metrics for the "early recognition" problem. Journal of Chemical Information and Modeling 47(2), 488–508 (2007)
18. Zhao, W., Hevener, K.E., White, S.W., Lee, R.E., Boyett, J.M.: A statistical framework to evaluate virtual screening. BMC Bioinformatics 10(1), 225 (2009)

Rate-Oriented Point-Wise Confidence Bounds for ROC Curves

Louise A.C. Millard[1,2], Meelis Kull[1], and Peter A. Flach[1,2]

[1] Intelligent Systems Laboratory, University of Bristol, United Kingdom
[2] MRC Integrative Epidemiology Unit, School of Social and Community Medicine,
University of Bristol, United Kingdom
{louise.millard,meelis.kull,peter.flach}@bristol.ac.uk

Abstract. Common approaches to generating confidence bounds around ROC curves have several shortcomings. We resolve these weaknesses with a new 'rate-oriented' approach. We generate confidence bounds composed of a series of confidence intervals for a consensus curve, each at a particular predicted positive rate (PPR), with the aim that each confidence interval contains new samples of this consensus curve with probability 95%. We propose two approaches; a parametric and a bootstrapping approach, which we base on a derivation from first principles. Our method is particularly appropriate with models used for a common type of task that we call rate-constrained, where a certain proportion of examples needs to be classified as positive by the model, such that the operating point will be set at a particular PPR value.

Keywords: Confidence bounds, rate-averaging, ROC curves, rate-constrained.

1 Introduction

ROC curves are informative visualisations of model performance that show the ranking performance at different regions of a ranking, or the performance of a scoring classifier at each possible choice of operating point. ROC curves are often used to determine if one model is better than other, and confidence bounds provide a measure of the uncertainty such that this can be determined, for a specified confidence level. In general when several independent sample ROC curves are generated, such as with m-fold cross validation, the variation between them can be used to estimate a confidence around the average (consensus) ROC curve. Several methods have been proposed to generate confidence bounds, mainly parametric approaches such as vertical [13] or threshold [5] averaging.

Vertical averaging is the most common approach, where the false positive rate is fixed and the mean and confidence interval across the true positive rate is calculated at each false positive rate value. Horizontal averaging is a similar approach that instead fixes the true positive rate and calculates the confidence interval across false positive rate values. However, these approaches have several shortcomings. Firstly, the false and true positive rates are metrics over which

T. Calders et al. (Eds.): ECML PKDD 2014, Part II, LNCS 8725, pp. 404–421, 2014.
© Springer-Verlag Berlin Heidelberg 2014

we have little control, such that it is difficult to set a threshold at a particular value. It is therefore preferable to evaluate a ROC curve with respect to a metric with which setting the threshold is simple in practice. Furthermore, vertical and horizontal averaging are not invariant to swapping the classes, such that if the x-axis and y-axis of ROC space become the false and true negative rate respectively, equivalent points will have different confidence bounds. Finally, depending on the distributional assumptions of points at each false (or true) positive rate value, the confidence bounds may not be constrained to the bounds of ROC space, such that $tpr \in [0, 1]$ and $fpr \in [0, 1]$ (where tpr and fpr are the true and false positive rates respectively).

Threshold-averaging is similar to vertical (and horizontal) averaging but instead fixes the score and averages over each cloud of points in ROC space with the same score. This has the advantage that we can easily use thresholds set at a particular score, classifying each example by whether its score is below or above this threshold value. However, how best to generate confidence bounds for a set points that are not constrained to a single dimension is not obvious. Fawcett et al [5] suggest averaging separately across false and true positive rates, but this creates a rectangular shaped bound for each score where a smoother bound would seem more natural.

To address these shortcomings of existing methods, we specify a set of properties we would like our confidence bounds to satisfy. Firstly, the generated confidence bounds should be invariant to swapping the classes, by which we mean that if the positive and negative classes are swapped such that the x-axis and y-axis of ROC space refer to the false and true negative rate respectively, of the original class labels, then the confidence bounds of these two ROC curves should be symmetrical about the line $tpr = 1 - fpr$ (the descending diagonal). Secondly, the confidence bounds should be constrained to sit within the bounds of ROC space at all points along the lower and upper confidence bounds.

Furthermore, there is a specific type of task in which we are particularly interested. A task may be constrained to a certain proportion of examples that should be classified as positive by the model, the predicted positive rate (PPR). We call these tasks *rate-constrained*, and these are common in many fields. For example, screening a database of customers to decide who should be targeted in a direct sales campaign, where time and monetary budgets mean it is only possible to approach a proportion of the potential customers. Furthermore, the PPR value, hence also the operating point it infers, may not be known precisely, such as the task described by Millard et al. [12], of ranking research articles for rapid reviews in epidemiology.

We suggest that when a task is rate-constrained, the consensus curve should be generated by averaging a set of sample ROC curves while fixing the rate, which we call *rate-averaging*. Furthermore, the comparison of several models should use confidence intervals also created at each PPR value, which we call a *rate-oriented* approach, such that they can be compared with respect to the PPR. We illustrate this with Figure 1, which shows two ROC curves and their consensus curve, created by vertical- (left) and rate-averaging (right). Each point on the

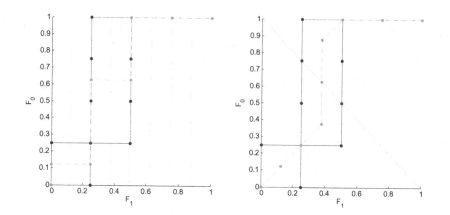

Fig. 1. Illustration of generating consensus curves (broken green curves) from two ROC curves. Left: vertical-averaging, right: rate-averaging. Dotted lines show false positive rate and PPR isometrics, in the left and right figures, respectively.

rate-averaged consensus curve gives the average performance of all sample ROC curves at a particular PPR value. In order for the confidence interval to give the uncertainty of a rate-averaged consensus curve, this should also be generated for each PPR value.

Our aim is to generate confidence intervals for a consensus curve at each PPR value, such that at significance level σ new samples generated from this consensus curve pass between the lower and upper confidence limits at a given PPR value, with probability $1 - \sigma$. The series of confidence intervals creates a *confidence bound* around the consensus curve. We call these *point-wise confidence bounds* in line with [10] in order to differentiate from the common meaning of ROC confidence bands, where the confidence refers to the proportion of whole curves sitting entirely inside the confidence band. Where we discuss methods that are solely used to generate a bound around the whole curve, we explicitly refer to these as bands.

Our main contribution is an approach to generate rate-oriented point-wise confidence bounds. We derive our approach from first principles and demonstrate its effectiveness experimentally.

2   Notation and Basic Definitions

We follow the notation of [7]. We assume a two-class classification problem with instance space $\mathcal{X}$. The positive and negative classes are denoted by 0 and 1, respectively. The learner outputs a score $s(x) \in [0, 1]$ for each instance $x \in \mathcal{X}$. The score densities (lower scores suggest positive class) and cumulative distributions are denoted by f_y and F_y for class $y \in \{0, 1\}$. Given a threshold at score t the true positive rate (also called sensitivity or positive recall) is $P(s(x) \leqslant t | y =$

$0) = F_0(t)$ and the false positive rate is $P(s(x) \leqslant t|y = 1) = F_1(t)$. The true negative rate, also called specificity or negative recall, is $1 - F_1(t)$.

The proportions of positives and negatives are denoted by π_0 and π_1 respectively. The score density of the mixed distribution is denoted by f and given by:

$$f(t) = \pi_0 \cdot f_0(t) + \pi_1 \cdot f_1(t) \tag{1}$$

The probability of a positive at score t is given by:

$$\pi_{0,t} = \frac{\pi_0 \cdot f_0(t)}{\pi_0 \cdot f_0(t) + \pi_1 \cdot f_1(t)} \tag{2}$$

The cumulative distribution of the mixed density distribution is denoted by F and given by:

$$F(t) = \pi_0 \cdot F_0(t) + \pi_1 \cdot F_1(t) \tag{3}$$

This is also the proportion of positive predictions at threshold t known as the predicted positive rate (PPR), which we abbreviate to the rate.

A ROC curve is a plot of true positive rate on the y-axis against false positive rate on the x-axis. A ROC table, such as that shown in Table 1, is a matrix with m rows and n columns, containing the results of independent tests using m samples, such as m-fold cross-validation.

Table 1. Example ROC table, with $m = 4$ samples and n columns, numbers of positive examples in each column pos_k

	k					
$y_{i,k}$	1	2	3	...	n-1	n
Sample 1	0	0	1	...	1	1
Sample 2	0	1	1	...	0	1
Sample 3	0	0	0	...	0	1
Sample 4	0	1	0	...	1	1
pos_k	4	2	2	...	2	0

Table 2. Example $S_{i,k}$ values (number of positive examples up to column k in a sample) for example ROC table(left)

$S_{i,k}$	$S_{i,1}$	$S_{i,2}$	$S_{i,3}$
Sample 1	1	2	2
Sample 2	1	1	1
Sample 3	1	2	3
Sample 4	1	1	2

Each cell contains the label $y_{i,k} \in [0, 1]$ of the example at position k along the ranking of sample i, where the examples of each sample are ranked by increasing score. A segment of consecutive positions in a ranking having the same score are assigned a fractional label to account for this – the average of the labels in this segment, calculated as $\frac{1}{1+q'-q} \sum_{j=q}^{q'} y_j$ where q and q' are, respectively, the start and end of the position range with equal score. The number of positives and negatives in a ranking are denoted by n_0 and n_1 respectively, such that $n = n_0 + n_1$.

The number of positives across samples at column k in the ROC table, denoted pos_k, is given in Equation 4 (and examples are given in Table 1). The number of positives up to position k of row i in the ROC table, which we refer to as the *true positive value* (as opposed to the true positive rate) and denote by $S_{i,k}$, is given in Equation 5 (and examples are given in Table 2 for the ROC table shown in Table 1).

$$pos_k = \sum_{i=1}^{m}(1 - y_{i,k}) \quad (4) \qquad\qquad s_{i,k} = \sum_{j=1}^{k}(1 - y_{i,j}) \quad (5)$$

The number of positives up to position k across all samples in the ROC table, denoted s_k, is given by:

$$s_k = \sum_{j=1}^{k} pos_j = \sum_{i=1}^{m} s_{i,k} \tag{6}$$

Recall (of the positive class) is the proportion of positive examples, correctly classified as positive, at a given point on the ROC curve (also known as the true positive rate). We specify this in terms of rates. The recall $tpr_{i,k}$ of sample i with operating point at position k is given by:

$$tpr_{i,k} = \frac{s_{i,k}}{n_0} \tag{7}$$

We denote an unsorted list of n items as $a_1, a_2 \ldots a_n$ and a sorted list as $a_{(1)}, a_{(2)} \cdots a_{(n)}$.

3 Generating Confidence Bounds

In this section we give our approach to generating rate-oriented point-wise confidence bounds. This includes a new approach to generating samples that uses the ROC curve, rather than the common approach of sampling from the score probability density function of each class (Section 3.2). We begin by describing a simple approach of inferring confidence bounds, used as a baseline in our experiments (Section 4).

3.1 Baseline Method

We use a simple parametric approach as a baseline method. This method is similar to previous approaches such as vertical-averaging, but we fix the rate rather than the false positive rate, in line with our aims. We calculate the mean and variance of recall across samples and, after making an assumption of the underlying distribution across the ROC points of each sample at each rate, calculate the 95% confidence intervals. Here we use positive recall as a distance measure along rate isometrics in ROC space, but any metric that varies linearly along rate isometrics could also be used (such as negative recall or accuracy).

The variance of mean sample recall at each position k along the ranking is given by:

$$\sigma_k^2 = \frac{1}{m \cdot (m-1)} \sum_{i=1}^{m}(tpr_{i,k} - \overline{tpr}_k)^2 \tag{8}$$

where m is the number of samples, $tpr_{i,k}$ is the recall for sample i at position k and $\overline{tpr}_k$ is the mean recall across the samples, at position k. The additional m in the denominator is because we need the variance of the sample mean.

In order to infer a confidence interval we need to assume a particular distribution across the recall at each position k. Assuming a normal distribution the confidence intervals are given by $\overline{tpr}_k \pm 1.96 \cdot \sigma_k$.

We also test this method using a beta distribution, which is bounded by $[0, 1]$ such that we can constrain our confidence intervals to the bounds of ROC space. To use the beta distribution we rescale, at each position k, each sample recall value from the range $max\left(0, \frac{r-\pi_1}{\pi_0}\right) \ldots min\left(1, \frac{r}{\pi_0}\right)$, where $r = \frac{k}{n}$ is the rate at k, to the range $0 \ldots 1$. We calculate the mean and standard deviation of these scaled recall values at each position k and use these to calculate the α and β parameters of the beta distribution. We find the lower and upper limits of the 95% confidence interval of this distribution, and then rescale this back to the original range.

3.2 Generating Sample ROC Curves

Given the score densities of each class, sample rankings can be generated using this distribution. For instance, for each example in the new ranking we can sample a score from the mixed distribution and then sample a label using the probabilities of each class at this score (shown in Table 3 left).

However, the score distribution is not determined by the ROC curve and hence may not be known. In this case we can sample using the ROC curve instead of the score densities, by sampling across the rate. The gradient on the ROC curve is the class likelihood ratio, from which we can calculate the class probabilities at this point on the curve, and then sample the label using this.

We do not need to know the scores because the rate also determines the order of the examples in the ranking, and the ROC curve determines the class probabilities at each rate. We call this the 'rate-first' approach, given in Table 3 (right).

Table 3. Two sampling approaches. Left. Score-first approach. Right: Rate-first approach.

Score-first:	Rate-first
Repeat n times:	Repeat n times:
$\quad$ Sample score $s_j \sim f$	$\quad$ Sample rate $r_j \sim uniform(0, 1)$
$\quad\quad$ Sample label $y_j \sim$ $bernoulli(\pi_{0,s_j})$	$\quad\quad \pi_{0,r_j} \leftarrow$ calculated from gradient at r_j on ROC curve
$\quad$ Rank labels by score s_j	$\quad\quad$ Sample label $y_j \sim bernoulli(\pi_{0,r_j})$
	$\quad$ Rank labels by rate r_j

3.3 Overview of Our Approaches

We assume a random process that generates ROC tables of size $n \cdot m$ from the usually unknown score densities. Let us denote by $S_{i,k}$ the random variable of the sum of the number of positives at position k. Formally, for any fixed true positive value s at this position, with n_0 and n_1 all fixed, we want to estimate:

$$p\left(S_{i,k} = s \mid S_{i,n} = n_0\right) = \frac{p\left(S_{i,k} = s, S_{i,n} = n_0\right)}{\sum_{s'} p\left(S_{i,k} = s', S_{i,n} = n_0\right)} \tag{9}$$

We condition on the class distribution to reflect the fact that a data sample has a finite number of examples with a certain number of each class. This also corresponds to the fact that ROC curves must pass through the points $(0,0)$ and $(1,1)$. We present two alternative methods, a parametric and a bootstrap approach. We derive the probability distribution across the number of positives up to a position, k, in a sample, and use this to infer these two approaches. We develop bootstrap approaches for cases where the distributional assumptions of the parametric approach are invalid.

Importantly, our approach is naturally invariant to swapping the classes. In ROC space, swapping the classes means that the x-axis becomes the false negative rate $(1 - tpr)$, and the y-axis becomes the true negative rate $(1 - fpr)$. The corresponding ROC curve in this 'swapped' ROC space is simply a line mirroring of the original ROC curve along the descending diagonal $(tpr = 1 - fpr)$. Furthermore, the rates are given by $r'(t) = \pi_0(1 - tpr) + \pi_1(1 - fpr)$. Therefore it follows that $r'(t) = 1 - r(t)$. Hence, for each set of points along a rate isometric in the original space, there is a corresponding rate isometric in the 'swapped' space along which this set of points also lie. The confidence bands along these corresponding rate isometrics will have equivalent confidence intervals.

3.4   Parametric Approach

We find the probability distribution across the number of positives from the first position to a position k in the ranking, $S_{i,k}$. We first derive an analytical solution (Theorem 1), and then provide an empirical version that can be used when only the ROC curve (and not the score densities) is available, as is usually the case. At this point we fix i as we refer only to a single sample, such that $S_{i,k}$ is denoted S_k and $S_{i,n}$ is denoted S_n.

Theorem 1. *Let the score densities, F_0 and F_1, and the number of examples of each class in the sample, n_0 and n_1, be fixed. Then:*

$$p(S_{i,k} = s, S_{i,n} = n_0)$$
$$= \int_0^1 [binom(s, k - 1, \pi_0^{<r}) \cdot (1 - \pi_0^{=r}) + binom(s - 1, k - 1, \pi_0^{<r}) \cdot (\pi_0^{=r})]$$
$$\cdot binom(n_0 - s, n - k, \pi_0^{>r}) \cdot p(R_k = r)dr \tag{10}$$

where

$$\pi_0^{<r} = \frac{\pi_0 F_0(t)}{\pi_0 F_0(t) + \pi_1 F_1(t)} \quad (11) \qquad \pi_0^{>r} = \frac{\pi_0(1 - F_0(t))}{\pi_0(1 - F_0(t)) + \pi_1(1 - F_1(t))} \quad (12)$$

$$\pi_0^{=r} = \frac{\pi_0 f_0(t)}{\pi_0 f_0(t) + \pi_1 f_1(t)} \quad (13)$$

$t = F^{-1}(r)$, $p(R_k = r) = beta(r, k, n - k + 1)$, R_k is the rate from which the example at position k was sampled and $binom(k_b, n_b, p_b)$ is the binomial distribution for k_b successes in n_b trials, with probability of success p_b, and $beta(x, a, b)$ is the probability of value x for beta distribution with $\alpha = a$ and $\beta = b$.

Proof. To compute the left hand side of Equation 9 it is sufficient to compute:

$$p(S_k = s, S_n = n_0) \quad (14)$$

The probability of $S = s$ and $S_n = n_0$ in the new sample depends on which rate it was sampled from, such that:

$$p(S_k = s, S_n = n_0) = \int_0^1 p(S_k = s, S_n = n_0 \mid R_k = r) \cdot p(R_k = r)dr \quad (15)$$

The order statistic states that when sampling n values uniformly within the range 0..1 and sorting these examples, the probability that an example at position k was sampled from a rate r is beta distributed with $\alpha = k$ and $\beta = n - k + 1$ [1]. Therefore, $p(R_k = r)$ of Equation 15 is the beta density.

The other component of Equation 15 is the probability of s positives up to a position k, given the example at this position is sampled from a particular rate r. There are two cases where value s is the number of positives up to a position k: 1) $s - 1$ positives occur before position k and the example at k is a positive, or 2) s positives occur before position k and the example at position k is a negative. In either case there must also be $n_0 - s$ positives after position k to ensure that the class distribution is correct.

The examples before position k can be sampled independently, with probability of a positive given by Equation 11. The examples after position k can also be sampled independently, with probability of a positive given by Equation 12. The independence between samples is valid because we are sampling a set of *unordered* examples, and this means that the probabilities of the set of examples before and after position k are binomially distributed, which infers:

$$p(S_k = s, S_n = n_0 | R_k = r)$$

$$= \left[p\left(\sum_{i=1}^{k-1}(1 - y_i) = s \right) p(y_k = 1) + p\left(\sum_{i=1}^{k-1}(1 - y_i) = s - 1 \right) p(y_k = 0) \right]$$

$$\cdot p\left(\sum_{i=k+1}^{n}(1 - y_i) = n_0 - s \right)$$

$$= \left[binom(s, k-1, \pi_0^{<r}) \cdot (1 - \pi_0^{=r}) + binom(s-1, k-1, \pi_0^{<r}) \cdot (\pi_0^{=r}) \right]$$

$$\cdot binom(n_0 - s, n - k, \pi_0^{>r})$$

$$(16)$$

Using Equation 16 in Equation 15 concludes the proof.

$\square$

To reiterate, a key point - while an example at position k has rate $r = \frac{k}{n}$ for this ROC table, we can imagine this table is sampled from a ROC curve of all possible examples. The rate from which it is sampled from this 'true' ROC curve is probabilistic, corresponding to $p(R_k = r)$ in Equation 10. The class probabilities used to generate this example are determined by the class distribution at the rate from which this example was sampled.

An important aspect of Theorem 1 is that the sampling probabilities before, at and after rate r (Equations 11 - 13) can be computed solely using the ROC curve. Recall from Section 3.2 that Equation 13 can be calculated from the gradient at r on the ROC curve. We can also infer the values of Equations 11 and 12 from the ROC curve. Equation 11 is equivalent to the average probability of sampling a positive across all rates before r, and this can be inferred from the gradient of the straight line from point $(0, 0)$ to the point at r on the ROC curve. Similarly, Equation 12 can be inferred from the gradient of the straight line from the ROC curve point at r to the point $(1, 1)$.

Theorem 1 gives the analytical calculation but we cannot use this directly in practice, as we have empirical ROC curves / ROC tables rather than the score densities. Firstly, our empirical ROC tables have discrete rates such that in the discrete case the integral of Equation 15 is changed to a summation. We implement this as an average of the joint probability, for a set of rates of the CDF of the beta distribution (the sampling distribution for this k) at each 0.01 interval:

$$p(S_k = s, S_n = n_0) = \sum_{t=1}^{99} p\left(S_k = s, S_n = n_0 \mid R_k = F_{beta}^{-1}(0.01 \cdot t) \right) \quad (17)$$

such that we sample the rates at each 0.01 interval of the CDF of the beta distribution (with $\alpha = k$ and $\beta = n - k + 1$). This CDF models the probability that an example at position k is sampled by each rate (according to the order statistic).

We also require discrete versions of Equations 11- 13 that can also be used with an empirical ROC table, and these are given in Equations 18- 20:

$$\pi_0^{<r} = \frac{1}{r \cdot n \cdot m} \left[S_{i,\lfloor r \cdot n \rfloor} + d \cdot pos_{\lceil r \cdot n \rceil} \right] \tag{18}$$

$$\pi_0^{=r} = \frac{1}{m} pos_{\lceil r \cdot n \rceil} \tag{19}$$

$$\pi_0^{>r} = \frac{1}{(1-r) \cdot n \cdot m} \left[n_0 - S_{i,\lceil r \cdot n \rceil} + (1-d) \cdot pos_{\lceil r \cdot n \rceil} \right] \tag{20}$$

where $d = r \cdot n - \lfloor r \cdot n \rfloor$ is the relative distance of the rate between positions $\lceil r \cdot n \rceil$ and $\lfloor r \cdot n \rfloor$.

The probabilities of each S_k value computed in Theorem 1, correspond to only a single row of the ROC table. We need the distribution across the number of positives up to position k of all samples in the ROC table. For each S_k value we need:

$$p \left(S_k = s \, | \forall i \in 1 \ldots m : \sum_{j=1}^{n} (1 - y_{i,j}) = n_0 \right) \tag{21}$$

Computing this exactly is computationally intractable, as for each possible s at a position k the probability is given as the summation of the probabilities of all possible combinations of values at position k that sum to this value. We instead approximate the confidence intervals using the estimated variance of this distribution. The mean and variance of the distribution of one sample up to position k are given by:

$$\mu_{1,k} = \sum_s p(S_k = s \, | S_n = n_0) \cdot s \tag{22}$$

$$\sigma_{1,k}^2 = \sum_s p(S_k = s \, | S_n = n_0) \cdot (s - \mu_{1,k})^2 \tag{23}$$

where 1 denotes that these functions correspond to a single sample. We assume each row is identically distributed such that the mean and variance of s at position k of the ROC table are given by:

$$\mu_k = \sum_{i=1}^{m} \mu_{i,k} = m \cdot \mu_{1,k} \quad (24) \qquad\qquad \sigma_k^2 = \sum_{i=1}^{m} \sigma_{i,k}^2 = m \cdot \sigma_{1,k}^2 \quad (25)$$

At each k we restrict to only the possible values of S_k, rescale these to between zero and one, and use a scaled beta distribution to model this distribution and estimate the confidence intervals. We calculate the mean and variance across S_k

values at each position k, where the S_k values have been rescaled to the range $[0, 1]$:

$$\mu_{k,\beta} = \frac{\mu_k - minS_k}{maxS_k - minS_k} \tag{26}$$

$$\sigma_{k,\beta}^2 = \frac{\sigma_k^2}{(maxS_k - minS_k)^2} \tag{27}$$

where $maxS_k = m \cdot maxS_{1,k}$ and $minS_k = m \cdot minS_{1,k}$ and:

$$minS_{1,k} = max(0, n_0 - n + k) \quad (28) \qquad\qquad maxS_{1,k} = min(k, n_0) \tag{29}$$

We use these to parameterise a beta distribution and infer a confidence interval, which we then rescale to the original scale.

3.5   Bootstrap Approach

We generate 2,000 bootstrapped ROC tables each with m samples. Each sample is generated independently using the rate-first sampling approach, as follows.
The rates are sampled uniformly and sorted:

$$r_1, r_2 \ldots r_n \xrightarrow{\;sort\;} r_{(1)}, r_{(2)} \cdots r_{(n)} \tag{30}$$

The probability distribution at each rate is found by:

$$\pi_{0,r} = \frac{1}{m} pos_{\lceil r \cdot n \rceil} \tag{31}$$

We then use this probability to generate a label at k:

$$l_k \sim binom(\pi_{0,r}) \tag{32}$$

In this way we generate a set of 2,000 bootstrap ROC tables (generating $2,000 \cdot m$ samples in total).

This sampling procedure does not ensure that each sample has the correct class distribution. This is needed so that the confidence intervals generated from these samples reflect that at rates 0 and 1 we are certain the curve passes through the points $(0, 0)$ and $(1, 1)$ in ROC space, respectively. A simple approach to restrict to a fixed class distribution discards all samples where the class distribution is not correct. However, this approach is only feasible when the number of examples is low, as otherwise samples are rarely generated with the correct class distribution and this method becomes too slow.

We propose another approach that can be used with a larger number of examples, where we adjust the rate and the number of true and false positives at each position in order to correct the class distribution. The rates of the bootstrap ROC tables are equally distributed along the ranking, as shown in Figure 2.

For each sample individually we adjust these rates and the true positive values at each position, by scaling each position according to a correction factor, a value for each sample and class that rescales the 'width' of each example in the ranking to correct the class distribution. This adjustment is illustrated in Figure 2, and shows how the effect is to stretch or narrow the examples along the ranking.

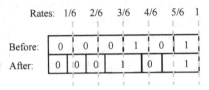

Fig. 2. Illustration of rate adjustment to correct class distribution

We use the bootstrapped ROC tables with the corrected true positive values, to estimate the confidence bound of the true ROC curve. For each ROC table, and at each position k along the ranking, we calculate the average recall across the samples:

$$\overline{tpr}_k = \frac{1}{m \cdot n_0} \sum_{i=1}^{m} s_{i,k} \tag{33}$$

Each position k in the ranking has a set of average recall values, one for each sample ROC table. This now corresponds to the probability density function we stated in Equation 9. The proportion of bootstrap ROC tables with recall value between $\overline{tpr}_k$ and $\overline{tpr}'_k$ gives an estimate of the probability that the recall at this position is between these values, given this sample has a particular class distribution.

The confidence interval for position k is obtained from the mean recall values, $\overline{tpr}_k$, of the bootstrapped ROC tables as follows. For each position k we take the $\overline{tpr}_k$ value of each ROC table, sort these values in ascending order, and select the 2.5% and 97.5% percentiles as the lower and upper endpoints of the 95% confidence interval. This gives a series of recall-rate pairs for the lower and upper limits of the confidence interval at each position k. A confidence bound can be created by interpolating between these points.

4   Experiments

Our experiments use a known ROC curve to generate samples for which we create confidence bounds, specified by normally distributed score density functions with mean 0 and 1 for the positive and negative class respectively, and a variance of 1. These score distributions, and the corresponding ROC curve are shown in Figure 3. Our tests use ROC tables with 10 samples and 50 examples per sample.

We evaluate whether the generated confidence intervals meet our aims, where at significance level σ new samples generated from this consensus curve pass between the lower and upper confidence limits at a given PPR value, with probability $1 - \sigma$. Given a single sample ROC table and its confidence bounds, we generate 1,000 new sample ROC tables from this sample. We count, at each rate, the number of consensus curves (of these samples) the confidence interval contains. A true 95% confidence interval at a given rate, should contain the consensus curve of new samples 95% of the time.

The results are shown in Figure 4. The results of the basic parametric approaches (Figures 4a and 4b) are highly variable. Our parametric approach (Figure 4c) reliably generates confidence bounds with close to 95% confidence, except at the extremes. This indicates that the assumption that the number of positives up to a particular position in the ranking is beta distributed is not valid in these regions.

Our bootstrap approaches are also much more effective compared to the baseline results. They are a little conservative, particularly at the extremes of the distribution, due to the nature of bootstrap sampling, where the variation between bootstraps may be too low to calculate strict confidence intervals (for instance, where the lower and upper bounds of the 95% limits are the same as those for the 94 or 96% limits). For example if a bootstrap sample contained only one value then the values at the 95% bounds would also be the same values as for the 1% or 100% limits. This also justifies the shape of the graph in Figure 4f, as where rates have a probability of a positive near to 1, there is little variation across samples.

Figure 5a shows an example ROC curve generated using our analytical approach, and the equivalent rate-recall curve is shown in Figure 5b.

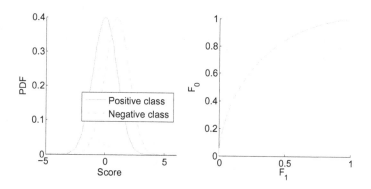

Fig. 3. Score probability densities for two classes (positive class: $\mu = 0$, $\sigma^2 = 1$; negative class: $\mu = 1$, $\sigma^2 = 1$), and corresponding ROC curve

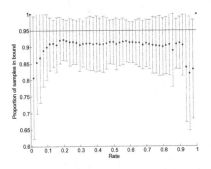

(a) Results of baseline with normal assumption

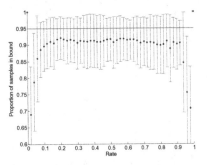

(b) Results of baseline with beta assumption

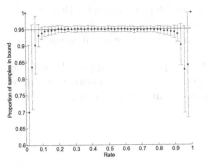

(c) Results of parametric approach

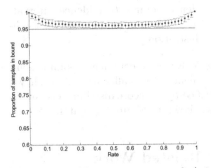

(d) Results of bootstrap approach (with discarding)

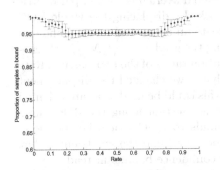

(e) Results of bootstrap approach (with adjustment)

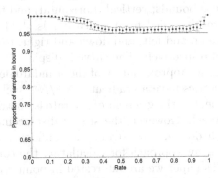

(f) Results of bootstrap approach (with discarding) for score distributions with: $\mu_0 = 0$, $\sigma_0^2 = 1$, $\mu_1 = 1$, $\sigma_1^2 = 0.2$

Fig. 4. Mean (variance) of the proportion of 1000 new samples (sampled from ROC table) within confidence interval at each rate, across 100 tests

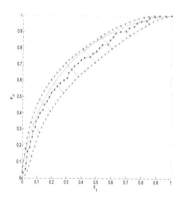 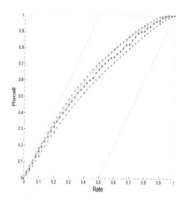

(a) Example ROC curve and confidence bounds (confidence intervals at a selection of rates shown for illustration)

(b) Equivalent rate-recall curve [12] and confidence bounds for ROC curve shown in Figure 5a. Grey lines indicate bounds of rate-recall space.

Fig. 5. Example confidence bounds generated with our parametric approach, and the equivalent rate-recall curve. Also shown are two curves: 1) The smooth true curve specified by the score distributions and 2) The consensus curve from the original sample (also shown in Figure 3 (right)).

5   Related Work

In the introduction we discussed two parametric approaches to generating confidence bounds; vertical (horizontal) and threshold averaging. A non-parametric approach, called fixed width bands [4, 11] works by displacing the whole ROC curve up and left, and down and right, to create an upper and lower confidence band respectively. The curve is displaced along the gradient $-\sqrt{(N^+/N^-)}$ (chosen as an approximation of the standard deviation ratios of the two classes). Rate isometrics have a gradient $-N^+/N^-$ such that if we changed the displacement gradient to the gradient of the rate isometric this could be used as a rate-oriented approach. However, the size of displacement is constant along the ROC curve which does not constrain the confidence bounds to ROC space. Furthermore, this is an approach for calculating the confidence around the whole curve, but in this paper we are interested in point-wise confidence bounds instead.

Other approaches include a non-parametric approach by Tilbury et al., which they derived from first principles [16], and the use of kernel estimation to estimate the continuous probability density functions of the scores of each class [6]. We also refer the reader to comparisons of various approaches, performed by Macskassy et al. [9, 10].

Early retrieval tasks are those where the top ranked examples are of most interest [2], and metrics used for this task weight the importance of an example by its position in the ranking. For instance, the rate-weighted AUC (rAUC) [12] is a general measure where the distribution of weights along the ranking can be chosen for the specific task at hand. Other metrics that are restricted to particular weight distributions include; discounted cumulative gain (DCG) and normalised discounted cumulative gain (NDCG) [8] in information retrieval, robust initial enhancement (RIE) [14], the Boltzmann-enhanced Discrimination of ROC (BEDROC) [17], concentrated ROC (CROC) [15] and sum of the log ranks (SLR) [18]. These metric all evaluate rankings with respect to the rate, such that when assessing tasks that use these metrics in ROC space, we suggest it is most appropriate to generate rate-averaged consensus curves with rate-oriented point-wise confidence bounds.

Rate-averaging has been previously used [3,12] to generate consensus curves, referred to as pooling in [3]. To our knowledge there is no approach in the literature to infer rate-oriented confidence bounds. [9] claims that rate-averaging makes the strong assumption that the rates are estimating the same point in ROC space, and this is not appropriate. However, other approaches make this similar assumptions across a different metric, such as the false positive rate in vertical-averaging.

6   Conclusions

We have described a new approach to generate confidence bounds, which we call rate-oriented point-wise confidence bounds. Our main aim was to address some important weaknesses of other existing methods. Calculating the consensus and confidences bounds at each rate is practical as rate is a measure over which we have control in practice. On the other hand, vertical (or horizontal) averaging fix the false positive rate (true positive rate) and average across the true positive rate (false positive rate), but these metrics are not under our control so are of little use in practice. Score-averaging creates confidence bounds around clouds of points, and how best to do this is an open problem. Rate-averaging does not have this problem because it constrains to a single dimension.

Our approach is also invariant to swapping the classes, and we suggest that this property is sensible when generating confidence bounds. The confidence of a point on the ROC curve should not depend on which class is labelled as positive. Furthermore, our bounds have the advantage that they are smooth, due to the sampling across rates we perform as part of our method.

Our secondary aim was to find appropriate bounds for assessing models used specifically for rate-constrained tasks. Using a rate-oriented approach ensured that the performance (and confidence interval) shown at a rate is an estimate for this particular rate.

In this paper we analytically derived the probability distribution of the number of positives up to each position in the ranking, and then used this to develop two methods, a parametric and a bootstrap approach. The parametric approach gave confidence bounds having very close to the 95% confidence, except at the extremes. The bootstrap approach did generate satisfactory bounds at the extreme but also had greater variance around the 95% confidence level. Therefore, we suggest that when the performance at the extremes of the ROC curve are of little importance, the parametric approach should be used, but where this is not the case the bootstrap approach can be used instead.

Acknowledgments. This work is supported by the REFRAME project granted by the European Coordinated Research on Long-term Challenges in Information and Communication Sciences & Technologies ERA-Net (CHIST-ERA), and funded by the Engineering and Physical Sciences Research Council in the UK. LACM is funded by a studentship from the UK Medical Research Council. This work was also supported by Medical Research Council grant MC_UU_12013/1-9.

References

1. Arnold, B.C., Balakrishnan, N., Nagaraja, H.N.: A first course in order statistics, vol. 54. SIAM (1992)
2. Berrar, D., Flach, P.: Caveats and pitfalls of ROC analysis in clinical microarray research (and how to avoid them). Briefings in Bioinformatics 13(1), 83–97 (2012)
3. Bradley, A.P.: The use of the area under the ROC curve in the evaluation of machine learning algorithms. Pattern Recognition 30(7), 1145–1159 (1997)
4. Campbell, G.: Advances in statistical methodology for the evaluation of diagnostic and laboratory tests. Statistics in Medicine 13(5-7), 499–508 (1994)
5. Fawcett, T.: ROC graphs: Notes and practical considerations for researchers. Machine Learning 31, 1–38 (2004)
6. Hall, P., Hyndman, R.J., Fan, Y.: Nonparametric confidence intervals for receiver operating characteristic curves. Biometrika 91(3), 743–750 (2004)
7. Hand, D.J.: Measuring classifier performance: A coherent alternative to the area under the ROC curve. Machine Learning 77(1), 103–123 (2009)
8. Järvelin, K., Kekäläinen, J.: IR evaluation methods for retrieving highly relevant documents. In: Proceedings of the 23rd Annual International ACM SIGIR Conference on Research and Development in Information Retrieval, pp. 41–48. ACM (2000)
9. Macskassy, S., Provost, F.: Confidence bands for ROC curves: Methods and an empirical study. In: Proceedings of the First Workshop on ROC Analysis in AI (2004)
10. Macskassy, S., Provost, F., Rosset, S.: Pointwise ROC confidence bounds: An empirical evaluation. In: Proceedings of the Workshop on ROC Analysis in Machine Learning (2005)
11. Macskassy, S.A., Provost, F., Rosset, S.: ROC confidence bands: An empirical evaluation. In: Proceedings of the 22nd International Conference on Machine Learning, ICML 2005, New York, NY, USA, pp. 537–544 (2005)
12. Millard, L.A.C., Flach, P.A., Higgins, J.P.T.: Rate-constrained ranking and the rate-weighted AUC. In: Calders, T., Esposito, F., Hüllermeier, E. (eds.) ECML/PKDD 2014, vol. 8725, pp. 383–398. Springer, Heidelberg (2014)
13. Provost, F.J., Fawcett, T., Kohavi, R.: The case against accuracy estimation for comparing induction algorithms. In: ICML, vol. 98, pp. 445–453 (1998)
14. Sheridan, R.P., Singh, S.B., Fluder, E.M., Kearsley, S.K.: Protocols for bridging the peptide to nonpeptide gap in topological similarity searches. Journal of Chemical Information and Computer Sciences 41(5), 1395–1406 (2001)
15. Joshua Swamidass, S., Azencott, C.-A., Daily, K., Baldi, P.: A CROC stronger than ROC: Measuring, visualizing and optimizing early retrieval. Bioinformatics 26(10), 1348–1356 (2010)

16. Tilbury, J.B., Van Eetvelt, W., Garibaldi, J.M., Curnsw, W.J., Ifeachor, E.C.: Receiver operating characteristic analysis for intelligent medical systems-a new approach for finding confidence intervals. IEEE Transactions on Biomedical Engineering 47(7), 952–963 (2000)

17. Truchon, J.-F., Bayly, C.I.: Evaluating virtual screening methods: good and bad metrics for the "early recognition" problem. Journal of Chemical Information and Modeling 47(2), 488–508 (2007)

18. Zhao, W., Hevener, K.E., White, S.W., Lee, R.E., Boyett, J.M.: A statistical framework to evaluate virtual screening. BMC Bioinformatics 10(1), 225 (2009)

A Fast Method of Statistical Assessment for Combinatorial Hypotheses Based on Frequent Itemset Enumeration

Shin-ichi Minato[1,2], Takeaki Uno[3], Koji Tsuda[4,5,2],
Aika Terada[6], and Jun Sese[6]

[1] Graduate School of Information Science and Technology,
Hokkaido University, Sapporo, 060-0814 Japan
[2] JST ERATO Minato Discrete Structure Manipulation System Project,
Sapporo, 060–0814 Japan
[3] National Institute of Informatics, Tokyo 101–8430, Japan
[4] Graduate School of Frontier Sciences, The University of Tokyo,
Kashiwa, 277–8561 Japan
[5] Computational Biology Research Center, National Institute of Advanced Industrial
Science and Technology, Tokyo, 135-0064 Japan
[6] Department of Computer Science, Ochanomizu University, Tokyo, 112-8610 Japan

Abstract. In many scientific communities using experiment databases, one of the crucial problems is how to assess the statistical significance (p-value) of a discovered hypothesis. Especially, combinatorial hypothesis assessment is a hard problem because it requires a multiple-testing procedure with a very large factor of the p-value correction. Recently, Terada et al. proposed a novel method of the p-value correction, called "Limitless Arity Multiple-testing Procedure" (LAMP), which is based on frequent itemset enumeration to exclude meaninglessly infrequent itemsets which will never be significant. The LAMP makes much more accurate p-value correction than previous method, and it empowers the scientific discovery. However, the original LAMP implementation is sometimes too time-consuming for practical databases. We propose a new LAMP algorithm that essentially executes itemset mining algorithm once, while the previous one executes many times. Our experimental results show that the proposed method is much (10 to 100 times) faster than the original LAMP. This algorithm enables us to discover significant p-value patterns in quite short time even for very large-scale databases.

1 Introduction

Discovering useful knowledge from large-scale databases has attracted considerable attention during the last decade. Such knowledge discovery techniques are widely utilized in many areas of experimental sciences, such as biochemistry, material science, medical science, etc. In those scientific communities using experiment databases, one of the crucial problems is how to assess the statistical significance (p-value) of a discovered hypothesis. The p-value-based assessment is

T. Calders et al. (Eds.): ECML PKDD 2014, Part II, LNCS 8725, pp. 422–436, 2014.
© Springer-Verlag Berlin Heidelberg 2014

one of the most important factors in the paper review process of academic journals in experimental sciences [11]. (Here are some related studies [8,26,18,15] on data mining algorithms considering p-values.)

For those scientific applications, detecting a combinatorial regulation of multiple factors is sometimes a very important issue. For example, it is well known that a key to generate iPS cells consists of the four factors of genes [16]. However, statistical assessment of a hypothesis for detected combinatorial effect is a hard problem because it requires a multiple-testing procedure with a very large factor of the p-value correction. This correction is necessary to avoid a false discovery caused by repetition of statistical tests. When we consider the combinations of j out of n hypotheses, the number of tested combinations increases exponentially as $O(n^j)$, and if we use a naive correction method in $j = 4$ or more, it is too conservative since almost all discoveries becomes extremely unlikely.

Recently, Terada et al. developed a novel p-value correction procedure, called "Limitless Arity Multiple-testing Procedure" (LAMP). Their paper [20] was published in *PNAS*, a leading journal in scientific community. This new procedure excludes meaninglessly infrequent hypotheses which will never be significant. The p-value correction factor calculated by LAMP is much more accurate than previous method, and it empowers the scientific discovery from the experiment databases. However, the original LAMP implementation is sometimes too time-consuming for practical databases, and a state-of-the-art algorithm has been desired.

The LAMP is based on the techniques of *frequent itemset mining*, to enumerate all frequent itemsets included in at least σ transactions of the database for a given threshold σ. Since the pioneering work by Agrawal et al. [1], various algorithms have been proposed to solve this problem [9,13,27]. Among those state-of-the-art algorithms, *LCM (Linear time Closed itemset Miner)*[24,22,23] by Uno et al. is known as one of the fastest algorithm, which is based on a depth-first traversal of a search tree for the combinatorial space.

In this paper, we propose a fast itemset enumeration algorithm to find the minimum support for satisfying the LAMP condition. Our new algorithm essentially executes itemset mining algorithm once, while the previous one executes many times. We show that LAMP condition is a kind of threshold function which is monotonically decreasing or increasing. We developed a general scheme to explore the maximum frequency satisfying a given threshold function. We successfully applied this new scheme to the LAMP condition. Those new procedures are implemented into the newest version of the LCM program. Our experimental results show that the proposed method is much (10 to 100 times) faster than the original LAMP. This algorithm enables us to discover significant p-value patterns in quite short time even for very large-scale databases.

In the rest of this paper, we first explain the preliminaries on frequent itemset mining algorithms. In Section 3, we then present the problem of statistical assessment for combinatorial hypotheses and the idea of LAMP. Section 4 describes our proposed methods for finding the minimum frequency for satisfying threshold function and the LAMP condition. Section 5 discusses efficient implementation, and Section 6 shows our experimental results, followed by the conclusion in Section 7.

2   Preliminary

Here we start with some basic definitions of itemset databases and frequent itemset mining.

Let $E = \{1, \ldots, n\}$ be a set of items. A subset of E is called an *itemset*. A *transaction database* is a database composed of *transactions* where a transaction is an itemset. A transaction database can include two or more identical transactions. For a transaction database $\mathcal{D}$, $|\mathcal{D}|$ denotes the number of transactions in $\mathcal{D}$, and $||\mathcal{D}||$ denotes the size of $\mathcal{D}$, that is the sum of the size of the transactions in $\mathcal{D}$, i.e., $||\mathcal{D}|| = \sum_{T \in \mathcal{D}} |T|$.

For an itemset X and a transaction database $\mathcal{D}$, an *occurrence* of X in $\mathcal{D}$ is a transaction including X. The *occurrence set* of X, denoted by $Occ(X)$ is the set of all occurrences of X in $\mathcal{D}$. The *frequency* of X is the number of occurrences of X, and is denoted by $frq(X)$. For a given constant number σ called *minimum support*, an itemset X is called *frequent*. The frequent itemset mining problem is to enumerate all frequent itemsets for given database $\mathcal{D}$ and threshold σ.

Without confusions, an item e also represents the itemset $\{e\}$, hence $frq(e)$, $X \cup e$ and $X \setminus e$ denote $frq(\{e\})$, $X \cup \{e\}$ and $X \setminus \{e\}$, respectively. Let $\kappa(\sigma)$ be the number of frequent patterns whose frequencies are no less than σ.

2.1   Frequent Itemset Mining Algorithm

The set system given by the set of frequent itemsets is anti-monotone, i.e., any subset Y of a frequent itemset X is always frequent. Thus, enumeration of frequent itemsets is done efficiently by hill-climbing algorithms, that start from the emptyset and recursively add items unless the itemset is infrequent. In particular, depth-first search type algorithms (backtracking) are known to be efficient [9]. In the backtracking way, we add an item e to the current itemset X, and explore all itemsets generated from $X \cup e$ before processing $X \cup e', e' \neq e$. To avoid duplicated solutions such that an itemset is output twice, backtracking adds only item $e > tail(X)$ where $tail(X)$ is the maximum item in X. Through this enumeration technique, any item Y is generated from the itemset $Y \setminus tail(X)$, thus no duplication occurs. The pseudo code of backtracking is written as follows.

ALGORITHM **BackTracking_Basic** (X)
1. **output** X
2. **for** each item $e > tail(X)$,
   if $frq(X \cup e) \geq \sigma$ **then call** **BackTracking_Basic** $(X \cup e)$

The most heavy computation in this algorithm is the computation of $frq(X \cup e)$ on Step 2 of the algorithm. There are several techniques for reducing this computation. Especially, *recursive database reduction* techniques, such as FP-tree representation of the database [12] and anytime database reduction [22], are quite efficient. These techniques can be applied only to depth-first type algorithm, this explains why BFS algorithms are slow compared to backtracking on real-world data [10]. There is one more important technique so called *equi-support* (see for example [24] written as *hypercube decomposition*). For an itemset

X, we call an item $e \notin X$ *addible* if $frq(X \cup e) = frq(X)$. $Eq(X)$ denotes the set of addible items e of X that satisfy $e > tail(X)$. We can see that any itemset $X \cup S, S \subseteq Eq(X)$ satisfies $frq(X) = frq(X \cup S)$, thus we output all these itemsets without the computation of their frequency. Using this observation, a more efficient algorithm can be written as follows. Duplications can be avoided by not generating recursive calls for $X \cup e$ with $e \in Eq(X)$.

ALGORITHM **BackTracking** (X)
1. **output** $X \cup S$ for all $S \subseteq Eq(X)$
2. **for** each item $e > tail(X)$, $e \notin Eq(X)$,
    if $frq(X \cup e) \geq \sigma$ **then call BackTracking** $(X \cup e)$

Fast implementations of backtracking find up to one million solutions in a second [10].

3   Statistical Assessment for Combinatorial Hypotheses

In this paper, we discuss a fast method of statistical assessment using the techniques of frequent itemset mining. First we assume the following experimental scenario.

Consider a scientific database including experimental results for a number of human gene samples, and each sample shows a set of expressions of targeted factors. Here we assume only one expression level (exist or not) for each factor. We also have another classification result for each sample whether it is positive or negative, for example, the gene sample is given by a patient of breast cancer or a normal person. Then we want to discover a combination of factors which is highly correlated to incidence of a breast cancer. This is a quite simple scenario and many similar cases may commonly appear in various areas of experimental sciences. Since we assume only binary values in the database, we can represent it as a transaction database $\mathcal{D}$ as shown in Fig. 1. Here we also assume a classifier $C : \mathcal{D} \to \{pos, neg\}$, which classify each transaction into a positive or a negative one. Suppose that the database has n items for the expressions of factors, m transactions in total, and m_p positive transactions.

Now we consider the assessment whether a given itemset has a strong correlation to appear in the positive transactions. Figure 2 shows a *contingency table* between the occurrence of the itemset X and the positive class of transactions. Here we note $\sigma = frq(X)$ and σ_p is a number of positive transactions in $Occ(X)$. We also show the value of each cell when $X = \{2, 3\}$ for the database shown in Fig. 1.

If the distribution of the contingency table is very biased, we may consider this is a kind of knowledge discovery because it is unlikely that it happened incidentally. The *p-value* represents the probability that such a biased distribution incidentally happens. In other words, it is the probability of a false discovery, and we can accept the statistical significance if the p-value is smaller than an arbitrary threshold α. ($\alpha = 0.05$ is often used.) The p-value is quite important in the paper review process of academic journals in experimental sciences.

$T \in \mathcal{D}$	$\mathcal{C}(T)$
2 3 4 5	**neg**
1 3 5	pos
2 3 4	**pos**
1 2 4 5	neg
1 2 3	**pos**
2 4	neg
1 2 3 5	**pos**
2 4 5	neg
1 3 4	neg

Fig. 1. An example of transaction database with positive-negative class

	pos	neg	all
$Occ(X)$	(σ_p) 3	$(\sigma - \sigma_p)$ 1	(σ) 4
$\overline{Occ(X)}$	$(m_p - \sigma_p)$ 1	$(m - m_p - \sigma + \sigma_p)$ 4	$(m - \sigma)$ 5
all	(m_p) 4	$(m - m_p)$ 5	(m) 9

Fig. 2. Contingency table ($X = \{2, 3\}$)

Fisher's exact test is one of the major methods for calculating the p-value for a given contingency table. This testing method assumes that each experimental result is independent and has an equal weight. Then, the p-value is calculated by counting all combinatorial cases to generate equally or more biased distributions. More exactly, the probability of generating a contingency table of Fig. 2 can be written as:

$$P(\sigma_p) = \left(\binom{m_p}{\sigma_p} \cdot \binom{m - m_p}{\sigma - \sigma_p} \right) / \binom{m}{\sigma}, \tag{1}$$

and the p-value is defined as $\displaystyle\sum_{\sigma'_p = \sigma_p}^{min(\sigma, \, m_p)} P(\sigma'_p)$, which is the total probabilities of all equally or more biased distributions.

3.1 P-value Correction in Multiple Tests

If we repeatedly calculate p-values for many different factors in a same database, we may more likely find a factor with an incidentally low p-value. For example, if we explore a family of hundred hypotheses each of which might be a false discovery in $p = 0.05$, then at least one false discovery can be found in $p = 0.994$. It is well known that such multiple tests may cause serious false positive problems [17]. Hence, a multiple testing correction must be used in order to avoid a false discovery. The *family-wise error rate (FWER)* indicates the probability that at least one false discovery happens in multiple tests. This rate increases at most linearly as the number of tests, which motivates the Bonferroni correction [7] that multiplies the raw p-value by the number of tests. In other words, we must compare the raw p-values with the adjusted threshold $\delta = \alpha/k$, where k is number of tests. The Bonferroni correction is a very conservative method, which likely causes false negative but hardly causes false positive, and often used in academic articles in experimental sciences.

When we consider the combinations of j out of n items, the number of tested combinations increases exponentially as $O(n^j)$, and if we naively use Bonferroni correction in $j = 4$ or more, most of the discoveries of combinatorial hypotheses becomes extremely unlikely. However, in many practical cases, not all the combinations occur in the databases, so the ideal size of hypothesis family would be much smaller than the naive combinatorial number. This is a motivation of the LAMP.

3.2   Idea of LAMP

Recently, Terada et al. developed a novel p-value correction procedure, called "Limitless Arity Multiple-testing Procedure" (LAMP) [20]. This new procedure is based on frequent itemset enumeration for checking the two principles:

(1) Meaninglessly infrequent itemsets which never be significant can be excluded from the number of hypotheses to be counted.
(2) Any different itemsets having completely the same occurrence set can be counted as one hypothesis in the family.

The principle (2) means that we may enumerate only the *closed* itemsets, and we can just use existing state-of-the-art algorithms of closed itemset mining [24,22,23]. Here we explain in detail how to check the principle (1).

Suppose an itemset X with a very low frequency σ ($< m_p$). Using Fisher's exact test, the raw p-value cannot be smaller than

$$f(\sigma) = \binom{m_p}{\sigma} / \binom{m}{\sigma}, \tag{2}$$

which means the most biased case of equation (1) that all $Occ(X)$ are classified into the positive class. Note that $f(\sigma)$ is monotonically decreasing, namely, the less frequent itemsets has the larger $f(\sigma)$. This observation means that all the infrequent itemsets satisfying $f(\sigma) > \delta$ can never be significant, regardless of the classification result. Thus, we can exclude such itemsets from the number of hypotheses to be counted. This is the key idea of the LAMP.

Let $\kappa_c(\sigma)$ be the number of all closed itemsets not less frequent than σ. Then, the adjusted threshold of Bonferroni correction can be written as $\delta = \alpha/\kappa_c(\sigma)$. Now our subject is to find the maximum frequency σ_{max} which satisfies:

$$f(\sigma_{max} - 1) > \frac{\alpha}{\kappa_c(\sigma_{max})} \quad \text{and} \quad f(\sigma_{max}) \leq \frac{\alpha}{\kappa_c(\sigma_{max} + 1)} . \tag{3}$$

3.3   Current Implementation of LAMP

In the above inequation of the LAMP condition, the left side $f(\sigma - 1)$ is monotonically decreasing and the right side $\alpha/\kappa_c(\sigma)$ is monotonically increasing in the range $1 \leq \sigma \leq m_p$. Thus, a naive idea is to enumerate all the itemsets in an order from the highest σ to the lowest one, and we may stop the enumeration

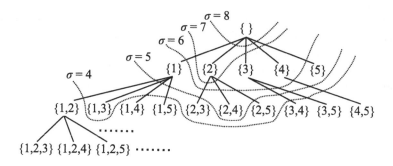

Fig. 3. An example of a breadth-first search

procedure when the current σ first satisfies the condition. This is well known as the breadth-first search approach used in the "Apriori-type" [1] frequent item-set mining algorithms. Figure 3 illustrates the execution steps of a breadth-first search on the same database in Fig. 1. This approach requires a large size of memories to store all the itemsets at the current frontier for each step, and in general, it is not very efficient for practical size of databases.

The current implementation of the original LAMP [19] is not using a breadth-first search due to memory limitation, but just calls the LCM algorithm repeatedly, as follows.

ALGORITHM **Original LAMP** (α)
1. $\sigma := max_{e \in E}(frq(e))$
2. **if** $\sigma < m_p$ **then** $\sigma := m_p$ // m_p: number of positive transactions.
3. $k := \kappa_c(\sigma)$ // call **LCM** algorithm to compute $\kappa_c(\sigma)$.
4. **if** $f(\sigma - 1) \le \alpha/k$ **then** $\sigma := \sigma - 1$; **goto** 3
5. **output** (σ, k)

This method requires less memory, but sometimes very time-consuming since the LCM may be called more than thousand times for practical cases, and it may take a day or more. The current LAMP has a bottleneck in computation time, and some efficient algorithms are desired.

One may consider that we can apply a sampling technique [6] to quickly esti-mate the number of frequent itemsets satisfying the LAMP condition. However, the result is used for statistical assessment, thus we must ensure the worst-case upper bound, but it is difficult to find a feasible bound by sampling-based meth-ods. Here we propose a fast and exact method of enumerating itemsets for the LAMP condition.

4   Proposed Algorithms

4.1   A Threshold Function for the LAMP Condition

Let $\theta(x, y) : (\mathbf{N} \times \mathbf{N}) \to \{true, false\}$ be a given threshold function such that $\theta(x, y) = true$ implies $\theta(x', y) = true$ for any $x' < x$, and that $\theta(x, y) = true$

implies $\theta(x, y') = true$ for any $y' > y$. In other words, $\theta(x, y)$ is monotonically decreasing for x and increasing for y. Now we consider a general scheme of the itemset mining problems, to find the largest σ satisfying $\theta(\sigma, \kappa(\sigma))$. Note that $\kappa(\sigma)$ is monotonically decreasing for $1 \leq \sigma \leq m$, thus $\theta(\sigma, \kappa(\sigma))$ is monotonically decreasing for the same $1 \leq \sigma \leq m$. We call the largest σ *maximum frequency* for the threshold function, and denote it by σ_{max}.

For examples of such threshold functions, the function for top-k mining can be defined as $\theta(x, y) = true$ iff $y \geq k$. We then explore the maximum frequency σ_{max} for satisfying $\theta(\sigma, \kappa(\sigma))$. It means that the k-th most frequent itemsets have the frequency σ_{max}.

Here we define a threshold function for the LAMP condition as follows.

$$\theta(x, y) = true \ \ \text{iff} \ \ f(x - 1) > \frac{\alpha}{y} \tag{4}$$

As discussed in Section 3, we can confirm that this threshold function is decreasing for x and increasing for y. We then explore the maximum frequency σ_{max} for satisfying $\theta(\sigma, \kappa(\sigma))$[1]. Hereafter we assume that $\theta(x, y)$ satisfies the above condition.

4.2   Support Increase Algorithm

For the computation of the maximum frequency for $\theta(\sigma, \kappa(\sigma))$, a natural way is to compute $\kappa(\sigma)$ by frequent itemset mining with the minimum support σ, for all possible candidates σ, one by one. This computation takes long time when σ is small since $\kappa(\sigma)$ is huge, thus we should compute $\theta(\sigma, \kappa(\sigma))$ in the decreasing order of σ. This is the basic scheme of the original LAMP [20] shown in the previous section. However, this needs long computation time since many frequent itemset mining processes are executed. In this paper, we propose a new algorithm that basically executes mining algorithm just once.

Suppose that we are given a function $\theta(\sigma, \kappa(\sigma))$. For a frequency σ, if we found some k and confirmed both $k \leq \kappa(\sigma)$ and $\theta(\sigma, k) = true$, then we get $\theta(\sigma, \kappa(\sigma)) = true$, since $\theta(x, y)$ is increasing for y. In such a case, we can see $\sigma_{max} \geq \sigma$, and in particular, $\sigma_{max} = \sigma$ if $\theta(\sigma + 1, \kappa(\sigma + 1)) = false$.

Suppose that we execute a backtracking algorithm for minimum support σ to check $\theta(\sigma, \kappa(\sigma))$, and during the mining process we found k frequent itemsets satisfying $\theta(\sigma, k)$. From the assumption of the function $\theta(x, y)$, we can then confirm that $\theta(\sigma, \kappa(\sigma)) = true$. We are then motivated to re-execute the backtracking with $\sigma := \sigma + 1$ to check $\theta(\sigma + 1, \kappa(\sigma + 1))$. However, in the current execution, we already found possibly many itemsets of frequency $\sigma + 1$, and in the past search process, we never missed such itemsets. This implies that the execution until the current iteration can be skipped. We just need to remove all itemsets of frequency σ from the set of past solutions that have already been

---

[1] Our implementation uses $\kappa_c(\sigma)$ instead of $\kappa(\sigma)$ for the LAMP condition in equation (2). The number of closed itemsets are also monotonically decreasing for the frequency, so we can use $\kappa_c(\sigma)$ as well. This is discussed in Section 5.2.

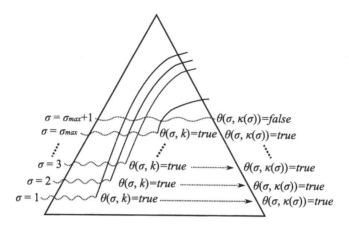

Fig. 4. The scheme of support increase algorithm

found, and can re-start the backtracking algorithm with $\sigma := \sigma + 1$. This implies that even if we start from very small σ having huge $\kappa(\sigma)$, we can increase it during the mining process for reducing the computation time.

A backtracking algorithm with this idea can be written as follows. Figure 4 illustrates the scheme of this algorithm. The algorithm start with $\sigma = 1$ and $S = \emptyset$ where S is a program variable for storing a set of frequent itemsets we already found. Let $S[\sigma]$ be the subset of itemsets in S whose frequency is σ.

ALGORITHM **SupportIncrease** (X)
global variable: σ, S (initialized $\sigma = 1, S = \emptyset$)
1. $S := S \cup \{X\}$
2. **if** $\theta(\sigma, |S|) = true$ **then** $S := S \setminus S[\sigma]$, $\sigma := \sigma + 1$; **go to** 2
3. **for** each item $e > tail(X)$,
    if $frq(X \cup e) \geq \sigma$ **then call SupportIncrease** $(X \cup e)$

Theorem 1. *Let σ^* and S^* denote the resulting values of σ and S after the execution of the algorithm* **SupportIncrease***. Then it holds that $\sigma^* = \sigma_{max} + 1$ and $|S^*| = \kappa(\sigma_{max} + 1)$.*

Proof. Since the algorithm never decreases σ, the itemsets of frequencies no less than σ^* cannot be missed, and are included in S^*. Since no itemset of frequency less than σ^* is in S, we have $|S^*| = \kappa(\sigma^*)$. We then prove $\sigma^* = \sigma_{max} + 1$ by contradiction.

Suppose that $\theta(\sigma^*, \kappa(\sigma^*)) = true$ holds. Let us consider the iteration in which the last itemset is inserted to S. Since $\theta(\sigma^*, \kappa(\sigma^*)) = true$ holds, the latter part of step 2 is not executed in this iteration, otherwise σ is increased so that $\theta(\sigma^*, \kappa(\sigma^*)) = false$ holds. Thus, we have $\sigma = \sigma^*$ in the iteration. After step 1, there is no more itemset of frequency no less than σ^* that has not been found, thus we have $|S| = \kappa(\sigma)$. This leads that $\theta(\sigma^*, \kappa(\sigma^*)) = false$, and contradicts to the assumption.

We next suppose that $\theta(\sigma^* - 1, \kappa(\sigma^* - 1)) = false$ holds. Let us consider the iteration in which σ is increased to σ^*. In the iteration, the latter part of step 2 is executed, and thus we have $\theta(\sigma^* - 1, |S|) = true$. Since $|S|$ is always no greater than $\kappa(\sigma)$, it implies that $\theta(\sigma^* - 1, k) = true$ holds for some $k \leq \kappa(\sigma^* - 1)$. This contradicts to the assumption.   □

The algorithm terminates in short time if $|S[\sigma]|$ is relatively small compared to $|S|$ on average. Particularly, if $|S[\sigma]| \leq q$ always holds, we can bound the number of iterations by $q \cdot \sigma_{max} + |S^*|$, since the algorithm removes at most $q \cdot \sigma_{max}$ itemsets from S. In the real-world data, we can naturally expect that $|S[\sigma]|$ is much smaller than $|S|$. In fact, we could confirm it in our computational experiments.

The computation of $|S[\sigma]|$ is done efficiently by using a heap that extracts the minimum frequency itemset from S. The computation then takes $O(|S[\sigma]| \log |S|)$ time. This computation time should be short than the computation time of an iteration, thus the total computation is expected not to increase much. However, if $|S|$ is large on average, or $\kappa(\sigma^*)$ is very large, the algorithm may take very long time. In the next section, we focus on such cases, and propose an efficient method for reducing the computation time.

5   Fast Implementation

The bottleneck of the computation of the algorithm in the previous section comes from the large size of the heap for S. This effects not only the computation time but also the memory usage. We here propose to use histogram counters instead of the heap. The histogram counters $s[\sigma]$ are prepared for keeping $|S[\sigma]|$, and an integer variable s is used for accumulating $|S|$. By using the histogram counters, we do not have to use the heap for storing the itemsets. The point is that we can compute $|S \setminus S[\sigma]|$ by just one subtract operation $s - s[\sigma]$. Another important point is the memory usage. The number of the histogram counters can be bounded by the number of transactions, thus it is only a linear factor to the input data size.

The use of histogram counters gives us one more advantage; we can use the equi-support technique. In each iteration with equi-support, we find many frequent itemsets, $2^{|Eq|}$ itemsets in exact, with the same frequency, at once. We can increase the counter for these in one step by adding $2^{|Eq|}$ to the counter. This saves the computation time of $2^{|Eq|} - 1$ iterations, thus the equi-support technique drastically shortens the total computation time. The algorithm is then written as follows.

ALGORITHM **EquiSupportIncrease** (X)
global variable: $\sigma, s, s[\ ]$ (initialized $\sigma = 1$, $s = 0$, $s[i] = 0$ for each i)
1. $s := s + 2^{|Eq(X)|}$, $s[frq(X)] := s[frq(X)] + 2^{|Eq(X)|}$
2. **if** $\theta(\sigma, s) = true$ **then** $s := s - s[\sigma]$, $\sigma := \sigma + 1$; **go to** 2
3. **for** each item $e > tail(X)$, $e \notin Eq(X)$,
   if $frq(X \cup e) \geq \sigma$ **then** call **EquiSupportIncrease** $(X \cup e)$

Theorem 2. *Let σ^* and s^* denote the resulting values of σ and s after the execution of the algorithm* **EquiSupportIncrease**. *Then it holds that $\sigma^* = \sigma_{max} + 1$ and $s^* = \kappa(\sigma_{max} + 1)$.*

Proof. We assume that the variable $\mathcal{S}$ is simultaneously computed as algorithm **SupportIncrease** during the execution of **EquiSupportIncrease**. Instead of insertion of itemsets X into $\mathcal{S}$, **EquiSupportIncrease** increases s and $s[frq(X)]$. Thus, in any iteration, $s = |\mathcal{S}|$ and $s[\sigma] = |\mathcal{S}[\sigma]|$ holds. This implies that the computation of σ in **EquiSupportIncrease** is the same as **SupportIncrease**, thus the statement holds.   □

Theorem 3. *Let P be the itemset enumerated by* **EquiSupportIncrease**. *The number of iterations needed to enumerate P by* **EquiSupportIncrease** *is equal to that by backtracking algorithms with equi-support technique.*   □

5.1   Calculating Family Size of the LAMP

Theorem 1 states that $|\mathcal{S}^*| = \kappa(\sigma_{max} + 1)$, namely, the histogram counters holds $\kappa(\sigma_{max} + 1)$ after the execution of the proposed algorithm. However, our final purpose of the LAMP is to know $\kappa(\sigma_{max})$, which is the hypothesis family size for the p-value correction. An easy way to know $\kappa(\sigma_{max})$ is calling LCM algorithm once again with σ_{max}. This is not so bad since calculating $\kappa(\sigma_{max})$ is not more time-consuming than finding σ_{max}, however, we can avoid such two-pass executions if we maintain not only $s[\sigma]$ but also $s[\sigma - 1]$ during the backtracking procedure. This can be done by a very small modification as shown below.

ALGORITHM **EquiSupportIncreaseLAMP** (X)
global variable: $\sigma, s, s[\ ]$ (initialized $\sigma = 1$, $s = 0$, $s[i] = 0$ for each i)
1. $s := s + 2^{|Eq(X)|}$, $s[frq(X)] := s[frq(X)] + 2^{|Eq(X)|}$
2. **if** $\theta(\sigma,\ s - s[\sigma - 1]) = true$ **then** $s := s - s[\sigma - 1]$, $\sigma := \sigma + 1$; **go to** 2
3. **for** each item $e > tail(X)$, $e \notin Eq(X)$,
    if $frq(X \cup e) \geq \sigma - 1$ **then** call **EquiSupportIncreaseLAMP** $(X \cup e)$

In this modification, the total number of backtracking will be a little increase because the condition of the recursive call in Step 3 is relaxed to one smaller frequency. This overhead is relatively small in the total computation time, and it is a reasonable cost for computing $\kappa(\sigma_{max})$ correctly.

5.2   Generalization to Other Patterns

Many patterns have been considered in the past researches of pattern mining. Our algorithm works on many of these patterns. The requirement is that the backtracking algorithm does work, i.e., the set of frequent patterns satisfies some monotone properties. For example, closed itemset mining accepts our algorithm, and also maximal frequent itemset mining also does. Closed itemsets have the anti-monotone property, and maximal frequent itemsets can be enumerated by backtracking. They can be solved by our implementation. Sequence pattern mining [25], frequent tree mining [4,5], frequent geograph mining [3], maximal motif mining [2], frequent graph mining [14], and other basic patterns also satisfy monotone property, so that we can construct a search tree in which parents have frequencies no smaller than their children, thereby backtracking works. They accept the counter implementation when the maximum possible frequency is not huge, and also Equi-support technique when it works. Our basic scheme of the algorithm is quite strong so that we can use it in many kinds of pattern mining problems.

6   Computational Experiments

We implemented our new LAMP algorithm by modifying LCM ver. 5 that is available on the author's website [21]. This is the latest version of LCM algorithm that won in FIMI04 competition of fast pattern mining implementations [10]. The fundamental issues of the implementation is described in [22]. We note that the modification is quite small such that we added/modified only up to 30 lines of C codes. We could not observe any relatively large difference of computational performance after the modification of the LCM. For the comparison, we also evaluated the original version of LAMP [19]. The original LAMP repeatedly calls the same LCM ver. 5, as shown in Section 3 of this paper.

Table 1 presents the specifications of the database instances used in our experiments. "yeast" and "breast cancer" are the real gene databases, which are also used in the original LAMP paper [20]. The others are well known benchmark datasets of FIMI competition and KDD CUP 2000, available from the FIMI repository [10]. The columns n and m show the numbers of items and transactions, respectively. Here we show the hypothesis family sizes of traditional Bonferroni correction if we limit the maximum arity (the number of items in a combination) up to 2, 3, and 4. We can see that the family size grows exponentially to the arity, and that it seems too large for meaningful knowledge discovery in practical size of databases.

Table 2 shows our experimental results of the new and original versions of LAMP. The columns m_p, σ_{max} and $\kappa_c(\sigma_{max})$ indicate the number of positive transactions, the maximum frequency for the LAMP condition, and the number of frequent closed itemsets (hypothesis family size by LAMP), respectively. Here we used the significance threshold $\alpha = 0.05$. Note that the datasets "yeast" and "breast cancer" have a positive-negative classification for each transaction. "mushroom" does not have such information but it has the two specific items which mean poisonous or edible mushroom, so we define the poisonous one as positive, and the rest of 117 items are assessed as the combinatorial hypotheses. For the other datasets, we did not specify a particular classification item, but we assumed the two cases that the positive transactions share 50% and 10% in the whole data. The experimental results show that the new LAMP is much faster than the original LAMP, as much as 10 to 100 times in many cases, and it can work well for practical size of databases with hundreds of items and thousands

Table 1. Specifications of databases

database	n	m	Bonferroni (family size)		
			max-arity: 2	max-arity: 3	max-arity: 4
yeast	102	6074	5253	176953	4426528
breast cancer	397	12773	79003	10428793	1029883108
mushroom(3-119)	117	8124	6903	267033	7680738
T10I4D100K	870	100000	378885	109751225	23816205920
BMS-WebView-1	497	59602	123753	20460993	2532110133
BMS-WebView-2	3340	77512	5579470	6209953450	5.182×10^{12}
BMS-POS	1657	515597	1373653	758258113	3.137×10^{11}

Table 2. Experimental Results of new and original LAMP

database	m_p	σ_{max}	$\kappa_c(\sigma_{max})$ (family size)	new LAMP time(sec)	orig. LAMP time(sec)
yeast	530	4	303	0.005	0.463
breast cancer	1129	8	3750336	36.538	86.315
mushroom(3-119)	3916	20	98723	0.740	141.327
T10I4D100K(m_p:50%)	50000	21	107080	3.092	799.738
T10I4D100K(m_p:10%)	10000	7	483300	5.714	820.756
BMS-WebView-1(m_p:50%)	29801	22	170660	12.349	122.303
BMS-WebView-1(m_p:10%)	5960	8	959435	33.351	248.055
BMS-WebView-2(m_p:50%)	38756	22	209016	1.813	229.406
BMS-WebView-2(m_p:10%)	7751	8	665411	5.655	246.278
BMS-POS(m_p:50%)	257799	31	74373743	580.321	78801.858
BMS-POS(m_p:10%)	51560	11	702878145	4513.052	50883.609

Intel Core i7-3930K 3.2GHz, 64GB Mem, 12MB Cache, OpenSuSE 12.1

of transactions. We can also see that the family sizes given by LAMP are often smaller than ones by the traditional Bonferroni with the max-arity up to only 3. This is very powerful in practical applications because the LAMP has no arity limit up to n, and this correction still guarantees the family-wise error rate bounded by α.

7 Conclusion

In this paper, we proposed a fast itemset enumeration algorithm to find the frequency threshold satisfying the LAMP condition. We developed a general scheme to explore the maximum frequency satisfying a monotonic threshold function. We successfully applied this new scheme to the LAMP condition. The procedure is implemented into the newest LCM program. Our experimental results show that the proposed method is much (10 to 100 times) faster than the original LAMP and that it can work well for practical size of experiment databases. The new enumeration algorithm solved the bottleneck of the LAMP for practical applications, and useful for various areas of experimental sciences.

We may have several kinds of future work. As well as multiple-testing correction, computing the p-values for a particular hypothesis is also time-consuming procedure. It will be useful if we can efficiently compute the p-values for many combinatorial hypotheses and can discover the best or top-k significant one. In this paper, we considered Fisher's exact test, however, there are some other types of the p-value calculation, such as χ-squared test and Mann-Whitney test, and we may consider different statistical models. In addition, here we assumed only the binary-valued databases, but extension to non binary-valued databases is also interesting problem. As described in Section 5.2, our basic scheme of the algorithm is quite strong so that we can use it in many kinds of pattern mining problems. Anyway, our result demonstrated that the state-of-the-art enumeration techniques of pattern mining can be a useful means to many kinds of statistical problems.

References

1. Agrawal, R., Imielinski, T., Swami, A.N.: Mining association rules between sets of items in large databases. In: Buneman, P., Jajodia, S. (eds.) Proc. of the 1993 ACM SIGMOD International Conference on Management of Data. SIGMOD Record, vol. 22(2), pp. 207–216 (1993)
2. Arimura, H., Uno, T.: A polynomial space and polynomial delay algorithm for enumeration of maximal motifs in a sequence. In: Deng, X., Du, D.-Z. (eds.) ISAAC 2005. LNCS, vol. 3827, pp. 724–737. Springer, Heidelberg (2005)
3. Arimura, H., Uno, T., Shimozono, S.: Time and space efficient discovery of maximal geometric graphs. In: Corruble, V., Takeda, M., Suzuki, E. (eds.) DS 2007. LNCS (LNAI), vol. 4755, pp. 42–55. Springer, Heidelberg (2007)
4. Asai, T., Abe, K., Kawasoe, S., Sakamoto, H., Arimura, H., Arikawa, S.: Efficient substructure discovery from large semi-structured data. IEICE Trans. on Information and Systems E87-D(12), 2754–2763 (2004)
5. Asai, T., Arimura, H., Uno, T., Nakano, S.-i.: Discovering frequent substructures in large unordered trees. In: Grieser, G., Tanaka, Y., Yamamoto, A. (eds.) DS 2003. LNCS (LNAI), vol. 2843, pp. 47–61. Springer, Heidelberg (2003)
6. Boley, M., Gärtner, T., Grosskreutz, H., Fraunhofer, I.: Formal concept sampling for counting and threshold-free local pattern mining. In: Proc. of 2010 SIAM International Conference on Data Mining (SDM 2010), pp. 177–188 (April 2010)
7. Bonferroni, C.: Teoria statistica delle classi e calcolo delle probabilita. Pubblicazioni del R Istituto Superiore di Scienze Economiche e Commerciali di Firenze (Libreria Internazionale Seeber, Florence, Italy), 8: Article 3–62 (1936)
8. Gallo, A., De Bie, T., Cristianini, N.: MINI: Mining informative non-redundant itemsets. In: Kok, J.N., Koronacki, J., Lopez de Mantaras, R., Matwin, S., Mladenič, D., Skowron, A. (eds.) PKDD 2007. LNCS (LNAI), vol. 4702, pp. 438–445. Springer, Heidelberg (2007)
9. Goethals, B.: Survey on frequent pattern mining (2003), http://www.cs.helsinki.fi/u/goethals/publications/survey.ps
10. Goethals, B., Zaki, M.J.: Frequent itemset mining dataset repository. In: Frequent Itemset Mining Implementations, FIMI 2003 (2003), http://fimi.cs.helsinki.fi/
11. Nature Publishing Group. Nature guide to authors: Statistical checklist, http://www.nature.com/nature/authors/gta/Statistical_checklist.doc

12. Han, J., Pei, J., Yin, Y.: Mining frequent patterns without candidate generation. In: Proc of the 2000 ACM SIGMOD International Conference on Management of Data, pp. 1–12 (2000)
13. Han, J., Pei, J., Yin, Y., Mao, R.: Mining frequent patterns without candidate generation: A frequent-pattern tree approach. Data Mining and Knowledge Discovery 8(1), 53–87 (2004)
14. Inokuchi, A., Washio, T., Motoda, H.: An apriori-based algorithm for mining frequent substructures from graph data. In: Zighed, D.A., Komorowski, J., Żytkow, J.M. (eds.) PKDD 2000. LNCS (LNAI), vol. 1910, pp. 13–23. Springer, Heidelberg (2000)
15. Low-Kam, C., Raissi, C., Kaytoue, M., Pei, J.: Mining statistically significant sequential patterns. In: Proc. of 13th IEEE International Conference on Data Mining (ICDM 2013), pp. 488–497 (2013)
16. Okita, K., Ichisaka, T., Yamanaka, S.: Generation of germline-competent induced pluripotent stem cells. Nature 448, 313–317 (2007)
17. van der Laan, M.J., Dudoit, S.: Multiple Testing Procedures with Applications to Genomics. Springer, New York (2008)
18. Tatti, N.: Maximum entropy based significance of itemsets. Knowledge and Information Systems 17(1), 57–77 (2008)
19. Terada, A., Okada-Hatakeyama, M., Tsuda, K., Sese, J.: LAMP limitless-arity multiple-testing procedure (2013), http://a-terada.github.io/lamp/
20. Terada, A., Okada-Hatakeyama, M., Tsuda, K., Sese, J.: Statistical significance of combinatorial regulations. Proceedings of National Academy of Sciences of United States of America 110(32), 12996–13001 (2013)
21. Uno, T.: Program codes of takeaki uno, http://research.nii.ac.jp/~uno/codes.htm
22. Uno, T., Kiyomi, M., Arimura, H.: LCM ver.2: Efficient mining algorithms for frequent/closed/maximal itemsets. In: Proc. IEEE ICDM 2004 Workshop FIMI 2004 (International Conference on Data Mining, Frequent Itemset Mining Implementations) (2004)
23. Uno, T., Kiyomi, M., Arimura, H.: LCM ver.3: Collaboration of array, bitmap and prefix tree for frequent itemset mining. In: Proc. Open Source Data Mining Workshop on Frequent Pattern Mining Implementations 2005 (2005)
24. Uno, T., Uchida, Y., Asai, T., Arimura, H.: LCM: An efficient algorithm for enumerating frequent closed item sets. In: Proc. Workshop on Frequent Itemset Mining Implementations, FIMI 2003 (2003), http://fimi.cs.helsinki.fi/src/
25. Wang, J., Han, J.: Bide: Efficient mining of frequent closed sequences. In: Proc. of 4th IEEE International Conference on Data Mining (ICDM 2004), pp. 79–90 (2007)
26. Webb, G.I.: Discovering significant patterns. Machine Learning 68(1), 1–33 (2007)
27. Zaki, M.J.: Scalable algorithms for association mining. IEEE Trans. Knowl. Data Eng. 12(2), 372–390 (2000)

Large-Scale Multi-label Text Classification — Revisiting Neural Networks

Jinseok Nam[1,2], Jungi Kim[1], Eneldo Loza Mencía[1],
Iryna Gurevych[1,2], and Johannes Fürnkranz[1]

[1] Department of Computer Science, Technische Universität Darmstadt, Germany
[2] Knowledge Discovery in Scientific Literature,
German Institute for Educational Research, Germany

Abstract. Neural networks have recently been proposed for multi-label classification because they are able to capture and model label dependencies in the output layer. In this work, we investigate limitations of BP-MLL, a neural network (NN) architecture that aims at minimizing pairwise ranking error. Instead, we propose to use a comparably simple NN approach with recently proposed learning techniques for large-scale multi-label text classification tasks. In particular, we show that BP-MLL's ranking loss minimization can be efficiently and effectively replaced with the commonly used cross entropy error function, and demonstrate that several advances in neural network training that have been developed in the realm of deep learning can be effectively employed in this setting. Our experimental results show that simple NN models equipped with advanced techniques such as rectified linear units, dropout, and AdaGrad perform as well as or even outperform state-of-the-art approaches on six large-scale textual datasets with diverse characteristics.

1 Introduction

As the amount of textual data on the web and in digital libraries is increasing rapidly, the need for augmenting unstructured data with metadata is also increasing. Time- and cost-wise, a manual extraction of such information from ever-growing document collections is impractical.

Multi-label classification is an automatic approach for addressing such problems by learning to assign a suitable subset of categories from an established classification system to a given text. In the literature, one can find a number of multi-label classification approaches for a variety of tasks in different domains such as bioinformatics [1], music [28], and text [9]. In the simplest case, multi-label classification may be viewed as a set of binary classification tasks that decides for each label independently whether it should be assigned to the document or not. However, this so-called *binary relevance* approach ignores dependencies between the labels, so that current research in multi-label classification concentrates on the question of how such dependencies can be exploited [23, 3]. One such approach is BP-MLL [32], which formulates multi-label classification problems as a neural network with multiple output nodes, one for each label. The output layer is able to model dependencies between the individual labels.

T. Calders et al. (Eds.): ECML PKDD 2014, Part II, LNCS 8725, pp. 437–452, 2014.
© Springer-Verlag Berlin Heidelberg 2014

In this work, we directly build upon BP-MLL and show how a simple, single hidden layer NN may achieve a state-of-the-art performance in large-scale multi-label text classification tasks. The key modifications that we suggest are (i) more efficient and more effective training by replacing BP-MLL's pairwise ranking loss with cross entropy and (ii) the use of recent developments in the area of deep learning such as rectified linear units (ReLUs), Dropout, and AdaGrad.

Even though we employ techniques that have been developed in the realm of deep learning, we nevertheless stick to single-layer NNs. The motivation behind this is two-fold: first, a simple network configuration allows better scalability of the model and is more suitable for large-scale tasks on textual data[1]. Second, as it has been shown in the literature [15], popular feature representation schemes for textual data such as variants of *tf-idf* term weighting already incorporate a certain degree of higher dimensional features, and we speculate that even a single-layer NN model can work well with text data. This paper provides an empirical evidence to support that a simple NN model equipped with recent advanced techniques for training NN performs as well as or even outperforms state-of-the-art approaches on large-scale datasets with diverse characteristics.

2   Multi-label Classification

Formally, multi-label classification may be defined as follows: $X \subset \mathbb{R}^D$ is a set of M instances, each being a D-dimensional feature vector, and L is a set of labels. Each instance $\mathbf{x}$ is associated with a subset of the L labels, the so-called *relevant* labels; all other labels are *irrelevant* for this example. The task of the learner is to learn a mapping function $f : \mathbb{R}^D \rightarrow 2^L$ that assigns a subset of labels to a given instance. An alternative view is that we have to predict an L-dimensional target vector $\mathbf{y} \in \{0, 1\}^L$, where $y_i = 1$ indicates that the i-th label is relevant, whereas $y_i = 0$ indicates that it is irrelevant for the given instance.

Many algorithms have been developed for tackling this type of problem. The most straightforward way is binary relevance (BR) learning; it constructs L binary classifiers, which are trained on the L labels independently. Thus, the prediction of the label set is composed of independent predictions for individual labels. However, labels often occur together, that is, the presence of a specific label may suppress or exhibit the likelihood of other labels.

To address this limitation of BR, pairwise decomposition (PW) and label powerset (LP) approaches consider label dependencies during the transformation by either generating pairwise subproblems [9, 20] or the powerset of possible label combinations [29]. Classifier chains [23, 3] are another popular approach that extend BR by including previous predictions into the predictions of subsequent labels. [7] present a large-margin classifier, RankSVM, that minimizes a ranking loss by penalizing incorrectly ordered pairs of labels. This setting can be used for multi-label classification by assuming that the ranking algorithm has to rank each relevant label before each irrelevant label.

---

[1] Deep NNs, in fact, scale well and work effectively by learning features from raw inputs which are usually smaller than hand-crafted features extracted from the raw inputs. However, in our case, the dimensions of raw inputs are relatively large where training deep NNs is costly.

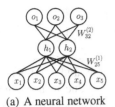

(a) A neural network

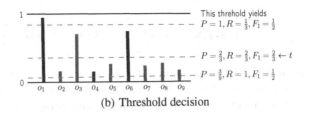

(b) Threshold decision

Fig. 1. (a) a neural network with a single hidden layer of two units and multiple output units, one for each possible label. (b) shows how threshold for a training example is estimated based on prediction output **o** of the network. Consider nine possible labels, of which o_1, o_4 and o_6 are relevant labels (blue) and the rest are irrelevant (red). The figure shows three exemplary threshold candidates (dashed lines), of which the middle one is the best choice because it gives the highest F1 score. See Section 3.3 for more details.

In order to make a prediction, the ranking has to be *calibrated* [9], i.e., a threshold has to be found that splits the ranking into relevant and irrelevant labels.

2.1 State-of-the-art Multi-label Classifiers and Limitations

The most prominent learning method for multi-label text classification is to use a BR approach with strong binary classifiers such as SVMs [24, 30] despite its simplicity. It is known that characteristics of high-dimensional and sparse data, such as text data, make decision problems linearly separable [15], and this characteristic suits the strengths of SVM classifiers well. Unlike benchmark datasets, real-world text collections consist of a large number of training examples represented in a high-dimensional space with a large amount of labels. To handle such datasets, researchers have derived efficient *linear* SVMs [16, 8] that can handle large-scale problems. However, their performance decreases as the number of labels grows and the label frequency distribution becomes skewed [19, 24]. In such cases, it is intractable to employ methods that minimize ranking errors among labels [7, 32] or that learn probability distributions of labels [11, 3].

3 Neural Networks for Multi-label Classification

In this section, we propose a neural network-based multi-label classification framework that is composed of a single hidden layer and operates with recent developments in neural network and optimization techniques, which allow the model to converge into good regions of the error surface in a few steps of parameter updates. Our approach consists of two modules (Fig. 1): a neural network that produces label scores (Sections 3.2–3.5), and a label predictor that converts label scores into binary (Section 3.3).

3.1 Rank Loss

The most intuitive objective for multi-label learning is to minimize the number of mis-ordering between a pair of relevant label and irrelevant label, which is called *rank loss*:

$$L(\mathbf{y}, f(\mathbf{x})) = w(\mathbf{y}) \sum_{y_i < y_j} \mathbb{I}\left(f_i(\mathbf{x}) > f_j(\mathbf{x})\right) + \frac{1}{2}\mathbb{I}\left(f_i = f_j\right) \tag{1}$$

where $w(\mathbf{y})$ is a normalization factor, $\mathbb{I}(\cdot)$ is the indicator function, and $f_i(\cdot)$ is a prediction score for a label i. Unfortunately, it is hard to minimize due to non-convex property of the loss function. Therefore, convex surrogate losses have been proposed as alternatives to rank loss [26, 7, 32].

3.2   Pairwise Ranking Loss Minimization in Neural Networks

Let us assume that we would like to make a prediction on L labels from D dimensional input features. Consider the neural network model with a single hidden layer in which F hidden units are defined and input units $\mathbf{x} \in \mathbb{R}^{D \times 1}$ are connected to hidden units $\mathbf{h} \in \mathbb{R}^{F \times 1}$ with weights $\mathbf{W}^{(1)} \in \mathbb{R}^{F \times D}$ and biases $\mathbf{b}^{(1)} \in \mathbb{R}^{F \times 1}$. The hidden units are connected to output units $\mathbf{o} \in \mathbb{R}^{L \times 1}$ through weights $\mathbf{W}^{(2)} \in \mathbb{R}^{L \times F}$ and biases $\mathbf{b}^{(2)} \in \mathbb{R}^{L \times 1}$. The network, then, can be written in a matrix-vector form, and we can construct a feed-forward network $f_\Theta : \mathbf{x} \to \mathbf{o}$ as a composite of non-linear functions in the range $[0, 1]$:

$$f_\Theta(\mathbf{x}) = f_o\left(\mathbf{W}^{(2)} f_h\left(\mathbf{W}^{(1)}\mathbf{x} + \mathbf{b}^{(1)}\right) + \mathbf{b}^{(2)}\right) \tag{2}$$

where $\Theta = \{\mathbf{W}^{(1)}, \mathbf{b}^{(1)}, \mathbf{W}^{(2)}, \mathbf{b}^{(2)}\}$, and f_o and f_h are *element-wise* activation functions in the output layer and the hidden layer, respectively. Specifically, the function $f_\Theta(\mathbf{x})$ can be re-written as follows:

$$\mathbf{z}^{(1)} = \mathbf{W}^{(1)}\mathbf{x} + \mathbf{b}^{(1)}, \quad \mathbf{h} = f_h\left(\mathbf{z}^{(1)}\right)$$

$$\mathbf{z}^{(2)} = \mathbf{W}^{(2)}\mathbf{h} + \mathbf{b}^{(2)}, \quad \mathbf{o} = f_o\left(\mathbf{z}^{(2)}\right)$$

where $\mathbf{z}^{(1)}$ and $\mathbf{z}^{(2)}$ denote the weighted sum of inputs and hidden activations, respectively. Our aim is to find a parameter vector Θ that minimizes a cost function $J(\Theta; \mathbf{x}, \mathbf{y})$. The cost function measures discrepancy between predictions of the network and given targets $\mathbf{y}$.

BP-MLL [32] minimizes errors induced by incorrectly ordered pairs of labels, in order to exploit dependencies among labels. To this end, it introduces a *pairwise error function* (PWE), which is defined as follows:

$$J_{PWE}(\Theta; \mathbf{x}, \mathbf{y}) = \frac{1}{|\mathbf{y}||\bar{\mathbf{y}}|} \sum_{(p,n) \in \mathbf{y} \times \bar{\mathbf{y}}} \exp(-(o_p - o_n)) \tag{3}$$

where p and n are positive and negative label index associated with training example $\mathbf{x}$. $\bar{\mathbf{y}}$ represents a set of negative labels and $|\cdot|$ stands for the cardinality. The PWE is the relaxation of the loss function in Equation 1 that we want to minimize.

As no closed-form solution exists to minimize the cost function, we use a gradient-based optimization method.

$$\Theta^{(\tau+1)} = \Theta^{(\tau)} - \eta \nabla_{\Theta^{(\tau)}} J(\Theta^{(\tau)}; \mathbf{x}, \mathbf{y}) \tag{4}$$

The parameter Θ is updated by adding a small step of negative gradients of the cost function $J(\Theta^{(\tau)}; \mathbf{x}, \mathbf{y})$ with respect to the parameter Θ at step τ. The parameter η, called the learning rate, determines the step size of updates.

3.3 Thresholding

Once training of the neural network is finished, its output can be used to rank labels, but additional measures are needed in order to split the ranking into relevant and irrelevant labels. For transforming the ranked list of labels into a set of binary predictions, we train a multi-label threshold predictor from training data. This sort of thresholding methods are also used in [7, 32]

For each document $\mathbf{x}_m$, labels are sorted by the probabilities in decreasing order. Ideally, if NNs successfully learn a mapping function f_Θ, all correct (positive) labels will be placed on top of the sorted list and there should be a large margin between the set of positive labels and the set of negative labels. Using F_1 score as a reference measure, we calculate classification performances at every pair of successive positive labels and choose a threshold value t_m that produces the best performance (Figure 1 (b)).

Afterwards, we can train a multi-label thresholding predictor $\hat{\mathbf{t}} = T(\mathbf{x}; \theta)$ to learn $\mathbf{t}$ as target values from input pattern $\mathbf{x}$. We use linear regression with $\ell2$-regularization to learn θ

$$J(\theta) = \frac{1}{2M} \sum_{m=1}^{M} (T(\mathbf{x}_m; \theta) - t_i)^2 + \frac{\lambda}{2} \|\theta\|_2^2 \tag{5}$$

where $\mathbf{x}_m$ is m-th document in the train data, $T(\mathbf{x}_m; \theta) = \theta^T \mathbf{x}_m$, and λ is a parameter which controls the magnitude of the $\ell2$ penalty.

At test time, these learned thresholds are used to predict a binary output $\hat{y}_{kl}$ for label l of a test document $\mathbf{x}_k$ given label probabilities o_{kl}; $\hat{y}_{kl} = 1$ if $o_{kl} > T(\mathbf{x}_k; \theta)$, otherwise 0. Due to the fact that the resulting parameter θ might get biased to the training data, the control parameter λ needs to be tuned via cross-validation.

3.4 Ranking Loss vs. Cross Entropy

BP-MLL is supposed to perform better in multi-label problems since it takes label correlations into consideration than the standard form of NN that does not. However, we have found that BP-MLL does not perform as expected in our preliminary experiments, particularly, on datasets in textual domain.

Consistency w.r.t Rank Loss. Recently, it has been claimed that none of convex loss functions including BP-MLL's loss function (Equation 3) is consistent with respect to *rank loss* which is non-convex and has discontinuity [2, 10]. Furthermore, univariate surrogate loss functions such as *log loss* are rather consistent with rank loss [4].

$$J_{log}(\Theta; \mathbf{x}, \mathbf{y}) = w(\mathbf{y}) \sum_{l} \log \left(1 + e^{-\hat{y}_l z_l} \right)$$

where $w(\mathbf{y})$ is a weighting function that normalizes loss in terms of $\mathbf{y}$ and z_l indicates prediction for label l. Please note that the log loss is often used for logistic regression in which $\hat{y} \in \{-1, 1\}$ is the target and z_l is the output of a linear function

$z_l = \sum_k W_{lk}x_k + b_l$ where W_{lk} is a weight from input x_k to output z_l and b_l is bias for label l. A typical choice is, for instance, $w(\mathbf{y}) = (\|\mathbf{y}\|\|\bar{\mathbf{y}}\|)^{-1}$ as in BP-MLL. In this work, we set $w(\mathbf{y}) = 1$, then the log loss above is equivalent to *cross entropy* (CE), which is commonly used to train neural networks for classification tasks if we use *sigmoid* transfer function in the output layer, i.e. $f_o(z) = 1/(1 + \exp(-z))$, or simply $f_o(z) = \sigma(z)$:

$$J_{CE}(\Theta; \mathbf{x}, \mathbf{y}) = -\sum_l (y_l \log o_l) + (1 - y_l) \log(1 - o_l)) \tag{6}$$

where o_l and y_l are the prediction and the target for label l, respectively. Let us verify the equivalence between the *log loss* and the *CE*. Consider the log loss function for only label l.

$$J_{log}(\Theta; \mathbf{x}, y_l) = \log(1 + e^{-\dot{y}_l z_l}) = -\log\left(\frac{1}{1 + e^{-\dot{y}_l z_l}}\right) \tag{7}$$

As noted, $\dot{y}$ in the log loss takes either -1 or 1, which allows us to split the above equation as follows:

$$-\log\left(\frac{1}{1 + e^{-\dot{y}_l z_l}}\right) = \begin{cases} -\log(\sigma(z_l)) & \text{if } \dot{y} = 1 \\ -\log(\sigma(-z_l)) & \text{if } \dot{y} = -1 \end{cases} \tag{8}$$

Then, we have the corresponding CE by using a property of the sigmoid function $\sigma(-z) = 1 - \sigma(z)$

$$J_{CE}(\Theta; \mathbf{x}, y_l) = -(y_l \log o_l + (1 - y_l) \log(1 - o_l)) \tag{9}$$

where $y \in \{0, 1\}$ and $o_l = \sigma(z_l)$.

Computational Expenses. In addition to consistency with rank loss, CE has an advantage in terms of computational efficiency; computational cost for computing gradients of parameters with respect to PWE is getting more expensive as the number of labels grows. The error term $\delta_l^{(2)}$ for label l which is propagated to the hidden layer is defined as

$$\delta_l^{(2)} = \begin{cases} -\frac{1}{|\mathbf{y}||\bar{\mathbf{y}}|} \sum_{n \in \bar{\mathbf{y}}} \exp(-(o_l - o_n)) f_o'(z_l^{(2)}), & \text{if } l \in \mathbf{y} \\ \frac{1}{|\mathbf{y}||\bar{\mathbf{y}}|} \sum_{p \in \mathbf{y}} \exp(-(o_p - o_l)) f_o'(z_l^{(2)}), & \text{if } l \in \bar{\mathbf{y}} \end{cases} \tag{10}$$

Whereas the computation of $\delta_l^{(2)} = -y_l/o_l + (1 - y_l)/(1 - o_l) f_o'(z_l^{(2)})$ for the CE can be performed efficiently, obtaining error terms $\delta_l^{(2)}$ for the PWE is L times more expensive than one in ordinary NN utilizing the cross entropy error function. This also shows that BP-MLL scales poorly w.r.t. the number of unique labels.

Plateaus. To get an idea of how differently both objective functions behave as a function of parameters to be optimized, let us draw graphs containing cost function values. Note that it has been pointed out that the slope of the cost function as a function of the parameters plays an important role in learning parameters of neural networks [27, 12].

Consider two-layer neural networks consisting of $W^{(1)} \in \mathbb{R}$ for the first layer, $\mathbf{W}^{(2)} \in \mathbb{R}^{4 \times 1}$ for the second, that is, the output layer. Since we are interested in function values with respect to two parameters $W^{(1)}$ and $W_1^{(2)}$ out of 5 parameters, $\mathbf{W}_{\{2,3,4\}}^{(2)}$

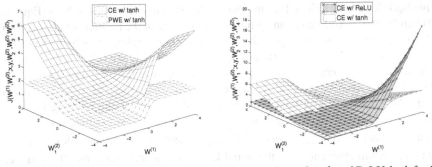

(a) Comparison of CE and PWE   (b) Comparison of tanh and ReLU, both for CE

Fig. 2. Landscape of cost functions and a type of hidden units. $W^{(1)}$ represents a weight connecting an input unit to a hidden unit. Likewise, $W_1^{(2)}$ denotes a weight from the hidden unit to output unit 1. The z-axis stands for a value for the cost function $J(W^{(1)}, W_1^{(2)}; \mathbf{x}, \mathbf{y}, W_2^{(2)}, W_3^{(2)}, W_4^{(2)})$ where instances $\mathbf{x}$, targets $\mathbf{y}$ and weights $W_2^{(2)}, W_3^{(2)}, W_4^{(2)}$ are fixed.

is set to a fixed value c. In this paper we use $c = 0.^2$ Figure 2 (a) shows different shapes of the functions and slope steepness. In figure 2 (a) both curves have similar shapes, but the curve for PWE has plateaus in which gradient descent can be very slow in comparison with the CE. Figure 2 (b) shows that CE with ReLUs, which is explained the next Section, has a very steep slope compared to CE with tanh. Such a slope can accelerate convergence speed in learning parameters using gradient descent. We conjecture that these properties might explain why our set-up converges faster than the other configurations, and BP-MLL performs poorly in most cases in our experiments.

3.5   Recent Advances in Deep Learning

In recent neural network and deep learning literature, a number of techniques were proposed to overcome the difficulty of learning neural networks efficiently. In particular, we make use of ReLUs, AdaGrad, and Dropout training, which are briefly discussed in the following.

Rectified Linear Units. *Rectified linear units* (ReLUs) have been proposed as activation units on the hidden layer and shown to yield better generalization performance [22, 13, 31]. A ReLU disables negative activation ($\text{ReLU}(x) = \max(0, x)$) so that the number of parameters to be learned decreases during the training. This sparsity characteristic makes ReLUs advantageous over the traditional activation units such as *sigmoid* and *tanh* in terms of the generalization performance.

---

^{2} The shape of the functions is not changed even if we set c to arbitrary value since it is drawn by function values in z-axis with respect to only $W^{(1)}$ and $W_1^{(2)}$.

Learning Rate Adaptation with AdaGrad. *Stochastic gradient descent* (SGD) is a simple but effective technique for minimizing the objective functions of NNs (Equation 4). When SGD is considered as an optimization tool, one of the problems is the choice of the learning rate. A common approach is to estimate the learning rate which gives lower training errors on subsamples of training examples [17] and then decrease it over time. Furthermore, to accelerate learning speed of SGD, one can utilize momentum [25].

Instead of a fixed or scheduled learning rate, an *adaptive learning rate* method, namely AdaGrad, was proposed [6]. The method determines the learning rate at iteration τ by keeping previous gradients $\Delta_{1:\tau}$ to compute the learning rate for each dimension of parameters $\eta_{i,\tau} = \eta_0 / \sqrt{\sum_{t=1}^{\tau} \Delta_{i,t}^2}$ where i stands for an index of each dimension of parameters and η_0 is the initial learning rate and shared by all parameters. For multi-label learning, it is often the case that a few labels occur frequently, whereas the majority only occurs rarely, so that the rare ones need to be updated with larger steps in the direction of the gradient. If we use AdaGrad, the learning rates for the frequent labels decreases because the gradient of the parameter for the frequent labels will get smaller as the updates proceed. On the other hand, the learning rates for rare labels remain comparatively large.

Regularization using Dropout Training. In principle, as the number of hidden layers and hidden units in a network increases, its expressive power also increases. If one is given a large number of training examples, training a larger networks will result in better performance than using a smaller one. The problem when training such a large network is that the model is more prone to getting stuck in local minima due to the huge number of parameters to learn. Dropout training [14] is a technique for preventing overfitting in a huge parameter space. Its key idea is to decouple hidden units that activate the same output together, by randomly dropping some hidden units' activations. Essentially, this corresponds to training an ensemble of networks with smaller hidden layers, and combining their predictions. However, the individual predictions of all possible hidden layers need not be computed and combined explicitly, but the output of the ensemble can be approximately reconstructed from the full network. Thus, dropout training has a similar regularization effect as ensemble techniques.

4   Experimental Setup

We have shown the reason why the structure of NNs needs to be reconsidered in the previous Sections. In this Section, we describe evaluation measures to show how effectively NNs perform by combining recent development in learning neural networks based on the fact that the *univariate* loss is consistent with respect to rank loss on large-scale textual datasets.

Evaluation Measures. Multi-label classifiers can be evaluated in two groups of measures: bipartition and ranking. Bipartition measures operate on classification results, i.e. a set of labels assigned by classifiers to each document, while ranking measures operate on the ranked list of labels. In order to evaluate the quality of a ranked list, we consider several ranking measures [26]. Given a document $\mathbf{x}$ and associated label information $\mathbf{y}$, consider a multi-label learner $f_\theta(\mathbf{x})$ that is able to produce scores for each label. These scores, then, can be sorted in descending order. Let $r(l)$ be the rank of a label l in the sorted list of labels. We already introduced *Rank loss*, which is concerned primarily in this work, in Section 3.1. *One-Error* evaluates whether the topmost ranked label with the highest score is a positive label or not: $\mathbb{I}\left(r^{-1}(1)(f_\theta(x)) \notin \mathbf{y}\right)$ where $r^{-1}(1)$ indicates the index of a label positioning on the first place in the sorted list. *Coverage* measures on average how far one needs to go down the ranked list of labels to achieve recall of 100%: $\max_{l_i \in \mathbf{y}} r(l_i) - 1$. *Average Precision* or AP measures the average fraction of labels preceding relevant labels in the ranked list of labels: $\frac{1}{|\mathbf{y}|} \sum_{l_i \in \mathbf{y}} \frac{|\{l_j \in \mathbf{y}|r(l_j) \leq r(l_i)\}|}{r(l_i)}$.

For bipartition measures, Precision, Recall, and F_1 score are conventional methods to evaluate effectiveness of information retrieval systems. There are two ways of computing such performance measures: *Micro-averaged* measures and *Macro-averaged* measures[3][21].

$$P_{micro} = \frac{\sum_{l=1}^{L} tp_l}{\sum_{l=1}^{L} tp_l + fp_l}, R_{micro} = \frac{\sum_{l=1}^{L} tp_l}{\sum_{l=1}^{L} tp_l + fn_l}, F_{1-micro} = \frac{\sum_{l=1}^{L} 2tp_l}{\sum_{l=1}^{L} 2tp_l + fp_l + fn_l}$$

$$P_{macro} = \frac{1}{L} \sum_{l=1}^{L} \frac{tp_l}{tp_l + fp_l}, R_{macro} = \frac{1}{L} \sum_{l=1}^{L} \frac{tp_l}{tp_l + fn_l}, F_{1-macro} = \frac{1}{L} \sum_{l=1}^{L} \frac{2tp_l}{2tp_l + fp_l + fnl}$$

Datasets. Our main interest is in large-scale text classification, for which we selected six representative domains, whose characteristics are summarized in Table 1. For Reuters21578, we used the same training/test split as previous works [30]. Training and test data were switched for RCV1-v2 [18] which originally consists of 23,149 train and 781,265 test documents. The EUR-Lex, Delicious and Bookmarks datasets were taken from the MULAN repository.[4] Except for Delicious and Bookmarks, all documents are represented with *tf-idf* features with cosine normalization such that length of the document vector is 1 in order to account for the different document lengths.

In addition to these standard benchmark datasets, we prepared a large-scale dataset from documents of the German Education Index (GEI).[5] The GEI is a database of links to more than 800,000 scientific articles with metadata, e.g. title, authorship, language of an article and index terms. We consider a subset of the dataset consisting of approximately 300,000 documents which have an abstract as well as the metadata. Each document has multiple index terms which are carefully hand-labeled by human experts with respect to the content of the articles. We processed plain text by removing stopwords and stemming each token. To avoid the computational bottleneck from a large number of labels, we chose the 1,000 most common labels out of about 50,000. We then randomly split the dataset into 90% for training and 10% for test.

---

[3] Note that scores computed by micro-averaged measures might be much higher than that by macro-averaged measures if there are many rarely-occurring labels for which the classification system does not perform well. This is because macro-averaging weighs each label equally, whereas micro-averaged measures are dominated by the results of frequent labels.

[4] http://mulan.sourceforge.net/datasets.html

[5] http://www.dipf.de/en/portals/portals-educational-information/
german-education-index

Table 1. Number of documents (D), size of vocabulary (D), total number of labels (L) and average number of labels per instance (C) for the six datasets used in our study

Dataset	M	D	L	C
Reuters-21578	10789	18637	90	1.13
RCV1-v2	804414	47236	103	3.24
EUR-Lex	19348	5000	3993	5.31
Delicious	16105	500	983	19.02
Bookmarks	87856	2150	208	2.03
German Education Index	316061	20000	1000	7.16

Algorithms. Our main goal is to compare our NN-based approach to BP-MLL. NN_A stands for the single hidden layer neural networks which have *ReLUs* for its hidden layer and which are trained with SGD where each parameter of the neural networks has their own learning rate using *AdaGrad*. NN_{AD} additionally employs *Dropout* based on the same settings as NN_A. T and R following BP-MLL indicate *tanh* and *ReLU* as a transfer function in the hidden layer. For both NN and BP-MLL, we used 1000 units in the hidden layer over all datasets. [6] As Dropout works well as a regularizer, no additional regularization to prevent overfitting was incorporated. The base learning rate η_0 was also determined among $[0.001, 0.01, 0.1]$ using validation data.

We also compared the NN-based algorithms to binary relevance (BR) using SVMs (Liblinear) as a base learner, as a representative of the state-of-the-art. The penalty parameter C was optimized in the range of $[10^{-3}, 10^{-2}, \ldots, 10^2, 10^3]$ based on either average of micro- and macro-average F_1 or rankloss on validation set. BR_B refers to linear SVMs where C is optimized with bipartition measures on the validation dataset. BR models whose penalty parameter is optimized on ranking measures are indicated as BR_R. In addition, we apply the same thresholding technique which we utilize in our NN approach (Section 3.3) on a ranked list produced by BR models (BR_R). Given a document, the distance of each predicted label to the hyperplane is used to determine the position of the label in the ranked list.

5 Results

We evaluate our proposed models and other baseline systems on datasets with varying statistics and characteristics. We first show experimental results that confirm that the techniques discussed in Section 3.5 actually contribute to an increased performance of NN-based multi-label classification, and then compare all algorithms on the six above-mentioned datasets in order to get an overall impression of their performance.

Better Local Minima and Acceleration of Convergence Speed. First we intend to show the effect of ReLUs and AdaGrad in terms of convergence speed and rank loss. The left part of Figure 3 shows that all three results of AdaGrad (red lines) show a lower

---

[6] The optimal number of hidden units of BP-MLL and NN was tested among 20, 50, 100, 500, 1000 and 4000 on validation datasets. Usually, the more units are in the hidden layer, the better performance of networks is. We chose it in terms of computational efficiency.

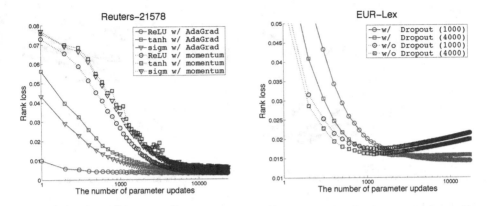

Fig. 3. (*left*) effects of AdaGrad and momentum on three types of transfer functions in the hidden layers in terms of rank loss on Reuters-21578. The number of parameter updates in *x*-axis corresponds to the number of evaluations of Eq. (4). (*right*) effects of dropout with two different numbers of hidden units in terms of rank loss on EUR-Lex.

rank loss than all three versions of momentum. Moreover, within each group, ReLUs outperform the versions using tanh or sigmoid activation functions. That NNs with ReLUs at the hidden layer converge faster into a better weight space has been previously observed for the speech domain [31].[7] This faster convergence is a major advantage of combining recently proposed learning components such as ReLUs and AdaGrad, which facilitates a quicker learning of the parameters of NNs. This is particularly important for the large-scale text classification problems that are the main focus of this work.

Decorrelating Hidden Units While Output Units Remain Correlated. One major goal of multi-label learners is to minimize rank loss by leveraging inherent correlations in a label space. However, we conjecture that these correlations also may cause overfitting because if groups of hidden units specialize in predicting particular label subsets that occur frequently in the training data, it will become harder to predict novel label combinations that only occur in the test set. Dropout effectively fights this by randomly dropping individual hidden units, so that it becomes harder for groups of hidden units to specialize in the prediction of particular output combinations, i.e., they decorrelate the hidden units, whereas the correlation of output units still remains. Particularly, a subset of output activations $\mathbf{o}$ and hidden activations $\mathbf{h}$ would be correlated through $\mathbf{W}^{(2)}$.

We observed overfitting across all datasets except for Reuters-21578 and RCV1-v2 under our experimental settings. The right part of Figure 3 shows how well Dropout prevents NNs from overfitting on the test data of EUR-Lex. In particular, we can see that with increasing numbers of parameter updates, the performance of regular NNs eventually got worse in terms of rank loss. On the other hand, when dropout is employed, convergence is initially slower, but eventually effectively prevents overfitting.

---

[7] However, unlike the results of [31], in our preliminary experiments adding more hidden layers did not further improve generalization performance.

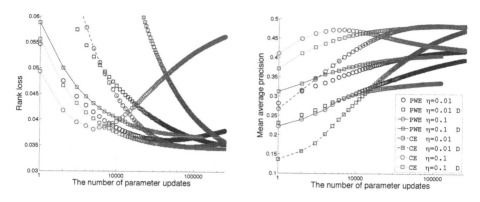

Fig. 4. Rankloss (left) and mean average precision (right) on the German Education Index test data for the different cost functions. η denotes the base learning rate and D indicates that Dropout is applied. Note that x-axis is in log scale.

Limiting Small Learning Rates in BP-MLL. The learning rate strongly influences convergence and learning speed [17]. As we have already seen in the Figure 2, the slope of PWE is less steep than CE, which implies that smaller learning rates should be used. Specifically, we observed PWE allows only smaller learning rate 0.01 (blue markers) in contrast with CE that works well a relatively larger learning rate 0.1 (red markers) in Figure 4. In the case of PWE with the larger learning rate (green markers), interestingly, dropout (rectangle markers in green) makes it converge towards much better local minima, yet it is still worse than the other configurations. It seems that the weights of BP-MLL oscillates in the vicinity of local minima and, indeed, converges *worse* local minima. However, it makes learning procedure of BP-MLL slow compared to NNs with CE making bigger steps for parameter updates.

With respect to Dropout, Figure 4 also shows that for the same learning rates, networks without Dropout converge much faster than ones working with Dropout in terms of both rank loss and MAP. Regardless of the cost functions, overfitting arises over the networks without Dropout and it is likely that overfitting is avoided effectively as discussed earlier.[8]

Comparison of Algorithms. Table 3 shows detailed results of all experiments with all algorithms on all six datasets, except that we could not obtain results of BP-MLL on EUR-Lex within a reasonable time frame. In an attempt to summarize the results, Table 2 shows the average rank of each algorithm in these six datasets according to all ranking an bipartition measures discussed in Section 4 [9].

We can see that although BP-MLL focuses on minimizing pairwise ranking errors, thereby capturing label dependencies, the single hidden layer NNs with cross-entropy

---

[8] A trajectory for PWE $\eta = 0.1$ is missing in the figure because it got 0.2 on the rankloss measure which is much worse than the other configurations.

[9] The Friedman test is passed for $\alpha = 1\%$ except for micro and macro recall ($\alpha = 10\%$) [5]. Nemenyi's critical distance between the average ranks, for which a statistical difference can be detected, is 4.7 for $\alpha = 5\%$ (4.3 for $\alpha = 10\%$).

Table 2. Average ranks of the algorithms on ranking and bipartition measures

Eval. measures	Ranking				Bipartition					
	rankloss	oneError	Coverage	MAP	miP	miR	miF	maP	maR	maF
Average Ranks										
NN$_A$	2.2	2.4	2.6	2.2	**2.0**	6.0	2.4	**1.8**	5.6	**2.0**
NN$_{AD}$	**1.2**	**1.4**	**1.2**	**1.6**	**2.0**	5.8	**1.8**	2.0	5.6	2.2
BP-MLL$_{TA}$	5.2	7.2	6.0	6.4	7.0	3.2	7.0	6.2	**2.0**	5.6
BP-MLL$_{TAD}$	4.1	6.0	4.4	5.9	7.4	**2.8**	7.4	7.2	3.2	7.0
BP-MLL$_{RA}$	5.9	6.7	5.6	6.4	5.2	3.2	4.6	5.6	3.8	4.8
BP-MLL$_{RAD}$	4.0	6.0	3.6	5.6	5.6	3.6	5.4	5.4	4.4	5.8
BR$_B$	7.4	3.3	6.9	4.3	3.2	6.8	4.6	4.4	6.8	5.6
BR$_R$	6.0	3.0	5.7	3.6	3.6	4.6	2.8	3.4	4.6	3.0

minimization (i.e., NN$_A$ and NN$_{AD}$) work much better not only on rank loss but also on other ranking measures. The binary relevance (BR) approaches show acceptable performance on ranking measures even though label dependency was ignored during the training phase. In addition, NN$_A$ and NN$_{AD}$ perform as good as or better than other methods on bipartition measures as well as on ranking measures.

We did not observe significant improvements by replacing hidden units of BP-MLL from tanh to ReLU. However, if we change the cost function in the previous setup from PWE to CE, significant improvements were obtained. Because BP-MLL$_{RAD}$ is the same architecture as NN$_{AD}$ except for its cost function,[10] we can say that the differences in the effectiveness of NNs and BP-MLL are due to the use of different cost functions. This also implies that the main source of improvements for NNs against BP-MLL is replacement of the cost function. Again, Figure 4 shows the difference between two cost functions more explicitly.

6   Conclusion

This paper presents a multi-label classification framework based on a neural network and a simple threshold label predictor. We found that our approach outperforms BP-MLL, both in predictive performance as well as in computational complexity and convergence speed. We have explored why BP-MLL as a multi-label text classifier does not perform well, and provided an empirical confirmation of a recent theoretical result that univariate losses might be more useful than bivariate losses for optimizing rank performance. Our experimental results showed the proposed framework is an effective method for the multi-label text classification task. Also, we have conducted extensive analysis to characterize the effectiveness of combining ReLUs with AdaGrad for fast convergence rate, and utilizing Dropout to prevent overfitting which results in better generalization.

Acknowledgments. The authors would like to thank the anonymous reviewers for their valuable comments. This work has been supported by the German Institute for Educational Research (DIPF) under the Knowledge Discovery in Scientific Literature

---

[10] For PWE we use *tanh* in the output layer, but *sigmoid* is used for CE because predictions **o** for computing CE with targets **y** needs to be between 0 and 1.

Table 3. Results on ranking and bipartition measures. Results for BP-MLL on EUR-Lex are missing because the runs could not be completed in a reasonably short time.

Eval. measures	Ranking				Bipartition					
	rankloss	oneError	Coverage	MAP	miP	miR	miF	maP	maR	maF
Reuters-21578										
NN_A	0.0037	0.0706	0.7473	0.9484	0.8986	0.8357	0.8660	**0.6439**	0.4424	0.4996
NN_{AD}	**0.0031**	0.0689	**0.6611**	0.9499	0.9042	0.8344	0.8679	0.6150	0.4420	0.4956
$BP\text{-}MLL_{TA}$	0.0039	0.0868	0.8238	0.9400	0.7876	0.8616	0.8230	0.5609	**0.4761**	0.4939
$BP\text{-}MLL_{TAD}$	0.0039	0.0808	0.8119	0.9434	0.7945	0.8654	0.8284	0.5459	0.4685	0.4831
$BP\text{-}MLL_{RA}$	0.0054	0.0808	1.0987	0.9431	0.8205	0.8582	0.8389	0.5303	0.4364	0.4624
$BP\text{-}MLL_{RAD}$	0.0063	0.0719	1.2037	0.9476	0.8421	0.8416	0.8418	0.5510	0.4292	0.4629
BR_B	0.0040	**0.0613**	0.8092	**0.9550**	**0.9300**	0.8096	0.8656	0.6050	0.3806	0.4455
BR_R	0.0040	**0.0613**	0.8092	**0.9550**	0.8982	**0.8603**	**0.8789**	0.6396	0.4744	**0.5213**
RCV1-v2										
NN_A	0.0040	0.0218	3.1564	0.9491	0.9017	0.7836	0.8385	0.7671	0.5760	0.6457
NN_{AD}	**0.0038**	**0.0212**	**3.1108**	**0.9500**	**0.9075**	0.7813	0.8397	**0.7842**	0.5626	0.6404
$BP\text{-}MLL_{TA}$	0.0058	0.0349	3.7570	0.9373	0.6685	0.7695	0.7154	0.4385	0.5803	0.4855
$BP\text{-}MLL_{TAD}$	0.0057	0.0332	3.6917	0.9375	0.6347	0.7497	0.6874	0.3961	0.5676	0.4483
$BP\text{-}MLL_{RA}$	0.0058	0.0393	3.6730	0.9330	0.7712	0.8074	0.7889	0.5741	0.6007	0.5823
$BP\text{-}MLL_{RAD}$	0.0056	0.0378	3.6032	0.9345	0.7612	0.8016	0.7809	0.5755	0.5748	0.5694
BR_B	0.0061	0.0301	3.8073	0.9375	0.8857	0.8232	**0.8533**	0.7654	0.6342	0.6842
BR_R	0.0051	0.0287	3.4998	0.9420	0.8156	**0.8822**	0.8476	0.6961	**0.7112**	**0.6923**
EUR-Lex										
NN_A	0.0195	0.2016	310.6202	0.5975	0.6346	0.4722	0.5415	0.3847	0.3115	0.3256
NN_{AD}	**0.0164**	**0.1681**	**269.4534**	**0.6433**	**0.7124**	0.4823	**0.5752**	**0.4470**	0.3427	0.3687
BR_B	0.0642	0.1918	976.2550	0.6114	0.6124	0.4945	0.5471	0.4260	**0.3643**	**0.3752**
BR_R	0.0204	0.2088	334.6172	0.5922	0.0329	**0.5134**	0.0619	0.2323	0.3063	0.2331
German Education Index										
NN_A	**0.0350**	0.2968	138.5423	**0.4828**	0.4499	0.4200	**0.4345**	0.4110	0.3132	**0.3427**
NN_{AD}	0.0352	**0.2963**	138.3590	0.4797	0.4155	0.4472	0.4308	0.3822	0.3216	0.3305
$BP\text{-}MLL_{TA}$	0.0386	0.8309	150.8065	0.3432	0.1502	**0.6758**	0.2458	0.1507	**0.5562**	0.2229
$BP\text{-}MLL_{TAD}$	0.0371	0.7591	139.1062	0.3281	0.1192	0.5056	0.1930	0.1079	0.4276	0.1632
$BP\text{-}MLL_{RA}$	0.0369	0.4221	143.4541	0.4133	0.2618	0.4909	0.3415	0.3032	0.3425	0.2878
$BP\text{-}MLL_{RAD}$	0.0353	0.4522	**135.1398**	0.3953	0.2400	0.5026	0.3248	0.2793	0.3520	0.2767
BR_B	0.0572	0.3052	221.0968	0.4533	**0.5141**	0.2318	0.3195	0.3913	0.1716	0.2319
BR_R	0.0434	0.3021	176.6349	0.4755	0.4421	0.3997	0.4199	**0.4361**	0.2706	0.3097
Delicious										
NN_A	0.0860	0.3149	396.4659	0.4015	**0.3637**	0.4099	0.3854	0.2488	0.1721	0.1772
NN_{AD}	**0.0836**	**0.3127**	389.9422	**0.4075**	0.3617	0.4399	**0.3970**	**0.2821**	0.1777	**0.1824**
$BP\text{-}MLL_{TA}$	0.0953	0.4967	434.8601	0.3288	0.1829	0.5857	0.2787	0.1220	0.2728	0.1572
$BP\text{-}MLL_{TAD}$	0.0898	0.4358	418.3618	0.3359	0.1874	0.5884	0.2806	0.1315	0.2427	0.1518
$BP\text{-}MLL_{RA}$	0.0964	0.6157	427.0468	0.2793	0.2070	**0.5894**	0.3064	0.1479	**0.2609**	0.1699
$BP\text{-}MLL_{RAD}$	0.0894	0.6060	411.5633	0.2854	0.2113	0.5495	0.3052	0.1650	0.2245	0.1567
BR_B	0.1184	0.4355	496.7444	0.3371	0.1752	0.2692	0.2123	0.0749	0.1336	0.0901
BR_R	0.1184	0.4358	496.8180	0.3371	0.2559	0.3561	0.2978	0.1000	0.1485	0.1152
Bookmarks										
NN_A	0.0663	0.4924	22.1183	0.5323	0.3919	0.3907	0.3913	0.3564	0.3069	0.3149
NN_{AD}	**0.0629**	**0.4828**	20.9938	**0.5423**	**0.3929**	0.3996	**0.3962**	**0.3664**	0.3149	**0.3222**
$BP\text{-}MLL_{TA}$	0.0684	0.5598	23.0362	0.4922	0.0943	0.5682	0.1617	0.1115	0.4743	0.1677
$BP\text{-}MLL_{TAD}$	0.0647	0.5574	21.7949	0.4911	0.0775	**0.6096**	0.1375	0.0874	**0.5144**	0.1414
$BP\text{-}MLL_{RA}$	0.0707	0.5428	23.6088	0.5049	0.1153	0.5389	0.1899	0.1235	0.4373	0.1808
$BP\text{-}MLL_{RAD}$	0.0638	0.5322	21.5108	0.5131	0.0938	0.5779	0.1615	0.1061	0.4785	0.1631
BR_B	0.0913	0.5318	29.6537	0.4868	0.2821	0.2546	0.2676	0.1950	0.1880	0.1877
BR_R	0.0895	0.5305	28.7233	0.4889	0.2525	0.4049	0.3110	0.2259	0.3126	0.2569

(KDSL) program, in part by the Volkswagen Foundation as part of the Lichtenberg-Professorship Program under grant No. I/82806, and in part by the German Research Foundation under grant 798/1-5.

References

[1] Bi, W., Kwok, J.T.: Multi-label classification on tree-and dag-structured hierarchies. In: Proceedings of the 28th International Conference on Machine Learning, pp. 17–24 (2011)

[2] Calauzènes, C., Usunier, N., Gallinari, P.: On the (non-)existence of convex, calibrated surrogate losses for ranking. In: Advances in Neural Information Processing Systems 25, pp. 197–205 (2012)

[3] Dembczyński, K., Cheng, W., Hüllermeier, E.: Bayes optimal multilabel classification via probabilistic classifier chains. In: Proceedings of the 27th International Conference on Machine Learning, pp. 279–286 (2010)

[4] Dembczyński, K., Kotłowski, W., Hüllermeier, E.: Consistent multilabel ranking through univariate losses. In: Proceedings of the 29th International Conference on Machine Learning, pp. 1319–1326 (2012)

[5] Demšar, J.: Statistical comparisons of classifiers over multiple data sets. The Journal of Machine Learning Research 7, 1–30 (2006)

[6] Duchi, J., Hazan, E., Singer, Y.: Adaptive subgradient methods for online learning and stochastic optimization. Journal of Machine Learning Research 12, 2121–2159 (2011)

[7] Elisseeff, A., Weston, J.: A kernel method for multi-labelled classification. In: Advances in Neural Information Processing Systems 14, pp. 681–687 (2001)

[8] Fan, R.E., Chang, K.W., Hsieh, C.J., Wang, X.R., Lin, C.J.: LIBLINEAR: A library for large linear classification. Journal of Machine Learning Research 9, 1871–1874 (2008)

[9] Fürnkranz, J., Hüllermeier, E., Loza Mencía, E., Brinker, K.: Multilabel classification via calibrated label ranking. Machine Learning 73(2), 133–153 (2008)

[10] Gao, W., Zhou, Z.H.: On the consistency of multi-label learning. Artificial Intelligence 199-200, 22–44 (2013)

[11] Ghamrawi, N., McCallum, A.: Collective multi-label classification. In: Proceedings of the 14th ACM International Conference on Information and Knowledge Management, pp. 195–200 (2005)

[12] Glorot, X., Bengio, Y.: Understanding the difficulty of training deep feedforward neural networks. In: Proceedings of the 13th International Conference on Artificial Intelligence and Statistics, JMLR W&CP, pp. 249–256 (2010)

[13] Glorot, X., Bordes, A., Bengio, Y.: Deep sparse rectifier neural networks. In: Proceedings of the 14th International Conference on Artificial Intelligence and Statistics, JMLR W&CP, pp. 315–323 (2011)

[14] Hinton, G.E., Srivastava, N., Krizhevsky, A., Sutskever, I., Salakhutdinov, R.R.: Improving neural networks by preventing co-adaptation of feature detectors. ArXiv preprint ArXiv:1207.0580 (2012)

[15] Joachims, T.: Text categorization with support vector machines: Learning with many relevant features. In: Nédellec, C., Rouveirol, C. (eds.) ECML 1998. LNCS, vol. 1398, pp. 137–142. Springer, Heidelberg (1998)

[16] Joachims, T.: Training linear svms in linear time. In: Proceedings of the 12th ACM SIGKDD International Conference on Knowledge Discovery and Data Mining, pp. 217–226 (2006)

[17] LeCun, Y., Bottou, L., Orr, G.B., Müller, K.R.: Efficient backprop. In: Neural Networks: Tricks of the Trade, pp. 9–48 (2012)

[18] Lewis, D.D., Yang, Y., Rose, T.G., Li, F.: RCV1: A new benchmark collection for text categorization research. Journal of Machine Learning Research 5, 361–397 (2004)

[19] Liu, T.Y., Yang, Y., Wan, H., Zeng, H.J., Chen, Z., Ma, W.Y.: Support vector machines classification with a very large-scale taxonomy. SIGKDD Explorations 7(1), 36–43 (2005)

[20] Loza Mencía, E., Park, S.H., Fürnkranz, J.: Efficient voting prediction for pairwise multilabel classification. Neurocomputing 73(7-9), 1164–1176 (2010)

[21] Manning, C.D., Raghavan, P., Schütze, H.: Introduction to Information Retrieval. Cambridge University Press (2008)

[22] Nair, V., Hinton, G.E.: Rectified linear units improve restricted boltzmann machines. In: Proceedings of the 27th International Conference on Machine Learning, pp. 807–814 (2010)

[23] Read, J., Pfahringer, B., Holmes, G., Frank, E.: Classifier chains for multi-label classification. Machine Learning 85(3), 333–359 (2011)

[24] Rubin, T.N., Chambers, A., Smyth, P., Steyvers, M.: Statistical topic models for multi-label document classification. Machine Learning 88(1-2), 157–208 (2012)

[25] Rumelhart, D.E., Hinton, G.E., Williams, R.J.: Learning representations by back-propagating errors. Nature 323(6088), 533–536 (1986)

[26] Schapire, R.E., Singer, Y.: BoosTexter: A boosting-based system for text categorization. Machine Learning 39(2/3), 135–168 (2000)

[27] Solla, S.A., Levin, E., Fleisher, M.: Accelerated learning in layered neural networks. Complex Systems 2(6), 625–640 (1988)

[28] Trohidis, K., Tsoumakas, G., Kalliris, G., Vlahavas, I.: Multi-label classification of music into emotions. In: Proceedings of the 9th International Conference on Music Information Retrieval, pp. 325–330 (2008)

[29] Tsoumakas, G., Katakis, I., Vlahavas, I.P.: Random k-labelsets for multilabel classification. IEEE Transactions on Knowledge and Data Engineering 23(7), 1079–1089 (2011)

[30] Yang, Y., Gopal, S.: Multilabel classification with meta-level features in a learning-to-rank framework. Machine Learning 88(1-2), 47–68 (2012)

[31] Zeiler, M.D., Ranzato, M., Monga, R., Mao, M.Z., Yang, K., Le, Q.V., Nguyen, P., Senior, A., Vanhoucke, V., Dean, J., Hinton, G.E.: On rectified linear units for speech processing. In: 2013 IEEE International Conference on Acoustics, Speech and Signal Processing (ICASSP), pp. 3517–3521 (2013)

[32] Zhang, M.L., Zhou, Z.H.: Multilabel neural networks with applications to functional genomics and text categorization. IEEE Transactions on Knowledge and Data Engineering 18, 1338–1351 (2006)

Distinct Chains for Different Instances: An Effective Strategy for Multi-label Classifier Chains

Pablo Nascimento da Silva[1], Eduardo Corrêa Gonçalves[1], Alexandre Plastino[1], and Alex A. Freitas[2]

[1] Fluminense Federal University, Institute of Computing, Niteroi RJ, Brazil
{psilva,egoncalves,plastino}@ic.uff.br
[2] University of Kent, School of Computing, UK
a.a.freitas@kent.ac.uk

Abstract. Multi-label classification (MLC) is a predictive problem in which an object may be associated with multiple labels. One of the most prominent MLC methods is the classifier chains (CC). This method induces q binary classifiers, where q represents the number of labels. Each one is responsible for predicting a specific label. These q classifiers are linked in a chain, such that at classification time each classifier considers the labels predicted by the previous ones as additional information. Although the performance of CC is largely influenced by the chain ordering, the original method uses a random ordering. To cope with this problem, in this paper we propose a novel method which is capable of finding a specific and more effective chain for each new instance to be classified. Experiments have shown that the proposed method obtained, overall, higher predictive accuracies than the well-established binary relevance, CC and CC ensemble methods.

Keywords: Multi-Label Classification, Classifier Chains, Classification.

1 Introduction

Multi-label classification (MLC) is the supervised learning problem of automatically assigning multiple labels to objects based on the features of these objects. An example of practical application is semantic scene classification [1], where the goal is to assign concepts to images. For instance, a photograph of the sun rising taken from a beach can be classified as belonging to the concepts "sky", "sunrise", and "ocean" at the same time. Other examples of important applications of MLC include text classification [16] (associating documents to various subjects), music categorization [17] (labeling songs with music genres or concepts) and functional genomics [3] (predicting the multiple biological functions of genes and proteins), just to name a few.

The multi-label classification problem can be formally defined as follows. Let $L = \{l_1, ..., l_q\}$ be a set of q class labels, where $q \geq 2$. Given a training set $D = \{(x_1, Y_1), (x_2, Y_2), ..., (x_N, Y_N)\}$ where each instance i is associated with a

T. Calders et al. (Eds.): ECML PKDD 2014, Part II, LNCS 8725, pp. 453–468, 2014.
© Springer-Verlag Berlin Heidelberg 2014

feature vector $x_i = \{(x_1, ..., x_d)\}$ and a subset of labels $Y_i \subseteq L$, the goal in the multi-label classification task is to learn a classifier $h(X) \to Y$ from D that, given an unlabeled instance $E = (x, ?)$, is capable of predicting its labelset Y.

MLC problems tend to be more challenging than traditional single-label classification problems (SLC), where objects can be associated with only a single target class label. This mainly occurs due to the existence of label correlations in most MLC problems. For instance, in scene classification, an image labeled as "ocean" is more likely to also be associated to labels such as "ship" or "beach", since these concepts are positively correlated. Similarly, an image associated to the label "desert" is less likely to also be associated to the label "snow", as these concepts are negatively correlated. Therefore, intuitively, it is expected that MLC methods which are able to identify and model label correlations should be more accurate. A large body of recent work [2,9,14,15,20,24,26,27] has primarily concentrated efforts to tackle this problem by using a wide range of different heuristics and statistical techniques.

Proposed in [14,15], the classifier chains method (CC) is one of the simplest and most prominent of such methods. The CC method involves the training of q single-label binary classifiers $y_1, y_2, ..., y_q$ where each one will be responsible for predicting a specific label in $\{l_1, l_2, ..., l_q\}$. These q classifiers are linked in a randomly-ordered chain $\{y_1 \to y_2 \to ... \to y_q\}$, such that, at classification time, each binary classifier y_j incorporates the labels predicted by the previous $y_1, ..., y_{j-1}$ classifiers as additional information. This is accomplished using a simple trick: in the training phase, the feature vector associated to each classifier y_j is extended with the binary values of the labels $l_1, ..., l_{j-1}$. Although it employs a simple approach to deal with label dependencies, CC has proved to be one of the best methods for multi-label learning in terms of both efficiency and predictive performance, having become a recommended benchmark algorithm [12,28].

However, there is a drawback in the CC approach even noted by its authors in [14,15]: the fact that the label ordering is decided at random. It is intuitive that an inadequate label ordering can potentially decrease accuracy, as the first binary classifiers could frequently output wrong predictions at classification time, thus resulting in significant error propagation along the chain. However, finding an optimized label sequence is a difficult problem because of the enormous search space of $q!$ different existing label permutations. In order to cope with this issue, the authors of CC suggest combining random orders via an ensemble of classifier chains (ECC) with the expectation that the effect of poorly ordered chains in predictive accuracy could be mitigated. Recently, other variations of the CC basic approach have been proposed in the literature [6,10,13,25], which are not based on ensembles. These novel techniques rely on the use of either statistical tests of correlation or heuristic search techniques (such as genetic algorithms and beam search) with the goal of finding a single sequence that leads to an improvement on the predictive accuracy of the CC model (i.e., an optimized label sequence). After being determined, this unique optimized chain should be used in the training and classification steps of the multi-label classifier chain model.

Nevertheless, none of the proposed CC variations have yet explored the idea of using a distinct label sequence for each new instance to be classified. In this concept, the aim is to construct a model which uses a specific label sequence tailored to each new instance at classification time. The main contribution of this paper is to demonstrate that this approach leads to a significant improvement in the predictive accuracy of the CC model. We propose a novel method called OOCC (One-to-One Classifier Chains) that addresses this problem by assigning a label sequence to a new instance in the test set based on the label sequences that perform well in training instances similar to the new instance, where such similar training instances are found using a conventional nearest neighbor (lazy learning) method. As a secondary contribution, this paper also aims at improving the fundamental understanding of the CC model. In this regard, we report the results of an experiment that, for the first time, investigated in depth the effect of different label sequences on the effectiveness of the CC method.

The remainder of this paper is organized as follows. Section 2 presents a brief overview on multi-label classification and discusses the original CC conceptual model, highlighting its main advantages and disadvantages. Section 3 presents an experiment that investigated the influence of the label sequence in the predictive accuracy of CC models. Section 4 is the main section of this work, where the OOCC algorithm is formalized and explained. Section 5 revises the related work. In Section 6, experimental results of OOCC and other MLC algorithms are presented. Conclusions and research directions are given in Section 7.

2   Multi-label Classification

2.1   Evaluation Measures

Several evaluation metrics have been proposed to evaluate multi-label classifiers [12,18,28]. This subsection presents the ones used in this paper. In the definitions throughout the text we adopted the following notation: n is number of test instances; q is the number of labels; Y_i and Z_i represents, respectively, the actual and the predicted labelset of the ith test instance.

The Exact Match (EM) measure, defined in Equation 1, assesses the proportion of instances that were fully correctly predicted in the test set. In Equation 1, $I(true) = 1$ and $I(false) = 0$.

$$EM = \frac{1}{n} \sum_{i=1}^{n} I(Y_i = Zi) \qquad (1)$$

The Accuracy (ACC) and F-Measure (FM) measures, respectively defined in Equations 2 and 3, are less strict than EM, providing the user with information about the proportion of correct predictions, meanwhile taking into consideration results that are partially correct.

$$ACC = \frac{1}{n} \sum_{i=1}^{n} \frac{|Z_i \cap Y_i|}{|Z_i \cup Y_i|} \qquad (2)$$

$$FM = \frac{1}{n} \sum_{i=1}^{n} \frac{2 \times |Z_i \cap Y_i|}{|Z_i| + |Y_i|} \qquad (3)$$

The Hamming Loss (HL) measure, defined in Equation 4, gives the average percentage of wrong label predictions to the total number of labels. The expression $Y_i \triangle Z_i$ represents the symmetric difference between Y_i and Z_i. Since HL is a loss function, its optimal value is zero.

$$HL = \frac{1}{n} \sum_{i=1}^{n} \frac{|Z_i \triangle Y_i|}{q} \qquad (4)$$

2.2   Basic Approaches for Multi-label Learning

Existing methods for MLC can be divided into two main categories: algorithm dependent or independent [18,28]. Algorithm dependent methods extend or adapt an existing single-label algorithm for the task of MLC. E.g., in [22] the authors developed a special topology and a new inference procedure for Bayesian networks so as to allow their use in multi-label problems.

By contrast, algorithm independent methods transform the multi-label problem into one or more single-label classification (SLC) problems. Then, any existing SLC algorithm can be directly applied by simply mapping its single label predictions into multi-label predictions. This enables abstraction from the underlying base algorithm, which is an important advantage since different classifiers achieve better performance in different application domains. There are a few distinct strategies to perform the transformation, being the binary relevance (BR) [11,15] approach the most commonly adopted. This method works by decomposing the multi-label problem into q independent single-label binary problems. In the training phase, one binary classifier is independently learned for each label. The labels of new instances are predicted by combining the outputs produced by each classifier.

The BR strategy presents some advantages: (i) it is simple and algorithm independent; (ii) it scales linearly with q; (iii) it can be easily parallelized. However, a serious disadvantage lies in that it is based on the assumption that all labels are independent. Each classifier works independently, disregarding the possible occurrence of relationships among labels. As a consequence, potentially important predictive information may be ignored.

2.3   Classifier Chains

The classifier chains model [14,15], briefly introduced in Section 1, represents a direct extension to the BR approach which is able to exploit label correlations. As with BR, the CC method involves the training of q single-label binary classifiers $y_1, y_2, ..., y_q$ where each one will be solely and respectively responsible for predicting a specific label in $\{l_1, l_2, ..., l_q\}$. The difference is that, in CC, these q classifiers are linked in a chain $\{y_1 \rightarrow y_2 \rightarrow ... \rightarrow y_q\}$. The first binary classifier in the chain, y_1, is trained using solely the attributes that compose the feature set X as its input attributes to predict the first label l_1. The second binary classifier, y_2, is trained using X augmented with l_1, which corresponds to the label associated to the classifier y_1. Each subsequent classifier y_j is trained using X augmented with the information of $j-1$ labels (the labels associated to the previous $j-1$ classifiers in the chain). Once the model is trained, the classification step should also be performed in a chained way. To predict the labelset of a new object, q binary classifications are needed, with the process beginning at y_1 and going along the chain. In this procedure, the classifier y_j predicts the relevance of label l_j, given the feature space augmented by the predictions carried out by the previous $j-1$ classifiers.

The CC conceptual model has many appealing properties. First, it is theoretically simple. While most MLC methods invest in complex probabilistic approaches to model label dependencies, CC adopts a quite straightforward strategy: it just passes label information between classifiers. It is also relatively efficient, since it scales linearly with q. Finally and more importantly, the method has proved to be highly effective. A comprehensive recent empirical study [12] comparing several state-of-the-art methods for MLC reported that CC is among the top best performing algorithms in terms of predictive performance. However, there is an important drawback in the basic CC approach: the label ordering is decided at random instead of being selected in a data-driven fashion. This issue is carefully investigated in the next section.

3   The Label Sequence Issue

In the basic CC model, the label sequence is decided at random. This has often been considered a major drawback, even noted by the authors of CC themselves, which deemed that if the first members of the chain have low accuracy (i.e., if they output many wrong predictions), error propagation will occur along the chain causing a significant decrease in predictive accuracy [14,15]. In a similar vein, [10,13] argued that different label orderings can lead to different results in terms of predictive accuracy mainly due to noisy data and finite sample effects. For example, if a label l_j is rare, then its binary model may be misestimated depending on the position of l_j in the chain. Nonetheless, the authors of [19] have a completely different belief. They consider that the effect of the chain order will often be very small when the number of features in the dataset is much higher than the number of labels (which corresponds to the most typical situation).

Nevertheless, [2] realized that "the effect of different orders on the prediction performance of the (CC) algorithm has not yet been studied in depth". Motivated by this consideration and by the conflicting views of [19] and [10,13,14,15], we decided to carry out an experiment to investigate the following questions:

1. Does finding a single optimized label sequence for the entire dataset indeed significantly improve the effectiveness of CC?
2. Does finding different optimized label sequences for distinct instances improve even more the effectiveness of CC?

The experiment consisted in assessing the predictive accuracy of CC considering all $q!$ label permutations of three benchmark datasets using the following single-label base algorithms: k-NN, C4.5, Naïve Bayes, and SMO [23]. The main goal is to observe the differences in predictive accuracy between the best (most accurate) chain and the worst chain for each of those base classifiers. If most of the differences are large, then there is evidence that the label sequence is actually important. In the experiment the predictive performance is determined in terms of the Accuracy measure (a brief note on results for other measures will be mentioned later).

The experiment was implemented in Java, within the MULAN tool [21], an open source platform for the evaluation of multi-label algorithms that works on the top of the WEKA API [7]. The datasets "emotions" ($q = 6$), "scene"($q = 6$) and "flags" ($q = 7$), obtained from the MULAN repository were used in this experiment. Since they have a small number of labels, it became feasible to build and test CC models for all possible label permutations. These models were evaluated by applying the holdout method with the use of the training and test parts supplied with the datasets.

Tables 1, 2 and 3, respectively present the results for the datasets "emotions", "scene" and "flags" in terms of Accuracy. These tables are divided into four main columns. The first indicates the name of the base algorithm (the acronym "x-NN" is used to refer to the k-NN algorithm configured with $k = x$). The second main column indicates the obtained Accuracy values when a unique chain is selected for the training and testing of all instances. It is divided into sub-columns {1},{2} and {3}, which respectively present the performance of the best label sequence, the performance of the worst label sequence and the difference in the Accuracy value between the best and the worst sequences. The third main column indicates the obtained Accuracy values when different label sequences are used for different instances. It is also divided into three sub-columns, labeled as {4}, {5} and {6}. Sub-column {4} presents the Accuracy value that is obtained when the best label sequence associated with each instance is selected. Sub-column {5} presents the computed Accuracy value when the worst sequence associated with each instance is selected. Sub-column {6} simply shows the difference between the values in {4} and {5}. Finally, the fourth main column presents the improvement in performance obtained when the best chain for each instance is selected in relation to the use of a unique best chain for the entire dataset (the best chain on average). In Sub-columns {3}, {6}, and {7}, the numbers between brackets in each cell denote the rank of the corresponding difference values.

The results revealed that: (i) using a single optimized label sequence indeed has a strong effect on predictive accuracy; and (ii) finding different optimized label sequences for distinct instances is even more effective. For example, consider the performance of the C4.5 algorithm in Table 3. Note that the difference in Accuracy between the model built with the best single chain (i.e., the best chain on average considering the entire dataset) and the model built with the worst single chain reached 12.65% (Sub-column {3}). However, the difference in the predictive performance between selecting the best chain for each instance and

Table 1. Results of the exhaustive experiment in terms of Accuracy values for the *emotions* dataset

Classifier	One chain for the dataset			One chain for each instance			Improvement
	Best {1}	Worst {2}	Diff {3}	Best {4}	Worst {5}	Diff {6}	{7}={4}−{1}
1-NN	0.4926	0.4926	0.0000 (7)	0.4926	0.4926	0.0000 (7)	0.0000 (7)
3-NN	0.5837	0.4983	0.0854 (5)	0.6885	0.3879	0.3006 (3)	0.1048 (5)
5-NN	0.5957	0.5314	0.0643 (3)	0.7227	0.3982	0.3245 (5)	0.1270 (4)
7-NN	0.6021	0.5307	0.0714 (4)	0.7244	0.3916	0.3328 (4)	0.1223 (3)
C4.5	0.5380	0.4059	0.1321 (1)	0.9059	0.0724	0.8335 (1)	0.3679 (1)
NB	0.5436	0.5184	0.0252 (6)	0.5840	0.4656	0.1184 (6)	0.0404 (6)
SMO	0.6167	0.4864	0.1303 (2)	0.7805	0.3426	0.4380 (2)	0.1638 (2)

Table 2. Results of the exhaustive experiment in terms of Accuracy values for the *scene* dataset

Classifier	One chain for the dataset			One chain for each instance			Improvement
	Best {1}	Worst {2}	Diff {3}	Best {4}	Worst {5}	Diff {6}	{7}={4}−{1}
1-NN	0.6368	0.6368	0.0000 (7)	0.6368	0.6368	0.0000 (7)	0.0000 (7)
3-NN	0.6785	0.6575	0.0210 (5)	0.7103	0.6315	0.0787 (5)	0.0318 (5)
5-NN	0.6898	0.6522	0.0376 (3)	0.7429	0.6196	0.1233 (3)	0.0531 (3)
7-NN	0.6819	0.6487	0.0332 (4)	0.7277	0.6116	0.1161 (4)	0.0458 (4)
C4.5	0.5993	0.5376	0.0617 (2)	0.8822	0.1564	0.7259 (1)	0.2829 (1)
NB	0.4415	0.4358	0.0057 (6)	0.4473	0.4309	0.0164 (6)	0.0058 (6)
SMO	0.6915	0.6069	0.0846 (1)	0.9034	0.3537	0.5497 (2)	0.2119 (2)

Table 3. Results of the exhaustive experiment in terms of Accuracy values for the *flags* dataset

Classifier	One chain for the dataset			One chain for each instance			Improvement
	Best {1}	Worst {2}	Diff {3}	Best {4}	Worst {5}	Diff {6}	{7}={4}−{1}
1-NN	0.5305	0.5305	0.0000 (7)	0.5305	0.5305	0.0000 (7)	0.0000 (7)
3-NN	0.6223	0.5159	0.1064 (5)	0.6964	0.4154	0.2810 (2)	0.0741 (5)
5-NN	0.6277	0.5343	0.0934 (4)	0.7487	0.4225	0.3262 (4)	0.1210 (4)
7-NN	0.6143	0.5102	0.1041 (3)	0.7394	0.4104	0.3290 (3)	0.1251 (3)
C4.5	0.6222	0.4957	0.1265 (1)	0.8089	0.2665	0.5424 (1)	0.1867 (1)
NB	0.5759	0.4873	0.0886 (6)	0.6291	0.4179	0.2112 (5)	0.0532 (6)
SMO	0.6068	0.5220	0.0848 (2)	0.7712	0.3751	0.3962 (6)	0.1644 (2)

selecting the worst chain for each instance is 54.24% (Sub-column {6}). More interestingly, note that the choice of the most accurate chain for each instance lead to an Accuracy value 18.67% higher than that achieved by the one obtained by the model built with the best single chain for the entire dataset (Sub-column {7}). The same characteristic can be also observed for the two other datasets (Tables 1 and 2) and nearly all base algorithms evaluated in the experiments, with the exception of 1-NN (for which there is no improvement in Sub-column {7} across Tables 1, 2 and 3).

Additionally, it is also evident that the different base (single-label) algorithms, due to their own characteristics, are affected to different degrees by the label ordering. The effect tends to be very large when the base algorithms are C.45 and SMO, but it can be rather small for Naïve Bayes. The k-NN presented large differences for some configurations of k and small differences for others.

We also ran the same experiment using the measures of Exact Match and Hamming Loss and the results were similar: they evidenced that building a model which uses a specific and more effective label sequence for each new instance at classification time can largely improve the predictive performance of CC. Motivated by this empirical finding, in the next section we propose a novel method that addresses this problem by assigning a label sequence to a new instance based on the label sequences that perform well in the training instances that are most similar to the new instance being classified.

4   One-to-One Classifier Chains (OOCC)

In this section we present a novel method called One-To-One Classifier Chains (OOCC), which assigns a label sequence to each new instance t in the test set based on the label sequences that perform well in training instances similar to t. The basic ideas of our OOCC method are as follows. First, we find the one or more label sequences that perform well for each training instance (see Subsection 4.1). Then we use a k-NN (k-nearest neighbors) algorithm to retrieve the k training instances that are most similar to the instance t being classified, and assign, to t, the label sequence that was found to perform best for the training instances. Due to the similarity between testing instance t and its nearest training instances, it is expected that an effective label sequence for instance t's nearest neighbors will also be an effective label sequence for instance t.

In order to measure the predictive accuracy associated with each candidate label sequence for a given training instance, we compute the quality function in Equation 5. This function (originally proposed in [6]) determines the quality of prediction performed by a CC model with regard to the instance t, by taking into account the measures of Exact Match, Accuracy and Hamming Loss.

$$Quality(t, CC) = \frac{(1 - HL) + ACC + EM}{3} \qquad (5)$$

The OOCC method modifies both the training and the classification steps of the original CC method. These changes are explained in the next subsections.

4.1   OOCC's Training Procedure

Algorithm 1 describes the algorithm used in the OOCC's training step. This algorithm produces as output an array named *bestChains*, which is responsible for storing the best label sequence(s) associated to each training instance at the

---

Algorithm 1. OOCC's training procedure

Input : D (training set), m (number of data partitions), r (number of label sequences)
Output: $bestChains$ (an array containing the best chains for each training instance)

1: Divide the training set D into m folds $\{D_1, D_2, ..., D_m\}$
2: $bestChains \rightarrow$ new Array(N)
3: **for all** folds $D_v \in D$ **do**
4:      $CCModels \leftarrow \emptyset$
5:      $D_{st} \leftarrow \{D - D_v\}$, where D_v = validation set, D_{st} = sub-training set
6:      $RS \leftarrow generateRandomSequences(r)$
7:      **for all** label sequences $l_s \in RS$ **do**
8:          $CC \leftarrow$ buildCC(D_{st}, l_s)
9:          $CCModels \leftarrow CCModels \cup CC$
10:     **end for**
11:     **for all** instances $I \in D_v$ **do**
12:         $bestQuality \leftarrow -1$
13:         **for all** classifier chain models $CC \in CCModels$ **do**
14:             $l_s \leftarrow$ label sequence associated to the model CC
15:             $curQuality \leftarrow Quality(I, CC)$
16:             **if** $curQuality > bestQuality$ **then**
17:                 $bestQuality \leftarrow curQuality$
18:                 $bestChains(I) \leftarrow \{l_s, curQuality\}$
19:             **else if** $curQuality = bestQuality$ **then**
20:                 $bestChains(I) \leftarrow bestChains(I) \cup \{l_s, curQuality\}$
21:             **end if**
22:         **end for**
23:     **end for**
24: **end for**
25: **return** $bestChains$

---

end of processing. First, the training dataset is partitioned into m distinct subsets (line 1), where m is a user-provided parameter. Each of the m subsets represents a different validation set (denoted as D_v within the algorithm specification). These validation sets are processed in turn in the FOR loop that encompasses lines 3 to 24. This FOR loop is divided into two phases: building CC models with random chains (lines 4 to 10) and selection of the best label sequences for each instance (lines 11 to 23).

The first phase works as follows. During each iteration, the data partition D_v is assigned r distinct random label sequences, where r is specified by the user. Next (lines 7 to 10), r CC models are induced, one for each distinct sequence, using a sub-training set D_{st} formed by the remainder $m - 1$ data partitions (i.e., all data partitions except D_v). These models are stored in the array $CCModels$.

Once the models are built, it becomes possible to identify the best label sequences associated to each instance I of the data partition D_v. This is done in the second phase of the OOCC's training procedure. In this phase, all r trained models contained in $CCModels$ are used to evaluate the Quality of each instance

Algorithm 2. OOCC's classification procedure

Input : D (training set), t (instance to be classified), k (number of neighbors)
Output: Z (the predicted labelset for instance t)

1: $NN \leftarrow$ find the k closest neighbors to t in D.
2: $S \leftarrow \emptyset$
3: $bestQuality \leftarrow -1$
4: **for all** neighbors $I \in NN$ **do**
5:     $chains \leftarrow$ label sequences stored in $bestChains(I)$
6:     $curQuality \leftarrow$ Quality of the label sequences stored in $bestChains(I)$
7:     **if** $curQuality > bestQuality$ **then**
8:         $bestQuality \leftarrow curQuality$
9:         $S \leftarrow chains$
10:    **else if** $curQuality = bestQuality$ **then**
11:        $S \leftarrow S \cup chains$
12:    **end if**
13: **end for**
14: **if** $S.size = 1$ **then**
15:    $l_s \leftarrow$ the label sequence stored in S
16: **else**
17:    $l_s \leftarrow$ randomly-choose a label sequence from S
18: **end if**
19: $CC \leftarrow$ buildCC(D, l_s)
20: $Z \leftarrow$ classify(t, CC)
21: **return** Z

I from D_v with the use of the function defined in Equation 5 (lines 14-15). The label sequence which achieves the highest value of Quality for an instance I must be stored in the output array $bestChains$, along with their associated Quality value (lines 16-21). Since for some instances, more than one label sequence may achieve the same best value of Quality, it is possible to store more than one label sequence for I.

4.2   OOCC's Classification Procedure

The OOCC method employs a lazy procedure to classify a new test instance t, which is described in the algorithm shown in Algorithm 2. This procedure can be divided into three phases which are explained below.

Phase 1 (line 1) consists in finding the k instances more similar to t in the training set, where k is a user-specified parameter. In Phase 2 (lines 3 to 13), the algorithm examines the best label sequences associated to each neighbor instance (which were found in the OOCC's training step and are stored in the $bestChains$ set). At the end of this phase, the highest-quality sequence(s) will be stored in the S set. Phase 3 (lines 14 to 21) actually performs the classification of t. First, an optimized label sequence l_s is selected from S. A CC model is induced using the training set D and l_s. This model is then used to classify t.

5   Related Work

The authors of the original CC method were the first to propose a method to address the label ordering issue. They suggested the use of an ensemble of classifier chains (ECC) [14,15] in order to cope with that issue. In this approach the individual classifiers vote and the output labelset for a new instance is determined based on the collection of votes.

The techniques proposed in [6,10,13,25] are based on the search for a single optimized label sequence rather than using an ensemble approach. In [25], the authors present the Bayesian Chain Classifier (BCC) algorithm. In this approach, the first step is to induce a maximum weighted spanning tree, according to the mutual dependence measure between each pair of labels. Then, different optimized sequences may be generated according to the selection of a distinct node as the root node. The algorithm presented in [10] tackles the label sequence optimization problem by performing a beam search over a tree in which every distinct path represents a different label permutation. Since the construction of a tree with $q!$ paths is infeasible even for moderate sizes of q, the algorithm employs a user adjustable input parameter b (beam width) to reduce the number of paths (at each level, only the top-b vertices in terms of predictive accuracy are maintained in the tree). The M2CC algorithm, described in [13], employs a double-Monte Carlo optimization technique to efficiently generate and evaluate a small population of distinct label sequences. The algorithm starts with a randomly chosen sequence, s_0. During the algorithm execution this sequence is modified with the aim of finding, at least, a local maximum of some payoff function (e.g.: Exact Match). Finally, the work of [6] proposes GACC – a genetic algorithm to solve the label sequence optimization problem. In this strategy, each chromosome represents a different label sequence and the fitness function is the same defined in Equation 5. The crossover operation works by transferring subchains of random length between pairs of individuals. The proposed GA follows the wrapper approach [5], evaluating the quality of an individual (candidate label sequence) by using the target MLC algorithm (i.e. the CC algorithm). All these proposals aim at finding a unique label sequence that is used to train a CC model for all instances in the training dataset. Differently, the OOCC method proposed in this work is capable of selecting a distinct and more effective chain for each instance of the training dataset.

The PCC algorithm, introduced in [4], represents a technique to improve CC through the use of inference optimization. The PCC's training step is identical to the CC's one: a label sequence is randomly chosen and used to train a CC model. However, its classification step works differently. According to the label sequence used in the training step, the PCC classifier aims at maximizing the posterior probability of the predicted labelset for each test instance. However, differently of our approach, the PCC requires a probabilistic single-label base classifier. Moreover, it has the disadvantage of employing an exhaustive search in the space of 2^q possible label combinations. Thus, its practical applications are restricted to problems where q is small.

6 Experiments

We implemented our OOCC method within the MULAN platform [21]. The proposed method was evaluated on nine distinct benchmark datasets, which were obtained from the Mulan repository. A holdout evaluation was performed to assess the predictive performance of the multi-label methods, by using the training and test parts that come with these datasets. We compared OOCC to the algorithms BR, CC and ECC. The WEKA's SMO implementation with default parameters was used as the base single-label classification algorithm for all evaluated methods, although other algorithms could have been used. The parameter values used in OOCC were $k = 5$, $m = 5$ and $r = 15$. For ECC, the number of members in the ensemble was set to 10.

The predictive performance of the algorithms was evaluated in terms of Accuracy, F-Measure, Hamming Loss and Exact Match. To determine whether the differences in performance for each measure are statistically significant, we ran the Friedman test and the Nemenyi post-hoc test, following the approach described in [8]. First, the Friedman test is executed with the null hypothesis that the performances of all methods are equivalent. Whenever the null hypothesis is rejected at the 95% confidence level, we ran the Nemenyi post-hoc multiple comparison test, which assesses if there is a statistically significant difference in the performances of each pair of methods.

The results for the measures of Accuracy, F-Measure, Hamming Loss and Exact Match are respectively presented in Tables 4, 5, 6 and 7. In these tables, N, d and q represent, respectively, the number of instances, attributes and labels for each dataset. The best results for each dataset are highlighted in bold type. The obtained rank for each method in each dataset is presented in parenthesis. In the lines right below Tables 4, 5 and 6, the symbol $\succ$ represents a significant difference between one or more methods, such that $\{a\} \succ \{b, c\}$ shows that the method a is significantly better than b and c.

The results presented in Tables 4 and 5 show that the OOCC performance is, in the majority of the datasets, superior to all other methods with respect

Table 4. Performance of BR, CC, ECC and OOCC in terms of Accuracy

Dataset (N, d, q)	Accuracy			
	BR	CC	ECC	OOCC
flags (194, 19, 7)	**0.5938 (1.0)**	0.5560 (4.0)	0.5748 (2.5)	0.5748 (2.5)
cal500 (502, 68, 174)	0.2017 (2.0)	0.1765 (4.0)	0.2007 (3.0)	**0.2113 (1.0)**
emotions (593, 72, 6)	0.4835 (4.0)	0.5202 (3.0)	0.5653 (2.0)	**0.5866 (1.0)**
birds (645, 300, 19)	0.5669 (3.0)	0.5623 (4.0)	**0.5682 (1.0)**	0.5672 (2.0)
genbase (662, 1186, 27)	**0.9908 (2.5)**	**0.9908 (2.5)**	**0.9908 (2.5)**	**0.9908 (2.5)**
medical (978, 1449, 45)	0.6990 (4.0)	0.7134 (3.0)	0.7161 (2.0)	**0.7220 (1.0)**
enron (1702, 1001, 53)	0.4063 (3.0)	0.4053 (4.0)	**0.4501 (1.0)**	0.4129 (2.0)
scene (2407, 294, 6)	0.5711 (4.0)	0.6598 (3.0)	0.6654 (2.0)	**0.6702 (1.0)**
yeast (2417, 103, 14)	0.5018 (3.0)	0.4892 (4.0)	0.5333 (2.0)	**0.5429 (1.0)**
rank sums	26.5	31.5	18.0	**14.0**

$\{OOCC\} \succ \{BR, CC, ECC\}, \{ECC\} \succ \{BR, CC\}, \{BR\} \succ \{CC\}$

Table 5. Performance of BR, CC, ECC and OOCC in terms of F-Measure

Dataset (N, d, q)	F-Measure			
	BR	CC	ECC	OOCC
flags (194, 19, 7)	**0.7139 (1.0)**	0.6764 (4.0)	0.7020 (2.0)	0.6927 (3.0)
cal500 (502, 68, 174)	0.3297 (2.0)	0.2919 (4.0)	0.3251 (3.0)	**0.3375 (1.0)**
emotions (593, 72, 6)	0.5556 (4.0)	0.5979 (3.0)	0.6429 (2.0)	**0.6627 (1.0)**
birds (645, 300, 19)	**0.6061 (1.0)**	0.5976 (4.0)	0.6039 (2.5)	0.6039 (2.5)
genbase (662, 1186, 27)	0.9940 (2.5)	0.9940 (2.5)	0.9940 (2.5)	0.9940 (2.5)
medical (978, 1449, 45)	0.7273 (4.0)	0.7409 (3.0)	0.7436 (2.0)	**0.7472 (1.0)**
enron (1702, 1001, 53)	0.5152 (4.0)	0.5110 (3.0)	**0.5575 (1.0)**	0.5197 (2.0)
scene (2407, 294, 6)	0.5985 (4.0)	0.6761 (3.0)	0.6870 (2.0)	**0.6883 (1.0)**
yeast (2417, 103, 14)	0.6101 (3.0)	0.5904 (4.0)	0.6361 (2.0)	**0.6436 (1.0)**
rank sums	25.5	30.5	19.0	**15.0**

$$\{OOCC\} \succ \{BR, CC, ECC\}, \{ECC\} \succ \{BR, CC\}, \{BR\} \succ \{CC\}$$

to the Accuracy and F-Measure metrics. Note that the obtained rank sums are always smaller (indicating a better result) for the OOCC method. Actually, the Friedman test reported a significant difference between the methods. The Nemenyi post-hoc test indicated that OOCC is significantly better than CC, BR and ECC for both Accuracy and F-Measure, at the 95% confidence level.

The results in Table 6 indicate that ECC obtained the best results in terms of Hamming Loss, being significantly superior to all other methods. For this measure, the OOCC method performed fairly well, as it is significantly better than CC and equivalent to BR. Finally, Table 7 presents the results regarding the Exact Match measure. Although the Friedman and Nemenyi tests indicated that no statistically significant differences exist between the Exact Match values achieved by the four methods, it is possible to observe that the OOCC algorithm has the smallest rank sum (i.e., the best overall result), having obtained the best results for five of the nine evaluated datasets.

Our empirical results indicate that the proposed OOCC method exhibits a very competitive performance, obtaining results significantly superior to the other evaluated methods, according to two of the four evaluated measures of pre-

Table 6. Performance of BR, CC, ECC and OOCC in terms of Hamming Loss

Dataset (N, d, q)	Hamming Loss			
	BR	CC	ECC	OOCC
flags (194, 19, 7)	**0.2637 (1.0)**	0.3011 (4.0)	0.2813 (2.0)	0.2835 (3.0)
cal500 (502, 68, 174)	**0.1375 (1.0)**	0.1527 (3.0)	0.1458 (2.0)	0.1635 (4.0)
emotions (593, 72, 6)	0.2145 (3.0)	0.2376 (4.0)	0.2137 (2.0)	**0.2063 (1.0)**
birds (645, 300, 19)	0.0658 (3.0)	0.0668 (4.0)	**0.0595 (1.0)**	0.0619 (2.0)
genbase (662, 1186, 27)	0.0007 (2.5)	0.0007 (2.5)	0.0007 (2.5)	0.0007 (2.5)
medical (978, 1449, 45)	0.0117 (4.0)	0.0115 (3.0)	**0.0111 (1.5)**	**0.0111 (1.5)**
enron (1702, 1001, 53)	0.0572 (2.0)	0.0585 (4.0)	**0.0512 (1.0)**	0.0573 (3.0)
scene (2407, 294, 6)	0.1144 (3.0)	0.1154 (4.0)	**0.1026 (1.0)**	0.1116 (2.0)
yeast (2417, 103, 14)	**0.1997 (1.0)**	0.2109 (4.0)	0.2024 (3.0)	0.2014 (2.0)
rank sums	20.5	32.5	**16.0**	21.0

$$\{ECC\} \succ \{BR, CC, OOCC\}, \{OOCC\} \succ \{CC\}, \{BR\} \succ \{CC\}$$

dictive accuracy. It is also worth noting that the original CC method performed rather poorly in terms of Accuracy, F-Measure and Hamming Loss, presenting a performance significantly inferior to ECC, OOCC and even to the baseline BR method (the CC method performed better than BR only in terms of Exact Match, however without statistical significance). This confirms that the use of a single randomly-generated label sequence seems to be an ineffective approach for multi-label chain classifiers, reinforcing the importance of either using an ensemble or searching for an optimized label sequence.

Table 7. Performance of BR, CC, ECC and OOCC in terms of Exact Match

Dataset (N, d, q)	Exact Match			
	BR	CC	ECC	OOCC
flags (194, 19, 7)	**0.1538 (1.5)**	0.1231 (3.5)	0.1231 (3.5)	**0.1538 (1.5)**
cal500 (502, 68, 174)	0.0000 (2.5)	0.0000 (2.5)	0.0000 (2.5)	0.0000 (2.5)
emotions (593, 72, 6)	0.2525 (4.0)	0.2822 (3.0)	0.3267 (2.0)	**0.3465 (1.0)**
birds (645, 300, 19)	0.4630 (3.5)	**0.4722 (1.5)**	**0.4722 (1.5)**	0.4630 (3.5)
genbase (662, 1186, 27)	**0.9799 (2.5)**	**0.9799 (2.5)**	**0.9799 (2.5)**	**0.9799 (2.5)**
medical (978, 1449, 45)	0.6140 (4.0)	0.6326 (3.0)	0.6357 (2.0)	**0.6465 (1.0)**
enron (1702, 1001, 53)	0.1209 (4.0)	0.1313 (2.0)	**0.1503 (1.0)**	0.1295 (3.0)
scene (2407, 294, 6)	0.4908 (4.0)	0.6112 (2.0)	0.6012 (3.0)	**0.6162 (1.0)**
yeast (2417, 103, 14)	0.1603 (4.0)	0.1952 (3.0)	0.2148 (2.0)	**0.2399 (1.0)**
rank sums	30.0	23.0	20.0	**17.0**

No significance differences according to the Friedman test

7   Conclusions and Future Work

The classifier chains approach has become one of the most influential methods for multi-label classification. It is distinguished from other methods by its simple and effective approach to exploit label dependencies. However, the basic CC model suffers from an important drawback: it decides the label sequence at random. The main contribution of this paper was the proposal of a novel multi-label classifier chain method called One-to-One Classifier Chains (OOCC), which is capable of finding, at classification time, a specific and more accurate label sequence for each new instance in the test set. The OOCC method was compared against the well-established BR, CC and ECC methods. The obtained results show that OOCC significantly outperformed all these three methods in terms of Accuracy and F-Measure. In terms of Hamming Loss, OOCC significantly outperformed CC and was significantly outperformed by ECC. There was no significant difference among the four methods in terms of the Exact Match measure.

Additionally, we contributed to a better understanding of the underlying principles of the CC method by reporting the results of a study that evidenced that: (i) finding a single optimized label sequence has a strong effect on predictive accuracy; (ii) finding different optimized label sequences for distinct instances is even more effective; and (iii) the different base (single-label) algorithms, due to their own characteristics, are affected to different degrees by the label ordering.

For future research, we first intend to perform a detailed analysis on the sensivity of the results to the parameters r, m and k. The main goal is to determine the best set of parameters for the OOCC algorithm, using the training set to optimize the parameters. We also intend to evaluate other approaches to determine the label sequence of a new instance to be classified (which may not be necessarily based on the Quality measure). Finally, we plan to compare OOCC against some of the methods described in Section 5 and to develop an ensemble version of the proposed OOCC method.

Acknowledgments. This work was supported by CAPES research grant BEX 1642/14-6 (Eduardo Corrêa Gonçalves), CNPq and FAPERJ research grants (Alexandre Plastino) and CAPES DS scholarship (Pablo Nascimento da Silva).

References

1. Boutell, M.R., Luo, J., Shen, X., Brown, C.M.: Learning Multi-Label Scene Classification. Pattern Recognition 37(9), 1757–1771 (2004)
2. Cherman, E.A., Metz, J., Monard, M.C.: Incorporating Label Dependency into the Binary Relevance Framework for Multi-label Classification. Expert Systems with Applications 39(2), 1647–1655 (2012)
3. Clare, A.J., King, R.D.: Knowledge discovery in multi-label phenotype data. In: Siebes, A., De Raedt, L. (eds.) PKDD 2001. LNCS (LNAI), vol. 2168, p. 42. Springer, Heidelberg (2001)
4. Dembczynski, K., Cheng, W., Hüllermeier, E.: Bayes Optimal Multilabel Classification via Probabilistic Classifier Chains. In: 27th Intl. Conf. on Machine Learning (ICML 2010), Haifa, pp. 279–286 (2010)
5. Freitas, A.A.: Data Mining and Knowledge Discovery with Evolutionary Algorithms. Natural Computing Series. Springer (2002)
6. Gonçalves, E.C., Plastino, A., Freitas, A.A.: A Genetic Algorithm for Optimizing the Label Ordering in Multi-Label Classifier Chains. In: IEEE 25th Intl. Conf. on Tools with Artificial Intelligence (ICTAI 2013), Herndon, pp. 469–476 (2013)
7. Hall, M., Frank, E., Holmes, G., Pfahringer, B., Reutemann, P., Witten, I.H.: The WEKA Data Mining Software: an Update. ACM SIGKDD Exploration Newsletter 11(1), 10–18 (2009)
8. Japkowicz, N., Shah, M.: Evaluating Learning Algorithms: A Classification Perspective. Cambridge University Press (2011)
9. Li, N., Zhou, Z.-H.: Selective ensemble of classifier chains. In: Proceedings of the 11th International Workshop on Multiple Classifier Systems (MCS 2013), Nanjing, pp. 146–156 (2013)
10. Kumar, A., Vembu, S., Menon, A.K., Elkan, C.: Beam Search Algorithms for Multilabel Learning. Machine Learning 92(1), 65–89 (2013)
11. Luaces, O., Díez, J., Barranquero, J., Coz, J.J., Bahamonde, A.: Binary Relevance Efficacy for Multilabel Classification. Progress in Artificial Intelligence 1(4), 303–313 (2012)
12. Madjarov, G., Kocev, D., Gjorgjevikj, D., Džeroski, S.: An Extensive Experimental Comparison of Methods for Multi-label Learning. Pattern Recognition 45(9), 3084–3104 (2012)

13. Read, J., Martino, L., Luengo, D.: Efficient Monte Carlo Methods for Multidimensional Learning with Classifier Chains. Pattern Recognition 47(3), 1535–1546 (2014)
14. Read, J., Pfahringer, B., Holmes, G., Frank, E.: Classifier chains for multi-label classification. In: Buntine, W., Grobelnik, M., Mladenić, D., Shawe-Taylor, J. (eds.) ECML PKDD 2009, Part II. LNCS, vol. 5782, pp. 254–269. Springer, Heidelberg (2009)
15. Read, J., Pfahringer, B., Holmes, G., Frank, E.: Classifier Chains for Multi-label Classification. Machine Learning 85(3), 333–359 (2011)
16. Schapire, R.E., Singer, Y.: BoosTexter: A Boosting-based System for Text Categorization. Machine Learning 39(2-3), 135–168 (2000)
17. Trohidis, K., Tsoumakas, G., Kalliris, G., Vlahavas, I.P.: Multi-Label Classification of Music into Emotions. In: 9th Intl. Conf. on Music Information Retrieval (ISMIR 2008), Philadelphia, pp. 325–330 (2008)
18. Tsoumakas, G., Katakis, I., Vlahavas, I.: Mining Multi-Label Data. In: Data Mining and Knowledge Discovery Handbook, pp. 667–685. Springer, US (2010)
19. Tenenboim-Chekina, L., Rokach, L., Shapira, B.: Identification of Label Dependencies for Multi-label Classification. In: 2nd Intl. Workshop on Learning from Multi-Label Data (MLD 2010), Haifa, pp. 53–60 (2010)
20. Sheng, V.S., Ling, C.X.: Roulette sampling for cost-sensitive learning. In: Kok, J.N., Koronacki, J., Lopez de Mantaras, R., Matwin, S., Mladenič, D., Skowron, A. (eds.) ECML 2007. LNCS (LNAI), vol. 4701, pp. 724–731. Springer, Heidelberg (2007)
21. Tsoumakas, G., Xioufis, E.S., Vilcek, J., Vlahavas, I.P.: MULAN: A Java Library for Multi-Label Learning. JMLR 12, 2411–2414 (2011)
22. van der Gaag, L., de Waal, P.R.: Multi-dimensional Bayesian Network Classifiers. In: 3rd European Workshop on Probabilistic Graphical Models (PGM 2006), Prague, pp. 107–114 (2006)
23. Witten, I.H., Frank, E., Hall, M.A.: Data Mining: Practical Machine Learning Tools and Techniques: Practical Machine Learning Tools and Techniques, 3rd edn. Elsevier Science (2011)
24. Yua, Y., Pedryczb, W., Miao, D.: Multi-label Classification by Exploiting Label Correlations. Expert Systems with Applications 41(6), 2989–3004 (2014)
25. Zaragoza, J.H., Sucar, L.E., Morales, E.F., Bielza, C., Larrañaga, P.: Bayesian Chain Classifiers for Multidimensional Classification. In: 22nd Intl. Joint Conf. on Artificial Intelligence (IJCAI 2011), Barcelona, pp. 2192–2197 (2011)
26. Zhang, M.-L., Zhang, K.: Multi-label Learning by Exploiting Label Dependency. In: 16th ACM SIGKDD Intl. Conf. on Knowledge Discovery and Data Mining (KDD 2010), Washington, D.C., pp. 999–1008 (2010)
27. Zhang, M.-L., Zhou, Z.-H.: ML-KNN: A Lazy Learning Approach to Multi-label Learning. Pattern Recognition 40(7), 2038–2048 (2007)
28. Zhang, M.-L., Zhou, Z.-H.: A Review On Multi-Label Learning Algorithms. IEEE Transactions on Knowledge and Data Engineering, 99(preprints) (2013)

A Unified Framework for Probabilistic Component Analysis

Mihalis A. Nicolaou[1], Stefanos Zafeiriou[1], and Maja Pantic[1,2]

[1] Department of Computing, Imperial College London, UK
[2] EEMCS, University of Twente, Netherlands (NL)
{mihalis,s.zafeiriou,m.pantic}@imperial.ac.uk

Abstract. We present a unifying framework which reduces the construction of probabilistic component analysis techniques to a mere selection of the latent neighbourhood, thus providing an elegant and principled framework for creating novel component analysis models as well as constructing probabilistic equivalents of deterministic component analysis methods. Under our framework, we unify many very popular and well-studied component analysis algorithms, such as Principal Component Analysis (PCA), Linear Discriminant Analysis (LDA), Locality Preserving Projections (LPP) and Slow Feature Analysis (SFA), some of which have no probabilistic equivalents in literature thus far. We firstly define the Markov Random Fields (MRFs) which encapsulate the latent connectivity of the aforementioned component analysis techniques; subsequently, we show that the projection directions produced by all PCA, LDA, LPP and SFA are also produced by the Maximum Likelihood (ML) solution of a single joint probability density function, composed by selecting one of the defined MRF priors while utilising a simple observation model. Furthermore, we propose novel Expectation Maximization (EM) algorithms, exploiting the proposed joint PDF, while we generalize the proposed methodologies to arbitrary connectivities via parametrizable MRF products. Theoretical analysis and experiments on both simulated and real world data show the usefulness of the proposed framework, by deriving methods which well outperform state-of-the-art equivalents.

Keywords: Unifying Framework, Probabilistic Methods, Component Analysis, Dimensionality Reduction, Random Fields.

1 Introduction

Unification frameworks in machine learning provide valuable material towards the deeper understanding of various methodologies, while also they form a flexible basis upon which further extensions can be easily built. One of the first attempts to unify methodologies was made in [17]. In this seminal work, models such as Factor Analysis (FA), Principal Component Analysis (PCA), mixtures of Gaussian clusters, Linear Dynamic Systems, Hidden Markov Models and Independent Component Analysis were unified as variations of unsupervised learning under a single basic generative model.

T. Calders et al. (Eds.): ECML PKDD 2014, Part II, LNCS 8725, pp. 469–484, 2014.
© Springer-Verlag Berlin Heidelberg 2014

Deterministic Component Analysis (CA) unification frameworks proposed in previous works, such as [1], [10], [4], [23] and [21], provide significant insights on how CA methods such as Principal Component Analysis, Linear Discriminant Analysis, Laplacian Eigenmaps and others can be jointly formulated as, e.g., least squares problems under mild conditions or general trace optimisation problems. Nevertheless, while several probabilistic equivalents of, e.g. PCA have been formulated (c.f., [22] [16]), to this date no unification framework has been proposed for *probabilistic* component analysis. Motivated by the latter, in this paper we propose the *first* unified framework for probabilistic component analysis. Based on Markov Random Fields (MRFs), our framework unifies *all* component analysis techniques whose corresponding deterministic problem is solved as a trace optimisation problem without domain constraints for the parameters, such as Principal Component Analysis (PCA), Linear Discriminant Analysis (LDA), Locality Preserving Projections (LPP) and Slow Feature Analysis (SFA). Our framework provides further insight on component analysis methods from a probabilistic perspective. This entails providing probabilistic explanations for the data at hand with explicit variance modelling, as well as reduced complexity compared to the deterministic equivalents. These features are especially useful in case of methods for which no probabilistic equivalent exists in literature so far, such as LPP. Furthermore, under our framework one can generate *novel* component analysis techniques by merely combining products of MRFs with arbitrary connectivity.

The rest of this paper is organised as follows. We initially introduce previous work on CA, highlighting the properties of the proposed framework (Sec. 2). Subsequently, we formulate the joint complete-data Probability Density Function (PDF) of observations and latent variables. We show that the Maximum Likelihood (ML) solution of this joint PDF is co-directional to the solutions obtained via deterministic PCA, LDA, LPP and SFA, by changing only the prior latent distribution (Sec. 3), which, as we show, models the latent dependencies and thus determines the resulting CA technique. E.g, when using a fully connected MRF, we obtain PCA. When choosing the product of a fully connected MRF and an MRF connected only to within-class data, we derive LDA. LPP is derived by choosing a locally connected MRF, while finally, SFA is produced when the joint prior is a linear Markov-chain. Based on the aforementioned PDF we subsequently propose Expectation Maximization (EM) algorithms (Sec. 4). Finally in Sec. 5, utilising both synthetic and real data, we demonstrate the usefulness and advantages of this family of probabilistic component analysis methods.

2 Prior Art and Novelties

An important contribution of our paper lies in the proposed unification of probabilistic component techniques, giving rise to the first framework that reduces the construction of probabilistic component analysis models to the design of a appropriate prior, thus defining only the latent neighbourhood. Nevertheless, other novelties arise in methods generated via our framework. In this section, we review the state-of-the-art in deterministic and probabilistic PCA, LDA, LPP and SFA.

While doing so, we highlight novelties and advantages that our proposed framework entails wrt. each alternative formulation. Throughout this paper we consider, a zero mean set of F-dimensional observations of length T, represented by the matrix $\mathbf{X} = [\mathbf{x}_1, \ldots, \mathbf{x}_T]$. All CA methods discover an N-dimensional latent space $\mathbf{Y} = [\mathbf{y}_1, \ldots, \mathbf{y}_T]$ which preserves certain properties of $\mathbf{X}$.

2.1 Principal Component Analysis (PCA)

The deterministic model of PCA finds a set of projection bases $\mathbf{W}$, with the latent space $\mathbf{Y}$ being the projection of the training set $\mathbf{X}$ (i.e., $\mathbf{Y} = \mathbf{W}^T\mathbf{X}$)). The optimization problem is as follows

$$\mathbf{W}_o = \arg \max_{\mathbf{W}} \text{tr} \left[\mathbf{W}^T \mathbf{S} \mathbf{W} \right], \text{ s.t. } \mathbf{W}^T\mathbf{W} = \mathbf{I} \qquad (1)$$

where $\mathbf{S} = \frac{1}{T}\sum_{i=1}^{T} \mathbf{x}_i\mathbf{x}_i^T$ is the total scatter matrix and $\mathbf{I}$ the identity matrix. The optimal N projection basis $\mathbf{W}_o$ are recovered (the N eigenvectors of $\mathbf{S}$ that correspond to the N largest eigenvalues). Probabilistic PCA (PPCA) approaches were independently proposed in [16] and [22]. In [22] a probabilistic generative model was adopted as:

$$\mathbf{x}_i = \mathbf{W}\mathbf{y}_i + \boldsymbol{\epsilon}_i, \ \mathbf{y}_i \sim \mathcal{N}(0, \mathbf{I}), \ \boldsymbol{\epsilon}_i \sim \mathcal{N}(0, \sigma^2\mathbf{I}) \qquad (2)$$

where $\mathbf{W} \in \mathbb{R}^{F \times N}$ is the matrix that relates the latent variable $\mathbf{y}_i$ with the observed samples $\mathbf{x}_i$ and $\boldsymbol{\epsilon}_i$ is the noise which is assumed to be an isotropic Gaussian model. The motivation is that, when $N < F$, the latent variables will offer a more parsimonious explanation of the dependencies arising in observations.

2.2 Linear Discriminant Analysis (LDA)

Let us now further assume that our data $\mathbf{X}$ is further separated into K disjoint classes $\mathcal{C}_1, \ldots, \mathcal{C}_K$ with $T = \sum_{c=1}^{K} |\mathcal{C}_c|$. The Fisher's Linear Discriminant Analysis (LDA) finds a set of projection bases $\mathbf{W}$ s.t. [26]

$$\mathbf{W}_o = \arg \min_{\mathbf{W}} \text{tr} \left[\mathbf{W}^T \mathbf{S}_w \mathbf{W} \right], \text{ s.t. } \mathbf{W}^T\mathbf{S}\mathbf{W} = \mathbf{I} \qquad (3)$$

where $\mathbf{S}_w = \sum_{c=1}^{K} \sum_{\mathbf{x}_i \in \mathcal{C}_c} (\mathbf{x}_i - \boldsymbol{\mu}_{\mathcal{C}_i})(\mathbf{x}_i - \boldsymbol{\mu}_{\mathcal{C}_i})^T$ and $\boldsymbol{\mu}_{\mathcal{C}_i}$ the mean of class i. The aim is to find the latent space $\mathbf{Y} = \mathbf{W}^T\mathbf{X}$ such that the within-class variance is minimized in a whitened space. The solution is given by the eigenvectors of $\mathbf{S}_w$ corresponding to the $N - K$ smallest eigenvalues of the whitened data. [1]

Several probabilistic latent variable models which exploit class information have been recently proposed (c.f., [14,29,8]). In [14,29] another two related attempts were made to formulate a PLDA. Considering $\mathbf{x}_i$ to be the i-th sample of the c-th class, the generative model of [14] can be described as:

$$\mathbf{x}_i = \mathbf{F}\mathbf{h}_c + \mathbf{G}\mathbf{w}_{ic} + \boldsymbol{\epsilon}_{ic}, \ \mathbf{h}_c, \mathbf{w}_{ic} \sim \mathcal{N}(0, \mathbf{I}), \ \boldsymbol{\epsilon}_{ic} \sim \mathcal{N}(0, \boldsymbol{\Sigma}) \qquad (4)$$

---

[1] We adopt this formulation of LDA instead of the equivalent of maximizing the trace of the between-class scatter matrix [2], since this facilitates our following discussion on Probabilistic LDA alternatives.

where $\mathbf{h}_c$ represents the class-specific weights and $\mathbf{w}_{ic}$ the weights of each individual sample, with $\mathbf{G}$ and $\mathbf{F}$ denoting the corresponding loadings. Regarding [29], the probabilistic model is as follows:

$$\mathbf{x}_i = \mathbf{F}_c \mathbf{h}_c + \boldsymbol{\epsilon}_{ic}, \quad \mathbf{h}_c, \mathbf{F}_{ic} \sim \mathcal{N}(\mathbf{0}, \mathbf{I}), \quad \boldsymbol{\epsilon}_{ic} \sim \mathcal{N}(\mathbf{0}, \boldsymbol{\Sigma}) \tag{5}$$

We note that the two models become equivalent when choosing a common $\mathbf{F}$ (Eq. 5) for all classes while also disregarding the matrix $\mathbf{G}$. In this case, the ML solution is given by obtaining the eigenvectors corresponding to the largest eigenvalues of $\mathbf{S}_w$. Hence, the solution is vastly different than the one obtained by deterministic LDA (which keeps the smallest ones, Eq. 3), resembling more to the solution of problems which retain the maximum variance. In fact, when learning a different $\mathbf{F}_c$ per class, the model of [29] reduces to applying PPCA per class. To the best of our knowledge the only probabilistic model where the ML solution is closely related to that of deterministic LDA is [8]. The probabilistic model is defined as follows: $\mathbf{x} \in \mathcal{C}_i$, $\mathbf{x}|\mathbf{y} \sim \mathcal{N}(\mathbf{y}, \boldsymbol{\Phi}_w)$, $\mathbf{y} \sim \mathcal{N}(\mathbf{m}, \boldsymbol{\Phi}_b)$, $\mathbf{V}^T \boldsymbol{\Phi}_b \mathbf{V} = \boldsymbol{\Psi}$ and $\mathbf{V}^T \boldsymbol{\Phi}_w \mathbf{V} = \mathbf{I}$, $\mathbf{A} = \mathbf{V}^{-T}$, $\boldsymbol{\Phi}_w = \mathbf{A}\mathbf{A}^T$ $\boldsymbol{\Phi} = \mathbf{A}\boldsymbol{\Psi}\mathbf{A}^T$, where the observations are generated as:

$$\mathbf{x}_i = \mathbf{A}\mathbf{u}, \quad \mathbf{u} \sim \mathcal{N}(\mathbf{V}, \mathbf{I}), \quad \mathbf{v} \sim \mathcal{N}(\mathbf{0}, \boldsymbol{\Psi}). \tag{6}$$

The drawback of [8] is the requirement for all classes to contain the same number of samples. As we show, we overcome this limitation in our formulation.

2.3   Locality Preserving Projections (LPP)

Locality Preserving Projections (LPP) is the linear alternative to Laplacian Eigenmaps [13]. The aim is to obtain a set of projections $\mathbf{W}$ and a latent space $\mathbf{Y} = \mathbf{W}^T \mathbf{X}$ which preserves the locality of the original samples. First, let us define a set of weights that represent locality. Common choices for the weights are the heat kernel $u_{ij} = e^{-\frac{||\mathbf{x}_i - \mathbf{x}_j||^2}{\gamma}}$ or a set of constant weights ($u_{ij} = 1$ if the i-th and the j-th vectors are adjacent and $u_{ij} = 0$ otherwise, while $u_{ij} = u_{ji}$). LPP finds a set of projection basis matrix $\mathbf{W}$ by solving the following problem:

$$\begin{aligned} \mathbf{W}_o &= \arg\min_{\mathbf{W}} \sum_{i,j=1}^{T} \sum_{n=1}^{N} u_{ij} ||\mathbf{w}_n^T \mathbf{x}_i - \mathbf{w}_n^T \mathbf{x}_j||^2 \\ &= \arg\min_{\mathbf{W}} \operatorname{tr}\left[\mathbf{W}^T \mathbf{X} \mathbf{L} \mathbf{X}^T \mathbf{W}\right] \text{ s.t. } \mathbf{W}^T \mathbf{X} \mathbf{D} \mathbf{X}^T \mathbf{W} = \mathbf{I} \end{aligned} \tag{7}$$

where $\mathbf{U} = [u_{ij}]$, $\mathbf{L} = \mathbf{D} - \mathbf{U}$ and $\mathbf{D} = \operatorname{diag}(\mathbf{U}\mathbf{1})$ (where $\operatorname{diag}(\mathbf{a})$ is the diagonal matrix having as main diagonal vector $\mathbf{a}$ and $\mathbf{1}$ is a vector of ones). The objective function with the chosen weights w_{ij} results in a heavy penalty if the neighbouring points $\mathbf{x}_i$ and $\mathbf{x}_j$ are mapped far apart. Therefore, its minimization ensures that if $\mathbf{x}_i$ and $\mathbf{x}_j$ are near, then the projected features $\mathbf{y}_i = \mathbf{W}^T \mathbf{x}_i$ and $\mathbf{y}_j = \mathbf{W}^T \mathbf{x}_i$ are near, as well. To the best of our knowledge no probabilistic models exist for LPPs. In the following (Sec. 3, 4), we show how a probabilistic version of LPPs arises by choosing an appropriate prior over the latent space $\mathbf{y}_i$.

2.4 Slow Feature Analysis

Now let us consider the case that the columns of $\mathbf{x}_i$ are samples of a time series of length T. The aim of Slow Feature Analysis (SFA) is, given T sequential observation vectors $\mathbf{X} = [\mathbf{x}_1 \ldots \mathbf{x}_T]$, to find an output signal representation $\mathbf{Y} = [\mathbf{y}_1 \ldots \mathbf{y}_T]$ for which the features change slowest over time [25], [9]. By assuming again a linear mapping $\mathbf{Y} = \mathbf{W}^T\mathbf{X}$ for the output representation, SFA minimizes the *slowness* for these values, defined as the variance of the first derivative of $\mathbf{Y}$. Formally, $\mathbf{W}$ of SFA is computed as

$$\mathbf{W}_o = \arg\min_{\mathbf{W}} \mathrm{tr} \left[\mathbf{W}^T \dot{\mathbf{X}}\dot{\mathbf{X}}\mathbf{W} \right], \text{ s.t. } \mathbf{W}^T\mathbf{S}\mathbf{W} = \mathbf{I}, \tag{8}$$

where $\dot{\mathbf{X}}$ is the first derivative matrix (usually computed as the first order difference i.e., $\dot{\mathbf{x}}_j = \mathbf{x}_j - \mathbf{x}_{j-1}$). An ML solution of SFA was recently proposed in [24], by incorporating a Gaussian linear dynamic system prior over the latent space $\mathbf{Y}$. The proposed generative model is

$$\begin{aligned} P(\mathbf{x}_t|\mathbf{W}, \mathbf{y}_t, \sigma_x) &= \mathcal{N}(\mathbf{W}^{-1}\mathbf{y}_t, \sigma_x^2\mathbf{I}) \\ P(\mathbf{y}_t|\mathbf{y}_{t-1}, \lambda_{1:N}, \sigma_{1:N}) &= \textstyle\prod_{n=1}^{N} P(y_{n,t}|y_{n,t-1}, \lambda_n, \sigma_n^2) \end{aligned} \tag{9}$$

with $P(y_{n,t}|y_{n,t-1}, \lambda_n, \sigma_n^2) = \mathcal{N}(\lambda_n y_{n,t-1}, \sigma_n^2)$ and $P(y_{n,1}|\sigma_{n,1}^2) = \mathcal{N}(0, \sigma_{n,1}^2)$. As we will show, SFA is indeed a special case of our general model.

Summarizing, in the following sections we formulate a unified, probabilistic framework for component analysis which: (1) incorporates PCA as a special case, (2) produces a Probabilistic LDA which (i) has an ML solution for the loading matrix $\mathbf{W}$ with similar direction to the deterministic LDA (Eq. 3) and (ii) does not make assumptions regarding the number of samples per class (as in [8]), (3) provides the first, to the best of our knowledge, probabilistic model that explains LPP, (4) naturally incorporates SFA as a special case, (5) provides variance estimates not only for observations but also per latent dimension (differentiating our approach from existing probabilistic CA (e.g., PPCA, PLDA), and (6) provides a straightforward framework for producing novel component analysis techniques.

3 A Unified ML Framework for Component Analysis

In this section, we will present the proposed Maximum Likelihood (ML) framework for probabilistic component analysis and show how PCA, LDA, LPP and SFA can be generated within this framework, also proving equivalence with known deterministic models. Firstly, to ease computation, we assume the generative model for the i-th observation, $\mathbf{x}_i$, is defined as

$$\mathbf{x}_i = \mathbf{W}^{-1}\mathbf{y}_i + \boldsymbol{\epsilon}_i, \ \boldsymbol{\epsilon}_i \sim N(\mathbf{0}, \sigma_x^2\mathbf{I}). \tag{10}$$

In order to fully define the likelihood we need to define a prior distribution on the latent variables $\mathbf{y}$. We will prove that by choosing one of the priors defined below and subsequently taking the ML solution wrt. parameters, we end up generating the aforementioned family of probabilistic component models. The priors, parametrised by $\beta = \{\sigma_{1:N}, \lambda_{1:N}\}$, are as follows (see also Fig. 1).

- An MRF with full connectivity - each latent node $\mathbf{y}_i$ is connected to all other latent nodes $\mathbf{y}_j, j \neq i$.

$$
\begin{aligned}
P(\mathbf{Y}|\beta) &= \tfrac{1}{Z} \exp \left\{ -\tfrac{1}{2} \sum_{n=1}^{N} \sum_{i=1}^{T} \tfrac{1}{T-1} \sum_{j=1, j \neq i}^{T} \tfrac{1}{\sigma_n^2} (y_{n,i} - \lambda_n y_{n,j})^2 \right\} \\
&\approx \tfrac{1}{Z} \exp \left\{ -\tfrac{1}{2} \sum_{n=1}^{N} \sum_{i=1}^{T} \tfrac{1}{T} \sum_{j=1}^{T} \tfrac{1}{\sigma_n^2} (y_{n,i} - \lambda_n y_{n,j})^2 \right\} \qquad (11) \\
&= \tfrac{1}{Z} \exp \left\{ -\tfrac{1}{2} \left(\operatorname{tr} \left[\boldsymbol{\Lambda}^{(1)} \mathbf{Y} \mathbf{Y}^T \right] + \operatorname{tr} \left[\boldsymbol{\Lambda}^{(2)} \mathbf{Y} \mathbf{M} \mathbf{Y}^T \right] \right) \right\},
\end{aligned}
$$

where $\mathbf{M} \triangleq -\tfrac{1}{T} \mathbf{1} \mathbf{1}^T$, $\boldsymbol{\Lambda}^{(1)} \triangleq \left[\delta_{mn} \tfrac{\lambda_n^2 + 1}{\sigma_n^2} \right]$, $\boldsymbol{\Lambda}^{(2)} \triangleq \left[\delta_{mn} \tfrac{\lambda_n}{\sigma_n^2} \right]$.

- A product of two MRFs. In the first, each latent node $\mathbf{y}_i$ is connected only to other latent nodes in the same class ($\mathbf{y}_j, j \in \tilde{\mathcal{C}}_i$). In the second, each latent node ($\mathbf{y}_i$) is connected to all other latent nodes ($\mathbf{y}_j, j \neq i$).

$$
\begin{aligned}
P(\mathbf{Y}|\beta) &= \tfrac{1}{Z} \exp \left\{ -\tfrac{1}{2} \sum_{n=1}^{N} \sum_{i=1}^{T} \tfrac{1}{|\mathcal{C}_i|} \sum_{j \in \tilde{\mathcal{C}}_i} \tfrac{\lambda_n}{\sigma_n^2} (y_{n,i} - y_{n,j})^2 \right\} \\
&\quad \exp \left\{ -\tfrac{1}{2} \sum_{n=1}^{N} \sum_{i=1}^{T} \tfrac{1}{T-1} \sum_{j=1}^{T} \tfrac{(1-\lambda_n)^2}{\sigma_n^2} (y_{n,i} - y_{n,j})^2 \right\} \qquad (12) \\
&= \tfrac{1}{Z} \exp \left\{ -\tfrac{1}{2} \left(\operatorname{tr} \left[\boldsymbol{\Lambda}^{(1)} \mathbf{Y} \mathbf{M}_c \mathbf{Y}^T \right] + \operatorname{tr} \left[\boldsymbol{\Lambda}^{(2)} \mathbf{Y} \mathbf{M}_t \mathbf{Y}^T \right] \right) \right\},
\end{aligned}
$$

where $\mathbf{M}_c \triangleq \mathbf{I} - \operatorname{diag}[\mathbf{C}_1, \ldots, \mathbf{C}_C]$, $\mathbf{C}_c \triangleq \tfrac{1}{|\mathcal{C}_c|} \mathbf{1}_c \mathbf{1}_c^T$, $\mathbf{M}_t \triangleq \mathbf{I} + \mathbf{M}$, $\boldsymbol{\Lambda}^{(1)} \triangleq \left[\delta_{mn}(\tfrac{\lambda_n}{\sigma_n^2}) \right]$, $\boldsymbol{\Lambda}^{(2)} \triangleq \left[\delta_{mn} \tfrac{(1-\lambda_n)^2}{\sigma_n^2} \right]$, while $\tilde{\mathcal{C}}_i = \{j : \exists \mathcal{C}_l \text{ s.t. } \{\mathbf{x}_j, \mathbf{x}_i\} \in \mathcal{C}_l, i \neq j\}$.

- A product of two MRFs. In the first, each latent node $\mathbf{y}_i$ is connected to all other latent nodes that belong in $\mathbf{y}_i$'s neighbourhood (symmetrically defined as $\mathcal{N}_i^s = \mathcal{N}_j^s = \{i \in \mathcal{N}_j \cup j \in \mathcal{N}_i\}$). In the second, we only have individual potentials per node.

$$
\begin{aligned}
P(\mathbf{Y}|\beta) &= \tfrac{1}{Z} \exp \left(-\tfrac{1}{2} \sum_{n=1}^{N} \sum_{i=1}^{T} \tfrac{1}{|\mathcal{N}_i^s|} \sum_{j \in \mathcal{N}_i^s} \tfrac{\lambda_n}{\sigma_n^2} (y_{n,i} - y_{n,j})^2 \right) \\
&\quad \exp \left(-\tfrac{1}{2} \sum_{n=1}^{N} \sum_{i=1}^{T} \tfrac{(1-\lambda_n)^2}{\sigma_n^2} y_{n,i}^2 \right) \qquad (13) \\
&= \tfrac{1}{Z} \exp \left\{ -\tfrac{1}{2} \left(\operatorname{tr} \left[\boldsymbol{\Lambda}^{(1)} \mathbf{Y} \tilde{\mathbf{L}} \mathbf{Y}^T \right] + \operatorname{tr} \left[\boldsymbol{\Lambda}^{(2)} \mathbf{Y} \tilde{\mathbf{D}} \mathbf{Y}^T \right] \right) \right\}
\end{aligned}
$$

where $\tilde{\mathbf{L}} = \mathbf{D}^{-1} \mathbf{L}$ and $\tilde{\mathbf{D}} = \mathbf{I}$ ($\mathbf{L}$ and $\mathbf{D}$ are defined in Sec. 2.3 referring to LPPs). $\boldsymbol{\Lambda}^{(1)}$ and $\boldsymbol{\Lambda}^{(2)}$ are defined as above.

- A linear dynamical system prior over the latent space.

$$
\begin{aligned}
P(\mathbf{Y}|\beta) &= \tfrac{1}{Z} \exp \left\{ -\sum_{n=1}^{N} \left(\tfrac{1}{2\sigma_{n,1}^2} y_{n,1}^2 + \tfrac{1}{2\sigma_n^2} \sum_{t=2}^{T} [y_{n,t} - \lambda_n y_{n,t-1}]^2 \right) \right\} \\
&\approx \tfrac{1}{Z} \exp \left\{ -\tfrac{1}{2} \left(\operatorname{tr} \left[\boldsymbol{\Lambda}^{(1)} \mathbf{Y} \mathbf{K}_1 \mathbf{Y}^T \right] + \operatorname{tr} \left[\boldsymbol{\Lambda}^{(2)} \mathbf{Y} \mathbf{Y}^T \right] \right) \right\}
\end{aligned} \qquad (14)
$$

where $\mathbf{K}_1 = \mathbf{P}_1 \mathbf{P}_1^T$ and $\mathbf{P}_1$ is a $T \times (T-1)$ matrix with elements $p_{ii} = 1$ and $p_{(i+1)i} = -1$ (the rest are zero). The approximation holds when $T \to \infty$. Again, $\boldsymbol{\Lambda}^{(1)}$ and $\boldsymbol{\Lambda}^{(2)}$ are defined as above.

In all cases the partition function Z is defined as $Z = \int P(\mathbf{Y}) d\mathbf{Y}$. The motivation behind choosing the above latent priors was given by the influential analysis made in [7] where the connection between (deterministic) LPP, PCA and LDA was explored. A further piece of the puzzle was added by the recent work [24] where the linear dynamical system prior (Eq. 14) was used in order to provide a derivation of SFA in a ML framework. By formulating the appropriate priors for these models we unify these subspace methods in a single probabilistic framework of a linear generative model along with a prior of the form

$$P(\mathbf{Y}) \propto \exp\left\{-\tfrac{1}{2}\left(\operatorname{tr}\left[\boldsymbol{\Lambda}^{(1)}\mathbf{Y}\mathbf{B}^{(1)}\mathbf{Y}^T\right] + \operatorname{tr}\left[\boldsymbol{\Lambda}^{(1)}\mathbf{Y}\mathbf{B}^{(2)}\mathbf{Y}^T\right]\right)\right\}. \tag{15}$$

The differentiation amongst these models lies in the neighbourhood over which the potentials are defined. In fact, the varying neighbouring system is translated into the matrices $\mathbf{B}^{(1)}$ and $\mathbf{B}^{(2)}$ in the functional form of the potentials, essentially encapsulating the latent covariance connectivity. E.g., for Eq. 11, $\mathbf{B}^{(1)} = \mathbf{I}$ and $\mathbf{B}^{(2)} = \mathbf{M}$, for Eq. 12, $\mathbf{B}^{(1)} = \mathbf{M}_c$ and $\mathbf{B}^{(2)} = \mathbf{M}_t$, for Eq. 13, $\mathbf{B}^{(1)} = \tilde{\mathbf{L}}$ and $\mathbf{B}^{(2)} = \tilde{\mathbf{D}}$ and finally for Eq. 14, $\mathbf{B}^{(1)} = \mathbf{K}$ and $\mathbf{B}^{(2)} = \mathbf{I}$. In the following we will show that ML estimation using these potentials is equivalent to the deterministic formulations of PCA, LDA and LPP. SFA is a special case for which it was already shown in [24] that a potential of the form of Eq. 14 with an ML framework produces a projection with the same direction as Eq. 8.

Adopting the linear generative model in Eq. 10, the corresponding conditional data (observation) probability is a Gaussian,

$$P(\mathbf{x}_t|\mathbf{y}_t, \mathbf{W}, \sigma_x^2) = \mathcal{N}(\mathbf{W}^{-1}\mathbf{y}_t, \sigma_x^2). \tag{16}$$

Having chosen a prior of the form described in Eq. 15 (e.g., as defined in Eq. 11,12,13,14) we can now derive the likelihood of our model as follows:

$$P(\mathbf{X}|\Psi) = \int \prod_{t=1}^{T} P(\mathbf{x}_t|\mathbf{y}_t, \mathbf{W}, \sigma^2) P(\mathbf{Y}|\sigma_{1:N}^2, \lambda_{1:N}) d\mathbf{Y}, \tag{17}$$

where the model parameters are defined as $\Psi = \{\sigma_x^2, \mathbf{W}, \sigma_{1:N}^2, \lambda_{1:N}\}$. In the following we will show that by substituting the above priors in Eq. 17 and maximising the likelihood we obtain loadings $\mathbf{W}$ which are co-directional (up to a scale ambiguity) to deterministic PCA, LDA and LPPs and SFA. Firstly, by substituting the general prior (Eq. 15) in the likelihood (Eq. 17), we obtain

$$\begin{array}{c} P(\mathbf{X}|\Psi) = \int \prod_{t=1}^{T} P(\mathbf{x}_t|\mathbf{y}_t, \mathbf{W}, \sigma^2)\tfrac{1}{Z}\exp \\ \left\{-\tfrac{1}{2}\left(\operatorname{tr}\left[\boldsymbol{\Lambda}^{(1)}\mathbf{Y}\mathbf{B}^{(1)}\mathbf{Y}^T\right] + \operatorname{tr}\left[\boldsymbol{\Lambda}^{(2)}\mathbf{Y}\mathbf{B}^{(2)}\mathbf{Y}^T\right]\right)\right\} d\mathbf{Y}. \end{array} \tag{18}$$

In order to obtain a zero-variance limit ML solution, we map $\sigma_x \to 0$

$$\begin{array}{c} P(\mathbf{X}|\Psi) = \int \prod_{t=1}^{T} \delta(\mathbf{x}_t - \mathbf{W}^{-1}\mathbf{y}_t)\tfrac{1}{Z}\exp \\ \left\{-\tfrac{1}{2}\left(\operatorname{tr}\left[\boldsymbol{\Lambda}^{(1)}\mathbf{Y}\mathbf{B}^{(1)}\mathbf{Y}^T\right] + \operatorname{tr}\left[\boldsymbol{\Lambda}^{(2)}\mathbf{Y}\mathbf{B}^{(2)}Y^T\right]\right)\right\} d\mathbf{Y} \end{array} \tag{19}$$

By completing the integrals and taking the log, we obtain the conditional log-likelihood:

$$\begin{array}{c} L(\Psi) = \log P(\mathbf{X}|\theta) = -\log Z + T\log|\mathbf{W}| - \tfrac{1}{2} \\ \operatorname{tr}\left[\boldsymbol{\Lambda}^{(1)}\mathbf{W}\mathbf{X}\mathbf{B}^{(1)}\mathbf{X}^T\mathbf{W}^T + \boldsymbol{\Lambda}^{(2)}\mathbf{W}\mathbf{X}\mathbf{B}^{(2)}\mathbf{X}^T\mathbf{W}^T\right] \end{array} \tag{20}$$

where Z is a constant term independent of $\mathbf{W}$. By maximising for $\mathbf{W}$ we obtain

$$T\mathbf{W}^{-T} - \left(\mathbf{\Lambda}^{(1)}\mathbf{WXB}^{(1)}\mathbf{X}^T + \mathbf{\Lambda}^{(2)}\mathbf{WXB}^{(2)}\mathbf{X}^T\right) = \mathbf{0},$$
$$\mathbf{I} = \mathbf{\Lambda}^{(1)}\mathbf{WXB}^{(1)}\mathbf{X}^T\mathbf{W}^T + \mathbf{\Lambda}^{(2)}\mathbf{WXB}^{(2)}\mathbf{X}^T\mathbf{W}^T. \tag{21}$$

It is easy to prove that since $\mathbf{\Lambda}^{(1)}, \mathbf{\Lambda}^{(2)}$ are diagonal matrices, the $\mathbf{W}$ which satisfies Eq. 21 simultaneously diagonalises (up to a scale ambiguity) $\mathbf{XB}^{(1)}\mathbf{X}^T$ and $\mathbf{XB}^{(2)}\mathbf{X}^T$. By substituting the $\mathbf{B}$ matrices as defined above in Eq. 21, we now consider all cases separately. For PCA, by utilising Eq. 11, Eq. 21 is reformulated as $\mathbf{WXX}^T\mathbf{W}^T = \left[\mathbf{\Lambda}^{(1)}\right]^{-1}$ hence $\mathbf{W}$ is given by the eigenvectors of the total scatter matrix $\mathbf{S}$. For LDA (Eq. 12), Eq. 21 is reformulated as $\mathbf{\Lambda}^{(1)}\mathbf{WXMX}^T\mathbf{W}^T + \mathbf{\Lambda}^{(2)}\mathbf{WXX}^T\mathbf{W}^T = \mathbf{I}$. Thus, $\mathbf{W}$ is given by the directions that simultaneously diagonalise $\mathbf{S}$ and $\mathbf{S}_w$. For LPP (Eq. 13), Eq. 21 yields $\mathbf{\Lambda}^{(1)}\mathbf{WXLX}^T\mathbf{W}^T + \mathbf{\Lambda}^{(2)}\mathbf{WXD}^T\mathbf{X}^T\mathbf{W}^T = \mathbf{I}$, therefore $\mathbf{W}$ is given by the directions that simultaneously diagonalise $\mathbf{X}\tilde{\mathbf{L}}\mathbf{X}^T$ and $\mathbf{X}\tilde{\mathbf{D}}\mathbf{X}^T$. Finally, for SFA, by utilising Eq. 14, Eq. 21 becomes $\mathbf{\Lambda}^{(1)}\mathbf{WXKX}^T\mathbf{W}^T + \mathbf{\Lambda}^{(2)}\mathbf{WXX}^T\mathbf{W}^T = \mathbf{I}$, and $\mathbf{W}$ is given by the directions that simultaneously diagonalise $\mathbf{XKX}^T$ and $\mathbf{XX}^T$.

The above shows that the ML solution following our framework is equivalent to the deterministic models of PCA, LDA, LPP and SFA. The direction of $\mathbf{W}$ does not depend of σ_n^2 and λ_n, which can be estimated by optimizing Eq. 20 with regards to these parameters. In this work we will provide update rules for σ_n and λ_n using an EM framework. As we observe, the ML loading $\mathbf{W}$ does not depend on the exact setting of λ_n, so long as they are all different. If $0 < \lambda_n < 1, \forall n$, then larger values of λ_n correspond to more expressive (in case of PCA), more discriminant (in LDA), more local (in LPP) and slower latents (in case of SFA). This corresponds directly to the ordering of the solutions from PCA, LDA, LPP and SFA. To recover exact equivalence to LDA, LPP, SFA another limit is required that corrects the scales. There are several choices, but a natural one is to let $\sigma_n^2 = 1 - \lambda_n^2$. This choice in case of LDA and SFA fixes the prior covariance of the latent variables to be one ($\mathbf{W}^T\mathbf{XXW} = \mathbf{I}$) and it forces $\mathbf{W}^T\mathbf{XDXW} = \mathbf{I}$ in case of LPP. This choice of σ_n has been also discussed in [24] for SFA. We note that in case of PCA, we should set σ_n to be analogous to the corresponding eigenvalue of the covariance matrix, since otherwise the method will result to a *minor* component analysis.

4   A Unified EM Framework for Component Analysis

In the following we propose a unified EM framework for component analysis. This framework can treat all priors with undirected links (such as Eq. 11, Eq. 12 and Eq. 13). The EM of the prior in Eq. 14 contains only directed links with no loops, and thus can be solved (without any approximations) similarly to the EM of a linear dynamical system [3]. If we treat the SFA links as undirected, we end up with an autoregressive component analysis (see Section 4.1).

In order to perform EM with an MRF prior we adopt the simple and elegant mean field approximation theory [15,5,27], which essentially allows computationally favourable factorizations within an EM framework. Let us consider a generalisation of the priors we defined in Sec. 3 to $\mathcal{M}$ MRFs:

$$P(\mathbf{Y}|\beta) = \prod_{\mu \in \mathcal{M}} \frac{1}{Z^\mu} \exp\{Q^\mu\} \tag{22}$$

$$Q^\mu = -\sum_{n=1}^{N} \frac{f_\mu(\lambda_n)}{2\sigma_n^2} \frac{1}{c} \sum_{i \in \omega_i} \frac{1}{c_j^\mu} \sum_{j \in \omega_j^\mu} (y_{n,i} - \phi_\mu(\lambda_n) y_{n,j})^2$$

where c and c_j are normalisation constants, while f_μ and ϕ_μ are functions of λ_n. Without loss of generality and for clarity of notation, we assume that $c = 1$, $c_j^\mu = |\omega_j^\mu|$ and $\omega_i^\mu = [1, \ldots, T]$. Furthermore, we now assume the linear model

$$\mathbf{x}_i = \mathbf{W} \mathbf{y}_i + \epsilon_i, \epsilon_i \sim \mathcal{N}(0, \sigma_x^2). \tag{23}$$

For clarity, the set of parameters associated with the prior (i.e. energy function) are denoted as $\beta = \{\sigma_{1:N}, \lambda_{1:N}\}$, the parameters related to the observation model $\theta = \{\mathbf{W}, \sigma_x\}$, while the total parameter set is denoted as $\Psi = \{\theta, \beta\}$. In agreement with [5], we replace the marginal distribution $P(\mathbf{Y}|\beta)$ by the mean-field

$$P(\mathbf{Y}|\beta) \approx \prod_{i=1}^{T} P(\mathbf{y}_i | \mathbf{m}_i^{\mathcal{M}}, \beta^{\mathcal{M}}). \tag{24}$$

Since different CA models have different latent connectivities (and thus different MRF configurations), the mean-field influence on each latent point $\mathbf{y}_i$ now depends on the model-specific connectivity via $\mathbf{m}_i^{\mathcal{M}}$, a function of $\mathbb{E}[\mathbf{y}_j]$. After calculating the normalising integral for the priors Eq. 11-13 and given the mean-field, it can be easily shown that Eq. 22 follows a Gaussian distribution,

$$P(\mathbf{y}_i | \mathbf{m}_i^{\mathcal{M}}, \beta) = \mathcal{N}(\mathbf{m}_i^{\mathcal{M}}, \mathbf{\Sigma}^{\mathcal{M}}), \tag{25}$$

$$\mathbf{m}_i^{\mathcal{M}} = \sum_{\mu \in \mathcal{M}} \left(\frac{f_\mu(\lambda_n)\phi_\mu(\lambda_n)}{F^M(\lambda_n)} \boldsymbol{\mu}_{\omega_j^\mu} \right) = \sum_{\mu \in \mathcal{M}} \Lambda^\mu \boldsymbol{\mu}_{\omega_j^\mu}, \tag{26}$$

$$\mathbf{\Sigma}^{\mathcal{M}} = \left[\delta_{mn} \frac{\sigma_n^2}{F^M(\lambda_n)} \right] \tag{27}$$

with $\boldsymbol{\mu}_{\omega_j^\mu} = \frac{1}{|\omega_j^\mu|} \sum_{j \in \omega_j^\mu} \mathbb{E}[\mathbf{y}_{n,j}]$ and $F^M(\lambda_n) = \sum_{\mu \in \mathcal{M}} f_\mu(\lambda_n)$.

Therefore, by simply replacing the parametrisation of the priors we defined in Eq. 11 (PCA), 12 (LDA) and 13 (LPP) (see also Tab. 1) for the mean and

Table 1. MRF configuration for PCA, LDA and LPP

$\mathcal{M} = \{\alpha, \beta\}$	$F^{\mathcal{M}} = \sum_\mu f_\mu$	f_α	ϕ_α	ω_j^α	f_β	ϕ_β	ω_j^β
PCA (11)	1	1	λ_n	$\{1 \ldots T\} \setminus \{i\}$			
LDA (12)	$\lambda_n + (1-\lambda_n)^2$	λ_n	1	$\bar{\mathcal{C}}_i$	$(1-\lambda_n)^2$	1	$\{1 \ldots T\} \setminus \{i\}$
LPP (13)	$\lambda_n + (1-\lambda_n)^2$	λ_n	1	$\mathcal{N}_i^s$	$(1-\lambda_n)^2$	0	$\{1\}$

variance (Eq. 26 and Eq. 27), we obtain the distribution for each CA method we propose. The means $\mathbf{m}_i^{\mathcal{M}}$ for PCA, LDA and LPP are obtained as

$$\mathbf{m}_i^{(\text{PCA})} = \mathbf{\Lambda}\boldsymbol{\mu}_{-i}, \mathbf{m}_i^{(\text{LDA})} = \mathbf{\Lambda}^{(\alpha)}\boldsymbol{\mu}_{-i} + \mathbf{\Lambda}^{(\beta)}\boldsymbol{\mu}_{\tilde{\mathcal{C}}_i}, \mathbf{m}_i^{(\text{LPP})} = \mathbf{\Lambda}^{(\alpha)}\boldsymbol{\mu}_{\mathcal{N}_i^s} \quad (28)$$

and the variances $\boldsymbol{\Sigma}^{\mathcal{M}}$ as

$$\boldsymbol{\Sigma}^{(\text{PCA})} = \left[\delta_{mn}\sigma_n^2\right], \boldsymbol{\Sigma}^{(\text{LDA})} = \boldsymbol{\Sigma}^{(\text{LPP})} = \left[\delta_{mn}\left(\frac{\sigma_n^2}{\lambda_n + (1-\lambda_n)^2}\right)\right] \quad (29)$$

where $\boldsymbol{\mu}_{-i} = \frac{1}{T-1}\sum_{j\neq i}^{T}\mathbb{E}^{\mathcal{M}}[\mathbf{y}_j]$ is the mean, $\boldsymbol{\mu}_{\tilde{\mathcal{C}}_i} = \frac{1}{|\tilde{\mathcal{C}}_i|}\sum_{j\in\tilde{\mathcal{C}}_i}\mathbb{E}^{\mathcal{M}}[\mathbf{y}_j]$ the class mean, and $\boldsymbol{\mu}_{\mathcal{N}_i^s} = \frac{1}{|\mathcal{N}_i^s|}\sum_{j\in\mathcal{N}_i^s}^{T}\mathbb{E}^{\mathcal{M}}[\mathbf{y}_j]$ the neighbourhood mean. Furthermore, $\mathbf{\Lambda} = [\delta_{mn}\lambda_n]$, $\mathbf{\Lambda}^{(\alpha)} = \left[\delta_{mn}\left(\frac{\lambda_n}{\lambda_n + (1-\lambda_n)^2}\right)\right]$ and $\mathbf{\Lambda}^{(\beta)} = \left[\delta_{mn}\left(\frac{(1-\lambda_n)^2}{\lambda_n + (1-\lambda_n)^2}\right)\right]$.

In order to complete the expectation step, we infer the first order moments of the latent posterior, defined as

$$P(\mathbf{y}_i|\mathbf{x}_i, \mathbf{m}_i^{\mathcal{M}}, \Psi^{\mathcal{M}}) = \frac{P(\mathbf{x}_i|\mathbf{y}_i, \theta^{\mathcal{M}})P(\mathbf{y}_i|\mathbf{m}_i^{\mathcal{M}}, \beta^{\mathcal{M}})}{\int_{\mathbf{y}_i} P(\mathbf{x}_i|\mathbf{y}_i, \theta^{\mathcal{M}})P(\mathbf{y}_i|\mathbf{m}_i^{\mathcal{M}}, \beta^{\mathcal{M}})d\mathbf{y}_i}. \quad (30)$$

Since the posterior is a product of Gaussians[2], we have

$$P(\mathbf{y}_i|\mathbf{x}_i, \mathbf{m}_i^{\mathcal{M}}, \Psi^{\mathcal{M}}) = \mathcal{N}(\mathbf{y}_i|(\mathbf{W}^T\mathbf{x}_i + \boldsymbol{\Sigma}^{\mathcal{M}^{-1}}\mathbf{m}_i^{\mathcal{M}})\mathbf{A}, \sigma_x^{\mathcal{M}^2}\mathbf{A}) \quad (31)$$

with $\mathbf{A} = (\mathbf{W}^T\mathbf{W} + (\hat{\boldsymbol{\Sigma}}^{\mathcal{M}})^{-1})^{-1}$ and $\hat{\boldsymbol{\Sigma}}^{\mathcal{M}} = \left[\delta_{mn}(\Sigma_{mn}^{\mathcal{M}}/\sigma_x^{\mathcal{M}^2})\right]$. Therefore $\mathbb{E}^{\mathcal{M}}[\mathbf{y}_i]$ is equal to the mean, and $\mathbb{E}^{\mathcal{M}}[\mathbf{y}_i\mathbf{y}_i^T] = \sigma_x^{\mathcal{M}^2}\mathbf{A} + \mathbb{E}[\mathbf{y}_i]\mathbb{E}[\mathbf{y}_i]^T$.

Having recovered the first order moments, we move on to the maximisation step. In order to maximize the marginal log-likelihood, $\log P(\mathbf{X}|\Psi^{\mathcal{M}})$, we adopt the usual EM bound [17], $\int_{\mathbf{Y}} P(\mathbf{Y}|\mathbf{X}, \Psi^{\mathcal{M}}) \log P(\mathbf{X}, \mathbf{Y})d\mathbf{Y}$. By adopting the approximation proposed in [5], the complete-data likelihood is factorised as

$$P(\mathbf{Y}, \mathbf{X}|\Psi^{\mathcal{M}}) \approx \prod_{i=1}^{T} P(\mathbf{x}_i|\mathbf{y}_i, \theta^{\mathcal{M}})P(\mathbf{y}_i|\mathbf{m}_i^{\mathcal{M}}, \beta^{\mathcal{M}}). \quad (32)$$

and therefore, the maximisation term (EM bound) becomes

$$\sum_{i=1}^{T}\int_{\mathbf{y}_i} P(\mathbf{y}_i|\mathbf{x}_i, \mathbf{m}_i^{\mathcal{M}}, \Psi^{\mathcal{M}}) \log P(\mathbf{x}_i, \mathbf{y}_i|\Psi^{\mathcal{M}})d\mathbf{y}_i. \quad (33)$$

As can be seen the likelihood can be separated due to the logarithm for estimating $\theta^{\mathcal{M}} = \{\mathbf{W}^{\mathcal{M}}, \sigma_x^{\mathcal{M}}\}$ and $\beta = \{\sigma_{1:N}^{\mathcal{M}}, \lambda_{1:N}^{\mathcal{M}}\}$ as follows:

$$\theta^{\mathcal{M}} = \arg\max\left\{\sum_{i=1}^{T}\int_{\mathbf{y}_i} P(\mathbf{y}_i|\mathbf{x}_i, \mathbf{m}_i^{\mathcal{M}}, \Psi^{\mathcal{M}}) \log P(\mathbf{x}_i|\mathbf{y}_i, \theta^{\mathcal{M}})d\mathbf{y}_i\right\}. \quad (34)$$

$$\beta^{\mathcal{M}} = \arg\max\left\{\sum_{i=1}^{T}\int_{\mathbf{y}_i} P(\mathbf{y}_i|\mathbf{x}_i, \mathbf{m}_i^{\mathcal{M}}, \Psi^{\mathcal{M}}) \log P(\mathbf{y}_i|\mathbf{m}_i^{\mathcal{M}}, \beta^{\mathcal{M}})d\mathbf{y}_i\right\}. \quad (35)$$

---

[2] The result can be easily obtained by completing the square for $\mathbf{y}_i$.

Subsequently, we maximise the log-likelihoods wrt. the parameters, recovering the update equations. For θ, by maximising Eq. 34, we obtain

$$\mathbf{W}^{\mathcal{M}} = \left(\sum_{i=1}^{T} \mathbf{x}_i \mathbb{E}^{\mathcal{M}}[\mathbf{y}_i]^T \right) \left(\sum_{i=1}^{T} \mathbb{E}^{\mathcal{M}}[\mathbf{y}_i \mathbf{y}_i^T] \right)^{-1} \tag{36}$$

$$\sigma_x^{\mathcal{M}^2} = \frac{1}{FT} \sum_{i=1}^{T} \{ ||\mathbf{x}_i||^2 - 2\mathbb{E}^{\mathcal{M}}[\mathbf{y}_i]^T (\mathbf{W}^{\mathcal{M}})^T \mathbf{x}_i \\ + \mathrm{Tr}[\mathbb{E}^{\mathcal{M}}[\mathbf{y}_i \mathbf{y}_i^T] (\mathbf{W}^{\mathcal{M}})^T \mathbf{W}^{\mathcal{M}}] \}. \tag{37}$$

Similarly, by maximising Eq. 35 for β, we obtain:

$$\sigma_n^{\mathcal{M}^2} = \frac{F^{\mathcal{M}}(\lambda_n)}{T} \sum_{i=1}^{T} (\mathbb{E}^{\mathcal{M}}[y_{n,i}^2] - 2\mathbb{E}^{\mathcal{M}}[y_{n,i}] m_{n,i}^{\mathcal{M}} + m_{n,i}^{\mathcal{M}^2}) \tag{38}$$

where, as defined in Eq. 27, for PCA $F^{\mathcal{M}}(\lambda_n) = 1$, and for LDA and LPP $F^{\mathcal{M}}(\lambda_n) = \lambda_n + (1 - \lambda_n)^2$. For λ_n we choose the updates as described in Sec. 3. In what follows, we discuss some further points wrt. the proposed EM framework.

4.1   Further Discussion

Comparison to other Probabilistic Variants of PCA. It is clear that regarding the proposed EM-PCA, the updates for $\theta = \{\mathbf{W}, \sigma_x^2\}$ as well as the distribution of the latent variable $\mathbf{y}_i$ are the same with previously proposed probabilistic approaches [16],[22]. The only variation is the mean of $\mathbf{y}_i$, which in our case is shifted by the mean field, $\hat{\mathbf{\Sigma}}^{(PCA)^{-1}} \mathbf{m}_i^{(PCA)}$, while in addition, our method models per-dimension variance (σ_n). Note that in order to fully identify with the PPCA proposed in [22], we can set $\lambda_n = 0$ and $\sigma_n = 1$.

EM for SFA. The SFA prior in Eq. 14 allows for two interpretations of the SFA graphical model: both as an undirected MRF and a directed Dynamic Bayesian Network (DBN). Based on the undirected MRF interpretation, SFA trivially fits into the EM framework described in this section, leading to an autoregressive SFA model [19], able to learn bi-directional latent dependencies. When considering the SFA prior as a directed Markov chain, one can resort to exact inference techniques applied on DBNs. In fact, the EM for SFA can be straightforwardly reduced to solving a standard Linear Dynamical System (Chap. 13 [3]), while also enforcing diagonal transition matrices and setting $\sigma_n^2 = 1 - \lambda_n^2$.

Complexity. The proposed EM algorithm iteratively recovers the latent space preserving the characteristics enforced by the selected latent neighbourhood. Similarly to PCCA [16,22], for $N \ll T, F$ the complexity of each iteration is bounded by $O(TNF)$, unlike deterministic models ($\mathcal{O}(T^3)$). This is due to the covariance appearing only in trace operations, and is of high value for our proposed models, especially in case where no other probabilistic equivalent exists.

Probabilistic LDA Classification. We can exploit the probabilistic nature of the proposed EM-LDA in order to probabilistically infer the most likely class

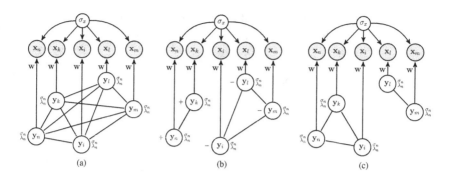

Fig. 1. MRF connectivities used for PCA, LDA and LPP under our unifying framework, with shaded nodes representing observations. (a) Fully connected MRF (PCA), (b) within-class connected MRF (LDA), and (c) a locally connected MRF (LPP).

assignment for unseen data. Instead of using the inferred projection, we can essentially utilise the log-likelihood of the model. In more detail, we can estimate the marginal log-likelihood for each test point $\mathbf{x}^*$ being assigned to each class c:

$$\arg_c \max \left\{ \log P(\mathbf{x}^* | \mathbf{m}^{\mathcal{M}_c}, \Psi^{\mathcal{M}}) \right\} \tag{39}$$

where by adopting the usual EM bound (as shown in Eq. 33) this boils down to

$$\arg_c \max \int_{\mathbf{y}_i^*} P(\mathbf{y}_i^* | \mathbf{x}_i^*, \mathbf{m}^{\mathcal{M}_c}, \Psi^{\mathcal{M}}) \log P(\mathbf{x}_i^*, \mathbf{y}_i^* | \Psi^{\mathcal{M}}) d\mathbf{y}_i^* \tag{40}$$

where $P(\mathbf{y}_i^* | \mathbf{x}_i^*, \mathbf{m}^{\mathcal{M}}, \Psi^{\mathcal{M}})$ is estimated as in Eq. 31, by utilising the inferred model parameters ($\Psi^{\mathcal{M}}$) along with the class model. Note that since the posterior mean given $\mathbf{x}_i$ depends on all *other* observations excluding i (Eq. 28), we only need to store the class mean estimated as a weighted average of all training data and all training data in class c, as

$$\mathbf{m}^{\mathcal{M}_c} = \Lambda^{(\alpha)} \frac{1}{T} \sum_{j=1}^{T} \mathbb{E}^{\mathcal{M}}[\mathbf{y}_j] + \Lambda^{(\beta)} \frac{1}{|\mathcal{C}_c|} \sum_{j \in \mathcal{C}_c} \mathbb{E}^{\mathcal{M}}[\mathbf{y}_j] \tag{41}$$

This is in contrast to traditional methods where all the (projected) training data have to be kept. Furthermore, during evaluation, we only need to estimate the likelihood of each test datum's assignment to each class ($\mathcal{O}(|C|)$), rather than compare each test datum to the entire training set ($\mathcal{O}(T)$).

5   Experiments

As proof of concept, we provide experiments both on synthetic and real-world data. We aim to (i) experimentally validate the equivalence of the proposed probabilistic models to other models belonging in the same class, and (ii) experimentally evaluate the performance of our models against others in the same class.

Synthetic Data. We demonstrate the application of our proposed probabilistic CA techniques on a set of synthetic data (see Fig. 2), generated utilising the Dimensionality Reduction Toolbox. In more detail, we compare the corresponding deterministic formulations of PCA, LDA and LLE to the proposed probabilistic models. The aim is mainly to qualitatively illustrate the equivalence of the proposed methods (by observing that the probabilistic projections match the deterministic equivalents). Furthermore, the variance modelling per latent dimension in our EM-LDA is clear in $\mathbb{E}[\mathbf{y}]$ of LDA (Fig. 2, Col. 3). This will prove beneficial prediction-wise, as we show in the following section.

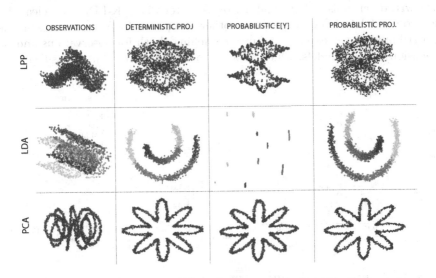

Fig. 2. Synthetic experiments on deterministic LLE, LDA and PCA (2nd col.) compared to the proposed probabilistic methods ($\mathbb{E}[y]$ in 3rd col., projections in 4th col.)

Real Data: Face Recognition via EM-LDA. One of the most common applications of LDA is face recognition. Therefore, we utilise various databases in order to verify the performance of our proposed EM-LDA. In more detail, we utilise the popular Extended Yale B database [6], as well as the PIE [20] and AR databases [12]. The experiments span a wide range of variability, such as various facial expressions, illumination changes, as well as pose changes. In more detail from the CMU PIE database [20] we used a total of 170 images near frontal images for each subject. For training, we randomly selected a subset consisting of 5 images per subject, while for testing the remaining images were used. For the extended Yale B database [6], we utilised a subset of 64 near frontal images per subject, where a random selection of 5 images per subject was used for training, while the rest of the images where used for testing. Regarding AR [12], we focus on facial expressions. We firstly randomly select 100 subjects. Subsequently, use the images which portray varying facial expressions from session 1, while using the corresponding images from session 2 for testing. In related experiments, we compared our EM-LDA against deterministic LDA, the Fukunaga-Koontz variant (FK-LDA) [28] and PLDA [14] (which has been shown to outperform other probabilistic methods such as [8] in [11]) under the presence of Gaussian noise. We used the gradients of each image pixel as features, since as we experimentally verified, this improved the results for all compared methods. The errors of each compared method for each database, accompanied by increasing Gaussian noise in the input, is shown in Fig. 3. Although PLDA offers a substantial improvement wrt. deterministic LDA and performs better than FK-LDA, it is clear that the proposed EM-LDA outperforms other compared LDA variants. This can be attributed to the explicit variance modelling (both for observations and per dimension) in our models, which appears to enable more robust classification.

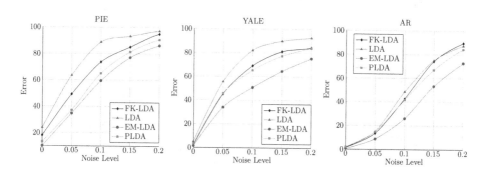

Fig. 3. Recognition error on PIE, YALE and AR under increasing Gaussian noise, comparing LDA, FK-LDA [28] the proposed EM-LDA and PLDA [14]

Real Data: Face Visualisation via EM-LPP. One of the typical applications of Neighbour Embedding methods is the visualisation of, usually high-dimensional, data at hand. In particular, LPPs have often been used in visualising faces, providing an intuitive understanding of the variance and structural properties of the data [16], [7]. In order to evaluate the proposed EM-LPP, which is to the best of our knowledge the first probabilistic equivalent to LPP [13], we experiment on the Frey Faces database [18], which contains 1965 images, captured as sequential frames of a video sequence. We apply a similar experiment to [7]. We firstly perturbed the images with random Gaussian noise, while subsequently we apply EM-LPP and LPP. The resulting space is illustrated in Fig. 4. It is clear that the deterministic LPP was unable to cope with the added Gaussian noise, failing to capture a meaningful data clustering. Note that the proposed EM-LPP was able to well capture the structure of the input data, modelling both pose and expression within the inferred latent space.

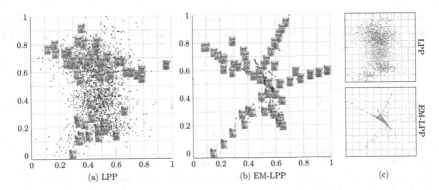

Fig. 4. Latent projections obtained by applying the proposed EM-LPP and LPP [13] to the Frey Faces database, with each image perturbed with random Gaussian noise

6   Conclusions

In this paper we introduced a novel, unifying probabilistic component analysis framework, reducing the construction of probabilistic component analysis models to selecting the proper latent neighbourhood via the design of the latent connectivity. Our framework can thus be used to introduce novel probabilistic component analysis techniques by formulating new latent priors as products of MRFs. We have shown specific priors which when used, generate probabilistic models corresponding to PCA, LPP, LDA and SFA, and by doing so, we introduced the first, favourable complexity-wise, probabilistic equivalent to LPP. Finally, by means of theoretical analysis and experiments, we have demonstrated various advantages that our proposed methods pose against existing probabilistic and deterministic techniques.

Acknowledgements. This work has been funded by the European Union's 7th Framework Programme [FP7/2007-2013] under grant agreement no. 288235 (FROG) and the EPSRC project EP/J017787/1 (4DFAB).

References

1. Akisato, K., Masashi, S., Hitoshi, S., Hirokazu, K.: Designing various multivariate analysis at will via generalized pairwise expression. JIP 6(1), 136–145 (2013)
2. Belhumeur, P., Hespanha, J., Kriegman, D.: Eigenfaces vs. fisherfaces: Recognition using class specific linear projection. IEEE TPAMI 19(7), 711–720 (1997)
3. Bishop, C.M.: Pattern Recognition and Machine Learning (Information Science and Statistics). Springer-Verlag New York, Inc., Secaucus (2006)
4. Borga, M., Landelius, T., Knutsson, H.: A unified approach to PCA, PLS, MLR and CCA (1997)
5. Celeux, G., Forbes, F., Peyrard, N.: EM procedures using mean field-like approximations for Markov model-based image segmentation. Pattern Recogn. 36(1), 131–144 (2003)

6. Georghiades, A., Belhumeur, P., Kriegman, D.: From few to many: Illumination cone models for face recognition under variable lighting and pose. IEEE TPAMI 23(6), 643–660 (2001)
7. He, X., Yan, S., Hu, Y., Niyogi, P., Zhang, H.: Face recognition using laplacianfaces. IEEE TPAMI 27(3), 328–340 (2005)
8. Ioffe, S.: Probabilistic Linear Discriminant Analysis. In: Leonardis, A., Bischof, H., Pinz, A. (eds.) ECCV 2006. LNCS, vol. 3954, pp. 531–542. Springer, Heidelberg (2006)
9. Klampfl, S., Maass, W.: Replacing supervised classification learning by slow feature analysis in spiking neural networks. In: Advances in NIPS, pp. 988–996 (2009)
10. Kokiopoulou, E., Chen, J., Saad, Y.: Trace optimization and eigenproblems in dimension reduction methods. Numer. Linear Algebra Appl. 18(3), 565–602 (2011)
11. Li, P., Fu, Y., Mohammed, U., Elder, J.H., Prince, S.J.: Probabilistic models for inference about identity. IEEE TPAMI 34(1), 144–157 (2012)
12. Martinez, A.M.: The AR face database. CVC Technical Report 24 (1998)
13. Niyogi, X.: Locality preserving projections. In: NIPS 2003, vol. 16, p. 153 (2004)
14. Prince, S.J.D., Elder, J.H.: Probabilistic linear discriminant analysis for inferences about identity. In: ICCV (2007)
15. Qian, W., Titterington, D.: Estimation of parameters in hidden markov models. Phil. Trans. of the Royal Society of London. Series A: Physical and Engineering Sciences 337(1647), 407–428 (1991)
16. Roweis, S.: EM algorithms for PCA and SPCA. In: NIPS 1998, pp. 626–632 (1998)
17. Roweis, S., Ghahramani, Z.: A unifying review of linear gaussian models. Neural Comput. 11(2), 305–345 (1999)
18. Roweis, S.T., Saul, L.K.: Nonlinear dimensionality reduction by locally linear embedding. Science 290(5500), 2323–2326 (2000)
19. Rue, H., Held, L.: Gaussian Markov random fields: Theory and applications. CRC Press (2004)
20. Sim, T., Baker, S., Bsat, M.: The CMU Pose, Illumination, and Expression Database. In: Proc. of the IEEE FG 2002 (2002)
21. Sun, L., Ji, S., Ye, J.: A least squares formulation for a class of generalized eigenvalue problems in machine learning. In: ICML 2009, pp. 977–984. ACM (2009)
22. Tipping, M.E., Bishop, C.M.: Probabilistic principal component analysis. Journal of the Royal Statistical Society, Series B 61, 611–622 (1999)
23. De la Torre, F.: A least-squares framework for component analysis. IEEE TPAMI 34(6), 1041–1055 (2012)
24. Turner, R., Sahani, M.: A maximum-likelihood interpretation for slow feature analysis. Neural Computation 19(4), 1022–1038 (2007)
25. Wiskott, L., Sejnowski, T.: Slow feature analysis: Unsupervised learning of invariances. Neural Computation 14(4), 715–770 (2002)
26. Yan, S., et al.: Graph embedding and extensions: A general framework for dimensionality reduction. IEEE TPAMI 29(1), 40–51 (2007)
27. Zhang, J.: The mean field theory in EM procedures for Markov random fields. IEEE Transactions on Signal Processing 40(10), 2570–2583 (1992)
28. Zhang, S., Sim, T.: Discriminant subspace analysis: A fukunaga-koontz approach. IEEE TPAMI 29(10), 1732–1745 (2007)
29. Zhang, Y., Yeung, D.-Y.: Heteroscedastic probabilistic linear discriminant analysis with semi-supervised extension. In: Buntine, W., Grobelnik, M., Mladenić, D., Shawe-Taylor, J. (eds.) ECML PKDD 2009, Part II. LNCS, vol. 5782, pp. 602–616. Springer, Heidelberg (2009)

Flexible Shift-Invariant Locality and Globality Preserving Projections

Feiping Nie, Xiao Cai, and Heng Huang*

Department of Computer Science and Engineering, University of Texas at Arlington,
Arlington, 76019, Texas, USA
feipingnie@gmail.com, xiao.cai@mavs.uta.edu, heng@uta.edu

Abstract. In data mining and machine learning, the embedding methods have commonly been used as a principled way to understand the high-dimensional data. To solve the out-of-sample problem, local preserving projection (LPP) was proposed and applied to many applications. However, LPP suffers two crucial deficiencies: 1) the LPP has no shift-invariant property which is an important property of embedding methods; 2) the rigid linear embedding is used as constraint, which often inhibits the optimal manifold structures finding. To overcome these two important problems, we propose a novel flexible shift-invariant locality and globality preserving projection method, which utilizes a newly defined graph Laplacian and the relaxed embedding constraint. The proposed objective is very challenging to solve, hence we derive a new optimization algorithm with rigorously proved global convergence. More importantly, we prove our optimization algorithm is a Newton method with fast quadratic convergence rate. Extensive experiments have been performed on six benchmark data sets. In all empirical results, our method shows promising results.

1 Introduction

In many data mining applications, it is highly desirable to map high-dimensional input data to a lower dimensional space, with a constraint that the data from similar classes will be projected to nearby locations in the new space. Thus, many data embedding methods have been developed. Depending on whether the label information is used, these methods can be classified into two categories, *i.e.*, unsupervised and supervised. A representative of unsupervised embedding methods is PCA [11], which aims at identifying a lower-dimensional space maximizing the variance among data. A representative of supervised embedding methods is LDA [4], which aims at identifying a lower dimensional space minimizing the inter-class similarity while maximizing the intra-class similarity simultaneously.

To discover the intrinsic manifold structure of the data, multiple nonlinear embedding algorithms have been recently proposed to use an eigen-decomposition for obtaining a lower-dimensional embedding of data lying on a non-linear manifold, such as Isomap [22], LLE [19], Laplacian Eigenmap [2], Local Tangent Space Alignment

---

* This work was partially supported by NSF IIS-1117965, IIS-1302675, IIS-1344152, DBI-1356628.

T. Calders et al. (Eds.): ECML PKDD 2014, Part II, LNCS 8725, pp. 485–500, 2014.
© Springer-Verlag Berlin Heidelberg 2014

(LTSA) [25] and Local Spline Embedding (LSE) [24]. However, many of them, such as Isomap and Laplacian Eigenmap, suffer from the out-of-sample problem, *i.e.* how to embed new points in relation to a previously specified configuration of fixed points. To deal with this problem, He *et al.* [7] developed the Locality Preserving Projections (LPP) method, in which the linear embedding function is used for mapping new data.Nie *et al.* [18] proposed a flexible linearization technique in which LPP and spectral regression [3] are two extreme cases.

Although LPP solved the out-of-sample problem, two crucial deficiencies exist in current LPP based methods. First, LPP has no shift-invariant property which is a basic property of subspace learning methods. The learned subspace (or the projection matrix) should be invariant when all training data points are shifted by the same constant vector. Second, in LPP, the rigid linear embedding is used as constraint, which often limits the search of optimal manifold structures.

To resolve these two important problems, we propose a novel flexible shift invariant locality and globality preserving projection (FLGPP) method. We reformulate the LPP objective using a correct Laplacian matrix which makes the new method shift invariant. Meanwhile, we show that the graph embedding methods are indeed locality and globality preserving projection methods, which were only considered as keeping the local geometrical structure. We relax the rigid linear embedding by allowing the error tolerance such that the data instances can be flexibly embedded. The proposed objective is very difficult to solve. As one important contribution of this paper, we derive a new optimization algorithm with proved global convergence. More importantly, we rigorously prove that our new optimization algorithm is a Newton method with fast quadratic convergence rate. To evaluate our method, we compare the new method to the LDA and LPP methods by performing them on six benchmark data sets. In all empirical results, our new FLGPP method shows promising results.

2   Locality Preserving Projections Revisit

2.1   Review of Related Graph Based Methods

Given n training data points $X = [x_1, \cdots, x_n] \in \mathbb{R}^{d \times n}$, where d is the data dimensionality and n is the number of data points, the graph based methods first construct a graph based on the data to encode the pairwise data similarities. With the graph affinity matrix $A \in \mathbb{R}^{n \times n}$, the Laplacian matrix is defined as $L = D - A$, where D is the diagonal matrix with the i-th diagonal element $D_{ii} = \sum_i A_{ij}$. L is positive semi-definite, and satisfies $L\mathbf{1} = \mathbf{0}$, where $\mathbf{1}$ is a vector having all elements as 1s, and $\mathbf{0}$ is a vector having all elements as 0s.

Traditional spectral clustering (or graph cut) [21,15] and Laplacian embedding (or graph embedding, manifold learning) [2] is to solve the following problem:

$$\min_{F^T Q F = I} Tr(F^T L F), \tag{1}$$

where Q would be D or the n by n identity matrix I. The optimal solution $F \in \mathbb{R}^{n \times m} (m < n)$ to Eq. (1) is the eigenvectors of $Q^{-1}L$ corresponding to the smallest eigenvalues.

The methods solving problem Eq. (1) only use the given training data, with no straightforward extension for out-of-sample examples. To handle the out-of-sample problem, a seminal work called Locality Preserving Projections (LPP) was proposed [7], which is to solve the following problem:

$$\min_{\substack{F^T Q F = I \\ X^T W = F}} Tr(F^T L F), \tag{2}$$

where $W \in \mathbb{R}^{d \times m}(m < d)$ is the projection matrix. This linearization method imposes a rigid constraint $X^T W = F$ on the problem Eq. (1), such that the data outside the training data can also be handled using the projection W. In LPP, only $Q = D$ is considered, thus the problem Eq. (2) can be written as:

$$\min_{W^T X D X^T W = I} Tr(W^T X L X^T W). \tag{3}$$

The optimal solution W to LPP is the eigenvectors of $(XDX^T)^{-1}XLX^T$ corresponding to the smallest eigenvalues. Many algorithms following this linearization method are also proposed for subspace learning and classifications in recent years.

2.2 Shift-Invariant Property

For subspace learning algorithms, shift invariance is a basic and important property. That is to say, the learned subspace (or the projection matrix W) should be invariant when every training data point x_i is shifted by the same constant vector c_i, i.e., X is shifted to $X + c\mathbf{1}^T$. For example, PCA, LDA and regularized least squares regression are all shift-invariant algorithms. We are going to demonstrate this observation.

PCA solves:

$$\min_{W^T W = I} Tr(W^T X L_t X^T W), \tag{4}$$

where $L_t = I - \frac{1}{n}\mathbf{1}\mathbf{1}^T$ is the centering matrix, which is a Laplacian matrix and satisfies $L_t \mathbf{1} = 0$. As a result, we have $(X + c\mathbf{1}^T)L_t(X + c\mathbf{1}^T)^T = XL_tX^T$, and thus the optimal solution W will not be changed when the training data are shifted by an arbitrary vector c.

LDA is to solve:

$$\max_{W} Tr((W^T X L_w X^T W)^{-1} W^T X L_b X^T W), \tag{5}$$

where L_w and L_b are another two Laplacian matrices satisfying $L_w \mathbf{1} = 0$ and $L_b \mathbf{1} = 0$. Obviously we have $(X + c\mathbf{1}^T)L_w(X + c\mathbf{1}^T)^T = XL_wX^T$ and $(X + c\mathbf{1}^T)L_b(X + c\mathbf{1}^T)^T = XL_bX^T$, and thus the optimal solution W is also invariant to an arbitrary shift vector c.

Ridge regression solves:

$$\min_{W, b} \left\| X^T W + \mathbf{1}b^T - Y \right\|_F^2 + \gamma \|W\|_F^2, \tag{6}$$

which has a closed form solution $W = (XL_tX^T + \gamma I)^{-1}XL_tY$. Thus, the optimal solution W of the ridge regression is also invariant to arbitrary shift vector c.

One can immediately observe that the original LPP algorithm does not satisfy the shift-invariant property. When the data points are shifted by a same constant vector, although the distribution of the data points is not changed, the learned subspace by LPP will be changed. This problem should be avoided for a subspace learning algorithm.

3   Shift-Invariant Locality Preserving Projections

The original LPP was derived from Eq. (1), in which the optimal solution is the eigen-vectors of $Q^{-1}L$ corresponding to the smallest eigenvalues. However, the smallest eigenvalue of $Q^{-1}L$ is 0 and the corresponding eigenvector is $\mathbf{1}$, which is usually dis-carded in practice. Thus the actual solutions are the eigenvectors of $Q^{-1}L$ correspond-ing the eigenvalues starting from the second smallest one, which is the solution to the following problem:

$$\min_{\substack{F^TQF=I \\ F^TQ\mathbf{1}=\mathbf{0}}} Tr(F^TLF). \tag{7}$$

Note that there is an additional constraint $F^TQ\mathbf{1} = \mathbf{0}$ in the problem. In the lineariza-tion method, when we use the linear constraint $X^TW = F$, the additional constraint $F^TQ\mathbf{1} = \mathbf{0}$ can not be satisfied since $W^TXQ\mathbf{1} \neq \mathbf{0}$. To fix this problem, we use the linear constraint with bias $X^TW + \mathbf{1}b^T = F$, where $b \in \mathbb{R}^{m \times 1}$ is the bias vector. With the additional constraint $F^TQ\mathbf{1} = \mathbf{0}$, we have $(W^TX + b\mathbf{1}^T)Q\mathbf{1} = \mathbf{0} \Rightarrow b = -\frac{1}{\mathbf{1}^TQ\mathbf{1}}W^TXQ\mathbf{1}$. Thus the linear constraint with bias in the linearization method is:

$$(I - \frac{1}{\mathbf{1}^TQ\mathbf{1}}\mathbf{1}\mathbf{1}^TQ)X^TW = F. \tag{8}$$

By imposing the linear constraint Eq.(8) to problem (1) or (7), the shift-invariant LPP is to solve the following problem [16]:

$$\min_{\substack{F^TQF=I \\ (I-\frac{1}{\mathbf{1}^TQ\mathbf{1}}\mathbf{1}\mathbf{1}^TQ)X^TW=F}} Tr(F^TLF). \tag{9}$$

Define

$$L_q = Q - \frac{1}{\mathbf{1}^TQ\mathbf{1}}Q\mathbf{1}\mathbf{1}^TQ, \tag{10}$$

then the problem (14) can be re-written as

$$\min_{W^TXL_qX^TW=I} Tr(W^TXLX^TW). \tag{11}$$

Note that L and L_q are Laplacian matrix satisfying $L\mathbf{1} = \mathbf{0}$ and $L_q\mathbf{1} = \mathbf{0}$, so we have $(X + c\mathbf{1}^T)L(X + c\mathbf{1}^T)^T = XLX^T$ and $(X + c\mathbf{1}^T)L_q(X + c\mathbf{1}^T)^T = XL_qX^T$. Therefore, the optimal solution W of the problem (11) is invariant to arbitrary shift vector c.

From the above analysis we know that the constraint $W^TXQX^TW = I$ (Q is a di-agonal matrix such as D or I) will make the learned subspace does not satisfy the basic shift invariance property. The correct constraint should be $W^T X L_q X^T W = I$. There are many works following LPP used the constraint $W^T X D X^T W = I$, so this issue should be pointed out. Although this issue could be alleviated if we centralize the data such that the mean of the training data is zero, the users who are not aware of this issue may not always perform this preprocessing when they apply this kind of algorithms. Therefore, it is worth to emphasizing that the correct constraint $W^T X L_q X^T W = I$ instead of the $W^T X Q X^T W = I$ should be used in subspace learning algorithm design.

4   Flexible Locality and Globality Preserving Embedding

4.1   Local and Global Viewpoints of The Graph Based Methods

It was known that the graph based data mining methods capture the local geometrical structure in training data. We will show that the graph based methods Laplacian embedding (solving Eq. (7)) and shift-invariant LPP (solving Eq. (11)) can capture both of local and global geometrical structure in training data.

Under the constraints in the problem (7), and according to Eq. (10), we know $Tr(F^T L_q F)$ is a constant. So problem (7) is equivalent to the following problem:

$$\min_{\substack{F^T Q F = I \\ F^T Q 1 = 0}} \frac{Tr(F^T L F)}{Tr(F^T L_q F)}. \tag{12}$$

Note that the following two equations hold:

$$Tr(F^T L F) = \sum_{i=1}^{n} \sum_{j=1}^{n} A_{ij} \left\| f_i - f_j \right\|^2,$$

$$Tr(F^T L_q F) = \sum_{i=1}^{n} Q_{ii} \left\| f_i - \bar{f} \right\|^2, \tag{13}$$

where $\bar{f} = \sum_{i=1}^{n} Q_{ii} f_i / \sum_{i=1}^{n} Q_{ii}$ is the weighted mean of $f_i|_1^n$. When $Q = I$, $Tr(F^T L_q F)$ is the variance of the n embedded data points $f_i|_1^n$. When $Q = D$, $Tr(F^T L_q F)$ is the weighed variance of the n embedded data points $f_i|_1^n$ with the weight D_{ii} for the i-th embedded data point f_i.

Thus, from Eq. (13), we can conclude that solving the problem (12) is to minimize the Euclidean distances between local data pairs in the embedded space and also to maximize the (weighted) variance of the total data points in the embedded space at the same time, which provides us a new understanding on the Laplacian embedding methods.

Similarly, problem (11) is equivalent to the following problem

$$\min_{W^T X L_q X^T W = I} \frac{Tr(W^T X L X^T W)}{Tr(W^T X L_q X^T W)}. \tag{14}$$

Thus, solving the problem (11) is to minimize the Euclidean distances between local data pairs in the projected subspace and also to maximize the (weighted) variance of the total data points in the projected subspace at the same time. That is to say, although the algorithm LPP is called "locality preserving", it can preserve both of the locality and globality structure in the training data.

If we use the orthogonal constraint instead of the constraint $W^T X L_q X^T W = I$, the problem (14) becomes the trace ratio LPP problem [16], which can be efficiently solved by an iterative algorithm with quadratic convergence rate [23,10]:

$$\min_{W^T W = I} \frac{Tr(W^T X L X^T W)}{Tr(W^T X L_q X^T W)}. \tag{15}$$

4.2 Locality and Globality Preserving Projections with Flexible Constraint

Traditional linearization method imposes a constraint $X^T W = F$ to learn the projection matrix W. Because F in the original problems (*e.g.* Eq. (1)) usually is nonlinear, imposing the constraint that F must be exactly equal to the linear model $X^T W$ is too rigid in practice. In this paper, we propose to use a flexible constraint $\left\| X^T W - F \right\|_F^2 \le \delta$ instead of the rigid constraint $X^T W = F$ in the linearization method. With this flexible linearization constraint and motivations inspired by Eq. (12), we propose the Flexible Locality and Globality Preserving Projections (FLGPP), which is to solve :

$$\min_{\substack{F, W^T W = I \\ \|X^T W - F\|_F^2 \le \delta}} \frac{Tr(F^T L F)}{Tr(F^T L_q F)}. \tag{16}$$

The problem (16) is equivalent to

$$\min_{F, W^T W = I} \frac{Tr(F^T L F)}{Tr(F^T L_q F)} + \lambda \left\| X^T W - F \right\|_F^2, \tag{17}$$

where $\lambda > 0$ is the Lagrangian multiplier coefficient. We propose to solve a similar problem to Eq. (17) for the FLGPP as follows:

$$\min_{F, W^T W = I} \frac{Tr(F^T L F) + \gamma \left\| X^T W - F \right\|_F^2}{Tr(F^T L_q F)}. \tag{18}$$

This new objective is very difficult to optimize, because there are two variables W and F to be solved. Moreover, the non-convex objective function is a ratio of two terms, meanwhile there is a non-convex constraint in the problem, which makes the optimization procedure more challenging. In next section, as one important contribution of this paper, we will propose an effective algorithm to solve the proposed objective, and also prove the algorithm converges to the global optimal solution with quadratic convergence rate, even though the problem is not convex.

5   New Optimization Algorithm

5.1   Proposed Algorithm

Denote $N = (L - \lambda L_q + \gamma I)^{-1}$ and define a function $g(\lambda)$ as follows:

$$g(\lambda) = \min_{F, W^T W = I} Tr F^T (L - \lambda L_q) F + \gamma \left\| X^T W - F \right\|_F^2 \qquad (19)$$

Eq. (19) can be written as:

$$g(\lambda) = \min_{F, W^T W = I} Tr(F^T N^{-1} F) \\ + \gamma Tr(W^T X X^T W) - 2\gamma Tr(W^T X F) \qquad (20)$$

From Eq. (20) we know N should be positive definite to guarantee the objective function is convex w.r.t. F, otherwise the objective function in Eq. (19) is not bounded. Suppose N is positive definite, by setting the derivative of Eq. (20) w.r.t. F to zero, we have

$$F = \gamma N X^T W \qquad (21)$$

Substituting F into Eq. (20), we have

$$g(\lambda) = \min_{W^T W = I} Tr W^T X (I - \gamma N) X^T W \qquad (22)$$

The optimal solution W consists of the m eigenvectors of $X(I - \gamma N)X^T$ corresponding to the smallest eigenvalues.

If we have an initial value λ_0 satisfying the following two conditions: $N_0 = (L - \lambda_0 L_q + \gamma I)^{-1}$ is positive definite (*i.e.*, the smallest eigenvalues of N_0 is larger than 0) and $g(\lambda_0) \leq 0$ (*i.e.*, the sum of the m smallest eigenvalues of $X(I - \gamma N_0)X^T$ is not larger than 0), we will have the algorithm to solve the proposed objective. The detailed algorithm to solve the problem (18) is described in Algorithm 1.

In the following subsections, we will prove our algorithm converges to the global optimal solution and provide the approach to find a λ_0 to satisfy the above two conditions.

5.2   Convergence Analysis of Our Algorithm

Denote

$$J(F, W) = \frac{Tr(F^T L F) + \gamma \left\| X^T W - F \right\|_F^2}{Tr(F^T L_q F)} \qquad (23)$$

Assume $\lambda^* = J(F^*, W^*)$ is the global optimal value of the objective function in Eq. (18). Denote

$$h(F, W; \lambda) = Tr F^T (L - \lambda L_q) F + \gamma \left\| X^T W - F \right\|_F^2 \qquad (24)$$

then $g(\lambda) = \min_{F, W^T W = I} h(F, W; \lambda)$.

Similar to the standard trace ratio problem [17], we have the following results.

Algorithm 1. Algorithm to solve the problem (18)

Input: X, Positive semi-definite matrices L and L_q, γ, m.

Initialize λ_0 such that $N_0 = (L - \lambda_0 L_q + \gamma I)^{-1}$ is positive definite and $g(\lambda_0) \leq 0$.

Let $t = 1$.

repeat

  1. Calculate $N_{t-1} = (L - \lambda_{t-1} L_q + \gamma I)^{-1}$.

  2. Calculate W_t, in which the columns are the m eigenvectors of $X(I - \gamma N_{t-1})X^T$ corresponding to the smallest eigenvalues.

  3. Calculate $F_t = \gamma N_{t-1} X^T W_t$.

  4. Calculate $\lambda_t = \frac{Tr(W_t^T X(I - \gamma N_{t-1})X^T W_t)}{\gamma Tr(W_t^T X N_{t-1} L_q N_{t-1} X^T W_t)} + \lambda_{t-1}$.

  5. Let $t = t + 1$.

until Converge

Output: F, W.

Lemma 1. *The below three equations hold:*

$$g(\lambda) = 0 \Rightarrow \lambda = \lambda^* \tag{25}$$

$$g(\lambda) > 0 \Rightarrow \lambda < \lambda^* \tag{26}$$

$$g(\lambda) < 0 \Rightarrow \lambda > \lambda^* \tag{27}$$

Proof: Since $\lambda^* = J(F^*, W^*)$ is the global optimal value, $\forall F, W^T W = I$, we have $J(F, W) \geq \lambda^*$. So $h(F^*, W^*; \lambda^*) = 0$ and $h(F, W; \lambda^*) \leq 0$. Thus $\min\limits_{F, W^T W = I} h(F, W; \lambda^*) = 0$, that is, $g(\lambda^*) = 0$. Similarly we can get Eq. (25).

If $\lambda \geq \lambda^*$, then

$$\begin{aligned}
g(\lambda) &= \min_{F, W^T W = I} h(F, W; \lambda) \leq h(F^*, W^*; \lambda) \\
&= g(\lambda^*) + (\lambda^* - \lambda)Tr(F^{*T} L_p F^*) \\
&= (\lambda^* - \lambda)Tr(F^{*T} L_p F^*) \\
&\leq 0,
\end{aligned} \tag{28}$$

which concludes Eq. (26).

If $\lambda \leq \lambda^*$, then

$$\begin{aligned}
g(\lambda) & \tag{29} \\
&= \min_{F, W^T W = I} h(F, W; \lambda) \\
&= \min_{F, W^T W = I} h(F, W; \lambda^*) + (\lambda^* - \lambda)Tr(F^T L_p F) \\
&\geq \min_{F, W^T W = I} h(F, W; \lambda^*) + \min_F (\lambda^* - \lambda)Tr(F^T L_p F) \\
&= g(\lambda^*) + (\lambda^* - \lambda)\min_F Tr(F^T L_p F) \\
&= 0,
\end{aligned}$$

which concludes Eq. (27). $\square$

Theorem 1. *In each iteration of Algorithm 1, the value of the objective function in Eq. (18) will not increase.*

Proof: According to Step 3 in Algorithm 1, $F_t = \gamma N_{t-1} X^T W_t$, and notice $N_{t-1} = (L - \lambda_{t-1} L_q + \gamma I)^{-1}$ according to Step 1, hence we have

$$
\begin{aligned}
J(F_t, W_t) &= \frac{Tr(F_t^T L F_t) + \gamma \left\| X^T W_t - F_t \right\|_F^2}{Tr(F_t^T L_q F_t)} \\
&= \frac{Tr W_t^T X (I - \gamma N_{t-1}) X^T W_t}{\gamma Tr W_t^T X N_{t-1} L_q N_{t-1} X^T W_t} + \lambda_{t-1} \\
&= \lambda_t .
\end{aligned}
\tag{30}
$$

Thus, $\lambda_t = J(F_t, W_t) \geq J(F^*, W^*) = \lambda^*$. According to Eq. (26) in Lemma 1, $g(\lambda_t) \leq 0$. On the other hand, according to the condition of λ_0, we have $g(\lambda_0) \leq 0$. Therefore, for $t \geq 0$, $g(\lambda_t) \leq 0$.

According to Steps 2 and 3, $\{F_{t+1}, W_{t+1}\}$ are the optimal solutions to $g(\lambda_t)$, so $g(\lambda_t) = h(F_{t+1}, W_{t+1}; \lambda_t)$. Therefore, for $t \geq 0$, we have

$$
g(\lambda_t) \leq 0
\tag{31}
$$
$$
\Rightarrow h(F_{t+1}, W_{t+1}; \lambda_t) \leq 0
$$
$$
\Rightarrow \frac{Tr(F_{t+1}^T L F_{t+1}) + \gamma \left\| X^T W_{t+1} - F_{t+1} \right\|_F^2}{Tr(F_{t+1}^T L_q F_{t+1})} \leq \lambda
$$
$$
\Rightarrow J(F_{t+1}, W_{t+1}) \leq J(F_t, W_t) ,
$$

which completes the proof.

□

Note that $J(F_t, W_t)$ has lower bound, thus the Algorithm 1 will converge.

Theorem 2. *The Algorithm 1 converges to the global optimal solution.*

Proof: According to Step 4 in Algorithm 1,

$$
\lambda_{t+1} = \frac{Tr W_{t+1}^T X (I - \gamma N_t) X^T W_{t+1}}{\gamma Tr W_{t+1}^T X N_t L_q N_t X^T W_{t+1}} + \lambda_t .
\tag{32}
$$

Note that $\lambda_{t+1} = \lambda_t$ in the convergence. Therefore

$$
\lambda_{t+1} = \frac{Tr W_{t+1}^T X (I - \gamma N_t) X^T W_{t+1}}{\gamma Tr W_{t+1}^T X N_t L_q N_t X^T W_{t+1}} + \lambda_{t+1}
\tag{33}
$$
$$
\Rightarrow \frac{Tr W_{t+1}^T X (I - \gamma N_t) X^T W_{t+1}}{\gamma Tr W_{t+1}^T X N_t L_q N_t X^T W_{t+1}} = 0
$$
$$
\Rightarrow Tr W_{t+1}^T X (I - \gamma N_t) X^T W_{t+1} = 0
$$
$$
\Rightarrow g(\lambda_t) = 0.
$$

According to Eq. (25) in Lemma 1, $\lambda_t = \lambda^*$. Therefore, the converged solution of Algorithm 1 is the global optimal solution.   □

To study the convergence rate of our algorithm, we prove the following theorem.

Theorem 3. *The Algorithm 1 is a Newton's method to find the root of* $g(\lambda) = 0$.

Proof: Denote the i-th smallest eigenvalue of $X(I - \gamma N_t)X^T$ by $\beta_i(\lambda_t)$ and the corresponding eigenvector by $w_i(\lambda_t)$. According to the definition of eigenvalues and eigenvectors, we have:

$$(X(I - \gamma N_t)X^T - \beta_i(\lambda_t)I)w_i(\lambda_t) = 0 \tag{34}$$

$$\Rightarrow \frac{\partial(X(I - \gamma N_t)X^T - \beta_i(\lambda_t)I)w_i(\lambda_t)}{\partial \lambda_t} = 0$$

$$\Rightarrow (-\gamma X N_t L_q N_t X^T - \beta_i'(\lambda_t)I)w_i(\lambda_t) +$$
$$(X(I - \gamma N_t)X^T - \beta_i(\lambda_t)I)w_i'(\lambda_t) = 0$$

$$\Rightarrow w_i^T(\lambda_t)(-\gamma X N_t L_q N_t X^T - \beta_i'(\lambda_t)I)w_i(\lambda_t) +$$
$$w_i^T(\lambda_t)(X(I - \gamma N_t)X^T - \beta_i(\lambda_t)I)w_i'(\lambda_t) = 0$$

$$\Rightarrow w_i^T(\lambda_t)(-\gamma X N_t L_q N_t X^T - \beta_i'(\lambda_t)I)w_i(\lambda_t) = 0$$

$$\Rightarrow \beta_i'(\lambda_t) = -\gamma w_i^T(\lambda_t) X N_t L_q N_t X^T w_i(\lambda_t)$$

From Eq. (22) we know, $g(\lambda) = \min_{W^T W = I} tr W^T X(I - \gamma N)X^T W$, so $g(\lambda_t) = \sum_{i=1}^{m} \beta_i(\lambda_t)$. Then we have:

$$g'(\lambda_t) = \sum_{i=1}^{m} \beta_i'(\lambda_t)$$

$$= \sum_{i=1}^{m} -\gamma w_i^T(\lambda_t) X N_t L_q N_t X^T w_i(\lambda_t)$$

$$= -\gamma Tr(W_{t+1}^T X N_t L_q N_t X^T W_{t+1}) . \tag{35}$$

According to Step 4 in Algorithm 1, we have:

$$\lambda_{t+1} = \frac{Tr(W_{t+1}^T X(I - \gamma N_t)X^T W_{t+1})}{\gamma Tr(W_{t+1}^T X N_t L_q N_t X^T W_{t+1})} + \lambda_t$$

$$= \lambda_t - \frac{g(\lambda_t)}{g'(\lambda_t)} . \tag{36}$$

Thus the iterative procedure of Algorithm 1 is essentially a Newton's method to find the root of $g(\lambda) = 0$.

$\square$

It is well known the rate of convergence of Newton's method is quadratic convergence under mild conditions, which is very fast to converge in practice. In our experiments, we find that the Algorithm 1 indeed converges very fast, and always converges within 5-20 iterations.

5.3   Approach to Find An Initial λ_0

Lemma 1 can be used to find a feasible λ_0 that satisfies the following two conditions: $N_0 = (L - \lambda_0 L_q + \gamma I)^{-1}$ is positive definite and $g(\lambda_0) \leq 0$.

Algorithm 2. Find a feasible value λ_0

Initialize F and W such that $W^T W = I$. Let:
$$\lambda_{\min} = 0 \text{ and } \lambda_{\max} = \frac{Tr(F^T L F) + \gamma \| X^T W - F \|_F^2}{Tr(F^T L_q F)}.$$
repeat
  Let $\lambda_0 = \frac{\lambda_{\min} + \lambda_{\max}}{2}$, $N_0 = (L - \lambda_0 L_q + \gamma I)^{-1}$.
  if the smallest eigenvalue of N_0 is not larger than 0 **then**
    $\lambda_{\max} \leftarrow \lambda_0$.
  end if
  if the sum of the m smallest eigenvalues of $X(I - \gamma N_0)X^T$ is larger than 0 **then**
    $\lambda_{\min} \leftarrow \lambda_0$.
  end if
until N_0 is positive definite and the sum of the m smallest eigenvalues of $X(I - \gamma N_0)X^T$ is not larger than 0

We apply bisection method to find such a λ_0. First, we evaluate the lower bound $\lambda_{\min}$ and upper bound $\lambda_{\max}$ of such a λ_0. According to Lemma 1, $g(\lambda_0) \leq 0$ indicates $\lambda_0 \geq \lambda^*$. If L and L_q are positive semi-definite, $\lambda^* \geq 0$, so we can set the initial lower bound $\lambda_{\min} = 0$.[1] Randomly initialize F and W such that $W^T W = I$, we have $J(F, W) \geq \lambda^*$, so we can set the initial upper bound $\lambda_{\max} = J(F, W)$. With the initial lower and upper bounds $\lambda_{\min}$ and $\lambda_{\max}$, we can use the bisection method to find a feasible λ_0 satisfying the two conditions. If N_0 is not positive, then the current λ_0 is too large, we update the upper bound $\lambda_{\max}$ with the current λ_0. If $g(\lambda_0) > 0$, then $\lambda_0 < \lambda^*$, which indicates the current λ_0 is too smaller, we update the lower bound $\lambda_{\min}$ with the current λ_0. The detailed approach is described in Algorithm 2.

It is worth noting that similar method can also be used to find an initial λ_0 for solving a different problem in [9], such that the algorithm in [9] is applicable with any parameter combination. We have updated the code for [9] in the author's website.

5.4 Shift Invariance of The Algorithm

It can be easily verified that $(I - \gamma N)\mathbf{1} = \mathbf{0}$, so we have $(X + c\mathbf{1}^T)(I - \gamma N)(X + c\mathbf{1}^T)^T = X(I - \gamma N)X^T$. Thus, according to the Algorithm 1, the optimal solution W to the problem (18) is invariant to arbitrary shift vector c.

6 Experiment

We evaluate the performance of the proposed flexible shift-invariant locality and globality preserving projection (FLGPP) on six benchmark data sets with the comparison to

---

[1] If the symmetric matrix L is not positive and L_q is positive, we can set $\lambda_{\min}$ to the smallest eigenvalue σ of $L_q^{-1} L$ since it can be verified $\lambda^* \geq \sigma$. We can also evaluate the smallest eigenvalue of L and the largest eigenvalue of L_q using the Gershgorin circle theorem to avoid computing the eigenvalue.

four related supervised embedding approaches, including multi-class Linear discriminant analysis (LDA), locality preserving projection (LPP), shift-invariant locality preserving projection (SILPP) in §3 as well as trace ratio locality preserving projection (TLPP) in §4.1.

Table 1. The summary of six benchmark datasets used in the experiments

data name	# classes(k)	image size	# data point(n)	# training per class
AT&T [20]	40	28×23	400	4
UMIST [6]	20	112×92	575	6
BINALPHA [1]	36	20×16	1404	6
COIL20 [14]	20	32×32	1440	12
YALEB [5]	31	24×21	1984	8
AR [12]	120	32×24	840	3

6.1 Data Descriptions

We use six image benchmark data sets in our experiments, because these data typically have high dimensionality.

AT&T [8] data set has 40 distinct subjects and each subject has 10 images. We downsampled each image (standard procedure to reduce the misalignment effect) to the size of 28×23. The training number per class is 4.

UMIST faces are for multiview face recognition. This data set contains 20 persons and totally there are 575 images. All these images of UMIST database are cropped and resized into 112×92 images. The training number per class is 6.

Binary Alpha data set contains binary digits of 0 through 9 and capital A through Z with size 20×16. There are 39 examples of each class. We randomly select 6 images per class as the training data.

Columbia University Image Library (COIL-20) data set [13] consists of color images of 20 objects where the images of the objects were taken at pose intervals of 5 degree, form the front view with 0 degree. Thus, there are 72 poses per objects. The images are converted to gray-scale image and they are normalized to the size of 32×32 pixels in our experiment. We randomly pick up 12 images for each object to do the training.

Yale database B data set [5] contains single light source images of 38 subjects (10 subjects in original database and 28 subjects in extended one) under 576 viewing conditions (9 poses $\times$ 64 illumination conditions). We fixed the pose. Thus, for each subject, we obtained 64 images under different lighting conditions. The facial areas were cropped into the final images for matching [5]. The size of each cropped image in our experiments is 24×21 pixels, with 256 gray levels per pixel. Because there is a set of images which are corrupted during the image acquisition [5], we have 31 subjects. We randomly select 64 illumination conditions for all 31 subjects to create the experimental dataset with 1984 images and randomly pick up 8 images per subject to do the training.

AR face database contains 120 people with different facial expressions, lighting conditions and occlusions. Each person has 26 different images, and the image resolution is 50×40. We random select 7 images per person and downsample the each image to the size of 32×24 to obtain the experimental dataset with 840 images. Then, we randomly select 3 per class as the training dataset.

We summarize the six data sets that used in our experiments in Table 1, and some image samples of the data sets are shown in Figure 1.

Fig. 1. Examples of the six data sets used in our experiments. From the first row to the sixth row: AT&T, UMIST, BINALPHA, COIL20, YALEB, AR.

6.2   Experiment Setup

In the training step, we firstly build the graph using the strategy described in next paragraph. Based on the same graph structure, five different embedding methods are conducted for a pre-defined reduced dimension. After we get the projection matrices for different methods, in the testing step, we use the simple k-NN (k=1) classifier (a simple classifier can avoid introducing any bias) to classify the testing data in the embedded space. In each experiment, we randomly select several data point per class for training and the rest are used as for testing. The average classification accuracy rates and standard deviations are reported over 50 random splits.

Regarding the graph construction, since we are discussing supervised embedding methods, we utilize the label information of the training data to build the graph. To be specific, $w_{i,j} = 1$, if i-th training data point and j-th training data point belong to the same class; $w_{i,j} = -1$, otherwise. We also remove the self-loop, $i.e.$ let $w_{i,j} = 0$, if $i = j$. The regularization parameter in FLGPP is set to 0.1 in all the experiments. We record the average classification accuracy rate V.S. the different reduced dimensions for all the methods. For multi-class LDA, we only record its performance up to $C - 1$, where C is the number of classes.

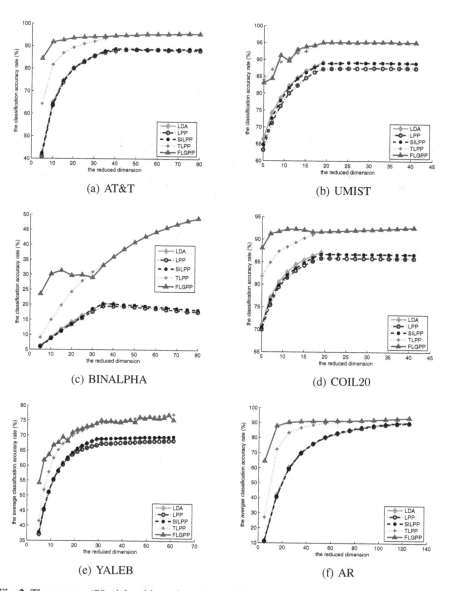

Fig. 2. The average (50 trials with random data split) classification accuracy of k-NN method on the embedded data by five different embedding approaches

Table 2. The average classification accuracy rate $\pm$ standard deviation on six benchmark datasets among all the reduced dimension from 5 to $C - 1$

data name	MLDA	LPP	SILPP	TLPP	FLGPP
AT&T	87.70 ± 2.06	87.24 ± 1.96	87.83 ± 2.49	93.48 ± 1.72	$\mathbf{94.28 \pm 1.91}$
UMIST	88.00 ± 2.63	86.24 ± 2.91	87.72 ± 2.70	94.07 ± 2.08	$\mathbf{94.22 \pm 2.07}$
BINALPHA	18.33 ± 1.52	17.63 ± 1.55	18.63 ± 1.44	30.89 ± 1.92	$\mathbf{31.38 \pm 3.24}$
COIL20	86.43 ± 1.53	85.16 ± 1.69	85.95 ± 1.49	91.03 ± 1.27	$\mathbf{92.26 \pm 1.07}$
YALEB	68.76 ± 5.25	66.74 ± 6.22	68.87 ± 5.67	73.75 ± 2.05	$\mathbf{74.05 \pm 2.56}$
AR	89.15 ± 1.18	88.68 ± 1.35	89.23 ± 1.31	92.28 ± 1.22	$\mathbf{92.39 \pm 1.14}$

6.3 Experiment Results

Fig. 2 shows the average classification accuracy rate evaluated by 1-NN v.s. the number of the reduced dimension on six datasets over 50 random data split. From Fig. 2 we clearly observe that the performance of our proposed FLGPP method consistently outperforms that of the other embedding approaches, especially when the reduced dimension is low. When the reduced dimension becomes larger, the performance of FLGPP and TLPP become similar. But they still beat the other three methods largely. Table 2 demonstrates the mean $\pm$ standard deviation of the best classification accuracy rate among all the reduced dimensions from 5 to $C - 1$ for different algorithms.

7 Conclusion

In this paper, we proposed a novel flexible shift-invariant locality and globality preserving projection (FLGPP) method. A refined graph Laplacian was formulated and used to preserve the shift-invariant property. Meanwhile, the relaxed linear embedding was introduced to allow the error tolerance, such that the flexible embedding results can reach the more optimal manifold structures. Because the proposed new objective is very difficult to solve, we derived a new optimization algorithm with rigorously proved global convergence. Moreover, we proved the new algorithm is a Newton method with the quadratic convergence rate. We evaluated our FLGPP method on six benchmark data sets. In all empirical results, our new method is consistently better than the related methods.

References

1. Belhumeur, P., Hespanha, J., Kriegman, D.: Eigenfaces vs. fisherfaces: Recognition using class specific linear projection. IEEE Transactions on Pattern Analysis and Machine Intelligence 19(7), 711–720 (1997)
2. Belkin, M., Niyogi, P.: Laplacian eigenmaps for dimensionality reduction and data representation. Neural Computation 15(6), 1373–1396 (2003)
3. Cai, D.: Spectral Regression: A Regression Framework for Efficient Regularized Subspace Learning. PhD thesis, Department of Computer Science, University of Illinois at Urbana-Champaign (May 2009)

4. Fisher, R.A.: The use of multiple measurements in taxonomic problems. Annals Eugen. 7, 179–188 (1936)
5. Georghiades, A.S., Belhumeur, P.N., Kriegman, D.J.: From few to many: Illumination cone models for face recognition under variable lighting and pose. IEEE Trans. Pattern Anal. Mach. Intell. 23(6), 643–660 (2001)
6. Graham, D.B., Allinson, N.M.: Characterizing virtual eigensignatures for general purpose face recognition. in face recognition: From theory to applications. NATO ASI Series F, Computer and Systems Sciences 163, 446–456 (1998)
7. He, X., Niyogi, P.: Locality preserving projections. In: NIPS (2003)
8. http://www.cl.cam.ac.uk/research/dtg/attarchive/facedatabase.html
9. Huang, Y., Xu, D., Nie, F.: Semi-supervised dimension reduction using trace ratio criterion. IEEE Trans. Neural Netw. Learning Syst. 23(3), 519–526 (2012)
10. Jia, Y., Nie, F., Zhang, C.: Trace ratio problem revisited. IEEE Transactions on Neural Networks 20(4), 729–735 (2009)
11. Jolliffe, I.T.: Principal Component Analysis, 2nd edn. Springer, New York (2002)
12. Martinez, A.: The ar face database. CVC Technical Report, 24 (1998)
13. Nayar, S., Nene, S., Murase, H.: Real-time 100 object recognition system. In: Proceedings of the 1996 IEEE International Conference on Robotics and Automation, vol. 3, pp. 2321–2325. IEEE (1996)
14. Nene, S.A., Nayar, S.K., Murase, H.: Columbia object image library (COIL-20), Technical Report CUCS-005-96. Columbia University (1996)
15. Ng, A.Y., Jordan, M.I., Weiss, Y.: On spectral clustering: Analysis and an algorithm. In: NIPS, pp. 849–856 (2001)
16. Nie, F., Xiang, S., Song, Y., Zhang, C.: Orthogonal locality minimizing globality maximizing projections for feature extraction. Optical Engineering 48, 017202 (2009)
17. Nie, F., Xiang, S., Zhang, C.: Neighborhood minmax projections. In: IJCAI, pp. 993–998 (2007)
18. Nie, F., Xu, D., Tsang, I.W.-H., Zhang, C.: Flexible manifold embedding: A framework for semi-supervised and unsupervised dimension reduction. IEEE Transactions on Image Processing 19(7), 1921–1932 (2010)
19. Roweis, S.T., Al, E.: Nonlinear dimensionality reduction by locally linear embedding. Science 290, 2323–2326 (2000)
20. Samaria, F.S., Harter, A.C.: Parameterisation of a stochastic model for human face identification. In: 2nd IEEE Workshop on Applications of Computer Vision, pp. 138–142 (1994)
21. Shi, J., Malik, J.: Normalized cuts and image segmentation. IEEE Transactions on PAMI 22(8), 888–905 (2000)
22. Tenenbaum, J.B., de Silva, V., Langford, J.C.: A global geometric framework for nonlinear dimensionality reduction. Science 290(5500), 2319–2323 (2000)
23. Wang, H., Yan, S., Xu, D., Tang, X., Huang, T.S.: Trace ratio vs. ratio trace for dimensionality reduction. In: CVPR (2007)
24. Xiang, S., Nie, F., Zhang, C., Zhang, C.: Nonlinear dimensionality reduction with local spline embedding. IEEE Transactions on Knowledge and Data Engineering 21(9), 1285–1298 (2009)
25. Zhang, Z., Zha, H.: Principal manifolds and nonlinear dimension reduction via local tangent space alignment. SIAM Journal of Scientific Computing 26, 313–338 (2005)

Interactive Knowledge-Based Kernel PCA*

Dino Oglic[1], Daniel Paurat[1], and Thomas Gärtner[1,2]

[1] University of Bonn, Bonn, Germany
[2] Fraunhofer IAIS, Sankt Augustin, Germany
{dino.oglic,daniel.paurat,thomas.gaertner}@uni-bonn.de

Abstract. Data understanding is an iterative process in which domain experts combine their knowledge with the data at hand to explore and confirm hypotheses. One important set of tools for exploring hypotheses about data are visualizations. Often, however, traditional, unsupervised dimensionality reduction algorithms are used for visualization. These tools allow for interaction, i.e., exploring different visualizations, only by means of manipulating some technical parameters of the algorithm. Therefore, instead of being able to intuitively interact with the visualization, domain experts have to learn and argue about these technical parameters. In this paper we propose a knowledge-based kernel PCA approach that allows for intuitive interaction with data visualizations. Each embedding direction is given by a non-convex quadratic optimization problem over an ellipsoid and has a globally optimal solution in the kernel feature space. A solution can be found in polynomial time using the algorithm presented in this paper. To facilitate direct feedback, i.e., updating the whole embedding with a sufficiently high frame-rate during interaction, we reduce the computational complexity further by incremental up- and down-dating. Our empirical evaluation demonstrates the flexibility and utility of this approach.

Keywords: Interactive visualization, kernel methods, dimensionality reduction.

1 Introduction

We investigate a variant of kernel principal component analysis (PCA) which allows domain experts to directly interact with data visualizations and to add domain-knowledge and other constraints in an intuitive way. Data visualization is an important part of knowledge discovery and at the core of data understanding and exploration tasks (see, e.g., **(author?)** [22]). Its importance for data science has been recognized already by **(author?)** [24]. While knowledge discovery is inherently interactive and iterative, most data visualizations are inherently static (we survey several methods in Section 2). Switching methods and changing algorithmic parameters allow for some interaction with the visualization but this interaction is rather indirect and only feasible for machine learning experts rather than domain experts. As data science and its tools are, however, getting more and more widespread, the need arises for tools that allow domain experts to directly interact with the data.

---

* Parts of this work have been presented in workshops [18, 19].

T. Calders et al. (Eds.): ECML PKDD 2014, Part II, LNCS 8725, pp. 501–516, 2014.
© Springer-Verlag Berlin Heidelberg 2014

Our interactive and knowledge-based variant of kernel PCA (described in detail in Section 3) incorporates different forms of supervision either as soft- or hard constraints. In particular, it allows for the placement of 'control points', the addition of must-link and cannot-link constraints, as well as being able to incorporate known class labels. Here, the motivation behind constraints is that the domain experts can choose one low dimensional embedding from the many possible ones not by tuning parameters but by dragging chosen data points in the embedding or grouping them by similarity and class labels whereby all related data points automatically and smoothly adjust their location accordingly. Similar to kernel PCA, each embedding direction/dimension corresponds to a function in the underlying reproducing kernel Hilbert space. For each direction, we propose to find a function which (i) maximizes the variance along this direction, (ii) has unit norm, (iii) is as orthogonal to the other functions as possible, and (iv) adheres to the knowledge-based constraints as much as possible.

The optimization problem (analyzed in detail in Section 4) derived from the above formulation is to maximize a—typically non-convex—quadratic function over an ellipsoid subject to linear equality constraints. Unconstrained quadratic optimization problems over spheres have been investigated by (author?) [11] who generally suggested two approaches: transforming the problem to a quadratic and then linear eigenvalue problem or reducing the problem to solving a one-dimensional secular equation. While both approaches have cubic complexity and the first approach is more elegant, the second one is numerically much more stable. To solve the quadratic function over ellipsoid subject to linear constraints, we extend the approach of (author?) [11] to this setting.

In order to allow a direct interaction with the embedding, i.e., updating the whole embedding with a sufficiently high frame-rate, the cubic complexity of the just mentioned approaches is, however, not sufficient. To overcome this, we observe that in an interactive set up it is hardly ever the case that the optimization problem has to be solved from scratch. Instead, consecutive optimization problems will be strongly related and indeed we show (in Section 5) that consecutive solutions differ only in rank-one updates which allows for much more fluent and natural interaction.

Our experiments focus on demonstrating the flexibility and usability of our approach in an interactive knowledge discovery setting (in Section 6). We show first that small perturbations of the location of control points only lead to small perturbations of the embedding of all points. This directly implies that it is possible to smoothly change the embedding without sudden and unexpected jumps (large changes) of the visualization. We then show that by appropriate placement of control points, knowledge-based kernel PCA can mimic other embeddings. In particular, we consider the sum of two different kernels and observe that by placing a few control points, the 2D kernel PCA embedding of either of the original kernels can be reasonably well recovered. In addition, we investigate the amount of information retained in low-dimensional embeddings. For that we embed benchmark semi-supervised classification data sets in 2D space and compare the predictive performance of learning algorithms in this embedding with other approaches. Last but not least, we show it is possible to discover structures within the data that do not necessarily exhibit high correlation with variance and, thus, remain hidden in the plain kernel PCA embedding.

2   Related Work

Interaction with traditional machine learning algorithms is hardly intuitive and forces domain experts to reason about parameters of the algorithm instead of data. To overcome this problem several tools for data visualization were designed with the ultimate goal to facilitate the understanding of the underlying model and the interaction with model parameters. One such tool facilitates the understanding of PCA by interpreting the influence of the movement of a point in the projection space along the principal directions on the coordinates of the corresponding instance in the input space [14]. The interaction with the PCA is, however, limited to feature weighting which allows domain experts to reason on the importance of particular features in the analyzed data set. In contrast, a user study conducted by (**author?**) [1] shows domain experts prefer to interact directly with a visualization by placing a few control points in accordance with the current understanding of data. In general, user studies report benefits of the interactive over the static approach in data visualization (see, e.g., (**author?**) [4, 14]).

In this paper we extend our previous work on interactive visualization in which we proposed a variant of supervised PCA [19] and provided a tool for interactive data visualization InVis [18]. In contrast to supervised PCA [19] which allowed interaction with a visualization only through the explicit placement of control points, the proposed knowledge-based kernel PCA allows interaction through a variety of soft and/or hard knowledge-based constraints (see Section 3.1, 3.2 or 3.3). Moreover, we relate the successive optimization problems arising during the interaction with rank-one updates and reduce the interaction complexity from cubic to quadratic in the number of data instances. The InVis tool, on the other hand, enables interaction with a visualization using the least square projections (LSP). As argued in our workshop paper [19], the LSP algorithm is in general not a good choice for data visualization and the same can be said about any purely supervised learning algorithm (e.g. linear discriminant analysis [13]). For instance, consider training a linear regression on a sparse high dimensional dataset. If it is trained using very few instances, the weight vector will only be non-zero over the union of their non-zero attributes and all instances having different non-zero attributes will be mapped to the origin—a pretty cluttered visualization.

From the interactive visualization perspective, the most related to our work are techniques developed by (**author?**) [8] and (**author?**) [16]. The proposed techniques allow placement of control points in a projection space, but the placement induces a different form of interaction. Namely, movement of points is interpreted as a similarity feedback (similar to must and cannot link constraints) and as such incorporated into the model parameters. Our approach, on the other hand, focuses on the explicit placement of control points enabling structural exploration of the data by observing how the embedding reacts to a placement of a selected control point. Both approaches [8, 16] perform interaction by incorporating expert feedback into the probabilistic PCA, multidimensional scaling and generative topographic mapping. The means to incorporate the similarity feedback into the named unsupervised algorithms are, however, limited to feature weighting and heuristic covariance estimates. Moreover, probabilistic PCA and multi-dimensional scaling approaches are highly sensitive to outliers and produce visualizations with huge number of overlapping points for such data sets.

From the methodological perspective, our method is a spectral method for semi-supervised learning and closely related to the semi-supervised kernel PCA [25]. This method can be described as a relaxation of T-SVM and/or a generalization of kernel PCA. Three such generalizations are proposed by (**author?**) [25] and the closest to our approach is the least square kernel PCA which can be viewed as choosing a projection with the best L_2-fit over the labelled data among the projections having a constant variance.

Methods for dimensionality reduction are also related to data visualization, and in many cases used primarily for visualization purposes. There are several well known methods for dimensionality reduction, but the majority of them is unsupervised and unable to incorporate the gained knowledge into a lower dimensional data representation. Some of the well known spectral methods for dimensionality reduction are principal component analysis [15], metric multidimensional scaling [6], isomap [23], maximum variance unfolding [26] and locally linear embedding [20]. Here we note that these methods can be also viewed as instances of kernel PCA with a suitably defined kernel matrix [12].

3   Knowledge-Based Kernel PCA

In this section we present a variant of semi-supervised kernel PCA which incorporates various domain-knowledge constraints while still maximizing the variance of the data 'along' the set of unit norm functions defining the embedding. Our formulation can take into account a variety of hard and/or soft constraints, allowing flexible placement of control points in an embedding space, the addition of must-link and cannot-link constraints, as well as known class labels. The goal of the proposed algorithm is to allow domain experts to interact with a low-dimensional embedding and choose from the many possible ones not by tuning parameters but by dragging or grouping the chosen data points in the embedding whereby all related data points automatically and smoothly adjust their location accordingly. To make the visualization more flexible with respect to orthogonality of maximum-variance directions we replace the usual hard orthogonality constraint with a soft-orthogonality term in the objective function.

Let $X = \{x_1, ..., x_n\}$ be a sample from an instance space $\mathcal{X}$, $\mathcal{H}$ a reproducing kernel Hilbert space with kernel $k(\cdot, \cdot) : \mathcal{X} \times \mathcal{X} \to \mathbb{R}$ and $\mathcal{H}_X = \mathrm{span}\{k(x_i, \cdot) | x_i \in X\}$. We iteratively find the constant $\mathcal{H}_X$-norm maximum variance directions $f_1, ..., f_d$ by solving the following optimization problem

$$
\begin{aligned}
f_s = \operatorname*{argmax}_{f \in \mathcal{H}} \quad & \frac{1}{n} \sum_{i=1}^{n} (f(x_i) - \langle f, \mu \rangle)^2 + \rho \Omega(f, s) - \nu \sum_{s'=1}^{s-1} \langle f_{s'}, f \rangle^2 \\
\text{subject to} \quad & \|f\|_{\mathcal{H}_X} = r, \\
& \Upsilon(f, y_s) = 0,
\end{aligned}
\tag{1}
$$

where Ω is a soft and Υ is a hard constraint term, $r \in \mathbb{R}^+$, $\mu = \frac{1}{n} \sum_{i=1}^{n} k(x_i, \cdot)$ and $y_s \in \mathbb{R}^m$ ($m < n$) is chosen interactively. Here we note that Υ is a linear operator over a direction f evaluated at $x \in X$. Additionally, the term can be used to express a hard orthogonality over the computed directions.

First, let us see that the weak representer theorem [7, 21] holds. As an optimizer $f_s \in \mathcal{H}$ we can write $f_s = u_s + v_s$ with $u_s \perp v_s$ and $u_s \in \mathcal{H}_X$. For the computation of the first extreme variance direction, f_1, there is no soft orthogonality term in the optimization objective. Plugging the substitution into Eq. (1) we conclude, provided the theorem holds for the Ω term, that the optimization objective is independent of v_s, and the weak representer theorem holds in this case. For the computation of the s-th variance direction f_s ($s > 1$), we additionally have orthogonality terms $\langle f_s, f_{s'} \rangle = \langle u_s + v_s, f_{s'} \rangle = \langle u_s, f_{s'} \rangle$ ($s' < s$) which are also independent of v_s. The hard constraint term Υ is also independent of v_s as it holds that $f_s(x) = u_s(x)$ for all $x \in X$. Therefore, the weak representer theorem holds for problem (1) and we can express an optimizer as $f_s = \sum_i \alpha_{si} k(x_i, \cdot)$ with $\alpha_{si} \in \mathbb{R}$.

Now, using the representer theorem we can rewrite terms from problem (1) as:

$$\langle f, \mu \rangle = \frac{1}{n} \sum_{i=1}^{n} \sum_{j=1}^{n} \alpha_i k(x_i, x_j) = \frac{1}{n} e^T K \alpha,$$

$$\sum_{s'=1}^{s-1} \langle f, f_{s'} \rangle^2 = \alpha^T K \left(\sum_{s'=1}^{s-1} \alpha_{s'} \alpha_{s'}^T \right) K \alpha,$$

with e denoting the vector of ones. If we ignore the hard constraint term (in Section 4.2 we show we can incorporate it into the objective) we can write problem (1) as:

$$\alpha_s = \underset{\alpha \in \mathbb{R}^n}{\arg\max} \quad \alpha^T K W K \alpha + \rho \Omega(\alpha, s) \tag{2}$$

$$\text{subject to} \quad \alpha^T K \alpha = r^2,$$

with $H_n = I_n - \frac{1}{n} e e^T$ and $W = \frac{1}{n} H_n - \nu \sum_{s'=1}^{s-1} \alpha_{s'} \alpha_{s'}^T$.

In the following sections we introduce several, quadratic and linear, knowledge-based constraints satisfying the weak representer theorem and enabling different forms of interaction with a low-dimensional embedding.

3.1 The Placement of Control Points

The most natural form of interaction with a low-dimensional embedding is the movement of control points across the projection space [1]. It is an exploratory form of interaction enabling domain experts to explore the structure of the data by observing how the embedding reacts to a movement of a selected control point. The placement of control points can be incorporated into the kernel PCA as a 'soft' constraint with

$$\Omega(f, s) = -\frac{1}{m} \sum_{i=1}^{m} \| f(x_i) - y_{si} \|^2,$$

where y_{si} denotes the coordinate 'along' the projection axis f_s of an example x_i. That the weak representer theorem holds for problem (1), including this Ω term, follows along the same lines as above and we are able to express Ω as

$$\Omega(\alpha, s) = -\frac{1}{m} \left(\alpha^T K_{[:n,:m]} K_{[:m,:n]} \alpha - 2 y_s^T K_{[:m,:n]} \alpha \right), \tag{3}$$

with $K_{[:m,:n]}$ denoting rows in the kernel matrix K corresponding to control points.

The alternative way is to treat the placements as hard constraints [19] and incorporate them into Υ term which can be written as

$$\Upsilon(f, \boldsymbol{y}_s) = K_{[:m,:n]}\boldsymbol{\alpha} - \boldsymbol{y}_s = 0. \tag{4}$$

Note that the soft constraint Ω is allowing some displacement which can lead to better visualizations if noise is to be expected in the positions of the control points.

3.2   Must-Link and Cannot-Link Constraints

Domain knowledge can also be expressed in terms of similarity between points and the knowledge-based term Ω can, for instance, be defined by pairing points which should or should not be placed close to each other. Squared distances between projections of paired points are then minimized for must-link pairs and maximized for cannot-link pairs, i.e.

$$\Omega(f, s) = -\frac{1}{|\mathscr{C}|} \sum_{(i,l)\in\mathscr{C}} y_{il}(f(\boldsymbol{x}_i) - f(\boldsymbol{x}_l))^2,$$

where $y_{il} = +1$ for a must-link and $y_{il} = -1$ for a cannot-link constraint, and $\mathscr{C}$ denotes the set of constraints. Analogously to the previous case, the weak representer theorem holds and the constraint can be written as

$$\Omega(\boldsymbol{\alpha}, s) = -\frac{1}{|\mathscr{C}|}\boldsymbol{\alpha}^T KLK\boldsymbol{\alpha}. \tag{5}$$

Here $\boldsymbol{\Delta}_{il} = \mathbf{e}_i - \mathbf{e}_l$ and $L = \sum_{(i,l)\in\mathscr{C}} y_{il}\boldsymbol{\Delta}_{il}\boldsymbol{\Delta}_{il}^T$ is a Laplacian matrix of the graph weighted with y_{il}.

3.3   Classification Constraints

In this case domain-knowledge is incorporated by providing positive and negative class labels for a small number of instances and the 'soft' knowledge-based term Ω can be written as

$$\Omega(f, s) = \langle f, \mu_\pm \rangle, \text{ with } \mu_\pm = \frac{1}{m}\sum_{i=1}^{m} y_i k(\boldsymbol{x}_i, \cdot).$$

Similar to previous cases, it can be checked that the weak representer theorem holds and the Ω term can be written as

$$\Omega(\boldsymbol{\alpha}, s) = \frac{1}{m}\boldsymbol{y}^T K_{[:m,:n]}\boldsymbol{\alpha}. \tag{6}$$

4   Optimization Problem

As stated in Section 3, it is possible to combine different knowledge-based constraints and for any such combination we are interested in the resulting optimization problem.

Each of the 'soft' knowledge-based constraints is either quadratic or linear and contributes with a quadratic or linear term to the optimization objective. Therefore, to compute our embedding, for any combination of 'soft' knowledge-based constraints, we have to solve the following optimization problem:

$$\underset{x\in\mathbb{R}^n}{\operatorname{argmax}} \; x^T W x - 2b^T x \quad \text{s. t. } x^T K x = r^2. \tag{7}$$

where $W \in \mathbb{R}^{n\times n}$ is a symmetric and $K \in \mathbb{R}^{n\times n}$ is a kernel matrix. Problem (7) is non-convex as it is a quadratic defined over an ellipsoid. If we also want to force the hard orthogonality or/and hard placement of control points we need to add an additional linear constraint

$$Lx = y, \tag{8}$$

where $L \in \mathbb{R}^{m\times n}$ is a rectangular constraint matrix and y is a coordinate placement along one of the projection axis or the zero vector (in hard orthogonality case).

In Section 4.1 we describe how to find the global optimizer for the non-convex problem (7) in a closed form.

4.1 Quadratic over a Hypersphere

We first transform problem (7) to optimize over a hypersphere instead of hyperellipsoid. To achieve this we decompose the positive definite matrix K and introduce a substitution $v = K^{\frac{1}{2}}x$. In the new optimization problem the symmetric matrix W is replaced with the symmetric matrix $C = K^{-\frac{1}{2}}WK^{-\frac{1}{2}}$ and the vector b with the vector $d = K^{-\frac{1}{2}}b$. Hence, after the transformation we are optimizing a quadratic over the hypersphere,

$$\underset{v\in\mathbb{R}^n}{\operatorname{argmax}} \; v^T C v - 2d^T v \quad \text{s. t. } v^T v = r^2. \tag{9}$$

To solve the problem we form the Lagrange function

$$\mathscr{L}(v, \lambda) = v^T C v - 2d^T v - \lambda(v^T v - r^2), \tag{10}$$

and set its derivatives to zero, i.e.

$$Cv = d + \lambda v, \quad v^T v = r^2. \tag{11}$$

As this is a non-convex problem a solution of the system in Eq. (11) is only a local optimum for problem (9). The following lemma, however, gives a criterion for distinguishing the global optimum of problem (9) from the solution set of the system in Eq. (11). Alternative and slightly more complex proofs for the same claim are given by (author?) [10] and (author?) [9]. Let us now denote the optimization objective of Eq. (9) with $\chi(v)$.

Lemma 1. *The maximum of the function $\chi(v)$ is attained at the tuple (v, λ) satisfying the stationary constraints (11) with the largest value of λ.*

Proof. Let $(\boldsymbol{v}_1, \lambda_1)$ and $(\boldsymbol{v}_2, \lambda_2)$ be two tuples satisfying the stationary constraints (11) with $\lambda_1 \geq \lambda_2$. Plugging the tuples into the first stationary constraint we obtain

$$C\boldsymbol{v}_1 = \lambda_1 \boldsymbol{v}_1 + \boldsymbol{d}, \tag{12}$$

$$C\boldsymbol{v}_2 = \lambda_2 \boldsymbol{v}_2 + \boldsymbol{d}. \tag{13}$$

Substracting (13) from (12) we have

$$C\boldsymbol{v}_1 - C\boldsymbol{v}_2 = \lambda_1 \boldsymbol{v}_1 - \lambda_2 \boldsymbol{v}_2. \tag{14}$$

Multiplying (14) first with $\boldsymbol{v}_1^T$ and then with $\boldsymbol{v}_2^T$ and adding the resulting two equations (having in mind that the matrix C is symmetric) we deduce

$$\boldsymbol{v}_1^T C \boldsymbol{v}_1 - \boldsymbol{v}_2^T C \boldsymbol{v}_2 = (\lambda_1 - \lambda_2)(r^2 + \boldsymbol{v}_1^T \boldsymbol{v}_2). \tag{15}$$

On the other hand, using the Cauchy-Schwarz inequality and (11) we deduce

$$\boldsymbol{v}_1^T \boldsymbol{v}_2 \leq \|\boldsymbol{v}_1\| \|\boldsymbol{v}_2\| = r^2. \tag{16}$$

Now, combining the results obtained in (15) and (16) with the initial assumption $\lambda_1 \geq \lambda_2$ we deduce

$$\boldsymbol{v}_1^T C \boldsymbol{v}_1 - \boldsymbol{v}_2^T C \boldsymbol{v}_2 \leq 2r^2(\lambda_1 - \lambda_2). \tag{17}$$

Finally, subtracting the optimization objectives for the two tuples and using (12) and (13) multiplied by $\boldsymbol{v}_1^T$ and $\boldsymbol{v}_2^T$, respectively, we prove

$$\chi(\boldsymbol{v}_1) - \chi(\boldsymbol{v}_2) = 2r^2(\lambda_1 - \lambda_2) - (\boldsymbol{v}_1^T C \boldsymbol{v}_1 - \boldsymbol{v}_2^T C \boldsymbol{v}_2) \geq 0,$$

where the last inequality follows from (17). □

Hence, instead of the original optimization problem (9) we can solve the system with two stationary equations (11) with maximal λ. **(author?)** [11] propose two methods for solving such problems. In the first approach, the problem is reduced to a quadratic eigenvalue problem and afterwards transformed into a linear eigenvalue problem. In the second approach the problem is reduced to solving a one-dimensional secular equation. The first approach is more elegant, as it allows us to compute the solution in a closed form. Namely, the solution to the problem (9) is given by [11]

$$\boldsymbol{v}^* = (C - \lambda_{max} \mathbf{I})^{-1} \boldsymbol{d},$$

where λ_{max} is the largest real eigenvalue of

$$\begin{bmatrix} C & -\mathbf{I} \\ -\frac{1}{r^2} \boldsymbol{d}\boldsymbol{d}^T & C \end{bmatrix} \begin{bmatrix} \boldsymbol{\gamma} \\ \boldsymbol{\eta} \end{bmatrix} = \lambda \begin{bmatrix} \boldsymbol{\gamma} \\ \boldsymbol{\eta} \end{bmatrix}.$$

Despite its elegance, the approach requires us to decompose a non-symmetric block matrix of dimension $2n$ and this is not a numerically stable task for every such matrix. Furthermore, the computed solution $\boldsymbol{v}^*$ highly depends on the precision up to which the optimal λ is computed and for an imprecise value the solution might not be on the hypersphere at all (for a detailed study refer to **(author?)** [11]).

For this reason, we rely on the secular approach in the computation of the optimal solution. In Section 5 we deal with an efficient algorithm for the computation of the parameter to a machine precision and here we describe how to derive the secular equation required to compute the optimal λ. In the first step the stationary constraint (11) is simplified by decomposing the symmetric matrix $C = P\Delta P^T$, i.e. $P\Delta P^T v = d + \lambda v$. Then, the resulting equation is multiplied with the orthogonal matrix P^T from the left and transformed into $\Delta t = \hat{d} + \lambda t$, with $\hat{d} = P^T d$ and $t = P^T v$. From the last equation we compute

$$t_i(\lambda) = \hat{d}_i / (\Delta_{ii} - \lambda) \quad (i = 1, 2, ..., n),$$

and substitute the computed t-vector into the second stationary constraint to form the secular equation

$$g(\lambda) = \sum_i t_i^2(\lambda) - r^2 = 0. \tag{18}$$

The optimal value of parameter λ is the largest root of the non-linear secular equation and the optimal solution to problem (9) is given by $v^* = P \cdot t(\lambda_{max})$. Moreover, the interval at which the root lies is known [11]. Namely, it must hold $\lambda_{max} \geq \Delta_{11}$, where Δ_{11} is the largest eigenvalue of the symmetric matrix C.

The complexity of both approaches (secular and eigenvalue) for a d-dimensional embedding is $O(dn^3)$, where n is the number of data instances. The cubic term arises from the eigendecompositions required to compute the solutions to problem (9) for each of the d variance direction.

In Section 4.2 we show how to transform the optimization problem (1) with additional hard constraints to optimize only problem (9).

4.2 Eliminating a Linear Constraint

For the optimization problems involving hard orthogonality or hard placement of control points we have an additional linear term (8) in the optimization problem (7). If the linear term is of rank $m < n$, we can eliminate it and transform the problem to optimize a quadratic over an $(n - m)$-dimensional hypersphere. The linear constraint term is first transformed from a constraint over a hyperellipsoid to a constraint over a hypersphere by replacing the matrix L (the linear constraint in Eq. (8)) with the matrix $LK^{-\frac{1}{2}}$. In the remainder of the section, we will denote the transformed constraint matrix with L.

In order to eliminate the linear constraint we do a QR factorization of the matrix $L^T = QR$, where $Q \in \mathbb{R}^{n \times n}$ is an orthogonal and $R = \begin{bmatrix} R \\ 0 \end{bmatrix} \in \mathbb{R}^{n \times m}$ is an upper-diagonal matrix. Substituting

$$Q^T v = \begin{bmatrix} z_1 \\ z_2 \end{bmatrix}, \quad \text{with } z_1 \in \mathbb{R}^m \text{ and } z_2 \in \mathbb{R}^{n-m},$$

linear and sphere constraints are transformed into

$$y = Lv = R^T(Q^Tv) = \overline{R}^Tz_1 \quad \Longrightarrow \quad z_1 = (\overline{R}^T)^{-1}y,$$
$$r^2 = v^Tv = z_1^Tz_1 + z_2^Tz_2 \quad \Longrightarrow \quad z_2^Tz_2 = r^2 - z_1^Tz_1 = \hat{r}^2.$$

As z_1 is a constant vector we can rewrite the objective in Eq. (9) as a quadratic over an $(n-m)$-dimensional sphere. Namely, the quadratic term can be rewritten as

$$v^TCv = v^TQQ^TCQQ^Tv = z_1^TFz_1 + 2z_2^TGz_1 + z_2^THz_2, \text{ where}$$

$$Q^TCQ = \begin{bmatrix} F & G^T \\ G & H \end{bmatrix} \text{ with } F \in \mathbb{R}^{m \times m}, G \in \mathbb{R}^{(n-m) \times m} \text{ and } H \in \mathbb{R}^{(n-m) \times (n-m)}.$$

On the other hand, the linear term is transformed into

$$d^Tv = d^TQQ^Tv = f_1^Tz_1 + f_2^Tz_2,$$

where $f_1 \in \mathbb{R}^m$ and $f_2 \in \mathbb{R}^{n-m}$ are blocks in the vector Q^Td. Denoting with $\hat{d} = f_2 - Gz_1$ and $\hat{z} = z_2$ we obtain

$$\operatorname*{argmax}_{\hat{z} \in \mathbb{R}^{n-m}} \hat{z}^TH\hat{z} - 2\hat{d}^T\hat{z} \quad \text{s. t. } \hat{z}^T\hat{z} = \hat{r}^2,$$

for which a closed form solution was given in Section 4.1.

5   Numerically Efficient Interaction

To shape an embedding interactively with the help of knowledge-based constraints it is required to solve the optimization problem (1) at each interaction step. In Section 4 we have described how to solve the arising optimization problem with complexity $O(dn^3)$, where n is the number of instances and d is the number of variance directions. In this section, we show how the proposed algorithm can be improved to enable a user interaction in $O(d^2n^2)$ time. To achieve this, we express the interaction in the form of rank-one updates of the original problem and review a linear time algorithm for solving secular equations arising in the process.

5.1   Efficient Formulation of the Interaction

Let us assume, without loss of generality, the algorithm is working with the classification constraints and the soft orthogonality. The optimization problem for this setting can be expressed in the form (9). Furthermore, assume a user, interested in a d-dimensional embedding, has provided labels y_k at an interaction step k. To compute embeddings interactively the algorithm needs to solve problem (9) for different interaction steps and for all d directions. We denote with C the symmetric matrix defining the quadratic term in the problem arising in a step k for a direction s. The variance term, independent of the provided labels and denoted with $\overline{C}$, can be decomposed prior to any interaction. The decomposition has complexity $O(n^3)$, but it is a one time cost paid prior to interaction steps. The linear term for a direction s at the step k is a function of a block of the kernel matrix and the label vector y_k and we denote it with a_k. Now, the first stationary constraint of the problem (9) for a direction $s + 1$ at the step k can be written as

$$C'x = \lambda x + a_k,$$

where the symmetric matrix C' is a rank-one update of the previous direction's quadratic term C for which the eigendecomposition is already computed. Reusing the decomposition, $C = U \Delta U^T$, we can rewrite the last equation as

$$C' = C - \mu \alpha_s \alpha_s^T = U \left(\Delta - \mu z z^T \right) U^T,$$

where α_s denotes the s-th direction vector and $z = U^T \alpha_s$. Let us denote the rank-one update to the diagonal matrix as

$$\Theta = \Delta - \mu z z^T \tag{19}$$

The complexity of a full eigendecomposition (see, e.g., **(author?)** [2, 3]) of the matrix Θ is $O(n^2)$ and it is computed by solving n secular equations (see Section 5.2), one for each of the eigenvalues of the matrix Θ. Rewriting the last equation using the substitution we get

$$U \Theta U^T x = \lambda x + a_k \implies \Theta \overline{x} = \lambda \overline{x} + \overline{a}, \text{ with } \overline{x} = U^T x \text{ and } \overline{a} = U^T a_k.$$

Now, using the decomposition $\Theta = V \Delta' V^T$ we transform the last problem into

$$\Delta' t = \lambda t + f, \text{ with } t = V^T \overline{x} \text{ and } f = V^T \overline{a}. \tag{20}$$

The second stationary constraint combined with Eq. (20) yields secular equation (18). Thus, the $(s + 1)$-th direction vector x_{s+1} is computed as

$$x_{s+1} = \overline{U} \left(\prod_{i=1}^{s} V_i \right) \cdot t(\lambda_{max}), \text{ with } t_i(\lambda) = \frac{\left(\overline{U} \prod_{i=1}^{s} V_i \right)^T a_k}{\delta_i^{(s+1)} - \lambda} \; (i = \overline{1, n}).$$

Note that $\overline{U}$ is the eigenvector matrix for the variance term $\overline{C}$ and $\delta_i^{(s+1)}$ is an eigenvalue of the quadratic term matrix for the direction $s + 1$ (see problems (1) and (9)).

Hence, to compute the directions at the interaction step k we need to perform $O(d^2)$ matrix vector multiplications, each incurring a quadratic cost, together with d quadratic time decompositions of Θ matrices. What remains to compute the data projection is a multiplication of direction vectors with the kernel matrix which is again of quadratic complexity. Therefore, the overall complexity of an interaction step is $O(d^2 n^2)$. In a similar fashion, it is possible to show the quadratic complexity of an interaction for other knowledge-based constraints.

5.2 Efficient and Stable Secular Solver

In this section, we review the state of the art in computation of the root of a secular equation in a given interval (δ_i, δ_{i+1}), with $\delta_1 \leq \delta_2 \leq ...\delta_{n-1} \leq \delta_n$. In particular, we are interested in finding a root of Eq. (18) and the equation [3]

$$\overline{g}(\lambda) = 1 + \mu \sum_{i=1}^{n} \frac{z_i^2}{\delta_i - \lambda},$$

whose roots are the eigenvalues of the perturbed matrix (19). In the discussion to follow we focus on finding the roots of $\overline{g}(\lambda)$ and digress to the secular function (18) only when the arguments are not eligible for it.

It is possible to compute a root of a secular equation with a linear number of flops [17]. An obvious choice for the root finder is a Newton method and, yet, it is not well suited for the problem. The tangent at certain points in the interval of interest crosses the x-axis outside this interval leading to incorrect solution or division by zero as the solution can converge to one of the poles δ_i. An efficient root finder, then, must overcome this issue and converge very quickly. The main idea behind the efficient root finder is to approximate a secular equation with a quadratic surrogate and find a zero of this surrogate. We go step-by-step through the procedure. First let us assume $\mu = 1$ and split the non-constant $\overline{g}$ terms into two functions

$$\psi_1(\lambda) = \sum_{k=1}^{i} \frac{z_k^2}{\delta_k - \lambda} \text{ and } \psi_2(\lambda) = \sum_{k=i+1}^{n} \frac{z_k^2}{\delta_k - \lambda}.$$

Then, each function ψ_k is approximated by a surrogate

$$h_k(x) = a_k + b_k/(\delta_{k-1+i} - x),$$

with constants computed such that at all candidate solutions x_j it holds [17]:

$$h_k(x_j) = \psi_k(x_j) \text{ and } h'_k(x_j) = \psi'_k(x_j).$$

Now, combining the surrogates for the particular terms we obtain the surrogate for the secular function $\overline{g}$,

$$\overline{h}(x) = c_3 + c_1/(\delta_i - x) + c_2/(\delta_{i+1} - x), \tag{21}$$

whose root is the next secular solution candidate. **(author?)** [3] proved the surrogate root finder converges to the desired root.

In contrast to the secular function $\overline{g}(\lambda)$, the secular equation (18) has only one surrogate [11]:

$$h(x) = p/(q - x)^2, h(x_j) = g(x_j) \text{ and } h'(x_j) = g'(x_j).$$

After computing the coefficients p and q we get the iteration step $\eta = x_{j+1} - x_j$ as

$$\eta = 2\frac{g(x_j) + r^2}{g'(x_j)} \left(1 - \frac{\sqrt{g(x_j) + r^2}}{r}\right).$$

For the initial solution $\delta_1 < x_0 < \lambda^*$ the convergence is monotonic [3], i.e. $\lambda^* > x_{j+1} > x_j$, where λ^* is the desired root of the secular equation.

Finally, we note that different roots of secular Eq. (18) belong to distinct closed intervals and can be, therefore, computed in parallel. This enables an efficient GPU implementation of the secular solver resulting in a significant speed-up to the presented rank-one update algorithm. Consequently, with a GPU implementation of the secular solver it is possible to increase the interaction frame rate and improve scalability.

6   Experiments

The best, and ultimately only true, way to evaluate an algorithm for interactive data visualization and exploration is via a study with real domain experts that are using a tool implementing that algorithm. In the absence of the study we performed a number of in silico experiments which aim at illustrating the utility and sensibility of our approach. In particular, we show: (i) a rendered embedding is robust under small changes in the placement of control points; (ii) the approach is flexible in choosing a low-dimensional embedding from the many possible ones; (iii) 'sufficient' amount of information is retained in a visualization; (iv) it is possible to detect structures that do not necessarily exhibit high correlation with variance and which are, therefore, obscured in a regular kernel PCA embedding. We study the properties $(i) - (iii)$ on benchmark data sets for semi-supervised learning [5] and generate an artificial data set to show the property (iv). In the experiments we use different kernels: Gaussian with a bandwidth equal to the median of pairwise distances between the instances, inverse Laplacian defined by the ϵ-neighbourhood graph, linear and polynomial kernel of degree three. All the reported results are averaged over ten runs.

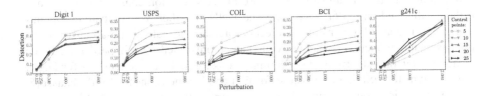

Fig. 1. Distortion of an embedding over the perturbation of a control point

How stable is our approach? In exploratory data visualization it should be possible to smoothly change the embedding by moving control point throughout the projection space. In other words, small perturbations of a control point should result in small perturbations of the overall embedding. We empirically verify the stability of the proposed method by moving a control point randomly throughout the projection space. We express the displacement of a control point as a fraction of the median of pairwise distances within the embedding. The distortion or the difference between the two embeddings is measured by the average displacement of a point between them and this value is scaled by the median pairwise distance between the points in the kernel PCA embedding. In Figure 1 we show the distortion as the perturbation increases across five benchmark data sets [5]. Results clearly indicate that the proposed method provides stable visual exploration of data.

How flexible is our approach? It is possible to generate different embeddings of the same dataset with kernel PCA using different kernels. To show the flexibility of the proposed method we set up an experiment with a sum of different kernels and show that the proposed method can choose the PCA embedding corresponding to a kernel by re-arranging control points accordingly. In particular, we combine a Gaussian and the inverse Laplacian kernel that produce geometrically very different PCA embeddings of the considered datasets. We again measure the distortion between the two embeddings.

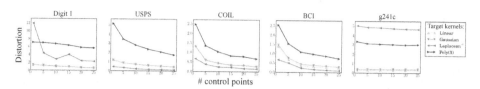

Fig. 2. Distortion between the target and the current embedding over the number of re-arranged control points. Results show, we can recover a target embedding with a small number of re-arranged control points.

The empirical results indicate that it is possible to recover the embeddings corresponding to PCA projection of each used kernel. Figure 2 shows the distortion between the current and the target embedding as the number of selected control points increases.

How informative is our approach? A satisfactory embedding should be able to retain a fair amount of information from the input space. To measure the amount of information the proposed method retains we simulate a classification task. We use the semi-supervised learning benchmark data set in which the small amount of labels is given to us. In building a classifier we use the classification constraints with a hard orthogonality and on top of it we apply the 1-NN classifier. We compare our method using only 1, 2 and 3 dimensional projections against the state-of-the-art unsupervised dimensionality reduction techniques with many more dimensions. In particular, dimensionality reduction algorithms use 38 dimensions for g241c, 4 dimensions for Digit1, 9 dimensions for USPS, 8 dimensions for BCI and 3 dimensions for COIL dataset [5]. The results indicate our algorithm is able to retain a satisfactory amount of information over the first 3 principal directions. We used Gaussian and inverse Laplacian kernel to compute the embeddings (see Figure 3).

Can we discover structures hidden by the plain PCA? We have created a 3D artificial data set to demonstrate the proposed approach is able to discover structures in data that do not exhibit high correlation with variance. Such structures remain hidden in the PCA projection and the proposed method is capable of detecting them by the appropriate placement of control points. In particular, we sample 3 plates of points from a 2D Gaussian distribution and embed these plates into the 3D space such that the z-axis coordinate for each plate is obtained by sampling from the normal distributions

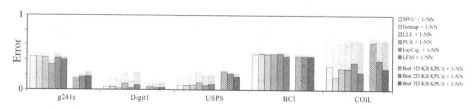

Fig. 3. A classification error comparison between the proposed method and unsupervised dimensionality algorithms with many more dimensions [5]. Supervision was done with either 10 (greyed out) or 100 labels. We point out that the experiment was conducted to show the amount of information retained by the visualization and not to compete these methods.

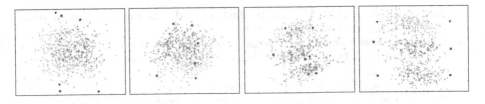

Fig. 4. Discovering cluster structures which are hidden in the kernel PCA embedding

with means at $1, 0$ and -1. We choose the variance for the z-axis sampling such that the cluster-plates barely touch each other. For the sample generated in this way the within-clusters variance is higher than the between-cluster variance and the cluster structure remains hidden in the kernel PCA projection (see the first picture in Figure 4). Moving the two most distant points of each cluster apart (a total of 6 displacements) we discover the cluster structure obscured by the plain kernel PCA (the last picture in Figure 4).

Scalability. As stated in Section 5, it is possible to implement the interaction using rank-one updates with $O(n^2)$ complexity. Our experiments indicate that it is possible to interact with a visualization with multiple frames per second for datasets with $\approx$ 1000 instances. The frame rate and scalability can be significantly improved with an efficient GPU implementation of the secular solver (see Section 5.2). Moreover, it is possible to use a kernel expansion over a subset of instances while searching for a suitable placement of a control point and upon such placement the embedding can be computed with a full kernel expansion.

7   Conclusion

We proposed a novel, knowledge-based variant of kernel PCA with the aim of making visual data exploration a more interactive endeavour, in which users can intuitively modify the embedding by placing 'control points', adding must-link and cannot-link constraints, and supplying known class labels. We showed that maximizing the variance of unit norm functions that are as orthogonal as possible and that adhere as much as possible to the user supplied knowledge can be formulated as a non-convex quadratic optimization problem over an ellipsoid with linear constraints. We gave an algorithm for computing the resulting embedding in cubic time and argued that the more typical situation in interactive data analysis corresponds, however, to a low-rank update of the previous solution. We then derived an algorithm for computing these low-rank updates in quadratic time.

Acknowledgement. Part of this work was supported by the German Science Foundation (DFG) under the reference number 'GA 1615/1-2'.

References

[1] Andrews, C., Endert, A., North, C.: Space to Think: Large High-resolution Displays for Sensemaking. In: Proceedings of the SIGCHI Conference on Human Factors in Computing Systems (CHI), pp. 55–64. ACM (2010)

[2] Arbenz, P.: Lecture Notes on Solving Large Scale Eigenvalue Problems, pp. 77–93. ETH Zürich (2012)

[3] Bunch, J.R., Nielsen, C.P., Sorensen, D.: Rank-one modification of the symmetric eigen-problem. Numerische Mathematik 31 (1978)

[4] Callahan, E., Koenemann, J.: A comparative usability evaluation of user interfaces for on-line product catalog. In: Proceedings of the 2nd ACM Conference on Electronic Commerce (EC), pp. 197–206. ACM (2000)

[5] Chapelle, O., Schölkopf, B., Zien, A. (eds.): Semi-Supervised Learning. MIT Press (2006)

[6] Cox, T.F., Cox, M.A.A.: Multidimensional Scaling. Chapman and Hall/CRC (2000)

[7] Dinuzzo, F., Schölkopf, B.: The representer theorem for Hilbert spaces: A necessary and sufficient condition. In: Proceedings of the Conference on Neural Information Processing Systems (NIPS), pp. 189–196 (2012)

[8] Endert, A., Han, C., Maiti, D., House, L., Leman, S., North, C.: Observation-level interaction with statistical models for visual analytics. In: IEEE VAST, pp. 121–130. IEEE (2011)

[9] Forsythe, G.E., Golub, G.H.: On the Stationary Values of a Second-Degree Polynomial on the Unit Sphere. Journal of the Society for Industrial and Applied Mathematics 13(4), 1050–1068 (1965)

[10] Gander, W.: Least Squares with a Quadratic Constraint. Numerische Mathematik 36, 291–308 (1981)

[11] Gander, W., Golub, G., von Matt, U.: A constrained eigenvalue problem. Linear Algebra and its Applications 114-115, 815–839 (1989)

[12] Ham, J., Lee, D.D., Mika, S., Schölkopf, B.: A Kernel View of the Dimensionality Reduction of Manifolds. In: Proceedings of the 21st International Conference on Machine Learning (2004)

[13] Izenman, A.J.: Linear Discriminant Analysis. Springer (2008)

[14] Jeong, D.H., Ziemkiewicz, C., Fisher, B.D., Ribarsky, W., Chang, R.: iPCA: An Interactive System for PCA-based Visual Analytics. Comput. Graph. Forum 28(3), 767–774 (2009)

[15] Jolliffe, I.T.: Principal Component Analysis. Springer (1986)

[16] Leman, S., House, L., Maiti, D., Endert, A., North, C.: Visual to Parametric Interaction (V2PI). PLoS One 8, e50474 (2013)

[17] Li, R.C.: Solving secular equations stably and efficiently (1993)

[18] Paurat, D., Gärtner, T.: Invis: A tool for interactive visual data analysis. In: Blockeel, H., Kersting, K., Nijssen, S., Železný, F. (eds.) ECML PKDD 2013, Part III. LNCS, vol. 8190, pp. 672–676. Springer, Heidelberg (2013)

[19] Paurat, D., Oglic, D., Gärtner, T.: Supervised PCA for Interactive Data Analysis. In: Proceedings of the Conference on Neural Information Processing Systems (NIPS) 2nd Workshop on Spectral Learning (2013)

[20] Roweis, S.T., Saul, L.K.: Nonlinear dimensionality reduction by locally linear embedding. Science 290(5500), 2323–2326 (2000)

[21] Schölkopf, B., Herbrich, R., Smola, A.J.: A generalized representer theorem. In: Helmbold, D., Williamson, B. (eds.) COLT/EuroCOLT 2001. LNCS (LNAI), vol. 2111, pp. 416–426. Springer, Heidelberg (2001)

[22] Shearer, C.: The CRISP-DM model: The new blueprint for data mining. Journal of Data Warehousing 5(4), 13–22 (2000)

[23] Tenenbaum, J.B., Silva, V.D., Langford, J.C.: A global geometric framework for nonlinear dimensionality reduction. Science 290(5500), 2319–2323 (2000)

[24] Tukey, J.W.: Mathematics and the picturing of data. In: Proceedings of the International Congress of Mathematicians, vol. 2, pp. 523–531 (1975)

[25] Walder, C., Henao, R., Mørup, M., Hansen, L.K.: Semi-Supervised Kernel PCA. Computing Research Repository (CoRR) abs/1008.1398 (2010)

[26] Weinberger, K.Q., Saul, L.K.: Unsupervised Learning of Image Manifolds by Semidefinite Programming. In: Proceedings of the IEEE Conference on Computer Vision and Pattern Recognition (CVPR), vol. 2, pp. 988–995 (2004)

A Two-Step Learning Approach for Solving Full and Almost Full Cold Start Problems in Dyadic Prediction

Tapio Pahikkala[1], Michiel Stock[2], Antti Airola[1], Tero Aittokallio[3],
Bernard De Baets[2] and Willem Waegeman[2]

[1] University of Turku and Turku Centre for Computer Science,
Joukahaisenkatu 3-5 B, FIN-20520, Turku, Finland
{firstname.surname}@utu.fi

[2] Department of Mathematical Modelling, Statistics and Bioinformatics, Ghent
University, Coupure links 653, B-9000 Ghent, Belgium
{firstname.surname}@UGent.be

[3] Institute for Molecular Medicine Finland (FIMM), P.O. Box 20 (Tukholmankatu
8), FI-00014 University of Helsinki, Helsinki, Finland
{firstname.surname}@fimm.fi

Abstract. Dyadic prediction methods operate on pairs of objects (dyads), aiming to infer labels for out-of-sample dyads. We consider the full and almost full cold start problem in dyadic prediction, a setting that occurs when both objects in an out-of-sample dyad have not been observed during training, or if one of them has been observed, but very few times. A popular approach for addressing this problem is to train a model that makes predictions based on a pairwise feature representation of the dyads, or, in case of kernel methods, based on a tensor product pairwise kernel. As an alternative to such a kernel approach, we introduce a novel two-step learning algorithm that borrows ideas from the fields of pairwise learning and spectral filtering. We show theoretically that the two-step method is very closely related to the tensor product kernel approach, and experimentally that it yields a slightly better predictive performance. Moreover, unlike existing tensor product kernel methods, the two-step method allows closed-form solutions for training and parameter selection via cross-validation estimates both in the full and almost full cold start settings, making the approach much more efficient and straightforward to implement.

Keywords: Dyadic prediction, pairwise learning, transfer learning, kernel ridge regression, kernel methods.

1 A Subdivision of Dyadic Prediction Methods

Many real-world machine learning problems can be naturally represented as pairwise learning or dyadic prediction problems, for which feature representations of two different types of objects (aka a dyad) are jointly used to predict a relationship between those objects. Amongst others, applications of that kind emerge

T. Calders et al. (Eds.): ECML PKDD 2014, Part II, LNCS 8725, pp. 517–532, 2014.
© Springer-Verlag Berlin Heidelberg 2014

in biology (e.g. predicting mRNA-miRNA interactions), medicine (e.g. design of personalized drugs), chemistry (e.g. prediction of binding between two types of molecules), social network analysis (e.g. link prediction) and recommender systems (e.g. personalized product recommendation).

For many dyadic prediction problems it is extremely important to implement appropriate training and evaluation procedures. [29] make in a recent Nature-review on dyadic prediction an important distinction between four main settings. Given t and d as the feature representations of the two types of objects, those four settings can be summarized as follows:

- **Setting A:** Both t and d are observed during training, as parts of separate dyads, but the label of the dyad (t, d) must be predicted.
- **Setting B:** Only t is known during training, while d is not observed in any dyad, and the label of the dyad (t, d) must be predicted.
- **Setting C:** Only d is known during training, while t is not observed in any dyad, and the label of the dyad (t, d) must be predicted.
- **Setting D:** Neither t nor d occur in any training dyad, but the label of the dyad (t, d) must be predicted (referred to as the full cold start problem).

Setting A is of all four settings by far the most studied setting in the machine learning literature. Motivated by applications in collaborative filtering and link prediction, matrix factorization and related techniques are often applied to complete partially observed matrices, where missing values represent (t, d) combinations that are not observed during training - see e.g. [15] for a review.

Settings B and C are very similar, and a variety of machine learning methods can be applied for these settings. From a recommender systems viewpoint, those settings resemble the cold start problem (new user or new item), for which hybrid and content-based filtering techniques are often applied – see e.g. [1, 10, 20, 35, 39] for a not at all exhaustive list. From a bioinformatics viewpoint, Settings B and C are often analyzed using graph-based methods that take the structure of a biological network into account – see e.g. [33] for a recent review. When the features of t are negligible or unavailable, while those of d are informative, Setting B can be interpreted as a multi-label classification problem (binary labels), a multivariate regression problems (continuous labels) or a specific multi-task learning problem. Here as well, a large number of applicable methods exists in the literature.

1.1   The Problem Setting Considered in this Article

Matrix factorization and hybrid filtering strategies are not applicable to Setting D. We will refer to this setting as the *full cold start* problem, which finds important applications in domains such as bioinformatics and chemistry – see experiments. Compared to the other three settings, Setting D has received less attention in the literature (with some exceptions, see e.g. [20, 23, 25, 28]), and it will be our main focus in this article. Furthermore, we will also investigate the transition phase between Settings C and D, when t occurs very few times in the training dataset, while $\mathbf{d}$ of the dyad $(\mathbf{d}, \mathbf{t})$ is only observed in the prediction phase. We refer to this setting as the *almost full cold start* problem.

Full and almost full cold start problems can only be solved by considering feature representations of dyads (aka side information in the recommender systems literature). Similar to several existing papers dealing with Setting D, we consider tensor product feature representations and their kernel duals. Such feature representations have been successfully applied in order to solve problems such as product recommendation [5, 28], prediction of protein-protein interactions [7, 14], drug design [13], prediction of game outcomes [26] and document retrieval [23]. For classification and regression problems a standard recipe exists of plugging pairwise kernels in support vector machines, kernel ridge regression (KRR), or any other kernel method. Efficient optimization approaches based on gradient descent [14, 23, 28] and closed form solutions [23] have been proposed. We compare KRR with a tensor product pairwise kernel (tensor KRR) both theoretically and experimentally to the two-step approach introduced in this paper.

1.2   Formulation as a Transfer Learning Problem

As discussed above, dyadic prediction is closely related to several subfields of machine learning. Further on in this article we decide to adopt a multi-task learning or transfer learning terminology, using $\mathbf{d}$ and $\mathbf{t}$ to denote the feature representations of instances and tasks, respectively. From this viewpoint, Setting C corresponds to a specific instantiation of a traditional transfer learning scenario, in which the aim is to transfer knowledge obtained from already learned *auxiliary tasks* to the *target task* of interest [27]. Stretching the concept of transfer learning even further, in the case of so-called *zero-data learning*, one arrives at Setting D, which is characterized by no available labeled training data for the target task [16]. If the target task is unknown during the training time, the learning method must be able to generalize to it "on the fly" at prediction time. The only available data here is coming from auxiliary training tasks.

We present a simple but elegant two-step approach to tackle these settings. First, a KRR model trained on auxiliary tasks is used to predict labels for the related target task. Next, a second model is constructed, using KRR on the target data, augmented by the predictions of the first phase. We show via spectral filtering that this approach is closely related to learning a pairwise model using a tensor product pairwise kernel. However, the two-step approach is much simpler to implement and it allows more heterogeneous transfer learning settings than the ordinary pairwise kernel ridge regression, as well as a more flexible model selection. Furthermore, it allows for a more efficient generalization to new tasks not known during training time, since the model built on auxiliary tasks does not need to be re-trained in such settings. In the experiments we consider three distinct dyadic prediction problems, concerning drug-target, newsgroup document similarity and protein functional similarity predictions. Our results show that the two-step transfer learning approach can be highly beneficial when there is no labeled data at all, or only a small amount of labeled data available for the target task, while in settings where there is a significant amount of labeled data available for the target task a single-task model suffices. In related work, [34] have recently proposed a similar two-step approach based on tree-based ensemble methods for biological network inference.

2 Solving Full and Almost Full Cold Start Problems via Transfer Learning

Adopting a multi-task learning methodology, the training set is assumed to consist of a set $\{\mathbf{x}_h\}_{h=1}^n$ of object-task pairs and a vector $\mathbf{y} \in \mathbb{R}^n$ of their real-valued labels. We assume that each training input can be represented as $\mathbf{x} = (\mathbf{d}, \mathbf{t})$, where $\mathbf{d} \in \mathcal{D}$ and $\mathbf{t} \in \mathcal{T}$ are the objects and tasks, respectively, and $\mathcal{D}$ and $\mathcal{T}$ are the corresponding spaces of objects and tasks. Moreover, let $D = \{\mathbf{d}_i\}_{i=1}^m$ and $T = \{\mathbf{t}_j\}_{j=1}^q$ denote, respectively, the sets of distinct objects and tasks encountered in the training set with $m = |D|$ and $q = |T|$. We say that the training set is *complete* if it contains every object-task pair with object in D and task in T exactly once. For complete training sets, we introduce a further notation for the matrix of labels $\mathbf{Y} \in \mathbb{R}^{m \times q}$, so that its rows are indexed by the objects in D and the columns by the tasks in T. In full and almost full cold start prediction problems, this matrix will not contain any target task information.

2.1 Kernel Ridge Regression with Tensor Product Kernels

Several authors (see [2, 4] and references therein) have extended KRR to involve task correlations via matrix-valued kernels. However, most of the literature concerns kernels for which the tasks are fixed at training time. An alternative approach, allowing the generalization to new tasks more straightforwardly, is to use the tensor product pairwise kernel [5, 7, 8, 12, 21, 23, 28], in which kernels are defined on object-task pairs

$$\Gamma(\mathbf{x}, \overline{\mathbf{x}}) = \Gamma\left((\mathbf{d}, \mathbf{t}), (\overline{\mathbf{d}}, \overline{\mathbf{t}})\right) = k\left(\mathbf{d}, \overline{\mathbf{d}}\right) g\left(\mathbf{t}, \overline{\mathbf{t}}\right) \tag{1}$$

as a product of the data kernel k and the task kernel g. Let

$$\mathbf{K} \in \mathbb{R}^{m \times m} \text{ and } \mathbf{G} \in \mathbb{R}^{q \times q} \tag{2}$$

be the kernel matrices for the data points and tasks, respectively. Then, the kernel matrix for the object-task pairs is, for a complete training set, the tensor product $\mathbf{\Gamma} = \mathbf{K} \otimes \mathbf{G}$, which is usually infeasible to use directly due to its large size. Tensor KRR seeks for a prediction function of type

$$f(\mathbf{x}) = \sum_{i=1}^n \alpha_i \Gamma(\mathbf{x}, \mathbf{x}_i),$$

where α_i are parameters that minimize the following objective function:

$$J(\boldsymbol{\alpha}) = (\mathbf{\Gamma}\boldsymbol{\alpha} - \mathbf{y})^{\mathrm{T}}(\mathbf{\Gamma}\boldsymbol{\alpha} - \mathbf{y}) + \lambda \boldsymbol{\alpha}^{\mathrm{T}} \mathbf{\Gamma}\boldsymbol{\alpha}, \tag{3}$$

Algorithm 1. Two-step kernel ridge regression

1: $\mathbf{C} \leftarrow \operatorname{argmin}_{\mathbf{C} \in \mathbb{R}^{m \times q}} \left\{ \|\mathbf{C}\mathbf{G} - \mathbf{Y}\|_F^2 + \lambda_t \operatorname{tr}(\mathbf{C}\mathbf{G}\mathbf{C}^{\mathrm{T}}) \right\}$
2: $\mathbf{z} \leftarrow \left(\mathbf{z}_L^{\mathrm{T}}, (\mathbf{C}_{\mathcal{U}}\mathbf{g})^{\mathrm{T}} \right)^{\mathrm{T}}$
3: $\mathbf{a} \leftarrow \operatorname{argmin}_{\mathbf{a} \in \mathbb{R}^m} \left\{ (\mathbf{K}\mathbf{a} - \mathbf{z})^{\mathrm{T}} (\mathbf{K}\mathbf{a} - \mathbf{z}) + \lambda_d \mathbf{a}^{\mathrm{T}} \mathbf{K}\mathbf{a} \right\}$
4: **return** $f_t(\cdot) = \sum_{i=1}^m a_i k(\mathbf{d}_i, \cdot)$

---

whose minimizer can be found by solving the following system of linear equations:

$$(\mathbf{\Gamma} + \lambda \mathbf{I}) \, \boldsymbol{\alpha} = \mathbf{y}. \tag{4}$$

Several authors have pointed out that, while the size of the above system is considerably large, its solution can be found efficiently via tensor algebraic optimization [2, 14, 19, 25, 30, 37]. Namely, the complexity scales roughly of order $O(|D|^3 + |T|^3)$ which is required by computing the singular value decomposition (SVD) of both the object and task kernel matrices, but the complexities can be scaled down even further by using sparse kernel matrix approximations.

However, the above computational short-cut only concerns the case in which the training set is complete. If some of the pairs are missing or if there are several occurrences of certain pairs, one has to resort, for example, to gradient descent based training approaches. While these approaches can also be accelerated via tensor algebraic optimization, they still remain considerably slower than the SVD-based approach. A serious short-coming of the approach is that when generalizing to new tasks, the whole training procedure needs to be re-done with the new training set that contains the union of the auxiliary data and the target data. If the amount of auxiliary data is large, as one would hope in order to expect any positive transfer to happen, this makes generalization to new tasks on-the-fly computationally impractical.

2.2 Two-Step Kernel Ridge Regression

Next, we present a two-step procedure for performing transfer learning. In the following, we assume that we are provided a training set in which every auxiliary task has the same labeled training objects. This assumption is fairly realistic in many practical settings, since one can carry out, for example, a preliminary completion step by using the extensive toolkit of missing value imputation or matrix completion algorithms. A newly given target task, in contrast, is assumed to have only a subset of the training objects labeled. That is, the training set consisting of both the auxiliary and the target tasks is incomplete, because of the missing object labels of the target task, ruling out the direct application of the SVD-based training. To cope with this incompleteness, we consider an approach of performing the learning in two steps, of which the first step is used for completing the training set for the target tasks part and the second step for building a model for the target task. A further benefit is that the first phase where a model is trained on auxiliary data needs to be performed only once, and the resulting model may be subsequently re-used when new target tasks appear.

Algorithm 2. Two-step with LOOCV-based automatic model selection

---

Require: $\mathbf{Y} \in \mathbb{R}^{m \times q}, \boldsymbol{\Phi} \in \mathbb{R}^{m \times d}, \boldsymbol{\Psi} \in \mathbb{R}^{q \times r}, \mathbf{g} \in \mathbb{R}^q, \mathbf{z}_{\mathcal{L}} \in \mathbb{R}^{|\mathcal{L}|}$ with $d \leq m$ and $r \leq q$.

1: $\mathbf{U}, \sqrt{\boldsymbol{\Sigma}}, \mathbf{V} \leftarrow \text{SVD}(\boldsymbol{\Phi})$, with $\mathbf{U} \in \mathbb{R}^{m \times d}$, $\mathbf{V} \in \mathbb{R}^{d \times d}$ $\triangleright \mathcal{O}(qr^2)$

2: $\mathbf{P}, \sqrt{\mathbf{S}}, \mathbf{Q} \leftarrow \text{SVD}(\boldsymbol{\Psi})$, with $\mathbf{P} \in \mathbb{R}^{q \times r}, \mathbf{Q} \in \mathbb{R}^{r \times r}$ $\triangleright \mathcal{O}(md^2)$

3: $e \leftarrow \infty$

4: **for** $\overline{\lambda_t} \in \{\text{Grid of parameter values}\}$ **do**

5:      **for** $j = 1, \ldots, q$ **do** $\widetilde{G}_{j,j} \leftarrow \mathbf{P}_j(\text{diag}((\mathbf{S} + \overline{\lambda_t}\mathbf{I})^{-1}) \odot \mathbf{P}_j^{\mathrm{T}})$ $\triangleright \mathcal{O}(qr)$

6:      $\overline{\mathbf{C}} \leftarrow \mathbf{YP}(\mathbf{S} + \overline{\lambda_t}\mathbf{I})^{-1}\mathbf{P}^{\mathrm{T}}$ $\triangleright \mathcal{O}(mqr)$

7:      **for** $i = 1, \ldots, m$ **and** $j = 1, \ldots, q$ **do** $\overline{R}_{i,j} \leftarrow Y_{i,j} - \left(\widetilde{G}_{j,j}\right)^{-1}\overline{C}_{i,j}$ $\triangleright \mathcal{O}(mq)$

8:      $\overline{e} \leftarrow \mathcal{E}(\overline{\mathbf{R}}, \mathbf{Y})$ $\triangleright$ Error between labels and LOO predictions

9:      **if** $\overline{e} < e$ **then** $\lambda_t, e, \mathbf{R}, \mathbf{C} \leftarrow \overline{\lambda_t}, \overline{e}, \overline{\mathbf{R}}, \overline{\mathbf{C}}$

10: $e \leftarrow \infty$

11: **for** $\overline{\lambda_d} \in \{\text{Grid of parameter values}\}$ **do**

12:      **for** $i = 1, \ldots, m$ **do** $\widetilde{K}_{i,i} \leftarrow \mathbf{U}_i(\text{diag}((\boldsymbol{\Sigma} + \overline{\lambda_d}\mathbf{I})^{-1}) \odot \mathbf{U}_i^{\mathrm{T}})$ $\triangleright \mathcal{O}(md)$

13:      $\overline{\mathbf{A}} \leftarrow \mathbf{U}(\boldsymbol{\Sigma} + \overline{\lambda_d}\mathbf{I})^{-1}\mathbf{U}^{\mathrm{T}}\mathbf{R}$ $\triangleright \mathcal{O}(mqd)$

14:      **for** $i = 1, \ldots, m$ **and** $j = 1, \ldots, q$ **do** $\overline{T}_{i,j} \leftarrow Y_{i,j} - \left(\widetilde{K}_{i,i}\right)^{-1}\overline{A}_{i,j}$ $\triangleright \mathcal{O}(mq)$

15:      $\overline{e} \leftarrow \mathcal{E}(\overline{\mathbf{T}}, \mathbf{Y})$ $\triangleright$ Error between labels and LOO predictions

16:      **if** $\overline{e} < e$ **then** $\lambda_d, e, \mathbf{T}, \mathbf{A} \leftarrow \overline{\lambda_d}, \overline{e}, \overline{\mathbf{T}}, \overline{\mathbf{A}}$

17: $\mathbf{z} \leftarrow \left(\mathbf{z}_{\mathcal{L}}^{\mathrm{T}}, (\mathbf{C}_{\mathcal{U}}\mathbf{g})^{\mathrm{T}}\right)^{\mathrm{T}}$

18: $\mathbf{a} \leftarrow \mathbf{U}(\boldsymbol{\Sigma} + \lambda_d\mathbf{I})^{-1}\mathbf{U}^{\mathrm{T}}\mathbf{z}$ $\triangleright \mathcal{O}(md)$

19: **return** $f_t(\cdot) = \sum_{i=1}^{m} \mathbf{a}_i k(\mathbf{d}_i, \cdot)$

Let $\mathcal{L} \subseteq D$ and $\mathcal{U} \subseteq D$ be the set of objects that are, respectively, labeled and unlabeled for the target task. Moreover, let $\mathbf{Y}$ now denote the matrix of labels for the auxiliary tasks and $\mathbf{z}_{\mathcal{L}} \in \mathbb{R}^{|\mathcal{L}|}$ the vector of known labels for the target task. Furthermore, in addition to the kernel matrices $\mathbf{K}$ and $\mathbf{G}$ for the training data points and auxiliary tasks defined in (2), let $\mathbf{g} \in \mathbb{R}^q$ denote the vector of task kernel evaluations between the target task and the auxiliary tasks, e.g. $\mathbf{g} = (g(\mathbf{t}, \mathbf{t}_1), \ldots, g(\mathbf{t}, \mathbf{t}_q))^{\mathrm{T}}$, where $\mathbf{t}$ is the target task and $\mathbf{t}_i$ the auxiliary tasks. Finally, let λ_t and λ_d be the regularization parameters for the first and the second learning steps, respectively. The two-step approach is summarized in Algorithm 1. The first training step is carried out by training a multi-output KRR model (line 1), in which a matrix $\mathbf{C}$ of parameters is estimated, and this model is used for predicting the labels indexed by $\mathcal{U}$ for the target task (line 2). The second step trains a single-output KRR for the targer task (line 3), in which a vector $\mathbf{a}$ of parameters is fitted to the data.

2.3 Computational Considerations and Model Selection

Let d and r denote the feature space dimensionalities of the object and task kernels, respectively. These dimensions can be reduced, for example, by the Nyström method in order to lower both the time and space complexities of kernel methods [32], and hence in the following we assume that $d \leq m$ and $r \leq q$. Let $\boldsymbol{\Phi} \in \mathbb{R}^{m \times d}$

and $\boldsymbol{\Psi} \in \mathbb{R}^{q \times r}$ be the matrices containing the feature representations of the train-ing objects and tasks in D and T, respectively, so that $\boldsymbol{\Phi}\boldsymbol{\Phi}^{\mathrm{T}} = \mathbf{K}$ and $\boldsymbol{\Psi}\boldsymbol{\Psi}^{\mathrm{T}} = \mathbf{G}$. Let $\boldsymbol{\Phi} = \mathbf{U}\sqrt{\boldsymbol{\Sigma}}\mathbf{V}^{\mathrm{T}}$ and $\boldsymbol{\Psi} = \mathbf{P}\sqrt{\mathbf{S}}\mathbf{Q}^{\mathrm{T}}$ be the SVDs of $\boldsymbol{\Phi}$ and $\boldsymbol{\Psi}$, respectively. Since the ranks of the feature matrices are at most the dimensions of the feature spaces, we can save both space and time by only computing the singular vectors that correspond to the nonzero singular values. That is, we compute the matrices $\mathbf{U} \in \mathbb{R}^{m \times d}$, $\mathbf{V} \in \mathbb{R}^{d \times d}$, $\mathbf{P} \in \mathbb{R}^{q \times r}$, and $\mathbf{Q} \in \mathbb{R}^{r \times r}$ via the economy sized SVD, requiring $\mathcal{O}(md^2 + qr^2)$ time. The outcomes of the first and second steps of the two-step KRR (e.g. the first and third lines of Algorithm 1) can be, respectively, written as $\mathbf{C} = \mathbf{Y}\widetilde{\mathbf{G}}$ and $\mathbf{a} = \widetilde{\mathbf{K}}\mathbf{z}$, where $\widetilde{\mathbf{G}} = (\mathbf{G} + \lambda_t\mathbf{I})^{-1} = \mathbf{U}(\boldsymbol{\Sigma} + \lambda_d\mathbf{I})^{-1}\mathbf{U}^{\mathrm{T}}$ and $\widetilde{\mathbf{K}} = (\mathbf{K} + \lambda_d\mathbf{I})^{-1} = \mathbf{U}(\boldsymbol{\Sigma} + \lambda_d\mathbf{I})^{-1}\mathbf{U}^{\mathrm{T}}$. Given that the above described SVD components are available, the computational complexity is dominated by the multiplication of the eigenvectors with the label matrix, which requires $\mathcal{O}(mqr)$ time if the matrix multiplications are performed in the optimal order.

We next present an automatic model selection and training approach for the two-step KRR that uses leave-one-out cross-validation (LOOCV) for selecting the values of both λ_t and λ_d. This is illustrated in Algorithm 2. It is well known that, for KRR, the LOOCV performance can be efficiently computed without training the model from scratch during each CV round (we refer to [24, 31] for details). Adapting this to the first step of the two-step KRR, the "leave-column-out" performance for the ith datum on the jth task (e.g. a CV in which each of the columns of $\mathbf{Y}$ are held out at a time to measure the generalization ability to new columns) can be obtained in constant time from $Y_{i,j} - \left(\widetilde{G}_{j,j}\right)^{-1} C_{i,j}$, given that the diagonal entries of $\widetilde{\mathbf{G}}$ and the dual variables $C_{i,j}$ are computed and stored in memory. Using the SVD components, both $\widetilde{G}_{j,j}$ and $C_{i,j}$ can be computed in $\mathcal{O}(r)$ time, which enables the efficient selection of the regularization parameter value with LOOCV. If the value is selected from a set of t candidates and LOOCV is computed for all data points and tasks, the overall complexity is $\mathcal{O}(mqrt)$. This is depicted in lines 4-9 of Algorithm 2, where the overline symbols denote temporary variables used in the search of the optimal candidate value and $\mathcal{E}$ denotes a prediction performance (such as the mean squared error, classification accuracy, or area under curve, for example).

By the definition of the two-step KRR, the second step consists of training a model using the predictions made during the first step as training labels, while the aim is to make good predictions of the true labels. Therefore, we select the regularization parameter value for the second step using LOOCV on a multi-output KRR model trained using the LOO prediction matrix $\mathbf{R}$ obtained from the first step as a label matrix. The second regularization parameter value is thus selected so that the error $\mathcal{E}(\overline{\mathbf{T}}, \mathbf{Y})$ between the LOO predictions made during the second step and the original labels $\mathbf{Y}$ is as small as possible. In contrast to the first step, the aim of the second step is to generalize to new data points, and hence the CV is done in the leave-row-out sense, which can again be efficiently computed as $Y_{i,j} - \left(\widetilde{K}_{i,i}\right)^{-1} A_{i,j}$, where $A_{i,j}$ are the model parameters of the multi-output KRR trained row-wise. This is done in lines 11-16 of Algorithm 2.

The overall computational complexity of the two-step KRR with automatic model selection is $\mathcal{O}(md^2 + qr^2 + mqrt + mqdt)$, where the first two terms denote the time required by SVD computations and the two latter the time spent for CV and grid search for the regularization parameter. The two-step KRR, in addition to enabling non-zero training sets for the target task, provides a very flexible machinery for CV and model selection. This is in contrast to the tensor KRR for which such short-cuts are not available to our knowledge, and there is no efficient closed form solution available for the almost full cold start settings. Note also that, while the above described method separately selects the regularization parameter values for the tasks and the data, the method is easy to modify so that it would select a separate regularization parameter value for each task and for each datum (e.g. altogether $m + q$ parameters), thus allowing considerably more degrees of freedom. However, the consideration of this variation is omitted due to the lack of space.

3   Theoretical Considerations

Here, we analyze the two-step learning approach by studying its connections to learning with pairwise tensor product kernels as in (1). These two approaches coincide in an interesting way for the full cold start problems (e.g. the special case in which there is no labeled data available for the target task). This, in turn, allows us to show the consistency of the two-step KRR via its universal approximation and spectral regularization properties.

The connection between the two-step and tensor KRR is characterized by the following result.

Proposition 1. *Let us consider a full cold start setting with a complete training set. Let $f_{\mathbf{t}}(\cdot)$ be a model trained with two-step KRR for the target task $\mathbf{t}$ and $f(\cdot, \cdot)$ be a model trained with an ordinary kernel least-squares regression (OKLS) on the object-task pairs with the following pairwise kernel function on $\mathcal{D} \times \mathcal{T}$:*

$$\varUpsilon\left((\mathbf{d}, \mathbf{t}), (\overline{\mathbf{d}}, \overline{\mathbf{t}})\right) = \left(k\left(\mathbf{d}, \overline{\mathbf{d}}\right) + \lambda_d \delta\left(\mathbf{d}, \overline{\mathbf{d}}\right)\right)\left(g\left(\mathbf{t}, \overline{\mathbf{t}}\right) + \lambda_t \delta\left(\mathbf{t}, \overline{\mathbf{t}}\right)\right) \qquad (5)$$

where δ is the delta kernel whose value is 1 if the arguments are equal and 0 otherwise. Then, $f_{\mathbf{t}}(\mathbf{d}) = f(\mathbf{t}, \mathbf{d})$ for any $\mathbf{d} \in \mathcal{D}$.

Proof. Writing the steps of the algorithm together and denoting $\widetilde{\mathbf{G}} = (\mathbf{G} + \lambda \mathbf{I})^{-1}$ and $\widetilde{\mathbf{K}} = (\mathbf{K} + \lambda \mathbf{I})^{-1}$, we observe that the model parameters $\mathbf{a}$ of the target task can also be obtained from the following closed form:

$$\mathbf{a} = \widetilde{\mathbf{K}} \mathbf{Y} \widetilde{\mathbf{G}} \mathbf{g}. \qquad (6)$$

The prediction for a datum $\mathbf{d}$ is $f_{\mathbf{t}}(\mathbf{d}) = \mathbf{k}^{\mathrm{T}} \mathbf{a}$, where $\mathbf{k} \in \mathbb{R}^m$ is the vector containing all kernel evaluations between $\mathbf{d}$ and the training data points.

The kernel matrix of $\varUpsilon$ for the full cold start setting can be expressed as: $\varUpsilon = (\mathbf{G} + \lambda_t \mathbf{I}) \otimes (\mathbf{K} + \lambda_d \mathbf{I})$. The OKLS problem with kernel $\varUpsilon$ being

$$\boldsymbol{\alpha} = \operatorname*{argmin}_{\boldsymbol{\alpha} \in \mathbb{R}^{mq}} \left\{ (\operatorname{vec}(\mathbf{Y}) - \varUpsilon\boldsymbol{\alpha})^{\mathrm{T}} (\operatorname{vec}(\mathbf{Y}) - \varUpsilon\boldsymbol{\alpha}) \right\},$$

its minimizer can be expressed as

$$\alpha = \Upsilon^{-1}\text{vec}(\mathbf{Y}) = \left((\mathbf{G} + \lambda_t\mathbf{I})^{-1} \otimes (\mathbf{K} + \lambda_d\mathbf{I})^{-1}\right)\text{vec}(\mathbf{Y})$$
$$= \text{vec}\left((\mathbf{K} + \lambda_d\mathbf{I})^{-1}\mathbf{Y}(\mathbf{G} + \lambda_t\mathbf{I})^{-1}\right) = \text{vec}\left(\widetilde{\mathbf{K}}\mathbf{Y}\widetilde{\mathbf{G}}\right). \qquad (7)$$

The prediction for the datum $\mathbf{d}$ is $(\mathbf{g} \otimes \mathbf{k})^{\mathrm{T}}\text{vec}\left(\widetilde{\mathbf{K}}\mathbf{Y}\widetilde{\mathbf{G}}\right) = \mathbf{k}^{\mathrm{T}}\widetilde{\mathbf{K}}\mathbf{Y}\widetilde{\mathbf{G}}\mathbf{g}$, where we make use of the rule $\text{vec}(\mathbf{MXN}) = (\mathbf{N}^{\mathrm{T}} \otimes \mathbf{N})\text{vec}(\mathbf{X})$ that holds for any conformable matrices $\mathbf{M}$, $\mathbf{X}$, and $\mathbf{N}$. $\qquad\qquad\qquad\qquad\qquad\qquad\qquad\qquad\square$

The kernel point of view allows us to consider the universal approximation properties of the learned knowledge transfer models. Recall the concept of universal kernel functions:

Definition 1. [36] *A continuous kernel k on a compact metric space $\mathcal{X}$ (i.e. $\mathcal{X}$ is closed and bounded) is called universal if the reproducing kernel Hilbert space (RKHS) induced by k is dense in $C(\mathcal{X})$, where $C(\mathcal{X})$ is the space of all continuous functions $f : \mathcal{X} \to \mathbb{R}$.*

The universality property indicates that the hypothesis space induced by an universal kernel can approximate any continuous function to be learned arbitrarily well, given that the available set of training data is large and representative enough, and the learning algorithm can efficiently find the approximation [36].

Proposition 2. *The kernel Υ on $\mathcal{D} \times \mathcal{T}$ defined in (5) is universal if the kernels k on $\mathcal{D}$ and g on $\mathcal{T}$ are both universal.*

Proof. We provide here a high-level sketch of the proof. The details are omitted due to lack of space but they can be easily verified from the existing literature. The RKHS of sums of reproducing kernels was characterized by [3] as follows: Let $\mathcal{H}(k_1)$ and $\mathcal{H}(k_2)$ be RKHSs over $\mathcal{X}$ with reproducing kernels k_1 and k_2, respectively. If $k = k_1 + k_2$ and $\mathcal{H}(k)$ denotes the corresponding RKHS, then $\mathcal{H}(k) = \{f_1 + f_2 : f_i \in \mathcal{H}(k_i), i = 1, 2\}$. Thus, if the object kernel is universal, the sum of the object and delta kernels is also universal and the same concerns the task kernel. The product of two universal kernels is also universal, as considered in our previous work [38]. $\qquad\qquad\qquad\qquad\qquad\qquad\qquad\qquad\square$

The full cold start setting with complete auxiliary training set allows us to consider the two-step approach from the spectral filtering regularization point of view [18], an approach that has recently gained some attention due to its ability to study various types of regularization approaches under the same framework. Continuing from (4), we observe that

$$\alpha = \varphi_\lambda(\Gamma)\text{vec}(\mathbf{Y}) = \mathbf{W}\varphi_\lambda(\Lambda)\mathbf{W}^{\mathrm{T}}\text{vec}(\mathbf{Y}),$$

where $\Gamma = \mathbf{W}\Lambda\mathbf{W}^{\mathrm{T}}$ is the eigen decomposition of the kernel matrix Γ and φ_λ is a filter function, parameterized by λ, such that if $\mathbf{v}$ is an eigenvector of Γ and σ is its corresponding eigenvalue, then $\Gamma\mathbf{v} = \varphi_\lambda(\sigma)\mathbf{v}$. The filter function corresponding to the Tikhonov regularization being $\varphi_\lambda(\sigma) = \frac{1}{\sigma+\lambda}$, and the ordinary least-squares approach corresponding to the $\lambda = 0$ case, several other learning approaches, such as spectral cut-off and gradient descent, can also be expressed as filter functions, but which cannot be expressed as a penalized empirical error minimization problem analogous to (3).

The eigenvalues of the kernel matrix obtained with the tensor product kernel on a complete training set can be expressed as the tensor product $\Lambda = \Sigma \otimes S$ of the eigenvalues Σ and S of the object and task kernel matrices. Now, instead of considering the two-step learning approach from the kernel point of view, one can also cast it into the spectral filtering regularization framework, resulting to the following filter function:

$$\varphi_\lambda(\sigma) = \frac{1}{(\sigma_1 + \lambda_t)(\sigma_2 + \lambda_d)} = \frac{1}{\sigma_1\sigma_2 + \lambda_d\sigma_1 + \lambda_t\sigma_2 + \lambda_t\lambda_d}, \tag{8}$$

where σ_1, σ_2 are the factors of σ, namely eigenvalues of K and G. This differs from the Tikhonov regularization only by the two middle terms in the denominator if one sets $\lambda = \lambda_t\lambda_d$. In the experiments, we observe that this difference is rather small also in practical cases, making the two-step learning approach a viable alternative for tensor KRR.

We assume Γ being bounded with $\kappa > 0$ such that $\sup_{x \in \mathcal{X}} \sqrt{\Gamma(x, x)} \leq \kappa$, indicating that the eigenvalues of kernel matrices are in $[0, \kappa^2]$. To further analyze the above filter functions, we follow [4, 6, 18] and say that a function $\varphi_\lambda : [0, \kappa^2] \to \mathbb{R}, 0 < \lambda \leq \kappa^2$, parameterized by $0 < \lambda \leq \kappa^2$, is an admissible regularizer if there exists constants $D, B, \gamma \in \mathbb{R}$ and $\bar{\nu}, \gamma_\nu > 0$ such that

$$\sup_{0 < \sigma \leq \kappa^2} |\sigma\varphi_\lambda(\sigma)| \leq D, \quad \sup_{0 < \sigma \leq \kappa^2} |\varphi_\lambda(\sigma)| \leq \frac{B}{\lambda}, \quad \sup_{0 < \sigma \leq \kappa^2} |1 - \sigma\varphi_\lambda(\sigma)| \leq \gamma,$$

$$\text{and} \quad \sup_{0 < \sigma \leq \kappa^2} |1 - \sigma\varphi_\lambda(\sigma)|\sigma^\nu \leq \gamma_\nu\lambda^\nu, \forall \nu \in (0, \bar{\nu}].$$

The admissibility, in turn, ensures that

$$R(\hat{f}^\lambda) - \inf_{f \in \mathcal{H}} R(f) = \mathcal{O}\left(n^{-\frac{\bar{\nu}}{2\bar{\nu}+1}}\right) \tag{9}$$

holds with high probability, where R denotes the expected prediction error with respect to some unknown probability measure $\rho(x, y)$ on the joint space $\mathcal{X} \times \mathbb{R}$ of inputs and labels that is, $R(f) = \int_{\mathcal{X} \times \mathbb{R}} (f(x) - y)^2 d\rho(x, y)$. We refer to [4, 6, 18] for a detailed consideration and further results. It is straightforward to see that, analogously to the Tikhonov regularization, the admissibility of the function (8) is confirmed by $D, B, \gamma, \gamma_\nu, \bar{\nu} = 1$ for arbitrary factorizations of $\lambda = \lambda_t\lambda_d$ and $\sigma = \sigma_1\sigma_2$ such that $\lambda_t, \lambda_d > 0$ and $\sigma_1, \sigma_2 \geq 0$. Thus, function (8) can be considered under the spectral filtering regularization framework with separate regularization parameter values for objects and tasks. The universality of the kernel ensures that $\inf_{f \in \mathcal{H}} R(f)$ in (9) is the error of the underlying regression function to be learned, and the admissibility of the regularizer ensures that $R(\hat{f}^\lambda)$ converges to it when the size of the training set approaches infinity. This, in turn, guarantees the consistency of the two-step KRR method.

4   Experiments

We compare different types of transfer learning settings in solving three dyadic prediction problems: drug-target, document similarity and protein similarity prediction. We simulate the full and almost full cold start problem as follows. In each experiment, one drug, document or protein is considered to be the target task in question, where the task is to predict the interactions of drugs or similarities of documents or proteins with respect to the target. Further, other tasks formed in the same way are provided as auxiliary information, leading to a full cold start or almost full cold start setting. The experiments are performed 100 times with different training/test set splits. The performances are averages over all repetitions and over all target tasks, and are measured using the concordance index [11] (C-index), also known as the pairwise ranking accuracy $\frac{1}{|\{(i,j)|y_i > y_j\}|} \sum_{y_i > y_j} H(\hat{y}_i - \hat{y}_j)$, where y_i denote the true and $\hat{y}_i$ the predicted labels, and H is the Heaviside step function. The regularization parameter selection is performed using LOOCV on the training data. For the two-step approach, we select the first regularization parameter via LOOCV on the auxiliary tasks, and the second one via LOOCV on the target task data augmented with predictions from the first step. The algorithms used in the experiments are implemented in the RLScore open source machine learning library[1].

The drug-target interaction prediction data[2] [9, 22] consists of 68 drug compounds and 442 protein targets. The kernel between the drugs is based on the 3D Tanimoto coefficient similarity, and the sequence similarity between the protein targets was computed using the normalized version of the Smith-Waterman score. Further, for each drug-protein pair we have a real-valued label, negative logarithm of the kinase disassociation constant K_d, that characterizes the interaction affinity between the drug and target in question. In each experiment, the task of interest corresponds to one of the drugs in the data set. The goal is to learn to predict for the given drug the K_d values for proteins unseen during the training phase. The performances are always computed over a testing set of 192 protein targets for a given task, i.e. we assess whether for a given target we can discriminate between proteins with more or less affinity for this drug.

For each task, we vary the number of available training proteins, from 5 to 250. In addition, we have available the training data for the 250 training proteins for the 67 auxiliary tasks. As summarized in Figure 1, we evaluate a number of different approaches:

- Single-task: KRR trained with data from the target task only
- Multi-task: both the target and auxiliary tasks have the same amount of training data available (multi-output learning leveraging task correlations, tackled with tensor KRR)
- Full cold start: tensor KRR with no data for the target task
- Almost full cold start: use a varying amount of data from the target task, and all the available data from auxiliary tasks (tackled with two-step KRR)

---

[1]   Available at https://github.com/aatapa/RLScore
[2]   http://users.utu.fi/aatapa/data/DrugTarget

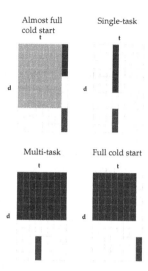

Fig. 1. Overview of the approaches investigated in this article. Green = training data of which the size is constant in the experiments. Blue = training data of which the number of objects varies over different experiments. Red = test data.

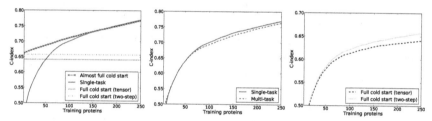

Fig. 2. Learning curves for the drug-target data. Left: target data increased, Middle: target and auxiliary data increased, Right: auxiliary data increased.

We do not consider tensor KRR in the almost full cold start experiment due to computational considerations. Unlike for the two-step KRR no closed-form solution exists for the method in this setting, and the iterative conjugate gradient based method has rather poor scalability.

In Figure 2, we present the results for the drug-target experiments. In Figure 2 (a) we present an experiment, where all the 67 auxiliary tasks have available the data for all 250 training proteins, and the amount of data available for the target task is varied. It can be seen that learning is possible even in the full cold start setting, where both two-step KRR and tensor KRR perform much better than randomly. The single-task approach begins to outperform the full cold start setting after the point when one has access to a bit more than 50 training proteins. Combining these two sources of information leads to the best performance up until 150 training proteins. However, once there is enough data available for the target task, there is no longer any positive transfer from the auxiliary tasks.

In Figure 2 (b) we consider the setting, where there is the same amount of data available for both the auxiliary tasks and the target tasks. This setting corresponds closely to the traditional multi-output regression problem, the exception being that only the label for the target task is of interest during testing. Here we can see that the multi-task method that uses the task correlation information fails to outperform the simple single-task approach, suggesting that on this type of data one requires significantly more data in the auxiliary tasks compared to the target tasks in order for it to be helpful for learning.

In Figure 2 (c) we consider the full cold start learning setting, while increasing the amount of data available for the auxiliary tasks. Here we observe that the simple two-step approach slightly outperforms tensor KRR, possibly due to the property that it allows regularizing the drugs and the targets separately. Both approaches generalize to the unknown target task, though the results are still much worse than when having significant amount of data for the target task.

Further, we experiment on the 20 Newsgroups data[3]. Here, given any target document, the goal is to predict the similarity of other documents with respect to it. This constitutes a three-level ordinal regression task, where documents from the same newsgroup as the target receive the highest rating, documents from similar newsgroups the second highest, and the rest the lowest rating. These similarities are assigned according to the taxonomy available at the data set web site. We use the bag-of-words feature representation. The number of target domain data ranges from 50 to 1500 documents (transfer learning, single-task, multi-task methods), and the number of auxiliary tasks and data available for each either ranges from 50 to 1500 documents (multi-task, full cold start learning), or stays fixed at 2000 documents (transfer learning).

The results are presented in Figure 3. For the transfer learning approaches, already 50 target domain documents suffices to reach the performance of the single-task method with 1500 documents. The multi-task learning setting does not outperform the single-task setting, and while learning is possible in the full cold start setting, some target task data is still required to reach a high predictive performance. Two-step learning slightly outperforms tensor KRR.

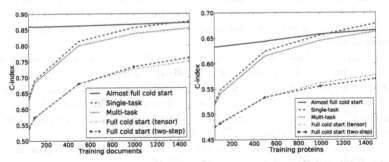

Fig. 3. Learning curves for 20 Newsgroups (left) and Uniprot (right)

---

[3] http://qwone.com/~jason/20Newsgroups/

The UniProt data was generated from all the protein amino acid sequences with all the gene ontology (GO) annotations of the Universal Protein Resource (UniProt) database. For the amino acid sequences we used the normalized spectrum kernel [17]. This kernel is a popular tool for comparing biological sequences without alignments. The normalized spectrum kernel is based on the number of k-mers two sequences have in common. In our experiments, k was set to three. Two proteins were labeled as 'similar in function' when they had at least one GO term in common, resulting in a binary classification problem. The experimental setup is the same as for the Newsgroup data, and the results, presented in Figure 3, are very similar, though at 1500 proteins the performance of the two-step method actually falls below that of the single-task approach.

In all experiments two-step KRR shows itself to be competitive compared to tensor KRR. Previously, [33] have in their overview article on dyadic prediction in the biological domain made the observation that in terms of predictive accuracy there does not seem to be a clear winner between the single-task and multi-task type of learning approaches. Based on our experimental results, a deciding factor on whether one may expect positive transfer from related tasks seems to be the amount of data available for the target task. The two-step method performs well in the almost full cold start settings with availability of a significant amount of auxiliary data and only very little data for the target task. But when there is enough data available for the target task, auxiliary data is no longer helpful.

Acknowledgments. We would like to thank the anonymous reviewers for their insightful comments.

References

[1] Adams, R.P., Dahl, G.E., Murray, I.: Incorporating side information into probabilistic matrix factorization using Gaussian processes. In: The 26th Conference on Uncertainty in Artificial Intelligence, pp. 1–9 (2010)

[2] Álvarez, M., Rosasco, L., Lawrence, N.: Kernels for vector-valued functions: a review. Foundation and Trends in Machine Learning 4(3), 195–266 (2012)

[3] Aronszajn, N.: Theory of reproducing kernels. Transactions of the American Mathematical Society 68 (1950)

[4] Baldassarre, L., Rosasco, L., Barla, A., Verri, A.: Multi-output learning via spectral filtering. Machine Learning 87(3), 259–301 (2012)

[5] Basilico, J., Hofmann, T.: Unifying collaborative and content-based filtering. In: 21st International Conference on Machine Learning, ICML 2004 (2004)

[6] Bauer, F., Pereverzev, S., Rosasco, L.: On regularization algorithms in learning theory. Journal of Complexity 23(1), 52–72 (2007)

[7] Ben-Hur, A., Noble, W.: Kernel methods for predicting protein-protein interactions. Bioinformatics 21(suppl. 1), 38–46 (2005)

[8] Bonilla, E.V., Agakov, F., Williams, C.: Kernel multi-task learning using task-specific features. In: The 11th International Conference on Artificial Intelligence and Statistics, AISTATS 2007, pp. 43–50 (2007)

[9] Davis, M.I., Hunt, J.P., Herrgard, S., Ciceri, P., Wodicka, L.M., Pallares, G., Hocker, M., Treiber, D.K., Zarrinkar, P.P.: Comprehensive analysis of kinase inhibitor selectivity. Nature biotechnology 29(11), 1046–1051 (2011)

[10] Fang, Y., Si, L.: Matrix co-factorization for recommendation with rich side information and implicit feedback. In: 2nd International Workshop on Information Heterogeneity and Fusion in Recommender Systems, pp. 65–69 (2011)

[11] Gönen, M., Heller, G.: Concordance probability and discriminatory power in proportional hazards regression. Biometrika 92(4), 965–970 (2005)

[12] Hayashi, K., Takenouchi, T., Tomioka, R., Kashima, H.: Self-measuring similarity for multi-task gaussian process. In: ICML Workshop on Unsupervised and Transfer Learning, JMLR Proceedings, vol. 27, pp. 145–154 (2012)

[13] Jacob, L., Vert, J.: Protein-ligand interaction prediction: an improved chemogenomics approach. Bioinformatics 24(19), 2149–2156 (2008)

[14] Kashima, H., Kato, T., Yamanishi, Y., Sugiyama, M., Tsuda, K.: Link propagation: A fast semi-supervised learning algorithm for link prediction. In: SIAM International Conference on Data Mining (SDM 2009), pp. 1099–1110 (2009)

[15] Koren, Y., Bell, R., Volinsky, C.: Matrix factorization techniques for recommender systems. Computer 42(8), 30–37 (2009)

[16] Larochelle, H., Erhan, D., Bengio, Y.: Zero-data learning of new tasks. In: 23rd National Conference on Artificial Intelligence (AAAI 2008), pp. 646–651 (2008)

[17] Leslie, C., Eskin, E., Noble, W.S.S.: The spectrum kernel: a string kernel for SVM protein classification. In: Pacific Symposium on Biocomputing, pp. 564–575 (2002)

[18] Lo Gerfo, L., Rosasco, L., Odone, F., De Vito, E., Verri, A.: Spectral algorithms for supervised learning. Neural Computation 20(7), 1873–1897 (2008)

[19] Martin, C.D., Van Loan, C.F.: Shifted Kronecker product systems. SIAM Journal on Matrix Analysis and Applications 29(1), 184–198 (2006)

[20] Menon, A., Elkan, C.: A log-linear model with latent features for dyadic prediction. In: The 10th IEEE International Conference on Data Mining (ICDM), pp. 364–373 (2010)

[21] Oyama, S., Manning, C.: Using feature conjunctions across examples for learning pairwise classifiers. In: Boulicaut, J.-F., Esposito, F., Giannotti, F., Pedreschi, D. (eds.) ECML 2004. LNCS (LNAI), vol. 3201, pp. 322–333. Springer, Heidelberg (2004)

[22] Pahikkala, T., Airola, A., Pietilä, S., Shakyawar, S., Szwajda, A., Tang, J., Aittokallio, T.: Toward more realistic drug-target interaction predictions. Briefings in Bioinformatics (in press, 2014), doi:10.1093/bib/bbu010

[23] Pahikkala, T., Airola, A., Stock, M., De Baets, B., Waegeman, W.: Efficient regularized least-squares algorithms for conditional ranking on relational data. Machine Learning 93(2–3), 321–356 (2013)

[24] Pahikkala, T., Suominen, H., Boberg, J.: Efficient cross-validation for kernelized least-squares regression with sparse basis expansions. Machine Learning 87(3), 381–407 (2012)

[25] Pahikkala, T., Waegeman, W., Airola, A., Salakoski, T., De Baets, B.: Conditional ranking on relational data. In: Balcázar, J.L., Bonchi, F., Gionis, A., Sebag, M. (eds.) ECML PKDD 2010, Part II. LNCS, vol. 6322, pp. 499–514. Springer, Heidelberg (2010)

[26] Pahikkala, T., Waegeman, W., Tsivtsivadze, E., Salakoski, T., De Baets, B.: Learning intransitive reciprocal relations with kernel methods. European Journal of Operational Research 206(3), 676–685 (2010)

[27] Pan, S.J., Yang, Q.: A survey on transfer learning. IEEE Transactions on Knowledge and Data Engineering 22(10), 1345–1359 (2010)

[28] Park, S.-T., Chu, W.: Pairwise preference regression for cold-start recommendation. In: 3rd ACM Conference on Recommender Systems, pp. 21–28 (2009)

[29] Park, Y., Marcotte, E.M.: Flaws in evaluation schemes for pair-input computational predictions. Nature Methods 9(12), 1134–1136 (2012)

[30] Raymond, R., Kashima, H.: Fast and scalable algorithms for semi-supervised link prediction on static and dynamic graphs. In: Balcázar, J.L., Bonchi, F., Gionis, A., Sebag, M. (eds.) ECML PKDD 2010, Part III. LNCS, vol. 6323, pp. 131–147. Springer, Heidelberg (2010)

[31] Rifkin, R., Lippert, R.: Notes on regularized least squares. Tech. Rep. MIT-CSAIL-TR-2007-025, Massachusetts Institute of Technology, Cambridge, Massachusetts, USA (2007)

[32] Mika, B., Burges, S., Knirsch, C.J.C., Schölkopf, P., Rätsch, K.-R.M.G., Input, A.J.S.: space versus feature space in kernel-based methods. IEEE Transactions on Neural Networks 10(5), 1000–1017 (1999)

[33] Schrynemackers, M., Küffner, R., Geurts, P.: On protocols and measures for the validation of supervised methods for the inference of biological networks. Frontiers in Genetics 4, 262 (2013)

[34] Schrynemackers, M., Wehenkel, L., Babu, M.M., Geurts, P.: Classifying pairs with trees for supervised biological network inference (2014) (submitted manuscript)

[35] Shan, H., Banerjee, A.: Generalized probabilistic matrix factorizations for collaborative filtering. In: The 10th IEEE International Conference on Data Mining (ICDM), pp. 1025–1030 (2010)

[36] Steinwart, I.: On the influence of the kernel on the consistency of support vector machines. Journal of Machine Learning Research 2, 67–93 (2002)

[37] Van Loan, C.F.: The ubiquitous kronecker product. Journal of Computational and Applied Mathematics 123(1–2), 85–100 (2000)

[38] Waegeman, W., Pahikkala, T., Airola, A., Salakoski, T., Stock, M., De Baets, B.: A kernel-based framework for learning graded relations from data. IEEE Transactions on Fuzzy Systems 20(6), 1090–1101 (2012)

[39] Zhou, T., Shan, H., Banerjee, A., Sapiro, G.: Kernelized probabilistic matrix factorization: Exploiting graphs and side information. In: 12th SIAM International Conference on Data Mining, pp. 403–414 (2012)

Deterministic Feature Selection for Regularized Least Squares Classification

Saurabh Paul and Petros Drineas

Computer Science Department, Rensselaer Polytechnic Institute, Troy, NY, USA
pauls2@rpi.edu, drinep@cs.rpi.edu

Abstract. We introduce a deterministic sampling based feature selection technique for regularized least squares classification. The method is unsupervised and gives worst-case guarantees of the generalization power of the classification function after feature selection with respect to the classification function obtained using all features. We perform experiments on synthetic and real-world datasets, namely a subset of TechTC-300 datasets, to support our theory. Experimental results indicate that the proposed method performs better than the existing feature selection methods.

Keywords: Feature Selection, Sampling, Regularized Least Squares Classification.

1 Introduction

Regularized Least Squares Classifier (RLSC) is a simple classifier based on least squares and has a long history in machine learning [17,12,13,10,15,18,1]. RLSC has been known to perform comparably to the popular Support Vector Machines (SVM) [13,10,15,18]. RLSC can be solved by simple vector space operations and do not require quadratic optimization techniques like SVM. The main focus of this paper is on a deterministic feature selection technique for RLSC with provable guarantees. There exist numerous feature selection techniques, which work well empirically. There also exist randomized feature selection methods [6] with provable guarantees which work well empirically. But the randomized methods have a failure probability and have to be re-run multiple times to get accurate results. Also, a randomized algorithm may not select the same features in different runs. A deterministic algorithm will select the same features irrespective of how many times it is run. This becomes important in many applications. Unsupervised feature selection involves selecting features oblivious to the class or labels. In this work, we present a *new provably accurate* unsupervised feature selection technique for RLSC. We study a deterministic sampling based feature selection strategy for RLSC with provable non-trivial worst-case performance bounds. The number of features selected is proportional to the rank of the training set. The deterministic sampling-based feature selection algorithm performs better in practice when compared to existing methods of feature selection.

T. Calders et al. (Eds.): ECML PKDD 2014, Part II, LNCS 8725, pp. 533–548, 2014.
© Springer-Verlag Berlin Heidelberg 2014

2 Our Contributions

We introduce single-set spectral sparsification as a provably accurate deterministic feature selection technique for RLSC in an unsupervised setting. The number of features selected by the algorithm is independent of the number of features, but depends on the number of data-points. The algorithm selects a small number of features and solves the classification problem using those features. Recently, Dasgupta et al. [6] used a leverage-score based randomized feature selection technique for RLSC and provided worst case guarantees of the approximate classifier function to that using all features. We use a deterministic algorithm to provide worst-case generalization error guarantees. The deterministic algorithm does not come with a failure probability and the number of features required by the deterministic algorithm is lesser than that required by the randomized algorithm. The leverage-score based algorithm has a sampling complexity of $O\left(\frac{n}{\epsilon^2}\log\left(\frac{n}{\epsilon^2\sqrt{\delta}}\right)\right)$, whereas single-set spectral sparsification requires $O\left(n/\epsilon^2\right)$ to be picked, where n is the number of training points, $\delta \in (0,1)$ is a failure probability and $\epsilon \in (0,1/2]$ is an accuracy parameter. Like in [6], we also provide additive-error approximation guarantees for any test-point and relative-error approximation guarantees for test-points that satisfy some conditions with respect to the training set.

From an **empirical perspective**, we evaluate single-set spectral sparsification on synthetic data and 48 document-term matrices, which are a subset of the TechTC-300 [7] dataset. We compare the single-set spectral sparsification algorithm with leverage-score sampling, information gain, rank-revealing QR factorization (RRQR) and random feature selection. We do not report running time because feature selection is an offline task. The experimental results indicate that single-set spectral sparsification out-performs all the methods in terms of out-of-sample error for all 48 TechTC-300 datasets. We observe that a much smaller number of features is required by the deterministic algorithm to achieve good performance when compared to leverage-score sampling.

3 Background and Related Work

Notation. $\mathbf{A}, \mathbf{B}, \ldots$ denote matrices and $\boldsymbol{\alpha}, \mathbf{b}, \ldots$ denote column vectors; $\mathbf{e}_i$ (for all $i = 1 \ldots n$) is the standard basis, whose dimensionality will be clear from context; and $\mathbf{I}_n$ is the $n \times n$ identity matrix. The Singular Value Decomposition (SVD) of a matrix $\mathbf{A} \in \mathbb{R}^{n \times d}$ is equal to $\mathbf{A} = \mathbf{U\Sigma V}^T$, where $\mathbf{U} \in \mathbb{R}^{n \times d}$ is an orthogonal matrix containing the left singular vectors, $\mathbf{\Sigma} \in \mathbb{R}^{d \times d}$ is a diagonal matrix containing the singular values $\sigma_1 \geq \sigma_2 \geq \ldots \sigma_d > 0$, and $\mathbf{V} \in \mathbb{R}^{d \times d}$ is a matrix containing the right singular vectors. The spectral norm of $\mathbf{A}$ is $\|\mathbf{A}\|_2 = \sigma_1$. σ_{max} and σ_{min} are the largest and smallest singular values of $\mathbf{A}$. $\kappa_{\mathbf{A}} = \sigma_{max}/\sigma_{min}$ is the condition number of $\mathbf{A}$. $\mathbf{U}^\perp$ denotes any $n \times (n-d)$ orthogonal matrix whose columns span the subspace orthogonal to $\mathbf{U}$. A vector $\mathbf{q} \in \mathbb{R}^n$ can be expressed as: $\mathbf{q} = \mathbf{A}\boldsymbol{\alpha} + \mathbf{U}^\perp\boldsymbol{\beta}$, for some vectors $\boldsymbol{\alpha} \in \mathbb{R}^d$ and $\boldsymbol{\beta} \in \mathbb{R}^{n-d}$, i.e. $\mathbf{q}$ has one component along $\mathbf{A}$ and another component orthogonal to $\mathbf{A}$.

Matrix Sampling Formalism. We now present the tools of feature selection. Let $\mathbf{A} \in \mathbb{R}^{d \times n}$ be the data matrix consisting of n points and d dimensions, $\mathbf{S} \in \mathbb{R}^{r \times d}$ be a matrix such that $\mathbf{SA} \in \mathbb{R}^{r \times n}$ contains r rows of $\mathbf{A}$. Let $\mathbf{D} \in \mathbb{R}^{r \times r}$ be the diagonal matrix such that $\mathbf{DSA} \in \mathbb{R}^{r \times n}$ rescales the rows of $\mathbf{A}$ that are in $\mathbf{SA}$. The matrices $\mathbf{S}$ and $\mathbf{D}$ are called the sampling and re-scaling matrices respectively. We will replace the sampling and re-scaling matrices by a single matrix $\mathbf{R} \in \mathbb{R}^{r \times d}$, where $\mathbf{R} = \mathbf{DS}$ denotes the matrix specifying which of the r rows of $\mathbf{A}$ are to be sampled and how they are to be rescaled.

RLSC Basics. Consider a training data of n points in d dimensions with respective labels $y_i \in \{-1, +1\}$ for $i = 1, .., n$. The solution of binary classification problems via Tikhonov regularization in a Reproducing Kernel Hilbert Space (RKHS) using the squared loss function results in Regularized Least Squares Classification (RLSC) problem [13], which can be stated as:

$$\min_{\mathbf{x} \in \mathbb{R}^n} \|\mathbf{Kx} - \mathbf{y}\|_2^2 + \lambda \mathbf{x}^T \mathbf{Kx} \tag{1}$$

where $\mathbf{K}$ is the $n \times n$ kernel matrix defined over the training dataset, λ is a regularization parameter and $\mathbf{y}$ is the n dimensional $\{\pm 1\}$ class label vector. In matrix notation, the training data-set $\mathbf{X}$ is a $d \times n$ matrix, consisting of n data-points and d features ($d \gg n$). Throughout this study, we assume that $\mathbf{X}$ is a full-rank matrix. We shall consider the linear kernel, which can be written as $\mathbf{K} = \mathbf{X}^T \mathbf{X}$. Using the SVD of $\mathbf{X}$, the optimal solution of Eqn. 1 in the full-dimensional space is

$$\mathbf{x}_{opt} = \mathbf{V} \left(\mathbf{\Sigma}^2 + \lambda \mathbf{I} \right)^{-1} \mathbf{V}^T \mathbf{y} \tag{2}$$

The vector $\mathbf{x}_{opt}$ can be used as a classification function that generalizes to test data. If $\mathbf{q} \in \mathbb{R}^d$ is the new test point, then the binary classification function is:

$$f(\mathbf{q}) = \mathbf{x}_{opt}^T \mathbf{X}^T \mathbf{q}. \tag{3}$$

Then, $sign(f(\mathbf{q}))$ gives the predicted label (-1 or $+1$) to be assigned to the new test point $\mathbf{q}$.

Our goal is to study how RLSC performs when the deterministic sampling based feature selection algorithm is used to select features in an unsupervised setting. Let $\mathbf{R} \in \mathbb{R}^{r \times d}$ be the matrix that samples and re-scales r rows of $\mathbf{X}$ thus reducing the dimensionality of the training set from d to $r \ll d$ and r is proportional to the rank of the input matrix. The transformed dataset into r dimensions is given by $\tilde{\mathbf{X}} = \mathbf{RX}$ and the RLSC problem becomes

$$\min_{\mathbf{x} \in \mathbb{R}^n} \left\| \tilde{\mathbf{K}}\mathbf{x} - \mathbf{y} \right\|_2^2 + \lambda \mathbf{x}^T \tilde{\mathbf{K}}\mathbf{x}, \tag{4}$$

thus giving an optimal vector $\tilde{\mathbf{x}}_{opt}$. The new test point $\mathbf{q}$ is first dimensionally reduced to $\tilde{\mathbf{q}} = \mathbf{Rq}$, where $\tilde{\mathbf{q}} \in \mathbb{R}^r$ and then classified by the function,

$$\tilde{f} = f(\tilde{\mathbf{q}}) = \tilde{\mathbf{x}}_{opt}^T \tilde{\mathbf{X}}^T \tilde{\mathbf{q}}. \tag{5}$$

In subsequent sections, we will assume that the test-point $\mathbf{q}$ is of the form $\mathbf{q} = \mathbf{X}\alpha + \mathbf{U}^\perp \beta$. The first part of the expression shows the portion of the test-point that is similar to the training-set and the second part shows how much the test-point is novel compared to the training set, i.e. $\|\beta\|_2$ measures how much of $\mathbf{q}$ lies outside the subspace spanned by the training set.

Related Work. The work most closely related to ours is that of Dasgupta et al. [6] who used a leverage-score based randomized feature selection technique for RLSC and provided worst case bounds of the approximate classifier with that of the classifier for all features. The proof of their main quality-of-approximation results provided an intuition of the circumstances when their feature selection method will work well. The running time of leverage-score based sampling is dominated by the time to compute SVD of the training set i.e. $O\left(n^2d\right)$, whereas, for single-set spectral sparsification, it is $O\left(rdn^2\right)$. Single-set spectral sparsification is a slower and more accurate method than leverage-score sampling. Another work on dimensionality reduction of RLSC is that of Avron et al. [2] who used efficient randomized-algorithms for solving RLSC, in settings where the design matrix has a Vandermonde structure. However, this technique is different from ours, since their work is focused on dimensionality reduction using linear combinations of features, but not on actual feature selection.

4   Our Main Tool: Single-Set Spectral Sparsification

We describe the Single-Set Spectral Sparsification algorithm (**BSS**[1] for short) of [3] as Algorithm 1. Algorithm 1 is a greedy technique that selects columns one at a time. Consider the input matrix as a set of d column vectors $\mathbf{U}^T = [\mathbf{u}_1, \mathbf{u}_2, ..., \mathbf{u}_d]$, with $\mathbf{u}_i \in \mathbb{R}^\ell$ $(i = 1, .., d)$. Given ℓ and $r > \ell$, we iterate over $\tau = 0, 1, 2, ..r - 1$. Define the parameters $L_\tau = \tau - \sqrt{r\ell}, \delta_L = 1, U_\tau = \delta_U\left(\tau + \sqrt{\ell r}\right)$ and $\delta_U = \left(1 + \sqrt{\ell/r}\right)/\left(1 - \sqrt{\ell/r}\right)$. For $U, L \in \mathbb{R}$ and $\mathbf{A} \in \mathbb{R}^{\ell \times \ell}$ a symmetric positive definite matrix with eigenvalues $\lambda_1, \lambda_2, ..., \lambda_\ell$, define

$$\Phi(L, \mathbf{A}) = \sum_{i=1}^{\ell} \frac{1}{\lambda_i - L}; \quad \hat{\Phi}(U, \mathbf{A}) = \sum_{i=1}^{\ell} \frac{1}{U - \lambda_i}$$

as the lower and upper potentials respectively. These potential functions measure how far the eigenvalues of $\mathbf{A}$ are from the upper and lower barriers U and L respectively. We define $\mathcal{L}\left(\mathbf{u}, \delta_L, \mathbf{A}, L\right)$ and $\mathcal{U}\left(\mathbf{u}, \delta_U, \mathbf{A}, U\right)$ as follows:

$$\mathcal{L}\left(\mathbf{u}, \delta_L, \mathbf{A}, L\right) = \frac{\mathbf{u}^T\left(\mathbf{A} - (L + \delta_L)\mathbf{I}_\ell\right)^{-2}\mathbf{u}}{\Phi(L + \delta_L, \mathbf{A}) - \Phi(L, \mathbf{A})} - \mathbf{u}^T\left(\mathbf{A} - (L + \delta_L)\mathbf{I}_\ell\right)^{-1}\mathbf{u}$$

$$\mathcal{U}\left(\mathbf{u}, \delta_U, \mathbf{A}, U\right) = \frac{\mathbf{u}^T\left((U + \delta_U)\mathbf{I}_\ell - \mathbf{A}\right)^{-2}\mathbf{u}}{\hat{\Phi}(U, \mathbf{A}) - \hat{\Phi}(U + \delta_U, \mathbf{A})} + \mathbf{u}^T\left((U + \delta_U)\mathbf{I}_\ell - \mathbf{A}\right)^{-1}\mathbf{u}.$$

At every iteration, there exists an index i_τ and a weight $t_\tau > 0$ such that, $t_\tau^{-1} \leq \mathcal{L}\left(\mathbf{u}_{i_\tau}, \delta_L, \mathbf{A}, L\right)$ and $t_\tau^{-1} \geq \mathcal{U}\left(\mathbf{u}_{i_\tau}, \delta_U, \mathbf{A}, U\right)$. Thus, there will be at

---

[1] The name BSS comes from the authors Batson, Spielman and Srivastava.

Input: $\mathbf{U} = [\mathbf{u}_1, \mathbf{u}_2, ... \mathbf{u}_d]^T \in \mathbb{R}^{d \times \ell}$ with $\mathbf{u}_i \in \mathbb{R}^{\ell}$ and $r > \ell$.
Output: Matrices $\mathbf{S} \in \mathbb{R}^{r \times d}, \mathbf{D} \in \mathbb{R}^{r \times r}$.

1. Initialize $\mathbf{A}_0 = \mathbf{0}_{\ell \times \ell}$, $\mathbf{S} = \mathbf{0}_{r \times d}$, $\mathbf{D} = \mathbf{0}_{r \times r}$.
2. Set constants $\delta_L = 1$ and $\delta_U = \left(1 + \sqrt{\ell/r}\right) / \left(1 - \sqrt{\ell/r}\right)$.
3. **for** $\tau = 0$ to $r - 1$ **do**
   - Let $L_\tau = \tau - \sqrt{r\ell}$; $U_\tau = \delta_U \left(\tau + \sqrt{\ell r}\right)$.
   - Pick index $i_\tau \in \{1, 2, ..d\}$ and number $t_\tau > 0$ (See Section 4 for definitions of $\mathcal{U}, \mathcal{L}$)

$$\mathcal{U}\left(\mathbf{u}_{i_\tau}, \delta_U, \mathbf{A}_\tau, U_\tau\right) \leq \mathcal{L}\left(\mathbf{u}_{i_\tau}, \delta_L, \mathbf{A}_\tau, L_\tau\right).$$

   - Let $t_\tau^{-1} = \frac{1}{2}\left(\mathcal{U}\left(\mathbf{u}_{i_\tau}, \delta_U, \mathbf{A}_\tau, U_\tau\right) + \mathcal{L}\left(\mathbf{u}_{i_\tau}, \delta_L, \mathbf{A}_\tau, L_\tau\right)\right)$
   - Update $\mathbf{A}_{\tau+1} = \mathbf{A}_\tau + t_\tau \mathbf{u}_{i_\tau} \mathbf{u}_{i_\tau}^T$; set $\mathbf{S}_{\tau+1, i_\tau} = 1$ and $\mathbf{D}_{\tau+1, \tau+1} = 1/\sqrt{t_\tau}$.
4. **end for**
5. Multiply all the weights in $\mathbf{D}$ by $\sqrt{r^{-1}\left(1 - \sqrt{(\ell/r)}\right)}$.
6. Return $\mathbf{S}$ and $\mathbf{D}$.

Algorithm 1. Single-set Spectral Sparsification

most r columns selected after τ iterations. The running time of the algorithm is dominated by the search for an index i_τ satisfying

$$\mathcal{U}\left(\mathbf{u}_{i_\tau}, \delta_U, \mathbf{A}_\tau, U_\tau\right) \leq \mathcal{L}\left(\mathbf{u}_{i_\tau}, \delta_L, \mathbf{A}_\tau, L_\tau\right)$$

and computing the weight t_τ. One needs to compute the upper and lower potentials $\hat{\Phi}(U, \mathbf{A})$ and $\Phi(L, \mathbf{A})$ and hence the eigenvalues of $\mathbf{A}$. Cost per iteration is $O(\ell^3)$ and the total cost is $O(r\ell^3)$. For $i = 1, .., d$, we need to compute $\mathcal{L}$ and $\mathcal{U}$ for every $\mathbf{u}_i$ which can be done in $O(d\ell^2)$ for every iteration, for a total of $O(rd\ell^2)$. Thus total running time of the algorithm is $O(rd\ell^2)$. We present the following lemma for the single-set spectral sparsification algorithm.

Lemma 1. BSS [3]: *Given* $\mathbf{U} \in \mathbb{R}^{d \times \ell}$ *satisfying* $\mathbf{U}^T\mathbf{U} = \mathbf{I}_\ell$ *and* $r > \ell$, *we can deterministically construct sampling and rescaling matrices* $\mathbf{S} \in \mathbb{R}^{r \times d}$ *and* $\mathbf{D} \in \mathbb{R}^{r \times r}$ *with* $\mathbf{R} = \mathbf{DS}$, *such that, for all* $\mathbf{y} \in \mathbb{R}^{\ell}$:

$$\left(1 - \sqrt{\ell/r}\right)^2 \|\mathbf{U}\mathbf{y}\|_2^2 \leq \|\mathbf{R}\mathbf{U}\mathbf{y}\|_2^2 \leq \left(1 + \sqrt{\ell/r}\right)^2 \|\mathbf{U}\mathbf{y}\|_2^2.$$

We now present a slightly modified version of Lemma 1 for our theorems.

Lemma 2. *Given* $\mathbf{U} \in \mathbb{R}^{d \times \ell}$ *satisfying* $\mathbf{U}^T\mathbf{U} = \mathbf{I}_\ell$ *and* $r > \ell$, *we can deterministically construct sampling and rescaling matrices* $\mathbf{S} \in \mathbb{R}^{r \times d}$ *and* $\mathbf{D} \in \mathbb{R}^{r \times r}$ *such that for* $\mathbf{R} = \mathbf{DS}$, $\left\|\mathbf{U}^T\mathbf{U} - \mathbf{U}^T\mathbf{R}^T\mathbf{R}\mathbf{U}\right\|_2 \leq 3\sqrt{\ell/r}$

Proof. From Lemma 1, it follows, $\sigma_\ell \left(\mathbf{U}^T \mathbf{R}^T \mathbf{R} \mathbf{U} \right) \geq \left(1 - \sqrt{\ell/r} \right)^2, \sigma_1 \left(\mathbf{U}^T \mathbf{R}^T \mathbf{R} \mathbf{U} \right)$
$\leq \left(1 + \sqrt{\ell/r} \right)^2$. Thus, $\lambda_{max} \left(\mathbf{U}^T \mathbf{U} - \mathbf{U}^T \mathbf{R}^T \mathbf{R} \mathbf{U} \right) \leq \left(1 - \left(1 - \sqrt{\ell/r} \right)^2 \right) \leq$
$2\sqrt{\ell/r}$. Similarly, $\lambda_{min} \left(\mathbf{U}^T \mathbf{U} - \mathbf{U}^T \mathbf{R}^T \mathbf{R} \mathbf{U} \right) \geq \left(1 - \left(1 + \sqrt{\ell/r} \right)^2 \right) \geq 3\sqrt{\ell/r}$.
Combining these, we have $\left\| \mathbf{U}^T \mathbf{U} - \mathbf{U}^T \mathbf{R}^T \mathbf{R} \mathbf{U} \right\|_2 \leq 3\sqrt{\ell/r}$.

Note: Let $\epsilon = 3\sqrt{\ell/r}$. It is possible to set an upper bound on ϵ by setting the value of r. In the next section, we assume $\epsilon \in (0, 1/2]$.

5   Our Main Theorems

The following theorem shows the additive error guarantees of the generalization bounds of the approximate classifer with that of the classifier with no feature selection. The classification error bound of BSS on RLSC depends on the condition number of the training set and on how much of the test-set lies in the subspace of the training set.

Theorem 1. *Let $\epsilon \in (0, 1/2]$ be an accuracy parameter, $r = O\left(n/\epsilon^2\right)$ be the number of features selected by BSS. Let $\mathbf{R} \in \mathbb{R}^{r \times d}$ be the matrix, as defined in Lemma 2. Let $\mathbf{X} \in \mathbb{R}^{d \times n}$ with $d \gg n$, be the training set, $\tilde{\mathbf{X}} = \mathbf{R}\mathbf{X}$ is the reduced dimensional matrix and $\mathbf{q} \in \mathbb{R}^d$ be the test point of the form $\mathbf{q} = \mathbf{X}\boldsymbol{\alpha} + \mathbf{U}^\perp \boldsymbol{\beta}$. Then, the following hold:*

- *If $\lambda = 0$, then $\left| \tilde{\mathbf{q}}^T \tilde{\mathbf{X}} \tilde{x}_{opt} - \mathbf{q}^T \mathbf{X} x_{opt} \right| \leq \frac{\epsilon \kappa_{\mathbf{X}}}{\sigma_{max}} \|\boldsymbol{\beta}\|_2 \|\mathbf{y}\|_2$*
- *If $\lambda > 0$, then $\left| \tilde{\mathbf{q}}^T \tilde{\mathbf{X}} \tilde{x}_{opt} - \mathbf{q}^T \mathbf{X} x_{opt} \right| \leq 2\epsilon \kappa_{\mathbf{X}} \|\boldsymbol{\alpha}\|_2 \|\mathbf{y}\|_2 + \frac{2\epsilon \kappa_{\mathbf{X}}}{\sigma_{max}} \|\boldsymbol{\beta}\|_2 \|\mathbf{y}\|_2$*

Proof. We assume that $\mathbf{X}$ is a full-rank matrix. Let $\mathbf{E} = \mathbf{U}^T \mathbf{U} - \mathbf{U}^T \mathbf{R}^T \mathbf{R} \mathbf{U}$ and $\|\mathbf{E}\|_2 = \left\| \mathbf{I} - \mathbf{U}^T \mathbf{R}^T \mathbf{R} \mathbf{U} \right\|_2 = \epsilon \leq 1/2$. Using the SVD of $\mathbf{X}$, we define
$$\boldsymbol{\Delta} = \boldsymbol{\Sigma} \mathbf{U}^T \mathbf{R}^T \mathbf{R} \mathbf{U} \boldsymbol{\Sigma} = \boldsymbol{\Sigma} \left(\mathbf{I} + \mathbf{E} \right) \boldsymbol{\Sigma}. \tag{6}$$
The optimal solution in the sampled space is given by,
$$\tilde{x}_{opt} = \mathbf{V} \left(\boldsymbol{\Delta} + \lambda \mathbf{I} \right)^{-1} \mathbf{V}^T \mathbf{y} \tag{7}$$
It can be proven easily that $\boldsymbol{\Delta}$ and $\boldsymbol{\Delta} + \lambda \mathbf{I}$ are invertible matrices. We focus on the term $\mathbf{q}^T \mathbf{X} x_{opt}$. Using the SVD of $\mathbf{X}$, we get
$$\mathbf{q}^T \mathbf{X} x_{opt} = \boldsymbol{\alpha}^T \mathbf{X}^T \mathbf{X} x_{opt} + \boldsymbol{\beta} \mathbf{U}^{\perp T} \left(\mathbf{U} \boldsymbol{\Sigma} \mathbf{V}^T \right) x_{opt}$$
$$= \boldsymbol{\alpha}^T \mathbf{V} \boldsymbol{\Sigma}^2 \left(\boldsymbol{\Sigma}^2 + \lambda \mathbf{I} \right)^{-1} \mathbf{V}^T \mathbf{y} \tag{8}$$
$$= \boldsymbol{\alpha}^T \mathbf{V} \left(\mathbf{I} + \lambda \boldsymbol{\Sigma}^{-2} \right)^{-1} \mathbf{V}^T \mathbf{y}. \tag{9}$$

Eqn(8) follows because of the fact that $\mathbf{U}^{\perp T} \mathbf{U} = \mathbf{0}$ and by substituting x_{opt} from Eqn.(2). Eqn.(9) follows from the fact that the matrices $\boldsymbol{\Sigma}^2$ and $\boldsymbol{\Sigma}^2 + \lambda \mathbf{I}$ are invertible. Now,

$$\left| \mathbf{q}^T \mathbf{X} \mathbf{x}_{opt} - \tilde{\mathbf{q}}^T \tilde{\mathbf{X}} \tilde{\mathbf{x}}_{opt} \right| = \left| \mathbf{q}^T \mathbf{X} \mathbf{x}_{opt} - \mathbf{q}^T \mathbf{R}^T \mathbf{R} \mathbf{X} \tilde{\mathbf{x}}_{opt} \right|$$

$$\leq \left| \mathbf{q}^T \mathbf{X} \mathbf{x}_{opt} - \boldsymbol{\alpha}^T \mathbf{X}^T \mathbf{R}^T \mathbf{R} \mathbf{X} \tilde{\mathbf{x}}_{opt} \right| \qquad (10)$$

$$+ \left| \boldsymbol{\beta}^T \mathbf{U}^{\perp T} \mathbf{R}^T \mathbf{R} \mathbf{X} \tilde{\mathbf{x}}_{opt} \right| \qquad (11)$$

We bound (10) and (11) separately. Substituting the values of $\tilde{\mathbf{x}}_{opt}$ and $\boldsymbol{\Delta}$,

$$\boldsymbol{\alpha}^T \mathbf{X}^T \mathbf{R}^T \mathbf{R} \mathbf{X} \tilde{\mathbf{x}}_{opt} = \boldsymbol{\alpha}^T \mathbf{V} \boldsymbol{\Delta} \mathbf{V}^T \tilde{\mathbf{x}}_{opt}$$

$$= \boldsymbol{\alpha}^T \mathbf{V} \boldsymbol{\Delta} \left(\boldsymbol{\Delta} + \lambda \mathbf{I} \right)^{-1} \mathbf{V}^T \mathbf{y}$$

$$= \boldsymbol{\alpha}^T \mathbf{V} \left(\mathbf{I} + \lambda \boldsymbol{\Delta}^{-1} \right)^{-1} \mathbf{V}^T \mathbf{y}$$

$$= \boldsymbol{\alpha}^T \mathbf{V} \left(\mathbf{I} + \lambda \boldsymbol{\Sigma}^{-1} \left(\mathbf{I} + \mathbf{E} \right)^{-1} \boldsymbol{\Sigma}^{-1} \right)^{-1} \mathbf{V}^T \mathbf{y}$$

$$= \boldsymbol{\alpha}^T \mathbf{V} \left(\mathbf{I} + \lambda \boldsymbol{\Sigma}^{-2} + \lambda \boldsymbol{\Sigma}^{-1} \boldsymbol{\Phi} \boldsymbol{\Sigma}^{-1} \right)^{-1} \mathbf{V}^T \mathbf{y} \qquad (12)$$

The last line follows from Lemma 3 in Appendix, which states that $\left(\mathbf{I} + \mathbf{E} \right)^{-1} = \mathbf{I} + \boldsymbol{\Phi}$, where $\boldsymbol{\Phi} = \sum\limits_{i=1}^{\infty} (-\mathbf{E})^i$. The spectral norm of $\boldsymbol{\Phi}$ is bounded by,

$$\|\boldsymbol{\Phi}\|_2 = \left\| \sum_{i=1}^{\infty} (-\mathbf{E})^i \right\|_2 \leq \sum_{i=1}^{\infty} \|\mathbf{E}\|_2^i \leq \sum_{i=1}^{\infty} \epsilon^i = \epsilon/(1 - \epsilon). \qquad (13)$$

We now bound (10). Substituting (9) and (12) in (10),

$$\left| \mathbf{q}^T \mathbf{X} \mathbf{x}_{opt} - \boldsymbol{\alpha}^T \mathbf{X}^T \mathbf{R}^T \mathbf{R} \mathbf{X} \tilde{\mathbf{x}}_{opt} \right|$$

$$= \left| \boldsymbol{\alpha}^T \mathbf{V} \{ \left(\mathbf{I} + \lambda \boldsymbol{\Sigma}^{-2} + \lambda \boldsymbol{\Sigma}^{-1} \boldsymbol{\Phi} \boldsymbol{\Sigma}^{-1} \right)^{-1} - \left(\mathbf{I} + \lambda \boldsymbol{\Sigma}^{-2} \right)^{-1} \} \mathbf{V}^T \mathbf{y} \right|$$

$$\leq \left\| \boldsymbol{\alpha}^T \mathbf{V} \left(\mathbf{I} + \lambda \boldsymbol{\Sigma}^{-2} \right) \right\|_2 \left\| \mathbf{V}^T \mathbf{y} \right\|_2 \|\boldsymbol{\Psi}\|_2$$

The last line follows because of Lemma 4 and the fact that all matrices involved are invertible. Here,

$$\boldsymbol{\Psi} = \lambda \boldsymbol{\Sigma}^{-1} \boldsymbol{\Phi} \boldsymbol{\Sigma}^{-1} \left(\mathbf{I} + \lambda \boldsymbol{\Sigma}^{-2} + \lambda \boldsymbol{\Sigma}^{-1} \boldsymbol{\Phi} \boldsymbol{\Sigma}^{-1} \right)^{-1}$$

$$= \lambda \boldsymbol{\Sigma}^{-1} \boldsymbol{\Phi} \boldsymbol{\Sigma}^{-1} \left(\boldsymbol{\Sigma}^{-1} \left(\boldsymbol{\Sigma}^2 + \lambda \mathbf{I} + \lambda \boldsymbol{\Phi} \right) \boldsymbol{\Sigma}^{-1} \right)^{-1}$$

$$= \lambda \boldsymbol{\Sigma}^{-1} \boldsymbol{\Phi} \left(\boldsymbol{\Sigma}^2 + \lambda \mathbf{I} + \lambda \boldsymbol{\Phi} \right)^{-1} \boldsymbol{\Sigma}$$

Since the spectral norms of $\boldsymbol{\Sigma}, \boldsymbol{\Sigma}^{-1}$ and $\boldsymbol{\Phi}$ are bounded, we only need to bound the spectral norm of $\left(\boldsymbol{\Sigma}^2 + \lambda \mathbf{I} + \lambda \boldsymbol{\Phi} \right)^{-1}$ to bound the spectral norm of $\boldsymbol{\Psi}$. The spectral norm of the matrix $\left(\boldsymbol{\Sigma}^2 + \lambda \mathbf{I} + \lambda \boldsymbol{\Phi} \right)^{-1}$ is the inverse of the smallest singular value of $\left(\boldsymbol{\Sigma}^2 + \lambda \mathbf{I} + \lambda \boldsymbol{\Phi} \right)$. From perturbation theory of matrices [14] and (13), we get

$$\left| \sigma_i \left(\boldsymbol{\Sigma}^2 + \lambda \mathbf{I} + \lambda \boldsymbol{\Phi} \right) - \sigma_i \left(\boldsymbol{\Sigma}^2 + \lambda \mathbf{I} \right) \right| \leq \|\lambda \boldsymbol{\Phi}\|_2 \leq \epsilon \lambda.$$

Here, $\sigma_i(\mathbf{Q})$ represents the ith singular value of the matrix $\mathbf{Q}$. Also, $\sigma_i^2 \left(\boldsymbol{\Sigma}^2 + \lambda \mathbf{I} \right) = \sigma_i^2 + \lambda$, where σ_i are the singular values of $\mathbf{X}$.

$$\sigma_i{}^2 + (1 - \epsilon)\lambda \leq \sigma_i \left(\boldsymbol{\Sigma}^2 + \lambda\mathbf{I} + \lambda\boldsymbol{\Phi}\right) \leq \sigma_i{}^2 + (1 + \epsilon)\lambda.$$

Thus, $\left\|\left(\boldsymbol{\Sigma}^2 + \lambda\mathbf{I} + \lambda\boldsymbol{\Phi}\right)^{-1}\right\|_2 = 1/\sigma_{min}\left(\boldsymbol{\Sigma}^2 + \lambda\mathbf{I} + \lambda\boldsymbol{\Phi}\right) \leq 1/\left(\sigma^2{}_{min} + (1-\epsilon)\lambda\right)$
Here, σ_{max} and σ_{min} denote the largest and smallest singular value of $\mathbf{X}$. Since $\|\boldsymbol{\Sigma}\|_2 \|\boldsymbol{\Sigma}^{-1}\|_2 = \sigma_{max}/\sigma_{min} \leq \kappa_{\mathbf{X}}$, (condition number of $\mathbf{X}$) we bound (10):

$$\left|\mathbf{q}^T\mathbf{X}\mathbf{x}_{opt} - \boldsymbol{\alpha}^T\mathbf{X}^T\mathbf{R}^T\mathbf{R}\mathbf{X}\tilde{\mathbf{x}}_{opt}\right| \leq \frac{\epsilon\lambda\kappa_{\mathbf{X}}}{\sigma^2{}_{min} + (1 - \epsilon)\lambda} \left\|\boldsymbol{\alpha}^T\mathbf{V}\left(\mathbf{I} + \lambda\boldsymbol{\Sigma}^{-2}\right)^{-1}\right\|_2 \left\|\mathbf{V}^T\mathbf{y}\right\|_2 \tag{14}$$

For $\lambda > 0$, the term $\sigma^2{}_{min} + (1 - \epsilon)\lambda$ in Eqn.(14) is always larger than $(1 - \epsilon)\lambda$, so it can be upper bounded by $2\epsilon\kappa_{\mathbf{X}}$ (assuming $\epsilon \leq 1/2$). Also,

$$\left\|\boldsymbol{\alpha}^T\mathbf{V}\left(\mathbf{I} + \lambda\boldsymbol{\Sigma}^{-2}\right)^{-1}\right\|_2 \leq \left\|\boldsymbol{\alpha}^T\mathbf{V}\right\|_2 \left\|\left(\mathbf{I} + \lambda\boldsymbol{\Sigma}^{-2}\right)^{-1}\right\|_2 \leq \left\|\boldsymbol{\alpha}\right\|_2.$$

This follows from the fact, that $\left\|\boldsymbol{\alpha}^T\mathbf{V}\right\|_2 = \|\boldsymbol{\alpha}\|_2$ and $\|\mathbf{V}\mathbf{y}\|_2 = \|\mathbf{y}\|_2$ as $\mathbf{V}$ is a full-rank orthonormal matrix and the singular values of $\mathbf{I} + \lambda\boldsymbol{\Sigma}^{-2}$ are equal to $1 + \lambda/\sigma_i{}^2$; making the spectral norm of its inverse at most one. Thus we get,

$$\left|\mathbf{q}^T\mathbf{X}\mathbf{x}_{opt} - \boldsymbol{\alpha}^T\mathbf{X}^T\mathbf{R}^T\mathbf{R}\mathbf{X}\tilde{\mathbf{x}}_{opt}\right| \leq 2\epsilon\kappa_{\mathbf{X}} \|\boldsymbol{\alpha}\|_2 \|\mathbf{y}\|_2. \tag{15}$$

We now bound (11). Expanding (11) using SVD and $\tilde{\mathbf{x}}_{opt}$,

$$\begin{aligned}
\left|\boldsymbol{\beta}^T\mathbf{U}^{\perp T}\mathbf{R}^T\mathbf{R}\mathbf{X}\tilde{\mathbf{x}}_{opt}\right| &= \left|\boldsymbol{\beta}^T\mathbf{U}^{\perp T}\mathbf{R}^T\mathbf{R}\mathbf{U}\boldsymbol{\Sigma}\left(\boldsymbol{\Delta} + \lambda\mathbf{I}\right)\mathbf{V}^T\mathbf{y}\right| \\
&\leq \left\|\mathbf{q}^T\mathbf{U}^{\perp}\mathbf{U}^{\perp T}\mathbf{R}^T\mathbf{R}\mathbf{U}\right\|_2 \left\|\boldsymbol{\Sigma}\left(\boldsymbol{\Delta} + \lambda\mathbf{I}\right)^{-1}\right\|_2 \left\|\mathbf{V}^T\mathbf{y}\right\|_2 \\
&\leq \epsilon\left\|\mathbf{U}^{\perp}\mathbf{U}^{\perp T}\mathbf{q}\right\|_2 \left\|\mathbf{V}^T\mathbf{y}\right\|_2 \left\|\boldsymbol{\Sigma}\left(\boldsymbol{\Delta} + \lambda\mathbf{I}\right)^{-1}\right\|_2 \\
&\leq \epsilon\|\boldsymbol{\beta}\|_2 \|\mathbf{y}\|_2 \left\|\boldsymbol{\Sigma}\left(\boldsymbol{\Delta} + \lambda\mathbf{I}\right)^{-1}\right\|_2
\end{aligned}$$

The first inequality follows from $\boldsymbol{\beta} = \mathbf{U}^{\perp T}\mathbf{q}$; and the second inequality follows from Lemma 6 given in appendix. To conclude the proof, we bound the spectral norm of $\boldsymbol{\Sigma}\left(\boldsymbol{\Delta} + \lambda\mathbf{I}\right)^{-1}$. Note that from Eqn.(6), $\boldsymbol{\Sigma}^{-1}\boldsymbol{\Delta}\boldsymbol{\Sigma}^{-1} = \mathbf{I} + \mathbf{E}$ and $\boldsymbol{\Sigma}\boldsymbol{\Sigma}^{-1} = \mathbf{I}$,

$$\boldsymbol{\Sigma}\left(\boldsymbol{\Delta} + \lambda\mathbf{I}\right)^{-1} = \left(\boldsymbol{\Sigma}^{-1}\boldsymbol{\Delta}\boldsymbol{\Sigma}^{-1} + \lambda\boldsymbol{\Sigma}^{-2}\right)^{-1}\boldsymbol{\Sigma}^{-1} = \left(\mathbf{I} + \lambda\boldsymbol{\Sigma}^{-2} + \mathbf{E}\right)^{-1}\boldsymbol{\Sigma}^{-1}.$$

One can get a lower bound for the smallest singular value of $\left(\mathbf{I} + \lambda\boldsymbol{\Sigma}^{-2} + \mathbf{E}\right)^{-1}$ using matrix perturbation theory and by comparing the singular values of this matrix to the singular values of $\mathbf{I} + \lambda\boldsymbol{\Sigma}^{-2}$. We get, $(1 - \epsilon) + \frac{\lambda}{\sigma_i{}^2} \leq \sigma_i\left(\mathbf{I} + \mathbf{E} + \lambda\boldsymbol{\Sigma}^{-2}\right) \leq (1 + \epsilon) + \frac{\lambda}{\sigma_i{}^2}$

$$\begin{aligned}
\left\|\left(\mathbf{I} + \lambda\boldsymbol{\Sigma}^{-2} + \mathbf{E}\right)^{-1}\boldsymbol{\Sigma}^{-1}\right\|_2 &\leq \frac{\sigma^2{}_{max}}{\left((1 - \epsilon)\sigma^2{}_{max} + \lambda\right)\sigma_{min}} \\
&= \frac{\kappa_{\mathbf{X}}\sigma_{max}}{(1 - \epsilon)\sigma^2{}_{max} + \lambda} \\
&\leq \frac{2\kappa_{\mathbf{X}}}{\sigma_{max}} \tag{16}
\end{aligned}$$

We assumed that $\epsilon \leq 1/2$, which implies $(1 - \epsilon) + \lambda/\sigma^2{}_{max} \geq 1/2$. Combining these, we get,

$$\left|\boldsymbol{\beta}^T \mathbf{U}^{\perp T} \mathbf{R}^T \mathbf{R} \mathbf{X} \tilde{\mathbf{x}}_{opt}\right| \leq \frac{2\epsilon\kappa_{\mathbf{X}}}{\sigma_{max}} \|\boldsymbol{\beta}\|_2 \|\mathbf{y}\|_2. \tag{17}$$

Combining Eqns (15) and (17) we complete the proof for the case $\lambda > 0$. For $\lambda = 0$, Eqn.(14) becomes zero and the result follows.

Our next theorem provides relative-error guarantees to the bound on the classification error when the test-point has no-new components, i.e. $\boldsymbol{\beta} = \mathbf{0}$.

Theorem 2. *Let $\epsilon \in (0, 1/2]$ be an accuracy parameter, $r = O\left(n/\epsilon^2\right)$ be the number of features selected by BSS and $\lambda > 0$. Let $\mathbf{q} \in \mathbb{R}^d$ be the test point of the form $\mathbf{q} = \mathbf{X}\boldsymbol{\alpha}$, i.e. it lies entirely in the subspace spanned by the training set, and the two vectors $\mathbf{V}^T \mathbf{y}$ and $\left(\mathbf{I} + \lambda\mathbf{\Sigma}^{-2}\right)^{-1} \mathbf{V}^T \boldsymbol{\alpha}$ satisfy the property,*

$$\left\|\left(\mathbf{I} + \lambda\mathbf{\Sigma}^{-2}\right)^{-1} \mathbf{V}^T \boldsymbol{\alpha}\right\|_2 \left\|\mathbf{V}^T \mathbf{y}\right\|_2 \leq \omega \left\|\left(\left(\mathbf{I} + \lambda\mathbf{\Sigma}^{-2}\right)^{-1} \mathbf{V}^T \boldsymbol{\alpha}\right)^T \mathbf{V}^T \mathbf{y}\right\|_2$$

$$= \omega \left|\mathbf{q}^T \mathbf{X} \mathbf{x}_{opt}\right|$$

for some constant ω. If we run RLSC after BSS, then $\left|\tilde{\mathbf{q}}^T \tilde{\mathbf{X}} \tilde{\mathbf{x}}_{opt} - \mathbf{q}^T \mathbf{X} \mathbf{x}_{opt}\right| \leq 2\epsilon\omega\kappa_{\mathbf{X}} \left|\mathbf{q}^T \mathbf{X} \mathbf{x}_{opt}\right|$

The proof follows directly from the proof of Theorem 1 if we consider $\boldsymbol{\beta} = \mathbf{0}$.

6 Experiments

All experiments were performed in MATLAB R2013b on an Intel i-7 processor with 16GB RAM.

6.1 BSS Implementation Issues

The authors of [3] do not provide any implementation details of the **BSS** algorithm. Here we discuss several issues arising during the implementation.

Choice of Column Selection: At every iteration, there are multiple columns which satisfy the condition $\mathcal{U}\left(\mathbf{u}_i, \delta_U, \mathbf{A}_\tau, U_\tau\right) \leq \mathcal{L}\left(\mathbf{u}_i, \delta_L, \mathbf{A}_\tau, L_\tau\right)$. The authors of [3] suggest picking any column which satisfies this constraint. Instead of breaking ties arbitrarily, we choose the column $\mathbf{u}_i$ which has not been selected in previous iterations and whose Euclidean-norm is highest among the candidate set. Columns with zero Euclidean norm never get selected by the algorithm. In the inner loop of Algorithm 1, $\mathcal{U}$ and $\mathcal{L}$ has to be computed for all the d columns in order to pick a good column. This step can be done efficiently using a single line of Matlab code, by making use of matrix and vector operations.

Ill-conditioning: The second issue related to the implementation is ill conditioning. It is possible for $\mathbf{A}_\tau$ to be almost singular. At every iteration τ, we check the condition number of $\mathbf{A}_\tau$. If it is high, then we regularize $\mathbf{A}_\tau$ as follows : $\mathbf{A}_\tau = \mathbf{A}_\tau + \gamma\mathbf{I}$. We set $\gamma = 0.01$ in our experiments. Smaller values of γ resulted in large eigenvalues of $\mathbf{A}_\tau{}^{-1}$, which in turn, resulted in large values of t_τ causing bad-scaling of the columns of the input matrix.

6.2   Other Feature Selection Methods

In this section, we describe other feature-selection methods with which we compare BSS.

Rank-Revealing QR Factorization (RRQR): Within the numerical linear algebra community, subset selection algorithms use the so-called Rank Revealing QR (RRQR) factorization. Here we slightly abuse notation and state $\mathbf{A}$ as a short and fat matrix as opposed to the tall and thin matrix. Let $\mathbf{A}$ be a $n \times d$ matrix with $(n < d)$ and an integer k $(k < d)$ and assume partial QR factorizations of the form

$$\mathbf{AP} = \mathbf{Q} \begin{pmatrix} \mathbf{R}_{11} & \mathbf{R}_{12} \\ \mathbf{0} & \mathbf{R}_{22} \end{pmatrix},$$

where $\mathbf{Q} \in \mathbb{R}^{n \times n}$ is an orthogonal matrix, $\mathbf{P} \in \mathbb{R}^{d \times d}$ is a permutation matrix, $\mathbf{R}_{11} \in \mathbb{R}^{k \times k}, \mathbf{R}_{12} \in \mathbb{R}^{k \times (d-k)}, \mathbf{R}_{22} \in \mathbb{R}^{(d-k) \times (d-k)}$ The above factorization is called a RRQR factorization if $\sigma_{min}(\mathbf{R}_{11}) \geq \sigma_k(\mathbf{A})/p(k,d)$, $\sigma_{max}(\mathbf{R}_{22}) \leq \sigma_{min}(\mathbf{A})p(k,d)$, where $p(k,d)$ is a function bounded by a low-degree polynomial in k and d. The important columns are given by $\mathbf{A}_1 = \mathbf{Q} \begin{pmatrix} \mathbf{R}_{11} \\ \mathbf{0} \end{pmatrix}$ and $\sigma_i(\mathbf{A}_1) = \sigma_i(\mathbf{R}_{11})$ with $1 \leq i \leq k$. We perform feature selection using RRQR by picking the important columns which preserve the rank of the matrix.

Random Feature Selection: We select features uniformly at random without replacement which serves as a baseline method. To get around the randomness, we repeat the sampling process five times.

Leverage Score Sampling: We describe the leverage-score sampling of [6]. Let $\mathbf{U}$ be the top-k left singular vectors of the training set $\mathbf{X}$. We create a carefully chosen probability distribution of the form $p_i = \frac{\|\mathbf{U}_i\|_2^2}{n}$. for $i = 1, 2, ..., d$, i.e. proportional to the squared Euclidean norms of the rows of the left-singular vectors and select r rows of $\mathbf{U}$ in i.i.d trials and re-scale the rows with $1/\sqrt{p_i}$. We repeat the sampling process five times to get around the randomness. In our experiments, k was set to the rank of $\mathbf{X}$.

Information Gain (IG): The Information Gain feature selection method [16] measures the amount of information obtained for binary class prediction by knowing the presence or absence of a feature in a dataset. The method is a supervised strategy, whereas the other methods used here are unsupervised.

Table 1. Most frequently selected features using the synthetic dataset

$r = 80$	$k = 90$	$k = 100$
BSS	89, 88, 87, 86, 85	100, 99, 98, 97, 95
RRQR	90, 80, 79, 78, 77	100, 80, 79, 78, 77
Lvg-Score	73, 85, 84, 81, 87	93, 87, 95, 97, 96
IG	80, 79, 78, 77, 76	80, 79, 78, 77, 76
$r = 90$	$k = 90$	$k = 100$
BSS	88, 87, 86, 85, 84	100, 99, 98, 97, 95
RRQR	90, 89, 88, 87, 86	100, 90, 89, 88, 87
Lvg-Score	67, 88, 83, 87, 85	100, 97, 92, 48, 58
IG	90, 89, 88, 87, 86	90, 89, 88, 87, 86

6.3 Synthetic Data

We run our experiments on synthetic data where we control the number of relevant features in the dataset and demonstrate the working of Algorithm 1 on RLSC. We generate synthetic data in the same manner as given in [4]. The dataset has n data-points and d features. The class label y_i of each data-point was randomly chosen to be 1 or -1 with equal probability. The first k features of each data-point $\mathbf{x}_i$ are drawn from $y_i \mathcal{N}(-j, 1)$ distribution, where $\mathcal{N}(\mu, \sigma^2)$ is a random normal distribution with mean μ and variance σ^2 and j varies from 1 to k. The remaining $d - k$ features are chosen from a $\mathcal{N}(0, 1)$ distribution. Thus the dataset has k relevant features and $(d - k)$ noisy features. By construction, among the first k features, the kth feature has the most discriminatory power, followed by $(k - 1)th$ feature and so on. We set n to 30 and d to 1000. We set k to 90 and 100 and ran two sets of experiments.

Table 2. Out-of-sample error of TechTC-300 datasets averaged over ten ten-fold cross-validation and over 48 datasets for three values of r. The first and second entry of each cell represents the mean and standard deviation. Items in bold indicate the best results.

$r = 300$	$\lambda = 0.1$	$\lambda = 0.3$	$\lambda = 0.5$	$\lambda = 0.7$
BSS	**31.76 ± 0.68**	**31.46 ± 0.67**	**31.24 ± 0.65**	**31.03 ± 0.66**
Lvg-Score	38.22 ± 1.26	37.63 ± 1.25	37.23 ± 1.24	36.94 ± 1.24
RRQR	37.84 ± 1.20	37.07 ± 1.19	36.57 ± 1.18	36.10 ± 1.18
Randomfs	50.01 ± 1.2	49.43 ± 1.2	49.18 ± 1.19	49.04 ± 1.19
IG	38.35 ± 1.21	36.64 ± 1.18	35.81 ± 1.18	35.15 ± 1.17
$r = 400$	$\lambda = 0.1$	$\lambda = 0.3$	$\lambda = 0.5$	$\lambda = 0.7$
BSS	**30.59 ± 0.66**	**30.33 ± 0.65**	**30.11 ± 0.65**	**29.96 ± 0.65**
Lvg-Score	35.06 ± 1.21	34.63 ± 1.20	34.32 ± 1.2	34.11 ± 1.19
RRQR	36.61 ± 1.19	36.04 ± 1.19	35.46 ± 1.18	35.05 ± 1.17
Randomfs	47.82 ± 1.2	47.02 ± 1.21	46.59 ± 1.21	46.27 ± 1.2
IG	37.37 ± 1.21	35.73 ± 1.19	34.88 ± 1.18	34.19 ± 1.18
$r = 500$	$\lambda = 0.1$	$\lambda = 0.3$	$\lambda = 0.5$	$\lambda = 0.7$
BSS	**29.80 ± 0.77**	**29.53 ± 0.77**	**29.34 ± 0.76**	**29.18 ± 0.75**
Lvg-Score	33.33 ± 1.19	32.98 ± 1.18	32.73 ± 1.18	32.52 ± 1.17
RRQR	35.77 ± 1.18	35.18 ± 1.16	34.67 ± 1.16	34.25 ± 1.14
Randomfs	46.26 ± 1.21	45.39 ± 1.19	44.96 ± 1.19	44.65 ± 1.18
IG	36.24 ± 1.20	34.80 ± 1.19	33.94 ± 1.18	33.39 ± 1.17

We set the value of r, i.e. the number of features selected by BSS to 80 and 90 for all experiments. We performed ten-fold cross-validation and repeated it ten times. The value of λ was set to 0, 0.1, 0.3, 0.5, 0.7, and 0.9. We compared BSS with RRQR, IG and leverage-score sampling. The mean out-of-sample error was 0 for all methods for both $k = 90$ and $k = 100$. Table 1 shows the set of five most frequently selected features by the different methods for one such synthetic dataset across 100 training sets. The top features picked up by the different methods are the relevant features by construction and also have good discriminatory power. This shows that supervised BSS is as good as any other method in terms of feature selection and often picks more discriminatory features than the other methods. We repeated our experiments on ten different synthetic datasets and each time, the five most frequently selected features were from the set of relevant features. Thus, by selecting only 8%-9% of all features, we show that we are able to obtain the most discriminatory features along with good out-of-sample error using BSS.

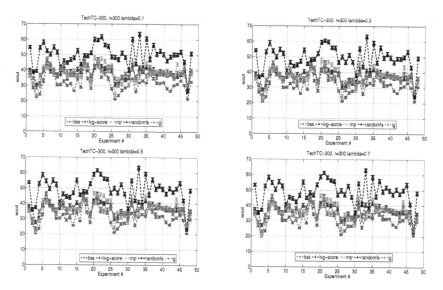

Fig. 1. Out-of-sample error of 48 TechTC-300 documents averaged over ten ten-fold cross validation experiments for different values of regularization parameter λ and number of features $r = 300$. Vertical bars represent standard deviation.

6.4   TechTC-300

We use the TechTC-300 data [7], consisting of a family of 295 document-term data matrices. The TechTC-300 dataset comes from the Open Directory Project (ODP), which is a large, comprehensive directory of the web, maintained by volunteer editors. Each matrix in the TechTC-300 dataset contains a pair of categories from the ODP. Each category corresponds to a label, and thus the resulting classification task is binary. The documents that are collected from the union of all the subcategories within each category are represented in the bag-of-words model, with the words constituting the features of the data [7]. Each data matrix consists of 150-280 documents, and each document is described with respect to 10,000-50,000 words. Thus, TechTC-300 provides a diverse collection of data sets for a systematic study of the performance of the RLSC using BSS. We removed all words of length at most four from the datasets. Next we grouped the datasets based on the categories and selected those datasets whose categories appeared at least thrice. There were 147 datasets, and we performed ten-fold

Table 3. A subset of the TechTC matrices of our study

id1_id2	id1	id2
1092_789236	Arts:Music:Styles:Opera	US Navy:Decommisioned Submarines
17899_278949	US:Michigan:Travel & Tourism	Recreation:Sailing Clubs:UK
17899_48446	US:Michigan:Travel & Tourism	Chemistry:Analytical:Products
14630_814096	US:Colorado:Localities:Boulder	Europe:Ireland:Dublin:Localities
10539_300332	US:Indiana:Localities:S	Canada:Ontario:Localities:E
10567_11346	US:Indiana:Evansville	US:Florida:Metro Areas:Miami
10539_194915	US:Indiana:Localities:S	US:Texas:Localities:D

cross validation and repeated it ten times on 48 such datasets. We set the values of the regularization parameter of RLSC to $0.1, 0.3, 0.5$ and 0.7. We do not report running times because feature selection is an offline task. We set r to 300, 400 and 500. We report the out-of-sample error for all 48 datasets. BSS consistently outperforms Leverage-Score sampling, IG, RRQR and random feature selection on all 48 datasets for all values of the regularization parameter. Table 2 and Fig 1 shows the results. The out-of-sample error decreases with increase in number of features for all methods. In terms of out-of-sample error, BSS is the best, followed by Leverage-score sampling, IG, RRQR and random feature selection. BSS is at least 3%-7% better than the other methods when averaged over 48 document matrices. From Fig 1 and 2, it is evident that BSS is comparable to the other methods and often better on all 48 datasets. Leverage-score sampling requires greater number of samples to achieve the same out-of-sample error as BSS (See Table 2, $r = 500$ for Lvg-Score and $r = 300$ for BSS). Therefore, for the same number of samples, BSS outperforms leverage-score sampling in terms of out-of-sample error. The out-of-sample error of supervised IG is worse than that of unsupervised BSS, which could be due to the worse generalization of the supervised IG metric. We also observe that the out-of-sample error decreases with increase in λ for the different feature selection methods.

Due to space constraints, we list the most frequently occurring words selected by BSS for the $r = 300$ case for seven TechTC-300 datasets over 100 training sets used in the cross-validation experiments. Table 3 shows the names of the seven TechTC-300 document-term matrices. The words shown in Table 4

Table 4. Frequently occurring terms of the TechTC-300 datasets of Table 3 selected by BSS

1092_789236	naval,shipyard,submarine,triton,music,opera,libretto,theatre
17899_278949	sailing,cruising,boat,yacht,racing,michigan,leelanau,casino
17899_48446	vacation,lodging,michigan,asbestos,chemical,analytical,laboratory
14630_814096	ireland,dublin,boulder,colorado,lucan,swords,school,dalkey
10539_300332	ontario,fishing,county,elliot,schererville,shelbyville,indiana,bullet
10567_11346	florida,miami,beach,indiana,evansville,music,business,south
10539_194915	texas,dallas,plano,denton,indiana,schererville,gallery,north

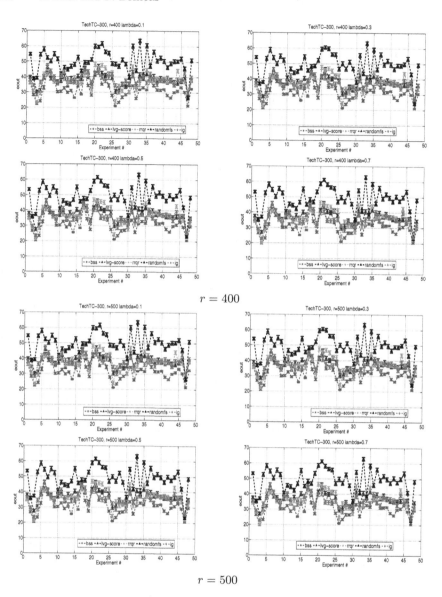

Fig. 2. Out-of-sample error of 48 TechTC-300 documents averaged over ten ten-fold cross validation experiments for different values of regularization parameter λ and number of features $r = 400$ and $r = 500$. Vertical bars represent standard deviation.

were selected in all cross-validation experiments for these seven datasets. The words are closely related to the categories to which the documents belong, which shows that BSS selects important features from the training set. For example, for the document-pair (1092_789236), where 1092 belongs to the category of "Arts:Music:Styles:Opera" and 789236 belongs to the category of "US:Navy:

Decommisioned Submarines", the BSS algorithm selects submarine, shipyard, triton, opera, libretto, theatre which are closely related to the two classes. Another example is the document-pair 10539_300332, where 10539 belongs to "US:Indiana:Localities:S" and 300332 belongs to the category of "Canada: Ontario: Localities:E". The top words selected for this document-pair are ontario, elliot, shelbyville, indiana, schererville which are closely related to the class values. Thus, we see that using only 2%-4% of all features we are able to select relevant features and obtain good out-of-sample error.

7 Conclusion

We present a provably accurate feature selection method for RLSC which works well empirically and also gives better generalization peformance than prior existing methods. The number of features required by BSS is of the order $O\left(n/\epsilon^2\right)$, which makes the result tighter than that obtained by leverage-score sampling. BSS has been recently used as a feature selection technique for k-means clustering [5], linear SVMs [11] and our work on RLSC helps to expand research in this direction. An interesting future work in this direction would be to include feature selection for non-linear kernels with provable guarantees.

Acknowledgements. We thank the reviewers for their insightful comments. SP and PD are supported by NSF CCF 1016501 and NSF IIS 1319280.

References

1. Agarwal, D.: Shrinkage estimator generalizations of proximal support vector machines. In: Proceedings of the Eighth ACM SIGKDD International Conference on Knowledge Discovery and Data Mining, pp. 173–182 (2002)
2. Avron, H., Sindhwani, V., Woodruff, D.: Sketching structured matrices for faster nonlinear regression. In: Advances in Neural Information Processing Systems, pp. 2994–3002 (2013)
3. Batson, J., Spielman, D., Srivastava, N.: Twice-ramanujan sparsifiers. In: Proceedings of the 41st Annual ACM STOC, pp. 255–262 (2009)
4. Bhattacharyya, C.: Second order cone programming formulations for feature selection. JMLR 5, 1417–1433 (2004)
5. Boutsidis, C., Magdon-Ismail, M.: Deterministic feature selection for k-means clustering. IEEE Transactions on Information Theory 59(9), 6099–6110 (2013)
6. Dasgupta, A., Drineas, P., Harb, B., Josifovski, V., Mahoney, M.: Feature selection methods for text classification. In: Proceedings of the 13th ACM SIGKDD International Conference on Knowledge Discovery and Data Mining, pp. 230–239 (2007)
7. Davidov, D., Gabrilovich, E., Markovitch, S.: Parameterized generation of labeled datasets for text categorization based on a hierarchical directory. In: Proceedings of the 27th Annual International ACM SIGIR Conference, pp. 250–257 (2004), http://techtc.cs.technion.ac.il/techtc300/techtc300.html

8. Demmel, J., Veselic, K.: Jacobi's method is more accurate than qr. SIAM Journal on Matrix Analysis and Applications 13(4), 1204–1245 (1992)

9. Drineas, P., Mahoney, M., Muthukrishnan, S.: Sampling algorithms for l2 regression and applications. In: Proceedings of the 17th Annual ACM-SIAM SODA, pp. 1127–1136 (2006)

10. Fung, G., Mangasarian, O.: Proximal support vector machine classifiers. In: Proceedings of the Seventh ACM SIGKDD International Conference on Knowledge Discovery and Data Mining, pp. 77–86 (2001)

11. Paul, S., Magdon-Ismail, M., Drineas, P.: Deterministic feature selection for linear svm with provable guarantees (2014), http://arxiv.org/abs/1406.0167

12. Poggio, T., Smale, S.: The mathematics of learning: Dealing with data. Notices of the AMS 50(5), 537–544 (2003)

13. Rifkin, R., Yeo, G., Poggio, T.: Regularized least-squares classification. Nato Science Series Sub Series III Computer and Systems Sciences 190, 131–154 (2003)

14. Stewart, G., Sun, J.: Matrix perturbation theory (1990)

15. Suykens, J., Vandewalle, J.: Least squares support vector machine classifiers. Neural Processing Letters 9(3), 293–300 (1999)

16. Yang, Y., Pedersen, J.: A comparative study on feature selection in text categorization. In: ICML, vol. 97, pp. 412–420 (1997)

17. Zhang, P., Peng, J.: SVM vs regularized least squares classification. In: Proceedings of the 17th International Conference on Pattern Recognition, vol. 1, pp. 176–179 (2004)

18. Zhang, T., Oles, F.: Text categorization based on regularized linear classification methods. Information Retrieval 4(1), 5–31 (2001)

8   Appendix

Lemma 3. *For any matrix* $\mathbf{E}$, *such that* $\mathbf{I} + \mathbf{E}$ *is invertible,* $(\mathbf{I} + \mathbf{E})^{-1} = \mathbf{I} + \sum_{i=1}^{\infty} (-\mathbf{E})^i$.

Lemma 4. *Let* $\mathbf{A}$ *and* $\tilde{\mathbf{A}} = \mathbf{A} + \mathbf{E}$ *be invertible matrices. Then* $\tilde{\mathbf{A}}^{-1} - \mathbf{A}^{-1} = -\mathbf{A}^{-1}\mathbf{E}\tilde{\mathbf{A}}^{-1}$.

Lemma 5. *Let* $\mathbf{D}$ *and* $\mathbf{X}$ *be matrices such that the product* $\mathbf{DXD}$ *is a symmetric positive definite matrix with matrix* $\mathbf{X}_{ii} = 1$. *Let the product* $\mathbf{DED}$ *be a perturbation such that,* $\|E\|_2 = \eta < \lambda_{min}(\mathbf{X})$. *Here* λ_{min} *corresponds to the smallest eigenvalue of* $\mathbf{X}$. *Let* λ_i *be the i-th eigenvalue of* $\mathbf{DXD}$ *and let* $\tilde{\lambda}_i$ *be the i-th eigenvalue of* $\mathbf{D}(\mathbf{X} + \mathbf{E})\mathbf{D}$. *Then,* $\left|\frac{\lambda_i - \tilde{\lambda}_i}{\lambda_i}\right| \leq \frac{\eta}{\lambda_{min}(\mathbf{X})}$.

The lemmas presented above are from matrix perturbation theory [14,8] and are used in the proof of our main theorem.

Lemma 6. *Let* $\epsilon \in (0, 1/2]$. *Then* $\left\|\mathbf{q}^T\mathbf{U}^{\perp}\mathbf{U}^{\perp T}\mathbf{R}^T\mathbf{R}\mathbf{U}\right\|_2 \leq \epsilon \left\|\mathbf{U}^{\perp}\mathbf{U}^{\perp T}\mathbf{q}\right\|_2$

The proof of this lemma is similar to Lemma 4.3 of [9].

Boosted Bellman Residual Minimization Handling Expert Demonstrations

Bilal Piot[1,2], Matthieu Geist[1,2], and Olivier Pietquin[3]

[1] Supélec, IMS-MaLIS Research group, France
{bilal.piot,matthieu.geist}@supelec.fr
[2] UMI 2958 (GeorgiaTech-CNRS), France
[3] University Lille 1, LIFL (UMR 8022 CNRS/Lille 1) - SequeL team, Lille, France
olivier.pietquin@univ-lille1.fr

Abstract. This paper addresses the problem of batch Reinforcement Learning with Expert Demonstrations (RLED). In RLED, the goal is to find an optimal policy of a Markov Decision Process (MDP), using a data set of fixed sampled transitions of the MDP as well as a data set of fixed expert demonstrations. This is slightly different from the batch Reinforcement Learning (RL) framework where only fixed sampled transitions of the MDP are available. Thus, the aim of this article is to propose algorithms that leverage those expert data. The idea proposed here differs from the Approximate Dynamic Programming methods in the sense that we minimize the Optimal Bellman Residual (OBR), where the minimization is guided by constraints defined by the expert demonstrations. This choice is motivated by the the fact that controlling the OBR implies controlling the distance between the estimated and optimal quality functions. However, this method presents some difficulties as the criterion to minimize is non-convex, non-differentiable and biased. Those difficulties are overcome via the embedding of distributions in a Reproducing Kernel Hilbert Space (RKHS) and a boosting technique which allows obtaining non-parametric algorithms. Finally, our algorithms are compared to the only state of the art algorithm, Approximate Policy Iteration with Demonstrations (APID) algorithm, in different experimental settings.

1 Introduction

This paper addresses the problem of batch Reinforcement Learning with Expert Demonstrations (RLED) where the aim is to find an optimal policy of a Markov Decision Process (MDP) only known through fixed sampled transitions, when expert demonstrations are also provided. Thus, RLED can be seen as a combination of two classical frameworks, Learning from Demonstrations (LfD) and batch Reinforcement Learning (RL). The LfD framework is a practical paradigm for learning from expert trajectories. A classical approach to LfD is to generalize the mapping between states and actions observed in the expert data. This can be done by a Supervised Learning method such as a multi-class classification algorithm [19]. However, those methods do not generalize well to regions of the state space that are not observed in the expert data, because they do not take

T. Calders et al. (Eds.): ECML PKDD 2014, Part II, LNCS 8725, pp. 549–564, 2014.
© Springer-Verlag Berlin Heidelberg 2014

into account the underlying dynamics of the MDP. To alleviate this, some recent methods [11,21] focus on querying the expert in some appropriate regions of the state space to improve the learning. However, this implies that the expert stays with the learning agent throughout the training process, which can reduce significantly the applicability of the technique. Therefore, the idea presented in [12] to overcome the limitation of conventional LfD methods is to use techniques from the batch Reinforcement Learning (RL) paradigm and combine them with LfD techniques. In batch Reinforcement Learning (RL), the goal is the same as in RLED, but without expert data. Usual techniques of batch RL have some difficulties to achieve good results from relatively little data. However, if some expert demonstrations are added to the set of sampled transitions, it is possible to improve significantly the results of the method [12]. Thus, a combination of expert data and non expert data offers the possibility to address the problem of learning optimal policies under realistic assumptions.

At our knowledge, there is only one algorithm taking into account expert data in order to find an optimal policy. This approach is Approximate Policy Iteration with Demonstrations (APID) [12], which consists in using expert demonstrations to define linear constraints that guide the optimization performed by Approximate Policy Iteration (API), a classical framework in RL. The practical algorithm APID is inspired by Least Squares Temporal Differences (LSTD) [6], where the choice of features is a key problem as it is a parametric method. Even if the optimization could be done in a Reproducing Kernel Hilbert Space (RKHS), which provides the flexibility of working with a non-parametric representation as claimed by [12], the choice of the kernel can be as difficult as the choice of the features. Therefore, we propose a method with no features choice in order to solve the RLED problem. Our method consists in the minimization of the norm of the Optimal Bellman Residual (OBR), guided by constraints defined by the expert demonstrations (see Sec 2.1). This minimization is motivated by the fact that if one is able to find a function with a small OBR, then this function is close to the optimal quality function. However, as far as we know, this technique is not used in RL for three reasons. First, the empirical norm of the OBR is biased. Second, it is not convex, so the minimization could lead to local minima. Third, it is not differentiable. Our contribution is to show how we can construct an empirical norm of the OBR which is not biased via the embedding of distributions in an RKHS (see Sec. 4.1) and how it is possible to minimize this non-convex and non-differentiable criterion via a boosting technique (see Sec. 4.2). In addition, boosting techniques are non-parametric methods which avoid choosing features.

In the proposed experiments (see Sec. 5), we compare our algorithms to the only state of the art algorithm APID, to an RL algorithm Least Square Policy Iteration (LSPI) [14], and to an LfD algorithm [19]. The first experiment is conducted on a generic task (randomly generated MDP called Garnet [3]) where expert demonstrations and non-expert transitions are provided and the aim is to find an optimal policy. The second experiment is realized on a classical LfD benchmark, the Highway problem.

2   Background and Notations

Let $(\mathbb{R}, |.|)$ be the real space with its canonical norm and X a finite set, $\mathbb{R}^X$ is the set of functions from X to $\mathbb{R}$. The set of probability distributions over X is noted Δ_X. Let Y be a finite set, Δ_X^Y is the set of functions from Y to Δ_X. Let $\alpha \in \mathbb{R}^X$, $p \in \mathbb{R}_+^*$ and $\nu \in \Delta_X$, we define the $\mathbf{L}_{p,\nu}$-norm of α, noted $\|\alpha\|_{p,\nu}$, by: $\|\alpha\|_{p,\nu} = (\sum_{x \in X} \nu(x)|\alpha(x)|^p)^{\frac{1}{p}}$. In addition, the infinite norm is noted $\|\alpha\|_\infty$ and defined as $\|\alpha\|_\infty = \max_{x \in X} |\alpha(x)|$. Let $x \in X$, $x \sim \nu$ means that x is sampled according to ν and $\mathbb{E}_\nu[\alpha] = \sum_{x \in X} \nu(x)\alpha(x)$ is the expectation of α under ν. Finally, $\delta_x \in \mathbb{R}^X$ is the function such that $\forall y \in X$, if $y \neq x$ then $\delta_x(y) = 0$, else $\delta_x(y) = 1$.

2.1   MDP, RL and RLED

In this section, we provide a very brief summary of some of the concepts and definitions from the theory of MDP and RL. For further information about MDP, the reader can be referred to [18]. Here, the agent is supposed to act in a finite MDP [1]. An MDP models the interactions of an agent evolving in a dynamic environment and is represented by a tuple $M = \{S, A, R, P, \gamma\}$ where $S = \{s_i\}_{1 \leq i \leq N_S}$ is the state space, $A = \{a_i\}_{1 \leq i \leq N_A}$ is the action space, $R \in \mathbb{R}^{S \times A}$ is the reward function, $\gamma \in]0, 1[$ is a discount factor and $P \in \Delta_S^{S \times A}$ is the Markovian dynamics which gives the probability, $P(s'|s, a)$, to reach s' by choosing the action a in the state s. A policy π is an element of A^S and defines the behavior of an agent. In order to quantify the quality of a policy π relatively to the reward R, we define the quality function. For a given MDP M and a given policy π, the quality function $Q^\pi \in \mathbb{R}^{S \times A}$ is defined as $Q^\pi(s, a) = \mathbb{E}_{s,a}^\pi[\sum_{t=0}^{+\infty} \gamma^t R(s_t, a_t)]$, where $\mathbb{E}_{s,a}^\pi$ is the expectation over the distribution of the admissible trajectories $(s_0, a_0, s_1, \pi(s_1), \dots)$ obtained by executing the policy π starting from $s_0 = s$ and $a_0 = a$. Moreover, the function $Q^* \in \mathbb{R}^{S \times A}$ defined as: $Q^* = \max_{\pi \in A^S} Q^\pi$ is called the optimal quality function. A policy π which has the following property: $\forall s \in S, \pi(s) \in \arg\max_{a \in A} Q^*(s, a)$ is said optimal with respect to R. Thus, it is quite easy to construct an optimal policy via the knowledge of the optimal quality function. For ease of writing, for each Q and each π, we define $f_Q^* \in \mathbb{R}^S$ such that $\forall s \in S, f_Q^*(s) = \max_{a \in A} Q(s, a)$ and $f_Q^\pi \in \mathbb{R}^S$ such that $\forall s \in S, f_Q^\pi(s) = Q(s, \pi(s))$. Q^π and Q^* are fixed points of the two following contracting operators T^π and T^* for the infinite norm:

$$\forall Q \in \mathbb{R}^{S \times A}, \forall (s, a) \in S \times A, \quad T^\pi Q(s, a) = R(s, a) + \gamma \mathbb{E}_{P(.|s,a)}[f_Q^\pi],$$

$$\forall Q \in \mathbb{R}^{S \times A}, \forall (s, a) \in S \times A, \quad T^* Q(s, a) = R(s, a) + \gamma \mathbb{E}_{P(.|s,a)}[f_Q^*].$$

The aim of Dynamic Programming (DP) is, given an MDP M, to find Q^* which is equivalent to minimizing a certain norm of the OBR defined as $T^* Q - Q$:

$$J_{DP}(Q) = \|T^* Q - Q\|,$$

---

[1] This work could be easily extended to measurable state spaces as in [9,12]; we choose the finite case for the ease and clarity of exposition.

where $\|.\|$ is a norm which can be equal to $\|.\|_\infty$ or $\|.\|_{\nu,p}$, $\nu \in \Delta_{S \times A}$ is such that $\forall (s,a) \in S \times A, \nu(s,a) > 0$ and $p \in \mathbb{R}_+^*$. Usual techniques of DP, such as Value Iteration (VI) or Policy Iteration (PI), do not directly minimize the criterion $J_{DP}(Q)$ but uses particular properties of the operators T^π and T^* to obtain Q^*. However, the motivation to minimize the norm of the OBR is clear as:

$$\|Q^* - Q\| \le \frac{C}{1-\gamma}\|T^*Q - Q\|,$$

where $C \in \mathbb{R}^*$ is a constant depending notably on the MDP (for details see the work of [16]). This means that if we are able to control the norm of the OBR, then we have found a function Q close to Q^*, which is the goal of DP.

Batch RL aims at estimating Q^* or finding an optimal policy when the model (the dynamics P and the reward function R) of the MDP M is known only through the RL data set noted D_{RL} which contains N_{RL} sampled transitions of the type (s_i, a_i, r_i, s_i') where $s_i \in S$, $a_i \in A$, $r_i = R(s_i, a_i)$ and $s_i' \sim P(.|s_i, a_i)$: $D_{RL} = (s_i, a_i, r_i, s_i')_{1 \le i \le N_{RL}}$. For the moment no assumption is made on the distribution $\nu_{RL} \in \Delta_{S \times A}$ from which the data are drawn, $(s_i, a_i) \sim \nu_{RL}$. In batch RLED, we suppose that we also have the expert data set D_E which contains N_E expert state-action couples of the the type $(s_j, \pi_E(s_j))$ where $s_j \in S$ and π_E is an expert policy which can be considered optimal or near-optimal: $D_E = (s_j, a_j = \pi_E(s_j))_{1 \le j \le N_E}$. The distribution from which the expert data are drawn is noted $\nu_E \in \Delta_S$, $s_j \sim \nu_E$.

3   A New Algorithm for the RLED Problem

Our problem consists in approximating Q^*. We saw in Sec. 2.1 that minimizing a certain norm of the OBR can lead us to a good approximation of the optimal quality function and of the optimal policy. However, the only available knowledge of the MDP lies in the sets of data D_{RL} and D_E. Thus for the set D_{RL}, we want to find a function $Q \in \mathbb{R}^{S \times A}$ that minimizes the empirical OBR:

$$J_{RL}(Q) = \frac{1}{N_{RL}} \sum_{i=1}^{N_{RL}} |T^*Q(s_i, a_i) - Q(s_i, a_i)|^p \underset{\text{def}}{=} \|T^*Q - Q\|_{D_{RL},p}^p,$$

where $p \ge 1$. For the expert set D_E, we would like to express that the action a_j is optimal, which means that $Q(s_j, a_j)$ is greater than $Q(s_j, a)$ where $a \in A \backslash a_j$. This can be expressed by the following large margin constraints:

$$\forall 1 \le j \le N_E, \quad \max_{a \in A}[Q(s_j, a) + l(s_j, a_j, a)] - Q(s_j, a_j) \le 0,$$

where $l \in \mathbb{R}_+^{S \times A \times A}$ is a margin function such that $\forall 1 \le j \le N_E$, $\forall a \in A \backslash a_j$, $l(s_j, a_j, a) > l(s_j, a_j, a_j)$. A canonical choice of margin function is $\forall 1 \le j \le N_E$, $\forall a \in A \backslash a_j$, $l(s_j, a_j, a_j) = 0, l(s_j, a_j, a) = 1$. This imposes that the function Q we are looking for is greater by a given amount determined by the margin function for the expert actions. If available, one prior knowledge can be used to structure the margin. Those constraints guide the minimization of J_{RL} as they impose a particular structure for the quality function we are looking for. Here, it is important to note that the constraints that guide the minimization of J_{RL} are compatible with this minimization as they are satisfied by the expert policy which is near optimal. Thus, we can think that those constraints help to accelerate and improve the minimization of J_{RL}.

However, notice that it is not the case if we use T^{π} in lieu of T^*, because the policy π can be completely different from the expert policy. This is a problem encountered in the APID algorithm, as they choose a Policy Iteration framework where there are several steps of the minimization of the norm of the Bellman Residual $\|T^{\pi}Q - Q\|$ guided by constraints on expert data (see Sec. 3.1).

As the expert policy might be suboptimal, the constraints can be violated by an optimal policy, that is why we smooth those constraints with slack variables:

$$\forall 1 \leq j \leq N_E, \quad \max_{a \in A}[Q(s_j, a) + l(s_j, a_j, a)] - Q(s_j, a_j) \leq \xi_j,$$

where ξ_j is a positive slack variable that must be the smallest possible. So the idea is to minimize:

$$J_{RL}(Q) + \frac{\lambda}{N_E} \sum_{j=1}^{N_E} \xi_j,$$

subject to $\max_{a \in A}[Q(s_j, a) + l(s_j, a_j, a)] - Q(s_j, a_j) \leq \xi_j, \quad \forall 1 \leq j \leq N_E.$

where λ determines the importance between the RL data and the expert data.

Following [20], as the slack variables are tight and positive, this problem is equivalent to minimizing:

$$J_{RLE}(Q) = J_{RL}(Q) + \lambda J_E(Q),$$

where: $J_E(Q) = \frac{1}{N_E} \sum_{j=1}^{N_E} \max_{a \in A}[Q(s_j, a) + l(s_j, a_j, a)] - Q(s_j, a_j)$. The minimization of the criterion $J_E(Q)$ is known in the literature and used in the LfD paradigm [19,13,17]. The minimization of $J_E(Q)$ can be seen as a score-based multi-class classification algorithm where the states s_j play the role of inputs and the actions a_j play the role of labels.

3.1   Comparaison to the APID Method

The APID method is couched in the API framework [5], which starts with an initial policy π_0. At the $k+1$-th iteration, given a policy π_k, the quality function Q^{π_k} is approximately evaluated by $\hat{Q}_k$. This step is called the approximate policy evaluation step. Then, a new policy π_{k+1} is computed, which is greedy with respect to $\hat{Q}_k$. In the APID algorithm, the policy evaluation step is realized by the following unconstrained minimization problem:

$$\hat{Q}_k = \operatorname*{argmin}_{Q \in \mathbb{R}^{S \times A}} J_{RL}^{\pi}(Q) + \lambda J_E(Q), \tag{1}$$

where: $J_{RL}^{\pi}(Q) = \frac{1}{N_{RL}} \sum_{i=1}^{N_{RL}} |T^{\pi}Q(s_i, a_i) - Q(s_i, a_i)|^p \overset{\text{def}}{=} \|T^{\pi}Q - Q\|_{D_{RL}, p}^p$. The difference between $J_{RL}(Q)$ and $J_{RL}^{\pi}(Q)$ is the use of the operator T^* in lieu of T^{π}. In the APID method, the policy evaluation step is slightly different from a classical API method which consists in the minimization of $J_{RL}^{\pi}(Q)$. This introduces an error in the evaluation step for the first iterations of the APID method, which can potentially slow down the learning process. Moreover, as the APID is an iterative method, the unconstrained problem Eq. (1) has to be resolved several times which can be time expensive. However, this problem is convex and easier to resolve than the minimization of J_{RLE}. In practice, in order to resolve the problem in Eq. (1), an LSTD-like algorithm [12] is used (this is the algorithm implemented in our experiments in order to represent the APID method). This method is by nature parametric and needs the choice of features or the choice of a kernel. Thus, the APID has as advantage the simplicity of minimizing $J_{RL}^{\pi}(Q) + \lambda J_E(Q)$, which is convex, but has also some drawbacks as it is an iterative and parametric method.

Our algorithm, which consists in the minimization of $J_{RLE}(Q) = J_{RL}(Q) + \lambda J_E(Q)$, avoids the APID drawbacks as it can be in practice a non parametric and non iterative method (see Sec. 4) but it is a non-convex criterion. Indeed, let us take a closer look at $J_{RL}(Q) = \|T^*Q - Q\|_{D_{RL}, p}^p$. This criterion is an empirical norm of the OBR, and the minimization of this criterion for solving the batch RL problem is an unused technique at our knowledge. There are several reasons to understand why the OBR minimization (OBRM) is usually not used. The first one is that this criterion is not convex in Q, so a direct minimization could lead us to local minima. The second reason is that this criterion is not differentiable because of the max operator, so we need to use sub-gradient techniques or generalized gradient techniques to minimize it. The third reason is that this technique is not directly inspired by a dynamic programming approach such as PI or VI.

In the next section, we exhibit the bias problem involved by J_{RL}. However, it is possible to construct two criterions $\hat{J}_{RL}$, which is a biased criterion but can be used in some specific conditions, and $\overline{J}_{RL}$ which converges in probability to $\|T^*Q - Q\|_{\nu, 2}^2$ when N_{RL} tends to infinity and ν is the distribution from which the data are drawn. $\overline{J}_{RL}$ is obtained thanks to the use of a distribution embedding in an RKHS. Finally, we show how we overcome the non-convex and non-differentiability difficulties via a boosting technique which allows obtaining non-parametric algorithms.

4   Practical Minimization of J_{RLE}

In this section, we present how the criterion J_{RLE} is minimized in order to obtain a practical and non-parametric algorithm. Here, we choose $p = 2$ and we suppose that the data (s_i, a_i) are drawn identically and independently (i.i.d) from a distribution $\nu \in \Delta_{S \times A}$ such that $\forall (s, a), \nu(s, a) > 0$. The i.i.d assumption is only done here to simplify the analysis. Indeed, this assumption could be relaxed using techniques that handle dependent data [25]. The i.i.d assumption for the data set D_{RL} in a batch RL setting is common in the literature [14,15,12].

First, let us take a closer look to the term $J_{RL}(Q) = \|T^*Q - Q\|^2_{D_{RL},2}$. If we only have the knowledge of the data set D_{RL}, we cannot compute $T^*Q(s_i, a_i)$, but we can compute an unbiased estimate $\hat{T}^*Q(s_i, a_i) = R(s_i, a_i) + \gamma \max_{a \in A} Q(s'_i, a)$. Then, our criterion becomes:

$$\hat{J}_{RL}(Q) = \frac{1}{N_{RL}} \sum_{i=1}^{N_{RL}} |\hat{T}^*Q(s_i, a_i) - Q(s_i, a_i)|^2 \underset{\text{def}}{=} \|\hat{T}^*Q - Q\|^2_{D_{RL},2}.$$

Unfortunately, this is a biased criterion. However, we have the following result:

Theorem 1.

$$\hat{J}_{RL}(Q) \underset{N_{RL} \to \infty}{\to} \|T^*Q - Q\|^2_{\nu,2} + \gamma^2 \sum_{(s,a) \in S \times A} \nu(s,a) \mathbb{E}_{P(.|s,a)}[(f^*_Q)^2 - \mathbb{E}_{P(.|s,a)}[|f^*_Q|]^2].$$

Proof. The proof follows the same line as the one of [2], where T^* replaces T^π. ∎

So, when the number of samples tends to infinity, the criterion $\hat{J}_{RL}(Q)$ tends to $\|T^*Q - Q\|^2_{\nu,2}$, which is what we want to minimize plus a term of variance $\gamma^2 \sum_{(s,a) \in S \times A} \nu(s,a) \mathbb{E}_{P(.|s,a)}[(f^*_Q)^2 - \mathbb{E}_{P(.|s,a)}[|f^*_Q|]^2]$. This term will favor functions which are smooth, but it is not controlled by a factor that we can choose such as in regularization theory. As pointed by [2], we have the same problem in the minimization of the Bellman residual in the PI framework. It is not satisfactory to present a criterion which has a bias even if it can work in some specific conditions such as when the MDP is deterministic (in that case $\hat{J}_{RL}(Q) \underset{N_{RL} \to \infty}{\to} \|T^*Q - Q\|^2_{\nu,2}$) or when the optimal action value function we are looking for is really smooth.

Thus, we also propose a criterion which does not have this bias. Several techniques have been used to get rid off the bias in the minimization of the Bellman Residual (see [2]), but here we are going to use the work developed by [15] where a conditional distribution is embbeded in a Reproducing Kernel Hilbert Space (RKHS), more appropriate for the considered batch setting.

4.1 RKHS Embbedings for MDP

Let us start with some notations relative to RKHS [4]. Let K be a positive definite kernel on a finite set X. The unique Hilbert space H with reproducing kernel K is denoted by H_K. Correspondingly the norm will be denoted by $\|.\|_K$ and the inner product will be denoted by $\langle .,.\rangle_K$.

Now, we can use the notion of distribution embeddings [23,15]. Given any probability distribution $P \in \Delta_X$ and a positive definite kernel K on X, a distribution embbeding of P in H_K is an element $\nu \in H_K$ such that:

$$\forall h \in H_K, \langle \nu, h\rangle_K = \mathbb{E}_P[h].$$

In our application, we want to find a distribution embbeding for the conditional distribution $P(.|s,a)$. Following the work done by [15], given the data set D_{RL}, a positive definite kernel K on $S \times A$, a positive definite kernel L on S, there is a way to estimate the element $\nu_{s,a} \in H_L$ such that $\langle \nu_{s,a}, f \rangle_K = \mathbb{E}_{P(.|s,a)}[f]$, for all $f \in H_L$. The estimation of $\nu_{s,a}$ is noted $\overline{\nu}_{s,a}$ and is such that:

$$\overline{\nu}_{s,a} = \sum_{i=1}^{N_{RL}} \overline{\beta}_i(s,a)L(s_i',.) \in H_L,$$

where $\overline{\beta}_i(s,a) = \sum_{j=1}^{N_{RL}} W_{ij}K((s_j,a_j),(s,a))$, and where $\mathbf{W} = (W_{ij})_{1 \le i,j \le N_{RL}} = (\mathbf{K} + \lambda_K N_{RL}\mathbf{I})^{-1}$ with $\mathbf{K} = (K((s_i,a_i),(s_j,a_j)))_{1 \le i,j \le N_{RL}}$, $\mathbf{I}$ the identity matrix of size N_{RL} and $\lambda_K \in \mathbb{R}_+$. In the case where S is finite, we can choose L to be the canonical dot product and in that case, we have $H_L = \mathbb{R}^S$ and $\forall Q \in S \times \mathbb{A}$, $f_Q^* \in H_L$. However if S is a measurable state space, the choice of L is not canonical.

Thus, if $f_Q^* \in H_L$ (which is the case when S is finite and L is the euclidian dot product) and if we define $\overline{T}^* Q(s_i,a_i) = R(s_i,a_i) + \gamma \sum_{j=1}^{N_{RL}} \overline{\beta}_j(s_i,a_i) \max_{a \in A} Q(s_j',a)$, we have that: $\overline{T}^* Q(s_i,a_i) = R(s_i,a_i) + \langle \overline{\nu}_{s_i,a_i}, f_Q^* \rangle$. So, if we define the following criterion:

$$\overline{J}_{RL}(Q) = \frac{1}{N_{RL}} \sum_{i=1}^{N_{RL}} |\overline{T}^* Q(s_i,a_i) - Q(s_i,a_i)|^2 \underset{\text{def}}{=} \|\overline{T}^* Q - Q\|_{D_{RL},2}^2,$$

we have the following Theorem:

Theorem 2. *Under some smoothness conditions of the MDP described in [15], the strict positivity of the Kernel L and by choosing $\lambda_K \underset{N_{RL} \to \infty}{\to} 0$ and $\lambda_K N_{RL}^3 \underset{N_{RL} \to \infty}{\to} \infty$, we have if $\|f_Q^*\|_L < \infty$:*

$$\overline{J}_{RL}(Q) \underset{N_{RL} \to \infty}{\overset{\nu}{\to}} \|T^* Q - Q\|_{\nu,2}^2.$$

Here, the convergence is in ν-Probability.

Proof. This comes directly from the Cauchy-Schwartz inequality and the Lemma 2.1 in [15].

$$\sup_{(s,a) \in S \times A} \|\nu_{s,a} - \overline{\nu}_{s,a}\|_L \underset{N_{RL} \to \infty}{\overset{\nu}{\to}} 0.$$

It is important to remark that we only need the coefficients of the form $(\overline{\beta}_j(s_i,a_i))_{1 \le i,j \le N_{RL}}$ in order to construct $\overline{J}_{RL}(Q)$. Thus we only need to compute the matrix product $\mathbf{B} = (B_{ij})_{1 \le i,j \le N_{RL}} = \mathbf{WK}$ because $B_{ij} = \sum_{k=1}^{N_{RL}} W_{ik}K((s_k,a_k),(s_j,a_j)) = \overline{\beta}_i(s_j,a_j)$. Finally, we can easily construct two criterions from the data set D_{RL}. One criterion, $\hat{J}_{RL}(Q)$, is naturally biased and the other one, $\overline{J}_{RL}(Q)$, converges in probability towards $\|T^* Q - Q\|_{\nu,2}^2$, with certain conditions of smoothness of the MDP. Those two criterions can take the same form if we rewrite

$\hat{T}^*Q(s_i, a_i) = R(s_i, a_i) + \gamma \sum_{j=1}^{N_{RL}} \hat{\beta}_j(s_i, a_i) \max_{a \in A} Q(s'_j, a)$ with $\hat{\beta}(s_i, a_i)_j = 1$ if $j = i$ and $\hat{\beta}_j(s_i, a_i) = 0$ otherwise. Thus the practical algorithms will consists in minimizing the two following criterions:

$$\hat{J}_{RLE}(Q) = \hat{J}_{RL} + \lambda J_E(Q),$$
$$\overline{J}_{RLE}(Q) = \overline{J}_{RL} + \lambda J_E(Q).$$

In order to realize it, we use a boosting technique.

4.2   Boosting

A boosting method is an interesting optimization technique: it minimizes directly the criterion without the step of choosing features, which is one of the major drawback of several RL methods. As presented by [10], a boosting algorithm is a projected sub-gradient descent [22] of a convex functional in a specific functions space (here $\mathbb{R}^{S \times A}$) which has to be a Hilbert space. The principle is to minimize a convex functional $F \in \mathbb{R}^H$ where H is a Hilbert space: $\min_{h \in H} F(h)$. This technique can be extended to non-smooth and non-convex functionals, yet the functional has to be Lipschitz in order to guarantee that the gradient of the functional exists almost everywhere [8]. For a Lipschitz and non smooth functional, the gradient can be calculated almost everywhere and if not the notion of generalized gradient is used (see [8] for details). To realize this minimization, we need to calculate the gradient $\partial_h F \in H$ and define $K \subset H$, a set of allowable directions (also called the restriction set) where the gradient is projected. Boosting algorithms use a projection step when optimizing over function space because the functions representing the gradient are often computationally difficult to manipulate and do not generalize well to new inputs [10]. In boosting literature, the restriction set corresponds directly to the set of hypotheses generated by a weak learner. The nearest direction k^*, which is the projection of the gradient $\partial_h F$, is defined by:

$$k^* = \underset{k \in K}{\operatorname{argmax}} \frac{\langle k, \partial_h F \rangle}{\|k\|},$$

where $\langle ., . \rangle$ is the inner product associated to the Hilbert space H and $\|.\|$ is the associated canonical norm. Then, the naive algorithm to realize the minimization of F is given by Algo. 1. More sophisticated boosting algorithms and their convergence proofs are presented by [10]. However, the naive approach is sufficient to obtain good results. For our specific problem, $H = \mathbb{R}^{S \times A}$, and $\langle ., . \rangle$ is the canonical dot product. The criterions which have to be minimized are $\overline{J}_{RLE}$ and $\hat{J}_{RLE}$. As those criterions have the same form (the only difference is the value of the coefficients $\hat{\beta}_j$ and $\overline{\beta}_j$), we present the boosting technique only for $\overline{J}_{RLE}$. Moreover, in our experiments, we choose the restriction set K to be weighted classification trees [7] from $\mathbb{R}^{S \times A}$ to $\{-1, 1\}$ where each k has the same norm (as it takes its values in $\{-1, 1\}$). Thus, our algorithm is given by Algo. 2. The output $Q_T = -\sum_{i=1}^{T} \xi_i k_i^*$ is a weighted sum of T classification

---

Algorithm 1. Naive boosting algorithm

Require: $h_0 \in \mathbb{R}^H$, $i = 0$, $T \in \mathbb{N}^*$ (number of iterations) and $(\xi_j)_{\{j \in \mathbb{N}\}}$ a family of learning rates.
1: While $i < T$ do
2: Calculate $\partial_{h_i} F$.
3: Calculate k_i^* associated to $\partial_{h_i} F$ (projection step).
4: $h_{i+1} = h_i - \xi_i k_i^*$
5: $i = i + 1$
6: end While, output h_T

Algorithm 2. Minimization of $\overline{J}_{RLE}$ with boosting

---

Require: $Q_0 \equiv 0$, $i = 0$, $T \in \mathbb{N}^*$ and $(\xi_j)_{\{j \in \mathbb{N}\}}$ a family of learning rates.
1: While $i < T$ do
2: Calculate k_i^* associated to $\partial_{Q_i} \overline{J}_{RLE}$. (projection step)
3: $Q_{i+1} = Q_i - \xi_i k_i^*$, $i = i + 1$
4: end While, output Q_T

---

trees: $\{k_i^*\}_{1 \leq i \leq T}$. Those T trees can be seen as the features of the problem which are automatically found by the boosting algorithm. The only step that we have to clarify is the calculation of k^*. For this particular choice of weak learners, we have the following Theorem that shows us how to compute it:

Theorem 3. *Calculating $k^* = \mathrm{argmax}_{k \in K} \langle k, \partial_Q \overline{J}_{RLE} \rangle$, where $Q \in \mathbb{R}^{S \times A}$, corresponds to training a weighted classification tree with the following training data set:*

$$D_C = \Big(((s_j, a_j), w_j, -1) \cup ((s_j, a_j^*), w_j, 1) \Big)_{\{1 \leq j \leq N_E\}}$$
$$\cup \, ((s_i, a_i), w_i, o_i)_{\{1 \leq i \leq N_{RL}\}} \cup ((s_p', a_p'), w_p, -o_p)_{\{1 \leq p \leq N_{RL}\}},$$

We recall that an element of a training data set of a weighted classification tree as the following form: (x, w, o) where x is the input, w the weight and o is the output. With $(s_j, a_j) \in D_E$, (s_i, a_i) corresponds to the first two elements in a sampled transition $(s_i, a_i, r_i, s_i') \in D_{RL}$, s_p' corresponds to the fourth element in a sampled transition $(s_p, a_p, r_p, s_p') \in D_{RL}$ and:

$$a_j^* = \mathrm{argmax}_{a \in A}[Q(s_j, a_j) + l(s_j, \pi_E(s_j), a)], \quad a_p' = \mathrm{argmax}_{a \in A} Q(s_p', a),$$

$$o_i = \mathrm{sgn}(Q(s_i, a_i) - \overline{T}^* Q(s_i, a_i)), \quad o_p = \mathrm{sgn}(\sum_{i=1}^{N_{RL}} \Big(Q(s_i, a_i) - \overline{T}^* Q(s_i, a_i) \Big) \beta_p(s_i, a_i)),$$

$$w_i = \frac{2}{N_P}|Q(s_i, a_i) - \overline{T}^* Q(s_i, a_i)|, \quad w_j = \frac{\lambda}{N_E},$$

$$w_p = \frac{2\gamma}{N_P}|\sum_{i=1}^{N_{RL}} \Big(Q(s_i, a_i) - \overline{T}^* Q(s_i, a_i) \Big) \beta_p(s_i, a_i)|.$$

Proof. Calculating $\partial_Q \overline{J}_{RLE}$ for a given $Q \in \mathbb{R}^{S \times A}$ is done as follows:

$$\partial_Q \max_{a \in A}\{Q(s_j, a) + l(s_j, \pi_E(s_j), a)\} = \delta_{(s_j, a_j^*)}, \quad \partial_Q Q(s_j, \pi_E(s_j)) = \delta_{(s_j, \pi_E(s_j))},$$

$$\partial_Q (\overline{T}^* Q(s_i, a_i) - Q(s_i, a_i))^2 = 2(Q(s_i, a_i) - \overline{T}^* Q(s_i, a_i))(\delta_{(s_i, a_i)} - \gamma \sum_{p=1}^{N_{RL}} \overline{\beta}_p(s_i, a_i)\delta_{(s_p', a_p')}),$$

$$\partial_Q \overline{J}_{RLE} = \frac{\sum_{1 \le j \le N_{RL}} \partial_Q (\overline{T}^* Q(s_i, a_i) - Q(s_i, a_i))^2}{N_{RL}} + \frac{\lambda \sum_{1 \le j \le N_E} \delta_{(s_j, a_j^*)} - \delta_{(s_j, \pi_E(s_j))}}{N_E}.$$

where $a_j^* = \mathrm{argmax}_{a \in A}[Q(s_j, a_j) + l(s_j, \pi_E(s_j), a)]$ and $a_p' = \mathrm{argmax}_{a \in A} Q(s_p', a)$. Obtaining k^* associated to $\partial_Q \overline{J}_{RLE}$ when K is the set of classification trees from $\mathbb{R}^{S \times A}$ to $\{-1, 1\}$ is done as follows. First, we calculate $\langle k, \partial_Q \overline{J}_{RLE} \rangle$:

$$\langle k, \partial_Q \overline{J}_{RLE} \rangle = \frac{2}{N_P} \sum_{i=1}^{N_{RL}} (Q(s_i, a_i) - \overline{T}^* Q(s_i, a_i))(k(s_i, a_i) - \gamma \sum_{p=1}^{N_{RL}} \overline{\beta}_p(s_i, a_i)k(s_p', a_p'))$$

$$+ \frac{\lambda}{N_E} \sum_{j=1}^{N_E} k(s_j, a_j^*) - k(s_j, \pi_E(s_j)).$$

To maximize $\langle k, \partial_Q \overline{J}_{RLE} \rangle$, we have to find a classifier k such that $k(s_j, a_j^*) = 1$, $k(s_i, a_i) = o_i = \mathrm{sgn}(Q(s_i, a_i) - \overline{T}^* Q(s_i, a_i))$, $k(s_j, \pi_E(s_j)) = -1$ and $k(s_p', a_p') = -o_p = \mathrm{sgn}(\sum_{i=1}^{N_{RL}} (Q(s_i, a_i) - \overline{T}^* Q(s_i, a_i))\beta_p(s_i, a_i))$ for a maximum of inputs while taking into consideration the weight factors for each input. The weight factors are the following $w_i = \frac{2}{N_P}|Q(s_i, a_i) - \overline{T}^* Q(s_i, a_i)|$, $w_p = \frac{2\gamma}{N_P}|\sum_{i=1}^{N_{RL}} (Q(s_i, a_i) - \overline{T}^* Q(s_i, a_i))\beta_p(s_i, a_i)|$, and $w_j = \frac{\lambda}{N_E}$ Thus, in order to obtain k^*, we train a classification tree with the following training set:

$$D_C = (((s_j, \pi_E(s_j)), w_j, -1) \cup ((s_j, a_j^*), w_j, 1))_{\{1 \le j \le N_E\}}$$
$$\cup ((s_i, a_i), w_i, o_i)_{\{1 \le i \le N_{RL}\}} \cup ((s_p', a_p'), w_p, -o_p)_{\{1 \le p \le N_{RL}\}}.$$

5   Experiments

In this section, we compare our algorithms (boosted minimization of $\hat{J}_{RLE}$ noted Residual1 and boosted minimization of $\overline{J}_{RLE}$ noted Residual2) to APID, LSPI and a classification algorithm noted Classif which is the boosted minimization of J_E as done by [19]. The comparison is performed on two different tasks. The first task is a generic task, called the Garnet experiment, where the algorithms are tested on several randomly constructed finite MDPs where there is a specific topology that simulates the ones encountered on real continuous MDP. The second experiment is realized on an LfD benchmark called the Highway problem. As the MDP are finite in our experiment, we choose a tabular representation for the parametric algorithms (LSPI, APID). For the boosted algorithms (Residual1, Residual2 and Classif), the features are automatically chosen by the algorithm so there is no features choice but we fix the number of weak learners, which are classification trees, to $T = 30$. The regularization parameter λ is fixed at 1 (the expert data and the non expert data are supposed to be of an equal importance), the learning rates are $\xi_i = \frac{1}{i+1}, i \in \mathbb{N}$ and the discount factor is $\gamma = 0.99$ in all of our experiments. Finally, the margin function is $\forall 1 \leq j \leq N_E, \forall a \in A \backslash a_j, l(s_j, a_j, a_j) = 0, l(s_j, a_j, a) = 1$, the Kernels K ans L are the canonical dot products in $\mathbb{R}^{S \times A}$ and in $\mathbb{R}^S$ and $\lambda_K = 10^{-5}$.

5.1   The Garnet Experiment

This experiment focuses on stationary Garnet problems, which are a class of randomly constructed finite MDPs representative of the kind of finite MDPs that might be encountered in practice [3]. A stationary Garnet problem is characterized by 3 parameters: $Garnet(N_S, N_A, N_B)$. The parameters N_S and N_A are the number of states and actions respectively, and N_B is a branching factor specifying the number of next states for each state-action pair. In this experiment, we choose a particular type of Garnets which presents a topological structure relative to real dynamical systems. Those systems are generally multi-dimensional state spaces MDPs where an action leads to different next states close to each other. The fact that an action leads to close next states can model the noise in a real system for instance. Thus, problems such as the highway simulator [13], the mountain car or the inverted pendulum (possibly discretized) are particular cases of this type of Garnets. For those particular Garnets, the state space is composed of d dimensions ($d = 3$ in this particular experiment) and each dimension i has a finite number of elements x_i ($x_i = 5$). So, a state $s = [s^1, s^2, .., s^i, .., s^d]$ is a tuple where each component s^i can take a finite value between 1 and x_i. In addition, the distance between two states s, s' is $\|s - s'\|^2 = \sum_{i=1}^{i=d}(s^i - s'^i)^2$. Thus, we obtain MDPs with a possible state space size of $\prod_{i=1}^d x_i$. The number of actions is $N_A = 5$. For each state action couple (s, a), we choose randomly N_B next states ($N_B = 5$) via a Gaussian distribution of d dimensions centered in s where the covariance matrix is the identity matrix of size d, I_d, multiply by a term σ (here $\sigma = 1$). σ allows handling the smoothness of the MDP: if σ is small the next states s' are close to s and if σ is large, the next states s' can be very far form each other and also from s. The probability of going to each next state s' is generated by partitioning the unit interval at $N_B - 1$ cut points selected randomly. We construct a sparse reward R by choosing $\frac{N_S}{10}$ states (uniform random choice without replacement) where $R(s, a) = 1$, elsewhere $R(s, a) = 0$. For each Garnet problem, it is possible to compute an expert policy $\pi_E = \pi^*$ which is optimal and the expert value function $V^{\pi_E} = f_{Q^*}^*$ via the policy iteration algorithm (as it is a finite MDP where the reward and the dynamics are perfectly known). In addition, we recall that the value function for a policy π is $V_R^\pi = f_{Q^\pi}^*$.

In this experiment, we construct 100 Garnets $\{G_p\}_{1 \leq p \leq 100}$ as explained before. For each Garnet G_p, we build 10 data sets $\{D_E^{p,q}\}_{1 \leq q \leq 10}$ composed of L_E trajectories of H_E expert demonstrations $(s_i, \pi_E(s_i))$ and 10 data sets $\{D_{RL}^{p,q}\}_{1 \leq q \leq 10}$ of L_R trajectories of H_R sampled transitions of the random policy (for each state, the action is uniformly chosen over the set of actions)

(s_i, a_i, s_i', r_i). Each trajectory begins from a state chosen uniformly over the state space, this uniform distribution is noted ρ. Then the RLED algorithms (APID, Residual1 and Residual2) are fed with the data sets $D_E^{p,q}$ and $D_{RL}^{p,q}$, LSPI is fed with $D_{RL}^{p,q}$ and the Classif algorithm is fed with $D_E^{p,q}$. Each algorithm outputs a function $Q_A^{p,q} \in \mathbb{R}^{S \times A}$ and the policy associated to $Q_A^{p,q}$ is $\pi_A^{p,q}(s) = \mathrm{argmax}_{a \in A} Q_A^{p,q}(s, a)$. In order to quantify the performance of a given algorithm, we calculate the criterion $T_A^{p,q} \frac{\mathbb{E}_\rho[V^{\pi_E} - V^{\pi_A^{p,q}}]}{\mathbb{E}_\rho[V^{\pi_E}]}$, where $V^{\pi_A^{p,q}}$ is calculated via the policy evaluation algorithm. The mean performance criterion T_A is $\frac{1}{1000} \sum_{p=1}^{100} \sum_{q=1}^{10} T_A^{p,q}$. We also calculate, for each algorithm, the variance criterion $\mathrm{std}_A^p = \left(\frac{1}{10} \sum_{q=1}^{10} (T_A^{p,q} - \frac{1}{10} \sum_{q=1}^{10} T_A^{p,q})^2 \right)^{\frac{1}{2}}$ and the resulting mean variance criterion is $\mathrm{std}_A = \frac{1}{100} \sum_1^{100} \mathrm{std}_A^p$. In Fig. 1(a), we plot the performance versus the length of the expert trajectories when $L_R = 300$, $H_R = 5$, $L_E = 5$ in order to see how the RLED algorithm manage to leverage the expert data. In Fig. 1(a), we see that the three RLED algorithms have quite the same performance. However, contrary to APID where the features are given by the user (the tabular representation has a size of 725 features), Residual1 and Residual2 manages to learn automatically 30 trees (which can be seen as features) in order to obtain the same performance as APID. The RLED algorithms outperforms the LSPI and Classif algorithm which was expected. We observe the same results in the experiments leaded by [12]. When the number of expert data grows, the RLED algorithms performance is getting better which shows that they are able to leverage those expert data. Besides, we observe that Residual1 and Residual2 have the same performance which shows that using an RKHS embedding is not critical in that case. In Fig. 1(b), we plot the performance versus the number of random trajectories when $H_R = 5$, $H_E = 50$, $L_E = 5$ in order to see the effects of adding non expert data on the RLED algorithms performance. In Fig. 1(b), we can observe that there is a gap between RLED algorithms and the Classif algorithm, and that LSPI does not manage to obtain the same results even when the number of data gets bigger. The gap between RLED and the Classif algorithm gets bigger as the number of RL data is growing which shows that RLED

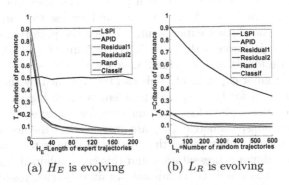

(a) H_E is evolving   (b) L_R is evolving

Fig. 1. Garnet Experiment

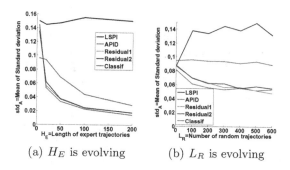

(a) H_E is evolving   (b) L_R is evolving

Fig. 2. Garnet Experiment

methods are able to use those data to improve their performance independently of the optimization technique and the parametrization used. In Fig. 2(a) and Fig. 2(b), we plot the mean variance.

5.2   RLED for the Highway

The Highway problem is used as benchmark in the LfD literature [1,24,13]. In this problem, the goal is to drive a car on a busy three-lane highway with randomly generated traffic. The car can move left and right, accelerate, decelerate and keep a constant speed (5 actions). The expert optimizes a handcrafted reward R which favorises speed, punishes off-road, punishes even more collisions and is neutral otherwise. This reward is quite sparse. We have 729 states corresponding to: 9 horizontal positions for the car, 3 horizontal and 9 vertical positions for the closest traffic car and 3 possible speeds. We compute π_E via the policy iteration algorithm as the dynamics P and the reward R of the car driving simulator are known (but unknown for the algorithm user). Here, we build 100 data sets $\{D_E^q\}_{1 \leq q \leq 100}$ composed of L_E trajectories of H_E expert demonstrations $(s_i, \pi_E(s_i))$ and 100 data sets $\{D_{RL}^q\}_{1 \leq q \leq 100}$ of L_R trajectories of H_R sampled transitions of the random policy. Each trajectory begins from a state chosen uniformly over the state space and we use the same criterion of performance as in the Garnet experiment. We plot the performance versus the length of expert trajectories with $L_E = 5$, $H_R = 5$ and $L_R = 50$. In Fig. 3(a), we observe that Residual1 and Residual2 have clearly better performances than APID in that particular experiment. This can be explained by the fact that the tabular representation for the Highway problem is much bigger than in the in the Garnets experiments (3645 features) and only few features are important. As our algorithms are non-parametric, they are able to select the necessary features in order to find a good policy. Here, the number of data is too small compared to the size of the tabular representation and this can explained why parametric algorithms such as LSPI and APID can not obtain satisfying results. Thus, this observation makes us believe that our algorithms are suited to scale up.

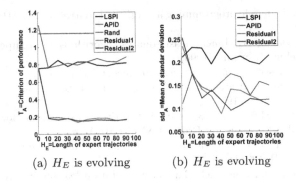

(a) H_E is evolving   (b) H_E is evolving

Fig. 3. Highway Experiment

6   Conclusion

In this paper, we present two new algorithms that minimize the empirical norm of the OBR guided by constraints define by expert demonstrations. These algorithms tackle the problem of RLED. Our algorithms are original in the sense that are not derived form the classical Approximate Dynamic Programming framework and manage to alleviate the difficulties inherent to the minimization of the OBR which are the non-convexity, the non-differentiability and the bias. Those drawbacks are overcome by the use of a distribution embbeding in an RKHS and a boosting technique which allows us to obtain non-parametric algorithms, avoiding the choice of features. We show also, in our experiments, that our algorithms perform well compared to the only state of the art algorithm APID, which is parametric. Finally, interesting perspective are to improve our algorithms by using better boosting algorithms, test our algorithms on large scale problems and to have a better understanding of our algorithm by a theoretical analysis.

Acknowledgements. The research leading to these results has received partial funding from the European Union Seventh Framework Programme (FP7/2007-2013) under grant agreement nŕ270780.

References

1. Abbeel, P., Ng, A.: Apprenticeship learning via inverse reinforcement learning. In: Proc. of ICML (2004)
2. Antos, A., Szepesvári, C., Munos, R.: Learning near-optimal policies with bellman-residual minimization based fitted policy iteration and a single sample path. Machine Learning (2008)
3. Archibald, T., McKinnon, K., Thomas, L.: On the generation of markov decision processes. Journal of the Operational Research Society (1995)
4. Aronszajn, N.: Theory of reproducing kernels. Transactions of the American Mathematical Society (1950)

5. Bertsekas, D.: Dynamic programming and optimal control, vol. 1. Athena Scientific, Belmont (1995)
6. Bradtke, S., Barto, A.: Linear least-squares algorithms for temporal difference learning. Machine Learning (1996)
7. Breiman, L.: Classification and regression trees. CRC Press (1993)
8. Clarke, F.: Generalized gradients and applications. Transactions of the American Mathematical Society (1975)
9. Farahmand, A., Munos, R., Szepesvári, C.: Error propagation for approximate policy and value iteration. In: Proc. of NIPS (2010)
10. Grubb, A., Bagnell, J.: Generalized boosting algorithms for convex optimization. In: Proc. of ICML (2011)
11. Judah, K., Fern, A., Dietterich, T.: Active imitation learning via reduction to iid active learning. In: Proc. of UAI (2012)
12. Kim, B., Farahmand, A., Pineau, J., Precup, D.: Learning from limited demonstrations. In: Proc. of NIPS (2013)
13. Klein, E., Geist, M., Piot, B., Pietquin, O.: Inverse reinforcement learning through structured classification. In: Proc. of NIPS (2012)
14. Lagoudakis, M., Parr, R.: Least-squares policy iteration. Journal of Machine Learning Research (2003)
15. Lever, G., Baldassarre, L., Gretton, A., Pontil, M., Grünewälder, S.: Modelling transition dynamics in mdps with rkhs embeddings. In: Proc. of ICML (2012)
16. Munos, R.: Performance bounds in l_p-norm for approximate value iteration. SIAM Journal on Control and Optimization (2007)
17. Piot, B., Geist, M., Pietquin, O.: Learning from demonstrations: Is it worth estimating a reward function? In: Blockeel, H., Kersting, K., Nijssen, S., Železný, F. (eds.) ECML PKDD 2013, Part I. LNCS, vol. 8188, pp. 17–32. Springer, Heidelberg (2013)
18. Puterman, M.: Markov decision processes: Discrete stochastic dynamic programming. John Wiley & Sons (1994)
19. Ratliff, N., Bagnell, J., Srinivasa, S.: Imitation learning for locomotion and manipulation. In: Proc. of IEEE-RAS International Conference on Humanoid Robots (2007)
20. Ratliff, N., Bagnell, J., Zinkevich, M.: Maximum margin planning. In: Proc. of ICML (2006)
21. Ross, S., Gordon, G., Bagnell, J.: A reduction of imitation learning and structured prediction to no-regret online learning. In: Proc. of AISTATS (2011)
22. Shor, N., Kiwiel, K., Ruszcaynski, A.: Minimization methods for non-differentiable functions. Springer (1985)
23. Sriperumbudur, B., Gretton, A., Fukumizu, K., Schölkopf, B., Lanckriet, G.: Hilbert space embeddings and metrics on probability measures. The Journal of Machine Learning Research (2010)
24. Syed, U., Bowling, M., Schapire, R.: Apprenticeship learning using linear programming. In: Proc. of ICML (2008)
25. Yu, B.: Rates of convergence for empirical processes of stationary mixing sequences. The Annals of Probability (1994)

Semi-supervised Learning
Using an Unsupervised Atlas[*]

Nikolaos Pitelis, Chris Russell, and Lourdes Agapito

University College London, London, United Kingdom
{n.pitelis,c.russell,l.agapito}@cs.ucl.ac.uk

Abstract. In many machine learning problems, high-dimensional datasets often lie on or near manifolds of locally low-rank. This knowledge can be exploited to avoid the "curse of dimensionality" when learning a classifier. Explicit manifold learning formulations such as LLE are rarely used for this purpose, and instead classifiers may make use of methods such as local co-ordinate coding or auto-encoders to implicitly characterise the manifold.

We propose novel manifold-based kernels for semi-supervised and supervised learning. We show how smooth classifiers can be learnt from existing descriptions of manifolds that characterise the manifold as a set of piecewise affine charts, or an atlas. We experimentally validate the importance of this smoothness vs. the more natural piecewise smooth classifiers, and we show a significant improvement over competing methods on standard datasets. In the semi-supervised learning setting our experiments show how using unlabelled data to learn the detailed shape of the underlying manifold substantially improves the accuracy of a classifier trained on limited labelled data.

1 Introduction

A fundamental challenge of machine learning lies in finding embeddings in high dimensional spaces that capture meaningful measures of distance. Bellman [3] coined the term curse of dimensionality to describe the problems that arise as the volume of the space grows exponentially with the number of dimensions and this in turn necessitates an exponentially larger number of observations to cover the space. However, in most applications, data is not uniformly distributed over the whole space, but instead lies on a locally low-dimensional topological structure. This key geometric intuition drives the use of manifolds in machine learning. By finding a compact representation which preserves the relevant topological structure of the data, manifold learning techniques avoid many of the statistical and computational difficulties that arise from high-dimensionality and provide meaningful low-dimensional representations.

In this work, we primarily target semi-supervised learning. We show how unsupervised knowledge of the data manifold can be exploited to learn Support

---

[*] This research was funded by the European Research Council under the ERC Starting Grant agreement 204871-HUMANIS.

T. Calders et al. (Eds.): ECML PKDD 2014, Part II, LNCS 8725, pp. 565–580, 2014.
© Springer-Verlag Berlin Heidelberg 2014

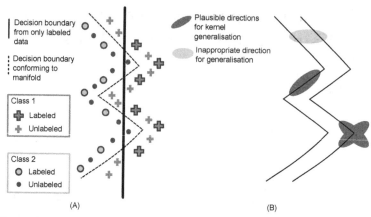

Fig. 1. Desirable properties of a learning algorithm, with respect to a manifold. *(A)* Knowledge of the underlying manifold structure of data can improve classification accuracy. Here unlabelled data can be used to discover the underlying shape of the manifold improving classification accuracy. *(B)* Given a manifold (**black**) an ideal classifier should generalise strongly in directions tangent to the manifold (**blue**) and generalise poorly with respect to directions orthogonal to the manifold (**pink, top**).

Vector Machine (SVM) kernels [24] that work in a set of low-dimensional charts associated with the manifold, avoiding the curse of dimensionality and exhibiting good generalisation to unseen data. Our formulation of manifold kernels is based on the mathematical definition of a manifold as an atlas [14]. Although our implementation makes use of the recent manifold learning technique of [18], it does not rely on the particular atlas using this method. In principle, given a soft-cost function for associating points with charts, it can be applied to any atlas either known a priori or discovered using a manifold learning technique that characterises the manifold as a set of parameterised charts that can be used to back-project points, such as [19].

Although manifold learning has shown much promise in finding embeddings that capture the intrinsic local low dimensionality of data, in practice the majority of such approaches have difficulty with the presence of noise and are unable to characterise closed manifolds such as the surface of a ball. [18] showed how any manifold, either closed or otherwise, could be approximated by an atlas of piecewise affine charts, and experimentally demonstrated their method's robustness to noise. Unfortunately, a good approximation of a smooth manifold as a piecewise affine manifold may require the use of a large number of charts (cf. figure 2). In addition, as a path on the manifold discontinuously jumps from the co-ordinate system of one chart to another, the use of many charts limits the generality of classifiers that can be learned from raw chart co-ordinates and encourages over-fitting.

In response to these difficulties, we present a new class of chart-based Mercer kernels suitable for use with SVMs that smoothly vary in the transition from one

Fig. 2. Approximations of a closed manifold of varying coarseness found using [18] with different minimum description length priors. The underlying manifold is shown in red, and local affine charts approximating the manifold are in green. The coloured dots shown are points sampled with Gaussian noise from the manifold, and their colour indicates which chart they belong to. As the approximation is refined the local affine approximations come closer to tangent planes of the manifold. However, more charts are required for these tighter approximations, and as such, classifiers trained directly on the raw charts that approximate the manifold well (e.g. rightmost) will have poor generalisation. To avoid this trade-off between generalisation, and a good local characterisation of the manifold, section 4 proposes kernels which smoothly vary along a path transitioning from one chart to another.

chart to another[1]. We experimentally verify our formulation and show that it outperforms a variety of existing methods including standard SVMs using: Linear and RBF kernels; standard manifold unwrapping followed by nearest neighbour (NN) and SVM classification using techniques such as [9, 18, 21, 30]; various forms of local co-ordinate coding (LCC) [13, 25, 32]; multi-class kernel-based classifiers [4, 8, 23]; and RBF kernels on the raw chart co-ordinates. We also present asymptotic speed-ups for our kernel computation, and show how a sparse approximation of it can be typically calculated in $O(n\sqrt{n})$ rather than the more usual $O(n^2)$ associated with Mercer kernels.

In using manifold learning as a preprocessing step before classification, we are conforming to the three tenets of manifold learning set out in [19]. Namely:

1. **The semi-supervised learning hypothesis:** The distribution of unlabelled data is informative and should be used to guide supervised classifiers.
2. **The unsupervised manifold learning hypothesis:** High-dimensional datasets often lie near locally low-rank manifolds.
3. **The manifold learning hypothesis for classification:** Data from different classes typically lies in different regions of the manifold and are often separated from one another by regions containing few samples.

Taken together these tenets give us an intuitive picture of supervised learning shown in figure 1. Note that the *strong generalisation* of a classifier in a particular direction, simply means that we expect the classifier response to vary slowly as we move in that direction, while *weak generalisation* refers to the fact that the classifier response may fluctuate quickly in that direction. As an SVM trained classifier is simply a weighted sum of kernel responses, such generalisation in a classifier can be encouraged by making the kernel responses behave in this manner.

---

[1] See [16] for an extensive discussion of the relationship between smoothness and the generalisation of classifiers.

As a complete method ours is a two-step approach:

1. **Unsupervised learning of the underlying manifold:** We approximate the manifold of data on the original space by fitting an atlas of low-dimensional overlapping affine charts.
2. **Supervised training of an** SVM: We propose a new family of Mercer Kernels for SVM-based supervised learning that make use of a soft assignment of datapoints to the underlying low-dimensional affine charts to generating the kernels.

Our contribution. As with earlier approaches that fuse manifold learning with supervised classification [2, 22], our two-stage approach has a natural application in semi-supervised learning. Unlabelled data can be used to generate a more detailed description of the manifold, which can then improve the trained classifier (see figure 1). In the experimental section, we provide an extensive evaluation that shows how these unsupervised manifolds can be used to substantially improve the generalisation of classifiers trained from limited data, where we outperform three other competing approaches: Eigenfunction [10], PVM [28], and AnchorGraphReg [15].

A further contribution of our work lies in the transition from learning on a single chart, found with a standard method like LLE, to learning on multiple charts. Most manifolds, such as the surface of a ball, cannot be expressed as a single chart without either folding or tearing the manifold. Learning kernels on manifolds that cannot be expressed as a single chart is currently a topic of interest. For instance [11] extended kernel-based algorithms to the Riemannian manifold of Symmetric Positive Definite (SPD) matrices. However, while they restricted both the type of manifold (SPD matrices), and the types of kernel considered, our work shows how any kernel defined over a local Euclidean space can be transformed into a kernel over any atlas.

2   Prior Work

While preprocessing a dataset with explicit manifold learning techniques such as [21, 29, 30], that explicitly find a single global mapping of the data lying in a high dimensional $\mathbb{R}^D$ to a lower dimensional $\mathbb{R}^d$, is an obvious way of avoiding the curse of dimensionality, with the exception of [30], such approaches have seen little use in practice. As argued by [18], this may well be because finding a single global mapping by aligning patches that capture local information, is an unnecessarily hard problem that should be avoided wherever possible. Such mappings are unable to capture the intrinsic structure of closed manifolds such as the surface of a ball, and as such methods typically try to preserve various metric properties of the local neighbourhood, they are vulnerable to noise, and a misestimation of the local neighbourhood can propagate throughout the manifold leading to degenerate solutions.

As an alternative to unwrapping a manifold, there has been much interest in local co-ordinate systems to characterise low dimensional subspaces. [19] made use of a variant of auto-encoders to characterise a manifold as a set of charts that were then fine tuned to improve classification accuracy. Local co-ordinate coding [27] and the quadratic variant local tangent based coding [26] approximate non-linear functions by interpolating between anchor points assumed to lie on a low-dimensional manifold.

The works [13] and [32] learnt a linear SVM over a set of full rank linear co-ordinates that smoothly vary from one cluster centre to another. While inspired by local coordinate coding, neither [13] nor [32] make the same manifold assumptions. Instead, they explicitly make use of a weighted concatenation of coordinate systems each of which spans the entire space, rather than focusing on local low-dimensional subspaces as in manifold learning.

Our work differs from previous approaches that fuse manifold learning with SVMs [2, 17, 22] both in the types of manifold that can be expressed –the previous approaches are based on Laplacian eigenmaps [1] that have difficulty with closed manifolds– and in the form taken. These previous methods alter their regularisation to penalise changes in classification response on the manifold, while we reshape kernels to generalise more in the direction of the manifold. As such our different approaches can be seen as complimentary descriptions of manifold constraints.

3   Learning a Manifold as an Atlas

The recent work of [18] formulated manifold learning as a problem of finding an atlas $\mathcal{A}$, defined as a set of overlapping charts $\mathcal{A} = \{c_1, c_2, \ldots c_n\}$, over points $\mathcal{X}$, such that each chart corresponds to an affine subspace of the original space $\mathbb{R}^D$ that accurately describe the local structure. This parametrization of the local transforms as affine spaces allows the efficient use of PCA to find local embeddings, without restricting the overall expressiveness of the atlas. Manifold learning is then formulated as a hybrid continuous/discrete optimisation that simultaneously estimates the affine mappings of charts and solving for a discrete labelling problem, that governs the assignment of points to charts. This objective takes the form of the minimisation of the cost:

$$C(\mathbf{z}) = \sum_{x \in \mathcal{X}} \left(\sum_{i \in \mathbf{z}_x} E_i(x) \right) + \lambda \mathrm{MDL}(\mathbf{z}), \qquad (1)$$

Where $\mathbf{z}_x$ is the set of charts associated with point x, and E_i is defined as in (4). Both subproblems, assigning points to charts and choosing the affine mappings, minimise the same cost –the reconstruction error associated with mapping points from their location in a chart back into the embedding space– subject to the spatial constraint that every point must belong to the *interior of one chart* – that is that each point and all its neighbours in a k-NNgraph should belong to the same chart[2]. Sparse solutions are encouraged by adding a *minimum description length* (MDL) prior[12] term to the energy that penalises the total number of

---

[2] Note that some points belong to more than one chart.

active charts used in an assignment. In practice, Atlas[18] is initialised by an excess of chart proposals in the form of random affine subspaces and alternates between assigning points to charts using the graph-cut [6] based optimisation of [20] and refitting the chart subspaces with PCA. Figure 2 illustrates the approximation of a closed manifold with an Atlas of locally affine subspaces using different MDL priors.

This manifold learning technique offers a set of attractive properties that we take advantage of in our chart-based approach to learning with SVMs. First, since the set of charts that characterise the atlas overlap, points may belong to more than one chart. Therefore, overlapping charts must explain some of the same data in the areas of overlap that connect neighbouring subspaces which results in implicit smoothness in the transition from one subspace to another. Furthermore, this method allows us to learn closed manifolds since it finds charts corresponding to affine subspaces on the original space $\mathbb{R}^D$ and does not require unwrapping into a lower dimensional space. In addition, this method is intrinsically adaptive in that the size of the region assigned to each chart is selected automatically in response to the amount of noise, the curvature of the manifold, and the sparsity of the data.

3.1   Formulation

More formally, each chart c_i contains a subset of points $X_i \subseteq \mathcal{X}$. We use $\mathbf{Z} = \{\mathbf{z}_1, \dots, \mathbf{z}_X\}$ to describe the labelling, where $\mathbf{z}_x$ refers to the assignment of charts to point x (the set of charts that point x belongs to).

We define the d-dimensional affine subspace associated with each chart c_i in terms of its mean μ_i, and an orthonormal matrix $\mathbf{C}_i$ which describes its principal directions of variance. Using x to refer to a datapoint in a feature space $\mathbb{R}^D$, we use $P_i^{\perp}(x) : \mathbb{R}^D \to \mathbb{R}^d$ to refer to a projection from the original feature space into a low rank subspace defined by chart c_i of the form:

$$P_i^{\perp}(x) = \mathbf{C}_i(x - \mu_i), \tag{2}$$

where μ_i is an offset corresponding to the mean of a subset of points used to define chart c_i, and $\mathbf{C}_i$ is the orthonormal matrix composed of the top d eigenvectors of the covariance matrix of the points X_i that belong to the chart, that projects from the embedding space into chart c_i.

We refer to the back-projection of point x into a low rank subspace of the original space as $P_i(x) : \mathbb{R}^d \to \mathbb{R}^D$

$$P_i(x) = \mathbf{C}_i^T P_i^{\perp}(x) + \mu_i, \tag{3}$$

and define the reconstruction error for point x belonging to chart c_i as the squared distance between a point and the back-projection of the closest vector on the chart c_i

$$E_i(x) = ||x - P_i(x)||_2^2. \tag{4}$$

4   Chart-Based Kernels

4.1   Definition

We associate with each chart c_i a unique Mercer kernel K^i defined over the projected space $\mathbb{R}^d$, and we define each element $K_{x,y}$ of the square kernel matrix K as:

$$K_{x,y} = \sum_{c_i \in \mathcal{A}} \exp\left(-\frac{E_i(x) + E_i(y)}{\gamma^2}\right) K_{x,y}^i, \tag{5}$$

where $E_i(x)$ is defined as in (4). This kernel can be understood as a natural softening of the obvious hard-assignment kernel H

$$H_{x,y} = \sum_{c_i \in \mathcal{A}} \Delta(x \in c_i)\Delta(y \in c_i)K_{x,y}^i, \tag{6}$$

where $\Delta(\cdot)$ is the indicator function that takes value 1 if $\cdot$ is true and 0 otherwise. This says that the inner product between two points is the same as a standard kernel defined over chart c_i if both points belong to c_i, and 0 otherwise.

In practice we consider two forms of local kernels K^i. Local linear kernels of the form

$$K_{x,y}^i = P_i^{\perp}(x) \cdot P_i^{\perp}(y) \tag{7}$$

and local Radial Basis Function (RBF) kernels of the form

$$K_{x,y}^i = \exp\left(-\frac{\|P_i^{\perp}(x) - P_i^{\perp}(y)\|_2^2}{\sigma^2}\right). \tag{8}$$

In the experimental section we compare against the hard-assignment kernel H and show the importance of our softening of the kernel response.

4.2   All such Kernels are Mercer

The proof follows directly by construction. We make use of two equivalent definitions of Mercer kernels. Namely: a kernel matrix is Mercer if and only if *(i)* it can be defined as a matrix of inner products over a Hilbert space; and equivalently a kernel is Mercer if and only if *(ii)* it is a positive semi-definite matrix.

We initially consider one of the kernels K^i. It follows from *(i)* that there must be some mapping $\phi_i(\cdot)$ from $\mathbb{R}^d$ to a Hilbert space such that

$$K_{x,y}^i = \langle \phi_i(x), \phi_i(y) \rangle. \tag{9}$$

We define

$$w_x^i = \exp\left(-\frac{\|x - P_i(x)\|_2^2}{\gamma^2}\right) \tag{10}$$

and linearly rescale the elements of the Hilbert space $\phi(x)$, by their weights w_x^i to induce a new kernel matrix $\bar{K}^i$

$$\bar{K}_{x,y}^i = \langle w_x^i \phi_i(x), w_y^i \phi_i(y) \rangle = w_x^i w_y^i K_{x,y}^i. \tag{11}$$

By *(i)* this kernel is Mercer. It follows that it is positive semi-definite, and consequently, the sum of all weighted kernels

$$K = \sum_{c_i \in \mathcal{A}} \bar{K}^i \tag{12}$$

is also positive semi-definite and therefore Mercer.□

While the weights w were chosen by analogy with the definition of a radial basis function, the proof holds for all choices of weight, and all choices of kernel. In the experimental section, we evaluate manifold variants of linear, local quadratic, and RBF kernels.

4.3   Efficient Approximation of the Kernel

In practice the majority of weights, w_x^i are close to zero and both w_y^i and $K_{x,y}^i$ are small[3]. As such, if w_x^i is small then the entire row $K_{x,-}^i$ and column $K_{-,x}^i$ can be safely set to 0, without altering the classification accuracy. To take advantage of this, we only compute explicitly the inner products $\langle \phi_i(x), \phi_i(y) \rangle$ if c_i is one of the closest subspaces for both x and y. This is equivalent to setting $w_x^i = 0$ if c_i is not one of the closest subspaces to x, and so the kernel remains Mercer.

We assume that the parameters of Atlas are chosen in such a way that for n datapoints Atlas will find $O(\sqrt{n})$ charts, each containing less than $O(\sqrt{n})$ points[4]. Then for each point, we must compute the distance to every subspace - which takes time $O(n\sqrt{n})$, and then for each subspace compute the local inner-products of all points assigned to it which again takes time $O(\sqrt{n}(\sqrt{n})^2) = O(n\sqrt{n})$ in total. Should these assumptions be violated the algorithm degrades naturally, with an overall run-time of $O(nm + \sum_i n_i^2)$, where m is the total number of charts, and n_i the number of points assigned to chart c_i.

Even with these modifications the asymptotic complexity of training an SVM using a cutting plane algorithm is $O(n^3)$. However for the datasets we consider the primary bottleneck lies in computation of the kernel matrix,and as such, a reduction in the complexity of computing the kernel has significant impact on run-time. See table 5 for a detailed breakdown of the run-time of the different components of our method vs. global RBF kernel. In practice, for all reported experiments we use the 10 closest subspaces in our approximation.

4.4   Integration with Efficient Primal Solvers

The restricted case in which $K_{x,y}^i = P_i^\perp(x) \cdot P_i^\perp(y)$ deserves special attention. In this case, we can solve the problem efficiently in the primal by taking as a

---

[3] $w_y^i K_{x,y}^i \leq 1$ in the case of an RBF kernel.

[4] These are sensible assumptions, and not just chosen to make asymptotic improvements possible. As the number of charts steadily increases Atlas will be able to approximate better any underlying manifold, while the fact that the number of charts grows sub-linearly means that Atlas should exhibit increasing robustness to sampling error.

feature vector for point x the concatenation of weighted projections $w_x^i P_i^\perp(x)$, and this allows the use of efficient schemes such as averaging stochastic gradient descent [5] that can exploit the sparsity of the training. As each feature vector is sparse with $O(\sqrt{n})$ non-zero components, computation of the inner products and sparse updates of the weight vector are $O(\sqrt{n})$ operations, making the overall run-time associated with a fixed number of passes over the training set $O(n\sqrt{n})$.

As a point of effectiveness, the linear kernel performs significantly better if we allow the SVM to learn a bias for each chart separately. As such, we train a standard linear SVM over the sparse feature vector

$$f_{\mathbf{x}} = \bigoplus_{c_i \in \mathcal{A}} w_x^i [1, \, P_i^\perp(x)]. \tag{13}$$

where $\bigoplus$ is the concatenation operator.

In the experimental section, we also explore the use of local quadratic kernels, both those without cross terms, in which the sparse feature vector takes the form

$$f_{\mathbf{x}} = \bigoplus_{c_i \in \mathcal{A}} w_x^i [1, \, P_i^\perp(x), \, P_i^\perp(x)^2] \tag{14}$$

with $P_i^\perp(x)^2$ being the *elementwise* square of $P_i^\perp(x)$, and those with cross-terms:

$$f_{\mathbf{x}} = \bigoplus_{c_i \in \mathcal{A}} w_x^i [1, \, P_i^\perp(x), \, l(P_i^\perp(x) \otimes P_i^\perp(x))] \tag{15}$$

where $l(P_i^\perp(x) \otimes P_i^\perp(x))$ is the vectorization of the lower triangular (inclusive of diagonals) component of the outer product matrix $P_i^\perp(x) \otimes P_i^\perp(x)$.

While in high-dimensional feature spaces, the use of quadratic features is largely unnecessary and incurs a substantial additional computation cost[5], in the local low-dimensional spaces of the manifold, the use of quadratic features incurs little overhead, and offers a noticeable improvement in discriminative performance.

5   Experiments

Semi-Supervised Learning: To illustrate the effectiveness of our approach in a semi-supervised situation, where the amount of labelled data is sparse relative to the total amount of data, we evaluate on MNIST by holding back the labels of a proportion of the training data. We generate a single Atlas over all training and test data, of local dimensionality 30, and calculate the classification error averaged over 20 trials varying the amount of labelled training data used from $\frac{1}{100}$th of the original training data (600 training samples) to $\frac{1}{2}$ of the data (30,000 training samples). As can be seen in figure 1, with sparse training data, AtlasRBF

---

[5] For example on MNIST, the raw feature vectors lie in a 784 dimensional space, while the quadratic features including cross terms lie in a 307,720 dimensional space.

Table 1. Classification performance on MNIST varying the proportion of labelled data. For all experiments, we use the same Atlas of local dimensionality 30, $\lambda = 100$, and $k = 2$, containing 835 charts. Zoom electronically to see standard deviation.

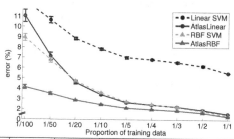

Training set ratio	1/100	1/50	1/20	1/10	1/5	1/4	1/3	1/2	1/1
Linear SVM	12.48 ± 0.33	10.65 ± 0.31	8.86 ± 0.13	7.85 ± 0.10	7.02 ± 0.12	6.83 ± 0.07	6.55 ± 0.10	6.20 ± 0.09	5.52
AtlasLin (eq. 13)	11.05 ± 0.60	7.15 ± 0.36	4.58 ± 0.16	3.41 ± 0.07	2.65 ± 0.08	2.44 ± 0.05	2.24 ± 0.06	1.94 ± 0.08	1.56
RBF SVM	8.95 ± 0.33	6.75 ± 0.24	4.73 ± 0.11	3.59 ± 0.07	2.73 ± 0.06	2.47 ± 0.06	2.20 ± 0.06	1.86 ± 0.05	1.41
AtlasRBF (eq. 12)	$\mathbf{4.13 \pm 0.20}$	$\mathbf{3.50 \pm 0.13}$	$\mathbf{2.87 \pm 0.06}$	$\mathbf{2.45 \pm 0.06}$	$\mathbf{2.12 \pm 0.05}$	$\mathbf{1.99 \pm 0.05}$	$\mathbf{1.87 \pm 0.05}$	$\mathbf{1.67 \pm 0.04}$	$\mathbf{1.31}$

drastically outperforms other methods –achieving significantly less than half the error of an RBF kernel at maximum sparsity (4.13% vs. 8.95% error)– while the performance of the efficient linear Atlas kernel approximately tracks that of the standard RBF kernel. In the limit, with full training data effectively covering the testing data, the performance of AtlasRBF and the RBF kernel almost converges, with AtlasRBF retaining a small edge (see table 1).

Table 2. Comparison with semi-supervised approaches. With 100 labelled points, the extreme sparsity of the training data required a simpler Atlas with fewer charts. For this, we set $\lambda = 1000$, resulting in an Atlas with 207 charts. The parameters γ, σ are the same as the experiments in tables 1 and 4.

Method	100 labelled points	1000 labelled points
RBF SVM	22.70 ± 1.35	7.58 ± 0.29
EigenFunction	21.35 ± 2.08	11.91 ± 0.62
PVM(hinge loss)	18.55 ± 1.59	7.21 ± 0.19
AnchorGraphReg	9.40 ± 1.07	6.17 ± 0.15
AtlasRBF	$\mathbf{8.10 \pm 0.95}$	$\mathbf{3.68 \pm 0.12}$

The majority of semi-supervised approaches can not be used on datasets as large as MNIST (see discussion in [15]). As such, we also follow the protocol of [15] and compare our generalisation performance trained with 100 and 1000 training samples against three other scalable approaches: Eigenfunction [10], PVM [28], and AnchorGraphReg [15], alongside RBF SVMs.

Supervised Learning. To validate our approach we tested our algorithm on standard classification datasets MNIST, USPS, SEMEION, and LETTER. In all cases we compare the results from our Atlas-based kernel SVMs with Linear SVMs and RBF-kernel SVMs on the original data. In addition, for MNIST, USPS and LETTER we show comparisons with state-of-the-art approaches that use different variants of local co-ordinate coding [13, 25, 32], as well as large margin multi-class kernel-based classifiers [4, 8, 23] (see Table 3). We use SEMEION to compare against the most recent manifold learning approaches followed by nearest neighbour classifier.

Datasets. The MNIST, USPS, and SEMEION datasets consist of grayscale images of handwritten digits '0' – '9'. Both USPS and SEMEION contain images of resolution 16×16 encoded as 256 dimensional binary feature vectors. USPS contains 7291 training and 2007 testing images while on SEMEION, following [30], we create 100 random splits of the data with 796 training and 797 testing images in each set and report average error. The MNIST dataset is substantially larger, with $60,000$ training and $10,000$ test Grey-scale images.

Our choice of datasets was driven by the desire to compare the performance of our approach against as many alternative methods as possible. The 4 datasets we selected are popular datasets, used by many authors and allow us to give scores from a wide variety of related methods and show that our approach provides improved performance.

Implementation. We first perform manifold learning using the *Atlas* algorithm [18] to approximate of the underlying manifold as an atlas of piecewise affine overlapping charts before running our efficient SVM learning approach using linear, quadratic, and RBF chart-based kernels.

[18] takes three parameters as an input: the local dimensionality d common for all charts, a weight $\lambda \in \{10^0, 10^1, \ldots, 10^5\}$ governing the strength of the MDL prior, the number of nearest neighbours $k \in \{2, 4, \ldots, 10\}$ and d the local dimensionality of the manifold. For LETTER a 16-dimensional dataset we take $d \in [5, 10]$, and for all other datasets, we search $d \in \{5, 10, 15, 20, 30\}$. LLE and LTSA also need the local dimensionality and the number of neighbours as an input, and we search over the same range of values as Atlas. For SMCE we finely tune its parameter λ so that its local dimensionality varies over the same range as other methods. For all SVM kernel methods $\sigma^{-1}, \gamma^{-1} \in \{2^{-1}2^{-2}, \ldots, 2^{-7}\}$, except on MNIST where a finer search of $\sigma^{-1} \in \{0.03, 0.031, \ldots 0.04\}$ was required to replicate the performance of an RBF kernel reported in http://yann.lecun.com/exdb/mnist/.

The parameter σ_r of a raw RBF kernel can be understood as a compromise between the two parameters γ and σ used in AtlasRBF in that it should be chosen to be somewhere close to γ preventing generalisation off the manifold, but also close to σ to allow generalisation on the manifold. Empirically, for the parameters selected, this is always the case: On AtlasRBF $\gamma > \sigma$, and the raw RBF $\sigma_r \in [\gamma, \sigma]$. For example on USPS $\gamma = 2^3$, $\sigma = 2^7$, while $\sigma_r = 2^5$.

In our experiments we used two SVM solvers: The primal linear solver *SvmAsgd* [5] combined with a *one-versus-all* merging of binary SVMs; *Lib*SVM [7] allows the use of a precomputed custom kernel such as our chart-based RBF kernel merged using the built-in implementation of the *one-versus-one* merging SVMs

Table 3. Supervised classification with efficient primal SVM solvers or 1-NN. Our chart-based linear and quadratic kernel SVMs outperform all single-chart manifold learning methods followed by SVMs as well as Atlas followed by 1-NN on all datasets. Variants of local coordinate coding with SVMs also perform worse than our method on LETTER and MNIST, while [32] has slightly lower error on USPS. Scores for Linear SVMs and manifold learning methods are from our experiments and scores for other methods are as reported elsewhere. SMCE failed to converge on LETTER. All manifold learning methods except Atlas required more than 30GB of ram on MNIST and failed to complete.

Nearest Neighbour and Efficient Primal Formulations

	USPS error (%)	LETTER error (%)	SEMEION error (%)	MNIST error (%)
Local coordinate based SVMs				
LL-SVM[13]	5.78	5.32	–	1.85
Linear SVM + G-OCC [32]	4.14	6.85	–	1.72
Linear SVM + C-OCC [32]	**3.94**	7.35	–	1.61
Linear SVM + LLC (512 anchor points) [25]	5.78	9.02	–	3.69
Linear SVM + LLC (4096 anchor points) [25]	4.38	4.12	–	2.28
Linear SVM + Tangent LLC (4096 points) [26]	–	–	–	1.64
Manifold Learning + Linear SVMs[5]				
Linear SVM on original space	8.42	35.75	7.40	5.55
SMCE[9]	6.88	–	9.04	–
LLE[21]	12.61	74.50	12.47	–
LTSA[31]	9.37	69.10	46.06	–
Manifold Learning + 1-NN classifier				
1-NN on original space	4.98	4.35	10.92	5.34
SMCE[9]	7.47	–	9.26	–
LLE[21]	6.83	19.03	9.41	–
wLTSA [30]	8.77	40.65	10.12	–
Atlas [18]	5.38	17.28	8.27	5.13
Primal Atlas SVMs (SvmAsgd)				
AtlasLinear - Hard Assignment (see eq. 6)	5.58	16.65	8.44	3.71
AtlasLinear (see eq. 13)	4.68	**3.13**	6.19	1.78
AtlasQuad (see eq. 14)	4.04	3.63	6.02	1.76
AtlasQuadCross (see eq. 15)	4.09	3.33	**5.48**	**1.46**

Comparison with standard Manifold learning. Looking at the results of tables 3 and 4, several themes can be seen. In general, the fusion of stock manifold learning techniques [9, 21, 30] with either linear or kernel SVMs is of limited value, and is perhaps more likely to hurt SVM scores than to improve them. In contrast, our Atlas kernels show substantial improvement over any baseline SVM approach (the only exception being the use of an RBF kernel on the already low dimensional dataset letter). Every type of our Atlas based kernels out-performs every use of stock manifold learning methods, both when used in conjuncture with a linear or kernel SVM, or as a nearest neighbour classifier.

Table 3 shows a comparison of the efficient methods on USPS, LETTER, MNIST, and SEMEION. On three of the four datasets, our approach, and particularly At-lasQuadCross, significantly outperforms all other methods. Note that, the local coordinate methods do not report scores on SEMEION. However, our efficient primal approach obtains substantially better scores than a standard RBF kernel (see table 4). In particular, on the LETTER dataset our approach to learning on an atlas halves the classification error of [32] and substantially improves on the classification error obtained with the coordinate coding approach of [13, 25].

Table 4. Classification performance on USPS, LETTER, SEMEION, and MNIST datasets. Our chart-based RBF kernel outperforms all other multi-class kernel based SVMs as well as all single-chart manifold learning methods followed by RBF SVMs. *Lib*SVM with an RBF kernel on the raw data achieved the best performance on LETTER, AtlasRBF is best on all other datasets. The comparison between AtlasRBF with soft and hard assignment shows the impact of our novel kernels. SMCE failed to converge on LETTER.

Kernel methods using cutting-plane type approaches

Method	USPS error (%)	LETTER error (%)	SEMEION error (%)	MNIST error (%)
Global SVMs				
MCVSVM[8]	4.24	2.42	–	1.44
SVM$_{struct}$ [23]	4.38	2.40	–	1.40
LaRank [4]	4.25	2.80	–	1.41
LibSVM on raw data[7]	4.53	**2.05**	6.41	1.41
Manifold Learning + RBF SVMs[7]				
SMCE[9]	6.18	–	8.68	–
LLE[21]	4.78	5.38	6.93	–
LTSA[31]	7.03	44.63	9.17	–
Atlas-Kernel SVMs (LibSVM)				
AtlasRBF - Hard Assignment (see eq. 6)	4.63	4.95	7.15	3.13
AtlasRBF (see eq. 12)	**3.68**	2.33	**5.14**	**1.31**

Table 4 shows that in comparison with RBF SVMs and the multi-class kernel-based SVMs of [4, 8, 23], we achieve substantial improvement in classification performance on USPS. Our AtlasRBF kernel outperforms all methods with the exception of the global RBF kernel SVM on the LETTER dataset. As LETTER is 16-dimensional, it does not allow for the advantages of the manifold learning methods to be fully employed, it is perhaps unsurprising that manifold learning is not only unnecessary, but also slightly detrimental, as we see higher errors for the LCC-based methods. As [18] allows the learning of a manifold of arbitrary dimension, we could learn the trivial 16-dimensional manifold, composed of a single chart, and where the projection matrix $P^{\perp}(x)$ is the identity function. In such cases our performance is identical to that of the RBF kernel. As such a result is uninformative, we instead cap the local manifold dimensionality at 10, when reporting our result. Our approach still achieves the second best performance and outperforms all other multi-class kernel-based methods.

Tables 3 and 4 clearly show the importance of forcing the classifier to vary smoothly, when generalising to the testing set. While the smooth AtlasRBF kernel consistently outperforms related work, the hard assignment kernels (6) of section 4.1 show that training a single kernel in each chart without soft assignment is noticeably worse than existing approaches. Table 5 shows the response of our Atlas-based approaches vs. global linear and RBF kernels to increasing levels of Gaussian noise. Our approach appears to behave better with respect to noise with lower classification errors. The parameters used are the same as in tables 3 and 4.

Chart Characterisation. Our results can also be used to validate the manifold learning tenets of [19]. Particularly table 4, where the improved results come from forcing an RBF kernel to conform to the manifold [6], clearly show unlabelled data is important and that a learned manifold can improve the performance of classifiers. Empirically, tenet 2 (that a manifold can be fitted to the data) also holds and explains the success of our approach. Tenet 3 states that different classes should lie on different areas of the manifold. This can be tested by seeing if different classes belong to different charts of the Atlas.

Although [18] learns the set of overlapping affine charts in a totally unsupervised manner, tenet 3 suggests that points which share similar statistical properties and are more likely to lie on the same subspace or chart would also share the same label information. In fact, on USPS, most of the 18 charts learned contain points from a single dominant class, where for the median chart 96% of the points assigned to it come from the same class. However, some charts contain two or three prevalent classes and around 10% of the data label differs from that of its interior chart. On MNIST 262 out of 835 charts contain data from the same class and 180 contain more than 10% points whose label differs from the dominant class. Similarly, for the median chart 98% come from the dominant class. In total 6.9% of the data does not belong to the dominant class of the interior of the chart it is assigned to. Along with providing empiric validation of tenet 3, the fact that 5-10% of the data does not reflect the dominant label of the chart provides some insight in the difference in performance between NN, linear and RBF kernels, and implicitly bounds the maximal error of any classifier trained on this Atlas.

Table 5. Extended analysis on USPS

(a) Run-time of various components. The top row shows the run-time of components, while the bottom row shows the accumulated time.

(b) Classification on USPS with increasing Gaussian noise.

	Init.	Atlas	Kernel	SVM train	SVM test
AtlasLinear	44.44	+10.29	+0.78	+0.78	+0.16
	44.44	54.73	55.51	56.29	56.45
AtlasRBF	44.44	+10.29	+28.99	+9.68	+2.53
	44.44	54.73	83.72	93.40	95.93
LibSVM	-	-	-	99.93	+95.47
	-	-	-	99.93	195.40

Noise	2%	5%	10%	15%	20%	30%
Linear SVM	8.72	8.82	9.07	9.82	10.21	11.36
AtlasLinear	5.08	5.83	6.03	5.93	6.78	11.46
RBF SVM	4.58	4.58	5.53	5.58	**6.33**	8.67
AtlasRBF	**4.04**	**4.14**	**4.48**	**5.08**	6.34	**7.57**

---

[6] In contrast, table 3 shows that the charts found can be used to raise the data into a high-dimensional space, where linear SVMs perform better.

6 Conclusion

We have presented a novel approach to supervised and semi-supervised learning via training a manifold on unlabelled data. We have shown superior performance to both RBF kernels and local co-ordinate based methods on standard datasets, and to manifold learning based nearest neighbour. As such it provides additional empiric validation of the tenets of manifold learning first proposed in [19]. Our method provides a principled way for Support Vector Machines to make use of unlabelled data in learning a kernel, and we intend to further explore the benefits of this.

References

[1] Belkin, M., Niyogi, P.: Laplacian eigenmaps and spectral techniques for embedding and clustering. Advances in Neural Information Processing Systems 14, 585–591 (2001)

[2] Belkin, M., Niyogi, P., Sindhwani, V.: On manifold regularization. AISTATS (2005)

[3] Bellman, R.: Dynamic Programming. Dover Publications (March 1957)

[4] Bordes, A., Bottou, L., Gallinari, P., Weston, J.: Solving multiclass support vector machines with larank. In: Proceedings of the 24th International Conference on Machine Learning, pp. 89–96. ACM (2007)

[5] Bottou, L.: Large-scale machine learning with stochastic gradient descent. In: Lechevallier, Y., Saporta, G. (eds.) Proceedings of the 19th International Conference on Computational Statistics (COMPSTAT 2010), pp. 177–187. Springer, Paris (2010), http://leon.bottou.org/papers/bottou-2010

[6] Boykov, Y., Kolmogorov, V.: An Experimental Comparison of Min-Cut/Max-Flow Algorithms for Energy Minimization in Vision. PAMI 26(9), 1124–1137 (2004)

[7] Chang, C.C., Lin, C.J.: LIBSVM: A library for support vector machines. ACM Transactions on Intelligent Systems and Technology 2, 27:1–27:27 (2011), software available at http://www.csie.ntu.edu.tw/~cjlin/libsvm

[8] Crammer, K., Singer, Y.: On the algorithmic implementation of multiclass kernel-based vector machines. J. Mach. Learn. Res. 2, 265–292 (2002), http://dl.acm.org/citation.cfm?id=944790.944813

[9] Elhamifar, E., Vidal, R.: Sparse manifold clustering and embedding. In: Advances in Neural Information Processing Systems, pp. 55–63 (2011)

[10] Fergus, R., Weiss, Y., Torralba, A.: Semi-supervised learning in gigantic image collections. In: Bengio, Y., Schuurmans, D., Lafferty, J., Williams, C.K.I., Culotta, A. (eds.) Advances in Neural Information Processing Systems 22, pp. 522–530 (2009)

[11] Jayasumana, S., Hartley, R., Salzmann, M., Li, H., Harandi, M.: Kernel methods on the riemannian manifold of symmetric positive definite matrices. In: CVPR IEEE (2013)

[12] Ladický, L., Russell, C., Kohli, P., Torr, P.H.: Inference methods for crfs with co-occurrence statistics. International Journal of Computer Vision 103(2), 213–225 (2013)

[13] Ladicky, L., Torr, P.: Locally linear support vector machines. In: Proceedings of the 28th International Conference on Machine Learning (ICML 2011), pp. 985–992 (2011)

[14] Lee, J.M.: Introduction to smooth manifolds, vol. 218. Springer (2012)

[15] Liu, W., He, J., Chang, S.F.: Large graph construction for scalable semi-supervised learning. In: Fürnkranz, J., Joachims, T. (eds.) Proceedings of the 27th ICML (ICML 2010), pp. 679–686. Omni Press, Haifa (2010), http://www.icml2010.org/papers/16.pdf

[16] von Luxburg, U., Bousquet, O.: Distance–based classification with lipschitz functions. The Journal of Machine Learning Research 5, 669–695 (2004)

[17] Melacci, S., Belkin, M.: Laplacian support vector machines trained in the primal. Journal of Machine Learning Research 12, 1149–1184 (2011)

[18] Pitelis, N., Russell, C., Agapito, L.: Learning a manifold as an atlas. In: IEEE Conference on Computer Vision and Pattern Recognition, CVPR (2013)

[19] Rifai, S., Dauphin, Y., Vincent, P., Bengio, Y., Muller, X.: The manifold tangent classifier. Advances in Neural Information Processing Systems 24, 2294–2302 (2011)

[20] Russell, C., Fayad, J., Agapito, L.: Energy based multiple model fitting for nonrigid structure from motion. In: 2011 IEEE Conference on Computer Vision and Pattern Recognition (CVPR), pp. 3009–3016. IEEE (2011)

[21] Saul, L., Roweis, S.: Think globally, fit locally: unsupervised learning of low dimensional manifolds. The Journal of Machine Learning Research 4, 119–155 (2003)

[22] Sindhwani, V., Niyogi, P.: Linear manifold regularization for large scale semisupervised learning. In: Proc. of the 22nd ICML Workshop on Learning with Partially Classified Training Data (2005)

[23] Tsochantaridis, I., Joachims, T., Hofmann, T., Altun, Y., Singer, Y.: Large margin methods for structured and interdependent output variables. Journal of Machine Learning Research 6(2), 1453 (2006)

[24] Vapnik, V.: The Nature of Statistical Learning Theory. Springer (1995)

[25] Wang, J., Yang, J., Yu, K., Lv, F., Huang, T., Gong, Y.: Locality-constrained linear coding for image classification. In: 2010 IEEE Conference on Computer Vision and Pattern Recognition (CVPR), pp. 3360–3367. IEEE (2010)

[26] Yu, K., Zhang, T.: Improved local coordinate coding using local tangents. In: Proc. of the Intl. Conf. on Machine Learning, ICML (2010)

[27] Yu, K., Zhang, T., Gong, Y.: Nonlinear learning using local coordinate coding. Advances in Neural Information Processing Systems 22, 2223–2231 (2009)

[28] Zhang, K., Kwok, J.T., Parvin, B.: Prototype vector machine for large scale semi-supervised learning. In: Proceedings of the 26th Annual ICML, ICML 2009, pp. 1233–1240. ACM, New York (2009), http://doi.acm.org/10.1145/1553374.1553531

[29] Zhang, T., Tao, D., Li, X., Yang, J.: Patch alignment for dimensionality reduction. IEEE Transactions on Knowledge and Data Engineering 21(9), 1299–1313 (2009)

[30] Zhang, Z., Wang, J., Zha, H.: Adaptive manifold learning. IEEE Transactions on Pattern Analysis and Machine Intelligence 34(2), 253–265 (2012)

[31] Zhang, Z., Zha, H.: Principal manifolds and nonlinear dimension reduction via local tangent space alignment. SIAM Journal of Scientific Computing 26, 313–338 (2002)

[32] Zhang, Z., Ladicky, L., Torr, P., Saffari, A.: Learning anchor planes for classification. In: Advances in Neural Information Processing Systems, pp. 1611–1619 (2011)

A Lossless Data Reduction for Mining Constrained Patterns in n-ary Relations

Gabriel Poesia and Loïc Cerf

Department of Computer Science, Universidade Federal de Minas Gerais
Belo Horizonte, Brazil
{gabriel.poesia,lcerf}@dcc.ufmg.br

Abstract. Given a binary relation, listing the itemsets takes exponential time. The problem grows worse when searching for analog patterns defined in n-ary relations. However, real-life relations are sparse and, with a greater number n of dimensions, they tend to be even sparser. Moreover, not all itemsets are searched. Only those satisfying some user-defined constraints, such as minimal size constraints. This article proposes to exploit together the sparsity of the relation and the presence of constraints satisfying a common property, the monotonicity w.r.t. one dimension. It details a pre-processing step to identify and erase n-tuples whose removal does not change the collection of patterns to be discovered. That reduction of the relation is achieved in a time and a space that is linear in the number of n-tuples. Experiments on two real-life datasets show that, whatever the algorithm used afterward to actually list the patterns, the pre-process allows to lower the overall running time by a factor typically ranging from 10 to 100.

1 Introduction

Given a binary relation, which generically represents objects having (or not) some Boolean properties, an *itemset* is a subset of the properties. It can be associated with the subset of all objects having all those properties. Those objects are called the *support* of the itemset. Mining the itemsets with their supports allow the discovery of correlations between arbitrary numbers of Boolean properties, between arbitrary number of objects and between the objects and the properties. For instance, mining a binary relation indicating whether a customer (an object) bought a food item (a property) can unveil interesting buying behaviors. The pattern ({Alice,Bob,Dave}, {bread,cheese,oil,salt}) indicates that the three customers in the support are the only ones who bought together the four food items in the itemset. The number of itemsets is exponential in the number of properties and so is the time to compute them.

To keep under control the size of that output, two techniques are classically used. First of all, the itemsets that are not *closed* can be removed from the output without any loss of information. Every non-closed itemset is, by definition, strictly included into another itemset with the exact same support. For instance, {bread,cheese,oil,salt} is not closed if Alice, Bob and Dave

T. Calders et al. (Eds.): ECML PKDD 2014, Part II, LNCS 8725, pp. 581–596, 2014.
© Springer-Verlag Berlin Heidelberg 2014

all bought some butter. If there is no other food item that they all bought, {bread,cheese,oil,salt,butter} is closed. With the support stored along with the closed itemset, the support of *any* itemset can easily be retrieved from the reduced collection of patterns: it is the support of the smallest superset that is closed. However the number of closed itemsets remains exponential in the number of objects or in the number of properties (whichever is smaller).

To further reduce the output, the sole *relevant* (closed) itemsets must be shown. The relevance is usually defined by the analyst as a conjunction of constraints that every output itemset must satisfy. For instance, knowing the subset W of customers who are women and the function p returning the price of the food item in argument, our analyst may want to take a look at the patterns (C, I) satisfying the following constraints:

$C_{\geq 8 \text{ women}}(C, I) \equiv |C \cap W| \geq 8$ for at least eight women in every pattern;

$C_{\leq 12 \text{ items}}(C, I) \equiv |I| \leq 12$ for at most twelve items in every pattern;

$C_{50\text{-min-area}}(C, I) \equiv |C \times I| \geq 50$ for at least fifty tuples in the cover of every pattern;

$C_{8\$\text{-max-price}}(C, I) \equiv \max_{i \in I} p(i) \leq 8$ for all items in every pattern having a price below 8\$;

$C_{4\$\text{-min-range-price}}(C, I) \equiv \max_{(i,i') \in I^2}(p(i) - p(i')) \geq 4$ for a price difference of at least 4\$ between the cheapest and the most expensive item in every pattern;

$C_{10\$\text{-min-total-price}}(C, I) \equiv \sum_{i \in I} p(i) \geq 10$ for at least 10\$ worth of items in every pattern.

Depending on the algorithm at work, some constraints can guide the search of the itemsets, i. e., regions of the pattern space are left unexplored because they do not contain any relevant itemset. Doing so, the relevant patterns can be discovered in a fraction of the time required to list every unconstrained pattern.

When dealing with "big data", whose growth in quantity is steeper than that of disk sizes, the first technique that is commonly applied is to simply identify irrelevant data that need not be stored. Constraints on itemsets play this role on the "big data output". But what about using the constraints before the actual extraction to reduce the input data? The binary relation is not "big". Nevertheless, because listing the constrained itemsets generally remains NP-hard, that simple idea can lead to a great reduction of the overall running time. This is especially true when a constraint allows, in a pre-processing step, to remove some tuples but cannot be used by the chosen algorithm to prune the pattern space (hence the need for a filter at the output). Notice that the removed tuples must be guaranteely useless, i. e., with or without them, the closed itemsets satisfying the constraints must be the same.

Such a pre-processing method has already been proposed for (not necessarily closed) itemset mining [3]. In this article, the closedness is taken into consideration. More challenging, the task is generalized toward n-ary relations. For example, our proposal can take advantage of some of the constraints listed above in the context of a ternary relations that encode whether customers buy items along time (e. g., the third element of a 3-tuple can be jan-14 or feb-14 or etc.).

That pre-process works at the level of *tubes*, i. e., one dimensional subspaces of the n-ary relation such as ({Alice}, {bread}, all months in which Alice bought bread), ({Alice}, all items Alice bought in jan-14, {jan-14}) and (all customers buying bread in jan-14, {bread}, {jan-14}). The Cartesian product of the n dimensions of a tube is called the *cover* of this tube. The less n-tuples in the cover of a tube, the more likely they can be seamlessly removed altogether from the relation thanks to a constraint that is *monotone w.r.t one dimension*, a property that this article introduces and that many common constraints happen to satisfy. Moreover, every n-tuple is in the cover of n tubes "oriented" in each of the n dimensions of the relation. As a consequence, emptying a tube makes it more likely that some of the orthogonal tubes can be emptied in a sequence. Indeed, their covers have just lost one n-tuple.

The pre-process therefore is effective as long as the relation contains tubes with small covers. It turns out that real-life n-ary relations often are sparse and even sparser for a greater n. In our example, a customer who ever bought an item usually did not buy it every month and the ternary relation is sparser than the binary one. Furthermore, the distribution, over all tubes, of the number of covered tuples often is skewed, i. e., most of the tubes cover few n-tuples. All those n-tuples, covered by the long tail of the distribution, are prone to be removed by the pre-process. Figure 1 shows such a distribution for one of the real-life ternary relations we used in our experiments. Each curve relates to one "orientation" for the tubes. In the log-scaled abscissa, those tubes were ordered in decreasing order of the number of 3-tuples they cover.

After presenting the related work in Sect. 2, Sect. 3 provides some definitions and formally defines the data-mining problem we consider. In Sect. 4, the pre-process is detailed and its correctness proved. Sect. 5 shows, on two real-life datasets that it frequently allows to solve the problem orders of magnitude faster. Finally, Sect. 6 briefly concludes.

2   Related Work

Given a binary relation, which can be seen as a Boolean matrix, the famous Apriori algorithm [1] mines itemsets under a minimal frequency constraint, i. e., a minimal number of rows in the support of the itemset. Apriori first considers the individual columns of the matrix and removes those with a number of present tuples that is below the frequency threshold. Indeed, such columns cannot be involved in any frequent itemset. This property of the frequency constraint has later been called *anti-monotonicity* by opposition to *monotonicity* [9].

[5] and [6] are among the early studies of the efficient extraction of patterns under both monotone and anti-monotone constraints. A monotone constraint on the rows of a pattern is anti-monotone when applied, instead, on its columns. This duality obviously vanishes when considering Boolean tensors, i. e., relations of higher arities. In this article, the expression *monotonicity w.r.t. one dimension* is coined. In the specific context of a binary relation, a constraint is monotone w.r.t. rows (respectively columns) if it only deals with rows

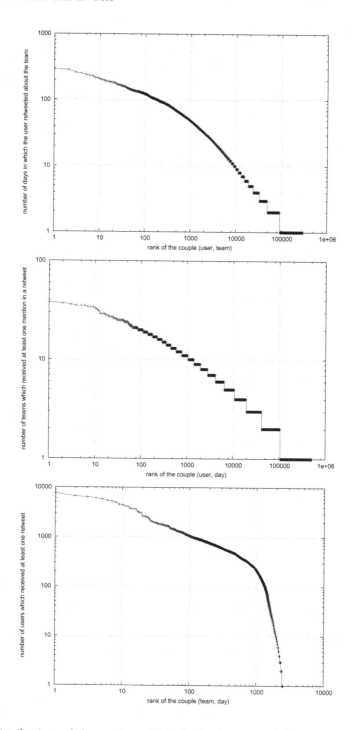

Fig. 1. Distributions of the number of 3-tuples in the cover of the tubes w.r.t. each of the three dimensions of the densest `Retweet` relation

(respectively columns) and is anti-monotone (respectively monotone) in the classical meaning of the word.

When constrained patterns are to be extracted from a binary relation, the individual rows (respectively columns) that do not satisfy the monotone (respectively anti-monotone) constraints can be removed. ExAnte [3] iteratively performs that reduction of a binary relation until a fixed point is reached, i.e., until no more row or column is removed. The algorithm presented in this article can be seen as a generalization of ExAnte toward n-ary relations ($n \geq 2$). To the best of our knowledge, it is the first attempt to pre-process an n-ary relation to speed up the subsequent search of constrained patterns.

Those "patterns" naturally generalize itemsets (along with their supports). More precisely, a pattern in an n-ary relation consists of n subsets of each of the n dimensions and only covers tuples present in the relation, i.e., the Cartesian product of the n subsets must be included in the relation. [11], [10] and [14] detail algorithms to mine such patterns in ternary relations. [7] and [13] directly tackle the search of patterns in arbitrary n-ary relations. All those algorithms actually enforce an additional maximality property, the closedness constraint, which is known to losslessly reduce the collection of patterns to the most informative ones [8]. Besides, they all are able to focus the search of the patterns on those having user-specified minimal numbers of elements in each of the dimensions. In fact, all of them but DATA-PEELER [7] *only* consider minimal size constraints. In contrast, DATA-PEELER can prune the search of the patterns with any *piecewise (anti)-monotone* constraint and, as a consequence, with any constraint that is *monotone w.r.t. one dimension* (as defined in this article).

3   Definitions and Problem Statement

All along the article, $\times$ denotes the Cartesian product and $\prod$ is used for the Cartesian product of an arbitrary number of sets. Given $n \in \mathbb{N}$ dimensions of analysis (i.e., n finite sets) $(D_i)_{i=1..n}$, the dataset is a relation $\mathcal{R} \subseteq \prod_{i=1}^{n} D_i$, i.e., a set of n-tuples. Table 1 represents such a relation $\mathcal{R}_E \subseteq \{\alpha, \beta, \gamma\} \times \{1, 2, 3, 4\} \times \{A, B, C\}$, hence a ternary relation. In this table, every '1' (resp. '0') at the intersection of three elements stands for the presence (resp. absence) of the related 3-tuple in $\mathcal{R}_E$. E.g., $(\alpha, 1, A) \in \mathcal{R}_E$ and $(\alpha, 1, C) \notin \mathcal{R}_E$.

Table 1. $\mathcal{R}_E \subseteq \{\alpha, \beta, \gamma\} \times \{1, 2, 3, 4\} \times \{A, B, C\}$

	A	B	C	A	B	C	A	B	C
1	1	1	0	0	0	1	0	1	0
2	0	0	0	1	1	0	1	1	0
3	1	0	1	0	1	1	1	0	0
4	1	0	0	1	0	0	0	1	0
		α			β			γ	

An *n-set* $(X_1, \cdots, X_n)$ consists of n subsets of each of the n dimensions, i.e., $(X_1, \cdots, X_n) \in \prod_{i=1}^{n} \mathcal{P}(D_i)$. For example, given the dimensions of $\mathcal{R}_E$, $(\{\alpha, \gamma\}, \{2, 4\}, \{B\})$ is an *n*-set, whereas $(\{\beta\}, \{4\}, \{A, \alpha\})$ is not because $\alpha \notin D_3$.

An *i-tube* (with $i \in \{1, \cdots, n\}$) is a special *n*-set: all its dimensions but the ith are singletons and its ith dimension contains all the elements in D_i that form, with the elements in the singletons, *n*-tuples present in the relation. Formally, $(T_1, \cdots, T_n) \in \prod_{i=1}^{n} \mathcal{P}(D_i)$ is an *i-tube* in $\mathcal{R} \subseteq \prod_{i=1}^{n} D_i$ if and only if:

$$\begin{cases} \forall j \in \{1, \cdots, i-1, i+1, \cdots, n\}, \ \exists t_j \in D_j \mid T_j = \{t_j\} \\ T_i = \{t_i \in D_i \mid (t_1, \cdots, t_{i-1}, t_i, t_{i+1}, \cdots, t_n) \in \mathcal{R}\} \end{cases}$$

$(\{\alpha\}, \{1, 3, 4\}, \{A\})$ is an example of a 2-tube in $\mathcal{R}_E$. $(\{\alpha\}, \{1, 3\}, \{A\})$ is not a 2-tube in $\mathcal{R}_E$ because 4 is not in its second dimension although $(\alpha, 4, A) \in \mathcal{R}_E$. $(\{\alpha\}, \{1, 2, 3, 4\}, \{A\})$ is not a 2-tube either because $(\alpha, 2, A) \notin \mathcal{R}_E$.

A *constraint* over an *n*-set is a predicate, i.e., a function that associates every *n*-set with a value of either true or false. The data mining task we consider is the extraction, from an *n*-ary relation, of *all* patterns ("pattern" will be defined in the next paragraph) satisfying a conjunction of constraints that are independent from the relation. The introduction of this article provides six examples of constraints. None of them depends on the relation, i.e., the sole *n*-set and, possibly, some external data (such as W and p in the introduction) are sufficient to evaluate the constraint.

A *pattern* in an *n*-ary relation is a natural generalization of a closed itemset and its support in a binary relation. It is an *n*-set (1) whose *cover* (the Cartesian product of its n dimensions) is included in the relation and (2) that is closed. The closedness is a property of maximality. It means that no element can be added to any of dimension of the pattern without breaking property (1). Formally, $(X_1, \cdots, X_n) \in \prod_{i=1}^{n} \mathcal{P}(D_i)$ is a pattern in $\mathcal{R} \subseteq \prod_{i=1}^{n} D_i$ if and only if:

$$\begin{cases} (1) \ \prod_{i=1}^{n} X_i \subseteq \mathcal{R} \\ (2) \ \forall (X_1', \cdots, X_n') \in \prod_{i=1}^{n} \mathcal{P}(D_i), \\ \quad \begin{cases} \forall i \in \{1, \cdots, n\}, X_i \subseteq X_i' \\ \prod_{i=1}^{n} X_i' \subseteq \mathcal{R} \end{cases} \Rightarrow \forall i \in \{1, \cdots, n\}, X_i = X_i' \end{cases}$$

For brevity, we sometimes write $\mathcal{C}_{\text{closed}}(X_1, \cdots, X_n, \mathcal{R})$ to mean that the *n*-set $(X_1, \cdots, X_n)$ satisfies property (2).

$(\{\beta, \gamma\}, \{2\}, \{A, B\})$ is an example of a pattern in $\mathcal{R}_E$ because (1) the four 3-tuples in $\{\beta, \gamma\} \times \{2\} \times \{A, B\}$ belong to $\mathcal{R}_E$ and (2) no element can be added to any of its dimensions without breaking property (1). $(\{\beta\}, \{2\}, \{A, B\})$ is not a pattern in $\mathcal{R}_E$ because it is not closed: it is "included" in $(\{\beta, \gamma\}, \{2\}, \{A, B\})$, which only covers 3-tuples that are present in $\mathcal{R}_E$. $(\{\alpha, \gamma\}, \{1\}, \{A\})$ is not a pattern either because $(\gamma, 1, A)$ can be formed by taking an element in each of its dimensions and $(\gamma, 1, A) \notin \mathcal{R}_E$.

The data mining task considered in this paper can now be formalized. Given a relation $\mathcal{R} \subseteq \prod_{i=1}^{n} D_i$ and a set $\mathcal{C}_{\text{all}}$ of constraints that are all independent from the relation, the problem is the computation of the following set $\mathcal{T}h(\mathcal{R}, \mathcal{C}_{\text{all}})$:

$$\left\{ (X_1, \cdots, X_n) \in \prod_{i=1}^{n} \mathcal{P}(D_i) \mid \begin{cases} (X_1, \cdots, X_n) \text{ is a pattern in } \mathcal{R} \\ \forall \mathcal{C} \in \mathcal{C}_{\text{all}}, \mathcal{C}(X_1, \cdots, X_n) \end{cases} \right\}$$

However, this work is not about a new algorithm to solve the problem. It is about computing a relation $\mathcal{R}' \subseteq \mathcal{R}$ that is as small as possible and yet guarantees that $Th(\mathcal{R}', \mathcal{C}_{\text{all}}) = Th(\mathcal{R}, \mathcal{C}_{\text{all}})$. In this way, the actual pattern miner potentially runs faster on $\mathcal{R}'$ and yet outputs the correct and complete collection of constrained patterns.

To shrink $\mathcal{R}$ into $\mathcal{R}'$, the algorithm proposed in this article exploits the constraints in $\mathcal{C}_{\text{all}}$ that are *monotone w.r.t. one dimension*. They are constraints that do not depend on any dimension of the n-set but one and that are monotone w.r.t. the inclusion order on this dimension, i. e., if an n-set satisfies a constraint that is monotone w.r.t. dimension i, then any n-set with a larger i^{th} dimension (w.r.t. set inclusion) satisfies it as well. Formally, a constraint $\mathcal{C}$ is *monotone w.r.t. dimension i* (with $i \in \{1, \cdots, n\}$) if and only if:

$$\forall (X_1, \cdots, X_n) \in \prod_{i=1}^{n} \mathcal{P}(D_i), \forall Y_i \subseteq D_i,$$
$$\mathcal{C}(X_1, \cdots, X_n) \Rightarrow \mathcal{C}(X_1, \cdots, X_{i-1}, X_i \cup Y_i, X_{i+1}, \cdots, X_n) \ .$$

Among the six constraints listed in the introduction, $\mathcal{C}_{\geq 8 \text{ women}}$ is monotone w.r.t. the customer dimension; $\mathcal{C}_{8\$\text{-max-price}}$, $\mathcal{C}_{4\$\text{-min-range-price}}$ and $\mathcal{C}_{10\$\text{-min-total-price}}$ are monotone w.r.t. the food item dimension. $\mathcal{C}_{\leq 12 \text{ items}}$ is not monotone: given an n-set that satisfies it, there exists another n-set with more than twelve items, including all those involved in the first n-set. $\mathcal{C}_{50\text{-min-area}}$ is not monotone w.r.t. one dimension either because it depends on two dimensions of the n-set.

4 Dataset Reduction

The reduction of the n-ary relation $\mathcal{R} \subseteq \prod_{i=1}^{n} D_i$, which is proposed in this article, is based on the removal of the n-tuples covered by an i-tube that does not verify a constraint that is monotone w.r.t. dimension i. This section first proves that this operation does not change the set of constrained patterns that is to be discovered. Then the actual algorithm is presented and its complexity is analyzed.

4.1 Fundamental Theorem

Let us first show that the ith dimension of a pattern necessarily contains a subset of the ith dimension of any i-tube that covers some of its tuples:

Lemma 1. *Given a pattern $(X_1, \cdots, X_n)$ in $\mathcal{R}$ and an i-tube $(T_1, \cdots, T_n)$ in $\mathcal{R}$ (with $i \in \{1, \cdots, n\}$), we have:*

$$\left(\prod_{j=1}^{n} X_j \right) \cap \left(\prod_{j=1}^{n} T_j \right) \neq \varnothing \Rightarrow X_i \subseteq T_i \ .$$

Proof. If $\left(\prod_{j=1}^n X_j\right) \cap \left(\prod_{j=1}^n T_j\right) \neq \varnothing$, then $\forall j \in \{1, \cdots, n\}$, $X_j \cap T_j \neq \varnothing$. By definition of the i-tube $(T_1, \cdots, T_n)$, $\forall j \neq i$, $|T_j| = 1$. As a consequence, $\forall j \neq i$, $T_j \subseteq X_j$ (1). Assume, by contradiction, $X_i \not\subseteq T_i$, i.e., $\exists e \in X_i \setminus T_i$. By (1) and the first property defining the pattern $(X_1, \cdots, X_n)$, $T_1 \times \cdots T_{i-1} \times \{e\} \times T_{i+1} \times \cdots \times T_n \subseteq \prod_{j=1}^n X_j \subseteq \mathcal{R}$ (2). By definition of the i-tube $(T_1, \cdots, T_n)$, $T_i = \{t_i \in D_i \mid (t_1, \cdots, t_{i-1}, t_i, t_{i+1}, \cdots, t_n) \in \mathcal{R}\}$. With (2), we therefore have $e \in T_i$, which contradicts the assumption.   □

From now on, $\mathcal{C}_i$ (with $i \in \{1, \cdots, n\}$) denotes the conjunction of all constraints in $\mathcal{C}_{\text{all}}$ that are monotone w.r.t. dimension i. Clearly, $\mathcal{C}_i$ is monotone w.r.t. dimension i. The following lemma states that whenever an i-tube violates $\mathcal{C}_i$, removing the n-tuples covered by this i-tube leads to a reduced relation that does not embed any pattern absent from the original relation.

Lemma 2. *Given $\mathcal{R}$, $\mathcal{C}_{\text{all}}$ and an i-tube $(T_1, \cdots, T_n)$ in $\mathcal{R}$ (with $i \in \{1, \cdots, n\}$), we have:*

$$\neg \mathcal{C}_i(T_1, \cdots, T_n) \Rightarrow \mathcal{T}h\left(\mathcal{R} \setminus \prod_{j=1}^n T_j, \mathcal{C}_{\text{all}}\right) \subseteq \mathcal{T}h(\mathcal{R}, \mathcal{C}_{\text{all}}) \ .$$

Proof. Let $(X_1, \cdots, X_n) \in \mathcal{T}h(\mathcal{R} \setminus \prod_{j=1}^n T_j, \mathcal{C}_{\text{all}})$.

By the first property defining the pattern $(X_1, \cdots, X_n)$ in $\mathcal{R} \setminus \prod_{j=1}^n T_j$, $\prod_{j=1}^n X_j \subseteq \mathcal{R} \setminus \prod_{j=1}^n T_j$. Because $\mathcal{R} \setminus \prod_{j=1}^n T_j \subseteq \mathcal{R}$ and by transitivity of $\subseteq$, $\prod_{j=1}^n X_j \subseteq \mathcal{R}$ (1).

Assume, by contradiction, $\neg \mathcal{C}_{\text{closed}}(X_1, \cdots, X_n, \mathcal{R})$. By definition of $\mathcal{C}_{\text{closed}}$,

$\exists (X_1', \cdots, X_n') \in \prod_{i=1}^n \mathcal{P}(D_i) \mid \begin{cases} \prod_{i=1}^n X_i' \subseteq \mathcal{R}(2) \\ \forall i \in \{1, \cdots, n\}, X_i \subseteq X_i'(3) \\ \exists i \in \{1, \cdots, n\} \mid X_i \subsetneq X_i'(4) \end{cases}$. We necessar-

ily have $\prod_{i=1}^n X_i' \not\subseteq (\mathcal{R} \setminus \prod_{i=1}^n T_i)$ otherwise, with (3) and (4), it would follow that $\neg \mathcal{C}_{\text{closed}}(X_1, \cdots, X_n, \mathcal{R} \setminus \prod_{j=1}^n T_j)$ what would contradict $(X_1, \cdots, X_n) \in \mathcal{T}h(\mathcal{R} \setminus \prod_{j=1}^n T_j, \mathcal{C}_{\text{all}})$. By adding (2) to that, we have $(\prod_{j=1}^n X_j') \cap (\prod_{j=1}^n T_j) \neq \varnothing$ and, by Lemma 1, $X_i' \subseteq T_i$. Because (3) imposes $X_i \subseteq X_i'$, we have, by transitivity of $\subseteq$, $X_i \subseteq T_i$. Therefore, by contraposition of the definition of the monotonicity w.r.t. dimension i that holds for $\mathcal{C}_i$, $\neg \mathcal{C}_i(T_1, \cdots, T_n) \Rightarrow \neg \mathcal{C}_i(X_1, \cdots, X_n)$. As a consequence, $(X_1, \cdots, X_n) \notin \mathcal{T}h(\mathcal{R} \setminus \prod_{j=1}^n T_j, \mathcal{C}_{\text{all}})$, a contradiction. Therefore, the assumption is wrong, i.e., $\mathcal{C}_{\text{closed}}(X_1, \cdots, X_n, \mathcal{R})$ (5).

Finally, because all constraints in $\mathcal{C}_{\text{all}}$ are independent from the relation, the fact that $(X_1, \cdots, X_n)$ satisfies them in $\mathcal{R} \setminus \prod_{j=1}^n T_j$ implies that it satisfies them as well in $\mathcal{R}$. Together with (1) and (5), we therefore have $(X_1, \cdots, X_n) \in \mathcal{T}h(\mathcal{R}, \mathcal{C}_{\text{all}})$.   □

One final lemma to state the opposite of Lemma 2, i.e., whenever an i-tube violates $\mathcal{C}_i$, removing the n-tuples covered by this i-tube leads to a reduced relation that embeds every pattern present in the original relation.

Lemma 3. *Given $\mathcal{R}$, $\mathcal{C}_{all}$ and an i-tube $(T_1, \cdots, T_n)$ in $\mathcal{R}$ (with $i \in \{1, \cdots, n\}$), we have:*

$$\neg \mathcal{C}_i(T_1, \cdots, T_n) \Rightarrow \mathcal{T}h(\mathcal{R}, \mathcal{C}_{all}) \subseteq \mathcal{T}h\left(\mathcal{R} \setminus \prod_{j=1}^{n} T_j, \mathcal{C}_{all}\right).$$

Proof. Let $(X_1, \cdots, X_n) \in \mathcal{T}h(\mathcal{R}, \mathcal{C}_{all})$.

By the first property defining the pattern $(X_1, \cdots, X_n)$ in $\mathcal{R}$, $\prod_{j=1}^{n} X_j \subseteq \mathcal{R}$ (1). Assume, by contradiction, $\prod_{j=1}^{n} X_j \not\subseteq \mathcal{R} \setminus \prod_{j=1}^{n} T_j$. With (1), we have $(\prod_{j=1}^{n} X_j) \cap (\prod_{j=1}^{n} T_j) \neq \varnothing$, i.e., Lemma 1 applies and $X_i \subseteq T_i$. By contraposition of the definition of the monotonicity w.r.t. dimension i that holds for $\mathcal{C}_i$, $\neg \mathcal{C}_i(T_1, \cdots, T_n) \Rightarrow \neg \mathcal{C}_i(X_1, \cdots, X_n)$. As a consequence, $(X_1, \cdots, X_n) \notin \mathcal{T}h(\mathcal{R}, \mathcal{C}_{all})$, a contradiction. Therefore, the assumption is wrong, i.e., $\prod_{j=1}^{n} X_j \subseteq \mathcal{R} \setminus \prod_{j=1}^{n} T_j$ (2).

$\mathcal{C}_{closed}(X_1, \cdots, X_n, \mathcal{R} \setminus \prod_{j=1}^{n} T_j)$ (3) directly follows from $\mathcal{R} \setminus \prod_{j=1}^{n} T_j \subseteq \mathcal{R}$ and, in a sequence, from the closedness of $(X_1, \cdots, X_n)$ in $\mathcal{R}$: $\forall (X'_1, \cdots, X'_n) \in$

$$\prod_{j=1}^{n} \mathcal{P}(D_j), \begin{cases} \forall j \in \{1, \cdots, n\}, X_j \subseteq X'_j \\ \prod_{j=1}^{n} X'_j \subseteq \mathcal{R} \setminus \prod_{j=1}^{n} T_j \subseteq \mathcal{R} \end{cases} \Rightarrow \forall j \in \{1, \cdots, n\}, X_j = X'_j.$$

Finally, because all constraints in $\mathcal{C}_{all}$ are independent from the relation, the fact that $(X_1, \cdots, X_n)$ satisfies them in $\mathcal{R}$ implies that it satisfies them as well in $\mathcal{R} \setminus \prod_{j=1}^{n} T_j$. Together with (2) and (3), we therefore have $(X_1, \cdots, X_n) \in \mathcal{T}h(\mathcal{R} \setminus \prod_{j=1}^{n} T_j, \mathcal{C}_{all})$. □

Finally, here is the theorem at the foundation of the data reduction proposed in this article.

Theorem 1. *Given $\mathcal{R}$, $\mathcal{C}_{all}$ and an i-tube $(T_1, \cdots, T_n)$ in $\mathcal{R}$ (with $i \in \{1, \cdots, n\}$), we have:*

$$\neg \mathcal{C}_i(T_1, \cdots, T_n) \Rightarrow \mathcal{T}h(\mathcal{R}, \mathcal{C}_{all}) = \mathcal{T}h\left(\mathcal{R} \setminus \prod_{j=1}^{n} T_j, \mathcal{C}_{all}\right).$$

Proof. The equality follows from Lemmas 2 and 3.

4.2   Algorithm

The obvious pre-process, which directly follows from Th. 1, would consider, one by one and for all $i \in \{1, \cdots, n\}$, every i-tube in $\mathcal{R}$. It would test whether the related $\mathcal{C}_i$ is satisfied and, if not, it would "empty" the i-tube. However, the removal of an n-tuple in an i-tube corresponds as well to the removal of this same n-tuple in every orthogonal j-tube (with $j \neq i$). Such a j-tube may have already been considered and was satisfying $\mathcal{C}_j$. However, since $\mathcal{C}_j$ is monotone w.r.t. dimension j, the j-tube may now violate $\mathcal{C}_j$ because it contains one element less in its jth dimension. In this way, the constraints that are monotone w.r.t. one dimension work in synergy with the constraints that are monotone w.r.t.

any other dimension. When one is effective, (i. e., allows to identify a tube to empty), it makes it more likely that the others become effective.

The following pseudo-code formalizes the pre-process. It enumerates, one by one, the n-tuples in the relation and checks all n tubes that cover each of the n-tuples. Whenever an i-tube is emptied because it violates C_i, the j-tubes $(j \neq i)$ that involve the removed n-tuples are rechecked. In this way, the pre-process only terminates when all i-tubes, for all $i \in \{1, \cdots, n\}$, are either empty or satisfy the related constraint C_i.

Data: relation $\mathcal{R} \subseteq \prod_{j=1}^{n} D_j$, set $\mathcal{C}_{\text{all}}$ of constraints that are all independent from $\mathcal{R}$
begin
    forall the $(t_1, \cdots, t_n) \in \mathcal{R}$ **do**
        forall the $i \in \{1, \cdots, n\}$ **do**
            CLEANTUBE$(\mathcal{R}, i, (t_1, \cdots, t_{i-1}, t_{i+1}, \cdots, t_n))$;

Algorithm 1. CLEANRELATION

Data: relation $\mathcal{R} \subseteq \prod_{j=1}^{n} D_j$, orientation of the tube $i \in \{1, \cdots, n\}$, elements in the singletons of the tube $(t_1, \cdots, t_{i-1}, t_{i+1}, \cdots, t_n)$
begin
    $T_i \leftarrow \{t_i \in D_i \mid (t_1, \cdots, t_{i-1}, t_i, t_{i+1}, \cdots, t_n) \in \mathcal{R}\}$;
    if $\neg C_i(\{t_1\}, \cdots, \{t_{i-1}\}, T_i, \{t_{i+1}\}, \cdots \{t_n\})$ **then**
        forall the $t_i \in T_i$ **do**
            $\mathcal{R} \leftarrow \mathcal{R} \setminus \{(t_1, \cdots, t_n)\}$;
            forall the $j \in \{1, \cdots, i-1, i+1, \cdots, n\}$ **do**
                CLEANTUBE$(\mathcal{R}, j, (t_1, \cdots, t_{j-1}, t_{j+1}, \cdots, t_n))$;

Procedure. CLEANTUBE()

In the pseudo-code, there may be no removal of n-tuples in an i-tube between two checks of this i-tube. To avoid that, the actual implementation does not directly execute the recursive calls of CLEANTUBE. Instead, the tubes in arguments of those calls are stored in a hash set (hence no duplicate). As long as the hash set is non-empty, a tube is retrieved from it and the related call of CLEANTUBE is made. Once the hash map is empty, the execution comes back to CLEANRELATION.

Also, the tubes are not actually computed from the set of all n-tuples whenever they are required. Instead, the n-ary relation is stored n times as the set all i-tubes $(i \in \{1, \cdots, n\})$.

4.3   Complexity Analysis

In the worst case scenario, CLEANRELATION's enumeration of the n-tuples does not identify any tube to empty but the last one. Then, every single n-tuple is removed one by one, hence $|\mathcal{R}|$ calls of the CleanTube function. In this scenario, the pre-process consists of four steps whose time complexities follow:

Storage of $\mathcal{R}$: $O(n|\mathcal{R}|)$ since every n-tuple in $\mathcal{R}$ is stored n times; it is the space complexity of the overall pre-process too (assuming no external data is required to verify or speed up the verification of some constraints);

Outer-most enumeration: $O(|\mathcal{R}| \sum_{i=1}^{n} \mathrm{check}(\mathcal{C}_i))$ where $\mathrm{check}(\mathcal{C}_i)$ denotes the cost of verifying whether one i-tube verifies the constraints in $\mathcal{C}_{\mathrm{all}}$ that are monotone w.r.t. dimension i;

Actual cleaning: $O(|\mathcal{R}| \sum_{i=1}^{n} \mathrm{check}(\mathcal{C}_i))$;

Output of the remaining n-tuples: $O(|\mathcal{R}|)$; the worst-case scenario for this step corresponds to no actual cleaning.

Overall, the pre-process has a $O(n|\mathcal{R}|)$ space complexity and a time complexity of $O(|\mathcal{R}| \sum_{i=1}^{n} \mathrm{check}(\mathcal{C}_i))$. Notice that, for common constraints, $\mathrm{check}(\mathcal{C}_i)$ is cheap. For instance, it is $O(1)$ for minimal size constraints (assuming every i-tube is stored in a container with a constant time access to its size) or minimal sum constraints over positive numbers (assuming the sums for each i-tube are stored and updated whenever an n-tuple in it is erased). It is $O(\log |D_i|)$ for a maximal, a minimal or a min-range constraint (using respectively max-heaps, min-heaps and both).

5   Experimental Study

CLEANRELATION is integrated to DATA-PEELER, which is free software[1]. It is implemented in C++ and compiled by GCC 4.7.2 with the O3 optimizations. Because it is a pre-process, any pattern extractor can work on the reduced relation it outputs. We received, from their respective authors, the implementations of CUBEMINER [11], TRIAS [10], DATA-PEELER [7], TRICONS [14] and CNS-MINER [13], i. e., all (exact) pattern extractor that handle ternary relations (or more in the cases of DATA-PEELER and CNS-MINER). Unfortunately, CNS-MINER never produced any output and we therefore decided to focus the experimental study on ternary relations where comparisons can be made. TRICONS did not work, either crashing or returning an incomplete output.

The remaining three algorithms are tested on a GNU/Linux[TM] system running on top of 3.10GHz cores and 12GB of RAM. CUBEMINER and DATA-PEELER, both implemented in C++, were compiled with GCC 4.7.2. TRIAS was compiled and interpreted by Oracle's JVM version 1.7.0_45. CUBEMINER and TRIAS can only prune the search space with minimal size constraints on some or all dimensions of the pattern. In contrast, DATA-PEELER's traversal of the

---

[1] It is available, under the terms of the GNU GPLv3, at
http://dcc.ufmg.br/~lcerf/en/prototypes.html#d-peeler.

pattern space can be guided by any number of piecewise (anti)-monotone constraints. That includes any constraint that is monotone w.r.t. one dimension. As a consequence, unless the sole minimal size constraints are desired, DATA-PEELER would be preferred. Fortunately, minimal size constraints are monotone w.r.t. one dimension. The remainder of this section compares the times CUBEMINER, TRIAS and DATA-PEELER take to list minimally sized patterns in ternary relations, with and without the pre-process

5.1   Retweet Dataset

The micro-blogging service Twitter is particularly popular in Brazil. Tweets about the Brazilian soccer championship were collected from January, 9th 2014 to April, 11th 2014 (92 days) and classified w.r.t. to the mentioned team(s) (supervised classification method, which is out of the scope of this paper). How many times a user is retweeted (i. e., other users "repeat" her tweets) is known to be a good measure of her influence [12]. 184,159 users were retweeted at least once during the considered period. A 3-dimensional tensor gives how many times each of them is retweeted (over all her messages) during a given day when writing about a given soccer team (among 29). That tensor contains 731,685 non-null values.

It is turned into a ternary relation by keeping the tuples relating to cells of the tensors with a high enough number of retweets. In the experiments on this dataset, the threshold is a variable. On the contrary, the minimal size constraints on the patterns are kept constant: at least two days, two teams and two users. Although those constraints are rather loose, the pre-process is efficient because the relation is very sparse. In the most challenging context, when one retweet is considered "influential enough", CLEANRELATION only keeps 263,413 out of the 731,685 3-tuples, a 64% reduction.

5.2   Distrowatch Dataset

DistroWatch^2 is a popular Web site that gathers comprehensive information about GNU/Linux^{TM}, BSD, and Solaris operating systems. Every distribution is described on a separate page. When a visitor loads a page, her country is known from the IP address. The logs of the Web server are turned into a ternary relation that gives for any time period (13 semesters from early 2004 to early 2010) and every page (describing 655 distributions), the countries that visited it more than 25 times. From that relation, we consider the extraction of all patterns involving at least four semesters, m distributions and m countries, where m is an integer variable ranging from 5 to 48.

The relation contains 150,834 3-tuples, a number that is comparable to those of the Retweet relations. However, it is considerably denser. Because of that, and even with strong minimal size constraints, some of the algorithms cannot mine the patterns in the relation that is not pre-processed. Those same algorithms

---

^{2} http://www.distrowatch.com

benefit a lot from the pre-process, with overall running times that become several orders of magnitude shorter. With $m = 10$, CLEANRELATION keeps 65,130 out of the 150,834 3-tuples, a 57% reduction.

5.3 Pre-processing Time

Figure 2 depicts DATA-PEELER's running times on the `Retweet` relations with and without the pre-processing step. The actual data reduction takes only a fragment of the time required by the subsequent extraction. Despite the loose constraints at work ("at least two elements in every dimension of the pattern"), the pre-process is effective. With it, DATA-PEELER lists all the constrained patterns in about one percent of the time it takes to process the non-reduced relation (for the exact same result).

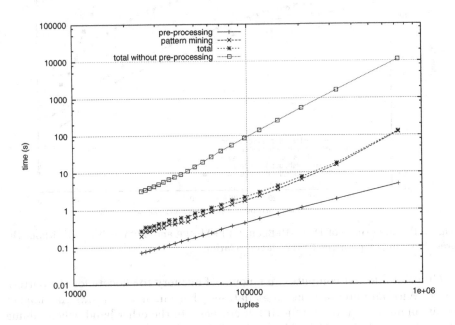

Fig. 2. Running times of CLEANRELATION and DATA-PEELER running with and without the pre-processing step on the `Retweet` relations

5.4 Time Gains over the Whole Task

Figures 3 and 4 show the time gains CLEANRELATION brings to all the three tested algorithms. 24 hours are not enough for CUBEMINER to directly mine the patterns in the `DistroWatch` dataset, even under the strongest considered constraints. However, in this same context but with the pre-process, it returns those patterns in 0.028s, i.e., at least three million times faster than without CLEANRELATION. DATA-PEELER, which is the fastest algorithm when no pre-process is used, remains the fastest when it is used. However DATA-PEELER benefits less from CLEANRELATION than CUBEMINER. The pre-process allows to divide the overall running time by a factor ranging between 2 and 4. TRIAS is 5 to 100 times faster when it mines the reduced relation rather than the original one. It is faster for TRIAS to compute from the reduced relation all patterns with at least four semesters, 16 distributions and 16 countries than to compute from the original relations the patterns with at least four semesters, 23 distributions and 23 countries.

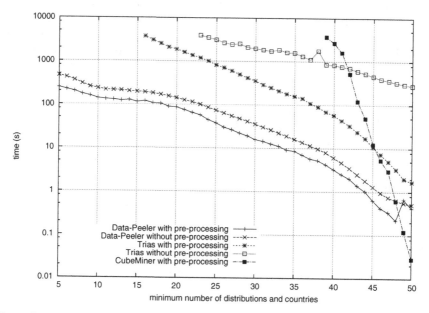

Fig. 3. Running times of DATA-PEELER, CUBEMINER and TRIAS with and without the pre-processing step on the `Distrowatch` relation

CUBEMINER first computes an amount of memory to allocate for the pattern space. With an unreduced `Retweet` dataset, that amount overflows the number of bits in an integer and CUBEMINER crashes. On the other hand, when mining the reduced relation, CUBEMINER is efficient. It even competes with DATA-PEELER for the sparsest versions of the dataset. Within a few hours, TRIAS manages to extract the constrained patterns only if the relation is very sparse. By preceding the call of TRIAS by the pre-process, the results dramatically improve. The running times are divided by about 50,000. DATA-PEELER is the only algorithm that allows to extract, in a reasonable time, the patterns in the densest versions of the dataset. The pre-processing step helps it a lot in those more challenging contexts. In the dataset encoding whether a user was retweeted at least once when writing about a team during a day, the ternary relation, which used to contain 731,685 3-tuples, is reduced to only 170,388 tuples ($\approx 23\%$ of the original size). In sequence, DATA-PEELER runs about 100 times faster on the reduced relation.

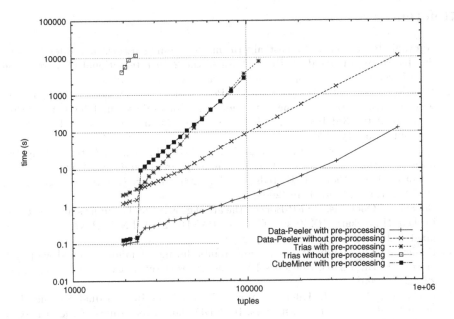

Fig. 4. Running times of DATA-PEELER, CUBEMINER and TRIAS with and without the pre-processing step on the Retweet relations

6   Conclusion

When searching for itemset-like patterns in an n-ary relation, constraints are, in practice, required. They specify some relevance criteria every pattern must satisfy, reduce the output to a manageable size and drastically lower the extraction times (if the algorithm can prune the search space with the constraints). In this article, we have identified a common property among constraints, the monotonicity w.r.t. one dimension, that allows to empty tubes (i. e., one-dimensional subspaces) of the relation while guaranteeing the presence of the same constrained patterns in the reduced data. Because an n-tuple belongs to n different tubes, constraints on the different dimensions of the pattern (e. g., minimal size constraints) work in synergy: emptying a tube makes it easier to empty the orthogonal tubes that used to contain the erased n-tuples. Once the fixed point reached, the actual pattern extraction, with any algorithm, takes place. Because real-life n-ary relations usually are sparse, the effectiveness of our pre-process can be impressive: in our experiments, the overall time to mine the patterns with the fastest algorithm, DATA-PEELER, is lowered by a factor typically ranging from 10 to 100 and it can reach millions for less efficient algorithms. In the same way that the idea behind ExAnte [3] was then applied along the search for the itemsets [2,4], we currently investigate an analog integration of the present proposal into DATA-PEELER.

References

1. Agrawal, R., Srikant, R.: Fast algorithms for mining association rules in large databases. In: VLDB 1994: Proceedings of the 20th International Conference on Very Large Data Bases, pp. 487–499. Morgan Kaufmann (1994)
2. Bonchi, F., Giannotti, F., Mazzanti, A., Pedreschi, D.: ExAMiner: Optimized level-wise frequent pattern mining with monotone constraints. In: ICDM 2003: Proceedings of the 3rd International Conference on Data Mining, pp. 11–18. IEEE Computer Society (2003)
3. Bonchi, F., Giannotti, F., Mazzanti, A., Pedreschi, D.: ExAnte: Anticipated data reduction in constrained pattern mining. In: Lavrač, N., Gamberger, D., Todorovski, L., Blockeel, H. (eds.) PKDD 2003. LNCS (LNAI), vol. 2838, pp. 59–70. Springer, Heidelberg (2003)
4. Bonchi, F., Goethals, B.: FP-Bonsai: the art of growing and pruning small FP-trees. In: Dai, H., Srikant, R., Zhang, C. (eds.) PAKDD 2004. LNCS (LNAI), vol. 3056, pp. 155–160. Springer, Heidelberg (2004)
5. Boulicaut, J.F., Jeudy, B.: Using constraints during set mining: should we prune or not? In: BDA 2000: Actes des 16ème Journées Bases de Données Avancées, pp. 221–237 (2000)
6. Bucila, C., Gehrke, J., Kifer, D., White, W.M.: DualMiner: a dual-pruning algorithm for itemsets with constraints. In: KDD 2002: Proceedings of the 8th ACM SIGKDD International Conference on Knowledge Discovery and Data Mining, pp. 42–51. ACM Press (2002)
7. Cerf, L., Besson, J., Robardet, C., Boulicaut, J.F.: Closed patterns meet n-ary relations. ACM Transactions on Knowledge Discovery from Data 3(1), 1–36 (2009)
8. Gallo, A., Mammone, A., Bie, T.D., Turchi, M., Cristianini, N.: From frequent itemsets to informative patterns. Tech. Rep. 123936, University of Bristol, Senate House, Tyndall Avenue, Bristol BS8 1TH, UK (December 2009)
9. Grahne, G., Lakshmanan, L.V.S., Wang, X.: Efficient mining of constrained correlated sets. In: ICDE 2000: Proceedings of the 16th International Conference on Data Engineering, pp. 512–521. IEEE Computer Society (2000)
10. Jaschke, R., Hotho, A., Schmitz, C., Ganter, B., Stumme, G.: TRIAS–an algorithm for mining iceberg tri-lattices. In: ICDM 2006: Proceedings of the 6th IEEE International Conference on Data Mining, pp. 907–911. IEEE Computer Society (2006)
11. Ji, L., Tan, K.L., Tung, A.K.H.: Mining frequent closed cubes in 3D data sets. In: VLDB'06: Proceedings of the 32nd International Conference on Very Large Data Bases, pp. 811–822. VLDB Endowment (2006)
12. Kwak, H., Lee, C., Park, H., Moon, S.: What is twitter, a social network or a news media? In: WWW 2010: Proceedings of the 19th International World Wide Web Conferences, pp. 591–600. ACM Press (2010)
13. Nataraj, R.V., Selvan, S.: Closed pattern mining from n-ary relations. International Journal of Computer Applications 1(9), 9–13 (2010)
14. Trabelsi, C., Jelassi, N., Ben Yahia, S.: Scalable mining of frequent tri-concepts from *Folksonomies*. In: Tan, P.-N., Chawla, S., Ho, C.K., Bailey, J. (eds.) PAKDD 2012, Part II. LNCS, vol. 7302, pp. 231–242. Springer, Heidelberg (2012)

Interestingness-Driven Diffusion Process Summarization in Dynamic Networks

Qiang Qu[1], Siyuan Liu[2], Christian S. Jensen[3], Feida Zhu[4], and Christos Faloutsos[5]

[1] Department of Computer Science, Aarhus University
[2] Heinz College, Carnegie Mellon University
[3] Department of Computer Science, Aalborg University
[4] School of Information Systems, Singapore Management University
[5] School of Computer Science, Carnegie Mellon University
qu@cs.au.dk, siyuan@cmu.edu, csj@cs.aau.dk, fdzhu@smu.edu.sg,
christos@cs.cmu.edu

Abstract. The widespread use of social networks enables the rapid diffusion of information, e.g., news, among users in very large communities. It is a substantial challenge to be able to observe and understand such diffusion processes, which may be modeled as networks that are both large and dynamic. A key tool in this regard is data summarization. However, few existing studies aim to summarize graphs/networks for dynamics. Dynamic networks raise new challenges not found in static settings, including time sensitivity and the needs for online interestingness evaluation and summary traceability, which render existing techniques inapplicable. We study the topic of dynamic network summarization: how to summarize dynamic networks with millions of nodes by only capturing the few most interesting nodes or edges over time, and we address the problem by finding interestingness-driven diffusion processes. Based on the concepts of diffusion radius and scope, we define interestingness measures for dynamic networks, and we propose OSNet, an online summarization framework for dynamic networks. We report on extensive experiments with both synthetic and real-life data. The study offers insight into the effectiveness and design properties of OSNet.

1 Introduction

The summarization of networks or graphs continues to be an important research problem due in part to the ever-increasing sizes of real-world networks. While most studies consider the summarization of static networks according to criteria such as compression ratio, network representation, minimum loss, and visualization friendliness [15,20], recent developments in social network mining and analysis as well as in location-based services [6, 13] and bioinformatics [20] give prominence to the study of a new kind of dynamic network [9, 10] that captures information diffusion processes in an underlying network. These developments offer new challenges to network summarization.

An information diffusion process in a network can be represented by a stream of timestamped pairs of nodes from the underlying network, where a timestamped pair indicates that information was sent from one node to the other at the given time. This stream can be modeled as a *dynamic network*. An example of an information diffusion

T. Calders et al. (Eds.): ECML PKDD 2014, Part II, LNCS 8725, pp. 597–613, 2014.
© Springer-Verlag Berlin Heidelberg 2014

process is the spread of news items among Twitter users by means of the network's "reply/(re)tweet" functionality.

While the network that may be created from a completed diffusion process by assembling the node pairs that represent the process is a static network, the summarization task for a diffusion process distinguishes itself from that of a static network. The critical difference lies in that, for a dynamic diffusion process, it is most valuable to capture each "interesting" development as the process evolves, in an online fashion. This problem is termed *dynamic network summarization* (DNS) and has many applications. We highlight several as follows.

In information visualization, massive dynamic networks are hard to visualize due to their size and evolution [14]. With DNS, it is possible to create online, time-labeled summaries in the form of "trajectories" such that it is possible to view the change in a diffusion process as it evolves. In social graph studies, DNS enables the identification of interesting dynamics in the form of "backbones" that describe how information propagates and that can help capture the evolving roles of different participants in diffusion processes. This is useful for tasks such as change detection [18] and trend mining [4]. In road traffic analysis, DNS can capture major traffic flows. Summaries for given periods can be projected onto the road network to detect traffic thoroughfares, provide better road planning services, or analyze how people move in a city [11].

One approach is to compute a summary from the evolving diffusion process periodically. Thus, the process is represented by a sequence of summaries of static networks. Each network aggregates edges and nodes in a time interval of size $\triangle t$ [14]. However, this approach is costly when networks are large. Further, parameter $\triangle t$ is fundamentally hard to set: if it is too small, performance deteriorates, while if it is too large, important diffusion dynamics may be missed. Even if given a $\triangle t$, most of the previous methods show difficulty in producing results that capture interesting dynamics, because their specific criteria and goals do not target dynamics.

As suggested by the application examples, DNS faces unique challenges.

(1) Time Sensitivity. Diffusion processes often represent vast, viral, and unpredictable processes, e.g., breaking news and bursty events [21]. As a result, the rate of diffusion can vary drastically over a short period of time. It is a difficult challenge to respond adaptively to the changing dynamics and to achieve timely summarizations.

(2) Online Interestingness Evaluation. A key challenge is to capture the most interesting nodes and edges in summarizations. Compared with traditional network summarization, interestingness evaluation in DNS assumes an extra degree of difficulty because of the partial view of the network at any time of the evaluation.

(3) Summary Traceability. An important goal is to enable a better understanding of the evolution of the diffusion process throughout its life cycle. A good summary should reveal the flow of the dynamics so that interesting developments can be traced.

To tackle the DNS problem, we propose **OSNet**, a framework for Online Summarization of Dynamic Networks that aims to produce concise, interestingness-driven summaries that capture the evolution of diffusion processes. Our contribution is fivefold: 1) Unlike previous proposals that apply optimization criteria in offline settings, we consider a setting where network summarization occurs online, as a diffusion process evolves. 2) Based on the concepts of propagation radius **proRadius** and propagation scope **proScope**, we formalize the problem of characterizing the interesting dynamics of an evolving diffusion process in a traceable manner. 3) We propose OSNet that encompasses online and incremental dynamic network summarization algorithms on a spreading tree model. In terms of entropy, OSNet archives the best summaries with respect to informativeness. 4) A generalization of OSNet is presented. 5) Extensive experiments are conducted with both synthetic and real-life datasets.

2 Problem Definition

The input to the problem is a stream of time ordered interactions (i.e., diffusion processes) on a network G. We define a network as a labeled graph $G = (V, E, l_G)$, where V is a set of nodes, $E \subset V \times V$ is a set of undirected edges, and l_G is a labeling function. Given a set $\Sigma = \{\varsigma_1, \varsigma_2, \ldots, \varsigma_k\}$ of labels, labeling function $l_G : V(G) \mapsto \Sigma$ maps nodes to labels.

A diffusion process on a network G, denoted by $\mathcal{D}(G)$, is a stream of time-ordered interactions. An interaction $x = (\delta, u, v, t) \in \mathcal{D}(G)$ indicates that a specific story is diffused from node u to node v at time $t \in \mathcal{T}$. A story is defined by a textual keyword list used to describe an event, such as breaking news in Twitter. The diffusion from u to v captures that node v receives the story from u. We also say that u is an infector of v while v is an infectee of u. We call time t the infection time of node v. Note that a diffusion process of a story can be initiated by different nodes that are regarded as seeds or roots. For each interaction x, we further define δ to be a three-tuple as a canonical identifier, i.e., $\delta = (storyID, v_r, t')$, where $storyID$ is the identity of the diffusing story, v_r represents the seed node starting the diffusion, and t' is the infection time of the infector u. The diffusion process from a seed over a time period forms a time-stamped graph, known as a network cascade C [9, 18] where each interaction is a directed edge from the infector to the infectee.

Definition 1. [Cascade C] *A cascade C is a directed graph $C = (V_C, E_C, l_{V_C}, l_{E_C})$, representing a diffusion process $\mathcal{D}(G) = \{x = (\delta, u_i, v_i, t_i)\}$ diffusing from a seed on a story during a time period $\mathcal{T}$. The node set is $V_C = \cup u_i + \cup v_i$, and the edge set is $E_C = \cup_{u_i, v_i \in x}(u_i, v_i)$. A node pair (u_i, v_i) for each x is considered as a directed edge from u_i to v_i. $l_{V_C} : V_C \mapsto \Sigma$ is node labeling function, and $l_{E_C} : E_C \mapsto \mathcal{T}$ is an edge labeling function.*

A network G with diffusion processes is termed a diffusion network or a dynamic network, which, for simplicity, we also denote by G. Given a diffusion network G, a set $\mathcal{I}(G) \in V(G)$ is given that contains the seed nodes from which a diffusion starts. The infection time of a seed v_r is given as t_{v_r}. We use $\deg^+(u)$ to denote the number of infectees of a node u in a cascade C.

Before we present the definition of interestingness, two measures are introduced to evaluate nodes in a dynamical process by i) how far the information can travel (Measure 1: depth) and ii) how many infectees a node can have (Measure 2: breadth). These two measures can be used for capturing the **interestingness** of a diffusion process for three reasons : 1) The two measures agree with intuition. 2) The two measures capture the cascade, enabling reconstruction with little more information. 3) The two measures offer a foundation for computing different properties of a cascade. In addition, we observe that other studies also suggest that the two measures can characterize diffusion processes [2, 22].

Measure 1. [Propagation Radius (proRadius)] *The propagation radius of a node v in a cascade C, denoted by $y(v)$, is the length of the path $l(v)$ from the root of C to v, $|l(v)|$. The maximum propagation radius of a node in C is the diameter of C: $d(C) = \max(y(v))$. Note that the propagation radius of the root is 0.*

Measure 2. [Propagation Scope (proScope)] *The **proScope**, $w(v) = deg^+(v)$, of a node v for a cascade C is the number of infectees of v in C.*

Definition 2. [Interestingness] *We represent a node v by a vector $(y(v), w(v))$ and use Equation 1 to quantify the total interestingness of the node. As the degree distribution of many networks follows a power-law, we use a log value of the **proScope**:*

$$\xi(v) = \alpha \log w(v) + (1 - \alpha)y(v), \tag{1}$$

*where $\alpha \in [0, 1]$ balances the two measures. We set $\log w(v) = 0$ if $w(v) = 0$. Note that cascades evolve over time as interactions arrive in the stream. We thus use $\xi_t(v)$ to denote the interestingness of a node v at time t, which is calculated using the values of **proScope** and **proRadius** of v at t.*

Definition 3. [Interesting Summary $S(C)$] *Given a cascade C and a threshold τ, an interesting summary $S(C)$ is a subgraph of C satisfying that for any node $v_i \in S(C)$, $\xi_t(v_i) > \tau$ holds; for two nodes u and v in $V(S(C))$, the edge $e' = (u, v)$ exists in $S(C)$ if and only if $e = (u, v)$ exists in C. Labels of the edges and nodes in $S(C)$ retain the labels they have in C.*

Definition 4. [Traceable Interesting Summary $\mathbf{S}(C)$] *Given an interesting summary $S(C) \subset C$, a traceable interesting summary $\mathbf{S}(C)$ is a super-graph of $S(C)$, denoted $S(C) \subset \mathbf{S}(C)$. A node v_i in C is in $\mathbf{S}(C)$ if: v_i is the seed, or $\xi_t(v_i) > \tau \vee (\exists v_j \in C, (v_i \in l(v_j) \wedge \xi_t(v_j) > \tau))$.*

As some nodes are removed from an interesting summary (Definition 3), remaining interesting nodes may become disconnected. Definition 4 includes the missed nodes on the paths from the seed to the remaining interesting nodes. A traceable interesting summary thus is possible to reveal the flow of dynamics and interesting developments can be traced throughout their life cycle. To explain the evolution in a traceable interesting summary, we next introduce the concepts **diffusion rise** and **diffusion decay**, defined by the notion of acceleration intensity. In the rest of the paper, we use a summary (summaries) to indicate a traceable interestingness summary (summaries) for simplicity.

Definition 5. [Acceleration Intensity ϱ] *Given a node v_i as an infector of a node v_j in a cascade C, the acceleration intensity is defined based on the diffusion path from v_i to v_j ($l(v_i, v_j)$) in C as*

$$\varrho(l(v_i, v_j)) = \frac{\xi_{t_j}(v_j) - \xi_{t_i}(v_i)}{|t_j - t_i|}, \tag{2}$$

where t_i and t_j are the infection times of v_i and v_j, respectively.

We can now define the rise and decay of a diffusion process: When $\varrho > 0$, the propagation process from v_i to v_j is a diffusion rise process; otherwise, it is a diffusion decay process.

The goal of the DNS problem is to better understand network dynamics. A summary thus needs to be informative with respect to the original data. There are several methods to evaluate informativeness. Among these, we propose to use Entropy. A review of Shannon Entropy and details are presented in Section 3.2. Here we denote the entropy of a traceable interesting summary $\mathbf{S}(C)$ by $H(\mathbf{S}(C))$. Recall that the entropy gains when its value decreases. We thus aim to find a summary with minimal entropy to achieve the best informativeness. The problem is stated as follows:

Problem Statement (Interestingness-driven Diffusion Process Compression). *Given a diffusion network G with seed sets $\cup \mathcal{I}(G)$, stories diffuse from each seed over time. The dynamic process is represented by a stream of interactions, which forms a set of cascades $\{\ldots, C_i, \ldots\}$. The output of the problem is a set of traceable interesting summaries $\mathbf{S}(G) = \{\ldots, S_i(C_i), \ldots\}$ $(|S_i(C_i)| > 0)$. The entropy $(H(\mathbf{S}(C)))$ of each summary $S_i(C_i)$, which reveals diffusion rise and decay, is minimized subject to the balancing parameter $0 \leq \alpha \leq 1$ of the aggregate score and the interestingness threshold $\tau \geq 0$.*

To solve the problem, two sub-problems have to be solved: i) How to model the dynamics on the top of graphs? Is the cascade model suitable? The diffusion processes we discuss are evolving over time. And all the cascades on a node are merged. This may cause problems for the summarization because the interestingness of a node is associated with time stamps and stories as node instances. This requires to design a labeling function to distinguish the node instances, which is ineffective. ii) How to set proper values for α and τ for different diffusion processes? Given an α in the range $[0, 1]$, each connected subgraph of a cascade C over time can be a summary, which yields a hard graph decomposition problem. On the other hand, the scale of a summary mostly depends on the threshold τ. A proper value is necessary because we intend to find all interesting developments. We proceed to develop the OSNet framework that encompasses new and incremental techniques capable of continuously summarizing dynamics based on a spreading tree model in step with the evolution of diffusion processes.

3 Our Method

3.1 Spreading Tree Model

Although network cascades can model diffusion processes. several issues of dynamics challenge the effectiveness of network cascades. First, the interactions on a node are merged in cascades [9]. However, in dynamic networks, a node may become interesting only at a specific interaction, which would require extra efforts in designing labels to distinguish different interactions and cascades. Furthermore, as cascades are directed graphs, there exist backward and forward edges or even cycles. This makes a cascade hard to interpret and navigate. Second, the cascade model is a graph model. Summary search can then be regarded as subgraph search. However, graph search is usually time-consuming since it involves isomorphism checking. Third, since cascades are merged into one directed graph, the graph search space grows exponentially, which makes the problem even harder. We propose to instead use a **Spreading Tree** model. First, spreading trees are constructed directly by interactions without any other efforts. The model distinguishes interactions and cascades by itself. Next, tree search is relatively efficient. Numerous proposals of efficient tree operations exist. Third, there are no backward and forward edges in spreading trees. The tree structure is not as complex as a cascade. The search space is proportional to the scale of the interactions.

Definition 6. **[Spreading Tree** T**]** *A spreading tree* $T = (v_r, V', E', l_{V'}, l_{E'})$*, is a rooted and labeled n-ary tree, where* $v_r \in V'$ *is the root,* V' *is a set of nodes,* $E' \subseteq V' \times V'$ *is a set of edges,* $l_{V'} : V' \mapsto \Sigma$ *is node labeling function, and* $l_{E'} : E' \mapsto \mathcal{T}$ *is an edge labeling function.*

Intuitively, a node represents a specific user in a network, and the node's label is the name of the user; an edge in a spreading tree connects an infector node with an infectee node, and the edge's label is the infection time of the infectee node. A non-root node has one infector. A non-leaf node has one or more infectees, and a leaf node has no infectees.

Given a diffusion network G, each seed $v_r \in V(G)$ forms the root of a spreading tree. When an interaction $x = (\delta, u, v, t) \in \mathcal{D}(G)$ arrives, the spreading tree for δ is updated by inserting a **new** node labeled v and an edge labeled t from an **existing** node labeled u to v. Note that both u and v are labels of the nodes. To find the existing node u, we search the tree in breadth-first order starting from the root until a node with label u and infection time $\delta.t'$ is found. Therefore, although multiple nodes have the same label, the three-tuple δ can determine from which node to insert the edge to the new infectee.

From the above, the spreading tree model achieves the following properties: 1) cascades can be equally modeled as spreading trees, such that the summarization on cascades equals the task on spreading trees; 2) the trees are separated by seeds; 3) a node can be duplicated in a spreading tree, which shows the model distinguishes node instances; 4) the size of trees is proportional to the scale of interactions; 5) infection occurs top-down, and diffusion always occurs from a parent node to a child node.

3.2 Parameter Relief

Although using fixed values for parameters is simple for implementation, two main issues demand better approaches. First, for a single diffusion process, prediction of the network statistics (arrival rate, number of infectees, propagating range, etc.) is usually difficult. Thus, it is hard to find parameter settings that can best capture the dynamics. Second, different diffusion processes vary substantially in range and scope. Thus, the same settings are not likely to work across different processes. Our study aims to provide a self-tuning mechanism that adapts to differences in the summarization.

Alpha Estimation. Recall that the entropy H of a random variable E with possible values $\{e_1, \ldots, e_n\}$ is defined as

$$H(E) = -\sum_i^n p(e_i) log_2 p(e_i), \tag{3}$$

where $p(e_i)$ is the probability mass function of outcome e_i. $H(E)$ is close to 0 if the distribution is highly skewed and informative.

We measure the entropy of a summary $\mathbf{S}(C)$ and aim to maximize the informativeness of $\mathbf{S}(C)$ to have the maximum possible information out of T. Given a set of continuous interactions $\mathcal{D}(G)$ by time t (denoted by $\mathcal{D}(G)_t$), the probability of diffusing story i from a node v_j is regarded as a conditional probability:

$$p_{(i,t)}(v_j) = p_{(i,t)}(v_i) \times \frac{\sum^{\mathcal{D}(G)_t} f_{(i,t)}(v_j, x)}{|\mathcal{D}(G)_t|},$$

where $p_{(i,t)}(v_i)$ is the probability of the node that infects v_j. Note that for a seed node, the probability of its infector is 1 in order to guarantee that a root is infected. The function $f_{(i,t)}(v_j, x)$ is an indicator that is 1 if v_j is an infectee in x when $storyID = i$ by time t, otherwise 0. Then we use the entropy $H_{p(i,t)}(\mathbf{S}(C))$ as an informativeness measure of a summary $\mathbf{S}(C)$ with respect to T:

$$H_{p(i,t)}(\mathbf{S}(C)) = - \sum_{j=1}^{|\mathbf{S}(C)|} p_{(i,t)}(v_j) \log p_{(i,t)}(v_j). \tag{4}$$

Thus, $\mathbf{S}(C)$ is the most informative by time t with respect to T if the value of its entropy $H_{p(i,t)}$ is minimized. Before we present the details of the estimation, Lemma 1 is introduced as a property of a summary's entropy.

Lemma 1. *If two summaries $\mathbf{S}(T)$ and $\mathcal{S}'(T)$ satisfy $d(\mathbf{S}(T)) > d(\mathcal{S}'(T))$, $V'(\mathcal{S}'(T) \setminus \mathbf{S}(T)) = \emptyset$, and $|l(v)| > d(\mathcal{S}'(T))$ where $v \in \mathbf{S}(T) \setminus \mathcal{S}'(T)$, then we have $H_{p(i,t)}(\mathbf{S}(T)) \le H_{p(i,t)}(\mathcal{S}'(T))$ holds.*

The proof is omitted due to the space limitation. It shows that the entropy is smaller for those summaries with greater depth. By Equation 1, to achieve the smallest entropy, we need to minimize α because a smaller α yields a higher weight for depth such that deep summaries are preferred. In the remainder of the section, we present the bounds on α followed by our estimation based on entropy.

Lemma 2. *The depth of a summary $\mathbf{S}(T)$ is bounded by the parameter α as $d(T) \ge \tau/(1-\alpha)$.*

Proof. By Measure 1, we have $\max(y(v)) = d(T), v \in \mathbf{S}(T)$. Given such a node v, we have $(1-\alpha)d(T) \ge \tau$ when we set $w(v) = 0$.

Theorem 1. *Let n as the maximal number of nodes in a summary $\mathbf{S}(T)$ with a threshold τ. The parameter α is bounded as*

$$\tau / \sqrt[d(T)]{\frac{n}{d(T)+1}} \le \alpha \le 1 - \tau/d(T). \tag{5}$$

Proof. Lemma 2 confirms the right part of Equation 5. The left part is achieved as follows. Similar to Lemma 2, the maximum fanout of $\mathbf{S}(T)$ is τ/α as a positive integer and larger than 1. Then we obtain $(\frac{\tau}{\alpha})^{d(T)} + \ldots + \frac{\tau}{\alpha} + 1 = \sum_i^{d(T)+1} (\frac{\tau}{\alpha})^i \le n$. Since $\sum_i^{d(T)+1} (\frac{\tau}{\alpha})^i \le (d(T)+1)(\frac{\tau}{\alpha})^{d(T)}$, the given $\mathbf{S}(T)$ has at most n nodes, i.e., $(\frac{\tau}{\alpha})^{d(T)} + \ldots + \frac{\tau}{\alpha} + 1 \le n$, if $(d(T)+1)(\frac{\tau}{\alpha})^{d(T)} \le n$. By transforming the inequality, the left part follows.

Several studies [9, 22] have shown that most diffusion processes are within 3 hops in social networks. Without loss of generality, we assume that the lower bound of the depth of $S(T)$ is 3. The bound yields the maximum lower-bounded α. As we know, the minimum α turns out to produce the most informative summaries. Thus, by Equation 5 we have the estimation for α as:

$$\alpha = \tau \sqrt[3]{\frac{4}{n}}, \tag{6}$$

to obtain the minimum entropy. The estimation is therefore able to facilitate summarization regardless of varying dynamics.

Threshold Selection. The goal is to find the most interesting developments of dynamics over time as summaries. This naturally requires OSNet to only focus on the small set of the interesting nodes and edges in a spreading tree T. Our goal is to find a proper threshold that can make the summarization converge fast and produce a small sized summary over time. However, the changes and differences of dynamics challenge the setting of such a threshold. Therefore, a selection mechanism adapting to the trends of dynamics (i.e., rise and fall) is necessary.

The idea of the proposed solution is to maintain a variable τ' for each spreading tree T, which is the maximum value (MAX) of $\xi_{t'}(v_i)$, $v_i \in T$ by time t'. During the summarization, we compare a new interestingness score $\xi_t(v_j)$ with τ': if $\xi_t(v_j) > \tau'$, then $\tau' = \xi_t(v_j)$, and v_j is inserted into the corresponding $S(T)$. If we have a value of τ' that is large enough, OSNet converges to a relatively steady state until there is a more interesting node, e.g., far away from the seed and with many infectees, to exhibit another rise of the diffusion. Thus, in a summary $S(T)$ based on MAX, the interesting nodes (by the first condition in Definition 3) in deeper levels always show diffusion rises from those in lower levels. From an interesting node to a node recovered for the next interesting node, the flow is always a diffusion decay.

Other methods than MAX would be possible, e.g., average value (AVG) of $\xi_{t'}(v_i)$ as $\sum_{v_i \in V} \xi_{t'}(v_i)/|V|$. We compare these alternatives experimentally in Section 4.

3.3 Algorithmic Framework and Details

Framework Overview. An overview of OSNet is shown in Figure 1. The input is a diffusion process $\mathcal{D}(G)$ that is captured by a set of indexed spreading trees. There are indexes on storyID and seeds, such that we can insert an interaction into a spreading tree T_i efficiently. By Equation 1, the interestingness-based operator is to evaluate the interestingness of nodes in spreading trees with two parameters, α and τ. We evaluate the interestingness of a node v when it infects new nodes (i.e., $w(v)$ increases). If v has $\xi_t(v) > \tau$, it is inserted into a summary $S(T_i)$. The summaries are also indexed in the same way as T. We thus insert v into $S(T_i)$ by searching storyID and seed. Once a node v is inserted into tree T, it is tagged with its branch such that a node cannot be reinserted into the summary $S(T_i)$. We only insert new nodes and edges into a tree over time, and it is not necessary to rebuild any part of T or $S(T)$.

When a node v of T_i is to be inserted into $S(T_i)$ at time t, there may be three cases: 1) $S(T_i)$ does not exist and v_i is not a seed ($v \notin \mathcal{I}(G)$); 2) $S(T_i)$ exists and the infector of v in T_i is already in $S(T_i)$; 3) $S(T_i)$ exists and the infector of v in T_i is not in $S(T_i)$. Cases 1) and 2) are straightforward. We can create a new tree for case 1); and for case 2),

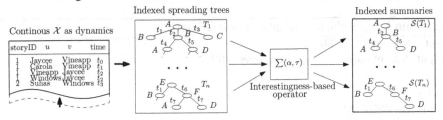

Fig. 1. Overview of the OSNet framework

we insert v as a child of its infector in $\mathbf{S}(T)$. In case 3), the insertion of v renders $\mathbf{S}(T_i)$ disconnected, and the process thus cannot be traced from the seed to v. A solution is to recover all the nodes in the path from the root to v_i. We call this problem the *Recovery Problem*.

Path Recovery. An efficient way in a tree-based data model to solve the *Recovery Problem* is to construct $\mathbf{S}(T)$ as a search tree. The basic idea is that all the siblings at each level of $\mathbf{S}(T)$ are ordered. The canonical ordering is based on timestamps of tree branches (edges) and node labels. If a node v_j gets infected from v_i at time t_i, v_j is inserted in the approach: the timestamps of edge labels of all the siblings on the left are no later than t_i, and the node labels on the left are not lexicographic larger than v_j. Lemma 3 presents the worst case search cost of the search tree.

Lemma 3. *Let a tree T have n nodes and fanout d. The worst case search cost when $d(T)$ is minimum is:*

$$O(\log_d^{(n(d-1)+1)}(\log_d^{(n(d-1)+1)}-1)\log_2 d).$$

The proof is omitted due to the space limitation.

Algorithm Details. There exist two essential components of OSNet depicted by Algorithm 1: 1) Constructing spreading trees (from lines 3 to 5); 2) Summarizing the most interesting dynamics into $\mathbf{S}(T)$ (from lines 8 to 10). Specifically, the following explains details. We allow users to terminate the summarization process through variable breakFlag in line 2. According to applications, one can also bound the size of $\mathbf{S}(T)$ to abort the algorithm. Note that we have no limitation on the size n. Once a new interaction $x(\delta, v_i, v_j, t)$ arrives (line 3), we call mapT in line 4 to retrieve the T of story δ. Next, branchOut in line 5 is an insertion, which inserts an infectee v_j from v_i with an edge labeled by t into T. We implement each $\mathbf{S}(T)$ as a search tree. From line 6, we summarize the updated node according to Equation 1. If the node's interestingness exceeds the threshold, it shows a diffusion rise, and the node is inserted into $\mathbf{S}(T)$. Parameters are automatically adjusted in line 7 based on discussion in Section 3.2.

Before inserting a node into $\mathbf{S}(T)$, we retrieve the path from the root in line 8 by iteratively pushing an infector (function Push) into list path. We then insert the missed nodes and edges into $\mathbf{S}(T)$ in line 10. These nodes show diffusion decays from the last interesting node, but rises to the next. Summaries are returned if necessary in line 11.

Algorithm 1. Algorithmic description of the OSNet

Input   : Network G, seed set $\mathcal{I}(G)$.
Output: A set of summarized spreading trees, $\mathbf{S}(G)$.

1 **begin**
2 | Threshold $\tau \leftarrow 0$; weight parameter $\alpha \leftarrow 0$;
| Boolean beakFlag $\leftarrow false$;
| List path $\leftarrow null$;
| Spreading tree set Set(T) rooted by seeds in $\mathcal{I}(G)$;
| **if** *beakFlag* $== false$ **then**
3 | | **if** $x(\delta, v_i, v_j, t) \leftarrow \mathcal{D}(G)[t_{ij}]$ *exists* **then**
4 | | | $T \leftarrow$ mapT(Set(T), δ)
| | | /* add x onto T. */
5 | | | $v_i \leftarrow$ Search(T, v_i) branchOut(v_i, v_j, t)
6 | | | **if** $\xi(v_i) > \tau$ **then**
7 | | | | $\tau \leftarrow \xi(v_i), \alpha \leftarrow \tau \sqrt[3]{\frac{4}{n}}$ /* retrieve path from T. */
8 | | | | **while** $v_i.getInfector(T) \notin \mathcal{I}(G)$ **do**
9 | | | | | path.Push($v_i.getinfector(T)$)
10 | | | | $\mathbf{S}(T) \leftarrow$ getST(Set(T), T) insertPath($\mathbf{S}(T)$, path)
11 | **return** *Set*(T)

---

4 Evaluation

4.1 Experimental Methodology and Settings

The study probes into three questions: 1) **Sense-making Evaluation**: Compared with the state-of-the-art, do summaries generated by OSNet make sense and achieve the goal of capturing interesting dynamics? 2) **Parameter Study**: Can we use fixed parameters? What are the effects of the parameters? Does OSNet converge fast, using MAX or AVG? 3) **Real-life Data**: How does OSNet work on real-life data?

The experiments on synthetic data are used to test whether our methods produce expected results in a controlled environment. We first generate an underlying structure G_0 containing 10,000 nodes. With a random seed set $\mathcal{I}(G_0)$, we then start the propagation for each seed in a breadth-first manner. The number of infectees of a node v obeys the following models to simulate different dynamics: I) Gaussian distribution (G); II) Poisson distribution (P); III) Zipf distribution (Z), which is an approximate power law probability distribution. We define the **modeled number of nodes** to be the number of nodes we choose for a dataset, and we require that their numbers of infectees obey one of the three distributions. To simulate continuous dynamics, we generate the interactions as a data stream with an arrival rate of 1 per millisecond.

Experiments were conducted on a 3.2 GHz Intel Core i5 with 16GB 1600 MHZ DDR3 main memory and running OSX 10.8.5. Algorithms were implemented in JDK 1.6.

4.2 Sense-Making Evaluation

We compare OSNet with several existing algorithms using synthetic data. To enable existing methods to support diffusion processes, we generate a graph sequence for each dataset, in which each graph aggregates all edges and nodes in a time interval Δt. Due to the space limitation, we only report results for several time intervals. Similar findings apply to other intervals. We compare our techniques against the following state-of-the-art algorithms:

- **DisSim-Alg:** This is a graph compression algorithm that abstracts a large graph into a smaller graph that contains approximately the same information. It is developed based on the notion of dissimilarity between the decompression graph and the original graph. We use an existing implementation [20] and set the weight of an edge to 1 if the adjacent nodes diffuse infection by time t: otherwise, edge weights are set to 0.
- **MDL-Alg:** MDL is a successful and popular technique for graph compression. We compare against a recent study by Navlakha et al. [15] where a graph is compressed and represented as a graph summary and a set of corrections. We use the original GREEDY algorithm that offers the best compression and lowest cost [15]. To enable cliques to be merged into a single supernode, we add self-edges to each node before applying the algorithm.

(**OSNet**): Figures 2 to 4 show an example from t_1 to t_3 using data generated by applying Zipf distribution. The infector as the central node of each group is labeled with a canonical identifier for ease of explanation. The node with identifier 0 is the seed of the propagation. In the figures, the red and darker nodes are the nodes that are already infected; the grey and lighter nodes are other nodes in the synthetic networks. To facilitate visualization, we remove the background nodes and edges in the underlying networks that are not involved in the diffusion process. Figures 5 to 7 present the summaries by OSNet from t_1 to t_3. The results show the incrementality of the summaries. The intuitively interesting nodes are captured, and the summaries are traceable and connected paths, such that we can spot the dynamics from the start to the nodes i) that infect nodes in great quantity; and ii) that are far from the seed. We observe that the summaries in Figures 5 and 6 are the same. Although in the original diffusion process from t_1 to t_2, the diffusion reached nodes 86 and 87 at t_2, the number of infectees is quite few. Compared with the other nodes in $S(T)$, 86 and 87 are thus not interesting enough to be summarized. This shows that from t_1 to t_2 the diffusion process is not rising according to Definition 5, and OSNet adapts to the changes in diffusion. In contrast at time t_3, both 108 and 109 have many infectees and they are far away from the seed 0. They again expedite the diffusion process such that we capture the two as interesting nodes. If we only summarize the two without including nodes 86 and 87, we loose the connections that allow us to interpret how information propagates. Thus, 86 and 87 are recovered and included. The findings show that OSNet is capable of finding a small set of connected interesting nodes that meaningfully capture the diffusion process.

(**DisSim-Alg**): We vary Δt to generate graph sequences and try various values for the internal compression ratio parameter. We report three of representatives at t_1 in Figure 8. The findings show that the summaries vary a lot w.r.t. compression ratio. Comparing (a) and (c) where (c) is with a higher compression ratio, the graph size of (c) is much smaller but it is with less information of the propagation because *DisSim-Alg* aims to minimize the dissimilarity according to edge weights. To maintain a smaller dissimilarity, some edges or superedges are removed (e.g., (c)). Figure 8(b) shows a summary

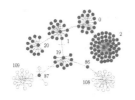

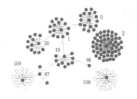

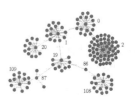

Fig. 2. A diffusion process $\mathcal{D}(G)$ at t_1

Fig. 3. The snapshot of $\mathcal{D}(G)$ at t_2

Fig. 4. The snapshot of $\mathcal{D}(G)$ at t_3

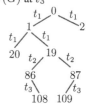

Fig. 5. The summary $\mathbf{S}(T)$ of $\mathcal{D}(G)$ at t_1

Fig. 6. The summary $\mathbf{S}(T)$ of $\mathcal{D}(G)$ at t_2

Fig. 7. $\mathbf{S}(T)$ at t_3 with recovery of 86 and 87

caused by using a smaller Δt that is larger than that of (a). This occurs because when the compression ratio is achieved, although new edges and nodes arrive, the algorithm only considers the dissimilarity and does not attempt further compression. As a result, the algorithm does not adapt to dynamics and capture traceable flows well.

(**MDL-Alg**): Figure 9 shows summaries obtained by *MDL-Alg* on the same diffusion process. *MDL-Alg* does not require the users to supply parameters. It computes the best cliques to merge in order to maintain a low cost. The findings show that *MDL-Alg* generates separate cliques, which makes little sense for diffusion process.

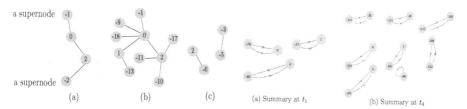

Fig. 8. Summaries by DisSim-Alg at t_1 **Fig. 9.** Summaries by MDL-Alg

4.3   Parameter Study and Understanding

(**Weight α on proScope**): We increase α from 0 to 1 in steps of 0.1. For synthetic data, the depth of the spreading trees is 100, and the modeled number of nodes is 1000. For Gaussian (G) datasets, we set the mean to 100 and the standard deviation to 20. The expect value for Poisson distribution (P) is 50. The maximum $deg^+(v)$ of the Zipf distribution(Z) is 200. Consequently, we have three datasets with $89,037$ (G), $45,306$ (P), and $36,892$ (Z) interactions, respectively. All the interactions are simulated as data streams with an arrival rate of 1 per millisecond. We set τ to 100, which means that the score of a node with expect out-degree $deg^+(v)$ is 100; and we set $\alpha = 1$. Figure 10

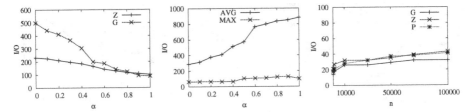

Fig. 10. I/O cost by varying α (a weight parameter)   **Fig. 11.** I/O cost by various strategies on τ selection   **Fig. 12.** I/O cost by varying n (the maximal number of nodes)

shows the I/O efficiency with respect to α. We count the I/O cost as the size of summaries, namely the number of interactions in $S(T)$. The I/O cost for the dataset P is 0 in the experiments, which means that no node in the propagation process gains a score that reaches 100. The findings show that the same fixed threshold does not work well across different datasets. For both G and Z in Figure 10, the I/O cost decreases as α increases. As we know, α controls the weight of **proScope**. Thus, when α is small, the **proRadius** becomes more important in Equation 1. As a result, nodes that are far away from a seed are more likely to be captured, which yields a higher I/O cost.

(**Threshold** τ): We compare our proposal that uses the current maximum score (MAX) against using the average historical score AVG. We use the same datasets as above. Figure 11 shows the findings on dataset G, which indicate that the I/O cost of MAX is much lower than that of AVG. And with a given α, the summarization with MAX converges faster to a relatively steady state than with AVG. MAX requires less updates on the summarized spreading trees than does AVG. Compared with the findings in Figure 10, the I/O cost increases as α increases when using AVG, because a larger value of α yields a larger score. This allows more nodes of a $\mathcal{D}(G)$ to be summarized, which increases the I/O cost. However for MAX, the cost remains almost the same when $\alpha < 0.6$ and it increases only slightly afterwards. We obtain similar results on the other two datasets.

By Equation 6, α never decreases because τ is based on the MAX strategy. This is beneficial for summarization for two reasons: i) With MAX, a larger α allows a bit more nodes to be summarized if diffusion rises; ii) a larger α decreases the influence of **proRadius** such that the summarization converges faster. This keeps OSNet from capturing too many nodes even when many are far away from seeds.

(**Maximum Possible Summary Size** n): We evaluate the effect of n in Equation 6 by varying n from 100 to 100,000. The findings in Figure 12 for all the three datasets show that the I/O cost increases as n increases. A larger n yields a smaller α by Equation 6. Figure 12 thus shows the same I/O cost trend as does Figure 10. However, the variation in Figure 12 is slight. For simplicity, we suggest to set n to be the (average) number of nodes of a T, which is also the maximum number of nodes that can be summarized in an $S(T)$.

4.4 Evaluation on a Real-life Social Network

We use data from Sina Weibo, a Chinese Twitter-like micro-blogging service platform (http://www.weibo.com) that has two important features that are not yet offered by

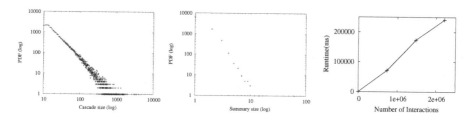

Fig. 13. Cascade size distribu- **Fig. 14.** Summary size distribu- **Fig. 15.** Runtime (including
tion of Weibo Data   tion of Weibo Data   data arrival time)

Fig. 16. A sample of diffusion processes   **Fig. 17.** A sample of summaries

Twitter: 1) A user can comment on any other user's tweets, which yields more user interactions; 2) The retweeting/forwarding chain is visible to the public, which is important for studying diffusion processes. Our dataset covers more than 1.8 million users, and we reconstruct the diffusion processes from their replies. There are 41,561 cascades (diffusion processes) with 2,211,221 interactions. We show that the probability density distribution (PDF) of the cascade size (Log-Log) as a property of the original data in Figure 13. The input is simulated as a stream with an arrival rate of 10 per millisecond, and **OSNet** outputs 8,647 summaries in which a seed has at least one infectee. Among the results, the summary with the most edges has 62 edges. The PDF of the summary size (Log-Log) is shown in Figure 14, which shows that most of the summaries are small. Figure 15 shows that the runtime of **OSNet** is proportional to the number of interactions in the dataset. Figure 16 shows a sample of diffusion processes represented by cascades. Figure 17 gives the corresponding **OSNet** summaries. Although the cascades in Figure 16 may merge on some nodes, the summaries are separated from each other w.r.t. stories. This is because **OSNet** models diffusion using spreading trees that naturally separate cascades. The summaries in Figure 17 show a vocabulary of patterns, which may be used for event classification or diffusion prediction based on diffusion processes.

5   Related Work

Statistical methods [3] are widely used to characterize properties of large graphs. However, most of these methods do not produce topological summaries, and their results are hard to interpret. Graph pattern mining [7] can be used for summarizing graphs, but usually yields overwhelmingly large numbers of patterns. Although constraint-based graph mining approaches [23, 24] are introduced to reduce the number of patterns, they only work for specific constraints. Further, summaries of diffusion processes are not inherently frequent.

Next, graph OLAP has been introduced to summarize large graphs [16,19]. However, most studies are designed for static network analyse and are limited to user-specified aggregation operations. Graph compression or simplification mainly focus on generating compact graph representations to simplify storage and manipulation. Much of the work has focused on lossless web graph compression [1, 17]. Most of these studies, however, only focus on reducing the number of bits needed to encode a link, and few compute topological summaries since the compressed representation is not a graph. Based on the MDL principle, Navlakha et al. [15] propose an error bounded representation that recreates the original graph within a bounded error. Toivonen et al. [20] merge nodes of a graph that share similar properties. Compared with these studies, our approach is developed to summarize diffusion processes.

As one of the attempts to consider time-evolving networks, Liu et al. [14] compress weighted time-evolving graphs, and they encode a dynamic graph by a three-dimensional array. The goal is to minimize the overall encoding cost of the graph. This is equivalent to compressing a sequence of static graphs according to time slices. Ferlež et al. [5] propose TimeFall in a principled MDL way to monitor network evolution, which clusters text in scientific networks and uses MDL to connect clusters. This class of studies are inherently distinct from ours in four aspects: 1) we use general networks and do not have assumptions on text processing; 2) OSNet takes as argument an interaction stream rather than a timestamped offline network; 3) we do not assume to be given a sequence of time-sliced graphs; 4) we aim to summarize diffusion processes. There are also studies on multiple social networks [12] and their temporal dynamics [8], including trend mining [4]. They focus on tasks different from ours.

6 Conclusion and Future Work

We studied the problem of dynamic network summarization and proposed an online, incremental summarization framework, OSNet, capable of simultaneously capturing the most intuitively interesting summaries that best represented network dynamics.

Several directions for future research are promising, including the development of techniques capable of exploiting the networks underlying diffusion networks, parallel processing of spreading trees, and summarization with structural network changes.

Acknowledgments. This research was supported in part by Geocrowd, funded by the European Commission as an FP7 Peoples Marie Curie Action under Grant Agreement Number 264994; the Singapore National Research Foundation; and the Pinnacle Lab at Singapore Management University. This material is based on work supported by the National Science Foundation under Grant No. CNS-1314632, and by the Army Research Laboratory under Cooperative Agreement Number W911NF-09-2-0053.

Any opinions, findings, and conclusions or recommendations expressed in this material are those of the author(s) and do not necessarily reflect the views of the National

Science Foundation or other funding parties. The U.S. Government is authorized to reproduce and distribute reprints for Government purposes notwithstanding any copyright notation here on.

References

1. Boldi, P., Vigna, S.: The webgraph framework I: compression techniques. In: WWW, pp. 595–602 (2004)
2. Cha, M., Mislove, A., Gummadi, K.P.: A measurement-driven analysis of information propagation in the Flickr social network. In: WWW, pp. 721–730 (2009)
3. Chakrabarti, D., Faloutsos, C.: Graph Mining: Laws, Tools, and Case Studies. Morgan & Claypool Publishers (2012)
4. Desmier, E., Plantevit, M., Robardet, C., Boulicaut, J.-F.: Trend mining in dynamic attributed graphs. In: ECML/PKDD, pp. 654–669 (2013)
5. Ferlež, J., Faloutsos, C., Leskovec, J., Mladenic, D., Grobelnik, M.: Monitoring network evolution using MDL. In: ICDE, pp. 1328–1330 (2008)
6. Hage, C., Jensen, C.S., Pedersen, T.B., Speicys, L., Timko, I.: Integrated data management for mobile services in the real world. In: VLDB, pp. 1019–1030 (2003)
7. Inokuchi, A., Washio, T., Motoda, H.: An apriori-based algorithm for mining frequent substructures from graph data. In: Zighed, D.A., Komorowski, J., Żytkow, J.M. (eds.) PKDD 2000. LNCS (LNAI), vol. 1910, pp. 13–23. Springer, Heidelberg (2000)
8. Kumar, R., Novak, J., Tomkins, A.: Structure and evolution of online social networks. In: SIGKDD, pp. 611–617 (2006)
9. Leskovec, J., McGlohon, M., Faloutsos, C., Glance, N.S., Hurst, M.: Patterns of cascading behavior in large blog graphs. In: SDM, pp. 551–556 (2007)
10. Lin, Y.-R., Sundaram, H., Kelliher, A.: Summarization of social activity over time: people, actions and concepts in dynamic networks. In: CIKM, pp. 1379–1380 (2008)
11. Liu, S., Liu, Y., Ni, L.M., Fan, J., Li, M.: Towards mobility-based clustering. In: SIGKDD, pp. 919–928 (2010)
12. Liu, S., Wang, S., Zhu, F., Zhang, J., Krishnan, R.: HYDRA: Large-scale social identity linkage via heterogeneous behavior modeling. In: SIGMOD Conference, pp. 51–62 (2014)
13. Liu, S., Yue, Y., Krishnan, R.: Adaptive collective routing using gaussian process dynamic congestion models. In: SIGKDD, pp. 704–712 (2013)
14. Liu, W., Kan, A., Chan, J., Bailey, J., Leckie, C., Pei, J., Kotagiri, R.: On compressing weighted time-evolving graphs. In: CIKM, pp. 2319–2322 (2012)
15. Navlakha, S., Rastogi, R., Shrivastava, N.: Graph summarization with bounded error. In: SIGMOD Conference, pp. 419–432 (2008)
16. Qu, Q., Zhu, F., Yan, X., Han, J., Yu, P.S., Li, H.: Efficient topological OLAP on information networks. In: Yu, J.X., Kim, M.H., Unland, R. (eds.) DASFAA 2011, Part I. LNCS, vol. 6587, pp. 389–403. Springer, Heidelberg (2011)
17. Raghavan, S., Garcia-molina, H.: Representing web graphs. In: ICDE, pp. 405–416 (2003)
18. Sun, J., Faloutsos, C., Papadimitriou, S., Yu, P.S.: GraphScope: parameter-free mining of large time-evolving graphs. In: SIGKDD, pp. 687–696 (2007)
19. Tian, Y., Hankins, R.A., Patel, J.M.: Efficient aggregation for graph summarization. In: SIGMOD Conference, pp. 567–580 (2008)
20. Toivonen, H., Zhou, F., Hartikainen, A., Hinkka, A.: Compression of weighted graphs. In: SIGKDD, pp. 965–973 (2011)
21. Xie, R., Zhu, F., Ma, H., Xie, W., Lin, C.: CLEar: A real-time online observatory for bursty and viral events. PVLDB 7(11) (2014)

22. Yang, J., Counts, S.: Predicting the speed, scale, and range of information diffusion in Twitter. In: ICWSM, pp. 355–358 (2010)
23. Zhu, F., Qu, Q., Lo, D., Yan, X., Han, J., Yu, P.S.: Mining top-k large structural patterns in a massive network. PVLDB 4(11), 807–818 (2011)
24. Zhu, F., Zhang, Z., Qu, Q.: A direct mining approach to efficient constrained graph pattern discovery. In: SIGMOD Conference, pp. 821–832 (2013)

Neural Gaussian Conditional Random Fields

Vladan Radosavljevic[1,*], Slobodan Vucetic[2], and Zoran Obradovic[2]

[1] Yahoo Labs, Sunnyvale, CA, USA
vladan@yahoo-inc.com
[2] Temple University, Philadelphia, PA, USA
{vucetic,zoran.obradovic}@temple.edu

Abstract. We propose a Conditional Random Field (CRF) model for structured regression. By constraining the feature functions as quadratic functions of outputs, the model can be conveniently represented in a Gaussian canonical form. We improved the representational power of the resulting Gaussian CRF (GCRF) model by (1) introducing an adaptive feature function that can learn nonlinear relationships between inputs and outputs and (2) allowing the weights of feature functions to be dependent on inputs. Since both the adaptive feature functions and weights can be constructed using feedforward neural networks, we call the resulting model Neural GCRF. The appeal of Neural GCRF is in conceptual simplicity and computational efficiency of learning and inference through use of sparse matrix computations. Experimental evaluation on the remote sensing problem of aerosol estimation from satellite measurements and on the problem of document retrieval showed that Neural GCRF is more accurate than the benchmark predictors.

Keywords: Gaussian conditional random fields, neural networks, graphical models.

1 Introduction

Learning from structured data is a frequently encountered problem in geoscience [1,2], computer vision [3,4], bioinformatics [5,6], and other areas where examples exhibit sequential [7,8], temporal [9,10], spatial [11], spatio-temporal [12,13], or some other dependencies. In such cases, the traditional unstructured supervised learning approaches could result in a weak model with low prediction accuracy [14]. Structured learning methods try to solve this problem by learning to simultaneously predict all outputs given all inputs. The structured approaches can exploit correlations among output variables, which often results in accuracy improvements over unstructured approaches that predict independently for each example. The benefits of structured learning grow with the strength of dependency between the examples and the data size.

In structured learning there is usually some prior knowledge about relationships among the outputs. Those relationships are application-specific and, very

---

* This study was conducted while the author was a postdoctoral associate at Temple University.

T. Calders et al. (Eds.): ECML PKDD 2014, Part II, LNCS 8725, pp. 614–629, 2014.
© Springer-Verlag Berlin Heidelberg 2014

often, they can be modeled by graphical models. The advantage of the graphical models is that one can make use of sparseness in the interactions between outputs and develop efficient learning and inference algorithms. In learning from structured data, the Markov Random Fields [2] and the Conditional Random Fields (CRF) [7] are among the most popular models. Originally, CRFs were designed for classification of sequential data [7] and have found many applications in areas such as computer vision [3] and computational biology [6].

Using CRF for regression is a less explored topic. Continuous Conditional Random Fields (CCRF) [8] is a ranking model that takes into account relationships among ranks of objects in document retrieval. With minor modifications, it can be used for structured regression problems. The Conditional State Space Model (CSSM) [15], an extension of the CRF to a domain with continuous multivariate outputs, was proposed for regression of sequential data. CSSM is an undirected model that makes no independence assumptions between outputs, which results in a more flexible framework. In [4] a conditional distribution of pixels given a noisy input image is modeled using the weighted quadratic factors obtained by convolving the image with a set of filters. Feature functions in [4] are specifically designed for image de-noising problems and are not readily applicable in regression. The Gaussian CRF for structured regression problems with feature functions constrained to quadratic form was introduced in [1]. The Sparse GCRF [10] is a variant of the GCRF model that incorporates $l1$ regularization in optimization function, thus enforcing sparsity in GCRF parameters. GCRF has recently been successfully utilized in a variety of real world applications. In the computational advertising field, GCRF significantly improved accuracy of click through rate estimation by taking into account relationship among advertisements [11]. An extension of GCRF to the non-Gaussian case using the copula transform was used in forecasting wind power [16]. In combination with decision trees, GCRF was successfully applied to short-term energy load forecasting [17], while in combination with support vector machines it was applied on automatic recognition of emotions from audio and visual features [18]. A tractable fully connected GCRF, which captures both long-range and short-range dependencies, was developed in [19] and was successfully applied on image de-noising and geoscience problems.

To improve expressive power of GCRF, we propose a Neural GCRF (NGCRF) regression model where CCRF and GCRF can be considered as special cases. In addition to using the existing unstructured predictors, the proposed NGCRF allows training of additional unstructured predictors simultaneously with other NGCRF parameters. This idea is motivated by the Conditional Neural Fields (CNF) [20,5] proposed for classification problems to facilitate modeling of complex relationships between inputs and outputs. Moreover, weights of NGCRF feature functions are themselves allowed to be nonlinear functions of inputs. In this way, NGCRF is able to capture non-homogeneous relationships among outputs and account for differing uncertainties in the unstructured predictors. We will show that learning and inference of NGCRF can be conducted efficiently through sparse matrix computations.

2 Gaussian Conditional Random Fields

Let us denote as $\mathbf{x} = (x_1, \ldots x_M)$ an M-dimensional vector of observations and as $\mathbf{y} = (y_1, \ldots y_N)$ an N-dimensional vector of real-valued output variables. The objective is to learn a non-linear mapping $f : \mathcal{R}^M \rightarrow \mathcal{R}^N$ that predicts the vector of output variables $\mathbf{y}$ as accurately as possible given all inputs $\mathbf{x}$. A CRF models a conditional distribution $P(\mathbf{y}|\mathbf{x})$, according to the associated graphical structure

$$P(\mathbf{y}|\mathbf{x}) = \frac{1}{Z(\alpha, \beta, \mathbf{x})} e^{\phi(\alpha, \beta, \mathbf{y}, \mathbf{x})}, \tag{1}$$

with energy function

$$\phi(\alpha, \beta, \mathbf{y}, \mathbf{x}) = \sum_{i=1}^{N} A(\alpha, y_i, \mathbf{x}) + \sum_{j \sim i} I(\beta, y_i, y_j, \mathbf{x}), \tag{2}$$

$A(\alpha, y_i, \mathbf{x})$ - association potential with parameters α,

$I(\beta, y_i, y_j, \mathbf{x})$ - interaction potential with parameters β,

$i \sim j$ - y_i and y_j are connected by an edge in the graph structure,

and the normalization function $Z(\alpha, \beta, \mathbf{x})$ defined as

$$Z(\alpha, \beta, \mathbf{x}) = \int_{\mathbf{y}} e^{\phi(\alpha, \beta, \mathbf{y}, \mathbf{x})} d\mathbf{y}. \tag{3}$$

The output y_i is associated with vector of observations $\mathbf{x} = (x_1, \ldots x_M)$ by a real-valued function called the association potential $A(\alpha, y_i, \mathbf{x})$, where α is a K-dimensional set of parameters. In general, A takes as input any appropriate combination of attributes from vector of observations $\mathbf{x}$. To model interactions among outputs, a real valued function called the interaction potential $I(\beta, y_i, y_j, \mathbf{x})$ is used, where β is an L dimensional set of parameters. Interaction potential represents the relationship between two outputs and in general can depend on inputs $\mathbf{x}$. Different applications can impose different interaction potentials. The larger the value of the interaction potential, the more related the two outputs are.

In CRF applications, A and I could be conveniently defined as linear combinations of a set of fixed features in terms of α and β, as in [7]

$$A(\alpha, y_i, \mathbf{x}) = \sum_{k=1}^{K} \alpha_k f_k(y_i, \mathbf{x}),$$

$$I(\beta, y_i, y_j, \mathbf{x}) = \sum_{l=1}^{L} \beta_l g_l(y_i, y_j, \mathbf{x}). \tag{4}$$

The use of feature functions is convenient because it allows us to model arbitrary relationships between inputs and outputs. In this way, any potentially relevant feature function could be included to the model and the learning algorithm can automatically determine their relevance.

Models with real valued outputs pose quite different challenges with respect to feature function complexity than in the discrete-valued case. Discrete valued models are always feasible, because Z is finite and defined as a sum over finitely many possible values of y. On the contrary, to have a feasible model with real valued outputs, Z must be integrable. Proving that Z is integrable in general might be difficult due to the complexity of association and interaction potentials.

2.1   Feature Functions

Construction of appropriate feature functions in CRF is a manual process that depends on prior beliefs of a practitioner about what features could be useful. The choice of features is often constrained to simple constructs to reduce the complexity of learning and inference from CRF.

If A and I are defined as quadratic functions of $\mathbf{y}$, $P(\mathbf{y}|\mathbf{x})$ becomes a multivariate Gaussian distribution such that learning and inference can be accomplished in a computationally efficient manner.

In the following, we describe the feature functions that led to Gaussian CRF. Let us assume we are given K unbiased unstructured predictors, $R_k(\mathbf{x})$, $k = 1, \ldots K$, that predict single output y_i taking into account $\mathbf{x}$ (in a special case, only corresponding x_i can be used as $\mathbf{x}$). To model the dependency between the prediction and output, we use quadratic feature functions

$$f_k(y_i, \mathbf{x}) = -(y_i - R_k(\mathbf{x}))^2, k = 1, \ldots K. \tag{5}$$

These feature functions follow the basic principle for association potentials in that their values are large when predictions and outputs are similar. To model the correlation among outputs, we use the quadratic feature function

$$g_l(y_i, y_j, \mathbf{x}) = -e_l(i, j, \mathbf{x})(y_i - y_j)^2,$$
$$e_l(i, j, \mathbf{x}) = \begin{cases} w_l(i, j, \mathbf{x}), & (i, j) \in G_l \\ 0, & (i, j) \notin G_l, \end{cases} \tag{6}$$

which imposes that outputs y_i and y_j have similar values if they are connected by an edge in the graph G_l. $w_l(i, j, \mathbf{x})$ represents the weight of an edge (i, j) in graph G_l. It should be noted that using multiple graphs G_l can facilitate modeling of different aspects of correlation between outputs (for example, spatial and temporal).

2.2   Multivariate Gaussian Model

Conditional distribution $P(\mathbf{y}|\mathbf{x})$ for the CRF model in Eq. (1), which uses quadratic feature functions defined in the previous section, can be represented as a multivariate Gaussian distribution. The resulting energy function of the GCRF model can be written as

$$\phi = -\sum_{i=1}^{N}\sum_{k=1}^{K}\alpha_k(y_i - R_k(\mathbf{x}))^2 - \sum_{i,j}\sum_{l=1}^{L}\beta_l e_l(i,j,\mathbf{x})(y_i - y_j)^2. \quad (7)$$

The energy function is a quadratic function in terms of $\mathbf{y}$. Therefore, $P(\mathbf{y}|\mathbf{x})$ can be transformed to a Gaussian form by representing ϕ as

$$\phi = -\frac{1}{2}(\mathbf{y} - \boldsymbol{\mu})^T\boldsymbol{\Sigma}^{-1}(\mathbf{y} - \boldsymbol{\mu}). \quad (8)$$

To transform $P(\mathbf{y}|\mathbf{x})$ to Gaussian form we determine $\boldsymbol{\Sigma}$ and $\boldsymbol{\mu}$ by matching Eq. (7) and (8)

$$\boldsymbol{\Sigma}_{i,j}^{-1} = 2\begin{cases}\sum_{k=1}^{K}\alpha_k + \sum_{n=1,n\neq j}^{N}\sum_l \beta_l e_l(i,n,\mathbf{x}), \ i = j \\ -\sum_l \beta_l e_l(i,j,\mathbf{x}), \ i \neq j, \end{cases} \quad (9)$$

$$\boldsymbol{\mu} = \boldsymbol{\Sigma}\mathbf{b}, \quad (10)$$

where $\mathbf{b}$ is a vector with elements

$$b_i = 2\sum_{k=1}^{K}\alpha_k R_k(\mathbf{x}). \quad (11)$$

If we calculate Z using the transformed exponent, we obtain

$$P(\mathbf{y}|\mathbf{x}) = \frac{1}{(2\pi)^{N/2}|\boldsymbol{\Sigma}|^{1/2}}e^{-\frac{1}{2}(\mathbf{y}-\boldsymbol{\mu})^T\boldsymbol{\Sigma}^{-1}(\mathbf{y}-\boldsymbol{\mu})}. \quad (12)$$

Therefore, the resulting conditional distribution is Gaussian with mean $\boldsymbol{\mu}$ and covariance $\boldsymbol{\Sigma}$. We observe that $\boldsymbol{\Sigma}$ is a function of parameters $\boldsymbol{\alpha}$ and $\boldsymbol{\beta}$, and interaction potential graphs G_l, while $\boldsymbol{\mu}$ is also a function of inputs $\mathbf{x}$. The resulting CRF is the Gaussian CRF (GCRF). In order for the model to be feasible, the conditional distribution has to be well defined. This means that we have to ensure that the precision matrix $\boldsymbol{\Sigma}^{-1}$ is positive semi-definite [1], which we will address in the following sections.

3   Neural Gaussian CRF

In this section we propose a new Neural Gaussian CRF model, which enhances GCRF and increases its representational power.

3.1   Neural GCRF Model

First, motivated by the recently proposed Conditional Neural Fields [20,5], we introduce the adaptive feature function defined as

$$f_a(y_i, \mathbf{x}) = -(y_i - R_a(\mathbf{w}, \mathbf{x}))^2, \tag{13}$$

where $R_a(\mathbf{w}, \mathbf{x})$ is a function of weights $\mathbf{w}$ that can be trained simultaneously with other GCRF parameters. In this way, $R_a(\mathbf{w}, \mathbf{x})$ can be trained directly with the goal of maximizing the log-likelihood such that it complements the existing predictors R_k. In this paper, we will assume that predictor $R_a(\mathbf{w}, \mathbf{x})$ is a feedforward neural network.

Second, as defined in Eq. (4), Gaussian CRF assigns weights $\boldsymbol{\alpha}$ and $\boldsymbol{\beta}$ to the feature functions. Considering that feature functions for the association potential are defined as squared errors of unstructured predictors, the role of weights $\boldsymbol{\alpha}$ and $\boldsymbol{\beta}$ is to measure their prediction uncertainty. Since it is likely that the quality of different predictors changes with $\mathbf{x}$, we enhance GCRF such that parameters α_k and β_l are replaced with the uncertainty functions $\alpha_k(\boldsymbol{\theta}_k, \mathbf{x})$ and $\beta_l(\boldsymbol{\psi}_l, \mathbf{x})$, where $\boldsymbol{\theta}_k$ and $\boldsymbol{\psi}_l$ are the parameters. We allow using feedforward neural networks for the uncertainty functions. By using the adaptive feature and uncertainty functions, we have

$$A(\boldsymbol{\theta}, y_i, \mathbf{x}) = -\sum_{k=1}^{K} \alpha_k(\boldsymbol{\theta}_k, \mathbf{x})(y_i - R_k(\mathbf{x}))^2 - \alpha_a(\boldsymbol{\theta}_a, \mathbf{x})(y_i - R_a(\mathbf{w}, \mathbf{x}))^2,$$

$$I(\boldsymbol{\psi}, y_i, y_j, \mathbf{x}) = -\sum_{l=1}^{L} \beta_l(\boldsymbol{\psi}_l, \mathbf{x})(y_i - y_j)^2. \tag{14}$$

In this way, $\alpha_k(\boldsymbol{\theta}_k, \mathbf{x})$ models the varying degree of importance of predictor R_k over different conditions. Similarly, $\beta_l(\boldsymbol{\psi}_l, \mathbf{x})$ models varying importance of correlation between outputs. As a result, $\boldsymbol{\Sigma}$ from Eq. (9) becomes dependent on inputs, thus allowing for error heteroscedasticity. Conditional distribution of the enhanced GCRF is Gaussian as in Eq. (12). Since both adaptive feature and uncertainty functions are assumed to be feedforward neural networks, we call the resulting model the Neural GCRF (NGCRF).

Let us analyze the feasibility condition for the NGCRF model. In order for the model to be feasible, the precision matrix $\boldsymbol{\Sigma}^{-1}$ has to be positive semi-definite. A common approach used in practice [21] is to enforce sufficient condition given by Gershgorin's circle theorem [22], which says that a symmetric matrix is positive definite if all diagonal elements are non-negative and if the matrix is diagonally dominant.

Definition 1. *A square matrix $\boldsymbol{\Sigma}^{-1}$ is diagonally dominant if the absolute value of each diagonal element is greater than the sum of absolute values of the non-diagonal elements in corresponding row $|\boldsymbol{\Sigma}_{i,i}^{-1}| > \sum_{j \neq i} |\boldsymbol{\Sigma}_{i,j}^{-1}|, \forall i$.*

Theorem 1. *If the values of functions $\boldsymbol{\alpha}$ and $\boldsymbol{\beta}$ in Eq (14) are always greater than 0, then the precision matrix $\boldsymbol{\Sigma}^{-1}$ that corresponds to NGCRF model defined by association and interaction potentials in Eq. (14) is diagonally dominant and hence positive definite.*

Proof. For each i, the absolute value of a diagonal element $\Sigma_{i,i}^{-1}$ of precision matrix Σ^{-1} can be represented as

$$
\begin{aligned}
|\Sigma_{i,i}^{-1}| &= |\sum_{k=1}^{K} \alpha_k(\boldsymbol{\theta}_k, \mathbf{x}) + \sum_{j \neq i} \sum_{l=1}^{L} \beta_l(\boldsymbol{\psi}_l, \mathbf{x})| \\
&= \sum_{k=1}^{K} \alpha_k(\boldsymbol{\theta}_k, \mathbf{x}) + \sum_{j \neq i} \sum_{l=1}^{L} \beta_l(\boldsymbol{\psi}_l, \mathbf{x}),
\end{aligned}
\tag{15}
$$

where we use the fact that values of $\boldsymbol{\alpha}$ and $\boldsymbol{\beta}$ are always greater than 0. Similarly, the absolute value of each off-diagonal element $\Sigma_{i,j}^{-1}$ equals

$$
|\Sigma_{i,j}^{-1}| = |\sum_{l=1}^{L} \beta_l(\boldsymbol{\psi}_l, \mathbf{x})| = \sum_{l=1}^{L} \beta_l(\boldsymbol{\psi}_l, \mathbf{x}).
\tag{16}
$$

Then, for each i we have

$$
|\Sigma_{i,i}^{-1}| - \sum_{j \neq i} |\Sigma_{i,j}^{-1}| = \sum_{k=1}^{K} \alpha_k(\boldsymbol{\theta}_k, \mathbf{x}) > 0.
\tag{17}
$$

which proves the theorem.   □

Therefore, one way to ensure that the NGCRF model is feasible is to impose the constraints $\alpha > 0$ and $\beta > 0$, which is analytically tractable [8,1], but is known to be conservative [21]. To analyze the effect of constraining $\alpha > 0$, we will assume that the interaction potential is not used (output variables are assumed to be conditionally independent). The prediction for each y_i becomes a weighted average of the unstructured predictors, where weights are positive values with their sum equal to 1. This constrains the range of outputs to $y_i \in [min(R_k(\mathbf{x})), max(R_k(\mathbf{x}))]$, which has negligible effect on NGCRF since we assumed that unstructured predictors are unbiased. In [21] it was empirically verified that constraint $\beta > 0$ reduces parameter search space more and more with decreasing sparsity and increasing number of parameters in beta functions. This leads to limited improvements when using NGCRF with constraint $\beta > 0$ on more dense graphs.

3.2   Learning and Inference of NGCRF

Learning. The learning task is to choose values of parameters $\boldsymbol{\theta}$, $\boldsymbol{\psi}$ and $\mathbf{w}$ to maximize the conditional log-likelihood on the set of training examples $\mathcal{D} = \{(\mathbf{x_t}, \mathbf{y_t}), t = 1 \dots T\}$

$$(\hat{\boldsymbol{\theta}}, \hat{\boldsymbol{\psi}}, \hat{\mathbf{w}}) = \underset{\boldsymbol{\theta}, \boldsymbol{\psi}, \mathbf{w}}{\operatorname{argmax}} (\mathcal{L}(\boldsymbol{\theta}, \boldsymbol{\psi}, \mathbf{w}))$$

$$\text{where } \mathcal{L}(\boldsymbol{\theta}, \boldsymbol{\psi}, \mathbf{w}) = \sum_{t=1}^{T} \log P(\mathbf{y_t}|\mathbf{x_t}). \tag{18}$$

By setting $\boldsymbol{\alpha}$ and $\boldsymbol{\beta}$ to be greater than 0, learning becomes a constrained optimization problem. To convert it to unconstrained optimization, we adopt a technique used in [8,1] that applies the exponential transformation of the functions to guarantee that their values are positive. We apply an exponential transformation on $\boldsymbol{\alpha}$ and $\boldsymbol{\beta}$

$$\alpha_k = e^{u_k(\boldsymbol{\theta_k}, \mathbf{x})}, \text{ for } k = 1, \dots K,$$
$$\alpha_a = e^{u_a(\boldsymbol{\theta_a}, \mathbf{x})}, \tag{19}$$
$$\beta_l = e^{v_l(\boldsymbol{\psi_l}, \mathbf{x})}, \text{ for } l = 1, \dots L.$$

where u_k and v_l are differentiable functions with respect to parameters $\boldsymbol{\theta_k}$ and $\boldsymbol{\psi_l}$.

All the parameters are learned by a gradient-based optimization. To apply the gradient-based method for learning, we need to find the gradient of the conditional log-likelihood. The derivatives of $\mathcal{L}$ with respect to θ, ψ, and $\mathbf{w}$ are

$$\frac{\partial \mathcal{L}}{\partial \theta_k} = \frac{\partial \mathcal{L}}{\partial \alpha_k} \frac{\partial \alpha_k}{\partial u_k} \frac{\partial u_k}{\partial \theta_k},$$
$$\frac{\partial \mathcal{L}}{\partial \psi_l} = \frac{\partial \mathcal{L}}{\partial \beta_l} \frac{\partial \beta_l}{\partial v_l} \frac{\partial v_l}{\partial \psi_l}, \tag{20}$$
$$\frac{\partial \mathcal{L}}{\partial \mathbf{w}} = \frac{\partial \mathcal{L}}{\partial R_a} \frac{\partial R_a}{\partial \mathbf{w}}.$$

The gradient of $\mathcal{L}$ with respect to θ and ψ has three components. The first components are $\partial \mathcal{L}/\partial \alpha_k$ and $\partial \mathcal{L}/\partial \beta_l$. The expression for $\partial \mathcal{L}/\partial \alpha_k$ is

$$\frac{\partial \mathcal{L}}{\partial \alpha_k} = -\frac{1}{2} (\mathbf{y} - \boldsymbol{\mu})^T \frac{\partial \boldsymbol{\Sigma}^{-1}}{\partial \alpha_k} (\mathbf{y} - \boldsymbol{\mu}) + (\frac{\partial \mathbf{b}^T}{\partial \alpha_k} - \boldsymbol{\mu}^T \frac{\partial \boldsymbol{\Sigma}^{-1}}{\partial \alpha_k})(\mathbf{y} - \boldsymbol{\mu})$$
$$+ \frac{1}{2} Tr(\boldsymbol{\Sigma} \frac{\partial \boldsymbol{\Sigma}^{-1}}{\partial \alpha_k}). \tag{21}$$

To calculate $\partial \mathcal{L}/\partial \beta_l$, we use $\partial \mathbf{b}/\partial \beta_l = 0$ and obtain

$$\frac{\partial \mathcal{L}}{\partial \beta_l} = -\frac{1}{2} (\mathbf{y} + \boldsymbol{\mu})^T \frac{\partial \boldsymbol{\Sigma}^{-1}}{\partial \beta_l} (\mathbf{y} - \boldsymbol{\mu}) + \frac{1}{2} Tr(\boldsymbol{\Sigma} \frac{\partial \boldsymbol{\Sigma}^{-1}}{\partial \beta_l}). \tag{22}$$

From Eq. (19), the second components are $\partial \alpha_k/\partial u_k = \alpha_k$ and $\partial \beta_l/\partial v_l = \beta_l$. The third components depend on the chosen functions u_k and v_l. The gradient of $\mathcal{L}$ with respect to $\mathbf{w}$ depends on the functional form of R_a. Since $\boldsymbol{\Sigma}^{-1}$ does not depend on R_a, $\partial \mathcal{L}/\partial R_a$ becomes

$$\frac{\partial \mathcal{L}}{\partial R_a} = 2\boldsymbol{\alpha_a}^T (\mathbf{y} - \boldsymbol{\mu}). \tag{23}$$

Algorithm 1.. Learning of NGCRF Parameters

Input: x, $R_k(\mathbf{x})$, **y.**
1. Initialize $\boldsymbol{\theta_k}$, $\boldsymbol{\psi_l}$.
2. Estimate $\boldsymbol{\theta_k}$, $\boldsymbol{\psi_l}$ by applying gradient based approach and Eq. (21) and (22), without taking into account R_a.
3. Initialize $\boldsymbol{\theta_a}$.
4. Learn predictor R_a using Eq. (23).
repeat
    Apply gradient based optimization to estimate all parameters.
until Convergence

We observe that an update for the adaptive model R_a is proportional to the difference between true output and the mean of the NGCRF model. This means that R_a will be updated only if NGCRF is not able to predict the output correctly and R_a will be updated more aggressively when the error is larger. This justifies our hypothesis that R_a will work as a complement of the existing non-structured models.

To ensure convergence, the iterative procedure presented in Algorithm 1 [23,20] is used for learning model parameters according to update formulas derived earlier in this section. To avoid overfitting, which is a common problem for maximum likelihood optimization, we added regularization terms for $\boldsymbol{\alpha}$, $\boldsymbol{\theta}$, β, ψ to the log-likelihood. In this way, we penalize large outputs of $\boldsymbol{\alpha}$ and β as well as large weights $\boldsymbol{\theta}$ and ψ.

Inference. The inference task is to find the outputs **y** for a given set of observations **x** and estimated parameters $\hat{\boldsymbol{\alpha}}$ and $\hat{\beta}$ such that the conditional probability $P(\mathbf{y}|\mathbf{x})$ is maximized. The NGCRF model is Gaussian and, therefore, the maximum a posteriori estimate of **y** is obtained as the expected value $\boldsymbol{\mu}$ of the NGCRF distribution

$$\hat{\mathbf{y}} = \underset{\mathbf{y}}{\arg\max} P(\mathbf{y}|\mathbf{x}) = \boldsymbol{\mu} = \boldsymbol{\Sigma}\mathbf{b}, \tag{24}$$

while $\boldsymbol{\Sigma}$ is a measure of uncertainty of the point estimate.

3.3   Complexity

If the size of the training set is N and the learning takes I iterations, the straightforward matrix computation results in $\mathcal{O}(IN^3)$ time to train the model. The main cost of computation is matrix inversion, since during the gradient-based optimization we need to find $\boldsymbol{\Sigma}$ as an inverse of $\boldsymbol{\Sigma}^{-1}$. However, this is the worst case performance. Since matrix $\boldsymbol{\Sigma}^{-1}$ is typically very sparse (it depends on the imposed neighborhood structure), the training time can be decreased to $\mathcal{O}(IN^2)$ by using sparse matrix apparatus or even to $\mathcal{O}(IN)$ if we do not consider interaction potential [21]. During inference, we need to compute $\boldsymbol{\mu}$, which takes $\mathcal{O}(N)$ time. As we eventually need to calculate the trace of the matrix, only the elements that correspond to the main diagonal should be stored. Therefore, memory requirements depend only on the imposed neighborhood structure.

4   Experiments

To demonstrate the strength of the NGCRF model, we applied it on two real-world structured regression applications. The experimental results indicate that NGCRF improves prediction accuracy by efficiently utilizing information from structured data.

4.1   The NGCRF Model for Document Retrieval

In this application the objective is to retrieving the most relevant documents with respect to the given query. In order to make a comparison to the GCRF method, we replicated the experimental setup from [8]. We obtained query-document data from OHSUMED dataset from LETOR [24], which is a standard data source used in document retrieval research (the same dataset was used in [8]). The OHSUMED dataset contains search queries, where each query is associated with a number of relevant documents. There are 106 queries, 348,566 documents and a total of 16,140 query-"relevant document" pairs. From the NGCRF perspective, each query-"set of relevant documents" represents an example $(\mathbf{x}, \mathbf{y})$. Each component of $\mathbf{y}$ represents a relevance of the corresponding document to a query, while $\mathbf{x}$ contains extracted features. Features $\mathbf{x}$ were used to construct $K = 25$ unstructured predictors $R_k(\mathbf{x})$ that predict document relevance for a given query. The outputs of unstructured predictors are available in OHSUMED (more details are in [24]). OHSUMED considers three levels of relevance - highly, partially and not relevant (each component in $\mathbf{y}$ can take values 2, 1, or 0 respectively). In addition, OHSUMED contains information about similarity between documents i and j, $w(i, j, \mathbf{x})$, which was determined based on similarity of their contents. Having this setup, the goal is to estimate relevance of each document in the database for a given query.

Benchmark Methods. As benchmark methods we use the following (all parameters were set using a small validation set)

Unstructured retrieval by neural network (NN) We trained NN with five hidden units to predict relevance of documents for a given query. The inputs to NN were outputs of unstructured predictors.

Structured retrieval by baseline GCRF We trained GCRF to predict relevance of documents. As unstructured predictors we used R_k, which are readily available in OHSUMED. GCRF also utilized relationship among documents by incorporating weights $w(i, j, \mathbf{x})$ from OHSUMED into the interaction potential.

Structured retrieval by GCRF+NN We trained a GCRF model using unstructured predictors R_k from OHSUMED and pre-trained NN. We call this model GCRF+NN.

RankSVM State-of-the-art retrieval method [25], which predictions are available as a part of OHSUMED.

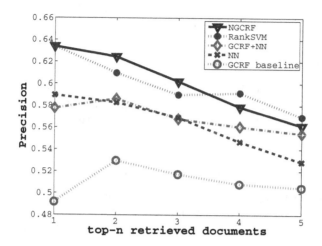

Fig. 1. Comparison of retrieval performance in terms of precision when top-n documents are retrieved

The NGCRF Model. We trained the NGCRF model where unstructured predictors were R_k, α was a function of unstructured predictors, β a function of similarity between documents, and adaptive NN was a function of R_k.

Evaluation. In our experiments, for each method we averaged results over 5 fold cross validation data sets provided in OHSUMED. As an evaluation measure, we used *precision@n*, which represents a percentage of relevant publications in top-n publications retrieved ($n = 1 \ldots 5$ in our experiments). To fetch top-n relevant publications we retrieved those publications which corresponded to the n largest predictions. In Figure 1 we see that NN and GCRF+NN outperform baseline GCRF, which can be explained by the ability of NN to capture nonlinearity in feature space. Furthermore, if we allow NN to be adaptive, we see that NGCRF outperforms all other alternatives. We see that NGCRF is comparable to state-of-the-art retrieval method RankSVM, which is specifically designed for ranking problems (while NGCRF has general applicability) and which also used R_k and $w(i, j, \mathbf{x})$ as its inputs.

4.2   The NGCRF Model for AOD Prediction

We evaluated the proposed Neural GCRF model on a high impact regression problem from remote sensing, namely, prediction of aerosol optical depth (AOD) in the atmosphere from satellite measurements. AOD is a measure of aerosol light extinction integrated vertically through the atmosphere. AOD prediction is important because one of the main challenges of today's climate research is to characterize and quantify the effect of aerosols on Earth's radiation budget [26].

We considered data from MODIS, an instrument aboard NASA's Terra satellites [27]. We used ground-based data obtained from the AERONET [28], which is a global remote sensing network of radiometers that measure AOD several times per hour at specific geographic locations. The data can be obtained from the official MODIS website of NASA [29].

We extracted satellite-based attributes that are used as inputs to domain-based deterministic prediction algorithms [27]. In addition, we extracted information about the location of each data point (longitude and latitude) and a quality of observation (QA) assigned to each point provided by domain scientist. Data quality index was provided at four levels from the lowest quality QA=0 to the highest quality QA=3. We collected 28,374 data points distributed over the entire globe at 217 AERONET sites during the years 2005 and 2006.

Benchmark Methods. Here we list benchmark methods that we compared NGCRF to.

Deterministic prediction algorithm C005 The primary benchmark for comparison with our CRF predictors was the most recent version of the MODIS operational algorithm, called C005 [27]. This is a deterministic algorithm that retrieves AOD from MODIS observations relying on the domain knowledge. It is based on the inversion of physical forward models developed by the domain scientists.

Statistical prediction by a neural network As a baseline statistical algorithm we used a neural network model trained to predict AERONET AOD from all MODIS attributes excluding location and quality flag. It has been shown previously that neural networks achieve comparable accuracy to C005 on the AOD prediction problem [30]. The neural network has a hidden layer with 10 nodes and an output layer with one node. In nested 5-cross-validation experiments we trained 5 neural networks. When tested on 2006 data, we used a single network trained on the entire training set.

Structured prediction by GCRF The aerosol data are characterized by strong spatial and temporal dependencies that a CRF is able to exploit by defining interactions among outputs using feature functions. Given a data set that consists of satellite observations and ground-based AOD measurements, a statistical prediction model (R_a) can be trained to use satellite observations as attributes and predict the labels which are ground-based AODs. The deterministic AOD prediction models (DP) are based on solid physical principles and tuned by domain scientists. To model the association potential, i.e the dependency between the predictions and output AOD, we introduce the following two feature functions,

$$f_1(y_i, \mathbf{x}_i) = - (y_i - DP(\mathbf{x}_i))^2,$$
$$f_a(y_i, \mathbf{x}_i) = - (y_i - R_a(\mathbf{x}_i))^2. \tag{25}$$

To model the interaction potential we introduce feature function

$$g_1(y_i, y_j, \mathbf{x}) = -(y_i - y_j)^2. \tag{26}$$

Table 1. RMSE and FRAC of C005, NN, GCRF and NGCRF on data with four quality flags

	C005	NN	GCRF+NN	NGCRF
RMSE	0.123	0.112 ± 0.002	0.105 ± 0.0006	0.102 ± 0.0008
FRAC	0.65	0.68 ± 0.03	0.71 ± 0.005	0.74 ± 0.007

This interaction potential will reflect correlation between spatio-temporal data examples i and j (closer examples are given larger weight). The learned parameter β represents the level of spatio-temporal correlation of neighboring outputs (large β indicates that spatio-temporal correlation is large). We partitioned data into four subsets corresponding to quality flags QA=0, 1, 2, and 3. We determined eight a parameters corresponding to C005 and NN predictions over these subsets. To model interaction potential we defined spatial-temporal neighbors as a pair of observations where temporal distance $temporalDist(i, j)$ is less than 7 days and spatial distance $spatialDist(i, j)$ is less than 50km. This choice is based on previous studies of aerosol dynamics by geoscientists. We multiply feature g with weights $w(i, j, \mathbf{x})$, that are products of Gaussians

$$
w(i, j, \mathbf{x}) = \begin{cases} e^{- \frac{spatialDist(i,j)^2}{2\sigma_s^2} - \frac{temporalDist(i,j)^2}{2\sigma_t^2}} & , i \sim j \\ 0, otherwise \end{cases} \tag{27}
$$

where $\sigma_s = 50$ and $\sigma_t = 10$ were determined using a small validation set.

The NGCRF Model. Here we use similar attributes as in the previous section but in the spirit of the proposed NGCRF model. Instead of defining manual partitions of the dataset, we use all observations as inputs to the $\boldsymbol{\alpha}$ functions. We define $\boldsymbol{\alpha}$ as an exponential function of linear combinations of observations. To incorporate potential bias, one observation is a vector with all ones.

$$
\alpha_k(\boldsymbol{\theta}, \mathbf{x}^{(i)}) = e^{\sum \theta_t x_t^{(i)}}, \tag{28}
$$

where $\mathbf{x}_1^{(i)}$ is a vector with all ones, $\mathbf{x}_{2,3,4,5}^{(i)}$ are quality flags. As an adaptive model R_a we used NN defined in previous sections. Its weight α_a follows the definition in Eq. (28).

To model spatio-temporal correlation, we use spatial and temporal distance between i and j as two observations for the β function. Similar to Eq. (28) we define β as

$$
\beta(\boldsymbol{\psi}, \mathbf{x}^{(i,j)}) = e^{\sum \psi_l x_l^{(i,j)}}, \tag{29}
$$

where $\mathbf{x}_1^{(i,j)}$ is a vector with all ones, $\mathbf{x}_2^{(i,j)}$ represents spatial distance between i and j and $\mathbf{x}_3^{(i,j)}$ represents their temporal proximity.

Evaluation. To evaluate proposed methods, we trained the models on 2005 data and used 2006 data for testing. There are many possible measures that could be used to assess AOD prediction accuracy. Given vector $\mathbf{t} = (t_1, t_2, \ldots t_N)^T$ of N outcome values and vector $\mathbf{y} = (y_1, y_2, \ldots y_N)^T$ of the corresponding predictions, we measure the root mean squared error (RMSE). We also report accuracy on the domain specific measure called *the fraction of successful predictions* (FRAC) that penalizes errors on small AOD more than errors on large AOD [27]

$$FRAC = \frac{I}{N} \times 100\%, \qquad (30)$$

where I is the number of predictions that satisfy $|y_i - t_i| \leq 0.05 + 0.15 t_i$.

RMSE error of the four models is presented in Table 1, where smaller numbers mean more accurate predictions. FRAC accuracy of these four models is also shown in Table 1, where larger numbers correspond to better predictions. We can see that in our experiments NN was more accurate than the operational C005 algorithm. GCRF showed an improvement in accuracy over both NN and C005 by taking advantage of a combination of models and spatio-temporal correlation in data. NGCRF achieves even better accuracy by utilizing nonlinear weights, an adaptive statistical model, and learning instead of assuming the level of correlation between points. Although NGCRF is a non-convex approach, it has only slightly larger variance in predictions than GCRF+NN.

The obtained results provide strong evidence that adaptive structured learning approaches can be successfully applied to AOD prediction, where even a small improvement of prediction accuracy results in huge uncertainty reduction in many geophysical studies that rely on AOD predictions [26].

5   Conclusion

Structured learning, as a fairly new research area in machine learning, has great success in classification, but its application on regression problems has not been explored sufficiently. In this article we proposed a method to adaptively combine the outputs of powerful non-structured regression models such as neural networks and a variety of correlated knowledge sources into a single prediction model by utilizing possible correlation among outcome variables. It is worth pointing to differences between our NGCRF model and the GCRF model proposed in [4]. The GCRF in [4] models a conditional distribution of pixels given a noisy input image using the weighted quadratic factors obtained by convolving the image with a set of filters. GCRF is designed for image de-noising problems, while NGCRF can be applied to general regression problems. By taking a closer look at GCRF we find that features in Eq. (5) and (6) are represented in GCRF, while GCRF does not model the adaptive component of NGCRF in Eq. (13). The proposed NGCRF is also readily applicable to other regression applications, where there is a need for knowledge integration and exploration of structure in outputs.

Acknowledgment. This work is supported in part by DARPA Grant FA9550-12-1-0406 negotiated by AFOSR, and NSF grant IIS-1117433.

References

1. Radosavljevic, V., Vucetic, S., Obradovic, Z.: Continuous conditional random fields for regression in remote sensing. In: European Conference on Artificial Intelligence (ECAI), pp. 809–814 (2010)
2. Solberg, A.H.S., Taxt, T., Jain, A.K.: A markov random field model for classification of multisource satellite imagery. IEEE Transactions on Geoscience and Remote Sensing 34(1), 100–113 (1996)
3. Kumar, S., Hebert, M.: Discriminative random fields: A discriminative framework for contextual interaction in classification. In: Proceedings Ninth IEEE International Conference on Computer Vision, vol. 2, pp. 1150–1157 (2003)
4. Tappen, M.F., Liu, C., Adelson, E.H., Freeman, W.T.: Learning gaussian conditional random fields for low-level vision. In: IEEE Conference on Computer Vision and Pattern Recognition (2007)
5. Peng, J., Bo, L., Xu, J.: Conditional neural fields. In: Advances in Neural Information Processing Systems 22, pp. 1419–1427 (2009)
6. Liu, Y., Carbonell, J., Klein-Seetharaman, J., Gopalakrishnan, V.: Comparison of probabilistic combination methods for protein secondary structure prediction. Bioinformatics 20(17), 3099–3107 (2004)
7. Lafferty, J., McCallum, A., Pereira, F.: Conditional random fields: Probabilistic models for segmenting and labeling sequence data. In: Proceedings International Conference on Machine Learning (2001)
8. Qin, T., Liu, T., Zhang, X., Wang, D., Li, H.: Global ranking using continuous conditional random fields. Neural Information Processing Systems (2008)
9. Grbovic, M., Vucetic, S.: Tracking concept change with incremental boosting by minimization of the evolving exponential loss. In: Gunopulos, D., Hofmann, T., Malerba, D., Vazirgiannis, M. (eds.) ECML PKDD 2011, Part I. LNCS, vol. 6911, pp. 516–532. Springer, Heidelberg (2011)
10. Wytock, M., Kolter, Z.: Sparse gaussian conditional random fields: Algorithms, theory, and application to energy forecasting. In: Dasgupta, S., Mcallester, D. (eds.) Proceedings of the 30th International Conference on Machine Learning (ICML 2013). JMLR Workshop and Conference Proceedings, vol. 28, pp. 1265–1273 (May 2013)
11. Xiong, C., Wang, T., Ding, W., Shen, Y., Liu, T.Y.: Relational click prediction for sponsored search. In: Proceedings of the Fifth ACM International Conference on Web Search and Data Mining, WSDM 2012, pp. 493–502. ACM, New York (2012)
12. Grbovic, M., Li, W., Xu, P., Usadi, A.K., Song, L., Vucetic, S.: Decentralized fault detection and diagnosis via sparse {PCA} based decomposition and maximum entropy decision fusion. Journal of Process Control 22(4), 738–750 (2012)
13. Djuric, N., Radosavljevic, V., Coric, V., Vucetic, S.: Travel speed forecasting by means of continuous conditional random fields. Transportation Research Record (2263), 131–139 (2011)
14. Neville, J., Gallagher, B., Eliassi-Rad, T., Wang, T.: Correcting evaluation bias of relational classifiers with network crossvalidation. Knowledge and Information Systems 30, 31–55 (2012)
15. Kim, M., Pavlovic, V.: Discriminative learning for dynamic state prediction. IEEE Transactions on Pattern Analysis and Machine Intelligence 31(10), 1847–1861 (2009)
16. Wytock, M., Kolter, J.: Large-scale probabilistic forecasting in energy systems using sparse gaussian conditional random fields. In: 2013 IEEE 52nd Annual Conference on Decision and Control (CDC), pp. 1019–1024 (December 2013)

17. Guo, H.: Modeling short-term energy load with continuous conditional random fields. In: Blockeel, H., Kersting, K., Nijssen, S., Železný, F. (eds.) ECML PKDD 2013, Part I. LNCS, vol. 8188, pp. 433–448. Springer, Heidelberg (2013)

18. Baltrušaitis, T., Banda, N., Robinson, P.: Dimensional affect recognition using continuous conditional random fields. In: IEEE Conference on Automatic Face and Gesture Recognition (2013)

19. Ristovski, K., Radosavljevic, V., Vucetic, S., Obradovic, Z.: Continuous conditional random fields for efficient regression in large fully connected graphs. In: des Jardins, M., Littman, M.L. (eds.) AAAI. AAAI Press (2013)

20. Do, T.M.T., Artieres, T.: Neural conditional random fields. In: Proceedings of the Thirteenth International Conference on Artificial Intelligence and Statistics, vol. 9, JMLR (May 2010)

21. Rue, H., Held, L.: Gaussian Markov Random Fields: Theory and Applications. Chapman & Hall/CRC Monographs on Statistics & Applied Probability. Taylor & Francis (2005)

22. Gerschgorin, S.: Uber die abgrenzung der eigenwerte einer matrix. Izv. Akad. Nauk. USSR Otd. Fiz.-Mat. Nauk 7, 749–754 (1931)

23. Nix, D.A., Weigend, A.S.: Learning local error bars for nonlinear regression. In: Tesauro, G., Touretzky, D.S., Leen, T.K. (eds.) Advances in Neural Information Processing Systems, vol. 7, pp. 489–495. MIT Press, Cambridge (1995)

24. Liu, T.Y., Xu, J., Qin, T., Xiong, W., Li, H.: Letor: Benchmark dataset for research on learning to rank for information retrieval. In: SIGIR 2007: Proceedings of the Learning to Rank Workshop in the 30th Annual International ACM SIGIR Conference on Research and Development in Information Retrieval (2007)

25. Joachims, T.: Optimizing search engines using clickthrough data. In: KDD 2002: Proceedings of the Eighth ACM SIGKDD International Conference on Knowledge Discovery and Data Mining. ACM, New York (2002)

26. Kaufman, Y.J., Tanre, D., Boucher, O.: A satellite view of aerosols in the climate system. Nature 419(6903), 215–223 (2002)

27. Remer, L.A., Kaufman, Y.: The modis aerosol algorithm, products and validation. Journal of the Atmospheric Sciences 62, 947–973 (2005)

28. Holben, B.N., Eck, T.F.: Aeronet: A federated instrument network and data archive for aerosol characterization. Remote Sensing of Environment 66, 1–16 (1998)

29. Official modis website, http://modis.gsfc.nasa.gov

30. Radosavljevic, V., Vucetic, S., Obradovic, Z.: A data-mining technique for aerosol retrieval across multiple accuracy measures. IEEE Geoscience and Remote Sensing Letters 7(2), 411–415 (2010)

Cutset Networks: A Simple, Tractable, and Scalable Approach for Improving the Accuracy of Chow-Liu Trees

Tahrima Rahman, Prasanna Kothalkar, and Vibhav Gogate

Computer Science Department
The University of Texas at Dallas
{tahrima.rahman,prasanna.kothalkar,
vibhav.gogate}@utdallas.edu

Abstract. In this paper, we present cutset networks, a new tractable probabilistic model for representing multi-dimensional discrete distributions. Cutset networks are rooted OR search trees, in which each OR node represents conditioning of a variable in the model, with tree Bayesian networks (Chow-Liu trees) at the leaves. From an inference point of view, cutset networks model the mechanics of Pearl's cutset conditioning algorithm, a popular exact inference method for probabilistic graphical models. We present efficient algorithms, which leverage and adopt vast amount of research on decision tree induction for learning cutset networks from data. We also present an expectation-maximization (EM) algorithm for learning mixtures of cutset networks. Our experiments on a wide variety of benchmark datasets clearly demonstrate that compared to approaches for learning other tractable models such as thin-junction trees, latent tree models, arithmetic circuits and sum-product networks, our approach is significantly more scalable, and provides similar or better accuracy.

1 Introduction

Learning tractable probabilistic models from data has been the subject of much recent research. These models offer a clear advantage over Bayesian networks and Markov networks: exact inference over them can be performed in polynomial time, obviating the need for unreliable, inaccurate approximate inference, not only at learning time but also at query time. Interestingly, experimental results in numerous recent studies [5,11,16,23] have shown that the performance of approaches that learn tractable models from data is similar or better than approaches that learn Bayesian and Markov networks from data. These results suggest that *controlling exact inference complexity* is the key to superior end-to-end performance.

In spite of these promising results, a key bottleneck remains: barring a few exceptions, algorithms that learn tractable models from data are computationally expensive, requiring several hours for even moderately sized problems (e.g., approaches presented in [11,16,23] need more than "10 hours" of CPU time for

T. Calders et al. (Eds.): ECML PKDD 2014, Part II, LNCS 8725, pp. 630–645, 2014.
© Springer-Verlag Berlin Heidelberg 2014

datasets having 200 variables and 10^4 examples). There are several reasons for this, with the main reason being the high computational complexity of *conditional independence tests*. For example, the LearnSPN algorithm of Gens and Domingos [11] and the ID-SPN algorithm of Rooshenas and Lowd [23] for learning tractable sum-product networks, spend a substantial amount of their execution time on partitioning the given set of variables into conditionally independent components. Other algorithms with strong theoretical guarantees, such as learning efficient Markov networks [13], and learning thin junction trees [1,4,19] also suffer from the same problem.

In this paper, we present a tractable class of graphical models, called *cutset networks*, which are rooted OR search trees with tree Bayesian networks (Chow-Liu trees) at the leaves. Each OR node (or sum node) in the OR tree represents conditioning over a variable in the model. Cutset networks derive their name from Pearl's cutset conditioning method [20]. The key idea in cutset conditioning is to condition on a subset of variables in the graphical model, called the cutset, such that the remaining network is a tree. Since, exact probabilistic inference can be performed in time that is linear in the size of the tree (using Belief propagation [20] for instance), the complexity of cutset conditioning is exponential in the cardinality (size) of the cutset. If the cutset is bounded, then cutset conditioning is tractable. However, note that unlike classic cutset conditioning, cutset networks can take advantage of determinism [3,12] and context-specific independence [2] by allowing different variables to be conditioned on at the same level in the OR search tree. As a result, they can yield a compact representation even if the size of the cutset is arbitrarily large.

The key advantage of cutset networks is that only the leaf nodes, which represent tree Bayesian networks, take advantage of conditional independence, while the OR search tree does not. As a result, to learn cutset networks from data, we do not have to run expensive conditional independence tests at any internal OR node. Moreover, the leaf distributions can be learned in polynomial time, using the classic Chow-Liu algorithm [6]. As a result, if we assume that the size of the cutset (or the height of the OR tree) is bounded by k, and given that the time complexity of the Chow-Liu algorithm is $O(n^2d)$, where n is the number of variables and d is the number of training examples, the optimal cutset network can be learned in $O(n^{k+2}d)$ time.

Although, the algorithm described above is tractable, it is infeasible for any reasonable k (e.g., 5) that we would like to use in practice. Therefore, to make our algorithm practical, we use splitting heuristics and pruning techniques developed over the last few decades for inducing decision trees from data [21,18]. The splitting heuristics help us quickly learn a reasonably good cutset network, without any backtracking, while the pruning techniques such as pre-pruning and reduced-error (post) pruning help us avoid over-fitting. To improve the accuracy further, we also consider mixtures of cutset networks, which generalize mixtures of Chow-Liu trees [17] and develop an expectation-maximization algorithm for learning them from data.

We conducted a detailed experimental evaluation comparing our three learning algorithms: learning cutset networks without pruning (CNet), learning cutset networks with pruning (CNetP), and learning mixtures of cutset networks (MC-Net), with six existing algorithms for learning tractable models from data: sum-product networks with direct and indirect interactions (ID-SPN) [23], learning tractable Markov networks with Arithmetic Circuits (ACMN) [16], mixtures of trees (MT) [17], stand alone Chow-Liu trees [6], learning sum-product networks (LearnSPN) [11] and latent tree models (LTM) [5]. We found that MCNet is the best performing algorithm in terms of test-set log likelihood score on 11 out of the 20 benchmarks that we experimented with. ID-SPN has the best test-set log likelihood score on 8 benchmarks while ACMN and CNetP are closely tied for the third-best performing algorithm spot. We also measured the time taken to learn the model for ID-SPN and our algorithms, and found that ID-SPN was the slowest algorithm. CNet was the fastest algorithm, while CNetP and MCNet were second and third fastest, respectively. Interestingly, if we look at learning time and accuracy as a whole, CNetP is the best performing algorithm, providing reasonably accurate results in a fraction of the time as compared to MCNet, ACMN and ID-SPN.

The rest of the paper is organized as follows. In section 2, we present background and notation. Section 3 provides the formal definition of cutset networks. Section 4 describes algorithms for learning cutset networks. We present experimental results in section 5 and conclude in section 6.

2   Notation and Background

We borrow notation from [17]. Let V be a set of n discrete random variables where each variable $v \in V$ ranges over a finite discrete domain Δ_v and let $x_v \in \Delta_v$ denote a value that can be assigned to v. Let $A \subseteq V$, then x_A denotes an assignment to all the variables in A. For simplicity, we often denote x_A as x and Δ_v as Δ. The set of domains is denoted by $\Delta_V = \{\Delta_i | i \in V\}$.

A probabilistic graphical model $\mathcal{G}$ is denoted by a triple $\langle V, \Delta_V, F \rangle$ where V and Δ_V are the sets of variables and their domains respectively, and F is a set of real-valued functions. Each function $f \in F$ is defined over a subset $V(f) \subseteq V$ of variables, called its scope. For Bayesian networks (cf. [20,7]) which are typically depicted using a directed acyclic graph (DAG), F is the set of conditional probability tables (CPTs), where each CPT is defined over a variable given its parents in the DAG. For Markov networks (cf. [14]), F is the set of potential functions. Markov networks are typically depicted using undirected graphs (also called the primal graph) in which we have a node in the graph for each variable in the model and edges connect two variables that appear together in the scope of a function.

A probabilistic graphical model represents the following joint probability distribution over V:

$$P(x) = \frac{1}{Z} \prod_{f \in F} f(x_{V(f)})$$

where x is an assignment to all variables in V, $x_{V(f)}$ denotes the projection of x on $V(f)$ and Z is a normalization constant. For Bayesian networks, $Z = 1$, while for Markov networks $Z > 0$. In Markov networks, Z is also called the *partition function*.

The two main inference problems in probabilistic graphical models are (1) posterior marginal inference: computing the marginal probability of a variable given evidence, namely computing $P(x_v|x_A)$ where x_A is the evidence and $v \in V \setminus A$; and (2) maximum-a-posteriori (MAP) inference: finding an assignment of values to all variables that has the maximum probability given evidence, namely computing $\arg\max_{x_B \in B} P(x_B|x_A)$ where x_A is the evidence and $B = V \setminus A$. Both problems are known to be NP-hard.

2.1 The Chow-Liu Algorithm for Learning Tree Distributions

A tree Bayesian network is a Bayesian network in which each variable has no more than one parent while a tree Markov network is a Markov network whose primal (or interaction) graph is a tree. It is known that both tree Bayesian and Markov networks have the same representation power and therefore can be used interchangeably.

The Chow-Liu algorithm [6] is a classic algorithm for learning tree networks from data. If $P(x)$ is a probability distribution over a set of variables V, then the Chow-Liu algorithm approximates $P(x)$ by a tree network $T(x)$. If $G_T = (V, E_T)$ is an undirected Markov network that induces the distribution $T(x)$, then

$$T(x) = \frac{\prod\limits_{(u,v) \in E_T} T(x_u, x_v)}{\prod\limits_{v \in V} T(x_v)^{deg(v)-1}}$$

where $deg(v)$ is the degree of vertex v or the number of incident edges to v. If G_T is a directed model such as a Bayesian network, then

$$T(x) = \prod_{v \in V} T(x_v|x_{pa(v)})$$

where $T(x_v|x_{pa(v)})$ is an arbitrary conditional probability distribution such that $|pa(v)| \leq 1$. The Kullback-Leibler divergence $KL(P,T)$ between $P(x)$ and $T(x)$ is defined as:

$$KL(P,T) = \sum_x P(x)\log\left(\frac{P(x)}{T(x)}\right)$$

In order to minimize $KL(P,T)$, Chow and Liu proved that each selected edge $(u,v) \in E_T$ has to maximize the total mutual information, $\sum_{(u,v) \in E_T} I(u,v)$. Mutual information, denoted by $I(u,v)$, is a measure of mutual dependence between two random variables u and v and is given by:

$$I(u,v) = \sum_{x_u \in \Delta_u} \sum_{x_v \in \Delta_v} P(x_u, x_v) \log\left(\frac{P(x_u, x_v)}{P(x_u)P(x_v)}\right) \tag{1}$$

To maximize (1), the Chow-Liu procedure computes the mutual information $I(u, v)$ for all possible pairs of variables in V and then finds the maximum weighted spanning tree $G_T = (V, E_T)$ such that each edge $(u, v) \in E_T$ is weighted by $I(u, v)$. The marginal distribution $T(u, v)$ of a pair of variables (u, v) connected by an edge is the same as $P(u, v)$.

Tree networks are attractive because: (1) learning both the structure and parameters of the distribution are tractable; (2) several probabilistic inference tasks can be solved in linear time; and (3) they have intuitive interpretations.

The time complexity of learning the structure and parameters of a tree network using the Chow-Liu algorithm is $O(n^2 d + n^2 \log(n))$ where n is the number of variables and d is the number of training examples. Since, in practice, $d > \log(n)$, for the rest of the paper, assume that the time complexity of the Chow-Liu algorithm is $O(n^2 d)$.

2.2 OR Trees

OR trees are rooted trees which are used to represent the search space explored during probabilistic inference by conditioning [20,9]. Each node in an OR tree is labeled by a variable v in the model. Each edge emanating from a node represents the conditioning of the variable v at that node by a value $x_v \in \Delta_v$ and is labeled by the marginal probability of the variable-value assignment given the path from the root to the node. For simplicity, we will focus on binary valued variables. For binary variables, assume that left edges represent the assignment of variable v to 0 and right edges represent $v = 1$. A similar representation can be used for multi-valued variables.

Any distribution can be represented using a OR Tree. In the worst-case, the tree will require $O(2^{n+1})$ parameters to specify the distribution. Figure 1 shows a probability distribution and a possible OR tree.

The distribution represented by an OR tree O is given by:

$$P(x) = \prod_{(v_i, v_j) \in path_O(x)} w(v_i, v_j) \tag{2}$$

where $path_O(x)$ is the path from the root to the unique leaf node $l(x)$ corresponding to the assignment x and $w(v_i, v_j)$ is the probability value attached to the edge between the OR nodes v_i and v_j.

3   Cutset Networks

Cutset Networks (CNets) are a hybrid of rooted OR trees and tree Bayesian networks, with an OR tree at the top and a tree Bayesian network attached to each leaf node of the OR tree. Formally a cutset network is a pair $C = (O, \mathbf{T})$ where O is a rooted OR tree and $\mathbf{T} = \{T_1, \dots, T_l\}$ is a collection of tree networks. The distribution represented by a cutset network is given by:

$$P(x) = \left(\prod_{(v_i, v_j) \in path_O(x)} w(v_i, v_j) \right) \left(T_{l(x)}(x_{V(T_{l(x)})}) \right) \tag{3}$$

a	b	c	P(a,b,c)
0	0	0	0.030
0	0	1	0.144
0	1	0	0.030
0	1	1	0.096
1	0	0	0.224
1	0	1	0.336
1	1	0	0.042
1	1	1	0.098

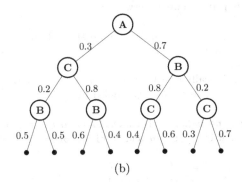

(a)   (b)

Fig. 1. (a) A probability distribution, (b) An OR tree representing the distribution given in (a). The left branch from a node represents conditioning by 0, whereas the right branch represents conditioning by 1.

where $path_O(x)$ is the path from the root to the unique leaf node $l(x)$ corresponding to the assignment x, $w(v_i, v_j)$ is the probability value attached to the edge between the OR nodes v_i and v_j and $T_{l(x)}$ is the tree Bayesian network associated with $l(x)$ and $V(T_{l(x)})$ is the set of variables over which $T_{l(x)}$ is defined. Fig. 2 shows an example cutset network.

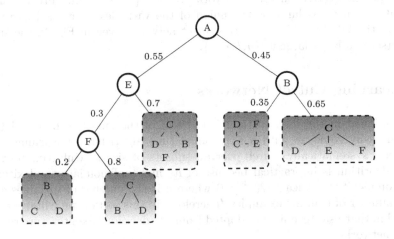

Fig. 2. A cutset network

Note that unlike classic cutset conditioning, cutset networks can take advantage of determinism [3] and context-specific independence [2] by branching on different variables at the same level (or depth) [12,13]. As a result, they can yield

---

Algorithm 1. LearnCNet

Input: Training dataset $\mathcal{D} = \{x^1, ..., x^d\}$, Variables V.
Output: A cutset network C

if Termination condition is satisfied **then**
   return ChowLiuTree($\mathcal{D}$)
end if
Heuristically select a variable $v \in V$ for splitting
Create a new node O_v labeled by v.
`/*`
`Each node has a left child` $O_v.left$`, a right child` $O_v.right$`, a left`
`probability` $O_v.lp$ `and a right probability` $O_v.rp$`.`
`*/`
Let $\mathcal{D}_{v=0} = \{x^i \in \mathcal{D} | x^i_v = 0\}$
Let $\mathcal{D}_{v=1} = \{x^i \in \mathcal{D} | x^i_v = 1\}$
$O_v.lp \leftarrow \frac{|\mathcal{D}_{v=0}|}{|\mathcal{D}|}$
$O_v.rp \leftarrow \frac{|\mathcal{D}_{v=1}|}{|\mathcal{D}|}$
$O_v.left \leftarrow$ LearnCNet($\mathcal{D}_{v=0}, V \setminus v$)
$O_v.right \leftarrow$ LearnCNet($\mathcal{D}_{v=1}, V \setminus v$)
return O_v

---

a compact representation, even if the size of the cutset[1] is arbitrarily large. For example, consider the cutset network given in Fig. 2. The left most leaf node represents a tree Bayesian network over $V \setminus \{a, e, f\}$ while the right most leaf node represents a tree Bayesian network over $V \setminus \{a, b\}$. Technically, the size of the cutset can be as large as the union of the variables mentioned at various levels in the OR tree. Thus, for the cutset network given in Fig. 2, the size of the cutset can be as large as $\{a, b, e, f\}$.

4   Learning Cutset Networks

As mentioned in the introduction, if the size of the cutset is bounded by k, we can easily come up with a polynomial time algorithm for learning cutset networks: systematically search over all subsets of size k. Unfortunately, this naive algorithm is impractical because of its high polynomial complexity; the time complexity is at least $\Omega(n^{k+2}d)$ where n is the number of variables and d is the number of training examples. Therefore, in this section we will present an algorithm that uses techniques adopted from the decision tree literature to learn cutset networks.

Simply put, given training data $\mathcal{D} = \{x^1, \ldots, x^d\}$ defined over a set V of variables, we can use the following recursive or divide-and-conquer approach to learn a cutset network from $\mathcal{D}$ (see Algorithm 1). Select a variable using the

---

[1] Given a graph $G = (V, E)$, $C \subseteq V$ is a cutset of G if the subgraph over $V \setminus C$ is a tree.

given *splitting heuristic*, place it at the root and make one branch for each of its possible values. Repeat the approach at each branch, using only those instances that reach the branch. If at any time, some pre-defined *termination condition* is satisfied, run the Chow-Liu algorithm on the remaining variables. It is easy to see that the optimal probability value, assuming that we are using the maximum likelihood estimation principle, attached to each branch in the OR tree is the fraction of the instances at the parent that actually reach the branch.

The two main choices in the above algorithm are which variable to split on and the termination condition. We discuss each of them in turn, next, followed by the algorithm to learn mixtures of cutset networks from data.

4.1 Splitting Heuristics

Intuitively, we should split on a variable that reduces the expected entropy (or the information content) of the two partitions of the data created by the split. The hope is that when the expected entropy is small, we will be able to represent it faithfully using a simple distribution such as a tree Bayesian network. Unfortunately, unlike traditional classification problems, in which we are interested in (the entropy of) a specific class variable, estimating the joint entropy of the data when the class variable is not known is a challenging task (we just don't have enough data to reliably measure the joint entropy). Therefore, we propose to approximate the joint entropy by the average entropy over individual variables. Formally, for our purpose, the entropy of data $\mathcal{D}$ defined over a set V of variables is given by:

$$\widehat{H}(\mathcal{D}) = \frac{1}{|V|} \sum_{v \in V} H_{\mathcal{D}}(v) \tag{4}$$

where $H_{\mathcal{D}}(v)$ is the entropy of variable v relative to $\mathcal{D}$. It is given by:

$$H_{\mathcal{D}}(v) = - \sum_{x_v \in \Delta_v} P(x_v) log(P(x_v))$$

Given a closed-form expression for the entropy of the data, we can calculate the information gain or the expected reduction in the entropy after conditioning on a variable v using the following expression:

$$Gain_{\mathcal{D}}(v) = \widehat{H}(\mathcal{D}) - \sum_{x_v \in \Delta_v} \frac{|\mathcal{D}_{x_v}|}{|\mathcal{D}|} \widehat{H}(\mathcal{D}_{x_v})$$

where $\mathcal{D}_{x_v} = \{x^i \in \mathcal{D} | x_v^i = x_v\}$.

From the discussion above, the splitting heuristic is obvious: select a variable that has the highest information gain.

4.2 Termination Condition and Post-Pruning

A simple termination condition that we can enforce is stopping when the number of examples at a node falls below a fixed threshold. Alternatively, we can also declare a node as a leaf node if the entropy falls below a threshold. Unfortunately, both of these criteria are highly dependent on the threshold used. A large threshold will yield shallow OR trees that are likely to underfit the data (high bias) while a small threshold will yield deep trees that are likely to overfit the data (high variance). To combat this, inspired by the decision tree literature [21,22], we propose to use reduced error pruning. In reduced error pruning, we grow the tree fully and post-prune in a bottom-up fashion. (Alternatively, we can also prune in a top-down fashion).

The benefits of pruning over using a fixed threshold are that it avoids the horizon effect (the thresholding method suffers from lack of sufficient look ahead). Pruning comes at a greater computational expense than threshold based stopped splitting and therefore for problems with large training sets, the expense can be prohibitive. For small problems, though, these computational costs are low and pruning should be preferred over stopped splitting. Moreover, pruning is an anytime method and as a result we can stop it at any time.

Formally, our proposed reduced error pruning for cutset networks operates as follows. We divide the data into two sets: training data and validation data. Then, we build a full OR tree over the training data, declaring a node as a leaf node using a weak termination condition (e.g., the number of examples at a node is less than or equal to 5). Then, we recursively visit the tree in a bottom up fashion, and replace a node and the sub-tree below it by a leaf node (namely, a Chow-Liu tree) if it increases the log-likelihood on the validation set.

We summarize the time and space complexity of learning (using Algorithm 1) and inference in cutset networks in the following theorem.

Theorem 1. *The time complexity of learning cutset networks is $O(n^2 ds)$ where s is the number of nodes in the cutset network, d is the number of examples and n is the number of variables. The space complexity of the algorithm is $O(ns)$, which also bounds the space required by the cutset networks. The time complexity of performing marginal and maximum-a-posteriori inference in a cutset network is $O(ns)$.*

Proof. The time complexity of computing the gain at each internal OR node is $O(n^2 d)$. Similarly, the time complexity of running the Chow-Liu algorithm at each leaf node is $O(n^2 d)$. Since there are s nodes in the cutset network, the overall time complexity is $O(n^2 ds)$. The space required to store an OR node is $O(1)$ while the space required to store a tree Bayesian network is $O(n)$. Thus, the overall space complexity is $O(max(n, 1)s) = O(ns)$. The time complexity of performing inference at each leaf Chow-Liu node is $O(n)$ while inference at each internal OR node can be done in constant time. Since the tree has s nodes, the overall inference complexity is $O(ns)$.

4.3   Mixtures of Cutset Networks

Similar to mixtures of trees [17], we define mixtures of cutset networks as distributions of the form:

$$P(x) = \sum_{i=1}^{k} \lambda_i C_i(x) \tag{5}$$

with $\lambda_i \geq 0$ for $i = 1, \ldots, k$, and $\sum_{i=1}^{k} \lambda_i = 1$. Each mixture component $C_i(x)$ is a cutset network and λ_i is its mixture co-efficient. At a high level, one can think of the mixture as containing a latent variable z which takes a value $i \in \{1, \ldots, k\}$ with probability λ_i.

Next, we present a version of the expectaton-maximization algorithm (EM) [10] for learning mixtures of cutset networks from data. The EM algorithm operates as follows. We begin with random parameters. At each iteration t, in the expectation-step (E-step) of the algorithm, we find the probability of completing each training example, using the current model. Namely, for each training example x^j and each component i, we compute

$$P^t(z = i | x^j) = \frac{\lambda_i^t C_i^t(x^j)}{\sum_{r=1}^{k} \lambda_r^t C_r^t(x^j)}$$

Then, in the maximization-step (M-step), we learn each mixture component i, using a weighted training set in which each example j has weight $P^t(z = i|x^j)$. This yields a new mixture component C_i^{t+1}. In the M-step, we also update the mixture co-efficients using the following expression:

$$\lambda_i^{t+1} = \frac{\sum_{j=1}^{d} P^t(z = i | x^j)}{d}$$

We can run EM until it converges or until a pre-defined bound on the number of iterations is exceeded. The quality of the local maxima reached by EM is highly dependent on the initialization used and therefore in practice, we typically run EM using several different initializations and choose parameter settings having the highest log-likelihood score. Notice that by varying the number of mixture components, we can explore interesting bias versus variance tradeoffs. Large k will yield high variance models and small k will yield high bias models.

We summarize the time and space complexity of learning and inference in mixtures of cutset networks in the following theorem.

Theorem 2. *The time complexity of learning mixtures of cutset networks is $O(n^2 dskt_{max})$ where s is the number of nodes in the cutset network, d is the number of examples, k is the number of mixture components, t_{max} is the maximum number of iterations for which EM is run and n is the number of variables. The space complexity of the algorithm is $O(nsk)$, which also bounds the space required by the mixtures of cutset networks. The time complexity of performing marginal and maximum-a-posteriori inference in a mixtures of cutset networks is $O(nks)$.*

Proof. Follows from Theorem 1

Table 1. Test-set Log-Likelihood. †indicates that the results of this algorithm are not available

DATASET	CNET	CNETP	MCNET	ID-SPN	ACMN	MT	CHOW-LIU	LEARN SPN	LTM
NLTCS	-6.10	-6.05	-6.00	-6.02	-6.00	-6.01	-6.76	-6.11	-6.49
MSNBC	-6.06	-6.05	-6.04	-6.04	-6.04	-6.07	-6.54	-6.11	-6.52
KDDCUP2000	-2.21	-2.19	-2.12	-2.13	-2.17	-2.13	-2.32	-2.18	-2.18
PLANTS	-13.37	-13.25	-12.78	-12.54	-12.80	-12.95	-16.51	-12.98	-16.39
AUDIO	-46.84	-41.97	-39.73	-39.79	-40.32	-40.08	-44.35	-40.50	-41.90
JESTER	-64.50	-55.26	-52.57	-52.86	-53.31	-53.08	-58.21	-53.48	-55.17
NETFLIX	-69.74	-58.72	-56.32	-56.36	-57.22	-56.74	-60.25	-57.33	-58.53
ACCIDENTS	-31.59	-30.66	-29.96	-26.98	-27.11	-29.63	-33.17	-30.04	-33.05
RETAIL	-11.12	-10.98	-10.82	-10.85	-10.88	-10.83	-11.02	-11.04	-10.92
PUMSB-STAR	-25.06	-24.28	-24.18	-22.40	-23.55	-23.71	-30.80	-24.78	-31.32
DNA	-109.79	-87.50	-85.82	-81.21	-80.03	-85.14	-87.70	-82.52	-87.60
KOSAREK	-11.53	-11.07	-10.58	-10.60	-10.84	-10.62	-11.52	-10.99	-10.87
MSWEB	-10.20	-10.12	-9.79	-9.73	-9.77	-9.85	-10.35	-10.25	-10.21
BOOK	-40.19	-37.51	-33.96	-34.14	-35.56	-34.63	-37.84	-35.89	-34.22
EACHMOVIE	-60.22	-57.71	-51.39	-51.51	-55.80	-54.60	-64.79	-52.49	†
WEBKB	-171.95	-161.58	-153.22	-151.84	-159.13	-156.86	-164.89	-158.20	-156.84
REUTERS-52	-91.35	-87.64	-86.11	-83.35	-90.23	-85.90	-96.85	-85.07	-91.23
20NEWSGROUP	-176.56	-161.68	-151.29	-151.47	-161.13	-154.24	-164.99	-155.93	-156.77
BBC	-300.33	-260.55	-250.58	-248.93	-257.10	-261.84	-261.41	-250.69	-255.76
AD	-16.31	-16.14	-16.68	-19.00	-16.53	-16.02	-16.67	-19.73	†

Table 2. Runtime Comparison(in seconds). Time-limit for each algorithm: 48 hours. †indicates that the algorithm did not terminate in 48 hours.

Dataset	Var#	Train	Valid	Test	CNet	CNetP	MCNet	ID-SPN	ACMN
NLTCS	16	16181	2157	3236	0.2	0.4	36.5	307.0	242.4
MSNBC	17	291326	38843	58265	13.0	29.2	2177.7	90354.0	579.9
KDDCup2000	64	180092	19907	34955	95.9	197.8	1988.0	38223.0	645.5
Plants	69	17412	2321	3482	6.5	10.5	135.0	10590.0	119.4
Audio	100	15000	2000	3000	17.2	19.6	187.0	14231.0	1663.9
Jester	100	9000	1000	4116	14.0	11.8	101.2	†	3665.8
Netflix	100	15000	2000	3000	25.2	22.6	224.4	†	1837.4
Accidents	111	12758	1700	2551	15.7	22.1	195.4	†	793.4
Retail	135	22041	2938	4408	18.9	27.6	104.7	2116.0	12.5
Pumsb-star	163	12262	1635	2452	30.1	41.8	233.8	18219.0	374.0
DNA	180	1600	400	1186	13.8	6.9	57.7	150850.0	39.9
Kosarek	190	33375	4450	6675	65.9	102.5	141.2	†	585.4
MSWeb	294	29441	32750	5000	208.6	365.8	642.8	†	286.3
Book	500	8700	1159	1739	129.1	204.2	154.4	125480.0	3035.0
EachMovie	500	4524	1002	591	90.7	133.4	204.8	78982.0	9881.1
WebKB	839	2803	558	838	169.7	228.7	160.4	†	7098.3
Reuters-52	889	6532	1028	1540	397.1	650.4	1177.2	†	2709.6
20Newsgroup	910	11293	3764	3764	695.2	935.8	1525.2	†	16255.3
BBC	1058	1670	225	330	206.7	223.9	70.2	4157.0	1862.2
Ad	1556	2461	327	491	365.8	594.3	155.4	285324.0	6496.4

5 Empirical Evaluation

The aim of our experimental evaluation is two fold: comparing the speed, measured in terms of CPU time, and accuracy, measured in terms of test-set log likelihood scores, of our methods with state-of-the-art methods for learning tractable models.

5.1 Methodology and Setup

We evaluated our algorithms as well as the competition on 20 benchmark datasets shown in Table 2. The number of variables in the datasets ranged from 16 to 1556, and the number of training examples varied from 1.6K to 291K examples. All variables in our datasets are binary-valued for a fair comparison with other methods, who operate primarily on binary-valued input. These datasets or a subset of them have also been used by [8,16,15,11,24].

We implemented three variations of our algorithms: (1) learning CNets without pruning (CNet), (2) learning CNets with pruning (CNetP) and (3) mixtures of CNets (MCNets). We compared their performance with the following learning algorithms from literature: learning sum-product networks with direct and indirect interactions (ID-SPN) [23], learning Markov networks using arithmetic circuits (ACMN) [16], learning mixture of trees (MT) [17], Chow-Liu trees [6],

Table 3. Head-to-head comparison of the number of wins (in terms of test-set log-likelihood score) achieved by one algorithm (row) over another (column), for all pairs of the 6 best performing algorithms used in our experimental study.

	CNetP	MCNet	ID-SPN	ACMN	MT	LearnSPN
CNetP	-	1	1	2	2	6
MCNet	19	-	11	13	15	18
ID-SPN	19	8	-	16	16	20
ACMN	18	5	3	-	8	15
MT	18	5	3	12	-	16
LearnSPN	14	2	0	5	·4	-

learning Sum-Product Networks (LearnSPN) [11] and learning latent tree models (LTM) [5]. Most of the results on the datasets (except the results on learning Chow-Liu models) were made available to us by [23]. They are part of the Libra toolkit available on Daniel Lowd's web page.

We smoothed all parameters using 1-laplace smoothing. For learning CNets without pruning, we stopped building the OR tree when the number of examples at the leaf node were fewer than 10 or the total entropy was smaller than 0.01. To learn MCNets, we varied the number of components from 5 to 40, in increments of 5 and ran the EM algorithm for 100 iterations or convergence whichever was earlier. For each iteration of EM, we could update both the structure and the parameters of the cutset network associated with each component. However, to speed up the learning algorithm, we chose to update just the parameters, utilizing the structure learned at the first iteration.

5.2   Accuracy

Table 1 shows the test-set log likelihood scores for the various benchmark networks while Table 3 shows head-to-head comparison of the six best performing algorithms namely CNetP, MCNet, ID-SPN, ACMN, MT and LearnSPN. Excluding the first two datasets where there are multiple winners, we can see that MCNet has the best log-likelihood score on 9 out of the remaining 18 benchmarks, while ID-SPN is the second best performing algorithm, with the best log-likelihood score on 7 out of the 18 benchmarks. In the head-to-head comparison, MCNet is better than CNetP on 19 benchmarks, ID-SPN on 11 benchmarks, ACMN on 13 benchmarks, MT on 15 benchmarks and LearnSPN on 18 benchmarks. CNetP is better than ID-SPN only on 1 benchmark, ACMN and MT on 2 benchmarks while it is better than LearnSPN on 6 benchmarks. A careful look at the datasets reveal that when the number of training examples is large, MCNet and to some extent CNetP are typically better than the competition. However, for small training set sizes, ID-SPN is the best performing algorithm. As expected, Chow-Liu trees and CNet are the worst-performing algorithms, the former underfits and the latter overfits.

MCNet is consistently better than CNetP which suggests that whenever possible *it is a good idea to use latent mixtures of simple models*. This conclusion can also be drawn from the performance of MT, which greatly improves the accuracy of Chow-Liu trees.

5.3 Learning Time

Table 2 shows the time taken by CNet, CNetP, MCNet, ID-SPN and ACMN to learn a model from data. We gave a time limit of 48 hours to all algorithms and ran all our timing experiments on a quad-core Intel i7, 2.7 GHz machine with 8GB of RAM. The fastest cutset network learners, in order, are: CNet, CNetP, and MCNet. On an average, ACMN is slower than MCNet. ID-SPN is the slowest algorithm. In fact, ID-SPN did not finish on 8 out of the 20 datasets in 48 hours (note that for the datasets on which ID-SPN did not finish in 48 hours, we report the test set log-likelihood scores from [23]). The best performing cutset network algorithm, MCNet, was faster than ID-SPN on all 20 datasets and ACMN on 14 datasets. If we look at the learning time and accuracy as a whole, CNetP is the best performing algorithm, providing reasonably accurate results in quick time.

6   Summary and Future Work

In this paper we presented cutset networks - a novel, simple and tractable probabilistic graphical model. At a high level, cutset networks are operational representation of Pearl's cutset conditioning method, with an OR tree modeling conditioning (at the top) and a tree Bayesian network modeling inference over trees at the leaves. We developed an efficient algorithm for learning cutset networks from data. Our new algorithm uses a decision-tree inspired learning algorithm for inducing the structure and parameters of the OR tree and the classic Chow-Liu algorithm for learning the tree distributions at the leaf nodes. We also presented an EM-based algorithm for learning mixtures of cutset networks.

Our detailed experimental study on a variety on benchmark datasets clearly demonstrated the power of cutset networks. In particular, our new algorithm that learns mixtures of cutset networks from data, was the best performing algorithm in terms of log-likelihood score on 55% of the benchmarks when compared with 5 other state-of-the-art algorithms from literature. Moreover, our new *one-shot algorithm*, which builds a cutset network using the information gain heuristic and employs reduced-error pruning is not only fast (as expected) but also reasonably accurate on several benchmark datasets. This gives us a spectrum of algorithms for future investigations: fast, accurate one-shot algorithms and slow, highly accurate iterative algorithms based on EM.

Future work includes learning polytrees having at most w parents at the leaves yielding w-cutset networks; using AND/OR trees or sum-product networks instead of OR trees yielding AND/OR cutset networks [9]; introducing structured latent variables in the mixture model; and merging identical sub-trees while learning to yield a compact graph-based representation.

Acknowledgements. This research was partly funded by ARO MURI grant W911NF-08-1-0242, by the AFRL under contract number FA8750-14-C-0021 and by the DARPA Probabilistic Programming for Advanced Machine Learning Program under AFRL prime contract number FA8750-14-C-0005. The views and conclusions contained in this document are those of the authors and should not be interpreted as representing the official policies, either expressed or implied, of DARPA, AFRL, ARO or the US government.

References

1. Bach, F., Jordan, M.: Thin junction trees. Advances in Neural Information Processing Systems 14, 569–576 (2001)
2. Boutilier, C., Friedman, N., Goldszmidt, M., Koller, D.: Context-specific independence in Bayesian networks. In: Proceedings of the Twelfth Conference on Uncertainty in Artificial Intelligence, pp. 115–123. Morgan Kaufmann, Portland (1996)
3. Chavira, M., Darwiche, A.: On probabilistic inference by weighted model counting. Artificial Intelligence 172(6-7), 772–799 (2008)
4. Chechetka, A., Guestrin, C.: Efficient principled learning of thin junction trees. In: Platt, J., Koller, D., Singer, Y., Roweis, S. (eds.) Advances in Neural Information Processing Systems 20. MIT Press, Cambridge (2008)
5. Choi, M.J., Tan, V., Anandkumar, A., Willsky, A.: Learning latent tree graphical models. Journal of Machine Learning Research 12, 1771–1812 (2011)
6. Chow, C.K., Liu, C.N.: Approximating discrete probability distributions with dependence trees. IEEE Transactions on Information Theory 14, 462–467 (1968)
7. Darwiche, A.: Modeling and reasoning with Bayesian networks. Cambridge University Press (2009)
8. Davis, J., Domingos, P.: Bottom-up learning of Markov network structure. In: Proceedings of the Twenty-Seventh International Conference on Machine Learning, pp. 271–278. ACM Press, Haifa (2010)
9. Dechter, R., Mateescu, R.: AND/OR search spaces for graphical models. Artificial Intelligence 171(2), 73–106 (2007)
10. Dempster, A.P., Laird, N.M., Rubin, D.B.: Maximum likelihood from incomplete data via the EM algorithm. Journal of the Royal Statistical Society, Series B 39, 1–38 (1977)
11. Gens, R., Domingos, P.: Learning the structure of sum-product networks. In: Proceedings of the Thirtieth International Conference on Machine Learning. JMLR: W&CP, vol. 28 (2013)
12. Gogate, V., Domingos, P.: Formula-Based Probabilistic Inference. In: Proceedings of the Twenty-Sixth Conference on Uncertainty in Artificial Intelligence, pp. 210–219 (2010)
13. Gogate, V., Webb, W., Domingos, P.: Learning efficient Markov networks. In: Proceedings of the 24th Conference on Neural Information Processing Systems, NIPS 2010 (2010)
14. Koller, D., Friedman, N.: Probabilistic Graphical Models: Principles and Techniques. MIT Press, Cambridge (2009)
15. Lowd, D., Davis, J.: Learning Markov network structure with decision trees. In: Proceedings of the 10th IEEE International Conference on Data Mining (ICDM), pp. 334–343. IEEE Computer Society Press, Sydney (2010)

16. Lowd, D., Rooshenas, A.: Learning Markov networks with arithmetic circuits. In: Proceedings of the Sixteenth International Conference on Artificial Intelligence and Statistics (AISTATS 2013), Scottsdale, AZ (2013)
17. Meila, M., Jordan, M.: Learning with mixtures of trees. Journal of Machine Learning Research 1, 1–48 (2000)
18. Mitchell, T.M.: Machine Learning. McGraw-Hill, New York (1997)
19. Narasimhan, M., Bilmes, J.: Pac-learning bounded tree-width graphical models. In: Proceedings of the Twentieth Conference on Uncertainty in Artificial Intelligence (2004)
20. Pearl, J.: Probabilistic Reasoning in Intelligent Systems: Networks of Plausible Inference. Morgan Kaufmann, San Francisco (1988)
21. Quinlan, J.R.: Induction of decision trees. Machine Learning 1, 81–106 (1986)
22. Quinlan, J.R.: C4.5: Programs for Machine Learning. Morgan Kaufmann, San Mateo (1993)
23. Rooshenas, A., Lowd, D.: Learning sum-product networks with direct and indirect interactions. In: Proceedings of the Thirty-First International Conference on Machine Learning. ACM Press, Beijing (2014)
24. Van Haaren, J., Davis, J.: Markov network structure learning: A randomized feature generation approach. In: Proceedings of the Twenty-Sixth National Conference on Artificial Intelligence. AAAI Press (2012)

Boosted Mean Shift Clustering

Yazhou Ren[1,3], Uday Kamath[2], Carlotta Domeniconi[3], and Guoji Zhang[4]

[1] School of Comp. Sci. and Eng., South China Uni. of Tech., Guangzhou, China
[2] BAE Systems Applied Intelligence, Mclean, VA, USA
[3] Department of Comp. Sci., George Mason University, Fairfax, VA, USA
[4] School of Sci., South China Uni. of Tech., Guangzhou, China
yazhou.ren@mail.scut.edu.cn, uday.kamath@baesystems.com,
carlotta@cs.gmu.edu, magjzh@scut.edu.cn

Abstract. Mean shift is a nonparametric clustering technique that does not require the number of clusters in input and can find clusters of arbitrary shapes. While appealing, the performance of the mean shift algorithm is sensitive to the selection of the bandwidth, and can fail to capture the correct clustering structure when multiple modes exist in one cluster. DBSCAN is an efficient density based clustering algorithm, but it is also sensitive to its parameters and typically merges overlapping clusters. In this paper we propose Boosted Mean Shift Clustering (BMSC) to address these issues. BMSC partitions the data across a grid and applies mean shift locally on the cells of the grid, each providing a number of intermediate modes (iModes). A mode-boosting technique is proposed to select points in denser regions iteratively, and DBSCAN is utilized to partition the obtained iModes iteratively. Our proposed BMSC can overcome the limitations of mean shift and DBSCAN, while preserving their desirable properties. Complexity analysis shows its potential to deal with large-scale data and extensive experimental results on both synthetic and real benchmark data demonstrate its effectiveness and robustness to parameter settings.

Keywords: Mean shift clustering, density-based clustering, boosting.

1 Introduction

Clustering aims to partition data into groups, so that points that are similar to one another are placed in the same cluster, and points that are dissimilar from each other are placed in different clusters. Clustering is a key step for many exploratory tasks. In the past decades, many clustering algorithms have been proposed, such as centroid-based clustering (e.g., k-means [21] and k-medoids [18]), distribution-based clustering (e.g., Expectation-Maximization with Gaussian mixture [8]), and density-based clustering (e.g., mean shift [4], DBSCAN [10] and OPTICS [1]).

Most of the existing clustering methods need the number of clusters in input, which is typically unknown in practice. The mean shift algorithm is an appealing and nonparametric clustering technique that estimates the number of clusters

T. Calders et al. (Eds.): ECML PKDD 2014, Part II, LNCS 8725, pp. 646–661, 2014.
© Springer-Verlag Berlin Heidelberg 2014

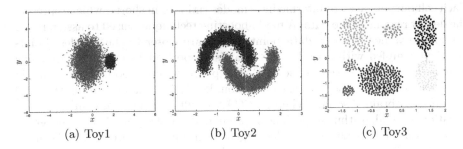

(a) Toy1   (b) Toy2   (c) Toy3

Fig. 1. Toy examples

directly from the data, and is able to find clusters with irregular shapes. It performs kernel density estimation, and iteratively locates the local maxima of the kernel mixture. Points that converge to the same mode are considered members of the same cluster [4]. The key parameter of mean shift is the kernel bandwidth. Its value can affect the performance of mean shift and is hard to set. Furthermore, mean shift may fail to find the proper cluster structure in the data when multiple modes exist in a cluster. As Fig. 1(b) shows, continuous dense regions exist in each cluster, possibly resulting in multiple modes detected by mean shift.

DBSCAN [10] is another popular density-based clustering method that does not require the number of clusters as input parameter. DBSCAN has the drawback of being sensitive to the choice of the neighborhood's radius (called *Eps*) [10]. DBSCAN tends to merge two clusters when an unsuitable *Eps* value is used, especially when the two clusters overlap, since the overlap may result in a contiguous high-density region, as shown in Fig. 1(a) and (c). (Experimental results on the three toy examples are presented in Section 5.)

Recently, a meta-algorithm known as Parallel Spatial Boosting Machine Learner (PSBML) has been introduced as a boosting algorithm for classification [17]. PSBML runs many classifiers in parallel on sampled data. The classifiers are organized in a two dimensional grid with a neighborhood structure. Data which are hard to classify are shared among the neighbor classifiers. PSBML is a robust algorithm that outperforms the underlying classifier in terms of accuracy and is less sensitive to parameter choice or noise [16]. The question we investigate in this research is whether the PSBML algorithm can be adapted to a clustering scenario to overcome the robustness issues related to parameter sensitivity as discussed above. The idea is to have a spatial grid framework as in PSBML, where a clustering algorithm such as mean shift runs at each node of the grid using local sampled data. A boosting process is applied to the local modes, which in turn are shared across the neighbors in the grid.

Specifically, we propose Boosted Mean Shift Clustering (BMSC) to address the aforementioned limitations of mean shift and DBSCAN. BMSC is an iterative and distributed version of mean shift clustering. Specifically, BMSC partitions the data across a grid, and applies mean shift locally on the cells of the grid.

Each cell outputs a set of intermediate modes (iModes in short), which represent the denser regions in the data. A mode-boosting technique is used to assign larger confidence values to those data points which are closer to the iModes. Points are then sampled with a probability that is proportional to the corresponding confidence. In successive iterations, BMSC progressively chooses data points in denser areas. Furthermore, at each iteration, DBSCAN is applied to partition all the iModes obtained so far. When DBSCAN results become stable in successive iterations, the algorithm stops. The accumulated iModes provide the "skeleton" of the data clusters and can be leveraged to group the entire data. The main contributions of this paper are summarized as follows:

- We introduce Boosted Mean Shift Clustering (BMSC) to overcome the disadvantages of mean shift and DBSCAN, while preserving their nonparametric nature. Our technique has the ability to identifying the essential structure (skeleton) of the clusters through the boosting of points around the modes.
- We present a complexity analysis to show the potential of BMSC to solve large-scale clustering tasks efficiently.
- Extensive experiments demonstrate the effectiveness and robustness of our proposed approach.

The rest of this paper is organized as follows. We review related work in Section 2 and introduce our methodology in Section 3. Section 4 presents the empirical evaluation and Section 5 discusses the experimental results. A final comment and conclusions are provided in Section 6 and 7, respectively.

2   Related Work

Mean shift [4,12] is a nonparametric feature space analysis technique that has been widely used in many machine learning applications, such as clustering [3], computer vision and image processing [4], and visual tracking [2]. It iteratively estimates the density of each point and computes the mean shift vector, which always points toward the direction of maximum increase in the density [4]. This defines a path leading to a stationary point (mode). The set of original data points that converge to the same mode defines a cluster. Mean shift uses a global fixed bandwidth, while the adaptive mean shift [5] sets different bandwidths for different data points. The convergence of mean shift procedure is guaranteed [4].

Density-based clustering methods [20] define a cluster as a set of points located in a contiguous region of high density, while points located in low-density areas are considered as noise or outliers. DBSCAN [10] is a popular clustering algorithm that relies on a density-based notion of clusters. It has only one parameter Eps, provided that the minimum number of points ($Minpts$) required to form a cluster is fixed. OPTICS [1] replaces the parameter Eps in DBSCAN with a maximum search radius and can be considered as a generalization of DBSCAN.

Adaboost [11] is the most popular boosting algorithm. It iteratively generates a distribution over the data in such a way that misclassified points by previous classifiers are more likely to be selected to train the next weak classifier. Adaboost is an ensemble algorithm that combines these weak classifiers to form a strong classifier that has shown to be more robust than the single classifier. Parallel spatial boosting machine learning (PSBML) [16] is a recent boosting algorithm which combines concepts from spatially structured parallel algorithms and machine learning boosting techniques. Both Adaboost and PSBML solve classification problems. The technique we introduce in this work is inspired by the PSBML framework. Unlike PSBML, though, our focus here is unsupervised learning, and in particular density-based clustering.

3   Boosted Mean Shift Clustering

Let $\mathcal{X} = \{\mathbf{x}_1, \mathbf{x}_2, \ldots, \mathbf{x}_n\}$ denote the data set, where n is the number of points and d is the dimensionality of each point $\mathbf{x}_i = (x_{i1}, x_{i2}, \ldots, x_{id})^T, i = 1, 2, \ldots, n$. A hard clustering $C = \{C_1, C_2, \ldots, C_{k^*}\}$ partitions $\mathcal{X}$ into k^* disjoint clusters, i.e., $C_i \cap C_j = \emptyset$ ($\forall i \neq j, i, j = 1, 2, \ldots, k^*$), and $\cup_{k=1}^{k^*} C_k = \mathcal{X}$.

3.1   Preliminary

In this section, we first give a brief review of the mean shift technique [4,5]. When using one global bandwidth h, the multivariate kernel density estimator with Kernel $K(\mathbf{x})$[1] is given by

$$\hat{f}(\mathbf{x}) = \frac{1}{nh^d} \sum_{i=1}^{n} K\left(\frac{\mathbf{x} - \mathbf{x}_i}{h}\right) \tag{1}$$

The *profile* of a kernel K is defined as a function $\kappa : [0, +\infty) \to \mathbb{R}$ such that $K(\mathbf{x}) = c \cdot \kappa(\|\mathbf{x}\|^2)$, where the positive constant c makes $K(\mathbf{x})$ integrate to one. Then, the sample point estimator (1) becomes

$$\hat{f}(\mathbf{x}) = \frac{c}{nh^d} \sum_{i=1}^{n} \kappa\left(\left\|\frac{\mathbf{x} - \mathbf{x}_i}{h}\right\|^2\right) \tag{2}$$

By taking the gradient of $\hat{f}(\mathbf{x})$ we obtain

$$\nabla \hat{f}(\mathbf{x}) = \frac{2c}{nh^{d+2}} \left[\sum_{i=1}^{n} g\left(\left\|\frac{\mathbf{x} - \mathbf{x}_i}{h}\right\|^2\right)\right] \times \underbrace{\left[\frac{\sum_{i=1}^{n} \mathbf{x}_i g(\|\frac{\mathbf{x} - \mathbf{x}_i}{h}\|^2)}{\sum_{i=1}^{n} g(\|\frac{\mathbf{x} - \mathbf{x}_i}{h}\|^2)} - \mathbf{x}\right]}_{\text{mean shift vector}} \tag{3}$$

where $g(x) = -\kappa'(x)$, provided that the derivative of κ exists. The first part of Eq. (3) is a constant, and the factor in bracket is the mean shift vector, which

---
[1] We use a Gaussian kernel in this paper.

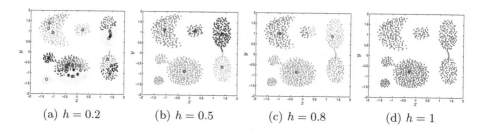

(a) $h = 0.2$ (b) $h = 0.5$ (c) $h = 0.8$ (d) $h = 1$

Fig. 2. Toy3: Clustering results of mean shift for different values of h

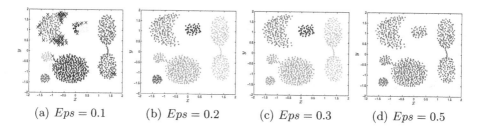

(a) $Eps = 0.1$ (b) $Eps = 0.2$ (c) $Eps = 0.3$ (d) $Eps = 0.5$

Fig. 3. Toy3: Clustering results of DBSCAN for different values of Eps

always points towards the direction of the greatest increase in density. Using the
mean shift vector, a sequence of estimation points $\{y_t\}_{t=1,2,\dots}$ is computed

$$y_{t+1} = \frac{\sum_{i=1}^{n} x_i g(\|\frac{y_t - x_i}{h}\|^2)}{\sum_{i=1}^{n} g(\|\frac{y_t - x_i}{h}\|^2)} \tag{4}$$

The starting point y_1 is one of the points x_i. The point that $\{y_t\}_{t=1,2,\dots}$ converges
to is considered as the mode of y_1. The points that converge to the same mode
are considered members of the same cluster. Please refer to [4] for more details.

3.2   The Algorithm

The performance of mean shift is sensitive to the choice of the bandwidth h.
To demonstrate this fact, Fig. 2 shows the clustering results of mean shift for
different values of h on a two-dimensional dataset containing clusters of different
shapes (called Toy3 in our experiments). The dark circles in Fig. 2 correspond
to the (global) modes generated by mean shift. For any given mode, the points
that converge to it are marked with the same color, and they define a cluster.
As shown in Fig. 2 (a), when the value $h = 0.2$ is used, mean shift finds several
modes and therefore detects a large number of clusters. Larger values of h lead
to fewer modes, and to the merging of separate clusters.

DBSCAN is another popular density-based clustering algorithm (refer to [10]
for more details) which is also sensitive to its input parameters and is likely
to merge overlapping clusters. Fig. 3 gives the results of DBSCAN on Toy3 for
different values of the parameter Eps. Here $Minpts$ is set to 4. Points of the

same color belong to the same cluster. In Fig. 3(a) the points marked as "×" are classified as noisy points by DBSCAN. More clusters and more noisy points are found by DBSCAN when *Eps*= 0.1. The detected noisy points are actually members of a cluster, and should not be considered as outliers. The larger *Eps* becomes, the more clusters are merged by DBSCAN. Eventually, for *Eps*= 0.5, DBSCAN detects only one cluster. The two rightmost clusters are also merged in Fig. 3(a), when *Eps*= 0.1 is used.

To overcome these limitations of mean shift and DBSCAN, while retaining their nonparametric nature, we propose the Boosted Mean Shift Clustering (BMSC) algorithm. We seek to capture the underlying group structure of the data by selecting the subset of data that provides the *skeleton* of the clusters. To achieve this goal, we iteratively compute modes relative to sampled data in a distributed fashion, and boost points proportionally to their distance from the modes. To achieve this goal, BMSC partitions the original data across a grid, and applies mean shift locally on the cells of the grid to search for the denser regions iteratively. The details are described below.

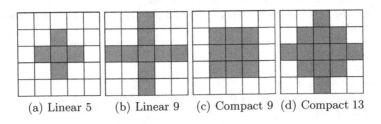

(a) Linear 5   (b) Linear 9   (c) Compact 9   (d) Compact 13

Fig. 4. Various neighborhood structures

Mode-boosting. BMSC first partitions the data uniformly across the cells of a two dimensional grid structure[2], as depicted in Fig. 4 [16,17,22] (Line 1 of Algorithm 1). The cells of the grid interact with the neighbors, where the neighborhood structure is user defined, as shown in Fig. 4[3]. Each cell applies a mean shift clustering algorithm on its local data. The mean shift algorithm outputs intermediate modes (iModes), which are located within dense regions (Line 6). Each cell uses its own iModes to assign confidence values to points assigned to the cell itself and to its neighbors. Specifically, given a set of iModes generated at $cell_j$, i.e., $iModes^{(j)} = \{iMode_1^{(j)}, \ldots, iMode_L^{(j)}\}$, we assign each local point (in $cell_j$ itself or in any neighboring cells) to the nearest iMode in $iModes$. For the points assigned to the same $iMode_l^{(j)}(l = 1, \ldots, L)$, we compute the confidence value of point i w.r.t. $cell_j$ as

$$conf_i^{(j)} = 1 - \frac{dis_i - min\_dis}{max\_dis - min\_dis} \qquad (5)$$

---

[2] The dimensionality of the grid affects the size of the neighborhood, and therefore the speed at which data is propagated through the grid. Note that the dimensionality d of the data can be arbitrary.

[3] In this paper, we use the 'Linear 5' structure.

Algorithm 1. Boosted Mean Shift Clustering

Input: $\mathcal{X}$, *width, height, h, Eps.*
Output: The final clustering result *cl_final.*
 1: INITIALIZEGRID($\mathcal{X}$, *width, height*); //Distribute $\mathcal{X}$ over $I = width \times height$ cells.
 2: $iModes \leftarrow \emptyset$;   //Initialize the set of intermediate modes.
 3: $counter \leftarrow 1$;
 4: **repeat**
 5:    **for** $j \leftarrow 1$ **to** I **do**
 6:       $newiModes \leftarrow$ MEANSHIFT($cellData_j, h$);
 7:       $iModes$.APPEND($newiModes$);
 8:    **end for**
 9:    CONFIDENCEASSIGNMENT(); // Assign confidence values to points in each cell via Eqs. (5) and (6).
10:    **for** $j \leftarrow 1$ **to** I **do**
11:       $CollectedData \leftarrow$ COLLECTNEIGHBORDATA(j) $\cup$ $cellData_j$;
12:       $cellData_j \leftarrow$ WEIGHTEDSAMPLING($CollectedData$);   // Update $cellData_j$.
13:    **end for**
14:    $[cl\_iModes, numberOfClustersDetected] \leftarrow$ DBSCAN($iModes, Eps$);
       //cl_iModes is the clustering result of $iModes$.
15:    **if** ($numberOfClustersDetected == lastnumberOfClustersDetected$) **then**
16:       $counter++$;
17:    **else**
18:       $counter \leftarrow 1$;
19:    **end if**
20: **until** $counter == 3$
21: $cl\_final \leftarrow$ DATAASSIGNMENT($\mathcal{X}$, $iModes, cl\_iModes$);   //Assign points in $\mathcal{X}$.
22: **return** cl_final.

---

where dis_i is the distance between point i and $iMode_l^{(j)}$, min_dis and max_dis are the minimum and maximum distances between the corresponding points and $iMode_l^{(j)}$, respectively. $conf_i^{(j)} \in [0,1]$. Intuitively, points near $iMode_l^{(j)}$ obtain larger confidence values, while those far away from $iMode_l^{(j)}$ are assigned smaller confidence values. Since a point $\mathbf{x}_i$ is a member of the neighborhood of multiple cells, an ensemble of confidence's assessments is obtained. We set the final confidence to the maximum confidence value obtained from any cell:

$$conf_i = \max_{j \in N_i} conf_i^{(j)} \tag{6}$$

where N_i is a set of indices of the neighbors of the cell to which point $\mathbf{x}_i$ belongs (Line 9).

The confidence values are used to select a sample of the data, locally at each cell, via a weighted sampling mechanism. Specifically, for each $cell_j$, all points in the cell and in its neighbors are collected. The larger the confidence value credited to a point $\mathbf{x}_i$ is (i.e., the closer $\mathbf{x}_i$ is to some iMode), the larger is the probability for $\mathbf{x}_i$ to be selected (Lines 11–12). As such, copies of points with larger confidence values will have higher probability of being selected, while points with low confidence will have a smaller probability of being selected. The sample size at each cell is kept constant at each epoch. Note that duplicate points may appear in a given cell, and the same point may appear in different cells.

Stopping Criterion. At each epoch, BMSC applies mean shift locally at each cell, thus generating a set of new iModes. We combine the new iModes with the iModes generated during the previous iterations, and apply DBSCAN on such updated set of iModes (Line 14). This process is repeated until the number of detected clusters by DBSCAN does not change for three consecutive iterations. At each iteration, the set of iModes produced so far gives a representation of the original data. A stable partition (i.e., a consistent number of detected clusters is obtained for three iterations) of iModes indicates a stable partition of the original data. In practice we have observed that the number of distinct points at each epoch quickly decreases, and BMSC always stops in less than 20 iterations in our experiments. The convergence of BMSC is empirically shown.

When BMSC stops, DBSCAN gives a partition of all the iModes. We then assign each original data point to the cluster to which its nearest iMode belongs. This gives the final clustering of the original data (Line 21). The pseudo-code of BMSC is given in Algorithm 1.

Computational Complexity. The computational complexity of estimating the density and computing the mean shift vector for one data point is $O(n)$, where n is the total number of data. Let T_1 be the maximum number of iterations it takes to compute the mode of any point in $\mathcal{X}$. Then the complexity of mean shift on the whole data is $O(T_1 n^2)$. The running time of BMSC is driven by the complexity of mean shift running on the cells of the grid and of DBSCAN running on the obtained iModes. BMSC applies mean shift locally on every cell, each with complexity $O(T_1 m^2)$, where $m = \frac{n}{I}$ and $I = width \times height$ is the number of cells in the spatial grid. The runtime complexity of DBSCAN on $\mathcal{X}$ is $O(n^2)$ and it can be reduced to $O(n \log n)$ if one uses R*-tree to process a range query [20]. Let s be the number of obtained iModes and T_2 be the number of iterations when BMSC stops, then the total computation complexity of BMSC is $O(T_2(IT_1 m^2 + s \log s))$. T_2 is empirically proved to be small, and $m \ll n$, $s \ll n$ when n is large. Thus, the computational complexity of BMSC is lower than mean shift and DBSCAN when dealing with large scale data. BMSC can be further speeded up with a parallel implementation. The complexity can be reduced to $O(T_2 \max\{T_1 m^2, s \log s\})$ with an I multi-thread process, which makes BMSC available for large-scale clustering tasks.

4   Experimental Setup

Datasets. We conducted experiments on three toy examples and ten real-world data sets to evaluate the performance of BMSC and comparing methods. Table 1 provides the characteristics of all the datasets used in our experiments. The toy examples are shown in Fig. 1. The two classes of toy

Table 1. Datasets used in the experiments

Data	#points	#features	#classes
Toy1	10000	2	2
Toy2	14977	2	2
Toy3	788	2	7
2D2K	1000	2	2
8D5K	1000	8	5
Letter	1555	16	2
Satimage	4220	36	4
Symbols	1020	398	6
KDD99	25192	38	22
Banknote	200	6	2
Chainlink	400	3	2
Image_Seg	990	19	3
Wall	5456	2	4

example 1 (Toy1), which were generated according to multivariate Gaussian distributions, consist of 8,000 and 2,000 points, respectively. The mean vector and the covariance matrix of the left class are (-10,0) and (10 0; 0 10), while those of the right class are (2,0) and (1 0; 0 1), respectively. Toy example 2 (Toy2, two-moons data), Banknote, and ChainLink contain two classes and are available at http://www.mathworks.com/matlabcentral/fileexchange/34412-fast-and-efficient-spectral-clustering. 400 points (200 points per class) of ChainLink were randomly chosen in the experiments. Toy3 (Aggregation data) was downloaded from http://cs.joensuu.fi/sipu/datasets/ and was used in [13]. 2D2K and 8D5K were two datasets used in [23] and were downloaded from http://strehl.com/. Symbols is a UCR time series data [19]. NSL-KDD data set retains all the important statistical characteristics of KDDCup-99. A subset which contains 25192 instances of 22 classes (1 normal class and 21 attack types) was downloaded from http://nsl.cs.unb.ca/NSL-KDD/ and the 38 numerical features were used for the experiments. Letter, Satimage, Image_Seg (Image segmentation) and Wall (Wall-following robot navigation data) are all available from the UCI repository (http://archive.ics.uci.edu/ml/index.html). The letters 'A' and 'B' were selected from the Letter database. The first 4 classes of Satimage and the first three classes of Image_Seg were used in our experiments. For each dataset, each feature was normalized to have zero mean value and unit variance.

Evaluation Criteria. We chose Rand Index (RI) [14], Adjusted Rand Index (ARI) [14], and Normalized Mutual Information (NMI) [23] as evaluation criteria since the label information of data are known. The label information is only used to measure the clustering results, and is not used during the clustering process. Both RI and NMI range from 0 to 1, while ARI belongs to [-1,1]. A value 1 of RI/ARI/NMI indicates a perfect clustering result.

Experimental Settings. As shown in Algorithm 1, BMSC requires four parameters in input: *width*, *height*, the bandwidth h for mean shift, and *Eps* for DBSCAN. We used a 5×5 spatial grid for Toy1, Toy2, and KDD99, and a 3×3 grid for all the other data sets. We set h to the average distance between each point and its k-th nearest neighbor, where $k = \alpha \sqrt{n\_Data}$. Here n_Data is the size of the sample assigned to a cell of a grid for BMSC, or the size of the whole dataset when we run mean shift on the entire collection. The value of α also affects the bandwidth; a larger α value corresponds to a larger global bandwidth h. α is always set to 0.5 for both BMSC and mean shift in our experiments. We set $Eps = 0.5$ for BMSC and DBSCAN on the three 2-dimensional toy datasets. Sensitivity analysis of parameters α and *Eps* is discussed in Section 5.4. To set the *Eps* parameter on real data, we consider the iModes generated after the first epoch of BMSC, and compute the distance to the 4-th nearest iMode [10] for each of the iModes. We then choose the median value of all such distances as the value of *Eps* for DBSCAN in all the successive iterations. When DBSCAN is run on the whole data, the 4-th nearest neighbor distances are computed with respect to the entire collection of data, and *Eps* is again set to the median of those values. The *MinPts* value of DBSCAN is always set to four in our experiments, as done in [10].

Besides mean shift and DBSCAN [7,10], we also performed comparisons against several other clustering algorithms: OPTICS [1,6], k-means [21], LAC [9], Aver-l (average-linkage clustering) [15], and EM (with a Gaussian mixture) [8]. OPTICS is a density based clustering algorithm which creates an augmented ordering of the data representing its clustering structure, and then retrieves DBSCAN clusters as the final clustering result. When OPTICS uses DBSCAN to extract clusters, the parameters were set as in DBSCAN itself. Both DBSCAN and OPTICS may output noisy clusters. k-means, LAC, Aver-l, and EM require the number of clusters in input, which we set equal to the number of classes in the data. LAC requires an additional parameter (weight of the regularization term; see [9] for details), which we set to 0.2 throughout our experiments. Mean shift, DBSCAN, OPTICS, and Aver-l are deterministic for fixed parameter values. For the remaining methods, the reported values are the average of 20 independent runs. One-sample t-test and paired-samples t-test were used to assess the statistical significance of the results at 95% significance level.

5   Results and Analysis

5.1   Results on Toy Examples

To illustrate the effectiveness of BMSC, we first conducted experiments on the three toy datasets. Fig. 5 shows the data selected by BMSC at different epochs on Toy1 in one independent run. For this run, BMSC stops at the fourth iteration. '(#5529)' in Fig. 5(a) means that at this epoch 5529 points are selected. The number of (distinct) points in each iteration greatly decreases, and points around the densest regions are more likely to survive. At successive iterations, the data becomes better separated, even though the original two classes overlap. Fig. 6

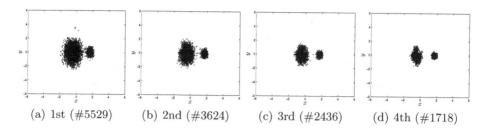

(a) 1st (#5529)   (b) 2nd (#3624)   (c) 3rd (#2436)   (d) 4th (#1718)

Fig. 5. Data selected by BMSC on Toy1 at different epochs

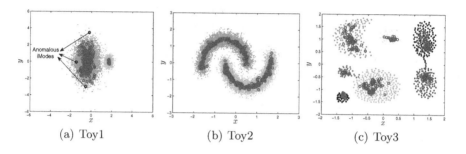

(a) Toy1   (b) Toy2   (c) Toy3

Fig. 6. BMSC clustering results on the three toy examples

gives the final clustering results of BMSC on the three toy examples for a given run. The red circles represent the iModes accumulated in successive iterations when BMSC stops. Points with the same color belong to the same cluster. The iModes successfully capture the structure of the different clusters. As a result, BMSC achieves a close to perfect performance on these datasets. When two clusters overlap (like in Toy1), the corresponding iModes are still well separated, and therefore easy to partition. When performing DBSCAN on the set of iModes, some iModes may be classified as noise. For example, three such iModes occur in Fig. 6(a). iModes detected as anomalous are discarded and not used to partition the whole data.

Table 2. Results on toy examples

Data		BMSC	MS	DBSCAN	OPTICS	k-means	LAC	Aver-l	EM
	RI	0.9955	0.3977	0.6792	0.6796	0.9294	0.9299	0.6798	**0.9978**
Toy1	ARI	0.9897	0.0755	-0.0009	-0.0004	0.8432	0.8443	-0.0001	**0.9949**
	NMI	0.9709	0.3846	0.0027	0.0018	0.7614	0.7626	0.0010	**0.9833**
	RI	**0.9995**	0.5380	0.5001	0.5001	0.8201	0.8185	0.8608	0.6584
Toy2	ARI	**0.9990**	0.0763	0.0000	0.0000	0.6401	0.6370	0.7216	0.3168
	NMI	**0.9970**	0.4495	0.0000	0.0000	0.5311	0.5280	0.6832	0.3675
	RI	0.9891	0.8697	0.2165	0.2165	0.9096	0.9006	**0.9971**	0.9063
Toy3	ARI	0.9686	0.5096	0.0000	0.0000	0.7061	0.6781	**0.9913**	0.7121
	NMI	0.9711	0.7925	0.0000	0.0000	0.8316	0.8115	**0.9869**	0.8379

Table 3. Results on real data (RI)

Data	BMSC	MS	DBSCAN	OPTICS	k-means	LAC	Aver-l	EM
2D2K	**0.9560**	0.9522	0.6962	0.4995	0.9531	0.9538	0.9250	0.9078
8D5K	**1.0000**	**1.0000**	0.9094	0.8164	0.9359	0.9448	**1.0000**	0.9488
Letter	**0.8928**	0.8909	0.6479	0.5133	0.8762	0.8767	0.8875	0.7054
Satimage	**0.8216**	0.8044	0.7305	0.5271	0.8030	0.8051	0.5370	0.7975
Symbols	**0.9081**	0.9007	0.8102	0.3922	0.8842	0.8773	0.6275	0.8193
KDD99	**0.7699**	0.7416	0.6843	0.6029	0.6777	0.6755	0.3966	0.7106
Banknote	**0.9694**	0.9510	0.8009	0.5934	0.9228	0.9261	0.4975	0.8892
Chainlink	**0.7475**	0.5626	0.5378	0.5264	0.5431	0.5410	0.5550	0.7350
Image_Seg	**0.9073**	0.8695	0.6978	0.7036	0.8091	0.8112	0.3461	0.7446
Wall	**0.7244**	0.7131	0.6609	0.5905	0.7055	0.7092	0.3656	0.6530

Table 4. Results on real data (ARI)

Data	BMSC	MS	DBSCAN	OPTICS	k-means	LAC	Aver-l	EM
2D2K	**0.9119**	0.9043	0.3922	0.0000	0.9062	0.9075	0.8499	0.8156
8D5K	**1.0000**	**1.0000**	0.6977	0.4010	0.8269	0.8446	**1.0000**	0.8558
LetterAB	**0.7856**	0.7817	0.2957	0.0265	0.7524	0.7533	0.7749	0.4109
Satimage	**0.5631**	0.5452	0.3508	0.0831	0.5234	0.5293	0.2242	0.5201
Symbols	**0.7042**	0.6645	0.4818	0.0071	0.6339	0.6186	0.2369	0.4566
KDD99	**0.4684**	0.3933	0.2979	0.0164	0.2229	0.2160	-0.0003	0.3154
Banknote	**0.9387**	0.9020	0.6011	0.1850	0.8456	0.8521	0.0000	0.7788
Chainlink	**0.4944**	0.1233	0.0737	0.0542	0.0865	0.0822	0.1110	0.4701
Image_Seg	**0.7843**	0.6775	0.2271	0.2078	0.6273	0.6218	0.0012	0.4812
Wall	0.2893	0.2437	0.1434	0.0532	0.3697	**0.3761**	0.0240	0.2656

Table 2 shows the results of the different algorithms on the toy examples using the three evaluation measures. In each row, the significantly best and comparable results are highlighted in boldface. On these datasets, BMSC improves upon mean shift, DBSCAN, and OPTICS by a large margin. As expected, EM gives the best performance on Toy1, which is a mixture of two Gaussians. Aver-l works quite well on Toy3. But both EM and Aver-l require the number of clusters in input, and their performance degrades on the other data. BMSC significantly outperforms all the comparing methods on Toy2, and it's the only approach that works well on all three toy datasets. The poor performance of k-means and LAC is mainly caused by the unbalanced data in Toy1, and the irregular shapes of clusters in Toy2 and Toy3.

5.2 Results on Real Data

This section evaluates the performance of the comparing methods on several real datasets. The RI and ARI values are shown in Tables 3 and 4, respectively. The best and comparable results are shown in boldface. In general, a better result on RI indicates a better result on ARI and NMI. But this is not always the case. Lets consider the Wall data for example. BMSC gives the best RI value, while LAC gives the best ARI value. BMSC, mean shift, and Aver-l do a perfect job in clustering 8D5K. For the remaining datasets, BMSC significantly outperforms all the other comparing methods. It's worth observing that, in terms of the ARI measure, BMSC outperforms mean shift by a considerable margin on Symbols, KDD99, Banknote, Chainlink, Imag_Seg, and Wall. Similar results were obtained on NMI and are not reported due to the limited space.

5.3   Performance Analysis of BMSC

The average number of iterations of BMSC and the average number of detected clusters by BMSC are shown in Table 5. The table shows that BMSC stops after a small number of iterations. Comparing the number of detected clusters (#clusters) and the actual number of classes (#classes), we can see that a larger number of clusters is detected by BMSC on KDD99, Chainlink, and Wall[4]. For this reason, lower values of RI and ARI are obtained on these three datasets, as shown in Tables 3 and 4. The number of clusters detected by BMSC is similar to the number of classes on Toy3, Satimage, Symbols, and Image_Seg. BMSC always detects a number of clusters that matches the number of classes on Toy1, Toy2, 2D2K, 8D5K, Letter, and Banknote datasets. This indicates that BMSC is capable of automatically finding a reasonable number of clusters.

Table 5. Performance Analysis of BMSC

Data	#iterations	#clusters	#classes
Toy1	3.30	2±0.00	2
Toy2	3.00	2±0.00	2
Toy3	4.45	6.80±0.42	7
2D2K	3.75	2±0.00	2
8D5K	3.15	5±0.00	5
Letter	3.35	2±0.00	2
Satimage	4.60	4.10±0.79	4
Symbols	5.50	6±1.49	6
KDD99	14.55	41.45±5.15	22
Banknote	3.05	2±0.00	2
Chainlink	4.15	4.65±1.23	2
Image_Seg	4.45	3.75±0.55	3
Wall	6.65	12.20±1.61	4

5.4   Sensitivity Analysis of the Parameters α and Eps

We tested the sensitivity of BMSC w.r.t. the parameters α and Eps on three datasets, namely Toy1, Toy2, and Satimage. We first tested the sensitivity of α which controls the kernel bandwidth in BMSC and mean shift. $Eps = 0.5$ was set for BMSC. The test range of α is $[0.05, 0.6]$ and Fig. 7 gives the results.

---

[4] This is not surprising since in practice there may not be a one-to-one correspondence between classes and clusters.

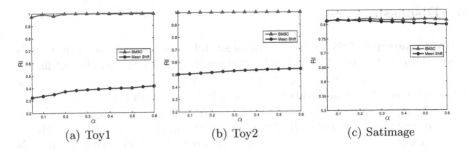

Fig. 7. Sensitivity analysis of the α parameter (RI)

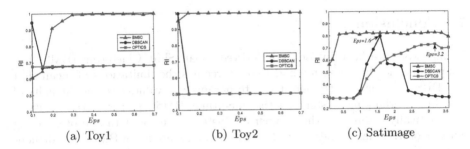

Fig. 8. Sensitivity analysis of parameter Eps (RI)

Mean shift performs slightly better as α increases on Toy1 and Toy2, and a general reduction in performance is observed with a larger α on Satimage. BMSC is stable and achieves good performance throughout. A larger α value leads to fewer modes found by mean shift, while a smaller α value results in more modes, thus generating more clusters in general. BMSC is more robust to the different values of α because α only affects the number of iModes generated by each cell, and such iModes are then linked together by DBSCAN.

We further compared BMSC against DBSCAN and OPTICS for different values of Eps. In these experiments, $\alpha = 0.5$ for BMSC. Tested ranges of Eps are $[0.1, 0.7]$ for Toy1 and Toy2, and $[0.1, 3.6]$ for Satimage. The results are given in Fig. 8. The performance of BMSC increases as the value of Eps increases, and it is stable for a wide range of Eps values. The main reason for this behavior is that BMSC applies DBSCAN on the iModes, rather than on the whole data. iModes of different clusters are well separated, even though the original clusters may overlap. DBSCAN works well for $Eps = 0.1$, and OPTICS always performs poorly on Toy1 and Toy2. On Satimage, the performance of both DBSCAN and OPTICS increases for larger Eps values, and reaches its peak at $Eps = 1.6$ and $Eps = 3.2$, respectively. After that, the performance drops. This shows that DBSCAN and OPTICS are sensitive to the choice of values for Eps, while our BMSC technique is robust to parameter settings. This provides insight to the superior performance of BMSC against DBSCAN and OPTICS obtained in the previous experiments.

6 Discussion

To overcome some of the limitations of mean shift, one may run mean shift on the whole data with a small bandwidth, thus generating a large number of modes. The modes can then be merged to obtain the final clustering results. But this approach has two major disadvantages: (1) Running mean shift on a large-scale data is of high complexity; and (2) With a fixed bandwidth, only one mode (the local maxima) will be found in dense areas. In contrast, our BMSC is able to find contiguous intermediate modes in dense areas. It achieves this through the partitioning of the data across a spatial grid and through the iterative process of collecting the iModes.

7 Conclusion

In this work we have introduced Boosted Mean Shift Clustering (BMSC), a nonparametric clustering method that overcomes the limitations of mean shift and DBSCAN, namely the sensitivity to parameters' values and the difficulty of handling overlapping clusters. At the same time, BMSC preserves the ability of automatically estimating the number of clusters from the data and of handling clusters of irregular shapes. The effectiveness and stability of BMSC are demonstrated through extensive experiments conducted on synthetic and real-world datasets. We are interested in extending the framework introduced here to other clustering methodologies, e.g. centroid-based, as well as to a semi-supervised scenario. As mentioned earlier in our discussion on computational complexity, BMSC can be easily parallelized via a multi-thread implementation (one thread per cell). We will proceed with such implementation in our future work and test the achieved speed-up for the big data clustering problems.

Acknowledgement. This paper is in part supported by the China Scholarship Council (CSC).

References

1. Ankerst, M., Breunig, M.M., Peter Kriegel, H., Sander, J.: OPTICS: Ordering points to identify the clustering structure. In: SIGMOD, pp. 49–60. ACM Press (1999)
2. Avidan, S.: Ensemble tracking. TPAMI 29(2), 261–271 (2007)
3. Cheng, Y.: Mean shift, mode seeking, and clustering. TPAMI 17(8), 790–799 (1995)
4. Comaniciu, D., Meer, P.: Mean shift: a robust approach toward feature space analysis. TPAMI 24(5), 603–619 (2002)
5. Comaniciu, D., Ramesh, V., Meer, P.: The variable bandwidth mean shift and data-driven scale selection. In: ICCV, pp. 438–445 (2001)
6. Daszykowski, M., Walczak, B., Massart, D.: Looking for natural patterns in analytical data. Part 2: Tracing local density with OPTICS. Journal of Chemical Information and Computer Sciences 42(3), 500–507 (2002)

7. Daszykowski, M., Walczak, B., Massart, D.L.: Looking for natural patterns in data. Part 1: Density based approach. Chemometrics and Intelligent Laboratory Systems 56(2), 83–92 (2001)
8. Dempster, A.P., Laird, N.M., Rubin, D.B.: Maximum likelihood from incomplete data via the EM algorithm. Journals of the Royal Statistical Society, Series B 39(1), 1–38 (1977)
9. Domeniconi, C., Gunopulos, D., Ma, S., Yan, B., Al-Razgan, M., Papadopoulos, D.: Locally adaptive metrics for clustering high dimensional data. DMKD 14(1), 63–97 (2007)
10. Ester, M., Kriegel, H.-P., Sander, J., Xu, X.: A density-based algorithm for discovering clusters in large spatial databases with noise. In: KDD, pp. 226–231 (1996)
11. Freund, Y., Schapire, R.E.: A decision-theoretic generalization of on-line learning and an application to boosting. Journal of Computer and System Sciences 55, 119–139 (1997)
12. Fukunaga, K., Hostetler, L.D.: The estimation of the gradient of a density function, with applications in pattern recognition. IEEE Transactions on Information Theory 21(1), 32–40 (1975)
13. Gionis, A., Mannila, H., Tsaparas, P.: Clustering aggregation. TKDD 1(1), 1–30 (2007)
14. Hubert, L., Arabie, P.: Comparing partitions. Journal of Classification 2(1), 193–218 (1985)
15. Jain, A.K., Murty, M.N., Flynn, P.J.: Data clustering: A review. ACM Computing Surveys 31(3), 264–323 (1999)
16. Kamath, U., Domeniconi, C., Jong, K.A.D.: An analysis of a spatial ea parallel boosting algorithm. In: GECCO, pp. 1053–1060 (2013)
17. Kamath, U., Kaers, J., Shehu, A., De Jong, K.A.: A spatial EA framework for parallelizing machine learning methods. In: Coello, C.A.C., Cutello, V., Deb, K., Forrest, S., Nicosia, G., Pavone, M. (eds.) PPSN 2012, Part I. LNCS, vol. 7491, pp. 206–215. Springer, Heidelberg (2012)
18. Kaufman, L., Rousseeuw, P.: Clustering by Means of Medoids. Faculty of Mathematics and Informatics (1987)
19. Keogh, E., Zhu, Q., Hu, B., Hao, Y., Xi, X., Wei, L., Ratanamahatana, C.A.: The UCR Time Series Classification/Clustering Homepage (2011), http://www.cs.ucr.edu/~eamonn/time_series_data/
20. Kriegel, H.-P., Kroger, P., Sander, J., Zimek, A.: Density-based clustering. DMKD 1(3), 231–240 (2011)
21. MacQueen, J.: Some methods for classification and analysis of multivariate observations. In: Proceedings of the 5th Berkeley Symposium on Mathematical Statistics and Probability, pp. 281–297. University of California Press (1967)
22. Sarma, J., Jong, K.: An analysis of the effects of neighborhood size and shape on local selection algorithms. In: Ebeling, W., Rechenberg, I., Voigt, H.-M., Schwefel, H.-P. (eds.) PPSN 1996. LNCS, vol. 1141, pp. 236–244. Springer, Heidelberg (1996)
23. Strehl, A., Ghosh, J.: Cluster ensembles - a knowledge reuse framework for combining multiple partitions. JMLR 3, 583–617 (2002)

Hypernode Graphs for Spectral Learning on Binary Relations over Sets*

Thomas Ricatte[1], Rémi Gilleron[2], and Marc Tommasi[2]

[1] SAP Research, Paris
[2] Lille University, LIFL and Inria Lille

Abstract. We introduce hypernode graphs as weighted binary relations between sets of nodes: a hypernode is a set of nodes, a hyperedge is a pair of hypernodes, and each node in a hypernode of a hyperedge is given a non negative weight that represents the node contribution to the relation. Hypernode graphs model binary relations between sets of individuals while allowing to reason at the level of individuals. We present a spectral theory for hypernode graphs that allows us to introduce an unnormalized Laplacian and a smoothness semi-norm. In this framework, we are able to extend spectral graph learning algorithms to the case of hypernode graphs. We show that hypernode graphs are a proper extension of graphs from the expressive power point of view and from the spectral analysis point of view. Therefore hypernode graphs allow to model higher order relations whereas it is not true for hypergraphs as shown in [1]. In order to prove the potential of the model, we represent multiple players games with hypernode graphs and introduce a novel method to infer skill ratings from game outcomes. We show that spectral learning algorithms over hypernode graphs obtain competitive results with skill ratings specialized algorithms such as Elo duelling and TrueSkill.

Keywords: Graphs, Hypergraphs, Semi Supervised Learning, Multiple Players Games.

1 Introduction

Graphs are commonly used as a powerful abstract model to represent binary relationships between individuals. Binary relationships between individuals are modeled by edges between nodes. This is for instance the case for social networks with the friendship relation, or for computer networks with the connection relation. The hypergraph formalism [2] has been introduced for modeling problems where relationships are no longer binary, that is when they involve more than two individuals. Hypergraphs have been used for instance in bioinformatics [11], computer vision [17] or natural language processing [3]. But, graphs and hypergraphs are limited when one has to consider relationships between sets of individual objects. A typical example is the case of multiple players games

---

* This work was supported by the French National Research Agency (ANR). Project Lampada ANR-09-EMER-007.

T. Calders et al. (Eds.): ECML PKDD 2014, Part II, LNCS 8725, pp. 662–677, 2014.
© Springer-Verlag Berlin Heidelberg 2014

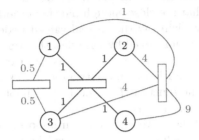

Fig. 1. A hypernode graph modeling 3 tennis games with 4 players. Each of the three hyperedges has one color and models a game for which players connected to the same long edge of a rectangle are in the same team.

where a game can be viewed as a relationship between two teams of multiple players. Other examples include relationships between groups in social networks or between clusters in computer networks. For these problems, considering both the group level and the individual level is a requisite. For instance for multiple players games, one is interested in predicting game outcomes for games between teams as well as in predicting player skills. Graphs and hypergraphs fail to model higher order relations considering both the individual level and the level of sets of individuals. This paper is a proposition to overcome this limitation.

A first contribution of this paper is to introduce a new class of undirected hypergraphs called *hypernode graphs* for modeling binary relationships between sets of individual objects. A relationship between two sets of individual objects is represented by a hyperedge which is defined to be a pair of disjoint hypernodes, where a hypernode is a set of nodes. Nodes in a hypernode of a hyperedge are given a non negative weight that represents the node contribution to the binary relationship. An example of hypernode graph is presented in Figure 1. There are four nodes that represent four tennis players and three hyperedges representing three games between teams: {1} against {3}, {1, 2} against {3, 4}, and {1, 4} against {2, 3}. For each hyperedge, each player has been given a weight which can be seen as the player's contribution. It can be noted that the hyperedge between singleton sets {1} and {3} can be viewed as an edge between nodes 1 and 3 with edge weight 0.5. Undirected graphs are shown to be hypernode graphs where hypernodes are singleton sets.

Given a hypernode graph modeling binary relationships between sets of individuals, an important task, as said above, is to evaluate individuals by means of node labelling or node scoring functions. The second contribution of this paper is to propose machine learning algorithms in the semi-supervised, batch setting on hypernode graphs for predicting node labels or node scores. To this aim, we develop a spectral learning theory for hypernode graphs. Similarly to the case of graph spectral learning, our approach relies on the homophilic assumption [4, Chapter 4] (also called assortative mixing assumption) which says that two linked nodes should have the same label or similar scores. For graphs, this assumption is reflected in the choice of smooth node functions for which linked nodes get values that are close enough. For hypernode graphs, we assume an additive model, and we will say that a real-valued node function over a hypernode graph is smooth if, for linked hypernodes, the weighted sum of function values over the two node sets are close enough. As an example, let us consider the blue hyperedge in Figure 1 between the two sets $\{1, 2\}$ and $\{3, 4\}$ and a real-valued node function f, the function f is said to be smooth over the hyperedge if $f(1) + f(2)$ is close to $f(3) + f(4)$.

For defining the smoothness, we introduce an unnormalized gradient for hypernode graphs. Then, we define an unnormalized Laplacian Δ for hypernode graphs by $\Delta = G^T G$ where G is the gradient. We show that the class of Laplacians of hypernode graphs is the class of symmetric positive semidefinite real-valued matrices M such that $\mathbf{1} \in \text{Null}(M)$, where $\text{Null}(M)$ denotes the null space of M and $\mathbf{1}$ is the vector full of 1's. Note that there exist hypernode graphs whose Laplacians do not match that of a graph (we can easily obtain extra-diagonal values that are positive as shown in Figure 2) whereas it has been proved in [1] that hypergraph Laplacians can be defined from graph Laplacians using adequate graph construction. The smoothness of a real-valued node function f on a hypernode graph can be characterized by the *smoothness semi-norm* defined by $\Omega(f) = f^T \Delta f$. We define the kernel of a hypernode graph to be the Moore-Penrose pseudoinverse [15] of its Laplacian. The spectral theory for hypernode graphs and its properties allow us to use spectral graph learning algorithms [16], [18], [20] for hypernode graphs.

We apply hypernode graph spectral learning to the rating of individual skills of players and to the prediction of game outcomes in multiple players games. We consider competitive games between two teams where each team is composed of an arbitrary number of players. Each game is modeled by a hyperedge and a set of games is represented by a hypernode graph. We define a skill rating function of players as a real-valued node function over the hypernode graph. And we show that finding the optimal skill rating function reduces to finding the real-valued function s^* minimizing $\Omega(s) = s^T \Delta s$, where Δ is the unnormalized Laplacian of the hypernode graph. The optimal individual skill rating function allows to compute the rating of teams and to predict game outcomes for new games. We apply this learning method on real datasets of multiple players games to predict game outcomes in a semi-supervised, batch setting. Experimental results show that we obtain very competitive results compared to specialized algorithms such as Elo duelling and TrueSkill.

Related Work. Hypernode graphs that we introduced can be viewed as an undirected version of directed hypergraphs popularized by [6] where a directed hyperedge consists in an oriented relation between two sets of nodes. As far as we know, this class of directed hypergraphs has not been studied from the machine learning point of view and no attempt was made to define a spectral framework for these objects. Hypernode graphs can also be viewed as an extension of hypergraphs. The question of learning with hypergraphs has been studied and, for an overview, we refer the reader to [1]. In this paper, the authors show that various formulations of the semi-supervised and the unsupervised learning problem on hypergraphs can be reduced to graph problems. For instance, the hypergraph Laplacian of [19] can be defined as a graph Laplacian by an adequate graph construction. To the best of our knowledge, no hypergraph Laplacian which cannot be reduced to a graph Laplacian has been defined so far. A very recent tentative to fully use the hypergraph structure was proposed by [8]. In this paper, the authors propose to use the hypergraph cut, and they introduce the total variation on a hypergraph as the Lovasz extension of the hypergraph cut. This allows to define a regularization functional on hypergraphs for defining semi-supervised learning algorithms.

2   Graphs and Hypernode Graphs

2.1   Undirected Graphs and Laplacians

In the following, we recall the commonly accepted definitions of undirected graphs and graph Laplacians. An *undirected graph* $\mathbf{g} = (V, E)$ is a set of nodes V with $|V| = n$ together with a set of undirected edges E with $|E| = p$. Each edge $e \in E$ is an unordered pair $\{i, j\}$ of nodes and has a non negative weight $w_{i,j}$. In order to define the smoothness of a real-valued node function f over a graph $\mathbf{g}$, we define the gradient function grad for f by, for every edge (i, j),

$$\operatorname{grad}(f)(i, j) = \sqrt{w_{i,j}}(f(j) - f(i)) \ .$$

We can note that $|\operatorname{grad}(f)(i, j)|$ is small whenever $f(i)$ is close to $f(j)$. Then, the smoothness of a real-valued node function f over a graph $\mathbf{g}$ is defined by

$$\Omega(f) = \sum_{i,j \in V^2} |\operatorname{grad}(f)(i, j)|^2 = f^T G^T G f \ ,$$

where G is the matrix of the linear mapping grad from $\mathbb{R}^n$ into $\mathbb{R}^p$. The symmetric matrix $\Delta = G^T G$ is called *undirected graph Laplacian*, which is also proved to be defined by $\Delta = D - W$ where D is the degree matrix of $\mathbf{g}$ and W the weight matrix of $\mathbf{g}$. $\Omega(f) = f^T \Delta f$ has been used in multiple works (see for example [20], [16]) to ensure the smoothness of a node labeling function f.

Additional information concerning the discrete analysis on graphs can be found in [18], which develop a similar theory with a normalized version of the gradient and Laplacian (G is replaced by $GD^{-1/2}$).

2.2   Hypernode Graphs

The following definition is our contribution to the modeling of binary relationships between sets of entities.

Definition 1. *A hypernode graph* $\mathbf{h} = (V, H)$ *is a set of nodes* V *with* $|V| = n$ *and a set of hyperedges* H *with* $|H| = p$. *Each hyperedge* $h \in H$ *is an unordered pair* $\{s_h, t_h\}$ *of two non empty and disjoint hypernodes (a hypernode is a subset* *of V). Each hyperedge $h \in H$ has a weight function w_h mapping every node i in $s_h \cup t_h$ to a positive real number $w_h(i)$ (for $i \notin s_h \cup t_h$, we define $w_h(i) = 0$). Each weight function w_h of $h = \{s_h, t_h\}$ must satisfy the* Equilibrium Condition *defined by*

$$\sum_{i \in t_h} \sqrt{w_h(i)} = \sum_{i \in s_h} \sqrt{w_h(i)} \ .$$

An example of hypernode graph is shown in Figure 1. The red hyperedge links the sets $\{1, 4\}$ and $\{2, 3\}$. The weights satisfy the Equilibrium condition which ensures that constant node functions have a null gradient as we will see in the next section. The green hyperedge is an unordered pair $\{\{1\}, \{3\}\}$ of two singleton sets with weights 0.5 for the nodes 1 and 3. It can be viewed as an edge between nodes 1 and 3 with edge weight 0.5. Indeed, when a hyperedge h is an unordered pair $\{\{i\}, \{j\}\}$ involving only two nodes, the Equilibrium Condition states that the weights $w_h(i)$ and $w_h(j)$ are equal. Thus, such a hyperedge can be seen as an edge with edge weight $w_{i,j} = w_h(i) = w_h(j)$. Therefore, a hypernode graph such that every hyperedge is an unordered pair of singleton nodes can be viewed as an undirected graph, and conversely.

2.3   Hypernode Graph Laplacians

In this section, we define the *smoothness* of a real-valued node function f over a hypernode graph with the gradient that we define now.

Definition 2. *Let* $\mathbf{h} = (V, H)$ *be a hypernode graph and f be a real-valued node function, the* (hypernode graph) *unnormalized gradient of* $\mathbf{h}$ *is a linear application, denoted by* grad, *that maps every real-valued node function f into a real-valued hyperedge function* grad(f) *defined, for every $h = \{s_h, t_h\}$ in H, by*

$$\mathrm{grad}(f)(h) = \sum_{i \in t_h} f(i)\sqrt{w_h(i)} - \sum_{i \in s_h} f(i)\sqrt{w_h(i)} \ ,$$

where an arbitrary orientation of the hyperedges has been chosen.

As an immediate consequence of the gradient definition and because of the Equilibrium Condition, the gradient of a constant node function is the zero-valued hyperedge function. Also, it can be noted that, for a hyperedge $h \in H$, $|\mathrm{grad}(f)(h)|^2$ is small when the weighted sum of the values $f(i)$ for nodes i in s_h is close to the weighted sum of the values $f(j)$ for nodes j in t_h. Thus, if we denote by $G \in \mathbb{R}^{p \times n}$ the matrix of grad, the smoothness of a real-valued node function f over a hypernode graph $\mathbf{h}$ is defined by $\Omega(f) = f^T G^T G f$.

Let $\mathbf{h}$ be a hypernode graph with unnormalized gradient G, the square $n \times n$ real valued matrix $\Delta = G^T G$ is defined to be the *unnormalized Laplacian* of the hypernode graph $\mathbf{h}$. It should be noted that, as in the graph case, the Laplacian Δ does not depend on the arbitrary orientation of the hyperedges used for defining the gradient. When the hypernode graph is a graph, the unnormalized hypernode graph Laplacian matches the unnormalized graph Laplacian. Last, we define the *hypernode graph kernel* of a hypernode graph **h** to be the Moore-Penrose pseudoinverse $\Delta^\dagger$ [15] of the hypernode graph Laplacian Δ .

2.4   Hypernode Graph Laplacians and Learning

We can characterize hypernode graph Laplacians by

Proposition 1. *The class of hypernode graph Laplacians is the class of symmetric positive semidefinite real-valued matrices M such that $\mathbf{1} \in \mathrm{Null}(M)$, where $\mathrm{Null}(M)$ denotes the null space of M.*

Proof. It is an immediate consequence of the definitions of the hypernode graph gradient and the hypernode graph Laplacian that a hypernode graph Laplacian is a symmetric positive semidefinite real-valued matrix, and that a constant function has a null gradient. For the other direction, let us consider a symmetric positive semidefinite real-valued matrix M such that $\mathbf{1} \in \mathrm{Null}(M)$. Then, consider a square root decomposition $M = G^T G$ of M. For each line of G, one can define a hyperedge $h = \{s_h, t_h\}$ with s_h the set of nodes with positive values in the line of G, t_h the set of nodes with negative values in the line of G, and weights equal to the square of values in the line of G. The Equilibrium condition is satisfied because $\mathbf{1} \in \mathrm{Null}(M)$ and it is easy to verify that the Laplacian of the resulting hypernode graph **h** is M.

As a consequence of the construction in the previous proof, it should be noted that there are several hypernode graphs with the same hypernode graph Laplacian because the square root decomposition is not unique. One can also find hypernode graphs whose Laplacian matches that of a graph. One can prove that this is not however the general case. For this, it suffices to consider a hypernode graph Laplacian with an extradiagonal term which is positive. For instance, consider the hypernode graph and its Laplacian matrix Δ in Figure 2, the Laplacian matrix has 1 as extradiagonal term, thus Δ is not a graph Laplacian.

As said in Proposition 1, hypernode graph Laplacians are positive semidefinite. This allows to leverage most of the spectral learning algorithms defined in [16] , [18], [20] from graphs to hypernode graphs. Note, however, that spectral hypernode graph learning can not be reduced to spectral graph learning since hypernode graph Laplacians are strictly more general than graph Laplacians.

2.5   Hypernode Graph Laplacians and Signed Graphs

In this section we present additional properties of hypernode graph Laplacians and kernels. As in the graph case, we have defined the kernel of a hypernode graph to be the Moore-Penrose pseudoinverse of its Laplacian. Because the pseudoinversion preserves semidefiniteness and symmetry, as a consequence of Proposition 1, one can show that the class of hypernode graph kernels is closed under the pseudoinverse operation. As a consequence, the class of hypernode graph kernels is equal to the class of hypernode graph Laplacians. It is worth noticing that the class of graph kernels is not closed by pseudoinversion.

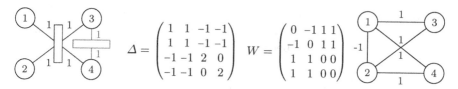

Fig. 2. From left to right : a hypernode graph, its Laplacian Δ, the pairwise weight matrix W, and the corresponding signed graph

It can also be shown that the class of hypernode graph Laplacians is closed by convex linear combination. This is an important property in the setting of learning from different sources of data. As graph kernels are hypernode graph kernels, it should be noted that the convex linear combination of graph kernels is a hypernode graph kernel, while it is not a graph kernel in general because the class of graph kernels is not closed by convex linear combination. This explains why problems for hypernode graphs can not be solved using graph constructions.

We have shown above that there does not exist in general a graph whose Laplacian is equal to the Laplacian of a given hypernode graph. Nevertheless, given a hypernode graph **h** and its Laplacian Δ, using Proposition 1, one can define a symmetric matrix W of possibly negative weights for pairs of nodes of **h** such that $\Delta = D - W$, where D is the degree matrix associated with W (the construction is illustrated in Figure 2). This means that, for every hypernode graph **h**, there is a unique signed graph with weight matrix W such that $D - W$ is the hypernode graph Laplacian of **h**. This result highlights the subclass of signed graphs whose Laplacian computed with the formula $D - W$ is positive semidefinite. This result also shows that homophilic relations between sets of nodes lead to non homophilic relations between nodes.

3   Hypernode Graph Model for Multiple Players Games

We consider competitive games between two teams where each team is composed of an arbitrary number of players. A first objective is to compute the skill ratings of individual players from game outcomes. A second objective is to predict a game outcome from a batch of games with their outcomes. For that, we will model games by hyperedges assuming that the performance of a team is the sum of the performances of its members as done by the team model proposed in [9].

3.1   Multiplayer Games

Let us consider a set of individual players $P = \{1, \ldots, n\}$ and a set of games $\Gamma = \{\gamma_1, \ldots, \gamma_p\}$ between two teams of players. Let us also consider that a player i contributes to a game γ_j with a non negative weight $c_j(i)$. We assume that each player has a skill $s(i)$ and that a game outcome can be predicted by comparing the weighted sum of the skills of the players of each of the two teams. More formally, given two teams of players $A = \{a_1, a_2, \ldots, a_\ell\}$ and $B = \{b_1, b_2, \ldots, b_k\}$ playing game γ_j, then A is predicted to be the winner if

$$\sum_{i=1}^{\ell} c_j(a_i) s(a_i) > \sum_{i=1}^{k} c_j(b_i) s(b_i) \ . \tag{1}$$

Equivalently, one can rewrite this inequality by introducing a non negative real number o_j on the right hand side such that

$$\sum_{i=1}^{\ell} c_j(a_i) s(a_i) = o_j + \sum_{i=1}^{k} c_j(b_i) s(b_i) \ . \tag{2}$$

The real number o_j quantifies the game outcome. In the case of a draw, the game outcome o_j is set to 0. Given a set of games, it may be impossible to assert that all constraints (1) can be simultaneously satisfied. Our goal is to estimate a skill rating function $s \in \mathbb{R}^n$ that respects the game outcomes in Γ as much as possible. We define the cost of a game γ_j with outcome o_j for a skill function s by

$$C_{\gamma_j}(s) = \| \sum_{i=1}^{\ell} c_j(a_i) s(a_i) - \sum_{i=1}^{k} c_j(b_i) s(b_i) - o_j \|^2 \ .$$

Consequently, given a set of games Γ and the corresponding game outcomes, the goal is to find a skill rating function s^* that minimizes the sum of the different costs, i.e. search for

$$s^* = \arg\min_s \sum_{\gamma_j \in \Gamma} C_{\gamma_j}(s) \ . \tag{3}$$

3.2   Modeling Games with Hypernode Graphs

We introduce the general construction by considering an example. Let us consider a game γ between two teams $A = \{1, 2\}$ and $B = \{3, 4\}$. Let us also assume that all the players contribute to the game with the same weight $c(1) = c(2) = c(3) = c(4) = 1$. Note that using uniform weights implies that the roles of the players inside a team are interchangeable and that equal skills for all players should lead to a draw. Such a game can be modeled by a hyperedge between sets of nodes $\{1, 2\}$ and $\{3, 4\}$ with weights equal to 1.

Now, let us suppose that A wins the game, then the skill rating function s must satisfy Equation (2), that is $s(1) + s(2) = o + s(3) + s(4)$ where $o > 0$ represents the outcome of the game γ. In order to model the game outcome in the hyperedge, we introduce a virtual player H that plays along with team B with a weight equal to 1 and we fix the skill rating function on H to be $s(H) = o > 0$. The virtual player is modeled by a node H, called outcome node, added to the set $\{3, 4\}$. Last, for the hyperedge to satisfy the equilibrium condition, we add a node Z, called lazy node, to the set $\{1, 2\}$. In this example, the weight of Z is set to 1. The skill $s(Z)$ of the lazy node Z is fixed to be 0 such as the equation

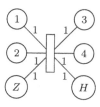

Fig. 3. Hyperedge h for a game γ between team $A = \{1,2\}$ and $B = \{3,4\}$. A wins and there is an additional outcome node H for the virtual player and an additional lazy node for the lazy virtual player. The node contributions are set to 1.

between skills can be rewritten as $s(1) + s(2) + s(Z) = s(3) + s(4) + s(H)$. And this equation is the definition of the *smoothness* of a node real valued function s over the hyperedge h with $s_h = \{1, 2, Z\}$ and $t_h = \{3, 4, H\}$ as represented in Figure 3 where s satisfies $s(H) = o$ and $s(Z) = 0$.

In the general case, let us consider a set of individual players $P = \{1, \ldots, n\}$, a set of games $\Gamma = \{\gamma_1, \ldots, \gamma_p\}$. Each game γ_j is between two teams (sets of players) A_j and B_j, the winning team is known as well as the game outcome o_j. Let us also consider that a player i contributes to a game γ_j with a non negative weight $c_j(i)$. We can define, for every game γ_j a hyperedge h_j as follows

1. The players of A_j define one of the two hypernodes of h_j. The weight of a player node i is defined to be $c_j(i)^2$,
2. do the same construction for the second team B_j,
3. add a *outcome node* H_j to the set of player nodes corresponding to the losing team. Its weight is set to 1,
4. add a *lazy node* Z_j to the set of player nodes corresponding to the winning team. Its weight is chosen in order to ensure the Equilibrium condition for the hyperedge h.

We define the hypernode graph $\mathbf{h} = (V, H)$ as the set of all hyperedges h_j for the games γ_j in Γ as defined above. Now, skill rating functions of players correspond to real-valued node functions over the hypernode graph. In order to model the game outcomes in the computation of the player skills, we fix the skill rating function values over the additional nodes for the virtual players. A skill function s over $\mathbf{h}$ must thus satisfy, for every lazy node Z_j, the function value $s(Z_j)$ is 0, and, for every outcome node H_j of game γ_j, the function value $s(Z_j)$ is the outcome o_j.

Formally, we assume a numbering of V such that $V = \{1, \ldots, N\}$ where N is the total number of nodes, the first n nodes are the player nodes followed by the t lazy nodes, then followed by the outcome nodes, that is, $V = \{1, \ldots, n\} \cup \{n+1, \ldots, n+t\} \cup \{n+t+1, \ldots, N\}$. Let Δ be the unnormalized Laplacian of $\mathbf{h}$, and let s be a real-valued node function on h, s can be seen as a real vector in $\mathbb{R}^N$ where the first n entries represent the skills of the n players. Then, it is easy to show that the *skill rating problem* (3) is equivalent to find the optimal vector s solving the optimization problem

$$\begin{aligned}
&\underset{s \in \mathbb{R}^N}{\text{minimize}} && s^T \Delta s \\
&\text{subject to} && \forall n + 1 \leq j \leq n + t,\ s(j) = 0 \text{ (for lazy nodes)} \\
& && \forall n + t + 1 \leq j \leq N,\ s(j) = o_j \text{ (for outcome nodes)}
\end{aligned} \tag{4}$$

3.3   Regularizing the Hypernode Graph

When the number of games is small, many players will participate to at most one game. Thus, in this case, the number of connected components can be quite large. The player skills in every connected component can be defined independently while satisfying the constraints. Thus, it will be irrelevant to compare player skills in different connected components. In order to solve this issue, we introduce in Equation (4) a regularization term based on the standard deviation of the players skills $\sigma(s_p)$, where $s_p = (s(1), \dots, s(n))$. This leads to the new formulation

$$\begin{aligned}
&\underset{s \in \mathbb{R}^N}{\text{minimize}} && s^T \Delta s + \mu \sigma(s_p)^2 \\
&\text{subject to} && \forall n + 1 \leq j \leq n + t,\ s(j) = 0 \text{ (for lazy nodes)} \\
& && \forall n + t + 1 \leq j \leq N,\ s(j) = o_j \text{ (for outcome nodes)},
\end{aligned} \tag{5}$$

where μ is a regularization parameter. Thus, we control the spread of s_p, avoiding to have extreme values for players participating in a small number of games.

In order to apply graph-based semi-supervised learning algorithms using hypernode graph Laplacians, we now show that the regularized optimization problem can be rewritten as an optimization problem for some hypernode graph Laplacian. For this, we will show that it suffices to add a regularizer node in the hypernode graph h. First, let us recall that if $\bar{s}$ is the mean of the player skills vector $s_p = (s(0), \dots, s(n))$, then, for all $q \in \mathbb{R}$, we have

$$\sigma(s_p)^2 = \frac{1}{n} \sum_{i=1}^{n} (s(i) - \bar{s})^2 \leq \frac{1}{n} \sum_{i=1}^{n} (s(i) - q)^2 .$$

Thus, in the problem 5, we can instead minimize $s^T \Delta s + \frac{\mu}{n} \sum_{i=1}^{n} (s(i) - q)^2$ over s and q. We now show that this can be written as the minimization of $r^T \Delta_\mu r$ for some vector r and well chosen hypernode graph Laplacian Δ_μ. For this, let us consider the $p \times N$ gradient matrix G of the hypernode graph h associated with the set of games Γ, and let us define the matrix G_μ by

$$G_\mu = \begin{pmatrix} \begin{array}{c|c} G & \begin{matrix} 0 \\ \vdots \\ 0 \end{matrix} \\ \hline \sqrt{\frac{\mu}{n}} B \end{array} \end{pmatrix} ,$$

where B is the $n \times (N + 1)$ matrix defined by, for every $1 \leq i \leq n$, $B_{i,i} = -1$, $B_{i,N+1} = 1$, and 0 otherwise. The matrix G_μ is the gradient of the hypernode graph $\mathbf{h}_\mu$ obtained from the hypernode graph $\mathbf{h}$ by adding a regularizer node R, an hyperedge between every player node and the regularizer node R with node weights μ/n (such a hyperedge can be viewed as an edge with edge weight μ/n). The construction is illustrated in Figure 4 with the hypernode graph reduced to a single hyperedge of Figure 3.

Let us denote by r the vector $(s(0), \ldots, s(N), q)$, then since $\Delta = G^T G$, we can write $r^T G_\mu^T G_\mu r = s^T \Delta s + \frac{\mu}{n} r B^T B r$. As $r B^T B r = \sum_i (s_i - q)^2$, if we denote by $\Delta_\mu = G_\mu^T G_\mu$ the $(N+1) \times (N+1)$ unnormalized Laplacian of the hypernode graph $\mathbf{h}_\mu$, we can finally rewrite the regularized problem (5) as

$$\underset{r \in \mathbb{R}^{N+1}}{\text{minimize}} \quad r^T \Delta_\mu r$$

$$\text{subject to} \quad \forall n + 1 \leq j \leq n + t, \; r(j) = 0 \text{ (for lazy nodes)} \tag{6}$$
$$\forall n + t + 1 \leq j \leq N, \; r(j) = o_j \text{ (for outcome nodes)}$$

3.4 Inferring Skill Ratings and Predicting Game Outcomes

We have shown that predicting skill ratings can be written as the optimization problem (6). It should be noted that it can also be viewed as a semi-supervised learning problem on the hypernode graph $\mathbf{h}_\mu$ because the question is to predict node scores (skill ratings) for player nodes when node scores for lazy nodes and outcome nodes are given. Using Proposition 1, we get that Δ_μ is a positive semidefinite real-valued matrix because it is a hypernode graph Laplacian. Therefore, we can use the semi-supervised learning algorithm presented in [20]. This algorithm was originally designed for graphs and solves exactly the problem (6) by putting hard constraints on the outcome nodes and on the lazy nodes. We denote this method by H-ZGL.

In order to predict skill ratings, another approach is to infer player nodes scores from lazy nodes scores and outcome nodes scores using a regression algorithm. For this, we consider the hypernode graph kernel $\Delta_\mu^\dagger$ (defined as the Moore-Penrose pseudoinverse of the Laplacian Δ_μ) and train a regression support vector machine. We denote this method by H-SVR.

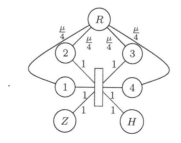

Fig. 4. Adding a regularizer node R to the hypergraph of Figure 3

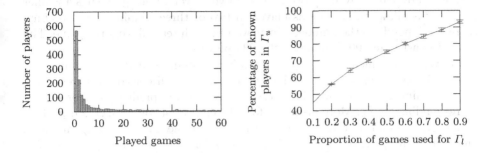

Fig. 5. [*left*] Distribution of the number of players against the number of played games; [*right*] Average percentage of players in Γ_u which are involved in some game in Γ_l

Using the two previous methods, we can infer skill ratings for players from a given set of games together with their outcomes. The inferred skill ratings can be used to predict game outcomes for new games. For this, we suppose that we are given a training set of games Γ_l with known outcomes together with a set of testing games Γ_u for which game outcomes are hidden. The goal is to predict game outcomes for the testing set Γ_u. Note that other works have considered similar questions in the online setting as in [9], [5] while we consider the batch setting. For the prediction of game outcomes, first we apply a skill rating prediction algorithm presented above given the training set Γ_l and output a skill rating function s^*. Then, for each game in Γ_u, we evaluate the inequality (1) with the skills defined by s^* and decide the winner. For every player which do not appear in the training set, the skill value is fixed a priori to the mean of known player skills.

Algorithm 1. Predicting game outcomes

Input: Training set of games Γ_l, set of testing games Γ_u

1: Build the regularized hypernode graph $\mathbf{h}_\mu$ as described in Sections 3.2 and 3.3
2: Compute an optimal skill rating s^* using H-ZGL or H-SVR.
3: Compute the mean skill $\tilde{s}$ among players in Γ_l
4: **for** each game in Γ_u **do**
5:    Assign skill given by s^* for players involved in Γ_l, and $\tilde{s}$ otherwise
6:    Evaluate the inequality (1) and predict the winner
7: **end for**

---

4 Experiments

4.1 Tennis Doubles

We consider a dataset of tennis doubles collected between January 2009 and September 2011 from ATP tournaments (World Tour, Challengers and Futures).

Tennis doubles are played by two teams of two players. Each game has a winner (no draw is allowed). A game is played in two or three winning sets. The final score corresponds to the number of sets won by each team during the game. The dataset consists in 10028 games with 1834 players.

In every experiment, we select randomly a training subset Γ_l of games and all remaining games define a testing subset Γ_u. We will consider different sizes for the training set Γ_l and will compute the outcome prediction error on the corresponding set Γ_u. More precisely, for a given proportion ρ varying from 10% to 90% , we build a training set Γ_l using ρ% of the games chosen randomly among the full game set, the remaining games form the test set Γ_u. We present in Figure 5 several statistics related to the Tennis dataset. It is worth noticing that many players have played only once. Therefore, the skill rating problem and the game outcome prediction problem become far more difficult to solve when few games are used for learning. Moreover, it should be noted that when the number of games in the training set is small, the number of players in the test set which are involved in a game of the training set is small. In this case many players will have a skill estimated to be the average skill.

Given a training set of games Γ_l and a test set Γ_u, we follow the experimental process described in Algorithm 1. For the definition of the hypergraph, we fix all player contributions in games to 1 because we do not have additional information than final scores. Thus the player nodes weights in every hyperedge are set to 1. In the optimization problem 6, the game outcomes o_j are defined to be the difference between the number of sets won by the two teams. This allows to take account of the score when computing player skills. In order to reduce the number of nodes, all lazy nodes are merged in a single one that is shared by all the hyperedges. We do the same for outcome nodes because score differences can be 1, 2 or 3. The resulting hypernode graph has at most 1839 nodes: at most 1834 player nodes, 1 lazy node, 3 outcome nodes, and 1 regularizer node.

To complete the definition of the hypernode graph $\mathbf{h}_\mu$ constructed from the game set Γ_l, it remains to fix the regularization node weights μ/n, i.e. fix the value of the regularization parameter μ. For this, assuming a Gaussian distribution for skill ratings and comparing expected values for the two terms $s^T \Delta s$ and $\mu\sigma(s_p)^2$, we can show that the value of μ/n should have the same order of magnitude than the average number of games played by a player. We fix the default value to be 16 for μ/n and use this default value in all experiments.

Given $\mathbf{h}_\mu$, following Algorithm 1, we apply the skill rating prediction algorithms H-ZGL and H-SVR. In order to compare our method, we also infer skill ratings using Elo Duelling and Trueskill [9][1] Then, we predict game outcomes from the inferred skill ratings. The results are given in Figure 6 (for each value of ρ, we repeat the experiment 10 times). It can be noted that Elo duelling performs poorly. Also, it can be noted that H-ZGL is significantly better than Trueskill whatever is the chosen proportion.

---

[1] TrueSkill and Elo implementations are from [7]. Results were double-checked using [14] and [13]. Parameters of Elo and TrueSkill are the default parameters of [7] ($K = 32$ for Elo, $\mu_0 = 25, \beta = 12.5, \sigma = 8.33$ and $\tau = 0.25$ for TrueSkill).

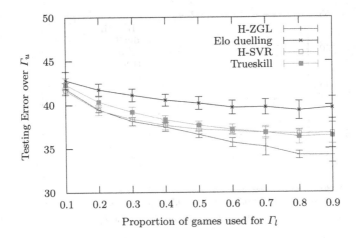

Fig. 6. Predictive error depending on the proportion of games used to build Γ_l

4.2   Xbox Title Halo2

The Halo2 dataset was generated by Bungie Studio during the beta testing of the XBox title Halo2. It has been notably used in [9] to evaluate the performance of the Trueskill algorithm. We consider the *Small Teams* dataset with 4992 players and 27536 games opposing up to 12 players in two teams which can have a different size. Each game can result in a draw or a win of one of the two teams. The proportion of draws is 22.8%. As reported in [9], the prediction of draws is challenging and it should be noted that Trueskill and our algorithm fail to outperform a random guess for the prediction of draw.

We again consider the experimental process described in Algorithm 1. As for the Tennis dataset, we fix all players contributions in games to 1. In the optimization problem 6, the game outcomes o_j are defined to be equal to 1 when the game has a winner and 0 otherwise because game scores in vary depending on the type of game. As above, we merge the lazy nodes into a single one and do the same for outcome nodes. The value of μ/n is again set to 16.

As for the Tennis dataset, we compare the skill rating algorithms H-ZGL, H-SVR, Elo Duelling and Trueskill. The number of prediction errors for game outcomes is computed assuming that a draw can be regarded as half a win, half a loss [12]. We present the experimental results in Figure 7. For a proportion of 10% of games in the training set, H-ZGL, H-SVR and Trueskill give similar results while with larger training sets, our hypernode graph learning algorithms outperform Trueskill. Contrary to the previous experiment, H-SVR performs better than H-ZGL. This result has however to be qualified given the fact that H-SVR depends on the soft margin parameter C whereas H-ZGL is strictly nonparametric, and we did not search for the better value of C.

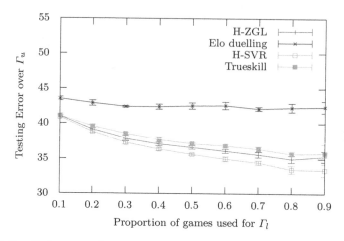

Fig. 7. Predictive error depending on the proportion of games used to build Γ_l

5 Conclusion

We have introduced hypernode graphs, defined a spectral theory for hypernode graphs, and presented an application to the problem of skill rating and game outcome prediction in multiple players games. This paper opens many research questions both from a theoretical perspective and from an applicatory perspective. First, the class of directed hypernode graphs should be investigated from a machine learning perspective. Second, following [8], it should be interesting to study the notion of cut for hypernode graphs. Third, we should define online learning algorithms for hypernode graphs following [10] which would be useful for large datasets for massive online games. Last, we are confident in the capability of our model to handle new applications in networked data.

References

1. Agarwal, S., Branson, K., Belongie, S.: Higher Order Learning with Graphs. In: Proc. of ICML, pp. 17–24 (2006)
2. Berge, C.: Graphs and hypergraphs. North-Holl Math. Libr. North-Holland, Amsterdam (1973)
3. Cai, J., Strube, M.: End-to-end coreference resolution via hypergraph partitioning. In: Proc. of COLING, pp. 143–151 (2010)
4. Easley, D., Kleinberg, J.: Networks, crowds, and markets: Reasoning about a highly connected world. Cambridge University Press (2010)
5. Elo, A.E.: The Rating of Chess Players, Past and Present. Arco Publishing (1978)
6. Gallo, G., Longo, G., Pallottino, S., Nguyen, S.: Directed hypergraphs and applications. Discrete Applied Mathematics 42(2-3), 177–201 (1993)
7. Hamilton, S.: PythonSkills: Implementation of the TrueSkill, Glicko and Elo Ranking Algorithms (2012), https://pypi.python.org/pypi/skills

8. Hein, M., Setzer, S., Jost, L., Rangapuram, S.S.: The Total Variation on Hypergraphs - Learning on Hypergraphs Revisited. In: Proc. of NIPS, pp. 2427–2435 (2013)
9. Herbrich, R., Minka, T., Graepel, T.: TrueSkill^{TM}: A Bayesian Skill Rating System. In: Proc. of NIPS, pp. 569–576 (2006)
10. Herbster, M., Pontil, M.: Prediction on a Graph with a Perceptron. In: Proc. of NIPS, pp. 577–584 (2006)
11. Klamt, S., Haus, U.-U., Theis, F.: Hypergraphs and Cellular Networks. PLoS Computational Biology 5(5) (2009)
12. Lasek, J., Szlávik, Z., Bhulai, S.: The predictive power of ranking systems in association football. International Journal of Applied Pattern Recognition 1(1), 27–46 (2013)
13. Lee, H.: Python implementation of Elo: A rating system for chess tournaments (2013), https://pypi.python.org/pypi/elo/0.1.dev
14. Lee, H.: Python implementation of TrueSkill: The video game rating system (2013), http://trueskill.org/
15. Penrose, R.: A generalized inverse for matrices. In: Proc. of Cambridge Philos. Soc., vol. 51, pp. 406–413. Cambridge University Press (1955)
16. Von Luxburg, U.: A tutorial on spectral clustering. Statistics and computing 17(4), 395–416 (2007)
17. Zhang, S., Sullivan, G.D., Baker, K.D.: The automatic construction of a view-independent relational model for 3-D object recognition. IEEE Transactions on Pattern Analysis and Machine Intelligence 15(6), 531–544 (1993)
18. Zhou, D., Huang, J., Schölkopf, B.: Learning from labeled and unlabeled data on a directed graph. In: Proc. of ICML, pp. 1036–1043 (2005)
19. Zhou, D., Huang, J., Schölkopf, B.: Learning with hypergraphs: Clustering, classification, and embedding. In: Proc. of NIPS, pp. 1601–1608 (2007)
20. Zhu, X., Ghahramani, Z., Lafferty, J., et al.: Semi-supervised learning using gaussian fields and harmonic functions. In: Proc. of ICML, vol. 3, pp. 912–919 (2003)

Discovering Dynamic Communities in Interaction Networks

Polina Rozenshtein, Nikolaj Tatti, and Aristides Gionis

Helsinki Institute for Information Technology, Aalto University
{firstname.lastname}@aalto.fi

Abstract. Online social networks are often defined by considering interactions over large time intervals, e.g., consider pairs of individuals who have called each other at least once in a mobilie-operator network, or users who have made a conversation in a social-media site. Although such a definition can be valuable in many graph-mining tasks, it suffers from a severe limitation: it neglects the precise time that the interaction between network nodes occurs.

In this paper we study *interaction networks*, where one considers not only the social-network topology, but also the exact time that nodes interact. In an interaction network an edge is associated with a time stamp, and multiple edges may occur for the same pair of nodes. Consequently, interaction networks offer a more fine-grained representation that can be used to reveal otherwise hidden dynamic phenomena in the network.

We consider the problem of discovering communities in interaction networks, which are dense and whose edges occur in short time intervals. Such communities represent groups of individuals who interact with each other in some specific time instances, for example, a group of employees who work on a project and whose interaction intensifies before certain project milestones. We prove that the problem we define is **NP**-hard, and we provide effective algorithms by adapting techniques used to find dense subgraphs. We perform extensive evaluation of the proposed methods on synthetic and real datasets, which demonstrates the validity of our concepts and the good performance of our algorithms.

Keywords: Community detection, graph mining, social-network analysis, dynamic graphs, time-evolving networks, interaction networks.

1 Introduction

Searching for communities in social networks is one of the most well-studied problems in social-network analysis. A number of different methods has been proposed, employing a diverse set of algorithmic tools, such as, agglomerative approaches, min-cut formulations, random walks, spectral methods, and more. Somewhat in contrast to this line of work, it has been observed that large networks are characterized by the lack of clear and well-defined communities [13,21].

T. Calders et al. (Eds.): ECML PKDD 2014, Part II, LNCS 8725, pp. 678–693, 2014.
© Springer-Verlag Berlin Heidelberg 2014

The lack of well-defined communities can be contributed to the high degree of interconnectivity, and the existence of overlapping communities. The phenomenon of overlapping communities is aggravated by the fact that community-detection methods typically ignore the time of interaction between network nodes, for instance, the same type of link can be used to represent friends in a hobby club and work colleagues.

On the other hand, as the amount of available data increases in volume and richness, it becomes possible to analyze not only the underlying topology but also the exact time of interactions. Analysis of such interaction events can reveal much more information about the structure and dynamics of the communities in the network. To be more concrete, consider the following examples.

Example 1: A group of researchers across many different European institutions are working on a large project. The members of the group go along with their everyday lives and other tasks, often unrelated to the project. However, once every few weeks or months, before deadlines of deliverables or project meetings, there is a lot of interaction among the group members.

Example 2: A group of twitter users is interested in technology products, in particular smartphones, and they are very active in blogging reviews and commenting the posts of each other. Their interaction is sparse, but it sustains over a long time, and it intensifies significantly after the release of a new product.

The main point of these two examples is that the communities are *not* isolated. Their members interact with each other, but they also interact with others outside the community. If one ignores the interaction dynamics and considers only the static social-network topology, the communities will be hidden and it will be impossible to discover them. It is only when considering the interaction time instances among the community members that it becomes possible to identify them: in both of the above examples, many interactions occur among the community members, but in a number of relatively short time intervals.

In this paper we formalize the idea exemplified above. We consider *interaction networks* for which we assume that all interaction events between the network nodes are known. Examples of such interaction networks include *call graphs* of telecommunications, *email communication networks*, *mention and commenting networks* in social media, *collaboration networks*, and more. Thus, interaction-network datasets are already abundant in many application domains.

In the context of interaction networks, we study the problem of discovering communities that are dense and whose edges occur in short time intervals. We prove that the problem we define is **NP**-hard, even though that the corresponding problem on static graphs is polynomially-time solvable. For the problem we define, we provide algorithms inspired by the literature of finding dense subgraphs. Our experiments demonstrate the effectiveness of the proposed algorithms, as well as the validity of our hypothesis. Namely, that it is possible to find communities that satisfy the requirements we set: dense interactions that occur within a number of short time intervals.

2   Preliminaries and Notation

An *interaction network* $G = (V, E)$ consists of a set of n nodes V and a set of m time-stamped interactions E between pairs of nodes

$$E = \{(u_i, v_i, t_i)\}, \text{ with } i = 1, \ldots, m, \text{ such that } u_i, v_i \in V \text{ and } t_i \in \mathbb{R}.$$

We consider that interactions are *undirected*. More than one interaction may take place between a pair of nodes, with different time stamps. Conversely, more than one interaction may take place at the same time, between different nodes.

For an interaction network $G = (V, E)$ we associate the set of edges $\pi(E)$ to be the pairs of nodes for which there is at least one interaction (one may think of π as "projecting" the edges of the interaction network along the time axis)

$$\pi(E) = \{(u, v) \in V \times V \mid (u, v, t) \in E \text{ for some } t\}.$$

Given an interaction network $G = (V, E)$, the network $\pi(G) = (V, \pi(E))$ is a standard graph with no time stamps on its edges. We refer to $\pi(G)$ as the *topology network* of G or as the *underlying network* of G.

Given an interaction network $G = (V, E)$ and a subset of nodes $W \subseteq V$, we define the *induced interaction network* $G(W) = (W, E(W))$, such that $E(W)$ consists of the interactions whose both end-points are contained in W,

$$E(W) = \{(u, v, t) \in E \mid u, v \in W\}.$$

We also consider time intervals $[s, f]$, where $s \in \mathbb{R}$ is the start point and $f \in \mathbb{R}$ is the end-point of the interval. We define the *span* of an interval to be its time duration, i.e., $\text{span}(T) = f - s$.

We define a *time-interval set* $\mathcal{T}$ to be a collection of non-overlapping time intervals, $\mathcal{T} = (T_1, \ldots, T_k)$. The span of $\mathcal{T}$ is the sum of individual spans,

$$\text{span}(\mathcal{T}) = \sum_{i=1}^{k} \text{span}(T_i).$$

Given an interaction network $G = (V, E)$ and a time interval $T = [s, f]$ we define the *spliced interaction network* $G(T) = (V, E(T))$, where $E(T)$ are the interactions that occur in T,

$$E(T) = \{(u, v, t) \in E \mid s \leq t \leq f\}.$$

The above notion can be extended in a straightforward manner, so as to define the spliced interaction network with respect to a set of time intervals $\mathcal{T} = (T_1, \ldots, T_k)$. This is achieved by collecting edges from individual time intervals, that is, $G(\mathcal{T}) = (V, E(\mathcal{T}))$, where $E(\mathcal{T}) = \bigcup_{i=1}^{k} E(T_i)$.

The concepts of *induced interaction network* and *spliced interaction network* provide two different ways to select subsets of interaction networks; one is based on subsets of nodes and the other is based on time intervals. The definition of dynamic communities, which is the central concept of our paper, relies on these two subset-selection strategies. In particular, for an interaction network $G = (V, E)$, a subset of nodes W, and a set of time intervals $\mathcal{T}$, we define a dynamic community $G(W, \mathcal{T})$ as the subgraph that consists of the nodes in W and the set of interactions among the nodes in W that occur within $\mathcal{T}$. In more formal terms, $G(W, \mathcal{T})$ is defined to be the spliced interaction network $H(\mathcal{T})$, where H is the induced interaction network $G(W)$.

To measure the quality of a dynamic community we rely on the notion of *density*. We recall the definition of density as defined for static graphs, e.g., for the topology network $\pi(G) = (V, \pi(E))$ of an interaction network $G = (V, E)$. We also review the *densest-subgraph problem* for static graphs.

Given a static graph $H = (V, F)$, i.e., the edges F do not have time stamps, the *density* of the graph $d(H)$ is twice the ratio of edges and the vertices,

$$d(H) = \frac{2\,|F|}{|V|}.$$

Problem 1 (Densest subgraph). Given a static graph $H = (V, F)$, find a subset of vertices W that maximizes the density $d(H(W))$.

Unlike the problem of finding the largest clique, which is **NP**-hard, finding the densest subgraph is polynomially-time solvable. Furthermore, there is a linear-time factor-2 approximation algorithm [2, 8]. The algorithm deletes iteratively a vertex with the lowest degree, obtaining a sequence of subgraphs. Among those subgraphs the algorithm returns the one with the highest density.

3   Dense Communities in Interaction Networks

Given an interaction network $G = (V, E)$ we aim to find a set of nodes W and a set of time intervals $\mathcal{T}$, such that the subgraph $G(W)$ is relatively dense within $\mathcal{T}$. To ensure that the time span of the subgraph $G(W)$ is short, we impose two types of constraints on the time-interval set $\mathcal{T}$: (i) constraints on the number of intervals of $\mathcal{T}$, and (ii) constraints on the total length of $\mathcal{T}$. We discuss these two constraints shortly. For the problem of finding dense dynamic communities, which we provide below, we also assume a *quality score* $q(W, \mathcal{T}; G)$ that measures the density of the community $G(W, \mathcal{T})$ in the interaction network G.

Problem 2. Assume that we are given a *quality score* $q(W, \mathcal{T}; G)$ that measures the quality of the community defined by nodes W and time interval span $\mathcal{T}$ in the interaction network G. Assume also we are given a budget K on the number of time intervals, and a budget B on the total time span. Our goal is to find a set of nodes W and a set of time intervals $\mathcal{T}$ that maximize

$$q(W, \mathcal{T}; G), \text{ such that } |\mathcal{T}| \leq K \text{ and } \mathrm{span}(T) \leq B.$$

The first constraint states that we can have at most K intervals while the second constraint requires that the total duration is at most B. Both constraints s are required: assuming that the quality score increases with the time span, if we drop the second constraint, then we can always choose the whole time span. Such a solution, however, does not capture the intuition of dynamic communities that we aim to discover. On the other hand, if we drop the first constraint, then we can pick individual edges by setting a time interval of duration 0 around each individual edge. Namely, the constraint on the number of intervals is necessary to impose time-continuity on the solutions found.

Regarding the score function used to assess the quality of a community, our proposed measure is the density of the topology network, after restricting to node set W and time-interval set $\mathcal{T}$

$$q(W, \mathcal{T}; G) = d(\pi(G(W, \mathcal{T}))),$$

that is, we count twice the number of interactions that occur between nodes of W within time intervals $\mathcal{T}$, and divide this number by $|W|$.

3.1   Complexity

We proceed to establish the complexity of the problem of finding a dense dynamic community in interaction networks (Problem 2).

Proposition 1. *The decision version of Problem 2 is **NP**-complete.*

Proof. We are given an interaction network G, budgets K, B, and a threshold σ, and we need to answer whether there is a node set W and a time-interval set $\mathcal{T}$, which satisfy the two budget constraints, and for which $q(W, \mathcal{T}; G) \geq \sigma$.

The problem is clearly in **NP**. To prove the hardness, we obtain a reduction from VERTEXCOVER. An instance of VERTEXCOVER specifies a graph H and budget ℓ, and asks whether there is a set $V' \subseteq V$, such that $|V'| \leq \ell$ and each edge of the graph is adjacent to at least one of the nodes of V'.

Consider graph $H = (U, F)$ with n nodes and m edges, and budget ℓ. Let us define an interaction network $G = (V, E)$. The node set V consists of U and $n+1$ additional auxiliary nodes, and the set of edges E is defined as follows: First we consider $n + 1$ distinct time points $t_0, \ldots, t_n$. At t_0 we consider interactions between all the auxiliary nodes, and between auxiliary nodes and each $v \in U$. We arbitrarily order the nodes in U and let v_i be the i-th node. At time t_i we connect v_i with all its neighbors in H.

Assume that there exists a solution W and $\mathcal{T}$, for Problem 2, with budgets $K = \ell+1$ and $B = 0$. We claim that W will contain all nodes and $\mathcal{T}$ will contain t_0 and the time points corresponding to the vertex cover of H.

Let us first prove that $W = V$ and $(t_0, t_0) \in \mathcal{T}$. Assume otherwise. Then, since the remaining time intervals have only edges between U, there must be at most $n(n - 1)/2$ edges, yielding density at most $n - 1$. Let us replace one of the selected time intervals with t_0 and reset W to be auxiliary nodes. This solution gives us a density of n, which is a contradiction.

Now we have established that t_0 is a part of $\mathcal{T}$. A straightforward calculation shows that it is always beneficial to add auxiliary nodes to W, if they are not part of a solution. Once this is shown, we can show further that adding any missing nodes from U also improves the density. Consequently, $W = V$.

Set $\sigma = 2(n(n + 1)/2 + n(n + 1) + m)/(2n + 1)$. The first two terms in the numerator correspond to the edges at t_0. The remaining m edges must come from the remaining time intervals. This is only possible if and only if the time intervals contain all edges from H, that is, the corresponding nodes cover every edge, which completes the reduction.   □

4 Algorithms for Discovering Communities

In this section we present the algorithm we propose for Problem 2. Since the problem is **NP**-hard, we propose an iterative method, which improves the solution by optimizing each one of the two components, the node set W and the time-interval set $\mathcal{T}$, in an alternating fashion, while keeping the other fixed.

Both of the objectives of our alternating optimization method give rise to interesting computational problems. One problem reduces to finding the densest subgraph, and the other is related to coverage, and it is even **NP**-hard. Next we formalize the two problems of our alternating optimization method.

Problem 3. Consider an interactive network $G = (V, E)$. Consider the problem of finding a dense dynamic community, with budgets K and B, and quality score q. Assume that a set of nodes W is provided as input. Find a time-interval set $\mathcal{T}$ that maximizes

$$q(W, \mathcal{T}; G), \text{ such that } |\mathcal{T}| \leq K \text{ and } \operatorname{span}(T) \leq B.$$

Problem 4. Consider the problem of finding a dense dynamic community on an interactive network $G = (V, E)$ with quality score q. Assume that a time-interval set $\mathcal{T}$ is given as input. Find a set of nodes W that maximizes $q(W, \mathcal{T}; G)$.

The proposed algorithm starts from an initial time interval set $\mathcal{T}_0$, and obtains a solution $(W, \mathcal{T})$ by iteratively solving the two problems defined above until convergence. Pseudocode of the method is given in Algorithm 1. As one may expect the iterative algorithm does not provide a guarantee for the quality of the solution that it returns. However, as it is stated by the following proposition, whose proof is straightforward, it has the desirable property that both of the alternating optimization problems return the correct component of the solution if they obtain as input the other component correctly.

Proposition 2. *Let $(W, \mathcal{T})$ be a solution to Problem 2 for a given interaction network G. Then (i) $\mathcal{T}$ is a solution to Problem 3 given G and W, and (ii) W is a solution to Problem 4 given G and $\mathcal{T}$.*

In the next two sections, 4.1 and 4.2, we present in detail our solution for the two subproblems of the iterative algorithm. In Section 4.3 we discuss the initialization of the algorithm.

Algorithm 1. Iterative algorithm for finding a dense dynamic community

---

1 $\mathcal{T}_0 \leftarrow$ initial sets of time intervals;
2 $i \leftarrow 0$;
3 **while** (convergence; $i{+}{+}$) **do**
4     $W_{i+1} \leftarrow$ solution to Problem 4 given $\mathcal{T}_i$;
5     $\mathcal{T}_{i+1} \leftarrow$ solution to Problem 3 given W_{i+1};
6 **return** $(W_i, \mathcal{T}_i)$;

4.1   Finding an Optimal Set of Nodes

We start with Problem 4 where the goal is to find an optimal set of nodes W given a set of time intervals $\mathcal{T}$. Assume that we are given a set of time intervals $\mathcal{T}$ and let $H = \pi(G(\mathcal{T}))$ be the topology network for the interactions that occur within $\mathcal{T}$ (i.e., the topology network of the interaction network spliced by $\mathcal{T}$). Note that

$$q(W, \mathcal{T}; G) = d(H(W)).$$

Consequently, finding the optimal set of nodes is equivalent to the densest-subgraph problem (Problem 1) on the (static) graph H. It follows that finding the optimal set of nodes W, given time-interval set $\mathcal{T}$, can be done in polynomial time. In our implementation, we use the linear-time algorithm of Charikar [8], which, as outlined in Section 2, offers a factor-2 approximation guarantee.

4.2   Finding an Optimal Set of Time Intervals

We now present our solutions for the second subproblem of the iterative algorithm, namely, finding an optimal set of time intervals for a given set of nodes. Unfortunately, even if this is a subproblem of the general community-discovery problem, it remains **NP**-hard. The proof of this claim is a simplified version of the proof of Proposition 1.

We view the problem of finding optimal time intervals as an instance of a *maximum-coverage with multiple budgets* (MCMB) problem.

Problem 5 (MCMB). Given a ground set $U = \{u_1, \ldots, u_m\}$ with weighted elements $w(u_i)$, a collection of subsets $\mathcal{S} = \{S_1, \ldots, S_k\}$, p cost functions c_i mapping each subset of $\mathcal{S}$ to a positive number, and n budget parameters B_i, find a subset $\mathcal{P} \subseteq \mathcal{S}$ maximizing

$$\sum_{u \in X} w(u), \text{ such that } X = \bigcup_{S \in \mathcal{P}} S, \text{ and } \sum_{S \in \mathcal{P}} c_i(S) \leq B_i, \text{ for all } i = 1, \ldots, p.$$

When $p = 1$, the problem is the standard budgeted maximum coverage. The problem is still **NP**-hard but there exists an approximation algorithm by Khuller et al. that achieves $(1 - 1/e)$ approximation ratio [17]. However this algorithm requires to enumerate all 3-subset collections, making it infeasible in practice.

The optimization problem can be also viewed as an instance of maximizing submodular function under multiple linear constraints. Kulik et al. presented a polynomial algorithm that achieves $(1 - \epsilon)(1 - 1/e)$ approximation ratio [18]. Unfortunately, this algorithm is not practical even for modest ϵ.

To see how finding a set of time intervals is related to maximum coverage, consider as ground set the set of edges $\pi(E(\mathcal{T}))$ (interactions that occur in $\mathcal{T}$ without the time stamps), and for each time interval $T \in \mathcal{T}$ create a subset S_T containing all edges whose corresponding interactions occur in T. There are two cost functions $c_1(T) = 1$ and $c_2(T) = \text{span}(T)$. The first budget constraint enforces the number of allowed time intervals to stay below K, while the second budget enforces the time-span constraint.

Thus, we need to solve the MCMB problem, defined above, with two budget constraints. We propose two solutions, both of which are inspired by the standard greedy approach for maximum coverage. The difference between the two proposed approaches is on how they try to satisfy the budget constraints. The first approach incorporates both budget constraints on the greedy step, while the second approach sets a parameter that controls the amount of violation of one constraint, and optimizes this parameter with binary search.

The standard greedy approach for maximum coverage is to select the set that has the best ratio of newly covered elements with respect to its cost. Motivated by this idea, we suggest the following greedy approach. Given a currently selected set of time intervals, say $\mathcal{T}$, we find the interval R that has the best ratio

$$\frac{q(W, \mathcal{T} \cup R, G) - q(W, \mathcal{T}, G)}{\max(x, y)}, \text{ where } x = \frac{1}{K - |\mathcal{T}|} \text{ and } y = \frac{\text{span}(R)}{B - \text{span}(\mathcal{T})}.$$

The numerator in the ratio is the number of new edges that can be covered with the new interval R. The denumerator is the maximum of two quantities, x and y, representing the two constraints on number of time intervals and time span, respectively. Both x and y are normalized so that they are equal to 1 if adding R will cap the corresponding constraint. By taking the maximum of the ratios we consider the constraint that is closer to be capped and penalize the ratio accordingly. The algorithm stops when one of the two constraints gets violated. We will refer to this approach as GREEDY.

Our second approach is based on the following observation. Assume that we are given a number α and consider optimizing

$$q(W, \mathcal{T}, G) - \alpha \cdot \text{span}(\mathcal{T}), \text{ such that } |\mathcal{T}| \leq K. \tag{1}$$

Note that we do not enforce any budget on the time span. If we set $\alpha = 0$, then the solution will contain the whole time. On the other hand, if we set α to be large, $\mathcal{T}$ will be just singular points. In fact, as it is shown in the following proposition, the time span of the optimal solution decreases as α increases.

Proposition 3. *Consider α_1 and α_2 with $\alpha_1 < \alpha_2$. Let $\mathcal{T}_1$ and $\mathcal{T}_2$ be the solutions of Equation (1) for α_1 and α_2, respectively. Then $\text{span}(\mathcal{T}_1) \geq \text{span}(\mathcal{T}_2)$.*

Proof. Define $\beta_i = \mathrm{span}(\mathcal{T}_i)$ and $d_i = q(W, \mathcal{T}_i, G)$. Due to optimality, we have

$$d_1 - \alpha_1\beta_1 \geq d_2 - \alpha_1\beta_2 \text{ and } d_2 - \alpha_2\beta_2 \geq d_1 - \alpha_2\beta_1.$$

By rearranging the terms we obtain $\alpha_1\beta_2 - \alpha_1\beta_1 \geq d_2 - d_1 \geq \alpha_2\beta_2 - \alpha_2\beta_1$. Rearranging the left and the right side gives us $(\alpha_1 - \alpha_2)(\beta_2 - \beta_1) \geq 0$. Since $\alpha_1 < \alpha_2$, we must have $\beta_1 \geq \beta_2$, which proves the proposition. □

Ideally, if we can solve Equation (1) optimally, we can use binary search to find the smallest α such that the time span of the solution does not exceed the budget. As we do not have an exact solver for Equation (1), we apply a greedy approach where in each step we find a single time interval that maximizes the score function. We then apply a binary search to find α that produces a feasible solution. We refer to this algorithm as BINARY.

4.3 Initialization

The quality of the solution discovered by the iterative algorithm depends on the set of time intervals $\mathcal{T}_0$ used as initial seed. Consider an optimal solution $(W, \mathcal{T})$, with $\mathcal{T} = (T_1, \ldots, T_K)$, which achieves density d^*. It follows that there is one single time interval $T \in \mathcal{T}$, for which the optimal set of nodes W has density at least d^*/K on $\pi(G(T))$. This observation motivates us to limit ourselves to consider only time interval sets of size 1. Assuming large computational power, one could test every possible time interval as a seed, consequently run the iterative algorithm, and return the best solution found out of all runs. There are $\mathcal{O}(m^2)$ such intervals, which is polynomial.

When running the algorithm $\mathcal{O}(m^2)$ times is expensive, we can select J random intervals, run the iterative algorithm for each of those random intervals, and return the best solution found out of all runs. In our experiments we evaluate the effect of the number of random seeds J to the quality of the solution found.

5 Experimental Evaluation

To evaluate the proposed methods we use several datasets: synthetic and real-world social communication networks. We describe our datasets in detail below.

Synthetic Data. We simulate activity on a network with a planted community. Different parameters for the planted community and the background noise are used, and the objective is to measure how the algorithms behave with respect to those parameters. The background network G is an Erdős-Rényi random graph, with expected degree being one of the model parameters. We plant a dense subgraph G', whose expected degree is a second model parameter. The length of whole time interval T is $|T| = 1000$ time units, while the interactions for the edges of G' can be covered by $k = 3$ arbitrary planted time intervals with total length of $|T'| = 100$ time units (10 times shorter than $|T|$). We randomly assign edges of G to time instances in T and edges of G' to time instances in T'.

Table 1. Characteristics of the two families of synthetic datasets. Planted community in Synthetic1 is a 5-clique. Planted community in Synthetic2 is an 8-node subgraph.

Name	$|V|$	$\text{Exp}[|E|]$	community avg degree	background avg degree
Synthetic1	100	200	4	$1-6$
Synthetic2	100	200	$2-7$	4

Table 2. Basic characteristics of real-world datasets. $|V|$: number of nodes; $|\pi(E)|$: number of edges of the topology network; $|E|$: number of interactions; $d(\pi(G))$: density of the whole topology network; $d^*(\pi(G))$: density of densest subgraph of the topology network.

Name	$|V|$	$|\pi(E)|$	$|E|$	$d(\pi(G))$	$d^*(\pi(G))$
Tumblr	1980	2454	7645	2.479	7.0
Students	883	2246	9865	5.087	11.292
Enron	1143	2019	6245	3.533	14.387

We test the ability of our algorithms to discover the planted communities in two settings. In the first setting (dataset family Synthetic1) we fix the planted subgraph and we vary the average degree of the background network. The objective is to test the robustness against background noise. In the second setting (dataset family Synthetic2) we fix the average degree of the background network and we vary the density of the planted subgraph. The characteristics of the synthetic datasets are given in Table 1.

Real-world Data. We use three datasets. The characteristics of these datasets are summarized in Table 2.

Tumblr: This is a subset of that Memetracker dataset,[1] which contains only quoting between Tumblr users. The subset covers three months: 02.2009–04.2009.

Students:[2] This dataset logs the activity in a student online community at University of California, Irvine. Nodes represent students and edges represent messages with ignored directions. We used a subset of the dataset that covers four months of communication from 2004-06-28 to 2004-10-26.

Enron:[3] This is the well-known dataset that contains the email communication of the senior management in a large company. It spans over 20 years from 1980.

Discovering Hidden Structure. We test the ability of our algorithms to detect the planted communities for different levels of background and in-community average degrees. We quantify the quality of our algorithms by measuring *precision* and *recall*, with respect to the ground-truth communities. We also report the

---

[1] http://snap.stanford.edu/data/memetracker9.html
[2] http://toreopsahl.com/datasets/#online_social_network
[3] http://www.cs.cmu.edu/~./enron/

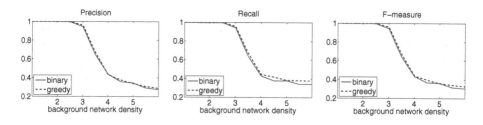

Fig. 1. Precision, recall and F-measure on Synthetic1, as a function of the background-network density. The planted community is a 5-clique.

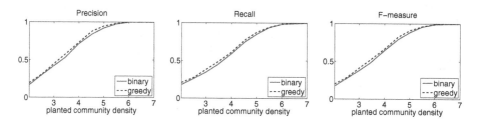

Fig. 2. Precision, recall and F-measure on Synthetic2, as a function of the density of a planted community of 8 nodes. The background-network density is set to 4.

F-measure, the harmonic mean of precision and recall. Results reported below are averages over $J = 1000$ independent runs.

Precision, recall and F-measure results for the two families of synthetic datasets are shown in Figures 1 and 2, respectively. Recall that datasets Synthetic1, contain a community based on a 5-clique. Both algorithms are able to discover this community correctly when the average degree of the underlying graph is smaller then the average degree of the planted community. Even when the community density is equal to the background-network density (around 4), the algorithms tend to keep high precision and recall. Precision and recall regrade at the same rate, indicating that with increase of background-network density the algorithms retrieve less nodes of planted community and more noisy nodes. Nevertheless, the measures do not drop very low, implying that the 5-clique spread over $k = 3$ short time intervals is distinguishable even within a dense background network.

The results on the second family of datasets (Synthetic2), shown in Figure 2, are similar. Both algorithms perform well when the background-network density is smaller than the planted-community density.

Effect of Random Seeds. Both of our algorithms are instances of Algorithm 1. In the experiments shown above we initialize the interval seed $\mathcal{T}_0$ with the whole time interval T spanned by the dataset. Starting from $\mathcal{T}_0 = \{T\}$ ensures that the subgraph we discover belongs to some dense structure in the topology network. However, if such a dense structure occurs in a scattered manner, the initialization $\mathcal{T}_0 = \{T\}$ may mislead. To overcome this problem and avoid dense structures

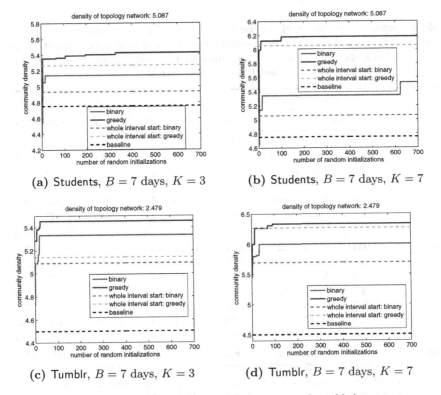

Fig. 3. Effect of random initializations on real-world datasets

that cannot be covered in the given time budget, we initialize Algorithm 1 with many random time intervals, and return the best solution found.

The improvement of performing random initializations is shown in Figure 3. The experiments are shown for Tumblr and Students. The figures show the best density discovered by our algorithms, when J independent random runs are performed. As expected, random initializations improve the performance of the algorithms. The most significant improvement is obtained for the Student dataset. We also experiment with a baseline that finds the densest subgraph over *all* possible intervals that satisfy the time budget B (no iterative process is followed). We see that our algorithms perform significantly better than this baseline.

Discovered Communities. Table 3 reports the densities of the communities discovered by our algorithms in the real-world datasets. We use $J = 200$ random initializations. We compare our algorithms with the same baseline as before: the densest subgraph over all intervals that satisfy the time budget B.

Overall, we observe that GREEDY and BINARY perform equally well, while in some settings BINARY yields denser communities than GREEDY.

Table 3. Densities of discovered subgraphs. The second column contains the number of allowed sets K and the column "budget" contains the time span budget B. For Tumblr and Students, B_1, B_2 and B_3 are equal to 1, 3 and 7 days, respectively. For Enron, B_1, B_2 and B_3 are 10, 30 and 120 days, respectively.

Name	K	budget = B_1			budget = B_2			budget = B_3		
		BINARY	GREEDY	BASE	BINARY	GREEDY	BASE	BINARY	GREEDY	BASE
Tumblr	1	3.818	3.818	3.866	4.0	4.0	4.0	4.5	4.5	4.5
	2	4.0	4.0	3.866	4.571	4.571	4.0	5.111	5.111	4.5
	3	4.6	4.285	3.866	5.2	5.2	4.0	5.5	5.4	4.5
	4	4.909	4.8	3.866	5.384	5.25	4.0	5.857	5.666	4.5
	5	5.166	5.111	3.866	5.666	5.5	4.0	6.0	5.866	4.5
	7	5.5	5.333	3.866	6.0	5.714	4.0	6.333	6.333	4.5
	10	6.181	5.818	3.866	6.428	6.181	4.0	6.8	6.666	4.5
Students	1	2.947	3.384	3.428	3.76	3.764	3.84	4.545	4.647	4.755
	2	3.5	3.3	3.428	4.32	4.133	3.84	5.225	5.125	4.755
	3	4.2	3.846	3.428	4.384	4.444	3.84	5.304	5.312	4.755
	4	4.0	4.0	3.428	4.545	4.615	3.84	5.642	5.368	4.755
	5	4.363	4.363	3.428	4.933	4.941	3.84	5.939	5.642	4.755
	7	4.625	4.545	3.428	5.210	5.185	3.84	6.108	6.0	4.755
	10	4.956	4.888	3.428	5.666	5.485	3.84	6.5	6.307	4.755
Enron	1	6.7272	6.7272	6.727	8.8	8.8	8.8	11.909	11.909	11.9
	2	8.875	8.4705	6.727	9.2222	9.625	8.8	13.047	11.913	11.9
	3	10.470	10.0	6.727	10.555	11.176	8.8	13.307	12.8	11.9
	4	11.058	10.736	6.727	11.904	12.2	8.8	13.642	13.047	11.9
	5	11.473	11.4	6.727	12.25	12.16	8.8	13.714	13.238	11.90
	7	12.370	12.16	6.727	12.666	13.0	8.8	13.931	13.857	11.9
	10	13.285	13.185	6.727	13.357	13.571	8.8	14.074	14.0	11.9

For fixed value of the time budget B, the density of the discovered community increases with K. For small values of K (1 to 3), the density of the communities discovered by our algorithm is equal, or in some cases slightly smaller, than the density of the communities discovered by the baseline. This behavior is expected, as the brute-force baseline tests all possible intervals, while our algorithms use only some random intervals for initialization. However, as the value of K increases, the algorithms take advantage of the provided flexibility to use many intervals effectively; for $K > 3$ both algorithms always outperform the baseline.

Furthermore, as we can see by contrasting Tables 2 and 3, the discovered communities are almost as dense as the densest subgraphs on the whole topology network, even though the time budget is significantly smaller than the time span of the dataset. For example, the densest subgraph of the over 20-year-large Enron dataset has average degree 14.387, while we were able to discover a subgraph with average degree 13.285 in a budget of 10 days, spanning 10 time intervals.

6   Related work

Community detection is one of the most studied problems in social-network analysis. A lot of research has been devoted to the case of static graphs, and the typical setting is to partitioning a graph into disjoint communities [9,12,26,30]; a thorough survey on such methods has been compiled by Fortunato [10].

Typically the term "dynamic graphs" refers to the model where edges are added or deleted. In this setting, once an edge is inserted in the graph it stays "alive" until the current time or until it is deleted. For example, this setting is used to model the process in which individuals establish friendship connections in a social network. On the contrary, our model intends to capture the continuous interaction between individuals. In the dynamic-graph setting, researchers have studied how networks evolve with respect to the arrival of new nodes and edges [19,20,31], the process of how groups and communities are formed [4], as well as methods for mining rules for graph evolution [5].

With respect to community detection in time-evolving graphs, the prominent line of work is to consider different graph snapshots, find communities in each snapshot separately (or by incorporating information from previous snapshots), and then establish correspondences among the communities in consecutive snapshots, so that it is possible to study how communities *appear, disappear, split, merge,* or *evolve.* A number of research papers follows this framework [3,14,22,24,29]. Similar recent works apply concepts of Laplacian dynamics [23] and frequent pattern mining [6] to ensure coherence and sufficiency of communities found in sequence of graph snapshots.

Many dynamic-graph studies are dedicated to the event-detection problem. The comprehensive tutorial by Akoglu and Faloutsos covers recent research on this topic.[4] The majority of the works focuses on how to compare different graph snapshots, and it aims to detect those snapshots that the graph structure changes significantly. The research tools developed in this area include novel metrics for graph similarity [25] and graph distance—see the survey of Gao et al. [11] and recent paper [28]—as well as extending scan-statistics methods for graphs [27], while a number of papers relies on matrix-decomposition methods [1,16].

To our knowledge, the approach that is best aligned with our problem setting, is presented by Bogdanov et al., for the problem of mining heavy subgraphs in time-evolving networks [7]. Yet, the two approaches are conceptually very distinct. First, the approach of Bogdanov et al. is still based on network snapshots, and thus sensitive to boundary quantization effects. Second, their concept of heavy subgraphs is based on edge weights, and their discovery problem maps to *prize collecting Steiner tree*, as opposed to a density-based objective.

Hu et al. propose a framework for mining frequent coherent dense subgraphs across a sequence of biological networks [15]. Their core concept is to construct a second-order graph, which represents co-activity of edges in the initial graph. As with the previous papers, Hu et al. work with network snapshots, which is quite a different model than the one we consider in this paper.

---

[4] http://www.cs.stonybrook.edu/~leman/icdm12/

In summary, in contrast to the existing work, in this paper we introduce a new point of view in the area of dynamic graphs, namely, we incorporate in our analysis point-wise interactions between the network nodes.

7 Concluding remarks

In this paper we considered the problem of finding dense dynamic communities in interaction networks, which are networks that contain time-stamped information regarding all the interactions among the network nodes. We formulated the community-discovery problem by asking to find a dense subgraph whose edges occur in short time intervals. We proved that the problem is **NP**-hard, and we provided effective algorithms inspired by methods for finding dense subgraphs.

Our paper is a step towards a more refined analysis of social networks, in which interaction information is taken into account and it is used to provide a more accurate description of communities and their dynamics in the network.

Our work opens many possibilities for future research. First we would like to extend the problem definition in order to discover many dense dynamic communities. This can be potentially achieved by asking to *cover* all (or a large fraction of) the interactions of the network with dense dynamic communities.

Second, we would like to incorporate additional information in our approach. As an example, think that the "smartphone community" discussed in the introduction, may use certain specialized vocabulary, brand names, or hashtags, which can provide additional clues for discovering the community. Our framework uses only time stamps of interactions; complementing our methods with additional information can potentially improve the quality of the results greatly.

Acknowledgements. This work was supported by Academy of Finland grant 118653 (ALGODAN).

References

1. Akoglu, L., Faloutsos, C.: Event detection in time series of mobile communication graphs. In: Army Science Conference (2010)
2. Asahiro, Y., Iwama, K., Tamaki, H., Tokuyama, T.: Greedily finding a dense subgraph. Journal of Algorithms 34(2) (2000)
3. Asur, S., Parthasarathy, S., Ucar, D.: An event-based framework for characterizing the evolutionary behavior of interaction graphs. TKDD 3(4) (2009)
4. Backstrom, L., Huttenlocher, D.P., Kleinberg, J.M., Lan, X.: Group formation in large social networks: membership, growth, and evolution. In: KDD (2006)
5. Berlingerio, M., Bonchi, F., Bringmann, B., Gionis, A.: Mining graph evolution rules. In: Buntine, W., Grobelnik, M., Mladenić, D., Shawe-Taylor, J. (eds.) ECML PKDD 2009, Part I. LNCS, vol. 5781, pp. 115–130. Springer, Heidelberg (2009)
6. Berlingerio, M., Pinelli, F., Calabrese, F.: Abacus: frequent pattern mining-based community discovery in multidimensional networks. Data Mining and Knowledge Discovery 27(3), 294–320 (2013)

7. Bogdanov, P., Mongiovì, M., Singh, A.K.: Mining heavy subgraphs in time-evolving networks. In: ICDM (2011)

8. Charikar, M.: Greedy approximation algorithms for finding dense components in a graph. In: Jansen, K., Khuller, S. (eds.) APPROX 2000. LNCS, vol. 1913, pp. 84–95. Springer, Heidelberg (2000)

9. Flake, G.W., Lawrence, S., Giles, C.L.: Efficient identification of web communities. In: KDD (2000)

10. Fortunato, S.: Community detection in graphs. Physics Reports 486 (2010)

11. Gao, X., Xiao, B., Tao, D., Li, X.: A survey of graph edit distance. Pattern Analysis and Applications 13(1) (2010)

12. Girvan, M., Newman, M.E.J.: Community structure in social and biological networks. PNAS 99 (2002)

13. Gleich, D.F., Seshadhri, C.: Vertex neighborhoods, low conductance cuts, and good seeds for local community methods. In: KDD (2012)

14. Greene, D., Doyle, D., Cunningham, P.: Tracking the evolution of communities in dynamic social networks. In: ASONAM (2010)

15. Hu, H., Yan, X., Huang, Y., Han, J., Zhou, X.J.: Mining coherent dense subgraphs across massive biological networks for functional discovery. Bioinformatics (2005)

16. Ide, T., Kashima, H.: Eigenspace-based anomaly detection in computer systems. In: KDD (2004)

17. Khuller, S., Moss, A., Naor, J.S.: The budgeted maximum coverage problem. Information Processing Letters 70(1) (1999)

18. Kulik, A., Shachnai, H., Tamir, T.: Maximizing submodular set functions subject to multiple linear constraints. In: SODA (2009)

19. Kumar, R., Novak, J., Tomkins, A.: Structure and evolution of online social networks. In: KDD (2006)

20. Leskovec, J., Backstrom, L., Kumar, R., Tomkins, A.: Microscopic evolution of social networks. In: KDD (2008)

21. Leskovec, J., Lang, K.J., Mahoney, M.W.: Empirical comparison of algorithms for network community detection. In: WWW (2010)

22. Lin, Y., Chi, Y., Zhu, S., Sundaram, H., Tseng, B.: Facetnet: A framework for analyzing communities and their evolutions in dynamic networks. In: WWW (2008)

23. Mucha, P.J., Richardson, T., Macon, K., Porter, M.A., Onnela, J.-P.: Community structure in time-dependent, multiscale, and multiplex networks. Science 328(5980), 876–878 (2010)

24. Palla, G., Barabási, A., Vicsek, T.: Quantifying social group evolution. Nature 446 (2007)

25. Papadimitriou, P., Dasdan, A., Garcia-Molina, H.: Web graph similarity for anomaly detection. Journal of Internet Services and Applications 1(1) (2010)

26. Pons, P., Latapy, M.: Computing communities in large networks using random walks. Journal of Graph Algorithms Applications 10(2) (2006)

27. Priebe, C.E., Conroy, J.M., Marchette, D.J., Park, Y.: Scan statistics on enron graphs. Computational & Mathematical Organization Theory 11(3) (2005)

28. Sricharan, K., Das, K.: Localizing anomalous changes in time-evolving graphs

29. Sun, J., Faloutsos, C., Papadimitriou, S., Yu, P.S.: Graphscope: parameter-free mining of large time-evolving graphs. In: KDD (2007)

30. van Dongen, S.: Graph Clustering by Flow Simulation. PhD thesis, University of Utrecht (2000)

31. Zhou, R., Liu, C., Yu, J.X., Liang, W., Zhang, Y.: Efficient truss maintenance in evolving networks. arXiv preprint arXiv:1402.2807 (2014)

Anti-discrimination Analysis
Using Privacy Attack Strategies

Salvatore Ruggieri[1], Sara Hajian[2], Faisal Kamiran[3], and Xiangliang Zhang[4]

[1] Università di Pisa, Italy
[2] Universitat Rovira i Virgili, Spain
[3] Information Technology University of the Punjab, Pakistan
[4] King Abdullah University of Science and Technology, Saudi Arabia

Abstract. Social discrimination discovery from data is an important task to identify illegal and unethical discriminatory patterns towards protected-by-law groups, e.g., ethnic minorities. We deploy privacy attack strategies as tools for discrimination discovery under hard assumptions which have rarely tackled in the literature: indirect discrimination discovery, privacy-aware discrimination discovery, and discrimination data recovery. The intuition comes from the intriguing parallel between the role of the *anti-discrimination authority* in the three scenarios above and the role of an *attacker* in private data publishing. We design strategies and algorithms inspired/based on Frèchet bounds attacks, attribute inference attacks, and minimality attacks to the purpose of unveiling hidden discriminatory practices. Experimental results show that they can be effective tools in the hands of anti-discrimination authorities.

1 Introduction

Discrimination refers to an unjustified distinction of individuals based on their membership, or perceived membership, in a certain group or category. Human rights laws prohibit discrimination on several grounds, such as sex, age, marital status, sexual orientation, race, religion or belief, membership of a national minority, disability or illness. Anti-discrimination authorities (equality enforcement bodies, regulation boards, consumer advisory councils) monitor, provide advice, and report on discrimination compliances based on investigations and inquiries. *Data* under investigation are studied by them with the main objective of *discrimination discovery*, which consists of unveiling contexts of discriminatory practices in a dataset of historical decision records. Discrimination discovery is a fundamental task in understanding past and current trends of discrimination, in judicial dispute resolution in legal trials, in the validation of micro-data or of aggregated data before they are publicly released. As an example of the last case, consider an employer noticing from public census data that the race or sex of workers act as proxy of the workers' productivity in his specific industry segment and geographical region. The employer may then use those visible

T. Calders et al. (Eds.): ECML PKDD 2014, Part II, LNCS 8725, pp. 694–710, 2014.
© Springer-Verlag Berlin Heidelberg 2014

traits of individuals, rather than their unobservable productivity, for driving (discriminatory) decisions in job interviews. Such a behavior, known as *statistical discrimination* [12], should be foreseen before data are publicly released.

Existing approaches for discrimination discovery [12,13] are designed with two assumptions: (1) the dataset under studying explicitly contains an attribute denoting the protected-by-law social group under investigation, and (2) the dataset has not been pre-processed prior to discrimination discovery. A first major source of complexity is to tackle the case that (1) does not hold – a problem known as *indirect discrimination discovery*, where indirect discrimination refers to apparently neutral practices that take into account personal attributes correlated with indicators of race, gender, and other protected grounds and that result in discriminatory effects on such protected groups. For example, even without race records of credit applicants, racial discrimination may occur in the practice of *redlining*: applicants living in a certain neighborhood are frequently denied, as most of people living in that neighborhood belong to the same ethnic minority. A second source of complexity, ignored in the literature so far, occurs when data contain attributes denoting protected groups but such data have been pre-processed to control the (privacy) risks of revealing confidential information, i.e., assumption (2) does not hold. If the anti-discrimination authority cannot be trusted, the original data cannot be accessed, and then discrimination discovery must be performed on the processed data. We name such a case *privacy-aware discrimination discovery*. A further case in which (2) may not hold occurs when data is pre-processed to hide discriminatory decisions to the anti-discrimination authority. Since the authority has to recover the original decisions as part of its investigation, we name such a case *discrimination data recovery*.

We follow the intriguing parallel between the role of the anti-discrimination authority in discrimination data analysis and the role of an *attacker* in privacy-preserving data publishing [1,4,5] – an unauthorized (possibly malicious) entity. Several attack strategies have been proposed in the literature, which model the reasonings of an attacker and its background knowledge. *Conceptually, the role of an anti-discrimination authority is similar to the one of an attacker.* In the case of indirect discrimination discovery, the authority has to infer personal data of individuals in the dataset under investigation, namely whether she belongs to a protected group or not (this step is necessary in order to measure the degree of discrimination in decisions). We substantiate this view by showing how combinatorial attacks based on Frèchet bounds inference [3] can be deployed to this purpose. In the case of privacy-aware discrimination discovery, the parallel is even more explicit: the anti-discrimination authority has to reason as an attacker to find out as much information as possible on the membership of individuals in the protected group. We will investigate a form of attribute inference attacks for discrimination discovery from a bucketized dataset [11]. Finally, in the case of discrimination data recovery the anti-discrimination authority has the objective of re-constructing original decisions from a perturbed dataset, which, again, is a typical task of privacy attackers. By exploiting an analogy with optimality attacks [14], we will devise an approach to reconstruct a dataset that has been decision

group	-	+	
protected	a	b	n_1
unprotected	c	d	n_2
	m_1	m_2	n

$p_1 = a/n_1$
$p_2 = c/n_2$
$p = m_1/n$
$RD = p_1 - p_2$

Fig. 1. Discrimination table

sanitized by means of the approach in [9]. The parallels highlighted open a new research direction consisting of applying the vast amount of methodologies and algorithms of privacy protection for discrimination data analysis.

This paper is organized as follows. Section 2 formalizes the three scenarios mentioned above. Section 3 recalls basic notions of discrimination analysis. The adaptation of privacy attack approaches and algorithms to each scenario is presented in Sections 4-6. Section 7 reports experimental results. Finally, conclusions report on related work and summarize our contributions.

2   Problem Scenarios

We assume two actors: a *data owner* and an *anti-discrimination authority*. The data owner releases to the anti-discrimination authority some data either in the form of micro-data, e.g., one or more relational or multidimensional tables, or in the form of aggregate data, e.g., one or more contingency tables. The anti-discrimination authority has access to additional information, called the *background knowledge*, that is exploited to unveil contexts of possible discrimination from the released data. The case when attributes to identify protected groups are part of the released data and data are without modification is known as *direct discrimination*. This is well-studied [12,13], and in this paper our main emphasis will be on the alternative case, consisting of one of the following scenarios.

Scenario I: Indirect discrimination discovery. The released data do not include attributes that explicitly identify protected-by-law groups. The task of the anti-discrimination authority is to unveil contexts of discrimination from the released data by exploiting background knowledge (e.g., correlations between attributes) to link the unknown attributes to attributes present in the data.

Scenario II: Privacy-aware discrimination discovery. The released data include attributes that explicitly identify protected-by-law groups, but the data were pre-processed by the data owner by applying a privacy-preserving inference control method to perturb such attributes. The anti-discrimination authority has the task of unveiling contexts of discrimination by exploiting background knowledge (e.g., aggregate counts on members of the protected group) and the awareness of the inference control algorithm used to pre-process the data.

Scenario III: Discriminatory data recovery. The released data were pre-processed by the data owner by applying a discrimination prevention inference control method that perturbed the data to hide discriminatory decisions. The task of the anti-discrimination authority is to reconstruct the original data by exploiting, again, background knowledge (e.g., amount of hidden discrimination) and the awareness of the inference control algorithm. Starting from the reconstructed dataset, standard direct discrimination discovery techniques can then be adopted to unveil contexts of discrimination.

3   Measures of Group Discrimination

A critical problem in the analysis of discrimination is precisely to quantify the degree of discrimination suffered by a given group (say, an ethnic group) in a given context (say, a geographic area and/or an income range) with respect to a decision (say, credit denial). To this purpose, several discrimination measures have been defined over a 4-fold contingency table, as shown in Fig. 1, where: the *protected* group is a social group which is suspected of being discriminated against; the *decision* is a binary attribute recording whether a benefit was granted (value "+") or not (value "-") to an individual; the *total population* denotes a context of possible discrimination, such as individuals from a specific city, job sector, income, or combination thereof.

We call the 4-fold contingency table of Fig. 1 a *discrimination table*. Different outcomes between groups are measured in terms of the proportion of people in each group with a specific outcome. Fig. 1 considers the proportions of benefits denied for the protected group (p_1), the unprotected group (p_2) and the overall population (p). Differences and rates of these proportions can model the legal principle of group under-representation of the protected group in positive outcomes or, equivalently, of over-representation in negative outcomes [12]. For space reasons, we restrict to consider only *risk difference* (RD = $p_1 - p_2$), which quantifies the marginal chance of the protected group of being given a negative decision. Once provided with a threshold α between "legal" and "illegal" degree of discrimination, we can isolate contexts of possible discrimination [13].

Definition 1 (α-protection). *A discrimination table is α-protective (w.r.t. the RD measure) if RD $\leq \alpha$. Otherwise, it is α-discriminatory.*

Direct discrimination discovery consists of finding α-discriminatory tables from a subset of past decision records. The original approach [13] performs a search in the space of discrimination tables of frequent (closed) itemsets. Fix a relational table whose attributes include GROUP, with values PROTECTED and UNPROTECTED, and DEC, with values + and -. An itemset is a set of items of the form $A = v$, where A is an attribute and $v \in dom(A)$, the domain of A. As usual in the literature, we write $A_1 = v_1, \ldots, A_k = v_k$ instead of $\{A_1 = v_1, \ldots, A_k = v_k\}$. Let **B** be an itemset without items over GROUP and DEC. The discrimination table associated to **B** regards the tuples in the cover of **B** as the total population. Therefore, n in Fig. 1 is the number of tuples satisfying **B** (i.e., its absolute support), and the cell values a, b, c and d are the counts of those also satisfying the cell coordinates. For instance, a is the support of the itemset "**B**, GROUP=PROTECTED, DEC=$-$".

	decision		
group	-	+	
protected	a	b	n_1
unprotected	c	d	n_2
	m_1	m_2	n

	rel. decision		
group	-	+	
g1	$\hat{a}$	$\hat{b}$	$\hat{n}_1$
g2	$\hat{c}$	$\hat{d}$	$\hat{n}_2$
	m_1	m_2	n

	rel. group	
group	g1	g2	
protected	e	f	n_1
unprotected	g	h	n_2
	$\hat{n}_1$	$\hat{n}_2$	n

Fig. 2. Indirect discrimination. Left: unknown contingency table. Center: known contingency table. Right: background knowledge contingency table.

4   Scenario I: Indirect Discrimination Discovery

The release of some aggregate data over a statistical database may lead to inferences on unpublished aggregates. In particular, the inference of bounds on entries in a 4-fold contingency table, given their marginals, trace back to the 1940's – and they are known as Fréchet bounds. They have been generalized to multidimensional contingency tables in the early 2000's [3]. We adopt an itemset based notation for contingency table cell entries. Let us denote by n_X the support of an itemset X in the dataset $\mathcal{R}$ under analysis: $n_X = |\{t \in \mathcal{R} | X \subseteq t\}|$. Consider now an itemset X of the form $A_1 = v_1, A_2 = v_2$, and Y of the form $A_2 = v_2, A_3 = v_3$. The itemset XY is $A_1 = v_1, A_2 = v_2, A_3 = v_3$ and the itemset $X \cap Y$ is $A_2 = v_2$. The Fréchet bounds for the support of XY are the following [3, Theorem 4]:

$$min\{n_X, n_Y\} \geq n_{XY} \geq max\{n_X + n_Y - n_{X \cap Y}, 0\} \qquad (1)$$

Let us exploit Fréchet bounds to model indirect discrimination discovery by means of background knowledge on attributes (cor-)related to membership to the protected group. Consider Fig. 2. Our problem is as follows: we want to derive bounds on a discrimination measure for an unknown contingency table (left) given a known/released contingency table (center) and some additional information contained in a background knowledge contingency table (right). The known contingency table shows data on an attribute that is related to the membership to the protected group through the background knowledge contingency table. The higher the correlation the closer the (known) discrimination measures for such an attribute are to the (unknown) discrimination measures for the protected group. The unknown value a can be decomposed into the number a_1 of individuals of the group g1 plus the number a_2 of individuals the group g2. Thus, $a_1 = n_{XY}$, where X is REL. GROUP=G1, DEC=- and Y is GROUP=PROTECTED, REL. GROUP=G1. The Fréchet bounds for a_1 yield:

$$min\{\hat{a}, e\} \geq a_1 \geq max\{\hat{a} + e - \hat{n}_1, 0\} = max\{e - \hat{b}, 0\}$$

and, with similar reasonings, those for a_2 yield: $min\{\hat{c}, f\} \geq a_2 \geq max\{\hat{c} + f - \hat{n}_2, 0\} = max\{f - \hat{d}, 0\}$. Therefore, for $a = a_1 + a_2$, we have the bounds:

$$min\{\hat{a}, e\} + min\{\hat{c}, f\} \geq a \geq max\{e - \hat{b}, 0\} + max\{f - \hat{d}, 0\} \qquad (2)$$

rel.	decision		
group	-	+	
g1	2	0	2
g2	6	18	24
	8	18	26

	rel. group		
group	g1	g2	
pro.	1	0	1
unp.	1	24	25
	2	24	26

Fig. 3. Sample known and background contingency tables

These bounds have an intuitive reading. Of the n_1 individuals in the protected group, e belong to group g1 and f belong to group g2. Consider the lower bounds. At most $min\{\hat{b}, e\}$ of those e (resp., $min\{\hat{d}, f\}$ of f) have a positive decision. Therefore, the number a is at least $max\{e - \hat{b}, 0\} + max\{f - \hat{d}, 0\}$. Consider now the upper bounds. At most e (resp., f) individuals of the protected group are in the g1 group (resp., g2 group), which, in turn, has at most $\hat{a}$ (resp., $\hat{c}$) negative decisions. Summarizing, the background knowledge necessary to derive the bounds for a consists of the distribution of the protected group into individuals of groups g1 and g2, namely values e and f in the background knowledge of Fig. 2. With similar means, one derives bounds for c:

$$min\{\hat{a}, g\} + min\{\hat{c}, h\} \geq c \geq max\{g - \hat{b}, 0\} + max\{h - \hat{d}, 0\}$$

Since n_1 and n_2 are in the background knowledge and m_1 is in the known contingency table, bounds for the proportions $p_1 = a/n_1$, $p_2 = c/n_2$, and $p = m_1/n$ can be readily computed. Finally, we derive a lower bound for RD:

$$RD \geq RDlb = \frac{max\{e - \hat{b}, 0\} + max\{f - \hat{d}, 0\}}{n_1} - \frac{min\{\hat{a}, g\} + min\{\hat{c}, h\}}{n_2}$$

Example 1. Consider the known and background knowledge tables in Fig. 3. The Frèchet bounds on a (number of protected individuals with negative decisions) and c (number of unprotected individuals with negative decisions) are:

$$1 = min\{2, 1\} + min\{6, 0\} \geq a \geq max\{1 - 0, 0\} + max\{0 - 18, 0\} = 1$$
$$7 = min\{2, 1\} + min\{6, 24\} \geq c \geq max\{1 - 0, 0\} + max\{24 - 18, 0\} = 7$$

We have $p_1 = 1/1$, $p_2 = 7/25 = 0.28$, and then $RD = p_1 - p_2 = 0.72$.

Notice that since Frèchet bounds are sharp [3], the bounds on discrimination measures are sharp as well. Although we described the case of a single attribute related to the protected attribute, the approach can be repeated for two or more related attributes, and the best bounds can be retained at each step. The overall approach is formalized in Algorithm 1, named *FrèchetDD* for Frèchet bounds-based Discrimination Discovery. The algorithm takes as input a relational table $\mathcal{R}$, background knowledge contingency tables $\mathcal{BK}$ and a threshold α for indirect discrimination discovery of α-discriminatory contingency tables. For each closed itemset **B**, the algorithm infers bounds ctu for its unknown contingency table. At the beginning (line 3), such bounds are the widest possible – from 0 to the

---

Algorithm 1. *FrèchetDD*($\mathcal{R}, \mathcal{BK}, \alpha$)

1: $\mathcal{C} \leftarrow \{$ frequent closed itemsets of $\mathcal{R}$ w/o GROUP and DEC items $\}$
2: **for** $\mathbf{B} \in \mathcal{C}$ **do**
3:     ctu $= ([0, n_\mathbf{B}], [0, n_\mathbf{B}], [0, n_\mathbf{B}], [0, n_\mathbf{B}])$
4:     $\mathcal{I} = \{A = v \mid$ no A-item is in $\mathbf{B}\}$
5:     **for** $A = v \in \mathcal{I}$ **do**
6:         **if** ctbg $= ct(\mathbf{B}, (\text{GROUP=PRO.}, \text{GROUP=UNPRO.}), (A = v, A \neq v)) \in \mathcal{BK}$ **then**
7:             ctk $= ct(\mathbf{B}, (A = v, A \neq v), (\text{DEC=-}, \text{DEC=+}))$
8:             ctu' $\leftarrow$ Frèchet bounds from ctk and ctbg
9:             ctu $\leftarrow min($ctu, ctu'$)$
10:         **end if**
11:     **end for**
12:     RDlb $\leftarrow$ RD lower bound from ctu
13:     **if** RDlb $\geq \alpha$ **then**
14:         output $\mathbf{B}$
15:     **end if**
16: **end for**

---

support $n_\mathbf{B}$ of $\mathbf{B}$. For every item $A = v$, where A is not already in $\mathbf{B}$ and such that a contingency table *ctbg* relating the protected group to $A = v$ in the context $\mathbf{B}$ is available in the background knowledge (line 6), the Frèchet bounds are calculated starting from such background contingency table and from a contingency table *ctk* that is computable from $\mathcal{R}$ (line 7), as described earlier in this section. The bounds are used to update *ctu* (line 9). After all items are considered, the final bounds *ctu* can be adopted for computing a lower bound on the discrimination measure at hand, RD in our case, to be checked against the threshold α (line 13). The computational complexity of Algorithm 1 is $O(|\mathcal{C}| \cdot |\mathcal{BK}|)$, i.e., the product of the size of closed itemsets by the size of the background knowledge.

A remarkable instance of indirect discrimination discovery is *redlining*, a practice banned in the U.S. consisting of denying credit on the basis of residence.

Example 2. Consider a released contingency table in Fig. 4 (right) regarding benefits granted and denied in a neighborhood specified by a ZIP code. In highly segregated cities, it may be very likely that specific neighborhoods, such as ZIP=100, are mostly populated by a specific race, say, a black minority. In such a case, the ZIP code acts as a proxy of the race of the population. Fig. 4 (left) shows the contingency table for the possibly discriminated group of black people living in the specific neighborhood ZIP=100. Entries of such a table may be unknown, due to the fact that the race of individuals is not recorded in the dataset. Fix the itemset X to ZIP=100,DEC=$-$, and Y to BLACK,ZIP=100. From the released contingency in Fig. 4 (right), we know that $n_X = \hat{a}$ and $n_{X \cap Y} = \hat{n}_1$. Assume now to have, as a background knowledge, the number n_Y of black people living in the neighborhood ZIP=100. Notice that, $n_Y = n_1$. Moreover, $n_{XY} = a$. The Frèchet bounds (1) are:

$$min\{\hat{a}, n_1\} \geq n_{XY} = a \geq max\{\hat{a} + n_1 - \hat{n}_1, 0\} = max\{n_1 - \hat{b}, 0\}$$

	decision				decision		
group	-	+		group	-	+	
black,zip=100	a	b	n_1	zip=100	$\hat{a}$	$\hat{b}$	$\hat{n}_1$
others	c	d	n_2	others	$\hat{c}$	$\hat{d}$	$\hat{n}_2$
	m_1	m_2	n		m_1	m_2	n

Fig. 4. Unknown (left) and known (right) contingency tables

Dividing by n_1, we get $min\{\hat{a}/n_1, 1\} \geq p_1 \geq max\{1 - \hat{b}/n_1, 0\}$. Since $c = m_1 - a$ and $n_2 = n - n_1$, bounds for $p_2 = c/n_2$ can be derived. The exact value $p = n_1/n$ is also known. Summarizing, ranges can be derived on all proportions in Fig. 1, and, *a fortiori*, on any discrimination measure based on them.

5   Scenario II: Privacy-aware Discrimination Discovery

In this scenario, the released dataset includes an attribute that explicitly identifies the protected group. However, since such an attribute is considered sensitive[1], data were pre-processed by the data owner using a privacy-preserving inference control method to diminish the correlation between such an attribute and other non-sensitive attributes. There could be different purposes for data sanitization: (1) to protect individuals' sensitive information; (2) to use data privacy as an excuse for hiding discriminatory practices. In both cases, the anti-discrimination authority has to unveil discrimination from the sanitized data.

There is a vast amount of privacy-preserving inference control methods. We investigate the scenario for one of the most popular ones, the *bucketization* method [15]. Bucketization disassociates the sensitive attributes from the non-sensitive attributes. The output of bucketization consists of two tables: a non-sensitive table (e.g., Fig. 5 left) and a sensitive table (e.g., Fig. 5 center). The non-sensitive table contains the entire non-sensitive attributes information, in addition to a group id GID (when tuples are partitioned into groups, a unique GID is assigned to each group). The sensitive table contains the sensitive values that appear in a specific group. Bucketization is a lossy join decomposition using the group id. For instance, tuple r_1 in group GID=1 has probability 25% of referring to a Muslim, Christian, Jewish, or Other individual, but it is impossible to determine which case actually holds. Thus, for the bucketized version $\mathcal{R}'$ of a dataset $\mathcal{R}$, the correlation between sensitive attribute and non-sensitive attributes is diminished. Note that in each group of our example table, every sensitive value is distinct and so the group size is equal to the parameter l in the l-diversity privacy model [11]. We assume that l is the cardinality of the attribute denoting protected and unprotected groups, e.g., the number of religions in our example.

---

[1] Protected group membership and private/sensitive information highly overlap [2], as e.g., for *religion, health status, genetic information* and *political opinions* attributes.

ID	Education	Job	Dec	GID
r_1	Bachelors	Engineer	-	1
r_2	Bachelors	Engineer	+	1
r_3	Doctorate	Engineer	+	1
r_4	Bachelors	Writer	+	1
r_5	Master	Engineer	+	2
r_6	Doctorate	Writer	+	2
r_7	Bachelors	Dancer	-	2
r_8	Master	Dancer	-	2
r_9	Master	Dancer	-	3
r_{10}	Master	Lawyer	+	3
r_{11}	Bachelors	Engineer	-	3
r_{12}	Bachelors	Dancer	-	3

GID	Religion
1	Muslim
1	Christian
1	Jewish
1	Other
2	Muslim
2	Christian
2	Jewish
2	Other
3	Muslim
3	Christian
3	Jewish
3	Other

	decision		
education=bachelors	-	+	
religion=muslim	a	b	3
religion$\neq$muslim	c	d	3
	4	2	6

Fig. 5. Non-sensitive (left) and sensitive (center) tables. Right: sample unknown c.t.

In this context, privacy-aware discrimination discovery can be formalized as the problem of deriving bounds on a discrimination measure for an unknown contingency table (see Fig. 2 left) given the bucketized dataset $\mathcal{R}'$. Consider a subset of n tuples from $\mathcal{R}'$ for which a contingency table has to be derived. The value m_1 is known (and also $m_2 = n - m_1$) because it consists of the number of tuples with negative decision. We assume that, as background knowledge, the number n_1 of tuples regarding protected group individuals is also known (and, a fortiori, $n_2 = n - n_1$). Starting from those aggregate values, bounds on cell values of the contingency table can be obtained by Frèchet bounds. Here, we propose to refine such bounds by exploiting the fact that in every bucket there is one and only one individual of the protected group. This yields the following bounds on a:

$$\Sigma_i \min\{1, n^i_-\} \geq a \geq n_1 - \Sigma_i \min\{1, n^i_+\} \tag{3}$$

where i ranges over group id's, n^i_- (resp., n^i_+) is the number of individuals with negative (resp., positive) decision with GID=i – this is available from the non-sensitive table. The bounds for c are easily derivable from those from a by noting that $c = m_1 - a$, since m_1 (the number of tuples with negative decision) is known. Similarly for $b = n_1 - a$, and for $d = n_2 - c$. Starting from them, bounds for p_1, p_2, p and discrimination measures defined over them can be computed.

Example 3. Consider the set of tuples from Fig. 5 (left) such that EDUCATION=BACHELORS. There are 6 such tuples: 4 with negative decision (r_1, r_7, r_{11}, r_{12}) and 2 with positive decision (r_2, r_4). Moreover, assume to know by background knowledge that $n_1 = 3$ out of the 6 tuples regard Muslims. This gives rise to the unknown contingency table in Fig. 5 (right). It turns out that $n^1_- = 1$, $n^2_- = 1$ and $n^3_- = 2$; and that $n^1_+ = 2$, $n^2_+ = 0$ and $n^3_+ = 0$. Therefore, we have:

$$\min\{1, 1\} + \min\{1, 1\} + \min\{1, 2\} = 3 \geq a \geq$$
$$2 = 3 - (\min\{1, 2\} + \min\{1, 0\} + \min\{1, 0\})$$

Frèchet bounds for Fig. 5 (right) would yield the strictly larger interval $\min\{4, 3\} = 3 \geq a \geq 1 = \max\{4 + 3 - 6, 0\}$. Since $a + c = 4$, we derive $2 \geq c \geq 1$. Thus, $p_1 = a/n_1 \in [2/3, 3/3]$, $p_2 = c/n_2 \in [1/3, 2/3]$ and then RD $= p_1 - p_2 \in [0, 2/3]$.

Algorithm 2. $PADD(\mathcal{R}', \mathcal{BK}, \alpha)$

1: $\mathcal{C} \leftarrow \{$ frequent closed itemsets of $\mathcal{R}'$ w/o GROUP and DEC items $\}$
2: **for** $\mathbf{B} \in \mathcal{C}$ **do**
3:    $n \leftarrow n_\mathbf{B}$
4:    $n_1 \leftarrow n_{\mathbf{B},\text{GROUP}=\text{PROTECTED}}$ // found in $\mathcal{BK}$
5:    $m_1 \leftarrow n_{\mathbf{B},\text{DEC}=-}$ // compute from $\mathcal{R}'$
6:    $a \in [a_l, a_u]$, with $a_u = min\{n_1, m_1, \Sigma_i min\{1, n_-^i\}\}$,
7:        $a_l = max\{n_1 + m_1 - n, 0, n_1 - \Sigma_i min\{1, n_+^i\}, lb(a)\}$ // $lb(a)$ found in $\mathcal{BK}$
8:    $c \in [c_l, c_u]$ with $c_u = m_1 - a_l$, $c_l = m_1 - a_u$
9:    RDlb $\leftarrow a_l/n_1 - c_u/(n - n_1)$
10:    **if** RDlb $\geq \alpha$ **then**
11:       output $\mathbf{B}$
12:    **end if**
13: **end for**

Given a bucketized dataset $\mathcal{R}'$ and background knowledge $\mathcal{BK}$, Algorithm 2, whose name is *PADD* for Privacy-Aware Discrimination Discovery, formalizes the search of itemsets $\mathbf{B}$ with a lower bound for RD greater or equal than α. We assume that $\mathcal{BK}$ may also include a further lower bound $lb(a)$ for a, obtained e.g., from answers to a survey or from allegations of discrimination against the data owner. The complexity of *PADD* is linear in the number of closed itemsets.

6   Scenario III: Discriminatory Data Recovery

To hide discrimination practices, data owners may apply discrimination prevention methods on datasets before publishing. For example, discrimination may be suppressed in the released data with minimal distortion of the decision attribute, i.e., by relabeling of some tuples to make the released dataset unbiased w.r.t. a protected group. Such discrimination prevention strategies are analogous to mechanisms of anonymization for data publication, where data anonymization is framed as a constrained optimization problem: produce the table with the smallest distortion that also satisfies a given set of privacy requirements. Such an attempt at minimizing information loss provides a loophole for attackers. The *minimality attack* [14] is one of the strategies to recover the private data from optimally anonymized data, given the non-sensitive information of individuals in the released dataset, the privacy policy, and the algorithm used for anonymization. The target of an anti-discrimination authority is precisely to reconstruct the original data from the released data, and then apply direct discrimination discovery techniques on the reconstructed data to unveil discrimination. In this sense, strategies such as minimality attacks can be readily re-proposed as a means in support of discrimination discovery.

We assume that the released dataset $\mathcal{R}'$ is changed minimally w.r.t. the original dataset $\mathcal{R}$ to suppress historical discriminatory practices. For instance, the *massaging* approach [9] changes a minimal number of tuples by promoting (from $-$ to $+$) or demoting (from $+$ to $-$) decision values. By "minimal" here it is

---

Algorithm 3. *DataRecovery(R′, DiscInt)*

1: $M \leftarrow 0.01 \cdot |protected\ group| \cdot |unprotected\ group| / |\mathcal{R}'|$
2: **for** 1 To $(DiscInt \cdot 100)$ **do**
3:     $(pr, dem) \leftarrow Rank(\mathcal{R}')$
4:     Change the decision of top M tuples of pr with DEC=- to DEC=+
5:     Change the decision of top M tuples of dem with DEC=+ to DEC=-
6:     $\mathcal{R}' \leftarrow \mathcal{R}'$ with new decision values of pr, dem
7: **end for**
8: **return** $\mathcal{R}'$

Algorithm 4. *Rank(R′)*

1: Learn a ranker L of DEC=+ using $\mathcal{R}'$ as training data
2: $pr \leftarrow$ unprotected group tuples in $\mathcal{R}'$ with DEC=-
            ordered descending w.r.t. the scores by L
3: $dem \leftarrow$ protected group tuples in $\mathcal{R}'$ DEC=+
            ordered ascending w.r.t. the scores by L
4: **return** (pr, dem)

---

meant that a number of changes is performed such that the RD measure for the released dataset is 0. We assume that the anti-discrimination authority knows, as background knowledge, the original value of RD, which we call *discrimination intensity (DiscInt)*. More realistically, such a value can be estimated on the basis of declarations made by individuals who claim to have been discriminated against. We exploit the observation proposed in [10] that discrimination affects the tuples close to the decision boundary of a classifier. To determine the decision boundary, we rank tuples of the protected and unprotected groups separately w.r.t. their positive decision probabilities accordingly to a classifier trained from $\mathcal{R}'$. We change the decision values of the tuples in the decision boundaries of the protected and unprotected groups to recover the original decision labels of $\mathcal{R}$. Algorithms 3 and 4 provide the pseudocode of this discriminatory data recovery process. Procedure *DataRecovery* takes as inputs the released data $\mathcal{R}'$ and the discrimination intensity *DiscInt*. The recovery is iteratively performed to recover one percent of released data in each step[2], rather than performing the entire data recovery in a single step. The reason is that the released data with altered attributes could lead to inaccurate calculation of probability scores. The gradual data recovery process improves the quality of data continuously, and thus provides more and more accurate probability scores.

Example 4. Let us assume that an employment bureau released its historical recruitment data as shown in Fig. 6. We, as an anti-discrimination authority,

---

[2] The number of modifications M at each step is determined as follow. Let n_1 (resp., n_2) be the size of the protected (resp., unprotected) group in $\mathcal{R}'$. A total of $M \cdot DiscInt \cdot 100$ tuples are demoted (resp., promoted) to move from RD = 0 to RD $= (M \cdot DiscInt \cdot 100)/n_1 + (M \cdot DiscInt \cdot 100)/n_2 = DiscInt$. By solving the equation, we get $M = 0.01 \cdot n_1 \cdot n_2/(n_1 + n_2)$ where $n_1 + n_2 = |\mathcal{R}'|$.

Sex	Ethnicity	Degree	Job Type	Dec	Prob
m	native	h.s.	board	+	98%
m	native	h.s.	board	+	98%
m	native	univ.	board	+	89%
m	*non-nat.*	*h.s.*	*health*	-	*47%*
m	non-nat.	univ.	health	-	30%

Sex	Ethnicity	Degree	Job Type	Dec	Prob
f	native	h.s.	board	+	93%
f	native	none	health	+	76%
f	*native*	*h.s.*	*edu.*	+	*51%*
f	non-nat.	univ.	edu.	-	2%
f	non-nat.	univ.	edu.	-	2%

Fig. 6. Sample job-application relation with positive decision probability scores

suspect of hidden discriminatory patterns in the released data due to complains about the biasness of this company w.r.t. sex of applicants. However, the bureau has changed minimally the original data to suppress historical discriminatory practices of the company. We have then to recover the discriminatory data. Assume the background knowledge that $DiscInt = 40\%$. We first calculate the positive decision probabilities for all tuples by adopting a probabilistic classifier (e.g., Naive Bayes), and then order the tuples of males and females w.r.t. these probability scores separately, as shown in Fig. 6. $DiscInt = 40\%$ in these 10 tuples implies that two tuples were relabeled for suppressing the sex discrimination, i.e., one male (resp. female) tuples was relabeled to negative (resp. positive) decision. The procedure $DataRecovery$ (steps 4, 5) selects tuples close to the decision boundaries as candidates for correction. In our example, those with $Prob$ around 50% (shown in red in Fig. 6) will have decision values changed: the male (resp., female) tuple is promoted from − to + (resp., demoted from + to −).

7   Experiments

In this section, we report experiments on three classical datasets available from the UCI Machine Learning repository (http://archive.ics.uci.edu/ml): *German credit*, which consists of 1000 tuples with attributes on bank account holders applying for credit; *Adult*, which contains 48848 tuples with census attributes on individuals; and *Communities and Crimes*, which contains 1994 tuples and describes the criminal behavior of different communities in the U.S.

Scenario I: Indirect Discrimination Discovery. We experimented the Frèchet bounds approach of Algorithm 1 on the *German credit* and *Adult* datasets. For the former dataset, the personal status attribute, denoting the protected group of non-single females, was removed before applying the algorithm. For the latter dataset, the same approach was taken for the protected group of non-Whites. Closed itemsets are computed by setting a minimum support threshold of 20, i.e., 2%, for *German credit* and of 48, i.e., 0.1%, for *Adult*. We simulate the availability of background knowledge contingency tables (*ctbg* in Algorithm 1) by computing them from the original dataset. In order to evaluate the impact of the size of the available background knowledge, only a random number ni of the items in the set $\mathcal{I}$ (see line 4 of Algorithm 1) are actually looked up. We experiment with $ni = 1$, i.e., the anti-discrimination authority has knowledge of only one related item, with $ni = 5$, and with an optimistic $ni = 30$. Fig. 7 (top)

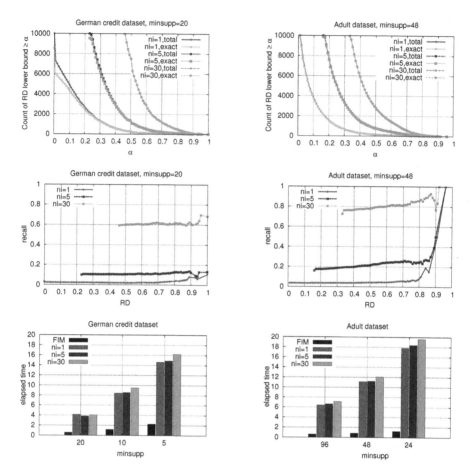

Fig. 7. Scenario I: precision (top), recall (middle), elapsed time (bottom) of *FrèchetDD*

shows the top 10K contingency tables w.r.t. the lower bound on the RD measure computed by Algorithm 1 for the *German credit* and the *Adult* datasets. The plots report the distributions of the contingency tables for which the lower bound is greater or equal than a given threshold α. It is shown the total number of such contingency tables (labels *total*) and the number of them for which the lower bound coincides with the upper bound (labels *exact*), namely, those for which Frèchet bounds are exact. Two facts can be concluded. First, if the inferred lower bound is higher than 0.3, then it is exact with high probability (95% or higher). Second, the higher is *ni* the higher are the inferred lower bounds. Fig. 7 (middle) shows the recall of the approach, namely the proportion of contingency tables with a given RD value of v that have been actually inferred a lower bound of v. The plots provide an estimate of the effectiveness of the indirect discrimination discovery approach for a given amount of background knowledge (*ni*). Finally,

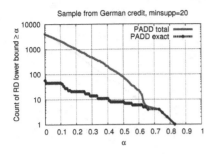
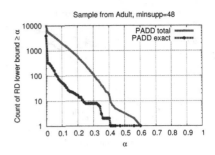

Fig. 8. Scenario II: precision of *PADD*

Fig. 7 (bottom) shows the elapsed times for the various experiments, including the time (denote by FIM) required for extracting closed itemsets. The time required by Algorithm 1 mainly depends on the number of closed itemsets, while the size of the background knowledge (the *ni* parameter) has a residual impact.

Scenario II: Privacy-aware discrimination discovery. We experimented with a subset of 200 tuples (resp., 14160) from the *German credit* (resp., *Adult*) dataset, randomly selected with a balanced distribution of the personal status (resp., race) attribute. Such a distribution is required to apply *l*-diversity data sanitization, where $l = 4$ (resp., $l = 2$) is the number of values of the personal status (resp., race) attribute, including the protected group of non-single females (resp., non-Whites). Tuples have been partitioned into groups of *l* elements with distinct personal status (resp., race). Fig. 8 shows the distributions of the RD lower bound for the top 10K contingency tables processed by Algorithm 2. The lower bound $lb(a)$ at line 7 is randomly generated in the interval from 0 to the actual value of *a*. Contrasting the plots with Fig. 7, we observe that the number and the exactness of the bounds inferred for RD is much lower than in the case of scenario I – notice that the plots in Fig. 8 are logscale in the y-axis. This is expected since the assumptions on the background knowledge exploitable in this scenario are much weaker than in scenario I. The anti-discrimination authority is assumed to know only the number of protected group individuals in the context under analysis as well as a lower bound on those with negative decision. In scenario I, correlation with groups whose decision value is precisely known is instead assumed. Nevertheless, scenario I and II are not mutually exclusive, and a hybrid approach could be applied to improve the inferred bounds.

Scenario III: Discriminatory data recovery. We conducted experiments on the *Adult* dataset, with protected group females, and on the *Crimes and Communities* dataset, with protected group blacks. As background knowledge, we assume to know that discrimination intensity is *DiscInt*=43% in *Crimes and Communities*, and *DiscInt*=19.45% in *Adult*. These numbers can be calculated from the original datasets. We proceeded with suppressing these differences by the method of massaging [9] before releasing the datasets. We then adopted the reverse engineering approach of Algorithm 3 to reconstruct the original data.

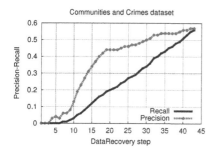
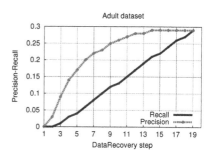

Fig. 9. Scenario III: performances of *DataRecovery*

The original dataset $\mathcal{R}$ can be used as ground truth for performance comparison. We measure the performances of Algorithm 3 by means of *Recall* and *Precision*. The *Recall* calculates how much massaged tuples were corrected, while the *Precision* measures how much corrected tuples were among those actually massaged. Algorithm 3 recovers data by iterations. In order to evaluate the performance at each iteration step, we compute recall and precision at the t-th step by $Recall = (\sum_{i=1}^{t} C_i)/(DiscInt \cdot |\mathcal{R}|)$ and $Precision = (\sum_{i=1}^{t} C_i)/(2 \cdot t \cdot M)$, respectively, where C_i is the number of tuples whose decision values are successfully corrected at the i-th step, and M is as in Algorithm 3. These sequential performance measures are shown in Fig. 9. The figure shows that our proposed method gives very promising results by reconstructing the *Adult* and the *Crimes and Communities* datasets (massaged to suppress 19.45% and 43% *DisctInt* resp.) with high precision and recall. We can observe that the method recovers the *Communities and Crimes* dataset with 59% precision and can assist the authorities to identify the suppressed discriminatory patterns. The recovery process is relatively less accurate over the *Adult* dataset due to a higher imbalance between protected and unprotected groups. Fig. 9 also shows the advantage of stepwise data recovery and refined probability score calculation. Our recovery algorithm continues to be more precise in the identification of perturbed tuples on the later recovery steps. This gradual and significant improvement in the performance can be attributed to the calculation of probability scores over the intermediary recovered and relatively corrected data.

8   Conclusions

Related Work. Discrimination analysis is a multi-disciplinary problem, involving sociological causes, legal argumentations, economic models, statistical techniques [12]. More recently, the issue of anti-discrimination has been considered from a data mining perspective. Some proposals are oriented to using data mining to measure and discover discrimination [13]; other proposals [6,9] deal with preventing data mining from becoming itself a source of discrimination. Summaries of contributions in discrimination-aware data mining are collected in [2,12]. The term privacy-preserving data mining (PPDM) was coined in

2000, although related work on inference control and statistical disclosure control (SDC) started in the 1970s. A detailed description of different PPDM and SDC methods can be found in [1, 5, 8]. Data are sanitized prior to publication and analysis (according to some privacy criterion). In some cases, however, an attacker can still re-identify sensitive information from the sanitized data using varying amounts of skill, background knowledge, and effort. Summaries of contributions and taxonomies of different privacy attacks strategies are collected in [1, 4]. Moreover, the problem of achieving simultaneous discrimination prevention and privacy protection in data publishing and mining was recently addressed in [7]. However, to the best of our knowledge, this is the first work that exploits tools from the privacy literature to the purpose of discovering discriminatory practices under hard conditions such as those of three scenarios considered.

Conclusion. The actual discovery of discriminatory situations and practices, hidden in a dataset of historical decision records, is an extremely difficult task. The reasons are as follows: First, there are a huge number of possible contexts may, or may not, be the theater for discrimination. Second, the features that may be the object of discrimination are not directly recorded in the data (scenario I). Third, the original data has previously been pre-processed due to privacy constraints (scenario II) or for hiding discrimination (scenario III). In this paper, we proposed new discrimination discovery methods inspired by the privacy attack strategies for the three scenarios above. The results of this paper can be considered a promising step towards the systematic application of techniques from the well explored area of privacy-preserving data mining to the emerging and challenging area of discrimination discovery.

References

1. Chen, B.C., Kifer, D., Le Fevre, K., Machanavajjhala, A.: Privacy-preserving data publishing. Foundations and Trends in Databases 2(1-2), 1–167 (2009)
2. Custers, B.H.M., Calders, T., Schermer, B.W., Zarsky, T.Z. (eds.): Discrimination and Privacy in the Information Society, Studies in Applied Philosophy, Epistemology and Rational Ethics, vol. 3. Springer (2013)
3. Dobra, A., Fienberg, S.E.: Bounds for cell entries in contingency tables given marginal totals and decomposable graphs. Proc. of the National Academy of Sciences 97(22), 11185–11192 (2000)
4. Domingo-Ferrer, J.: A survey of inference control methods for privacy-preserving data mining. In: Aggarwal, C.C., Yu, P.S. (eds.) Privacy-Preserving Data Mining. Advances in Database Systems, vol. 34, pp. 53–80. Springer (2008)
5. Fung, B.C.M., Wang, K., Chen, R., Yu, P.S.: Privacy-preserving data publishing: A survey of recent developments. ACM Comput. Surv. 42(4), Article 14 (2010)
6. Hajian, S., Domingo-Ferrer, J.: A methodology for direct and indirect discrimination prevention in data mining. IEEE Trans. on Knowledge and Data Engineering 25(7), 1445–1459 (2013)
7. Hajian, S., Domingo-Ferrer, J., Farràs, O.: Generalization-based privacy preservation and discrimination prevention in data publishing and mining. Data Mining and Knowledge Discovery, 1–31 (2014), doi:10.1007/s10618-014-0346-1

8. Hundepool, A., Domingo-Ferrer, J., Franconi, L., Giessing, S., Nordholt, E.S., Spicer, K., de Wolf, P.P.: Statistical Disclosure Control. Wiley (2012)

9. Kamiran, F., Calders, T.: Data preprocessing techniques for classification without discrimination. Knowledge and Information Systems 33, 1–33 (2012)

10. Kamiran, F., Karim, A., Zhang, X.: Decision theory for discrimination-aware classification. In: Proc. IEEE ICDM 2012, pp. 924–929 (2012)

11. Machanavajjhala, A., Kifer, D., Gehrke, J., Venkitasubramaniam, M.: L-diversity: Privacy beyond k-anonymity. ACM Trans. on Knowledge Discovery from Data 1(1), Article 3 (2007)

12. Romei, A., Ruggieri, S.: A multidisciplinary survey on discrimination analysis. The Knowledge Engineering Review, 1–57 (2014), doi:10.1017/S0269888913000039

13. Ruggieri, S., Pedreschi, D., Turini, F.: Data mining for discrimination discovery. ACM Trans. on Knowledge Discovery from Data 4(2), Article 9 (2010)

14. Wong, R.C.W., Fu, A.W.C., Wang, K., Pei, J.: Minimality attack in privacy preserving data publishing. In: Proc. of VLDB 2007, pp. 543–554 (2007)

15. Xiao, X., Tao, Y.: Anatomy: Simple and effective privacy preservation. In: Proc. of VLDB 2006, pp. 139–150 (2006)

Author Index